HANDBOOK OF PLASTICS, ELASTOMERS, AND COMPOSITES

HANDBOOK OF PLASTICS, ELASTOMERS, AND COMPOSITES

Charles A. Harper Editor in Chief

Technology Seminars, Inc.
Lutherville, Maryland

Third Edition

McGRAW-HILL

New York   St. Louis   San Francisco   Auckland   Bogotá
Caracas   Lisbon   London   Madrid   Mexico City   Milan
Montreal   New Delhi   San Juan   Singapore
Sydney   Tokyo   Toronto

Library of Congress Cataloging-in-Publication Data

Handbook of plastics, elastomers, and composites / Charles A. Harper,
   editor-in-chief. —3rd ed.
      p.   cm.
      Includes index.
      ISBN 0-07-026693-X
      1. Plastics—Handbooks, manuals, etc. 2. Elastomers—Handbooks,
manuals, etc. I. Harper, Charles A.
      TP1130.H36 1996
      668.4—dc20   96-20787
                                             CIP

McGraw-Hill

*A Division of The **McGraw·Hill** Companies*

4 5 6 7 8 9 0 DOC/DOC 9 0 9 8 7 6 5

ISBN 0-07-026693-X

*The sponsoring editor for this book was Zoe G. Faundotos, the editing super-
visor was Sheila Gillams, and the production supervisor was Don Schmidt.
This book was set in Times Roman. It was composed by North Market
Street Graphics.*

Printed and bound by R. R. Donnelley & Sons Company.

This book is printed on acid-free paper.

CONTENTS

Contributors   ix
Preface   xi

Chapter 1. Fundamentals of Plastics and Elastomers   1.1

1.1 Introduction / *1.1*
1.2 The Nature of Plastics / *1.1*
1.3 Polymer Structures and Polymerization Reactions / *1.2*
1.4 Plastic-Processing Methods and Design Guidelines / *1.5*
1.5 Plastic Properties / *1.13*
1.6 Thermosetting Plastics / *1.22*
1.7 Thermoplastics / *1.37*
1.8 Glass-Fiber-Reinforced Thermoplastics / *1.69*
1.9 Plastic Films and Tapes / *1.69*
1.10 Plastic Surface Finishing / *1.71*
1.11 Summary of Plastics / *1.72*
1.12 The Nature of Elastomers / *1.75*
References / *1.86*

Chapter 2. Laminates and Reinforced Plastic Materials and Processes   2.1

2.1 Laminates / *2.1*
2.2 Reinforced Plastics / *2.39*
2.3 Manufacturing Processes for Reinforced Plastics / *2.45*
2.4 Standards / *2.55*
References / *2.64*

Chapter 3. Advanced Composite Materials and Processes   3.1

3.1 Introduction / *3.1*
3.2 Material Systems / *3.4*
3.3 Composite Fabrication Techniques / *3.29*
3.4 Overview of Mechanics of Composite Materials / *3.52*
3.5 Design of Composites / *3.54*
3.6 Composite Testing / *3.64*
3.7 Safety Issues with Composite Materials / *3.69*
References / *3.72*

Chapter 4. Liquid and Low-Pressure Resin Systems   4.1

4.1 Introduction / *4.1*
4.2 Epoxies / *4.9*
4.3 Polyesters / *4.23*
4.4 Polyurethanes / *4.30*
4.5 Silicones / *4.37*
4.6 Cast Phenolics / *4.42*
4.7 Allylic Resins / *4.46*
4.8 Polybutadienes / *4.52*
4.9 Depolymerized Rubber / *4.56*
4.10 Polysulfide Rubber / *4.57*
4.11 Ethylene-Propylene Liquid Polymers / *4.58*
4.12 Cast Nylon / *4.58*
4.13 Acrylics / *4.60*
4.14 Vinyl Plastisols / *4.62*
4.15 Cyclic Thermoplastics / *4.63*
4.16 Expanding Monomer / *4.64*
4.17 Vinyl Esters / *4.64*
4.18 Cyanate Esters / *4.64*
4.19 Photopolymers / *4.67*
References / *4.67*

Chapter 5. Thermoplastic Elastomers   5.1

5.1 Introduction / *5.1*
5.2 TPEs in the Rubber and Plastics Industries / *5.2*
5.3 Comparison of TPEs and Thermoset Rubbers / *5.5*
5.4 Chemistry, Morphology, and Properties of TPEs / *5.8*
5.5 Generic Classes of TPEs / *5.11*
5.6 Processing of TPEs / *5.27*
5.7 Applications and Markets for TPEs / *5.37*
5.8 The Recovery and Recycle of TPEs / *5.39*
5.9 Acknowledgments / *5.40*
References / *5.40*

Chapter 6. Protective and Decorative Coatings   6.1

6.1 Introduction / *6.1*
6.2 Ecology / *6.3*
6.3 Surface Preparation / *6.4*
6.4 Coating Selection / *6.6*
6.5 Coating Materials / *6.8*
6.6 Application Methods / *6.27*
6.7 Curing / *6.41*
6.8 Summary / *6.43*
References / *6.45*

Chapter 7. Joining of Plastics, Elastomers, and Composites   7.1

7.1 Introduction / *7.1*
7.2 Design of Joints / *7.11*
7.3 Test Methods / *7.19*

7.4 Surface Preparation for Adhesive Bonding / *7.28*
7.5 Types of Adhesives / *7.50*
7.6 Selecting an Adhesive / *7.64*
7.7 Methods of Welding Polymeric Substrates / *7.69*
7.8 Methods of Mechanical Joining / *7.78*
7.9 Recommended Processes for Common Polymeric Substrates / *7.84*
7.10 Effect of Environment / *7.87*
7.11 Processing and Quality Control / *7.94*
References / *7.101*

Chapter 8. Plastics in Packaging   8.1

8.1 Packaging Plastics / *8.3*
8.2 Properties of Packaging Plastics / *8.17*
8.3 Mass Transfer in Polymeric Packaging Systems / *8.30*
References / *8.61*

Chapter 9. Elastomers and Engineering Thermoplastics for Automotive Applications   9.1

9.1 Introduction / *9.1*
9.2 Elastomers and Plastics / *9.2*
9.3 Why Use Elastomers? / *9.3*
9.4 Why Use Plastics? / *9.3*
9.5 The Importance of Material Selection / *9.4*
9.6 Engineering the Automobile / *9.5*
9.7 Automotive Elastomers / *9.12*
9.8 Automotive Engineering Thermoplastics / *9.25*
9.9 Selection Protocol for Automotive Elastomers and Plastics / *9.31*
9.10 Specification and Testing of Elastomers / *9.35*
9.11 Conclusion—A Look to the Future / *9.39*
References / *9.39*

Chapter 10. Design and Processing of Plastic Parts   10.1

10.1 Introduction / *10.1*
10.2 Design Procedure / *10.1*
10.3 Prototyping / *10.2*
10.4 Processes for Producing Plastic Parts / *10.3*
10.5 Assembly and Machining Guidelines / *10.19*
10.6 Postmolding Operations / *10.23*
10.7 Process-Related Design Considerations / *10.26*
10.8 Mold Construction and Fabrication / *10.31*
10.9 Summary / *10.33*

Chapter 11. Recycling of Plastic Materials   11.1

11.1 Introduction / *11.1*
11.2 Polyethylene Terephthalate (PET) Recycling / *11.14*
11.3 High-Density Polyethylene (HDPE) Recycling / *11.21*
11.4 Low-Density Polyethylene (LDPE) Recycling / *11.24*
11.5 Polystyrene (PS) Recycling / *11.26*

11.6 Polypropylene (PP) Recycling / *11.28*
11.7 Polyvinyl Chloride (PVC) Recycling / *11.30*
11.8 Recycling of Commingled Plastics / *11.32*
11.9 Recycling of Thermosets / *11.34*
11.10 Recycling of Automotive Plastics / *11.36*
11.11 Recycling of Other Thermoplastics / *11.38*
11.12 Sorting, Separation, and Compatibilization of Plastics / *11.39*
11.13 Thermal Recycling / *11.44*
11.14 Design Issues / *11.45*
11.15 The Future of Plastics Recycling / *11.48*
References / *11.50*
Further Reading / *11.54*

Appendix A. Glossary of Terms and Definitions   **A.1**

**Appendix B. Some Common Abbreviations Used in the Plastics
Industry**   **B.1**

Appendix C. Important Properties for Designing with Plastics   **C.1**

Appendix D. Electrical Properties of Resins and Compounds   **D.1**

**Appendix E. Sources of Specifications and Standards for Plastics
and Composites**   **E.1**

Index   **I.1**

CONTRIBUTORS

Leonard S. Buchoff   *Elastomeric Technologies, Inc., Hatboro, Pennsylvania.* (CHAP. 4)

Ruben Hernandez   *The School of Packaging, East Lansing, Michigan* (CHAP. 8)

John L. Hull   *Hull Corporation, Hatboro, Pennsylvania.* (CHAP. 10)

Carl P. Izzo   *Industry Consultant, Export, Pennsylvania.* (CHAP. 6)

Joseph F. Meier   *Industry Consultant, Export, Pennsylvania.* (CHAP. 1)

Stanley T. Peters   *Process Research, Mountain View, California* (CHAP. 3)

Edward M. Petrie   *ABB Transmission Technology Institute, Raleigh, North Carolina.* (CHAP. 7)

Charles P. Rader   *Advanced Elastomer Systems, L.P., Akron, Ohio* (CHAP. 5)

Ronald N. Sampson   *Technology Seminars, Pittsburgh, Pennsylvania.* (CHAP. 2)

Susan E. Selke   *The School of Packaging, East Lansing, Michigan* (CHAP. 11)

Ronald Toth   *Chrysler Corporation, Dearborn, Michigan* (CHAP. 9)

ABOUT THE EDITOR IN CHIEF

Charles A. Harper is an industrial consultant and president of Technology Seminars, Inc., based in Lutherville, Maryland. Previously, he was Manager of Materials Engineering for Westinghouse Electric Corporation in Baltimore, Maryland. He is a member of the Society of Plastics Engineers, IEEE, and the Society for the Advancement of Materials and Process Engineers (SAMPE). He is widely recognized for his teaching and writing activities, and for his leadership roles in numerous major professional societies.

PREFACE

In recent years, the development of new and improved polymers and their application in new and improved products have led to almost unlimited product opportunities. In fact, there are probably few who would not rate this area of product growth as one of the most important industry growth areas. The impact of polymers—plastics, elastomers, and composites—in all of their material forms has been little short of phenomenal. New polymers and improvements in established polymer groups regularly extend the performance limits of plastics, elastomers, and composites. These achievements in polymer and plastic technology offer major benefits and opportunities for the myriad of products in which they can be used.

With all of these achievements, however, a major impediment exists to the successful use of plastics, elastomers, and composites in products. This impediment is the lack of fundamental understanding of plastics, elastomers, and composites by product designers. Along with this lack of understanding is the absence of a useful consolidated source of information, data, and guidelines that can be practically used by product designers, most of whom do not "speak plastics." The usual practice is to use random supplier data sheets and data tables for guidance. It is, therefore, the object of this handbook to present, in a single source, all of the fundamental information required to understand the large number of materials and material forms, and to provide the necessary data and guidelines for optimal use of these materials and forms in the broad range of industry products. At the same time, this handbook will be invaluable to the plastics industry in acquainting its specialists with product requirements for which they must develop, manufacture, and fabricate plastic materials and forms.

This *Handbook of Plastics, Elastomers, and Composites* has been prepared as a thorough sourcebook of practical data for all ranges of interests. It contains an extensive array of property and performance data, presented as a function of the most important product variables. Further, it presents all important aspects of application guidelines, fabrication-method trade-offs, design, finishing, performance limits, and other important application considerations. It also fully covers chemical, structural, and other basic polymer properties. The handbook's other major features include thorough lists of standards and specifications sources, a completely cross-referenced and easy-to-use index, a comprehensive glossary, useful end-of-chapter reference lists, and several appendixes containing invaluable data and information for product engineers.

The chapter organization and coverage of the handbook is equally well suited for reader convenience. The opening chapter presents the fundamentals of plastics and elastomers and provides an overall foundation for the book, thus enabling readers to more fully understand the presentations in the following chapters.

The next six chapters deal with major plastic product forms that are so important in product design. One chapter each is devoted to laminates and conventional composites, advanced composites such as graphite-epoxy and others, liquid and low-pressure resin systems, thermoplastic elastomers, protective and decorative coatings, and adhesives. Each of these subjects requires special treatment for product design, and designers will find that these chapters provide excellent guidelines.

The following two chapters thoroughly cover the use of plastics and elastomers in two of the largest application fields, namely packaging and automotive. A special chapter is devoted to a clearly illustrated presentation of all of the important considerations for the design and fabrication of molded plastic products.

The final chapter is an excellent presentation on a subject of increasingly vital importance to all of those in all areas of plastics and elastomers—the recycling of waste products.

The result of these presentations is an extremely comprehensive and complete single reference and text—a must for the desk of anyone involved in any aspect of product design, development, or application of plastics, elastomers, and composites. This handbook will be invaluable for every reference library.

Charles A. Harper

CHAPTER 1
FUNDAMENTALS OF PLASTICS AND ELASTOMERS

Joseph F. Meier
Industry Consultant
Export, Pennsylvania

1.1 INTRODUCTION

The *Handbook of Plastics, Elastomers, and Composites* will attempt to address important application and use data and guidelines for the entire field of plastics and elastomers. This initial chapter covers the nature of plastics and elastomers, including the chemical nature of polymers, an overview of testing and the significance of standard tests, primary characteristics of individual polymer classes, processing and processed forms of plastics, and related subjects that are basic to the field of plastics and elastomers.

Subsequent chapters will cover in detail many important individual applications of plastics and elastomers that have developed since the last edition of this handbook.[1] Thermoplastic elastomers (Chap. 5) and plastic and elastomer adhesives (Chap. 7) are just two examples.

Because of the tremendous volume of data available and the limited space, references at the end of each chapter will direct the interested reader to more in-depth information.

1.2 THE NATURE OF PLASTICS

Practically stated, a plastic is an organic polymer, available in some resin form or some form derived from the basic polymerized resin. These forms can be liquid or pastelike resins for embedding, coating, and adhesive bonding; or they can be molded, laminated, or formed shapes, including sheet, film, or larger mass bulk shapes.

The number of basic plastic materials is large, and the list is increasing. In addition, the number of variations and modifications to these basic plastic materials is also quite large. Taken together, the resultant quantity of materials available is just too large to be completely understood and correctly applied by anyone other than those whose day-to-day work puts them in direct contact with a diverse selection of materials. The practice of mixing brand names, trade names, and chemical names of various plastics only makes the problem of understanding these materials more troublesome.

Another variable that makes it difficult for those not versed in plastics to understand and properly design with plastics is the large number of processes by which plastics can be fabricated. Fortunately, there is an organized pattern on which an orderly presentation of these variables can be based.

While there are numerous minor classifications for polymers, depending on how one wishes to categorize them, nearly all can be placed into one of two major classifications—thermosetting materials (or thermosets) and thermoplastic materials.[2] Likewise, foams, adhesives, embedding resins, elastomers, and so on, can be subdivided into the thermoplastic and thermosetting classifications.

1.2.1 Thermosetting Plastics

As the name implies, thermosetting plastics, or thermosets, are cured, set, or hardened into a permanent shape. Curing is an irreversible chemical reaction known as crosslinking, which usually occurs under heat. For some thermosetting materials, curing is initiated or completed at room temperature. Even here, however, it is often the heat of the reaction, or the exotherm, that actually cures the plastic material. Such is the case, for instance, with a room-temperature-curing epoxy or polyester compound.

The cross-linking that occurs in the curing reaction is brought about by the linking of atoms between or across two linear polymers, resulting in a three-dimensional rigidized chemical structure. One such reaction is shown in Fig. 1.1.[3] Although the cured part can be softened by heat, it cannot be remelted or restored to the flowable state that existed before curing. Continued heating for long times leads to degradation or decomposition.

1.2.2 Thermoplastics

Thermoplastics differ from thermosets in that they do not cure or set under heat as do thermosets. Thermoplastics merely soften, or melt when heated, to a flowable state, and under pressure they can be forced or transferred from a heated cavity into a cool mold. Upon cooling in a mold, thermoplastics harden and take the shape of the mold. Since thermoplastics do not cure or set, they can be remelted and then rehardened by cooling. Thermal aging, brought about by repeated exposure to the high temperatures required for melting, causes eventual degradation of the material and so limits the number of reheat cycles. Most of the common thermoplastic materials are discussed in detail in a following section of this chapter.

1.3 POLYMER STRUCTURES AND POLYMERIZATION REACTIONS

1.3.1 Polymer Structures

All polymers are formed by the creation of chemical linkages between relatively small molecules, or monomers, to form very large molecules, or polymers. As men-

Reaction A

One quantity of unsaturated acid reacts with two quantities of glycol to yield linear polyester (alkyd) polymer of n polymer units

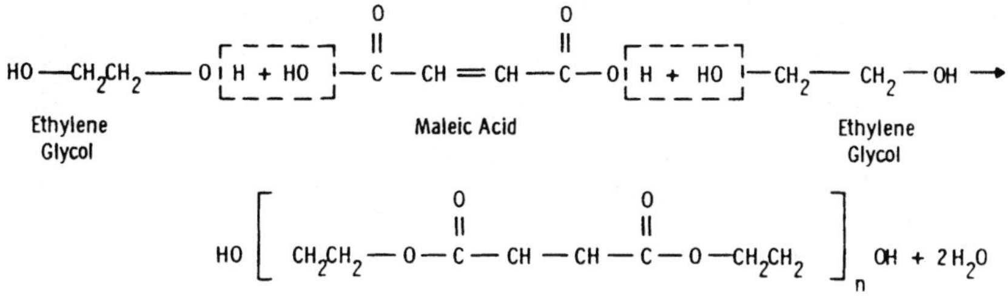

Ethylene
Glycol   Maleic Acid   Ethylene
                                                                                           Glycol

Ethylene Glycol Maleate Polyester

Reaction B

Polyester polymer units react (copolymerize) with styrene monomer in presence of catalyst and/or heat to yield styrene-polyester copolymer resin or, more simply, a cured polyester. (Asterisk indicates points capable of further cross-linking.)

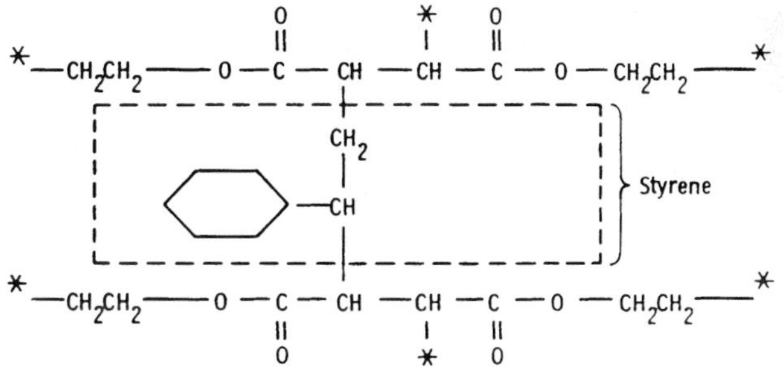

Styrene-Polyester Copolymer

FIGURE 1.1   Simplified diagrams showing how cross-linking reactions produce polyester resin (styrene-polyester copolymer resin) from basic chemicals.

tioned, if the chemical linkages form a rigid, cross-linked molecular structure, a thermosetting plastic results. If a somewhat flexible molecular structure is formed, either linear or branched, a thermoplastic results. Illustrations of these molecular structures are presented in Fig. 1.2.[1]

1.3.2   Polymerization Reactions

Polymerization reactions may occur in a number of ways; four common techniques are bulk, solution, suspension, and emulsion polymerization.[1,2] Bulk polymerization involves the reaction of monomers or reactants among themselves, without placing them in some form of extraneous media as is done in the other types of polymerization.

Solution polymerization is similar to bulk polymerization, except that whereas the solvent for the forming polymer in bulk polymerization is the monomer, the sol-

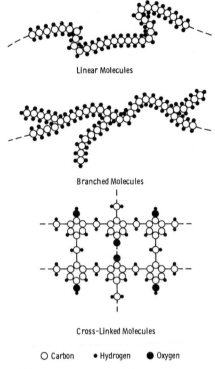

Linear Molecules

Branched Molecules

Cross-Linked Molecules

○ Carbon   • Hydrogen   ● Oxygen

FIGURE 1.2 Some possible molecular structures in polymers.

vent in solution polymerization is usually a chemically inert medium. The solvents used may be complete, partial, or nonsolvents for the growing polymer chains.

Suspension polymerization normally is used only for catalyst-initiated or free-radical addition polymerizations. The monomer is dispersed mechanically in a liquid, usually water, which is a nonsolvent for the monomer as well as for all sizes of polymer molecules that form during the reaction. The catalyst initiator is dissolved in the monomer, and it is preferable that it does not dissolve in the water so that it remains with the monomer. The monomer and the polymer that is formed from it stay within the beads of organic material dispersed in the phase. Actually, suspension polymerization is essentially a finely divided form of bulk polymerization. The main advantage of suspension polymerization over bulk is that it allows cooling of the exothermic polymerization reaction and maintains closer control over the chain-building process. By controlling the degree of agitation, monomer-to-water ratios, and other variables, it is also possible to control the particle size of the finished polymer, thus eliminating the need to re-form the material into pellets from a melt, as is usually necessary with bulk polymerization.

Emulsion polymerization is a technique in which addition polymerizations are carried out in a water medium containing an emulsifier (a soap) and a water-soluble initiator. Emulsion polymerization is much more rapid than bulk or solution polymerization at the same temperatures and produces polymers with molecular weights much greater than those obtained at the same rate in bulk polymerizations.

In emulsion polymerization, the monomer diffuses into micelles, which are small spheres of soap film. Polymerization occurs within the micelles. Soap concentration, overall reaction-mass recipe, and reaction conditions can be varied to provide control of the reaction rate and yield.

1.3.3   Raw-Material Sources of Major Plastic Materials

The primary raw-material sources of the major plastic and elastomer materials are natural gas, coal, and crude oil.[3,4]

1.4   PLASTIC-PROCESSING METHODS AND DESIGN GUIDELINES

Although most users of plastics buy parts from plastic processors, they should still have some knowledge of plastic processing, as such information can often be helpful in optimizing product design. Also, an increasing number of user companies are doing some in-house processing. For these reasons, some guideline information in plastic processing and some guidelines on the design of plastic parts are presented here. Further extensive and detailed guidelines are included in Chap. 9.

It should be mentioned that the information presented at this point applies broadly to all classes of plastics and types of processing. Most plastic suppliers will provide very specific data and guidelines for their individual products. This invaluable source of guidance is too often unused. It is strongly recommended that plastic suppliers be more fully utilized for product-design guidance. However, the information presented at this point will be valuable for making initial design and process decisions.

Table 1.1[5] explains the major ways in which plastic materials can be formed into parts, and the advantages, limitations, and relative cost of each processing method. In general, a plastic part is produced by a combination of cooling, heating, flowing, deformation, and chemical reaction. As noted previously, the processes differ, depending on whether the material is a thermoplastic or a thermoset.

The usual sequence of processing a thermoplastic is to heat the material so that it softens and flows, force the material in the desired shape through a die or in a mold, and chill the melt into its final shape. By comparison, a thermoset is typically processed by starting out with partially polymerized material, which is softened and activated by heating (either in or out of the mold), forcing it into the desired shape by pressure, and holding it at the curing temperature until final polymerization reaches the point where the part hardens and stiffens sufficiently to keep its shape when demolded.

The cost of the finished part depends on the material and the process used. A very rough estimate of the finished cost of a part can be obtained by multiplying the material cost by a factor ranging from 1.5 to 10. The cost factors shown in Table 1.1 are based on general industry experience.

Table 1.2 gives guidelines on part design for the various plastic-processing methods listed in Table 1.1. The design of a part frequently depends on the processing

TABLE 1.1 Descriptions and Guidelines for Plastic-Processing Methods[5]

Process	Description	Key advantages	Notable limitations	Cost factor*
Blow molding.........	An extruded tube (parison) of heated thermoplastic is placed between two halves of an open split mold and expanded against the sides of the closed mold by air pressure. The mold is open, and the part is ejected	Low tool and die costs; rapid production rates; ability to mold relatively complex hollow shapes in one piece	Limited to hollow or tubular parts; wall thickness and tolerances often hard to control	1.5–5, 2–3
Calendering.........	Doughlike thermoplastic mass is worked into a sheet of uniform thickness by passing it through and over a series of heated or cooled rolls. Calenders also are used to apply plastic covering to the back of other materials	Low cost; sheet materials are virtually free of molded-in stresses; i.e., they are isotropic	Limited to sheet materials; very thin films not possible	1.5–3, 2–5.5
Casting.........	Liquid plastic (usually thermoset except for acrylics) is poured into a mold (without pressure), cured, and removed from the mold. Cast thermoplastic films are made by depositing the material, either in solution or in hot-melt form, against a highly polished supporting surface	Low mold cost; ability to produce large parts with thick cross sections; good surface finish; suitable to low-volume production	Limited to relatively simple shapes; except for cast films, becomes uneconomical at high-volume production levels; most thermoplastics not suitable	1.5–3, 2–2.5
Compression molding.........	A thermoplastic or partially polymerized thermosetting resin compound, usually preformed, is placed in a heated mold cavity; the mold is closed, heat and pressure are applied, and the material flows and fills the mold cavity. Heat completes polymerization, and the mold is opened to remove the part. The process is sometimes used for thermoplastics, e.g., vinyl phonograph records	Little waste of material and low finishing costs; large, bulky parts are possible	Extremely intricate parts involving undercuts, side draws, small holes, delicate inserts, etc., not practical; very close tolerances difficult to produce	2–10, 1.5–3
Cold forming.........	Similar to compression molding in that material is charged into split mold; it differs in that it uses no heat—only pressure. Part is cured in an oven in a separate operation. Some thermoplastic sheet material and billets are cold-formed in process similar to drop hammer–die forming of metals. Shotgun shells are made in this manner from polyethylene billets	Ability to form heavy or tough-to-mold materials; simple; inexpensive; often has rapid production rate	Limited to relatively simple shapes; few materials can be processed in this manner	

1.6

Process		Characteristics	Limitations	
Extrusion	Thermoplastic or thermoset molding compound is fed from a hopper to a screw and barrel where it is heated to plasticity and then forwarded, usually by a rotating screw, through a nozzle having the desired cross-section configuration	Low tool cost; great many complex profile shapes possible; very rapid production rates; can apply coatings or jacketing to core materials, such as wire	Limited to sections of uniform cross section	2–5, 3–4
Filament winding	Continuous filaments, usually glass, in form of rovings are saturated with resin and machine-wound onto mandrels having shape of desired finished part. Once winding is completed, part and mandrel are placed in oven for curing. Mandrel is then removed through porthole at end of wound part	High-strength fiber reinforcements are oriented precisely in direction where strength is needed; exceptional strength/weight ratio; good uniformity of resin distribution in finished part	Limited to shapes of positive curvature; openings and holes reduce strength	5–10, 6–8
Injection molding	Thermoplastic or thermoset molding compound is heated to plasticity in cylinder at controlled temperature; then forced under pressure through a nozzle into sprues, runners, gates, and cavities of mold. The resin solidifies rapidly, the mold is opened and the part(s) ejected. In modified version of process—runnerless molding—the runners are part of mold cavity	Extremely rapid production rates, hence low cost per part; little finishing required; good dimensional accuracy; ability to produce relatively large, complex shapes; very good surface finish	High initial tool and die costs; not practical for small runs	1.5–5, 2–3
Laminating, high pressure	Material, usually in form of reinforcing cloth, paper, foil, etc., preimpregnated or coated with thermoset resin (sometimes a thermoplastic), is molded under pressure greater than 1,000 lb/in.^{2} into sheet, rod, tube, or other simple shape	Excellent dimensional stability of finished product; very economical in large production of parts	High tool and die costs; limited to simple shapes and cross-section profiles	2–5, 3–4
Matched-die molding	A variation of conventional compression molding, this process uses two metal molds having a close-fitting, telescoping area to seal in the plastic compound being molded and to trim the reinforcement. The reinforcement, usually mat or preform, is positioned in the mold, and the mold is closed and heated (pressures generally vary between 150 and 400 lb/in.^{2}). Mold is then opened and part lifted out	Rapid production rates; good quality and reproducibility of parts	High mold and equipment costs; parts often require extensive surface finishing, e.g., sanding	2–5, 3–4

TABLE 1.1 Descriptions and Guidelines for Plastic-Processing Methods[5] (*Continued*)

Process	Description	Key advantages	Notable limitations	Cost factor*
Rotational molding	A predetermined amount of powdered or liquid thermoplastic or thermoset material is poured into mold. Mold is closed, heated, and rotated in the axis of two planes until contents have fused to inner walls of mold. The mold is opened and part removed	Low mold cost; large hollow parts in one piece can be produced; molded parts are essentially isotropic in nature	Limited to hollow parts; in general, production rates are slow	1.5–5, 2–3
Slush molding	Powdered or liquid thermoplastic material is poured into a mold to capacity. Mold is closed and heated for a predetermined time to achieve a specified buildup of partially cured material on mold walls. Mold is opened, and unpolymerized material is poured out. Semifused part is removed from mold and fully polymerized in oven	Very low mold costs; very economical for small-production runs	Limited to hollow parts; production rates are very slow; limited choice of materials that can be processed	1.5–4, 2–3
Thermoforming	Heat-softened thermoplastic sheet is placed over male or female mold. Air is evacuated from between sheet and mold, causing sheet to conform to contour of mold. There are many variations, including vacuum snapback, plug assist, drape forming, etc.	Tooling costs generally are low; produces large parts with thin sections; often economical for limited production of parts	In general, limited to parts of simple configuration; limited number of materials to choose from; high scrap	2–10, 3–5
Transfer molding	Thermoset molding compound is fed from hopper into a transfer chamber where it is heated to plasticity. It is then fed by means of a plunger through sprues, runners, and gates of closed mold into mold cavity. Mold is opened and the part ejected	Good dimensional accuracy; rapid production rate; very intricate parts can be produced	Molds are expensive; high material loss in sprues and runners; size of parts is somewhat limited	1.5–5, 2–3
Wet lay-up or contact molding	Number of layers, consisting of a mixture of reinforcement (usually glass cloth) and resin (thermosetting), are placed in mold and contoured by roller to mold's shape. Assembly is allowed to cure (usually in an oven) without application of pressure. In modification of process, called spray molding, resin systems and chopped fibers are sprayed simultaneously from spray gun against mold surface; roller assist also is used. Wet lay-up parts sometimes are cured under pressure, using vacuum bag, pressure bag, or autoclave	Very low cost; large parts can be produced; suitable for low-volume production of parts	Not economical for large-volume production; uniformity of resin distribution very difficult to control; mainly limited to simple shapes	1.5–4, 2–3

* Material cost × factor = purchase price of a part: top figure is overall range, bottom is probable average cost.

TABLE 1.2 Guidelines on Part Design for Plastic-Processing Methods[3]

Design rules	Blow molding	Casting	Compression molding	Extrusion	Injection molding	Reinforced plastic molding			Rotational molding	Thermoforming	Transfer molding
						Wet lay-up (contact molding)	Matched-die molding	Filament winding			
Major shape characteristics	Hollow bodies	Simple configurations	Moldable in one plane	Constant cross-section profile	Few limitations	Moldable in one plane	Moldable in one plane	Structure with surfaces of revolution	Hollow bodies	Moldable in one plane	Simple configurations
Limiting size factor	M	M	ME	M	ME	MS	ME	WE	M	M	ME
Min inside radius, in	0.125	0.01–0.125	0.125	0.01–0.125	0.01–0.125	0.25	0.06	0.125	0.01–0.125	0.125	0.01–0.125
Undercuts	Yes	Yes[a]	NR[b]	Yes	Yes[a]	Yes	NR	NR	Yes[c]	Yes[a]	NR
Min draft, degrees	0	0–1	>1	NA[b]	<1	0	1	2–3	1		1
Min thickness, in	0.01	0.01–0.125	0.01–0.125	0.001	0.015	0.06	0.03	0.015	0.02	0.002	0.01–0.125
Max thickness buildup, in	>0.25	None	0.5	6	1	0.5	1	3	0.5	3	1
Inserts	NA	2–1	2–1	NA	2–1	2–1	2–1[d]	NR	NA	NA	2–1
Built-up cores	Yes	Yes	Yes	Yes	Yes	Yes	Yes	Yes	Yes	NR	Yes
Molded-in holes	Yes	Yes	No	Yes	Yes	Yes	Yes	Yes	Yes	Yes	Yes
Bosses	Yes	Yes	Yes	Yes[e]	Yes	Yes	Yes	No	Yes	No	Yes
Fins or ribs	Yes	Yes	Yes	Yes	Yes	Yes	No[f]	No[g]	Yes	Yes	Yes
Molded-in designs and nos	Yes	Yes	Yes	No	Yes	Yes	Yes	No	Yes	Yes	Yes
Overall dimensional tolerance, in./in.	±0.01	±0.001	±0.001	±0.005	±0.001	±0.02	±0.005	±0.005	±0.01	±0.01	±0.001
Surface finish[h]	1–2	2	1–2	1–2	1	4–5	4–5	5	2–3	1–3	1–2
Threads	Yes	Yes	Yes	No	Yes	No	No	No	Yes	No	Yes

M = material. ME = molding equipment. MS = mold size. WE = winding equipment.
[a] Special molds required.
[b] NR—not recommended; NA—not applicable.
[c] Only with flexible materials.
[d] Using premix: as desired.
[e] Only in direction of extrusion.
[f] Using premix: yes.
[g] Possible using special techniques.
[h] Rated 1 to 5:1 = very smooth, 5 = rough.

method selected to make the part. Also, of course, selection of the best processing method frequently is a function of the part design. Major plastic-processing methods and their respective design capabilities, such as minimum section thicknesses and radii and overall dimensional tolerances, are listed in Table 1.2. The basic purpose of this guide is to show the fundamental design limits of the many plastic-processing methods.

1.4.1 Plastic-Fabrication Processes and Forms

There are many plastic-fabrication processes, and a wide variety of plastics can be processed by each of these processes or techniques. Fabrication processes can be broadly divided into pressure processes and pressureless or low-pressure processes. Pressureless or low-pressure processes such as potting, casting, impregnating, encapsulating, and coating are reviewed in Chap. 4. Pressure processes are usually either thermoplastic-materials processes (such as injection molding, extrusion, and thermoforming) or thermosetting processes (such as compression molding, transfer molding, and laminating). There are exceptions to each, however, as mentioned later.

Compression Molding and Transfer Molding. Compression molding and transfer molding are the two major processes used for forming molded parts from thermosetting raw materials. The two can be carried out in the same type of molding press, but different types of molds are used. The thermosetting materials are normally molded by the compression or transfer process, but it is also possible to mold thermoplastics by these processes since the heated thermoplastics will flow to conform to the mold-cavity shape under suitable pressure. These processes are usually impractical for thermoplastic molding, however, since after the mold cavity is filled to its final shape, the heated mold would have to be cooled to solidify the thermoplastic part. Since repeated heating and cooling of this large mass of metal and the resultant long cycle time per part produced are both objectionable, injection molding is commonly used to process thermoplastics.

  Compression Molding. In compression molding, which is shown schematically in Fig. 1.3, the open mold is placed between the heated platens of the molding press, filled with a given quantity of molding material, and closed under pressure, causing the material to flow into the shape of the mold cavity. The actual pressure required depends on the molding material being used and the geometry of the mold. The mold is kept closed until the plastic material is suitably cured. Then the mold is

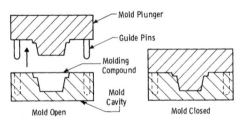

FIGURE 1.3 Simplified illustration of compression-molding process.

opened, the part ejected, and the cycle repeated. The mold is usually made of steel with a polished or plated cavity.

The simplest form of compression molding involves the use of a separate self-contained mold or die that is designed for manual handling by the operator. It is loaded on the bench, capped, placed in the press, closed, cured, and then removed for opening under an arbor press. The same mold in most instances (and with some structural modifications) can be mounted permanently into the press and opened and closed as the press itself opens and closes. The press must have a positive up-and-down movement under pressure instead of the usual gravity drop found in the standard hand press.

Transfer Molding. The transfer-molding sequence is shown in Fig. 1.4. The molding material is first placed in a heated pot, separate from the mold cavity. The hot plastic material is then transferred under pressure from the pot through the runners into the closed cavity of the mold.

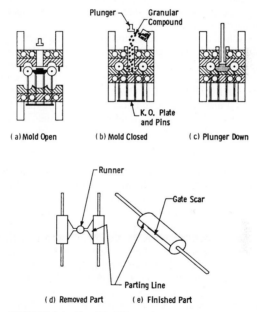

FIGURE 1.4 Simplified illustration of transfer-molding process.

The advantage of transfer molding lies in the fact that the mold proper is closed at the time the material enters. Parting lines that might give trouble in finishing are held to a minimum. Inserts are positioned and delicate steel parts of the mold are not subject to movement. Vertical dimensions are more stable than in straight compression. Also, delicate inserts can often be molded by transfer molding, especially with the low-pressure molding compounds.

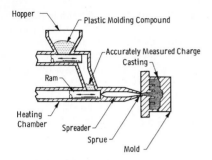

FIGURE 1.5 Simplified illustration of injection-molding process.

Injection Molding. Injection molding, shown schematically in Fig. 1.5, is the most practical process for molding thermoplastic materials. The operating principle is simple, but the equipment is not.

A material with thermoplastic qualities—one that is viscous at some elevated temperature and stable at room temperature without appreciable deterioration during the cycle—is maintained in a heated reservoir. This hot, soft material is forced from the reservoir into a cool mold. The mold is opened as soon as the material has cooled enough to hold its shape on demolding. The cycle speed is determined by the rapidity with which the temperature of the material used can be reduced, which in turn depends on the thermal conductivity of that material. Acrylics are slow performers, and styrenes are among the fastest.

The machine itself is usually a horizontal cylinder, the bore of which determines the capacity. Within the bore is a piston, that, when retracted, opens a hole in the top of the cylinder through which new material can be added to replace the charge shot into the mold. The cylinder is heated by electric bands that permit temperature variation along its length. Inside the exit end of the cylinder is a torpedo over which the hot material is forced just before coming out of the nozzle into the channels leading to the cavities. This gives the material a final churning and ensures thorough heating. The mold opens and closes automatically, and the whole cycle is controlled by timers.

Thermoset Injection Molding. Because of the chemical nature of the plastic materials, injection molding has traditionally been the primary molding method for thermoplastics, and compression and transfer molding have been the primary molding methods for thermosetting plastics. Because of the greater molding cycle speeds and lower molding costs in injection molding, thermoplastics have had a substantial molding cost advantage over thermosets. As a result, advances in equipment and in thermosetting molding compounds have resulted in a rapid transition to screw-injection, in-line molding. This has been especially prominent with phenolics, but other thermosets are also included to varying degrees. The growth in screw-injection molding of phenolics has been extremely rapid. The development of this technique allows the molder to automate further, reduce labor costs, improve quality, reduce rejects, and gain substantial overall molding cycle efficiency.

Extrusion and Pultrusion. The process of extrusion consists basically of forcing heated, melted plastic continuously through a die that has an opening shaped to produce a desired finished cross section. Normally it is used for processing thermoplastic materials, but it can also be used for processing thermosetting materials. The main application of extrusion is the production of continuous lengths of film, sheeting, pipe, filaments, wire jacketing, and other useful forms and cross sections. After the plastic melt has been extruded through the die, the extruded material is hardened by cooling, usually by air or water.

Extruded thermosetting materials are used increasingly in wire and cable coverings. The main object here is the production of shapes, parts, and tolerances not obtainable in compression or transfer molding. Pultrusion is a special, increasingly used technique for pulling resin-soaked fibers through an orifice, at it offers significant strength improvements. Any thermoset, B-stage, granular molding compound can be extruded, and almost any type of filler may be added to the compound. In fiber-filled compounds, the length of fiber is limited only by the cross-sectional thickness of the extruded piece.

A metered volume of molding compound is fed into the die feed zone, where it is slightly warmed. As the ram forces the compound through the die, the compound is heated gradually until it becomes semifluid. Before leaving the die, the extruded part is cured by controlling the time it takes to travel through a zone of increasing temperature. The cured material exits from the die at temperatures of 300 to 350°F and at variable rates.

Thermoforming.   Thermoforming is a relatively simple basic process that consists of heating a plastic sheet and forming it to conform to the shape of the mold, either by differential air pressure or by some mechanical means. By this processing technique, thermoplastic sheets can be converted rapidly and efficiently to a myriad of shapes, the thicknesses of which depend on the thickness of the film being used and the processing details of the individual operations. Although there are many variations of this process, they generally involve heating the plastic sheet and making it conform to the contour of a male or female form, either by air pressure or a matching set of male and female molds.

1.5 PLASTIC PROPERTIES

Much work has been done on the standardization of the measurements for properties of plastics and the analysis of plastic properties. An understanding of plastic performance, as indicated by standard tests, is especially important to the large percentage of nonchemically trained users of plastic materials. While more detailed presentations are made in subsequent chapters for special areas such as mechanical design and electrical design, and Chap. 8 covers in detail the types and sources of specifications and standards, some fundamental, general-use data and guidelines are presented at this point.

1.5.1 Standard Tests and Their Significance

Among the most widely used test procedures are those developed by the American Society for Testing and Materials (ASTM). These test procedures, divided into categories of performance, chemical, mechanical, thermal, analytical, optical, and electrical, are listed in Table 1.3. The significance of each test is given in Refs. 6 and 7. A cross reference of some important ASTM tests and federal test methods is presented in Table 1.4.[8]

1.5.2 Rating of Plastics by Property Comparisons

Subsequent sections of this chapter will present data on plastics as a function of the most important variables. It is frequently useful, however, to compare or rate plastics for a given property or characteristic. Any such data must be considered approximate, of course, because of the many possible variables involved. Some such comparative data approximations for electrical and mechanical characteristics are

TABLE 1.3 Widely Used ASTM Tests for Plastics[6,7]

Performance tests
    D1693, environmental stress cracking
    G23 (formerly E42), accelerated weathering
    Pipe tests (Commercial Standard CS 255-63)
    D794, permanent effect of heat
    D1435, outdoor weathering
    Weight loss on heating (D706, specification for cellulose acetate molding compounds)
    D570, water absorption

Chemical tests
    C619, chemical resistance of asbestos-fiber-reinforced thermosetting resins
    C581, chemical resistance of thermosetting resins of color in CIE 1931 system

Mechanical tests
    D790, flexural properties
    D1822, tensile impact
    D747, stiffness in flexure
    D256, Izod impact
    D638, tensile properties
    D785, Rockwell hardness
    D621, deformation under load
    D695, compressive properties of rigid plastics
    D732, shear strength

Thermal tests
    D648, deflection temperature
    D635, flammability (for self-supporting materials)
    D1238, flow rate (melt index) by extrusion plastometer
    D569, flow properties
    D1525, Vicat softening point
    D746, brittleness temperature

Analytical tests
    D792, specific gravity and density
    D1505, density by density-gradient technique

Optical tests
    E308, spectrophotometry and description
    D1003, haze

Electrical tests
    D618, conditioning procedures
    D495, arc resistance
    D149, dielectric strength
    D150, dielectric constant and dissipation factor
    D257, tests for electrical resistance, insulation resistance, volume resistivity, volume
        resistance

---

presented in Chaps. 2 and 3 of the *Handbook of Plastics, Elastomers, and Composites,* 2d edition. Data on some other important characteristics (specific gravity, water absorption, hardness, thermal conductivity, coefficient of expansion, and refractive index) are presented in Harper[8] and Grafton.[9] The relationship of the different hardness scales is illustrated in Fig. 1.6. Frequently it is desirable to compare plastics on a comparative cost basis for various property requirements. Such data are presented in Table 1.5[10] for a number of major plastics. However, due to events in the Persian Gulf crisis, volatile oil prices make actual prices subject to wide fluctuations depending on the spot price of oil.

TABLE 1.4 Cross Reference of ASTM and Federal Tests[8]

Test method	Federal Standard Method No.	ASTM No.
Abrasion wear (loss in weight)	1091	
Accelerated service tests (temperature and humidity extremes)	6011	D 756-56
Acetone extraction test for degree of cure of phenolics	7021	D 494-46
Arc resistance	4011	D 495-61
Bearing strength	1051	D 953-54, Method A
Bonding strength	1111	D 229-63T, pars 40–43
Brittleness temperature of plastics by impact	2051	D 746-64T
Compressive properties of rigid plastics	1021	D 695-63T
Constant-strain flexural fatigue strength	1061	
Constant-stress flexural fatigue strength	1062	
Deflection temperature under load	2011	D 648-56, Procedure 6(a)
Deformation under load	1101	D 621-64, Method A
Dielectric breakdown voltage and dielectric strength	4031	D 149-64
Dissipation factor and dielectric constant	4021	D 150-64T
Drying test (for weight loss)	7041	
Effect of hot hydrocarbons on surface stability	6062	
Electrical insulation resistance of plastic films and sheets	4052	
Electrical resistance (insulation, volume, surface)	4041	D 257-61
Falling-ball impact	1074	
Flame resistance	2023	
Flammability of plastics 0.050 in. and under in thickness	2022	D 568-61
Flammability of plastics over 0.050 in. in thickness	2021	D 635-63
Flexural properties of plastics	1031	D 790-63
Indentation hardness of rigid plastics by means of a durometer	1083	D 1706-61
Interlaminar and secondary bond shear strength of structural plastic laminates	1042	
Internal stress in plastic sheets	6052	
Izod impact strength	1071	D 256-56, Method A
Linear thermal expansion (fused-quartz tube method)	2031	D 696-44
Machinability	5041	
Mar resistance	1093	D 673-44
Mildew resistance of plastics, mixed culture method, agar medium	6091	
Porosity	5021	
Punching quality of phenolic laminated sheets	5031	D 617-44
Resistance of plastics to artificial weathering using fluorescent sunlamp and fog chamber	6024	D 1501-57T
Resistance of plastics to chemical reagents	7011	D 543-60T
Rockwell indentation hardness test	1081	D 785-62, Method A
Salt-spray test	6071	
Shear strength (double shear)	1041	
Shockproofness	1072	
Specific gravity by displacement of water	5011	
Specific gravity from weight and volume measurements	5012	
Surface abrasion	1092	D 1044-56
Tear resistance of film and sheeting	1121	D 1004-61
Tensile properties of plastics	1011	D 638-64T
Tensile properties of thin plastic sheets and films	1013	D 882-64T
Tensile strength of molded electrical insulating materials	1012	D 651-48
Tensile time—fracture and creep	1063	
Thermal-expansion test (strip method)	2032	
Warpage of sheet plastics	6054	D 1181-56
Water absorption of plastics	7031	D 570-63

TABLE 1.5 Properties and Prices per Unit Property of Selected Plastics[10]

Material	Price, cents/lb	D[a]	TM[b]	TYS[c]	FYS[d]	CS[e]	IS[f]	VR[g]	Cents/in.^{3}	Cents/TM[h]	Cents/TYS[i]	Cents/FYS[k]	Cents/CS[l]	Cents/IS[m]	Cents/VR[n]
LDPE	13.25	0.924		1.7					0.45		2.64				
HDPE	14.0	0.960	1.55	4.4			1.0		0.49	3.16	1.11			0.49	1.90
PE, GR[o]	41.0	1.28	9.00	11.0	11.0	6.0	3.5	1	1.90	2.11	1.73	1.73	3.16	1.15	0.69
Polypropylene (PP)	21.0	0.902		5.0			0.6	1	0.69		1.38			0.43	
PP copolymer	28.0	0.897	0.90	3.3			2.1		0.91	10.10	2.75			0.78	
PP, GR	71.0	1.22	9.00	9.0	11.0	7.0	4.0	1	3.14	3.49	3.49	2.85	4.49	0.78	3.14
Ionomer	45.0	0.940	0.50	2.0			14.0		1.53	30.60	7.64			0.11	
Poly(4-methylpentene-1)	105.0	0.83	2.10	4.0	9.5		0.8	1	3.15	15.00	7.89	3.31		3.94	3.15
Polystyrene (PS)	14.8	1.05	4.50	6.5	17.0		0.4	10	0.56	1.24	0.86	0.59		1.40	0.06
PS, GR	54.0	1.29	13.00	14.0	17.5	19.0	2.5	1	2.52	1.94	1.80	1.48	1.33	1.01	2.52
SAN	26.0	1.03	5.40	11.0	15.6		0.5		0.97	1.79	0.88	0.55		1.94	9.70
SAN, high-temp	23.0	1.08	5.00	9.8	16.5		0.5	0.1	0.90	1.80	0.92	0.58		1.80	
SAN, GR	51.0	1.35	18.00	18.0	22.0	23.0	3.0	1	2.49	1.38	1.38	1.13	1.18	0.83	2.49
Impact styrene	17.8	1.05	4.2	4.2	6.2		1.8	1	0.68	1.62	1.62	1.10		0.38	0.68
Impact styrene, GR	56.0	1.32	13.00	14.5	17.0	16.0	2.5	1	2.68	2.06	1.85	1.58	1.68	1.07	2.68
ABS	28.0	1.07	4.10	7.9	12.0	13.0	1.1	3.7	1.08	2.64	1.37	0.90	0.83	0.98	0.29
ABS, high-impact	38.0	1.04	3.10	5.9	9.6	7.8	6.0		1.43	4.62	2.43	1.49	1.83	0.24	
ABS, transparent	50.0	1.07	2.90	5.6	9.9	7.0	5.3		1.97	6.80	3.52	1.99	2.82	0.37	
ABS, GR	85.0	1.36	10.00	16.0	25.0	17.0	2.4		4.19	4.19	2.62	1.68	2.46	1.74	
ABS/PVC polyblend	40.0	1.21	3.30	6.0	10.2	7.4	12.5		1.75	5.30	2.92	1.72	2.37	0.14	
ABS/PC polyblend	55.0	1.12	3.70	8.2	14.3	11.7	11.3		2.23	6.04	2.72	1.56	1.91	0.19	
Acrylonitrile/styrene/acrylic	42.0	1.07	3.30	6.3	10.0		1.6		1.63	4.94	2.59	1.63		1.02	
Rigid PVC	36.5	1.35	4.07	8.3	11.0		0.4		1.79	4.40	2.16	1.62		4.48	
Rigid PVC, high-impact	40.0	1.35	3.65	5.7			15.0		1.95	5.33	3.42			0.13	
PVC/acrylic polyblend	50.0	1.30	2.75	5.0	8.7	6.2	12.0		2.35	8.55	4.70	2.70	3.79	0.20	
PTFE	325.0	2.20		3.0			3.0	100	25.90		86.40			8.64	0.26
PCTFE	490.0	2.13	1.80	6.0	9.0	7.4	NB[p]	10	37.80	210.00	63.00	42.00	51.00	NB	3.78
PMMA	45.5	1.18	4.00	9.5	15.0	18.0	0.3	0.01	1.95	4.88	2.05	1.30	1.53	6.50	195.00
PMMA, high-temp	65.5	1.16	4.65	10.0	12.0	18.0	0.3		2.75	5.92	2.75	2.29	1.53	9.20	
Phenylene oxide-based resin	59.0	1.06	3.55	9.6	13.5	16.4	5.0	10	2.68	7.54	2.79	1.98	1.63	1.49	0.27
Phenylene oxide-based resin, GR	80.0	1.27	12.00	17.0	20.0	13.9	1.4	10	3.91	3.26	2.30	1.95	3.24	2.79	0.39
Polysulfone	100.0	1.24	3.60	10.2	15.4	13.9	1.3	5	4.50	12.50	4.41	2.92	3.24	3.46	0.90
Polysulfone/ABS polyblend	85.0	1.14	2.90	7.5	11.0	21.0	8.0	1.5	3.51	12.10	4.69	3.19	3.55	0.44	2.34
Polysulfone, GR	149.0	1.38	15.00	19.0	23.5	21.0	2.5		7.46	4.97	3.92	3.17	3.55	2.98	
Polysulfone, high-temp	2,500.0	1.36		13.0	17.2	17.9			123.00		94.80	71.50	68.80		

1.16

Material	Cost	a	b	c	d	e	f	g	h	i	j	k	l	m	n
Acetal	65.0	1.42	5.20	10.0	14.1	18.0	1.4	0.1	3.34	6.44	3.34	2.35	1.86	2.39	33.40
Acetal copolymer	65.0	1.41	4.10	8.8		16.0	1.2	0.01	3.32	8.10	3.78		2.07	2.76	332.00
Acetal copolymer, high-impact	65.0	1.41	4.10	8.8	13.0	16.0	4.7	0.01	3.32	8.10	3.78	2.55	2.07	0.71	332.00
Acetal, GR	128.0	1.69	8.00	12.5	16.0	12.0	3.0	1	7.85	9.80	6.28	4.90	6.54	2.61	7.85
Cellulose acetate	52.0	1.30		6.4	10.0		1.2	0.001	2.44		3.81	2.18		2.04	2,440.00
Cellulose propionate	63.0	1.23		7.0	6.9		0.8	0.01	2.81		4.02	2.70		3.51	281.00
Cellulose acetate butyrate	62.0	1.20	3.50	4.5	12.5	10.5	4.4	0.01	2.70		6.00	3.95	3.09	0.61	270.00
Polycarbonate (PC)	75.0	1.20	3.43	9.0	14.2	13.3	13.0	1	3.25	9.30	3.61	2.60	5.51	0.25	3.25
PC, nonburning	150.0	1.35		9.9	25.0	19.0	1.0	3	7.34	21.40	7.42	5.17	3.48	7.34	2.45
PC, GR	120.0	1.52	17.00	20.0	12.0	12.0	3.5	0.15	6.61	3.89	3.30	2.64	2.68	1.89	44.00
Thermoplastic polyester	68.0	1.31		8.2	10.0	30.0	1.0	1	3.22		3.93	2.68	0.38	3.22	3.22
Phenolic, general-purpose	23.5	1.36	12.00	7.0	14.0	31.0	0.28	0.0001	1.15	0.96	1.64	1.15	2.49	4.11	11,500.00
DAP	125.0	1.71	14.00	8.0	17.5	25.0	0.50	0.01	7.72	5.51	9.65	5.51	1.25	15.44	772.00
Alkyl molding compound	39.0	2.21	20.00	7.0	7.0	7.0	3.0	0.0006	3.12	1.56	4.45	1.78	17.60	1.04	5,200.00
Polyurethane, GR	222.0	1.53	7.50	10.0	18.0		10.0	0.1	12.30	16.40	12.30	17.60		1.23	30.70
Nylon-6	75.0	1.13	4.40	11.8	16.5		0.9	0.01	3.07	6.98	2.60	1.70		3.41	310.00
Nylon-6/6	75.0	1.14		12.0	35.0	24.0	1.0	0.1	3.10		2.58	1.88	2.22	3.10	53.50
Nylon-6/6, GR	100.0	1.47	20.00	30.0		41.9	3.4	0.01	5.32	2.66	1.78	1.52	6.26	1.57	465.00
Nylon-6/10	120.0	1.07		8.5		34.0	0.6	0.92	4.65		5.48		0.53	7.75	28.50
Polyimide, GR	380.0	1.90	45.00	28.0	56.0	42.0	17.0		26.20	5.83	9.37	4.69	0.54	1.54	36,200.00
Urea formaldehyde	32.0	1.56	14.00	6.0	15.0		0.25	0.00005	1.81	1.29	3.02	1.21		7.24	16,300.00
Melamine formaldehyde	42.0	1.50	13.50	7.5	12.0		0.26	0.00014	2.28	1.69	3.04	1.90		8.78	
Cast steel	18.0	7.77	300.00	128.0					5.07	0.17	0.40				
Stainless steel	40.0	7.69	290.00	60.0					11.15	0.39	1.86				
Aluminum	50.0	2.67	102.00	13.0					4.85	0.48	3.73				
Magnesium	100.0	1.76	65.00	28.0					6.37	0.98	2.28				
Zinc alloy	30.00	7.09		25.0					7.70		3.08				
Copper	95.0	8.77	170.00	10.0					30.16	1.77	30.16				

a Density, g/cm^3.

b Tensile modulus, 10^6 lb/in^2.

c Tensile yield strength, 10^3 lb/in^2.

d Flexural yield strength, 10^3 lb/in^2.

e Compressive strength, 10^3 lb/in^2.

f Impact strength, ft-lb/in.

g Volume resistivity, 10^{16} Ω-cm.

h Tensile modulus, 10^{-6} cents/in^3 × in^2/lb.

i Tensile yield strength, 10^{-4} cents/in^3 × in^2/lb.

j Flexural yield strength, 10^{-4} cents/in^3 × in^2/lb.

k Compressive strength, 10^{-4} cents/in^3 × in^2/lb.

l Impact strength, cents/in.^{3} × in/ft-lb.

m Volume resistivity, 10^{-16} cents/in^3 Ω-cm.

o GR = glass-fiber-reinforced.

p NB = no break.

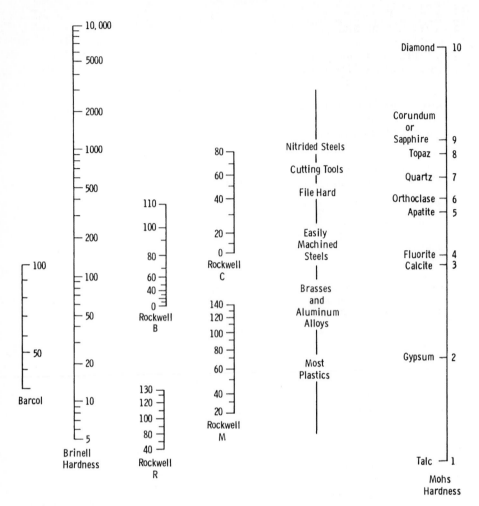

FIGURE 1.6   Comparison of hardness scales.

1.5.3   Flammability and Flame Retardancy of Plastics

Plastic usage is increasing both in volume and in new markets, and the issue of flammability is becoming a key requirement for continued growth in many areas, such as home furnishings, clothing, wire and cable, automobiles, and aircraft. Some major fires have been partially attributed to plastics,[11] and numerous deaths have been attributed to toxic smoke generated by burning plastics.

Strong feelings have been expressed that small-scale laboratory tests do not accurately predict results for large-scale fire tests.[12] Likewise some investigators feel that many descriptive terms for flammability rating, such as self-extinguishing (SE), tend to give the user an unjustified sense of security.[13,14]

In order to get products approved, a number of companies have been created that will measure flame test results to specific standards. While large-scale tests have been and are being developed, small-scale tests continue to be used widely. The available test methods are manifold and have been created by numerous sources for many applications.

A summary of some of the major flammability tests is given in Table 1.6.[15] Several updated tests aimed specifically at plastics in aircraft passenger compartments are described in ASTM E906[16] and the Federal Register.[17]

Flame Retardants. When plastics burn, heat from an external source pyrolyzes the solid plastic to produce gases and liquids that act as fuel for the fire. Fire-retardant chemicals affect the burning rate of a solid in one or more ways[18]:

1. By interfering with the combustion reactions
2. By making the products of pyrolysis less flammable
3. By reducing the transfer of heat from the flame to the solid
4. By reducing the rate of diffusion of pyrolysis products to the flame front

Since various plastics burn differently and often at different temperatures and rates, there is no single universal fire retardant. Almost every application requires either a different agent or different amounts of agent to obtain the desired flame retardancy with minimum effect on other properties.

A number of elements can act as flame retardants, but they must be incorporated in a structure that enables them to become active at the proper temperature. They include nitrogen, phosphorus, arsenic, antimony, bismuth, fluorine, bromine, chlorine, iodine, and boron. Of these, phosphorus, bromine, chlorine, and antimony are currently considered to be the most efficient.[18]

Table 1.7 shows the levels at which these elements are used as flame retardants in common polymers. Notice that in many cases several of these elements are used in

TABLE 1.6   Summary of Some of the Major Flammability Tests[15]

Test	Material
Ignition	
ASTM D2863, oxygen index test	All
UL hot-wire ignition test	Plastics
UL high-current arc ignition test	Plastics
UL high-voltage arc ignition test	Plastics
ASTM D2859, methenamine pill test	Carpets and floor coverings
Flame propagation	
ASTM D635, FTM 2021	Plastics sheet
ASTM E84, UL 723, NFPA 255, 25-ft tunnel test	All
UL 94, test for self-extinguishing polymers	Plastics sheet
MVSS302, horizontal burn test	All
FAA vertical test	All
FAA horizontal test	All
Fire endurance	
ASTM E119, UL 263, MFPA 251, fire endurance test	All
Bureau of Mines, flame penetration	Plastics foams
Heat contribution factory mutual calorimeter	All
Smoke generation	
ASTM D2843, Rohm & Hass XP2 smoke density chamber	Plastics
NBS chamber	All

TABLE 1.7 Flame Retardants for Plastics and Typical Composition Ranges[18]

Plastic	Typical flame retardants	Typical percentages of flame-retardant elements for equivalent retardance							
		Br	Cl	P	P + Br	P + Cl	Sb_2O_3 + Br	Sb_2O_3 + Cl	Sb_2O_3
ABS..............	Bromine and chlorine-containing organic additive compounds (such as chlorinated paraffins, brominated biphenyl, phosphate-containing aliphatic and aromatic compounds). Antimony trioxides, hydrated aluminum oxide, and zinc borate with halogenated and phosphate-containing organic compounds. Terpolymerization with halogen-containing monomers such as bis (2,3 dibromo propyl) fumerate	3	23					5 + 7	
Acrylics..........	Halogen- and phosphorus-containing organic compound. PVC to form alloy	16	20	5	1 + 3	2 + 4	7 + 5		
Cellulose acetate, cellulose butyrate, cellulose propionate	Halogenated and/or phosphate-containing organic plasticizers and halogenated compounds (such as brominated biphenyl)	13–15	24	2.5–2.5	1 + 9			12–15 + 9–12	
Epoxies...........	Halogenated and phosphate-containing compounds. Chlorinated brominated bisphenol A. Halogenated anhydrides (like chlorendic anhydride). Antimony trioxide, hydrated aluminum oxide with halogenated and phosphate-containing organic compounds		26–30	5–6	2 + 5	2 + 6		10 + 6	
Phenolic compounds	Halogen and/or phosphorus organic compounds (like chlorinated paraffins). Tris (2,3-dibromo propyl) phosphate. Hydrated aluminum oxide, zinc borate	16	16	6					
Polyamides........	Chlorinated and brominated biphenyls. Antimony trioxide with halogenated compounds	3.5–7	3.5–7	3.5				10 + 6	
Polycarbonate......	Brominated biphenyl, chlorinated paraffin. Tetrabromo bisphenol A. Antimony trioxide with halogenated compounds	4–5	10–15					7 + 7–8	

Polyesters........	Halogenated phosphate-containing plasticizers. Halogenated compounds (like chlorinated paraffin and brominated biphenyl). Halogen-containing intermediates (such as tetrachlorophthalic anhydride, tetrabromophthalic anhydride, and chlorendic anhydride). Antimony trioxide, hydrated aluminum oxide, hydrated zinc borate with halogenated and phosphate-containing organic compounds	12–15	26	5	2 + 6	1 + 15–20	2 + 8–9	1 + 16–18
Polyolefins........	Halogen-containing compounds (such as chlorinated paraffins and brominated biphenyl). Halogenation of polymer as chlorinated polyethylene. Antimony trioxide, hydrated aluminum oxide, zinc borate, or barium metaborate in conjunction with halogenated compounds	20	40	5	2.5 + 7	2.5 + 9	3 + 6	5 + 8
Polystyrene........	Halogen- and phosphate-containing compounds like tris (2,3-dibromo propyl) phosphate and brominated biphenyls. Halogen efficiency as flame retardant is increased by incorporating small amount of free radical initiators such as peroxides. Chemical bonding of halogenated compounds into the polystyrene chain. Antimony trioxide and hydrated aluminum oxide	4–5	10–15		0.2 + 3	0.5 + 5	7 + 7-8	7 + 7–8
Polyurethanes......	Halogenated phosphorus-containing plasticizers where flexibility is not a problem. Phosphorus-containing intermediates (such as phosphate, phosphite, phosphonates, and amino phosphonate polyols). Brominated isocyanates and halogen-containing prepolymers/polyols. Chlorine- and fluorine-containing compounds as blowing agents in foam. Antimony trioxide hydrated aluminum oxide, and zinc borate in conjunction with halogenated compound	12–14	18–20	1.5	0.5 + 4-7	1 + 10–15	2.5 + 2.5	4 + 4
Polyvinyl chloride..	Replace other plasticizer such as DOP, with halogen- and/or phosphorus-containing plasticizer. Antimony trioxides, hydrated aluminum oxide, and zinc borate		40	2–4				5–15

Further references: C. J. Hilado, "Flammability Handbook for Plastics," Technomic Publishers; J. W. Lyons, "The Chemistry and Uses of Fire Retardants," Interscience-Wiley.

1.21

combination to achieve a synergistic effect; that is, the combination is more effective than any one element used at the same level of loading.[18] A current summary on flame retardants and fire testing is given in Ref. 19. Often plastic parts, construction beams, and so on, are protected by use of intumescent materials, which foam and form a heat-resistant char.[20,21]

1.6 THERMOSETTING PLASTICS

Plastic materials included in the thermosetting plastic category and discussed separately in this section are alkyds, diallyl phthalates, epoxies, melamines, phenolics, polyesters, silicones, and ureas. A list of typical trade names and suppliers is given in Table 1.8. In general, unfilled thermosetting plastics tend to be harder, more brittle, and not as tough as thermoplastics. Thus it is common practice to add fillers to thermosetting materials. A wide variety of fillers can be used for varying product properties. For molded products, usually compression or transfer molding, mineral or cellulose fillers are often used as lower-cost, general-purpose fillers, and glass fiber fillers are often used for optimum strength or dimensional stability. There are always product and processing tradeoffs, but a general guide to the application of fillers is given in Seymour.[22] It should be added that filler form and filler surface treatment can also be major variables. Thus it is important to consider fillers along with the thermosetting material, especially for molded products. Other product forms may be filled or unfilled, depending on requirements.

1.6.1 Alkyds

Alkyds are available in granular, rope, and putty form, some suitable for molding at relatively low pressures, and at temperatures in the range of 300 to 400°F. They are formulated from polyester-type resins, in general reactions as shown in Fig. 1.1. Other possible monomers, aside from styrene, are diallyl phthalate and methyl methacrylate. Alkyd compounds are chemically similar to the polyester compounds but make use of higher-viscosity, or dry, monomers. Alkyd compounds often contain glass fiber filler but may include clay, calcium carbonate, or alumina, for example.

These unsaturated resins are produced through the reaction of an organic alcohol with an organic acid. The selection of suitable polyfunctional alcohols and acids permits selection of a large variation of repeating units. Formulating can provide resins that demonstrate a wide range of characteristics involving flexibility, heat resistance, chemical resistance, and electrical properties. Typical properties of alkyds are shown in Harper[1] and Ref. 23.

Alkyds are easy to mold and economical to use. Molding dimensional tolerances can be held to within ±0.001 in/in. Postmolding shrinkage is small, as shown in Fig. 1.7.[23] Their greatest limitation is in extremes of temperature (above 350°F) and humidity. Silicones and diallyl phthalates are superior here, silicones especially with respect to temperature and diallyl phthalates with respect to humidity.

1.6.2 Diallyl Phthalates (Allyls)

Diallyl phthalates, or allyls, are among the best of the thermosetting plastics with respect to high insulation resistance and low electrical losses, which are maintained

TABLE 1.8 Typical Trade Names and Suppliers of Thermosetting Plastics*

Plastic	Typical trade names and suppliers
Alkyd..............................	Plaskon (Allied Chemical) Durez (Hooker Chemical) Glaskyd (American Cyanamid)
Diallyl phthalates.....................	Dapon (FMC) Diall (Allied Chemical) Durez (Hooker Chemical)
Epoxies.............................	Epon (Shell Chemical) Epi-Rez (Celanese) D.E.R. (Dow Chemical) Araldite (Ciba) ERL (Union Carbide)
Melamines...........................	Cymel (American Cyanamid) Plaskon (Allied Chemical)
Phenolics............................	Bakelite (Union Carbide) Durez (Hooker Chemical) Genal (General Electric)
Polybutadienes........................	Dienite (Firestone) Ricon (Colorado Chemical Specialties)
Polyesters...........................	Laminac (American Cyanamid) Paraplex (Rohm and Haas) Selectron (PPG)
Silicones............................	DC (Dow Corning); see also Table 41
Ureas..............................	Plaskon (Allied Chemical) Beetle (American Cyanamid)

* These are but a few typical basic material suppliers. In addition, there are many companies which formulate filled plastic systems from these basic plastics, and which are sometimes the ultimate supplier to the end-product fabricator. See also App. C for expanded listing.

up to 400°F or higher, and in the presence of high-humidity environments. Also, diallyl phthalate resins are easily molded and fabricated.

There are several chemical variations of diallyl phthalate resins, but the two most commonly used are diallyl phthalate (DAP) and diallyl isophthalate (DAIP). The primary application difference is that DAIP will withstand somewhat higher temperatures than will DAP. Typical properties of DAP and DAIP molding compounds are shown in Harper[1] and Ref. 24.

The excellent dimensional stability of DAPs is demonstrated in Fig. 1.8,[25] which compares them to other plastic materials at various temperatures.

DAPs are extremely stable, having very low after-shrinkage, on the order of 0.1 percent, based on the MIL-M-14F dimensional stability test. The ultimate in electrical properties is obtained by the use of the synthetic fiber fillers. However, these materials are expensive and have high mold shrinkage and a strong, flexible flash that is extremely difficult to remove from the parts. Consequently, the largest commercial volume is in the short-glass-fiber-filled materials, which combine moldability in thin sections with extremely high tensile and flexural strengths.

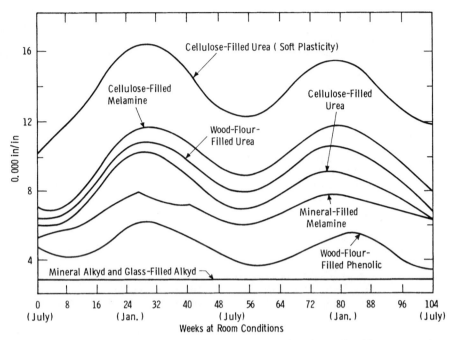

FIGURE 1.7 Stability of alkyds. Postmolding shrinkage variation of several molding compounds over a period of weeks.[23]

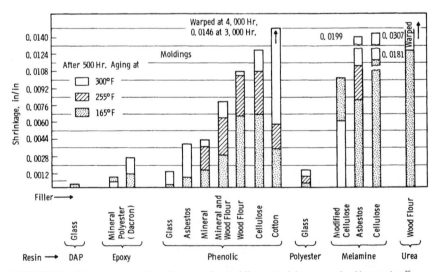

FIGURE 1.8 Shrinkage of various thermosetting molding materials as a result of heat aging.[25]

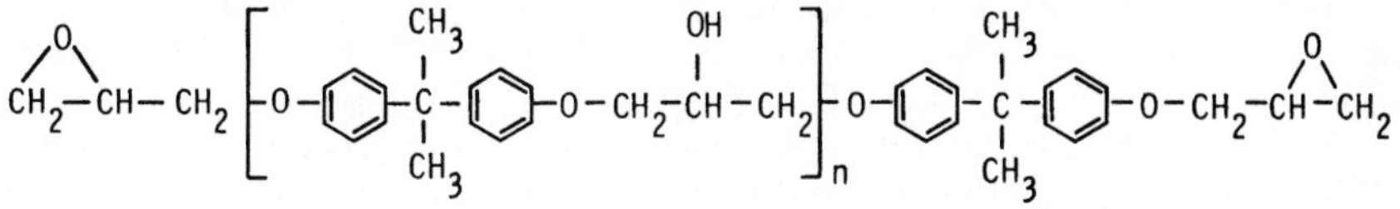

FIGURE 1.9 General structure of epoxy (diglycidyl ether of bisphenol A).

1.6.3 Epoxies

Types of Epoxies. Epoxy resins are characterized by the epoxide group (oxirane rings).[26,27] The most widely used resins are diglycidyl ethers of bisphenol A (Fig. 1.9). These are made by reacting epichlorohydrin with bisphenol A in the presence of an alkaline catalyst. By controlling operating conditions and varying the ratio of epichlorohydrin to bisphenol A, products of different molecular weights can be made. For liquid resins the n of Fig. 1.9 is generally less than 1; for solid resins n is 2 or greater. Solids with very high melting points have n values as high as 20.

Another class of epoxy resins is the novolacs, particularly the epoxy cresols (Fig. 1.10) and the epoxy phenol novolacs. These are produced by reacting a novolac resin, usually formed by the reaction of o-cresol or phenol and formaldehyde with epichlorohydrin. These highly functional materials are particularly recommended for transfer-molding powders, electrical laminates, and parts where superior thermal properties, high resistance to solvents and chemicals, and high reactivity with hardeners are needed.

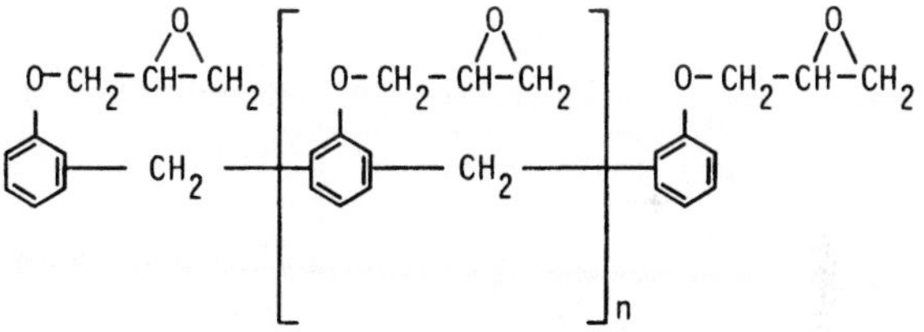

Epoxy Novolac

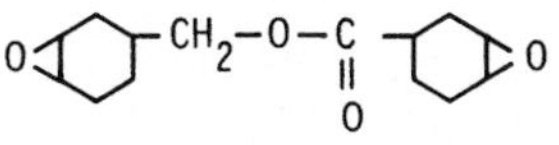

Cycloaliphatic Epoxy

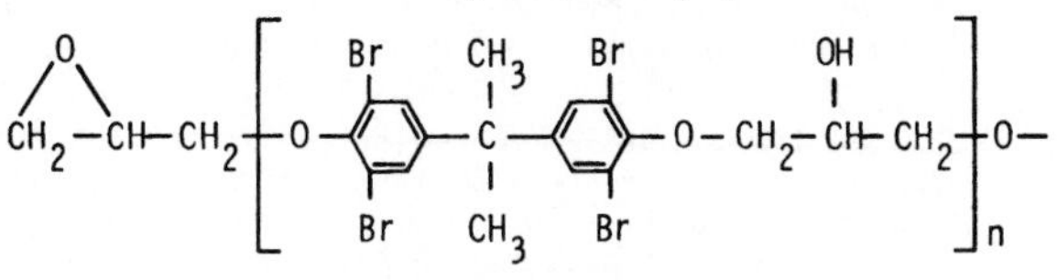

Brominated Epoxy

FIGURE 1.10 General structures of novolac and cycloaliphatic and brominated epoxies.

Another group of epoxy resins, the cycloaliphatics (Fig. 1.10), is particularly important when superior arc-track and weathering resistance are necessary requirements.

A distinguishing feature of cycloaliphatic resins is the location of the epoxy group(s) on a ring structure rather than on the aliphatic chain. Cycloaliphatics can be produced by the peracetic epoxidation of cyclic olefins and by the condensation of an acid such as tetrahydrophthalic anhydrite with epichlorohydrin, followed by dehydrohalogenation.

Epoxy resins must be cured with cross-linking agents (hardeners) or catalysts to develop desirable properties. The epoxy and hydroxyl groups are the reaction sites through which cross-linking occurs. Useful agents include amines, anhydrides, aldehyde condensation products, and Lewis acid catalysts. Careful selection of the proper curing agent is required to achieve a balance of application properties and initial handling characteristics.[26,27] Major types of curing agents are aliphatic amines, aromatic amines, catalytic curing agents, and acid anhydrides, as shown in Table 1.9.

TABLE 1.9 Curing Agents for Epoxy Resins

Curing-agent type	Characteristics	Typical materials
Aliphatic amines ..	Aliphatic amines allow curing of epoxy resins at room temperature, and thus are widely used. Resins cured with aliphatic amines, however, usually develop the highest exothermic temperatures during the curing reaction, and therefore the mass of material which can be cured is limited. Epoxy resins cured with aliphatic amines have the greatest tendency toward degradation of electrical and physical properties at elevated temperatures	Diethylene triamine (DETA) Triethylene tetramine (TETA)
Aromatic amines ..	Epoxies cured with aromatic amines usually have a longer working life than do epoxies cured with aliphatic amines. Aromatic amines usually require an elevated-temperature cure. Many of these curing agents are solid and must be melted into the epoxy, which makes them relatively difficult to use. The cured resin systems, however, can be used at temperatures considerably above those which are safe for resin systems cured with aliphatic amines	Metaphenylene diamine (MPDA) Methylene dianiline (MDA) Diamino diphenyl sulfone (DDS or DADS)
Catalytic curing agents..........	Catalytic curing agents also have a working life better than that of aliphatic amine curing agents and, like the aromatic amines, normally require curing of the resin system at a temperature of 200°F or above. In some cases, the exothermic reaction is critically affected by the mass of the resin mixture	Piperidine Boron trifluoride ethylamine complex Benzyl dimethyl-amine (BDMA)
Acid anhydrides...	The development of liquid acid anhydrides provides curing agents which are easy to work with, have minimum toxicity problems compared with amines, and offer optimum high-temperature properties of the cured resins. These curing agents are becoming more and more widely used	Nadic methyl anhydride (NMA) Dodecenyl succinic anhydride (DDSA) Hexahydrophthalic anhydride (HHPA) Alkendic anhydride

Aliphatic amine curing agents produce a resin-curing agent mixture that has a relatively short working life but that cures at room temperature or at low baking temperatures in relatively short time. Resins cured with aliphatic amines usually develop the highest exothermic temperatures during the curing reaction; thus the amount of material that can be cured at one time is limited because of possible cracking, crazing, or even charring of the resin system if too large a mass is mixed and cured. Also, physical and electrical properties of epoxy resins cured with aliphatic amines tend to degrade as the operating temperature increases. Epoxies cured with aliphatic amines find their greatest usefulness where small masses can be used, where room-temperature curing is desirable, and where the operating temperature required is below 100°C.

Epoxies cured with aromatic amines have a considerably longer working life than do those cured with aliphatic amines, but they require curing at 100°C or higher. Resins cured with aromatic amines can operate at a temperature considerably above the temperature necessary for those cured with aliphatic amines. However, aromatic amines are not so easy to work with as aliphatic amines, because of the solid nature of the curing agents and the fact that some (such as metaphenylene diamine) sublime when heated, causing stains and residue deposition.

Catalytic curing agents also have longer working lives than the aliphatic amine materials, and like the aromatic amines, catalytic curing agents normally require curing of the epoxy system at 100°C or above. Resins cured with these systems have good high-temperature properties as compared with epoxies cured with aliphatic amines. With some of the catalytic curing agents, the exothermic reaction becomes high as the mass of the resin mixture increases, as shown in Fig. 1.11.

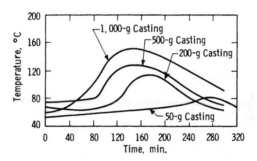

FIGURE 1.11 Exothermic curves as a function of resin mass for bisphenol epoxy and 5 percent piperidene curing agent (catalytic type) cured at 60°C.

Acid anhydride curing agents are particularly important for epoxy resins, especially the liquid anhydrides. The high-temperature properties of resin systems cured with these materials are better than those of resin systems cured with aromatic amines. Some anhydride-cured epoxy-resin systems retain most electrical properties to 150°C and higher, and are little affected physically, even after prolonged heat aging at 200°C. In addition, the liquid anhydrides are extremely easy to work with; they blend easily with the resins and reduce the viscosity of the resin system. Also, the working life of the liquid acid anhydride systems is long compared with that of mixtures of aliphatic amine and resin, and odors are slight. Amine promoters such as benzyl dimethylamine (BDMA) or DMP-30 are used to promote the curing of mixtures of acid anhydride and epoxy resin. The thermal stability of epoxies is improved by anhydride curing agents, as shown in Fig. 1.12.

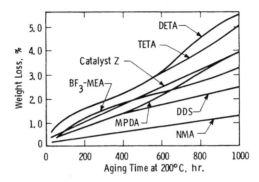

FIGURE 1.12 Weight-loss data at 200°C for Epon 828 cured with various curing agents. DETA, TETA—aliphatic amines; MPDA, DDS—aromatic amines; NMA—liquid acid anhydride.

Applications of Epoxies.   Epoxies are among the most versatile and most widely used plastics in the electronics field. This is primarily because of the wide variety of formulations possible and the ease with which these formulations can be made and utilized with minimal equipment requirements. Formulations range from flexible to rigid in the cured state and from thin liquids to thick pastes and molding powders in the uncured state. Conversion from uncured to cured state is made by use of hardeners or heat, or both. The largest application of epoxies is in embedding applications (potting, casting, encapsulating, and impregnating) in molded parts, and in laminated constructions such as metal-clad laminates for printed circuits and unclad laminates for various types of insulating and terminal boards. Molded parts have excellent dimensional stability, as shown in Fig. 1.8. The detailed properties of laminated products are given in Chap. 2. Typical properties of glass-fiber-filled and mineral-filled epoxy compounds are shown in Chap. 2 and in Appendix C.

1.6.4   Melamines and Ureas (Aminos)

As compared with alkyds, diallyl phthalates, and epoxies, which are polymers created by addition reactions and hence have no reaction by-products, melamines and ureas (also commonly referred to as aminos) are polymers that are formed by condensation reactions and do give off by-products. Another example of this type of reaction is the polymerization reaction, which produces phenolics. A typical condensation reaction is shown in Sec. 1.6.5 on phenolics. Melamines and ureas are a reaction product of formaldehyde with amino compounds containing NH_2 groups. Hence they are often also referred to as melamine formaldehydes and urea formaldehydes. Their general chemical structure is shown in Fig. 1.13.

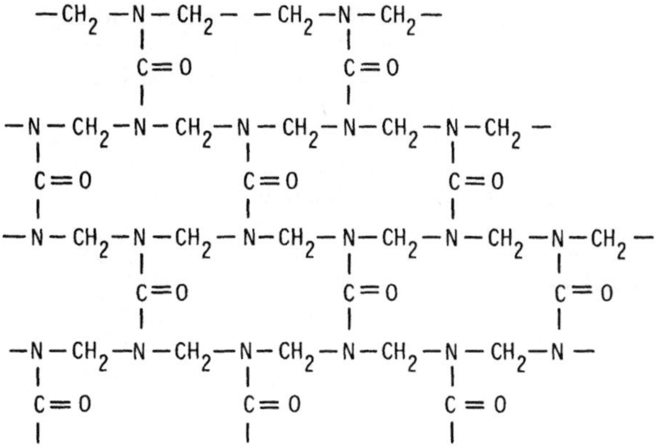

Urea Formaldehyde

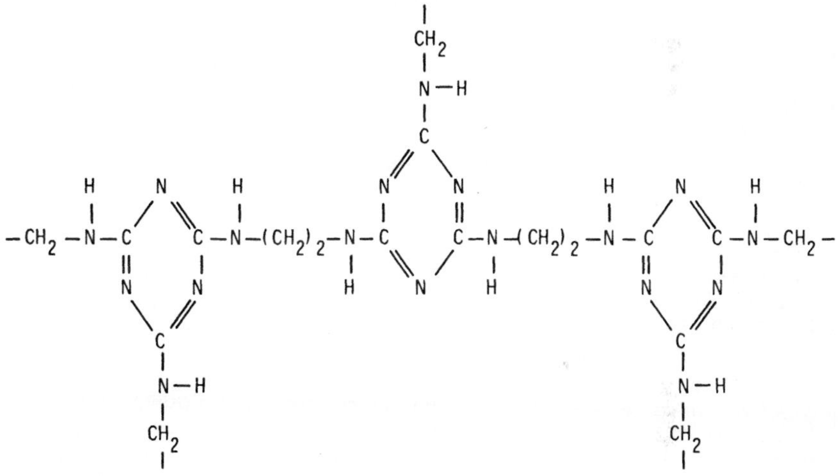

Melamine Formaldehyde

FIGURE 1.13 General chemical structures of urea formaldehyde and melamine formaldehyde.[28]

Amino resins have found applications in the fields of industrial and decorative laminating, adhesives, protective coatings, textile treatment, paper manufacture, and molding compounds. Their clarity permits products to be fabricated in virtually any color. Finished products having an amino-resin surface exhibit excellent resistance to moisture, greases, oils, and solvents; are tasteless and odorless; are self-extinguishing; offer excellent electrical properties; and resist scratching and marring. The melamine resins offer better chemical, heat, and moisture resistance than do the ureas.[28]

Amino molding compounds can be fabricated by economical molding methods. They are hard, rigid, and abrasion-resistant, and they have high resistance to deformation under load. These materials can be exposed to subzero temperatures without embrittlement. Under tropical conditions, the melamines do not support fungus growth.

Amino materials are self-extinguishing and have excellent electrical-insulation characteristics. They are unaffected by common organic solvents, greases and oils, and weak acids and alkalies. Melamines are superior to ureas in resistance to acids, alkalies, heat, and boiling water, and are preferred for applications involving cycling between wet and dry conditions or rough handling. Aminos do not impart taste or odor to foods.

Addition of alpha cellulose filler, the most commonly used filler for aminos, produces an unlimited range of light-stable colors and high degrees of translucency. Colors are obtained without sacrifice of basic material properties. Shrinkage characteristics with cellulose filler are shown in Fig. 1.7.

Melamines and ureas provide excellent heat insulation; temperatures up to the destruction point will not cause parts to lose their shape. Amino resins exhibit relatively high mold shrinkage and also shrink on aging. Cracks develop in urea moldings subjected to severe cycling between dry and wet conditions. Prolonged exposure to high temperature affects the color of both urea and melamine products.

A loss of certain strength characteristics also occurs when amino moldings are subjected to prolonged elevated temperatures. Some electrical characteristics are also adversely affected; the arc resistance of some industrial types, however, remains unaffected after exposure at 500°F.

Ureas are unsuitable for outdoor exposure. Melamines experience little degradation in electrical or physical properties after outdoor exposure, but color changes may occur.

Typical physical properties of amino plastics are shown in Table 1.10[29] and typical mechanical and electrical properties in Table 1.11.[29]

1.6.5 Phenolics

Like melamines and ureas, phenolic resin precursors are formed by a condensation reaction, the general nature of which is shown in Fig. 1.14. The general chemical structure of a cross-linked phenol-formaldehyde resin is given in Fig. 1.15.

Phenolics are among the oldest and best known general-purpose molding materials. They are also among the lowest in cost and the easiest to mold. An extremely large number of phenolic materials are available, based on the many resin and filler combinations, and they can be classified in many ways. One common way of classifying them is by type of application or grade. Typical properties for some of these common molding-material classifications are shown in Harper[8] and Grafton[9] and in Appendix C. In addition to molding materials, phenolics are used to bond friction materials for automotive brake linings, clutch parts, and transmission bands. They serve as binders for wood-particle board used in building panels and core material for furniture, as the water-resistant adhesive for exterior-grade plywood, and as the bonding agent for converting both organic and inorganic fibers into acoustical and thermal insulation pads, batts, or cushioning for home, industrial, and automotive applications. They are used to impregnate paper for electrical or decorative laminates and as special additives to tackify, plasticize, reinforce, or harden a variety of elastomers.

Although it is possible to obtain various molding grades of phenolics for various applications, as discussed, phenolics, generally speaking, are not equivalent to diallyl phthalates and epoxies in resistance to humidity and retention of electrical properties in extreme environments. Phenolics are, however, quite adequate for a

TABLE 1.10 Typical Physical Properties of Amino Compounds[29]

	Urea		Melamine					
	Alpha cellulose	Wood flour	Alpha cellulose	Wood flour	Alpha cellulose, modified	Rag	Asbestos	Glass fiber
Specific gravity	1.5	1.5	1.5	1.42	1.43	1.5	1.78	1.94–2.0
Density, g/in.³	24.6	24.6	24.6	23.8	23.5	24.6	29.2	31.8–32.8
Hardness, Rockwell E	94–97	95	110	94	....	100	90	
Shrinkage,* in./in.:								
Molding	0.006–0.009	0.006–0.014	0.008–0.009	0.007–0.008	0.006–0.008	0.003–0.004	0.005–0.007	0.002–0.004
Postmold	0.006–0.012	0.006–0.012	0.009–0.011	0.004–0.007	0.001–0.002	0.004–0.008	0.002–0.003	0.002–0.005
Deflection temp, °F, at 264 lb/in.²	266	270	361	266	266	310	266	400
Heat resistance, continuous, °F	170†	170	210†	250	250	250	300	300
Coefficient of thermal expansion, per °C × 10⁻⁶	22–36	30	20–57	32–50	34–36	25–30	21–43	12–25
Thermal conductivity, (cal)(cm)/(s²)(cm)(°C) × 10⁻⁴	10.1	10.1	10.1	8.4	....	10.6	13.1	
Water absorption, 24 h, at 23°C, %	0.4–0.8	0.7	0.3–0.5	0.34–0.6	0.3–0.6	0.3–0.6	0.13–0.15	0.09–0.3
Color possibilities	Unlimited	Brown, black	Unlimited	Brown	Brown	Limited	Brown	Natural gray

* Test specimen: 4-in. diam × ⅛-in disc.

† Based on no color change.

TABLE 1.11 Typical Mechanical and Electrical Properties of Amino Compounds[29]

	Urea		Melamine					
	Alpha cellulose	Wood flour	Alpha cellulose	Wood flour	Alpha cellulose modified	Rag	Asbestos	Glass fiber
Mechanical:								
Tensile strength, 1,000 lb/in.^{2}	5.5-7	5.5-10	7-8	5.7-6.5	5.5-6.5	8-10	5.5-6.5	5.9
Compressive strength, 1,000 lb/in.^{2}	30-38	25-35	40-45	30-35	24.5-26	30-35	25-30	20-29
Flexural strength, 1,000 lb/in.^{2}	11-18	8-16	12-15	6.5-9	11.5-12	12-15	7.4-10	13.2-24
Shear strength, 1,000 lb/in.^{2}	11-12	......	11-12	10-10.5	11.4-12.2	12-14	7-8	13.0-15.6
Impact strength, Izod ft-lb/in. of notch	0.24-0.28	0.25-0.35	0.30-0.35	0.25-0.35	0.30-0.42	0.55-0.90	0.30-0.40	0.5-6.0
Tensile modulus, 10^6 lb/in.^{2}	1.3-1.4	......	1.35	1.0	1.0	1.4	1.95	2.4
Flexural modulus, 10^6 lb/in.^{2}	1.4-1.5	1.3-1.6	1.1	1.0	1.1	1.4	1.8	
Electrical:								
Arc resistance, s	80-100	80-100	125-136	70-106	90-120	122-128	120-180	180-186
Dielectric strength, V/mil								
Short time: At 23°C	330-370	300-400	270-300	350-370	350-390	250-340	410-430	170-370
At 100°C	200-270	......	170-210	290-330	140-190	110-130	280-310	90-350*
Step by step: At 23°C	220-250	250-300	240-270	200-250	200-250	220-240	280-300	170-270
At 100°C	110-150	......	90-130	190-210	90-100	60-90	190-210	60-250*
Slow rate of rise: At 23°C	250-260	......	210-240	240-260	280-290	210-240	270-290	170-210
At 100°C	120-170	......	90-120	170-200	90	70-80	170-190	70-90
Dielectric constant: At 60 Hz	7.7-7.9	7.0-9.5	7.9-8.2	6.4-6.6	7.0-7.7	8.1-12.6	10.0-10.2	7.0-11.1
At 10^6 Hz	6.7-6.9	6.4-6.9	7.6-8.0	5.6-5.8	5.2-6.0	6.7-6.9	5.3-6.1	6.6-7.9
At 3 × 10, Hz	......	......	......	......	......	......	4.9	5.5
Dissipation factor: At 60 Hz	0.034-0.043	0.035-0.040	0.052-0.083	0.026-0.033	0.192	0.100-0.340	0.100	0.14-0.23
At 10^6 Hz	0.029-0.031	0.028-0.032	0.026-0.030	0.034-0.035	0.044-0.12	0.036-0.041	0.039-0.048	0.013-0.016
At 3 × 10^9 Hz	......	......	......	......	......	......	0.032	0.040
Dielectric loss factor: At 60 Hz	0.28-0.34	0.24-0.38	0.44-0.78	0.17-0.22	0.90-2.4	2.0-5.0	0.5-1.0	1.5-2.5
At 10^6 Hz	0.19-0.21	0.18-0.22	0.20-0.33.	0.20-0.21	0.19-0.28	0.24-0.26	0.21-0.31	0.09-0.19
Volume resistivity, Ω-cm	0.5-5.0 × 10^{11}	......	0.8-2.0 × 10^{12}	6-10 × 10^{11}	6 × 10^{06}	1.0-3.0 × 10^{11}	1.2 × 10^{12}	0.9-20 × 10^{11}
Surface resistivity, Ω	0.4-3.0 × 10^{11}	......	0.8-4.0 × 10^{11}	0.3-5.0 × 10^{12}	1.7 × 10^{12}	0.7-7.0 × 10^{11}	1.9 × 10^{13}	3.0-4.6 × 10^{12}
Insulation resistance, Ω	0.2-5.0 × 10^{11}	......	1.0-4.0 × 10^{10}	1.0-3.0 × 10^{11}	2.0-5.0 × 10^9	0.1-3.0 × 10^{10}	1.0-4.0 × 10^{10}	0.2-6.0 × 10^{10}

* At 50°C.

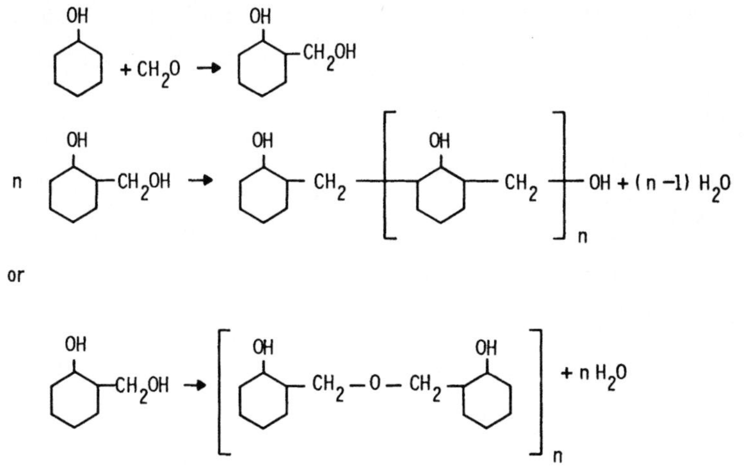

FIGURE 1.14 General nature of condensation reaction.

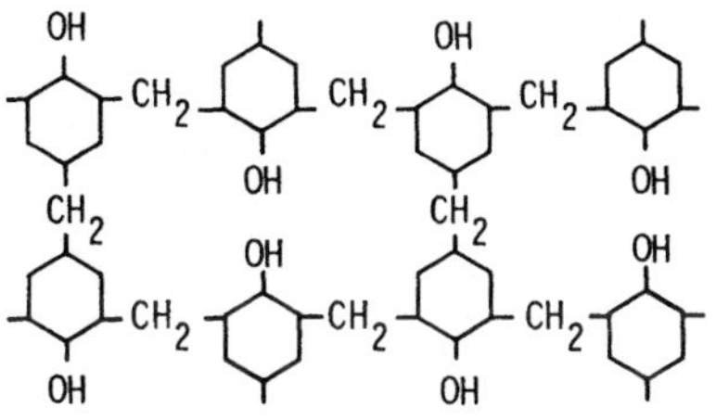

Phenol-Formaldehyde Resin

FIGURE 1.15 General chemical structure of phenol formaldehyde.

large percentage of electrical applications. Grades have been developed which yield considerable improvements in humid environments and at higher temperatures. The glass-filled, heat-resistant grades are outstanding in thermal stability up to 400°F and higher, with some being useful up to 500°F. Shrinkage in heat aging varies over a fairly wide range, depending on the filler used. Glass-filled phenolics are the more stable, as shown in Fig. 1.8. Martin[30] and Carswell[31] give considerable background on phenolic resin chemistry and applications.

1.6.6 Polybutadienes

Polybutadiene polymers that vary in 1,2 microstructure from 60 to 90 percent offer potential as moldings, laminating resins, coatings, and cast-liquid and formed-sheet products. These materials, which are essentially pure hydrocarbon, have outstanding electrical and thermal stability properties. Typical properties are shown in Table 1.12 and the chemical structure is as follows:

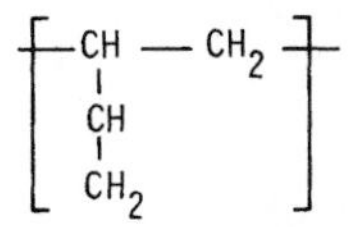

1, 2 - polybutadiene
microstructure

TABLE 1.12 Typical Properties of Mineral-Filled Ricon Molding Compound[32] (Moldable at Pressures of 300 lb/in^2 and above)

Formulations	
Ricon 150 (~70% 1,2 vinyl content), wt %	50.0
Polyethylene powder, wt %	50.0
Peroxide, wt %	4.0
Antioxidant, wt %	0.4
Vinyl Silane, wt %	1.2
Diatomaceous earth, wt %	50.0
Curing conditions at 340°F (171°C) and 500 lb/in^2, min	5
Physical properties	
Specific gravity at 4°C	1.17
Bulk factor	2.4
Mold shrinkage, in/in	0.030
Water absorption in 24 h, %	0.056
Spiral mold flow at 500 lb/in^2, in	43
Mechanical properties	
Impact strength (Izod), ft · lb/in notch	0.7–1.0
Tensile strength, lb/in^2	3000
Elongation, %	12
Flexural strength, lb/in^2	5000
Electrical properties	
Dielectric strength (short time), V/mil	530
Dielectric constant	
60 Hz	3.1
1 kHz	2.9
1 MHz	3.2
Dissipation factor	
60 Hz	0.0031
1 kHz	0.0030
1 MHz	0.0040
Volume resistivity, Ω · cm	1.4×10^{15}
Thermal stability	
Weight loss in 100 h at 400°F (205°C), wt %	1.64

Polybutadienes are cured by peroxide catalysts, which produce carbon-to-carbon bonds at the double bonds in the vinyl groups.[32] The final product is 100 percent hydrocarbon except where the starting polymer is the —OH or —COOH terminated variety. The nature of the resultant product may be more readily understood if the structure is regarded as polyethylene with a cross-link at every other carbon in the main chain.

Use of the high-temperature peroxides maximizes the opportunity for thermoplastic-like processing, because even the higher-molecular-weight forms become quite fluid at temperatures well below the cure temperature. Compounds can be injection-molded in an in-line machine with a thermoplastic screw.

1.6.7 Polyesters (Thermosetting)

Unsaturated, thermosetting polyesters are produced by addition polymerization reactions, as shown in simplified form in Fig. 1.1. Polyester resins can be formulated to have a range of physical properties from brittle and hard to tough and resistant to soft and flexible. Viscosities at room temperature may range from 50 to more than 25,000 cP. Polyesters can be used to fabricate a myriad of products by many techniques—open-mold casting, hand lay-up, spray-up, vacuum-bag molding, matched-metal-die molding, filament winding, pultrusion, encapsulation, centrifugal casting, and injection molding.[33]

By the appropriate choice of ingredients, particularly to form the linear polyester resin, special properties can be imparted. Fire retardance can be achieved through the use of one or more of the following: chlorendic anhydride, tetrabromophthalic anhydride, tetrachlorophthalic anhydride, dibromoneopentyl glycol, and chlorostyrene. Chemical resistance is obtained by using neopentyl glycol, isophthalic acid, hydrogenated bisphenol A, and trimethyl pentanediol. Weathering resistance can be enhanced by the use of neopentyl glycol and methyl methacrylate. Appropriate thermoplastic polymers can be added to reduce or eliminate shrinkage during curing and thereby minimize one of the disadvantages historically inherent in polyester systems.[33]

Thermosetting polyesters are widely used for moldings, laminated or reinforced structures, surface gel coatings, liquid castings, furniture products, and structures. Cast products include furniture, bowling balls, simulated marble, gaskets for vitrified-clay sewer pipe, pistol grips, pearlescent shirt buttons, and implosion barriers for television tubes.

By lay-up and spray-up techniques, large- and short-run items are fabricated. Examples include boats of all kinds—pleasure sailboats and powered yachts, commercial fishing boats and shrimp trawlers, small military vessels—dune buggies, all-terrain vehicles, custom auto bodies, truck cabs, horse trailers, motor homes, housing modules, concrete forms, and playground equipment.[33]

Molding is also performed with premix compounds, which are doughlike materials generally prepared by the molder shortly before they are to be molded by combining the premix constituents in a sigma-blade mixer or similar equipment. Premix, using conventional polyester resins, is used to mold automotive-heater housings and air-conditioner components. Low-shrinkage resin systems permit the fabrication of exterior automotive components such as fender extensions, lamp housings, hood scoops, and trim rails.

Wet molding of glass mats or preforms is used to fabricate such items as snack-table tops, food trays, tote boxes, and stackable chairs. Corrugated and flat paneling for room dividers, roofing and siding, awnings, skylights, fences, and the like is a very important outlet for polyesters. Pultrusion techniques are used to make fishing-rod stock and profiles from which slatted benches and ladders can be fabricated. Chemical storage tanks are made by filament winding.[33]

1.6.8 Silicones

Silicones are a family of unique synthetic polymers that are partly organic and partly inorganic.[34] They have a quartzlike polymer structure, being made up of alternating silicon and oxygen atoms rather than the carbon-to-carbon backbone, which is a characteristic of the organic polymers. Silicones have outstanding thermal stability.

Typically, the silicon atoms will have one or more organic side groups attached to them, generally phenyl (C_6H_5—), methyl (CH_3—), or vinyl ($CH_2 = CH$—) units. Other alkyl, aryl, and reactive organic groups on the silicon atom are also possible. These groups impart characteristics such as solvent resistance, lubricity and compatibility, and reactivity with organic chemicals and polymers.[34]

Silicone polymers may be filled or unfilled, depending on properties desired and application. They can be cured by several mechanisms, either at room temperature (by room-temperature vulcanization, or RTV) or at elevated temperatures. Their final form may be fluid, gel, elastomeric, or rigid.[34]

Some of the properties which distinguish silicone polymers from their organic counterparts are (1) relatively uniform properties over a wide temperature range, (2) low surface tension, (3) high degree of slip or lubricity, (4) excellent release properties, (5) extreme water repellency, (6) excellent electrical properties over a wide range of temperatures and frequencies, (7) inertness and compatibility, both physiologically and in electronic applications, (8) chemical inertness, and (9) weather resistance.

Flexible Silicone Resins. Flexible two-part, solvent-free silicone resins are available in filled and unfilled forms. Their viscosities range from 3000 cP to viscous thixotropic fluids of greater than 50,000 cP.[34] The polymer base for these resins is primarily dimethylpolysiloxane. Some vinyl and hydrogen groups attached to silicon are also present as part of the polymer.

These products are cured at room or slightly elevated temperatures. During cure there is little if any exotherm, and there are no by-products from the cure. The flexible resins have Shore A hardnesses of 0 to 60 and Bashore resiliencies of 0 to 80. Flexibility can be retained from −55°C or lower to 250°C or higher.

Flexible resins find extensive use in electrical and electronic applications where stable dielectric properties and resistance to harsh environments are important. They are also used in many industries to make rubber molds and patterns.

Rigid Silicone Resins. Rigid silicone resins exist as solvent solutions or as solvent-free solids. The most significant uses of these resins are as paint intermediates to upgrade thermal and weathering characteristics of organic coatings, as electrical varnishes, glass tape, and circuit-board coatings.

Glass cloth, asbestos, and mica laminates are prepared with silicone resins for a variety of electrical applications. Laminated parts can be molded under high or low pressures, vacuum-bag-molded, or filament-wound.

Thermosetting molding compounds made with silicone resins as the binder are finding wide application in the electronic industry as encapsulants for semiconductor devices. Inertness toward devices, stable electrical and thermal properties, and self-extinguishing characteristics are important reasons for their use. Typical properties are given in Ref. 35 and thermal stability is shown in Fig. 1.16.

Similar molding compounds containing refractory fillers can be molded on conventional thermoset equipment. Molded parts are then fired to yield a ceramic article. High-impact, long-glass-fiber-filled molding compounds are also available for use in high-temperature structural applications.

In general, silicone resins and composites made with silicone resins exhibit outstanding long-term thermal stabilities at temperatures approaching 300°C, and excellent moisture resistance and electrical properties.

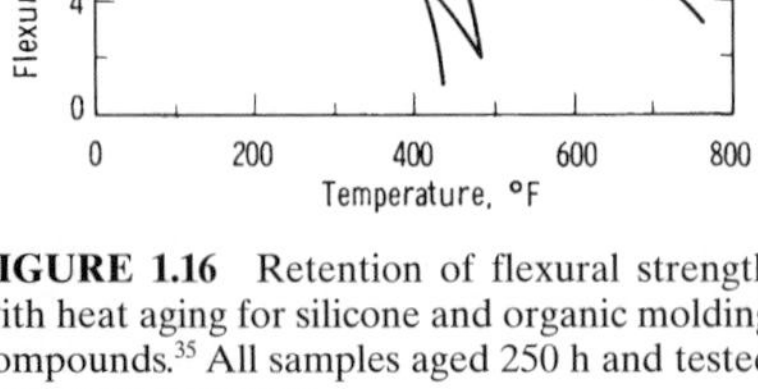

FIGURE 1.16 Retention of flexural strength with heat aging for silicone and organic molding compounds.[35] All samples aged 250 h and tested at room temperature.

1.7   THERMOPLASTICS

The general nature of thermoplastics has been discussed earlier in this chapter, and specific data for a variety of thermoplastics can be found in Ref. 36. In general, thermoplastic materials tend to be tougher and less brittle than thermosets so that they can be applied without the use of fillers. However, while some are very tough, others do tend to craze or crack easily, so each case must be considered on its individual merits. Traditionally, by virtue of their basic polymer structure, thermoplastics have been much less dimensionally and thermally stable than thermosetting plastics. Hence thermosets have offered a performance advantage, although the lower processing costs for thermoplastics have given the latter a cost advantage. However, three major trends tend to put both thermoplastics and thermosets on a performance-consideration basis. First, much has been done in the development of reinforced, fiber-filled thermoplastics, greatly increasing stability in many areas. Second, much has been achieved in the development of so-called engineering thermoplastics, or high-stability, higher-performance plastics, which can also be reinforced with fiber fillers to increase their stability further. Third, and countering these gains in thermoplastics, has been the development of lower-cost processing of thermosetting plastics, especially the screw-injection-molding technology. All these options should be considered in optimizing the design, fabrication, and performance of plastic parts.

A list of typical trade names and suppliers of the more common thermoplastics is given in Table 1.13, and a list of so-called advanced thermoplastic materials is presented in Table 1.14. Specific data for a variety of formulations is given in Ref. 36. Some of the more common classes of thermoplastics are discussed in this section.

1.7.1   ABS Plastics

ABS plastics are derived from acrylonitrile, butadiene, and styrene. They have the following general chemical structure:

Acrylonitrile   Butadiene   Styrene

Acrylonitrile-butadiene styrene copolymer

This class possesses hardness and rigidity without brittleness, at moderate costs. ABS materials have a good balance of tensile strength, impact resistance, surface hardness, rigidity, heat resistance, low-temperature properties, and electrical characteristics.[37] There are many ABS modifications and many blends of ABS with other thermoplastics. A list of trade names and suppliers is given in Ref. 36 with general data on a variety of ABS-based plastics.

Polymer Properties.   The most outstanding mechanical properties of ABS plastics are impact resistance and toughness. A wide variety of modifications are available for improved impact resistance, toughness, and heat resistance. Impact resistance does not fall off rapidly at lower temperatures. Stability under limited load is excellent, as shown in Fig. 1.17.[38] Heat-resistant ABS is equivalent to or better than

TABLE 1.13 Typical Trade Names and Suppliers of Thermoplastics*

Thermoplastic	Typical trade names and suppliers
ABS	Marbon Cycolac (Borg-Warner) Abson (B. F. Goodrich) Lustran (Monsanto)
Acetals	Delrin (E. I. du Pont) Celcon (Celanese)
Acrylics	Plexiglas (Rohm and Haas) Lucite (E. I. du Pont)
Aramids	Nomex (E. I. du Pont)
Cellulosics	Tenite (Eastman Chemical) Ethocel (Dow Chemical) Forticel (Celanese)
Ionomers	Surlyn A (E. I. du Pont) Bakelite (Union Carbide)
Low-permeability thermoplastics	Barex (Vistron/Sohio) NR-16 (E. I. du Pont) LPT (Imperial Chemical Industries)
Nylons (see also Aramids)	Zytel (E. I. du Pont) Plaskon (Allied Chemical) Bakelite (Union Carbide)
Parylenes	Parylene (Union Carbide)
Polyaryl ether	Arylon T (Uniroyal)
Polyaryl sulfone	Astrel (3M)
Polycarbonates	Lexan (General Electric) Merlon (Mobay Chemical)
Polyesters	Valox (General Electric) Celanex (Celanese) Celanar Film (Celanese) Mylar Film (E. I. du Pont) Tenite (Eastman Chemical)
Polyethersulfone	Polyethersulphone (Imperial Chemical Industries)

* See also App. C for expanded listing.

acetals, polycarbonates, and polysulfones in room-temperature creep at 3000 lb/in². The Izod impact strength at 75°F is in the range of 3 to 5 ft·lb/in notch. This value is gradually reduced to 1 ft·lb/in notch at −40°F. When impact failure does occur, the failure is ductile rather than brittle. The modulus of elasticity versus temperature is shown in Fig. 1.18.[39] Physical properties are little affected by moisture, which contributes greatly to the dimensional stability of ABS materials.

ABS Alloys and Electroplating Grades. Much work has been done to modify ABS plastics by alloying to improve certain properties, and by modifying to enhance adhesion of electroplated coatings.[36] ABS alloyed or blended with polycarbonate combines some of the best qualities of both materials, resulting in a thermoplastic that is easier to process, has high heat and impact resistance, and sells for considerably less than polycarbonate.

TABLE 1.13 Typical Trade Names and Suppliers of Thermoplastics* (*Continued*)

Thermoplastic	Typical trade names and suppliers
Polyethylenes, polypropylenes, and polyallomers...................	Alathon Polyethylene (E. I. du Pont) Petrothene Polyethene (U.S.I.) Hi-Fax Polyethylene (Hercules) Pro-Fax Polypropylene (Hercules) Bakelite Polyethylene and Polypropylene (Union Carbide) Tenite Polyethylene and Polypropylene (Eastman) Irradiated Polyolefin (Raychem)
Polyimides and polyamide-imides......	Vespel SP Polyimides (E. I. du Pont) Kapton Film (E. I. du Pont) Pyralin Laminates (E. I. du Pont) Keramid/Kinel (Rhodia) P13N (Ciba-Geigy) Torlon Polyamide-Imide (Amoco)
Polymethyl pentene..................	TPX (Imperial Chemical Industries)
Polyphenylene oxides................	Noryl (General Electric)
Polyphenylene sulfides..............	Ryton (Phillips Petroleum)
Polystyrenes.......................	Styron (Dow Chemical) Lustrex (Monsanto) Dylene (Koppers) Rexolite (American Enka)
Polysulfones......................	Ucardel (Union Carbide)
Vinyls............................	Pliovic (Goodyear Chemical) Diamond PVC (Diamond Alkali) Geon (B. F. Goodrich) Bakelite (Union Carbide)

TABLE 1.14 Advanced Thermoplastic Materials

Matrix systems	Abbreviation	Trade names and suppliers
Polyphenylene sulfide	PPS	Ryton (Phillips Petroleum)
Polysulfone	PSF	Udel (Union Carbide)
Polyetheretherketone	PEEK	APC (Imperial Chemical Industries)
Polyethersulfone	PES	Victrex (Imperial Chemical Industries)
Polyetherimide	PEI	Ultem (General Electric)
Polyamide-imide	PAI	Torlon (Amoco)
Polyetherketone	PEK	—(Imperial Chemical Industries)
Polyamide	PA	J-2 (E.I. du Pont)
Polyimide	PI	K-III (E.I. du Pont)
Polyarylene sulfide	PAS	PAS-2 (Phillips Petroleum)
Polyarylene ketone	—	HTA (Imperial Chemical Industries)
Polyetherketoneketone	PEKK	—(E. I. du Pont)

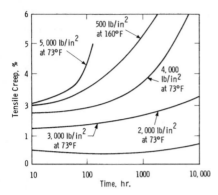

FIGURE 1.17 Tensile creep of ABS under various loads.[38]

The impact strength of the ABS-polycarbonate alloy, 10.7 ft·lb/in notch, is well above average for the high-impact engineering thermoplastics but still not as high as that of polycarbonate. However, unlike polycarbonate, the alloy does not have critical thicknesses with respect to notched impact strength. Its notched Izod value drops by only 2 to 4 ft·lb/in in the ⅛- to ¼-in range. The notched Izod value for a polycarbonate ⅛ in thick is about 16 ft·lb/in but only 3 to 4 ft·lb/in at thicknesses greater than ¼ in. The flexural modulus of the ABS-polycarbonate alloy is about 15 percent greater than that of polycarbonate alone. The alloy remains more rigid than polycarbonate up to about 200°F, as shown in Fig. 1.19.[40]

The 264-lb/in² heat-deflection temperature of the alloy is 245°F, and the 66-lb/in² value is 260°F. These values are 35 and 30°F, respectively, and are lower than those of polycarbonate. However, the maximum recommended continuous (no-load) temperature of the alloy is only 10°F lower than that of polycarbonate. The good creep resistance of polycarbonate, one of its biggest advantages and shown in Fig. 1.20,[40] is maintained after alloying.

In addition to the polycarbonate alloy, ABS can be alloyed with other plastics to obtain special properties. Furthermore, ABS has been modified to gain improved

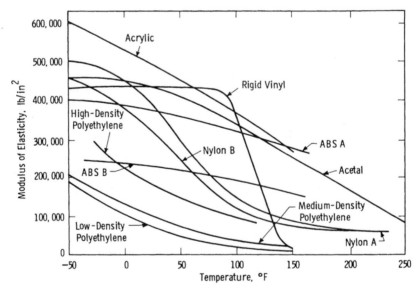

FIGURE 1.18 Modulus of elasticity of several thermoplastic materials as a function of temperature.[39]

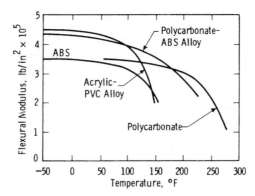

FIGURE 1.19 Flexural modulus of ABS-polycarbonate alloy and base materials.[40]

adhesion of electroplated metals. Advances have been so great in this area that ABS is perhaps the most widely used material for producing electroplated plastic parts. Electroplated ABS is used extensively for many electrical and mechanical products in many forms and shapes. Adhesion of the electroplated metal is excellent.

1.7.2 Acetals

Acetals are among the group of high-performance engineering thermoplastics that resembles nylon somewhat in appearance but not in properties. The general repeating chemical structural unit is:

Polyacetal Resin

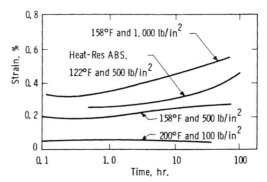

FIGURE 1.20 Creep resistance of ABS-polycarbonate alloy and heat-resistant ABS.[40]

Acetals are strong and rigid (but not brittle) and have good moisture, heat, and chemical resistance. There are two basic types of acetals: the homopolymers by du Pont and the copolymers by Celanese. Reference 36 further identifies these materials and gives typical properties. The homopolymers are harder, have higher resistance to fatigue, are more rigid, and have higher tensile and flexural strengths with lower elongation. The copolymers are more stable in long-term, high-temperature service and more resistant to hot water. Neither type of acetal is resistant to strong mineral acids, but the copolymers are resistant to strong bases. References made to acetals without identification of polymer type usually imply the homopolymer materials.

Polymer Properties.   The most outstanding properties of acetals are high tensile strength and stiffness, resilience, good recovery from deformation under load, and toughness under repeated impact. They exhibit excellent long-term load-carrying properties and dimensional stability and can be used for precision parts. Acetals have low static and dynamic coefficients of friction and are usable over a wide range of environmental conditions. The plastic surface is hard, smooth, and glossy. A fluorocarbon fiber-filled acetal, Delrin AF, is available and offers even better low-friction and resistance properties.

The modulus of elasticity as a function of temperature for acetals and several other thermoplastic materials is shown in Fig. 1.18.[39] The tensile yield strength of acetals is compared in Fig. 1.21[41] with that of some other thermoplastics. The deflection under load for Delrin acetal is compared in Fig. 1.22[41] with that of other thermoplastics.

Effect of Moisture.   Because acetals absorb a small amount of water, the dimensions of molded parts are affected by their water content. Figure 1.23[42] presents the relationship of dimensional changes with changes in moisture content. The dimensional change resulting from an environmental change is found by subtracting the "percent in length" found at the first humidity-temperature condition from the "percent in length" at the final conditions. For example, to go from 77°F (25°C) and 0 percent water (as-molded condition) to 100°F (38°C) and 100 percent humidity, a change of 0.45 percent will occur. Figure 1.24[42] shows the rate of water absorption of acetals at various conditions.

Effect of Space and Radiation.   Acetal resin will be stable in the vacuum of space under the same time-temperature conditions it can withstand in air.[42] Exposure to

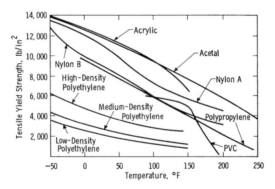

FIGURE 1.21   Tensile yield strength of several thermoplastic materials as a function of temperature.[41]

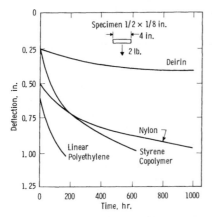

FIGURE 1.22 Deflection of several thermoplastic materials as a function of time at 90% relative humidity and 150°F.[41]

vacuum alone causes no loss of the engineering properties. The principal result is slight outgassing of small amounts of moisture and free formaldehyde. In a vacuum, as in air, prolonged exposure to elevated temperatures results in the liberation of increasing amounts of formaldehyde due to thermal degradation of the polymer.

Particulate radiation, such as the protons and electrons of the Van Allen radiation belts, is damaging to acetal resins and will cause loss of engineering properties.[41] For example, acetals should not be used in a radiation environment where the total electron dose is likely to exceed 1 Mrad. When irradiated with 2-MeV electrons, a 1-Mrad dose causes only slight discoloration while 2.3 Mrad causes considerable embrittlement. At 0.6 Mrad, however, acetals are still mechanically sound except for a moderate decrease in impact strength.

The regions of the electromagnetic spectrum that are most damaging to acetal resins are ultraviolet light and gamma rays.[42] In space, the deleterious effects of ultraviolet light are of prime consideration. This is due to the absence of the protective air atmosphere which normally filters out much of the sun's ultraviolet energy. Therefore the amount of ultraviolet light may be 10 to 100 times as intense in space as on earth.

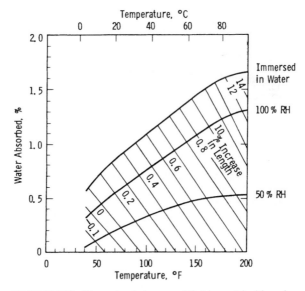

FIGURE 1.23 Dimensional changes of Delrin acetal with variations of temperature and moisture content.[42]

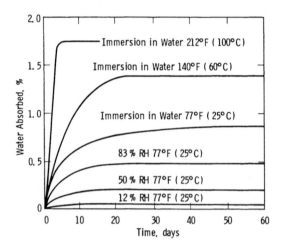

FIGURE 1.24   Rate of water absorption of Delrin acetal at various conditions.[42]

Fluorocarbon Fiber-Filled Acetal.   This is a modified acetal homopolymer developed to meet the need for a thermoplastic injection-molding material to be used in moving parts in which low friction and exceptional wear resistance are the principal requirements. This resin consists of oriented tetrafluoroethylene (TFE) fluorocarbon fibers uniformly dispersed in a matrix of acetal resin. The result is an injection-molding and extrusion resin that combines the strength, toughness, dimensional stability, and fabrication economy of acetals with the unusual surface and low frictional characteristics of the fluorocarbons.

The outstanding properties of TFE fiber-filled acetal are those associated with sliding friction. Bearings made from this material sustain high loads when operating at high speeds and show little wear. In addition, such bearings are essentially free of slipstick behavior because their static and dynamic coefficients are almost equal. Comparative properties of the filled and unfilled acetal are given in Refs. 36 and 43.

1.7.3   Acrylics

The general properties of acrylics are presented in Ref. 44. Trade names and suppliers are listed in Table 1.13. Acrylics are based on polymethyl methacrylate and their chemical structure is as follows:

$$\left[\begin{array}{c} H \;\; CH_3 \\ | \;\;\;\; | \\ -C-C- \\ | \;\;\;\; | \\ H \;\; COOCH_3 \end{array}\right]_n$$

Polymethyl methacrylate

Acrylics have exceptional optical clarity and basically good weather resistance, strength, electrical properties, and chemical resistance. They do not discolor or shrink after fabrication; they have low water-absorption characteristics, a slow burning rate, and will not flash-ignite. Acrylics are attacked by strong solvents, gasoline acetone, and other similar fluids.

Acrylics can be injection-molded, extruded, cast, vacuum- and pressure-formed, and machined, although molded parts for load bearing should be carefully analyzed, especially for long-term loading.

Optical Properties. Parts molded from acrylic powders in their natural state may be crystal clear and nearly optically perfect. The index of refraction ranges from 1.486 to 1.496. The total light transmittance is as high as 92 percent, and haze measurements average only 1 percent. Transmittance at various wavelengths is shown in Fig. 1.25.[45] Light transmittance and clarity can be modified by the addition of a wide range of transparent and opaque colors, most of these being formulated for long outdoor service.

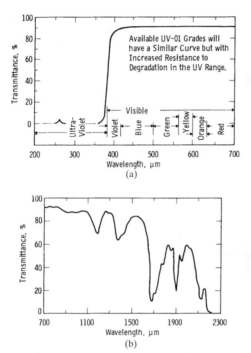

FIGURE 1.25 Percent transmittance as a function of wavelength for acrylic resin. Crystal, ⅛-in (3.2-mm) thickness, grades 129, 130, 140, 147, 148, molded only.[45] (*a*) Below 700 μm. (*b*) Above 700 μm.

Dimensional Stability and Aging. The amount of dimensional change for a given application is determined from the coefficient of linear thermal expansion of the material. In the case of an exterior glazing unit, which is 1 by 1 ft (30.4 by 30.4 cm) and is subjected to a temperature variation of 32 to 100°F (0 to 38°C), the space to be provided in the frame for expansion is calculated as follows (assuming no thermal expansion of the frame material):

$$\Delta L = \alpha L T$$

where ΔL = change in length, in or cm
α = coefficient of thermal expansion [= 4×10^{-5} in/in·°F (7.2×10^{-5} cm/cm·°C)]
L = initial length [= 12 in (30.4 cm)]
T = temperature variation [= 68°F (38°C)]

Therefore,

$$\Delta L = 4 \times 10^{-5} \times 12 \times 68 = 0.033 \text{ in}$$
$$= 7.2 \times 10^{-5} \times 30.4 \times 38 = 0.083 \text{ cm}$$

This indicates the need for an expansion space of approximately 0.0165 in (0.0419 cm) at each end for the direction in which this calculation was made (in this case the major length).

The moisture level in acrylics will depend on the relative humidity of the environment. Fig. 1.26[45] shows the equilibrium moisture content of acrylics versus the relative humidity in the surrounding air at room temperature.

When the relative humidity of the air changes from a lower value to a higher one, acrylics will absorb moisture, which in turn causes a slight dimensional expansion. Depending on the thickness of the part, the equilibrium moisture level may be reached in a couple of weeks or in months.

Fig. 1.27[45] shows the dimensional changes taking place in a ⅛-in (3.2-mm)-thick sheet as a function of time and the environmental relative humidity at room temperature. The flat horizontal part of the curves corresponds to the equilibrium conditions at the specific environmental relative humidity.

For example, assume that a ⅛-in (3.2-mm)-thick lighting panel as molded from dry resin is initially 24 in (61 cm) long. This panel, installed in a 50 percent relative-humidity atmosphere, will expand approximately 0.1 percent, or 0.024 in (0.61 mm).

Because of their chemical structure, acrylic resins are inherently resistant to discoloration and loss of light transmission. They are unsurpassed in this respect by any other transparent plastic. Their outstanding weatherability has been proved by their long-time performance in such products as automotive lenses, fluorescent street lights, outdoor signs, and boat windshields.

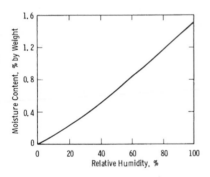

FIGURE 1.26 Equilibrium moisture content in acrylic resins versus percent relative humidity at 73°F.[45]

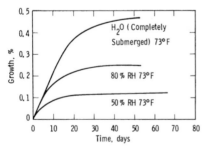

FIGURE 1.27 Dimensional changes versus time for acrylics under various environmental moisture conditions. Test samples ⅛ in (3.2 mm) thick.[45]

1.7.4 Cellulosics

Cellulosics are among the toughest of plastics. They are generally economical, basically good insulating materials. Their general structure is as follows:

Cellulose (natural polymer)

However, they are temperature-limited and are not as resistant to extreme environments as many other thermoplastics. The four most prominent industrial cellulosics are cellulose acetate, cellulose acetate butyrate, cellulose propionate, and ethyl cellulose. A fifth member of this group is cellulose nitrate. Cellulose materials are available in a great number of formulas and flows and are manufactured to offer a wide range of properties. They are formulated with a wide range of plasticizers for specific plasticized properties.

Cellulose butyrate, proprionate, and acetate provide a range of toughness and rigidity that is useful for many applications, especially where clarity, outdoor weatherability, and aging characteristics are needed. The materials are fast-molding plastics and can be provided with hard, glossy surfaces over the full range of color and texture.

Butyrate, propionate, and acetate are rated in that order in dimensional stability in relation to the effects of water absorption and plasticizers. The materials are slow-burning, although self-extinguishing forms of acetate are available. Special formulations of butyrate and propionate are serviceable outdoors for long periods. Acetate is generally considered unsuitable for outdoor uses. From an application standpoint, the acetates generally are used where tight dimensional stability, under anticipated humidity and temperature, is not required. Hardness, stiffness, and cost are lower than for butyrate or propionate. Butyrate is generally selected over propionate where weatherability, low-temperature impact strength, and dimensional stability are required. Propionate is often chosen for hardness, tensile strength, and stiffness, combined with good weather resistance.

Ethyl cellulose, best known for its toughness and resiliency at subzero temperatures, also has excellent dimensional stability over a wide range of temperature and humidity conditions. Alkalies or weak acids do not affect this material, but cleaning fluids, oils, and solvents are very harmful.

1.7.5 Fluorocarbons

Tetrafluoroethylene (TFE) and Fluorinated Ethylene Propylene (FEP). For practical purposes there are eight types of fluorocarbons, as summarized in Table 1.15 and discussed hereafter. Suppliers and trade names are also given in Table 1.15. Like other plastics, each type is available in several grades. The original, basic fluorocarbon, and perhaps still the most widely known one, is TFE fluorocarbon. It has the optimum of electrical and thermal properties and almost complete moisture resistance and chemical inertness, but it does have the disadvantage of cold flow or creep under mechanical loading. TFE hardness as a function of temperature is shown in Fig. 1.28.[46]

Stronger, filled modifications exist, as do newer, more cold-flow-resistant grades. FEP fluorocarbon is quite similar to TFE in most properties, except that its useful

TABLE 1.15   Structures, Trade Names, and Suppliers of Fluorocarbons

Fluorocarbon	Structure	Typical trade names and suppliers
TFE (tetrafluoroethylene)....	(structure: $-[-C-C-]_n$ with F, F on top and F, F on bottom)	Teflon TFE (E. I. du Pont) Halon TFE (Allied Chemical)
FEP (fluorinated ethylenepropylene).........	(structure of FEP, bracketed $_n$)	Teflon FEP (E. I. du Pont)
ETFE (ethylene-tetra-fluoroethylene copolymer)...	Copolymer of ethylene and TFE	Tefzel (E. I. du Pont)
PFA (perfluoroalkoxy).......	(structure of PFA with R_f*, bracketed $_n$)	Teflon PFA (E. I. du Pont)
CTFE (chlorotrifluoro-ethylene).................	(structure: $-[-C-C-]_n$ with Cl, F on top and F, F on bottom)	Kel-F (3M)
E-CTFE (ethylene-chlorotri-fluoroethylene copolymer)...	Copolymer of ethylene and CTFE	Halar E-CTFE (Allied Chemical)
PVF$_2$ (vinylidene fluoride)....	(structure: $-[-C-C-]_n$ with H, F on top and H, F on bottom)	Kynar (Pennsalt Chemicals)
PVF (polyvinyl fluoride)......	(structure: $-[-C-C-]_n$ with H, H on top and H, F on bottom)	Tedlar (E. I. du Pont)

* $R_f = C_n F_{2n-1}$.

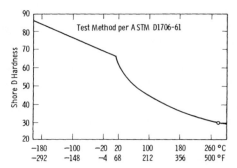

FIGURE 1.28 Hardness of TFE fluorocarbon as a function of temperature.[46]

temperature is limited to about 400°F. FEP is much more easily processed, and molded parts are possible with FEP, which might not be possible with TFE. Thermal-expansion curves for both are shown in Fig. 1.29.[47]

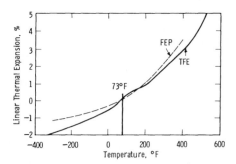

FIGURE 1.29 Linear thermal expansion of TFE and FEP fluorocarbons as a function of temperature.[47]

Ethylene-Tetrafluoroethylene (ETFE) Copolymer. ETFE, a high-temperature thermoplastic fluoropolymer, is readily processed by conventional methods, including extrusion and injection molding. As a copolymer of ethylene and tetrafluoroethylene, it is closely related to TFE. It has good thermal properties and abrasion resistance, and excellent impact strength, resistance to chemicals, and electrical-insulation characteristics.

ETFE can be described best as a rugged thermoplastic with an outstanding balance of properties. Mechanically, it is tough and has medium stiffness (200,000 lb/in²), excellent flex life, impact, cut-through, and abrasion resistance. The glass-fiber-reinforced compound has even higher tensile strength (12,000 lb/in²), stiffness (950,000 lb/in²), and creep resistance, but is still tough and impact-resistant.

Thermally, ETFE has a continuous temperature rating of 300°F (150°C). It can be used intermittently up to 200°C depending on exposure time, load, and environment. Glass-fiber-reinforced ETFE appears capable of useful service at 392°F (200°C).

ETFE is weather-resistant, inert to most solvents and chemicals, and hydrolytically stable. It has excellent resistance to high-energy radiation and is ultraviolet-resistant.

The dielectric constant of the fluoropolymer is 2.6 and the dissipation factor 0.0006, making it an excellent low-loss dielectric. These values are not affected by changes in temperature and other environmental conditions. Dielectric strength is high; resistivity is excellent.

The molding characteristics of the resin are excellent. It exhibits good melt flow, allowing the filling of thin sections (10 mil for small parts). Cycle times are equivalent to those of other thermoplastics.

It is the first fluoroplastic that can be reinforced—not merely filled—with glass fiber. Because the resin will bond to the fibers, strength, stiffness, creep resistance, heat-distortion temperature, and dimensional stability are enhanced. Electrical and chemical properties approach those of the unreinforced resin, while the coefficient of friction is actually lower. Tests indicate that the service temperature limit is approximately 400°F. The fluoropolymer, reinforced with glass, can be molded by conventional methods with rapid molding cycles.

Perfluoroalkoxy (PFA) Resins.   PFA resins, a class of melt-processible fluoroplastics, combine the ease and economics of thermoplastic processing with high-temperature performance in the range of TFE fluorocarbon resins. PFA resins resemble FEP fluorocarbon resins in having a branched polymer chain that provides good mechanical properties at melt viscosities much lower than those of TFE. However, the unique branch in PFA is longer and more flexible, leading to improvements in high-temperature properties, higher melting point, and greater thermal stability.

In use, PFA resins have the desirable properties typical of fluorocarbons, including resistance to virtually all chemicals, antistick, low coefficient of friction, excellent electrical characteristics, low smoke, excellent flammability resistance, ability to perform in temperature extremes, and excellent weatherability. Their strength and stiffness at high operating temperatures are at least equivalent to those of TFE, while their creep resistance appears to be better over a wide temperature range. They should perform successfully in the 500°F area.

Film and sheet are expected to find use as electrical insulations in flat cables and circuitry, and in laminates used in electrical and mechanical applications.

Chlorotrifluoroethylene (CTFE).   The CTFE resins, like the FEP materials, are melt-processible and can be injection-, transfer-, and compression-molded or screw-extruded. Compared with TFE, CTFE has greater tensile and compressive strength within its service-temperature range. However, at the temperature extremes, CTFE does not perform as well as TFE for parts such as seals. At the low end of the range, TFE has somewhat better physical properties; at the higher temperatures, CTFE is more prone to stress cracking and other difficulties.

The electrical properties of CTFE are generally excellent, but dielectric losses are higher than those of TFE. Chemical resistance is poorer than that of TFE, but radiation resistance is better. CTFE does not have the low-friction and bearing properties of TFE.

Ethylene-Chlorotrifluoroethylene (E-CTFE) Copolymer.   E-CTFE copolymer, a copolymer of ethylene and CTFE, is a strong, highly impact-resistant material that retains useful properties over a broad temperature range. E-CTFE is available in pellet and powder form and can be readily extruded, injection-, transfer-, and compression-molded, rotocast, and powder-coated. Its tensile strength, impact resistance, hardness, creep properties, and abrasion resistance at ambient temperature are comparable with those of nylon 6. E-CTFE retains its strength and impact resistance down to cryogenic temperatures.

E-CTFE also exhibits excellent ac loss properties. Its dielectric constant measures 2.5 to 2.6 over the frequency range of 10^2 to 10^6 Hz and is unaffected by temperature over the range of −80 to 300°F. Its dissipation factor is low, ranging from 0.0008 to 0.015 depending on temperature and frequency. It has a dielectric strength of 2000 V/mil in 10-mil wall thickness and a volume resistivity of 10^{16} Ω·cm.

Vinylidene Fluoride (PVF₂). PVF$_2$ is another melt-processable fluorocarbon capable of being injection- and compression-molded and screw-extruded. Its 20 percent lower specific gravity compared with that of TFE and CTFE and its good processing characteristics permit economy and provide excellent chemical and physical characteristics. The useful temperature range is from −80 to +300°F. Although it is stiffer and has higher resistance to cold flow than TFE, its chemical resistance, useful temperature range, antistick properties, lubricity, and electrical properties are lower. Polyvinylidene polymers have seen some use recently in piezoelectric devices as sensors.[48]

Polyvinyl Fluoride (PVF). PVF is manufactured as a film, the combination of excellent weathering and fabrication properties of which has made it widely accepted for surfacing industrial, architectural, and decorative building materials. It has outstanding weatherability, solvent resistance, chemical resistance, abrasion resistance, and color retention. It is supplied with a variety of surfaces for different bonding and antistick objectives.

1.7.6 Ionomers

Outstanding advantages of this polymer class are combinations of toughness, transparency, low-temperature impact, and solvent resistance.[49] Ionomers have high melt strength for thermoforming and extrusion-coating processes, and a broad processing-temperature range. There are resin grades for extrusion, films, injection molding, blow molding, and other thermoplastic processes.

Limitations of ionomers include low stiffness, susceptibility to creep, low heat-distortion temperature, and poor ultraviolet resistance unless stabilizers are added where these properties are important. Most ionomers are very transparent. In 60-mil sections, internal haze ranges from 5 to 25 percent. Light transmission ranges from 80 to 92 percent over the visible region, and in specific compositions high transmittance extends into the ultraviolet region.

Basically, commercial ionomers are nonrigid, unplasticized plastics. Outstanding low-temperature flexibility, resilience, high elongation, and excellent impact strength typify the ionomer resins.

Deterioration of mechanical and optical properties can occur when ionomers are exposed to ultraviolet light and weather. Some grades are available with ultraviolet stabilizers that provide up to 1 year of outdoor exposure with no loss in mechanical properties. Formulations containing carbon black provide ultraviolet resistance equal to that of black polyethylene.

Most ionomers have good dielectric characteristics over a broad frequency range. The combination of these electrical properties, high melt strength, and abrasion resistance qualifies these materials for insulation and jacketing of wire and cable. Their excellent mechanical and optical properties make them specially useful for thermoformed packaging and abrasion- and impact-resistant shoe parts. Typical materials are shown in Table 1.13, and properties are given in Ref. 44.

1.7.7 Low-Permeability Thermoplastics

Several grades of nitrile polymers have been found to offer outstanding low-permeability barrier properties against gases, odor, and flavor, some approaching the barrier properties of metal and glass. This strong characteristic, coupled with good transparency, good thermal and mechanical properties, and convenient processibility, makes this class of thermoplastics useful as containers for food and beverage products. Typical materials are shown in Table 1.13, and properties are given in Ref. 44.

Nylon 6

Nylon 6/6

Nylon 6/10

Nylon 11

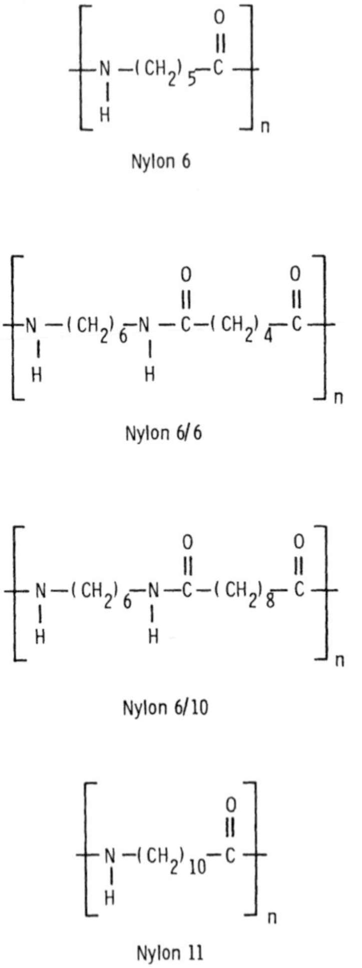

FIGURE 1.30 Chemical structures for four nylons.

1.7.8 Nylons

Also known as polyamides, nylons are strong, tough thermoplastics with good impact, tensile, and flexural strengths from freezing temperatures up to 300°F; excellent low-friction properties; and good electrical resistivities. The structures of four common nylons are shown in Fig. 1.30. Since all nylons absorb some moisture from environmental humidity, moisture-absorption characteristics must be considered in designing with these materials. They will absorb from 0.5 to nearly 2 percent moisture after 24-h water immersion. There are low-moisture-absorption grades, however; and hence moisture-absorption properties do not have to limit the use of nylons, especially for the lower-moisture-absorption grades. Typical materials and properties are shown in Ref. 36.

Regarding the identification of the various grades of nylon, certain nylons are identified by the number of carbon atoms in the diamine and dibasic acid used to produce that particular grade. For instance, nylon 6/6 is the reaction product of hexamethylenediamine and adipic acid, both of which are materials containing six carbon atoms in their chemical structure. Some common commercially available nylons are 6/6, 6, 6/10, 8, 11, and 12. Grades 6 and 6/6 are the strongest structurally; grades 6/10 and 11 have the lowest moisture absorption, best electrical properties, and best dimensional stability; and grades 6, 6/6, and 6/10 are the most flexible. Grades 6, 6/6, and 8 are heat-sealable, with nylon 8 being capable of cross-linking. Another grade, nylon 12, offers advantages similar to those of grades 6/10 and 11, but lower cost possibilities due to being more easily and economically processed. Also a high-temperature type of nylon exists. It is discussed separately hereafter.

In situ polymerization of nylon permits massive castings. Cast nylons are readily polymerized directly from the monomer material in the mold at atmospheric pressure. The method finds application where the size of the part required or the need for low tooling cost precludes injection molding. Cast nylon displays excellent bearing and fatigue properties as well as the other properties characteristic of other basic nylon formulations, with the addition of size and short-run flexibility advantages of the low-pressure casting process.

One special process exists in which nylon parts are made by compressing and sintering, thereby creating parts having exceptional wear characteristics and dimensional stability. Various fillers such as molybdenum disulfide and graphite can be incorporated into nylon to give special low-friction properties. Also, nylon can be reinforced with glass fibers, thus giving it considerable additional strength. These variations are further discussed in a later part of this chapter dealing with glass-fiber-reinforced thermoplastic materials.

Effect of Temperature on Nylon. One of the major advantages of nylon resins is that they retain useful mechanical properties over a range of temperatures from –60 to +400°F.

Both long-term and short-term effects of temperature must, however, be considered. In the short term, there are effects on such properties as stiffness and toughness. There is also the possibility of stress relief and its effect on dimensions. Of most concern in long-term applications at high temperature is gradual oxidative embrittlement, and for such cases the use of heat-stabilized resins is recommended.

The important consideration for design work with nylon resins is that exposure to high temperatures in air for a period of time will result in a permanent change in properties due to oxidation. The degree of change in properties depends on the temperature level, the time exposed, and the composition of the nylon used. The effect on the tensile strength, as measured at room temperature and 2.5 percent moisture content, is that a 25 percent strength reduction can occur in 3 months at 185°F, and 50 percent reduction can occur in 3 months at 250°F. Stabilized nylon does not change appreciably at 250°F aging.

High-temperature oxidation reduces the impact strength even more than the static-strength properties. For instance, the impact strength and elongation of nylon are reduced considerably after several days at 250°F.

A similar change in properties is encountered on exposure to high-temperature water for long periods of time. In this case a reaction with water takes place. There is no significant reaction up to 120°F. This has been confirmed by molecular-weight measurements on 15-year-old samples.

In boiling water, the tensile strength is slowly reduced until, after 2500 h, it levels off at 6000 lb/in^2 (tested at room temperature and 2.5 percent moisture content). The elongation drops rapidly after 1500 h; hence this time has been taken as the limit for the use of basic nylon. Some compositions are especially resistant to hot-water exposure, however.

Effect of Moisture on Dimensional Control. Freshly molded objects normally contain less than 0.3 percent of water, since only dry molding powder can be molded successfully. These objects will then absorb moisture when they are exposed to air or water. The amount of absorbed water will increase in any environment until an equilibrium condition based on relative humidity (RH) is reached. Equilibrium moisture contents for two humidity levels are approximately as follows[50]:

	Zytel 101	Zytel 31
50% RH air	2.5%	1.4%
100% RH air (or water)	8.5%	3.5%

These equilibrium moisture contents are not affected by temperature to any significant extent. Thus the final water content at equilibrium will be the same whether objects of nylon are exposed to water at room temperature or at boiling temperature.

The time required to reach equilibrium, however, is dependent on the temperature, the thickness of the specimen, and the amount of moisture present in the surroundings. Nylon exposed in boiling water will reach the equilibrium level, 8.5 percent, much sooner than nylon in cool water.

When nylon that contains some moisture is exposed to a dry atmosphere, the loss of water will be the reverse of the changes described, and it will take about the same length of time for a corresponding change to occur.

In the most common exposure, an environment of constantly varying humidity, no true equilibrium moisture content can be established. However, moldings of nylon will gradually gain in moisture content in such an environment until a balance is obtained with the midrange humidities. A slow cycling of moisture content near this value will then occur. In all but very thin moldings, the day-to-day or week-to-week variations in relative humidity will have little effect on the total moisture content. The long period changes, such as between summer and winter, will have some effect, depending on the thickness and the relative-humidity range. The highest average humidity for a month will not generally be above 70 percent. In cold weather, heated air may average as low as 20 percent relative humidity. Even at these extremes, the change in the moisture content of nylon is small in most cases, because of the very low rate of both absorption and desorption.

There are two significant dimensional effects that occur after molding. In some cases these oppose each other so that critical dimensions may change very little in a typical air environment. The first is a shrinkage in the direction of flow due to the relief of molded-in stresses. The second is an increase due to moisture absorption. In applications where dimensions are critical to performance and one of these effects predominates, it may be necessary to anneal or moisture-condition the parts to obtain the best performance. For example, a part that will be exposed to high temperatures might require annealing, and an object in water service might be moisture-conditioned to effect dimensional changes before use.

The magnitude of the dimensional change due to molded-in stresses depends on molding conditions and part geometry. This change in size can be determined in a given case only by measuring and annealing a few pieces. In general, long dimensions (in the direction of flow) will shrink, but the short dimensions will increase in an amount that is often too small to measure.

The effect of the moisture content on the dimensions of a molding can be predicted more accurately than the effect of annealing. The changes in dimension at various water contents for Zytel 101 and Zytel 31 are shown in Fig. 1.31. These data are for the annealed or stress-free condition. For most applications in air, the dimensions corresponding to those obtained in equilibrium with 50 percent relative-humidity air are usually chosen as the average size expected when a moisture balance is established. As shown, Zytel 101 will increase 0.006 in/in from the dry condition to equilibrium with 50 percent relative humidity. Zytel 31 under similar conditions will

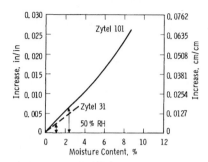

FIGURE 1.31 Changes in dimensions with moisture content for two nylon materials in stress-free, annealed conditions.[50]

increase 0.0025 in/in. In many applications these changes are small enough and occur so slowly that they do not affect the operation of the finished parts.

Stress-free objects of Zytel 101 and Zytel 31 will increase approximately 0.026 and 0.007 in/in, respectively, during the change from dry conditions to completely saturated with water. In water service, if dimensions are critical, the molding can be moisture-conditioned before assembly to allow for this change. Once a part is at or near saturation, little dimensional change can occur unless it is allowed to dry for long periods.

Aramid (High-Temperature, Nylon). In addition to conventional nylon molding resins, there is a high-temperature nylon now called aramid. This high-temperature nylon, Nomex, retains about 60 percent of its strength at temperatures of 475 to 500°F, which would melt conventional nylons. It has good dielectric strength (constant to 400°F and 95 percent relative humidity) and volume resistivity and a low dissipation factor. Fabricated primarily in sheet, fiber, and paper form, this material is being used in wrapped electrical-insulation constructions such as transformer coils and motor stators. It retains high tensile strength, resistance to wear, and electrical properties after prolonged exposure at temperatures up to 500°F.

1.7.9 Parylenes

Parylene is the generic name for members of a thermoplastic-polymer series developed by Union Carbide that are unique among plastics in that they are produced as thin films by vapor-phase polymerization.[51] The polymers are highly crystalline, high-molecular-weight, straight-chain compounds that are known as tough materials with excellent dielectric characteristics.

Parylene is extremely resistant to chemical attack, exceptionally low in trace-metal contamination, and compatible with all organic solvents used in the cleaning and processing of electronic circuits and systems. Although parylene is insoluble in most materials, it will soften in solvents having boiling points in excess of 150°C.

The basic member of this thermoplastic polymer family, parylene N or poly-para-xylylene, exhibits superior dielectric strength, exceptionally high surface and volume resistivities, and electrical properties that vary remarkably little with changes in temperature. Its chemical structure is as follows:

$$-\left[CH_2-\bigcirc-CH_2 \right]-$$

Since its melting point far exceeds that of many other thermoplastics, parylene N can even be used at temperatures exceeding 220°C in the absence of oxygen. Because parylene can be deposited in very thin coatings, heat generated by coated components is easier to control, and differences in thermal expansions are less of a problem than with conventional coatings.

The other commercially available member of the group, parylene C, is poly-mono-chloro-para-xylylene. Its chemical structure is as follows:

$$\left[CH_2 - \underset{}{\underset{}{\bigcirc}}^{Cl} - CH_2 \right]$$

Parylene C offers significantly lower permeability to moisture and gases, such as nitrogen or oxygen, while retaining excellent electrical properties.

The Parylene Process.   Unlike most plastics, parylene is not produced and sold as a polymer. It is not practical to melt, extrude, mold, or calender it as with other thermoplastics. Nor can it be applied from solvent systems, since it is insoluble in conventional solvents.

The advantages that characterize this rugged, intractable material are made possible by a fast, relatively simple vacuum-application system. While it is a unique system, it is often less cumbersome, less complex, and easier to use and maintain than many techniques devised for more conventional plastic-coating systems.

The parylene processor starts with a dimer rather than a polymer and polymerizes it on the surface of an object. To achieve this, the dimer must first go through a two-step heating process. The solid dimer is converted to a reactive vapor of the monomer. When passed over room-temperature objects, the vapor will rapidly coat them with polymer.

This process of polymerization on the object surface has much to recommend it, particularly when the goal is coating uniformity. Unlike dip or spray coating, condensation coating does not run off or sag. It is not "line of sight" as in vacuum metallizing; the vapor coats evenly—over edges, points, and internal areas. The vapor is pervasive, but it coats without bridging so that holes can be jacketed evenly. Masking tape can be used if it is desired that some areas not be coated. Moreover, with parylene, the object to be coated remains at or near room temperature, eliminating all risk of thermal damage. The coating thickness is controlled easily and very accurately simply by regulating the amount of dimer to be vaporized.

Deposition chambers of virtually any size can be constructed. Those currently in use range from 500 to 28,000 in³. Large parts (up to 5 ft long and 18 in high) can be processed in this equipment. The versatility of the process also enables the simultaneous coating of many small parts of varying configurations. Because of this, a time savings in labor requirements can readily be achieved.

Parylenes can also be deposited onto a cold condenser and then stripped off as a free film, or they can be deposited onto the surface of objects as a continuous, adherent coating in thicknesses ranging from 1000 Å (about 0.004 mil) to 3 mil or more. The deposition rate is normally about 0.5 µm/min (about 0.02 mil). On cooled substrates, the deposition rate can be as high as 1.0 mil/min.

Parylene Properties.   The material can be used at both elevated and cryogenic temperatures. The 1000-h service life for the N and the C members is 200 to 240°F. Corresponding 10-year service in air is limited to 140 to 175°F. Parylenes are excellent in having low gas permeability, low moisture-vapor transmission, and low-temperature ductility. Dimensional stability is reported as better than that of polycarbonate, and over-all barrier properties to most gases are reported superior to those of many other barrier films. Trade names and suppliers are listed in Table 1.13, and important properties of this unique material are listed in Tables 1.16 and 1.17.

TABLE 1.16   Thermal, Physical, and Mechanical Properties of Parylenes[51]

	Parylene N	Parylene C
Typical Thermal Properties		
Temp (melting), °C........................	405	280
Linear coefficient of expansion, 10^{-5}/°C........	6.9	3.5
Thermal conductivity, 10^{-4} cal/s/(cm²)(°C/cm).	$\sim$3	
Typical Physical and Mechanical Properties		
Tensile strength, lb/in.².....................	6,500	10,000
Yield strength, lb/in.².......................	6,100	8,000
Elongation to break, %.....................	30	200
Yield elongation, %........................	2.5	2.9
Density, g/cm³.............................	1.11	1.289
Coefficient of friction:		
Static..................................	0.29	0.25
Dynamic...............................	0.29	0.25
Water absorption, 24 h.....................	0.06 (0.029 in.)	0.01 (0.019 in.)
Index of refraction n_D 23°C.................	1.661	1.639

Data recorded following appropriate ASTM method.

TABLE 1.17   Film-Barrier Properties of Parylenes[51]

Polymer	Gas permeability, cm³-mil/100 in.², 24 h-atm (23°C)						Moisture-vapor transmission, g-mil/100 in.², 24 h, 37°C, 90% RH
	N_2	O_2	CO_2	H_2S	SO_2	Cl_2	
Parylene N.....	7.7	39.2	214	795	1,890	74	1.6
Parylene C.....	1.0	7.2	7.7	13	11	0.35	0.5
Epoxies........	4	5–10	8	...	...	...	1.8–2.4
Silicones........	...	50,000	300,000	...	...	...	4.4–7.9
Urethanes......	80	200	3,000	...	...	...	2.4–8.7

Data recorded following appropriate ASTM method.

1.7.10   Polyaryl Ether

Another material that can be classed as an engineering thermoplastic is polyaryl ether, available from Uniroyal, Inc., as Arylon T. A unique combination of four important, desirable properties makes polyaryl ether outstanding. The material has a high heat-deflection temperature of 300°F measured at 264 lb/in² on a ¼- by ¼- by 5-in bar; high impact strength (Izod impact strength of 8 ft·lb/in notch at 72°F and 2.5 ft·lb/in notch at –20°F); excellent chemical resistance, resisting organic solvents except chlorinated aromatics, esters, and ketones; and it appears to resist hydrolysis, withstanding prolonged immersion in boiling water. Fabricating advantages include a low shrinkage factor, precision molding of void-free parts, and insensitivity to moisture, requiring only nominal drying before molding. Markets for Arylon T include the automotive, appliance, and electrical industries.

1.7.11   Polyaryl Sulfone

Polyaryl sulfone offers the unique combination of thermoplasticity and retention of structurally useful properties at 500°F.[52] This material is supplied in both pellet and powder form.

Polyaryl sulfone consists mainly of phenyl and biphenyl groups linked by thermally stable ether and sulfone groups. It is distinguished from polysulfone polymers by the absence of aliphatic groups, which are subject to oxidative attack. This aromatic structure gives it excellent resistance to oxidative degradation and accounts for its retention of mechanical properties at high temperatures. The presence of ether-oxygen linkages gives the polymer chain flexibility to permit fabrication by conventional melt-processing techniques. Typical materials are shown in Table 1.13, and typical properties are given in Ref. 53.

Thermal Properties. Polyaryl sulfone is characterized by a very high heat-deflection temperature, 535°F at 264 lb/in^2, which is approximately 150°F higher than many other commercially available thermoplastics, as shown in Fig. 1.32. This is a consequence of its high glass-transition temperature, 550°F, rather than the effect of filler reinforcement or a crystalline melting point. At 500°F it maintains a tensile strength in excess of 4000 lb/in^2 and a flexural modulus of 250,000 lb/in^2. The resistance to oxidative degradation is indicated by the ability of polyaryl sulfone to retain its tensile strength after 2000-h exposure to 500°F air-oven aging.

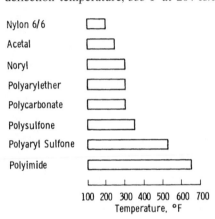

FIGURE 1.32 Approximate heat-deflection temperatures of some engineering thermoplastics at 264 lb/in^2.

Chemical Resistance. Polyaryl sulfone has good resistance to a wide variety of chemicals, including acids, bases, and common solvents. It is unaffected by practically all fuels, lubricants, hydraulic fluids, and cleaning agents used on or around electrical components. Highly polar solvents such as *N,N*-dimethylformamide, *N,N*-dimethylacetamide, and *N*-methylpyrrolidone are solvents for the material.

1.7.12   Polycarbonates

This group of plastics is also among those classified as engineering thermoplastics because of their high performance characteristics in engineering designs. The generalized chemical structure is as follows:

Polycarbonate

Polycarbonates are especially outstanding in impact strength, having strengths several times higher than other engineering thermoplastics. Polycarbonates are tough, rigid, and dimensionally stable and are available as transparent or colored parts. They are easily fabricated with reproducible results, using molding or machining techniques. An important molding characteristic is the low and predictable mold shrinkage, which sometimes gives polycarbonates an advantage over nylons and acetals for close-tolerance parts. As with most other plastics containing aromatic groups, radiation stability is high. The most commonly useful properties of polycarbonates are creep resistance, high heat resistance, dimensional stability, good electrical properties, self-extinguishing properties, product transparency, and the exceptional impact strength which compares favorably with that of some metals and exceeds that of many competitive plastics. In fact, polycarbonate is sometimes considered to be competitive with zinc and aluminum castings. Although such comparisons have limits, the fact that the comparisons are sometimes made in material selection for product design indicates the strong performance characteristics possible in polycarbonates.

In addition to their performance as engineering materials, polycarbonates are also alloyed with other plastics in order to increase the strength and rigidity of these plastics. Notable among the plastics with which polycarbonates have been alloyed are the ABS plastics. In addition to standard grades of polycarbonates, a special film grade exists for high-performance capacitors.[54] Typical properties are shown in Table 1.13.

Moisture-Resistance Properties. Oxidation stability on heating in air is good, and immersion in water and exposure to high humidity at temperatures up to 212°F have little effect on dimensions. Steam sterilization is another advantage that is attributable to the resin's high heat stability. However, if the application requires continuous exposure in water, the temperature should be limited to 140°F. Polycarbonates are among the most stable plastics in a wet environment, as shown in Figs. 1.33 and 1.34.[55,56]

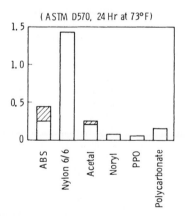

FIGURE 1.33 Water absorption of several thermoplastics.[55,56]

1.7.13 Polyesters

Thermoplastic polyesters have been and are currently used extensively in the production of film and fibers. These materials are denoted chemically as polyethylene terephthalate. In the past few years a new class of high-performance molding and extrusion grades of thermoplastic polyesters has been made available and is becoming increasingly competitive among plastics. These polymers are denoted chemically as poly(1,4-butylene terephthalate) and poly(tetramethylene terephthalate). These thermoplastic polyesters are highly crystalline, with a melting point of about 430°F. They are fairly translucent in thin molded sections and opaque in thick sections, but can be extruded into transparent thin film. Both unreinforced and reinforced formulations are extremely easy to process and can be molded in very fast cycles. Typical properties are shown in Ref. 57.

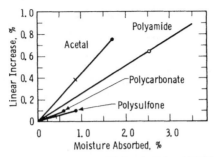

FIGURE 1.34 Dimensional changes of several thermoplastics due to absorbed moisture.[55,56]

The unreinforced resin offers the following characteristics: (1) hard, strong, and extremely tough; (2) high abrasion resistance, low coefficient of friction; (3) good chemical resistance, very low moisture absorption and resistance to cold flow; (4) good stress crack and fatigue resistance; (5) good electrical properties; and (6) good surface appearance. Electrical properties are stable up to the rated temperature limits.

The glass-reinforced polyester resins are unusual in that they are the first thermoplastics that can compare with, or are better than, thermosets in electrical, mechanical, dimensional, and creep properties at elevated temperatures (approximately 300°F), while having superior impact properties.

The glass-fiber concentration usually ranges from 10 to 30 percent in commercially available grades. In molded parts the glass fibers remain slightly below the surface so that finished items have a very smooth surface finish as well as an excellent appearance.

Unreinforced resins are primarily used in housings requiring excellent impact, and in moving parts such as gears, bearings, and pulleys, packaging applications, and writing instruments. The flame-retardant grades are primarily aimed at television, radio, and electrical and electronics parts as well as business-machine and pump components. Reinforced resins are being used in automotive, electrical and electronic, and general industrial areas, replacing thermosets, other thermoplastics, and metals. Electrical and mechanical properties coupled with low finished-part cost are enabling reinforced thermoplastic polyesters to replace phenolics, alkyds, DAP, and glass-reinforced thermoplastics in many applications.

1.7.14 Polyethersulfone

Polyethersulfone is a relatively new high-temperature engineering thermoplastic with an outstanding long-term resistance to creep at temperatures up to 150°C.[58] It is also capable of being used continuously under load at temperatures of up to about 180°C and, in some low-stress applications, up to 200°C. Other grades are capable of operating at temperatures above 200°C and for specialized adhesive and lacquer applications.

The polyethersulfone chemical structure shown here gives an amorphous polymer, which possesses only bonds of high thermal and oxidative stability.

$$\left[\!\!\left[\bigcirc \!\!-\! SO_2 \!\!-\! \bigcirc \!\!-\! O \right]\!\!\right]_n$$

While the sulfone group confers high-temperature performance, the ether linkage contributes toward practical processing by allowing mobility of the polymer chain when in the melt phase.

Polyethersulfone exhibits low creep. A constant stress of 3000 lb/in^2 at 20°C for 3 years produces a strain of 1 percent, while a stress of 6500 lb/in^2 results in a strain of only 2.6 percent over the same period of time. Higher modulus values are obtained with polyethersulfone at 150°C than with polysulfone, phenylene oxide-based resins, or polycarbonate at considerably lower temperatures.

Although its load-bearing properties are reduced above 150°C, polyethersulfone can still be considered for applications at temperatures up to 180°C. It remains form-stable to above 200°C and has a heat-deflection temperature of 203°C at 264 lb/in^2.

Polyethersulfone is especially resistant to acids, alkalies, oils, greases, and aliphatic hydrocarbons and alcohols. It is attacked by ketones, esters, and some halogenated and aromatic hydrocarbons. Available materials are identified in Table 1.13.

1.7.15 Polyethylenes, Polypropylenes, and Polyallomers

This large group of polymers is basically divided into the three separate polymer groups listed under this heading; all belong to the broad chemical classification known as polyolefins. Polyethylene and polypropylene can be considered as the first two members of a large group of polymers based on the ethylene structure. Their structures are as follows:

$$\left[\begin{matrix} H & H \\ | & | \\ -C - C - \\ | & | \\ H & H \end{matrix} \right]_n \qquad\qquad \left[\begin{matrix} H & H \\ | & | \\ -C - C - \\ | & | \\ H & CH_3 \end{matrix} \right]_n$$

Polyethylene   Polypropylene

Molecular changes beyond these two structures give quite different polymers and properties and are covered separately in other parts of this chapter. The chemical changes result from the replacement of the methyl group (—CH$_3$) in polypropylene with substituents such as chlorine (polyvinyl chloride), —OH (polyvinyl alcohol), F (polyvinyl fluoride), and —CN (polyacrylonitrile). There are many categories or types even within each of the three polymer groups discussed in this section. Although property variations exist among these three polymer groups and among the subcategories within these groups, there are also many similarities. The differences or unique features of each are discussed in this section. The similarities are, broadly speaking, appearance, general chemical characteristics, and electrical properties. The differences are more notably in physical and thermal-stability properties. Basically, polyolefins are all waxlike in appearance and extremely inert chemically, and they exhibit decreases in physical strength at somewhat lower temperatures than the higher-performance engineering thermoplastics. Polyethylenes were the first of these materials developed and, hence, for some of the original types, have the weakest mechanical properties. The later-developed polyethylenes, polypropylenes, and polyallomers offer improvements. The unique features of each of these three polymer groups are outlined in the following paragraphs. Typical materials are shown in Table 1.13, and typical properties are given in Ref. 36.

Polyethylenes. Polyethylenes are among the best-known plastics and come in three main classifications based on density: low, medium, and high. These density ranges are 0.910 to 0.925, 0.925 to 0.940, and 0.940 to 0.965, respectively. These three density grades are also sometimes known as types I, II, and III. All polyethylenes are relatively soft, and hardness increases as density increases. Generally, the higher the density, the better are the dimensional stability and physical properties, particularly as a function of temperature. The thermal stability of polyethylenes ranges from 190°F for the low-density material up to 250°F for the high-density material. Toughness is maintained to low negative temperatures.

Polypropylenes. Polypropylenes are chemically similar to polyethylenes but have somewhat better physical strength at a lower density. The density of polypropylenes is among the lowest of all plastic materials, ranging from 0.900 to 0.915. Polypropylenes offer more of a balance of properties than a single unique property, with the exception of flex-fatigue resistance. These materials have an almost infinite life under flexing, and they are thus often said to be "self-hinging." Use of this characteristic is widespread as plastic hinges. Polypropylenes are perhaps the only thermoplastics surpassing all others in combined electrical properties, heat resistance, rigidity, toughness, chemical resistance, dimensional stability, surface gloss, and melt flow, at a lower cost than that of most others.

Because of their exceptional quality and versatility, polypropylenes offer outstanding potential in the manufacture of products through injection molding. Mold shrinkage is significantly less than that of other polyolefins; uniformity in and across the direction of flow is appreciably greater. Shrinkage is therefore more predictable, and there is less susceptibility to warpage in flat sections.

Polyallomers. Polyallomers are also polyolefin-type thermoplastic polymers produced from two or more different monomers, such as propylene and ethylene, which would produce a propylene-ethylene polyallomer. The monomers, or base chemical materials, are similar to those of polypropylene or polyethylene. Hence, as was mentioned, and as would be expected, many properties of polyallomers are similar to those of polyethylenes and polypropylenes. Having a density of about 0.9, they, like polypropylenes, are among the lightest plastics.

Polyallomers have a brittleness temperature as low as −40°F and a heat-distortion temperature as high as 210°F at 66 lb/in². The excellent impact strength plus exceptional flow properties of polyallomer provide wide latitude in product design. Notched Izod impact strengths run as high as 12 ft·lb/in notch.

Although the surface hardness of polyallomers is slightly less than that of polypropylenes, resistance to abrasion is greater. Polyallomers are superior to linear polyethylene in flow characteristics, moldability, softening point, hardness, stress-crack resistance, and mold shrinkage. The flexural-fatigue-resistance properties of polyallomers are as good as or better than those of polypropylenes.

Cross-Linked Polyolefins. While polyolefins have many outstanding characteristics, they, like all thermoplastics to some degree, tend to creep or cold-flow under the influence of temperature, load, and time. In order to improve this and some other properties, considerable work has been done on developing cross-linked polyolefins, especially polyethylenes. The cross-linked polyethylenes offer thermal-performance improvements of up to 25°C or more. Cross-linking has been achieved primarily by chemical means and by ionizing radiation. Products of both types are available. Radiation-cross-linked polyolefins have gained particular prominence in a heat-shrinkable form. This is achieved by cross-linking the extruded or molded polyolefin using high-energy electron-beam radiation, heating the irradiated material above its crystalline melting point to a rubbery state, mechanically stretching to an expanded form (up to four or five times the original size), and cooling the stretched material. Upon further heating, the material will return to its original size, tightly shrinking onto the object around which it has been placed. Heat-shrinkable boots, jackets, and tubing are widely used. Also, irradiated polyolefins, sometimes known as irradiated polyalkenes, are important materials for certain wire and cable jacketing applications.

1.7.16 Polyimides and Polyamide-Imides

Among the commercially available plastics generally considered as having high heat resistance, polyimides can be used at the highest temperatures, and they are the strongest and most rigid. Polyimides have a useful operating range to about 900°F (482°C) for short durations and 500 to 600°F (260 to 315°C) for continuous service in air. Prolonged exposure at 500°F (260°C) results in moderate (25 to 30 percent) loss of original strength and rigidity.

These materials, which can be used in various forms including moldings, laminates, films, coatings, and adhesives, have high mechanical properties, wear resistance, chemical and radiation inertness, and excellent dielectric properties over a broad temperature range. Materials and sources are listed in Table 1.13, and properties are given in Ref. 36. The thermal stability is compared with that of other engineering plastics in Fig. 1.32.

Chemical Structures. Polyimides are heterocyclic polymers, having a noncarbon atom of nitrogen in one of the rings in the molecular chains.[36] The atom is nitrogen and it is in the inside ring as follows:

The fused rings provide chain stiffness essential to high-temperature strength retention. The low concentration of hydrogen provides oxidative resistance by preventing thermal degradative fracture of the chain.

The other resins considered as members of this family of polymers are the poly(amide-imide)s. These compositions contain aromatic rings and the characteristic nitrogen linkages, as follows:

$$\left[\begin{array}{c} \\ -R-\overset{H}{\underset{|}{N}}-\overset{}{\underset{\overset{||}{O}}{C}}-\overset{}{\overset{\overset{O}{||}}{\underset{\underset{||}{O}}{\underset{C}{}}}}\hspace{-2mm}\overset{}{\underset{C}{}}\hspace{-2mm}N- \\ \end{array}\right]_n$$

There are two basic types of polyimides: (1) condensation and (2) addition resins. The condensation polyimides are based on a reaction of an aromatic diamine with an aromatic dianhydride. A tractable (fusible) polyamic acid intermediate produced by this reaction is converted by heat to an insoluble and infusible polyimide, with water being given off during the cure. Generally, the condensation polyimides result in products having high void contents that detract from inherent mechanical properties and result in some loss of long-term heat-aging resistance.

The addition polyimides are based on short, preimidized polymer-chain segments similar to those comprising condensation polyimides. These prepolymer chains, which have unsaturated aliphatic end groups, are capped by termini that polymerize thermally without the loss of volatiles. The addition polyimides yield products that have slightly lower heat resistance than the condensation polyimides.

The condensation polyimides are available as either thermosets or thermoplastics, and the addition polyimides are available only as thermosets. Although some of the condensation polyimides technically are thermoplastics, which would indicate that they can be melted, this is not the case, since they have melting temperatures that are above the temperature at which the materials begin to decompose thermally.

Properties.   Polyimides and polyamide-imides exhibit some outstanding properties due to their combination of high-temperature stability up to 500 to 600°F continuously and to 900°F for intermediate use, excellent electrical and mechanical properties that are also relatively stable from low negative temperatures to high positive temperatures, dimensional stability (low cold flow) in most environments, excellent resistance to ionizing radiation, and very low outgassing in high vacuum. They have very low coefficients of friction, which can be further improved by use of graphite or other fillers. Materials and properties are shown in Ref. 36. Polyamide-imides and polyimides have very good electrical properties, though not as good as those of TFE fluorocarbons, but they are much better than TFE fluorocarbons in mechanical and dimensional-stability properties. This provides advantages in many high-temperature electronic applications. All these properties also make polyamide-imides and polyimides excellent material choices in extreme environments of space and temperature. These materials are available as solid (molded and machined) parts, films, laminates, and liquid varnishes and adhesives. Since the data are relatively similar, except for the form factor, the data presented are for solid polyimides unless indicated otherwise. Films are quite similar to Mylar, except for improved high-temperature capabilities.

1.7.17   Polymethylpentene

Another thermoplastic based on the ethylene structure, polymethylpentene, has special properties due to its combination of transparency and relatively high melting point. This polymer has four combined properties of (1) a high crystalline melting point of 464°F, coupled with useful mechanical properties at 400°F and retention of form stability to near melting; (2) transparency with a light-transmission value of 90 percent in comparison with 88 to 92 percent for polystyrene and 92 percent for acrylics; (3) a density of 0.83, which is close to the theoretical minimum for thermoplastics materials; and (4) excellent electrical properties with power factor, dielectric constant (2.12) and volume resistivity of the same order as PTFE fluorocarbon. Material and source are identified in Table 1.13, and properties are given in Ref. 36. Applications for polymethylpentene have been developed in the field of lighting and in the automotive, appliance, and electrical industries.

1.7.18 Polyphenylene Oxides

A patented process for oxidative coupling of phenolic monomers is used in formulating Noryl phenylene oxide-based thermoplastic resins.[59] The basic phenylene oxide structure is as follows:

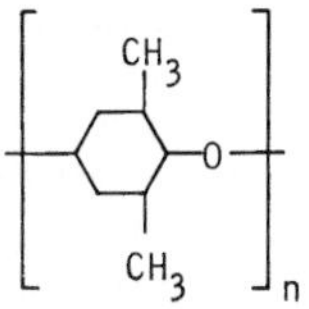

This family of engineering materials is characterized by outstanding dimensional stability at elevated temperatures, broad temperature-use range, outstanding hydrolytic stability, and excellent dielectric properties over a wide range of frequencies and temperatures. Several grades are available that have been developed to provide a choice of performance characteristics to meet a wide range of engineering-application requirements.

Among their principal design advantages are (1) excellent mechanical properties over temperatures from below −40°F to above 300°F; (2) self-extinguishing, nondripping characteristics; (3) excellent dimensional stability with low creep, high modulus, and low water absorption; (4) good electrical properties; (5) excellent resistance to aqueous chemical environments; (6) ease of processing with injection-molding and extrusion equipment; and (7) excellent impact strength. Properties are shown in Ref. 36. Thermal stability and moisture absorption are compared with those of other engineering thermoplastics in Figs. 1.32 and 1.33, respectively. The supplier is identified in Table 1.13.

1.7.19 Polyphenylene Sulfides

A crystalline polymer, polyphenylene sulfide (PPS), has a symmetrical, rigid backbone chain consisting of recurring para-substituted benzene rings and sulfur atoms. Its chemical structure is as follows:

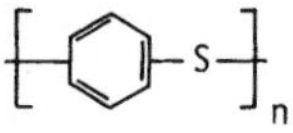

This chemical structure is responsible for the high melting point (550°F), outstanding chemical resistance, thermal stability, and nonflammability of the polymer. There are no known solvents below 375 to 400°F. The polymer is characterized by high stiffness and good retention of mechanical properties at elevated temperatures, which provide utility in coatings as well as in molding compounds. Polyphenylene sulfide is available in a variety of grades suitable for slurry coating, fluidized-bed coating, flocking, electrostatic spraying, as well as injection and compression molding.[60] This material is identified in Table 1.13, and properties of unfilled and glass-filled material are shown in Ref. 36.

At normal temperatures the unfilled polymer is a hard material with high tensile and flexural strengths. Substantial increases in these properties are realized by the addition of fillers, especially glass. Tensile strength and flexural modulus decrease with increasing temperature, leveling off at about 250°F, with good tensile strength and rigidity retained up to 500°F. With increasing temperature there is a marked increase in elongation and a corresponding increase in toughness.

The mechanical properties of PPS are unaffected by long-term exposure in air at 450°F. Molded items have applications where chemical resistance and high-temperature properties are of prime importance. For injection-molding applications, a 40 percent glass-filled grade is recommended. Coatings of PPS require a baking operation. Nonstick formulations can be prepared when a combination of hardness, chemical inertness, and release behavior is required.

Good adhesion to aluminum requires grit blasting and degreasing treatment. Good adhesion to steel is obtained by grit blasting and degreasing, followed by treatment at 700°F in air. Polyphenylene sulfide adheres well to titanium and to bronze after the metal surface has been degreased.

1.7.20 Polystyrenes

Commercial polystyrene is produced by continuous bulk, suspension, and solution polymerization techniques or by combining various aspects of these techniques.[1,61] Its structure is as follows:

$$\left[\begin{array}{c} \underset{\underset{H}{\overset{H}{|}}}{\overset{H}{\underset{|}{C}}}-\underset{\bigcirc}{\overset{H}{\underset{|}{C}}} \end{array}\right]_n$$

The polymerization is a highly exothermic, free-radical reaction. The homopolymer is characterized by its rigidity, sparkling clarity, and ease of processibility; however, it tends to be brittle. Impact properties are improved by copolymerization or grafting polystyrene chains to unsaturated rubbers such as polybutadiene. Rubber levels typically range from 3 to 12 percent. Commercially available impact-modified polystyrene is not as transparent as the homopolymers, but it has a marked increase in toughness.

The versatility of the styrene polymerization processes allows manufacturers to produce products with a wide variety of properties by varying the molecular-weight characteristics, additives, plasticizer content, and rubber levels. Heat resistance ranges from 170 to 200°F. Polystyrenes with tensile elongations from near zero to over 50 percent are produced. Various melt viscosities are also available.

Since properties can be varied so extensively, polystyrene is used in sheet and profile extrusion, thermoforming, injection and extrusion blow molding, heavy and thin-wall injection molding, direct-injection foam-sheet extrusion, biaxially oriented sheet extrusion, injection molding and extrusion of structural foam, and rotational molding. Polystyrenes can be printed; painted; vacuum-metallized and hot-stamped; sonic, solvent, adhesive, and spin welded; and screwed, nailed, and stapled. Polystyrenes are most attractive when considered on a cost-performance comparison with other thermoplastics. Limitations of polystyrene include poor weatherability, loss of clarity with impact modification, limited heat resistance, and flammability. These polymers are identified in Table 1.13, and properties are shown in Ref. 36.

Polystyrenes represent an important class of thermoplastic materials in the electronics industry because of very low electrical losses. Mechanical properties are adequate within operating-temperature limits, but polystyrenes are temperature-limited with normal temperature capabilities below 200°F. Polystyrenes can, however, be cross-linked to produce a higher-temperature material, as noted in Table 1.13.

Cross-linked polystyrenes are actually thermosetting materials and hence do not remelt, even though they may soften. The improved thermal properties, coupled with the outstanding electrical properties, hardness, and associated dimensional stability, make cross-linked polystyrenes the leading choice of dielectric for many high-frequency-radar-band applications.

Conventional polystyrenes are essentially polymerized styrene monomer alone. By varying manufacturing conditions or by adding small amounts of internal and external lubrication, it is possible to vary such properties as ease of flow, speed of setup, physical strength, and heat resistance. Conventional polystyrenes are frequently referred to as normal, regular, or standard polystyrenes.

Since conventional polystyrenes are somewhat hard and brittle and have low impact strength, many modified polystyrenes are available. Modified polystyrenes are materials in which the properties of elongation and resistance to shock have been increased by incorporating into their composition varying percentages of elastomers, as was described. Hence these types are frequently referred to as high-impact, high-elongation, or rubber-modified polystyrenes. The so-called superhigh-impact types can be quite rubbery. Electrical properties are usually degraded by these rubber modifications.

Polystyrenes are subject to stresses in fabrication and forming operations, and often require annealing to minimize such stresses for optimized final-product properties. Parts can usually be annealed by exposing them to an elevated temperature approximately 5 to 10°F lower than the temperature at which the greatest tolerable distortion occurs.

Polystyrenes generally have good dimensional stability and low mold shrinkage, and are easily processed at low costs. They have poor weatherability and are chemically attacked by oils and organic solvents. Resistance is good, however, to water, inorganic chemicals, and alcohols.

1.7.21　Polysulfones

In the natural and unmodified form, polysulfone is a rigid, strong thermoplastic,[36] which can be molded, extruded, or thermoformed (in sheets) into a wide variety of shapes. Characteristics of special significance to the design engineer are their heat-deflection temperature of 345°F at 264 lb/in^2 and long-term use temperature of 300 to 340°F. This is compared with some other engineering thermoplastics in Fig. 1.32. These polymers are identified in Table 1.13 and properties are shown in Ref. 36.

Thermal gravimetric analyses show polysulfone to be stable in air up to 500°C. This excellent thermal resistance of polysulfones, together with outstanding oxidation resistance, provides a high degree of melt stability for molding and extrusion.

Some flexibility in the polymer chain is derived from the ether linkage, thus providing inherent toughness. Polysulfone has a second, low-temperature glass transition at −150°F, similar to other tough, rigid thermoplastic polymers. This minor glass transition is attributable to the ether linkages. The linkages connecting the benzene rings are hydrolytically stable in polysulfones. These polymers therefore resist hydrolysis and aqueous acid and alkaline environments.

Polysulfone is produced by the reaction between the sodium salt of 2,2-bis(4-hydroxyphenol) propane and 4,4′-dichlorodiphenyl sulfone.[62] The sodium phenoxide end groups react with methyl chloride to terminate the polymerization. The molecular weight of the polymer is thereby controlled and thermal stability is assisted.

1.7.22 Vinyls

Vinyls are structurally based on the ethylene molecule through substitution of a hydrogen atom with a halogen or other group. Materials are identified in Table 1.13, and properties are shown in Ref. 36. Basically, the vinyl family is comprised of seven major types. These are polyvinyl acetals, polyvinyl acetate, polyvinyl alcohol, polyvinyl carbazole, polyvinyl chloride, polyvinyl chloride-acetate, and polyvinylidene chloride.

Polyvinyl acetals consist of three groups, namely polyvinyl formal, polyvinyl acetal, and polyvinyl butyral. These materials are available as molding powders, sheet, rod, and tube. Fabrication methods include molding, extruding, casting, and calendering.

Polyvinyl chloride (PVC) is perhaps the most widely used and highest-volume type of the vinyl family. PVC and polyvinyl chloride-acetate are the most commonly used vinyls for electronic and electrical applications.

Vinyls are basically tough and strong. They resist water and abrasion and are excellent electrical insulators. Special tougher types provide high wear resistance. Excluding some nonrigid types, vinyls are not degraded by prolonged contact with water, oils, foods, common chemicals, or cleaning fluids such as gasoline or naphtha. Vinyls are affected by chlorinated solvents.

Generally, vinyls will withstand continuous exposure to temperatures ranging up to 130°F; flexible types, filaments, and some rigids are unaffected by even higher temperatures. Some of these materials, in some operations, may be health hazards. These materials also are slow-burning, and certain types are self-extinguishing; but direct contact with an open flame or extreme heat must be avoided.

PVC is a material with a wide range of rigidity or flexibility. One of its basic advantages is the way it accepts compounding ingredients. For instance, PVC can be plasticized with a variety of plasticizers to produce soft, yielding materials to almost any desired degree of flexibility. Without plasticizers, it is a strong, rigid material that can be machined, heat-formed, or welded by solvents or heat. It is tough, with high resistance to acids, alcohol, alkalies, oils, and many other hydrocarbons. It is available in a wide range of colors. Typical uses include profile extrusions, wire and cable insulation, and foam applications. It also is made into film and sheets.

PVC raw materials are available as resins, latexes, organosols, plastisols, and compounds. Fabrication methods include injection, compression, blow or slush molding, extruding, calendering, coating, laminating, rotational and solution casting, and vacuum forming.

1.8 GLASS-FIBER-REINFORCED THERMOPLASTICS

Basically, thermoplastic molding materials are developed and can be used without fillers, as opposed to thermosetting molding materials, which are more commonly used with fillers incorporated into the compound. This is primarily because shrinkage, hardness, brittleness, and other important processing and use properties necessitate the use of fillers in thermosets. Thermoplastics, on the other hand, do not suffer from the same shortcomings as thermosets and hence can be used as molded products without fillers. However, thermoplastics do suffer from creep and dimensional-stability problems, especially under elevated temperature and load conditions. Because of this shortcoming, most designers find difficulty in matching the techniques of classical stress-strain analysis with the nonlinear, time-dependent strength-modulus properties of thermoplastics. Glass-fiber-reinforced thermoplastics (FRTPs) help to simplify these problems. For instance, 40 percent glass-fiber-reinforced nylon outperforms its unreinforced version by exhibiting two and one-half times greater tensile and Izod impact strengths, four times greater flexural modulus, and only one-fifth of the tensile creep.

Thus FRTPs fill a major materials gap in providing plastic materials that can be reliably used for strength purposes, and which in fact can compete with metal die castings. Strength is increased with glass-fiber reinforcement, as are stiffness and dimensional stability. The thermal expansion of the FRTPs is reduced, creep is substantially reduced, and molding precision is much greater.

The dimensional stability of glass-reinforced polymers is invariably better than that of the nonreinforced materials. Mold shrinkages of only a few mils per inch are characteristic of these products. Low moisture absorption of reinforced plastics ensures that parts will not suffer dimensional increases under high-humidity conditions. Also, the characteristic low coefficient of thermal expansion is close enough to that of such metals as zinc, aluminum, and magnesium that it is possible to design composite assemblies without fear that they will warp or buckle when cycled over temperature extremes. In applications where part geometry limits maximum wall thickness, reinforced plastics almost always afford economies for similar strength or stiffness over their unreinforced equivalents. A comparison of some important properties for unfilled and glass-filled (20 and 30 percent) thermoplastics is given in Ref. 36.

Chemical resistance is essentially unchanged, except that environmental stress-crack resistance of such polymers as polycarbonate and polyethylene is markedly increased by glass reinforcement.

1.9 PLASTIC FILMS AND TAPES

1.9.1 Films

Films are thin sections of the same polymers described previously in this chapter. Most films are thermoplastic in nature because of the great flexibility of this class of resins. Films can be made from most thermoplastics.

Films are made by extrusion, casting, calendering, and skiving. Certain of the materials are also available in foam form. The films are sold in thicknesses from 0.5 to 10 mil. Thicknesses in excess of 10 mil are more properly called sheets.

1.9.2 Tapes

Tapes are films slit to some acceptable width and are frequently coated with adhesives. The adhesives are either thermosetting or thermoplastic. The thermoset adhesives consist of rubber, acrylic, silicones, and epoxies, whereas the thermoplastic adhesives are generally acrylic or rubber. Tackifying resins are generally added to increase the adhesion. The adhesives all deteriorate with storage. The deterioration is marked by loss of tack or bond strength and can be inhibited by storage at low temperature.

1.9.3 Film Properties

Films differ from similar polymers in other forms in several key properties, but are identical in all others. Since an earlier section of this chapter described in detail most of the thermoplastic resins, this section will be limited to film properties. The properties of common films are presented in Ref. 63. To aid in the selection of the proper films, the most important features are summarized in Table 1.18.

Films differ from other polymers chiefly in improved electric strength and flexibility. Both these properties vary inversely with the film thickness. Electric strength is also related to the method of manufacture. Cast and extruded films have higher electric strength than skived films. This is caused by the greater incidence of holes in the latter films. Some films can be oriented, which improves their physical properties substantially. Orientation is a process of selectively stretching the films, thereby reducing the thickness and causing changes in the crystallinity of the polymer. This process is usually accomplished under conditions of elevated temperature, and the benefits are lost if the processing temperatures are exceeded during service.

Most films can be bonded to other substrates with a variety of adhesives. Those films which do not readily accept adhesives can be surface-treated for bonding by chemical and electrical etching. Films can also be combined to obtain bondable surfaces. Examples of these combined films are polyolefins laminated to polyester films and fluorocarbons laminated to polyimide films.

TABLE 1.18 Film-Selection Chart

Film	Cost	Thermal stability	Dielectric constant	Dissipation factor	Strength	Electric strength	Water absorption	Folding endurance
Cellulose	Low	Low	Medium	Medium	High	Medium	High	Low
FEP fluorocarbon	High	High	Low	Low	Low	High	Very low	Medium
Polyamide	Medium	Medium	Medium	Medium	High	Low	High	Very high
PTFE polytetrafluoroethylene	High	High	Low	Low	Low	Low	Very low	Medium
Acrylic	Medium	Low	Medium	Medium	Medium	Low	Medium	Medium
Polyethylene	Low	Low	Low	Low	Low	Low	Low	High
Polypropylene	Low	Medium	Low	Low	Low	Medium	Low	High
Polyvinyl fluoride	High	High	High	High	High	Medium	Low	High
Polyester	Medium	Medium	Medium	Low	High	High	Low	Very high
Polytrifluorochloroethylene	High	High	Low	Low	Medium	Medium	Very low	Medium
Polycarbonate	Medium	Medium	Medium	Medium	Medium	Low	Medium	Low
Polyimide	Very high	High	Medium	Low	High	High	High	Medium

1.10   PLASTIC SURFACE FINISHING

While the greatest majority of plastic parts can be and often are used either with their as-molded natural-colored surface or with colors obtained by use of precolored resins, color concentrate, or dry powder molded into the resin, increasing quantities of plastics are being surface-finished after molding to provide color or metallization. Some important points related to painting and plating are presented in the following sections.

1.10.1   Painting of Plastics

Painting of plastics frequently yields lower-cost coloring than precolored resin or molded-in coloring.[64] Plastics are often difficult to paint, and proper consideration must be given to all the important factors involved. In Harper[1] (Tables 37 and 38) a selection guide to paints for plastics is presented and application ratings are given for various paints. Some important considerations related to painting plastics are given in the following.

Heat-Distortion Point and Heat Resistance.   This determines whether a bake-type paint can be used and, if so, the maximum baking temperature the plastic can tolerate.

Solvent Resistance.   The susceptibility of the plastic to solvent attack dictates the choice of paint system. Some softening of the substrate is desirable to improve adhesion, but a solvent that attacks the surface aggressively and results in cracking or crazing obviously must be avoided.

Residual Stress.   Molding operations often produce parts with localized areas of stress. Application of coating to these areas may swell the plastic and cause crazing. Annealing of the part before coating will minimize or eliminate the problem. Often it can be avoided entirely by careful design of the molded part to prevent locked-in stress.

Mold-Release Residues.   Excessive amounts of mold-release agents often cause surface-finishing adhesion problems. To assure satisfactory adhesion, the plastic surface must be rinsed or otherwise cleaned to remove the release agents.

Plasticizers and Other Additives.   Most plastics are compounded with plasticizers and chemical additives. These materials usually migrate to the surface and may eventually soften the coating, destroying adhesion. A coating should be checked for short- and long-term softening or adhesion problems for the specific plastic formulation on which it will be used.

Other Factors.   Stiffness or rigidity, dimensional stability, and coefficient of expansion of the plastic are factors that affect the long-term adhesion of the coating. The physical properties of the paint film must accommodate those of the plastic substrate.

1.10.2   Plating on Plastics

The advantages of metallized plastics in many industries, coupled with major advances in both platable plastic materials and plating technology, have resulted in a continuing and rapid growth of metallized plastic parts. Some of the major problems have been adhesion of plating to plastic, differential expansion between plastics and metals, failure of plated part in thermal cycling, heat distortion and warpage of plastic parts during plating and in system use, and improper design for plating. The major plastics which are plated and their characteristics for plating are identified in Table 1.19.[65] Improvements are being made continuously, especially in ABS and polypropylene, which yield generally lower product costs. Thus the guidelines of Table 1.19 should be reviewed at any given time and for any given application. Aside from the commercial plastics described in Table 1.19, excellent plated plastics can be obtained with other plastics. Notable is the plating of TFE fluorocarbon, where otherwise unachievable electrical products of high quality are reproducibly made. Examples are corona-free capacitors and low-loss high-frequency electronic components.[66]

Design Considerations.   Proper design is extremely important in producing a quality plated-plastic part, and some important design considerations are presented in Ref. 67.

Appearance.   Because most plated-plastic parts now being produced are decorative (such as washer end caps, escutcheons) rather than functional (such as copper-plated conductive plastic automotive distributor parts), appearance is extremely critical. For a smooth, even finish, one-piece or integral parts should be designed. Mechanical welds are difficult to plate. If they are necessary, they should be hidden on a noncritical surface. Gates should be hidden on noncritical surfaces or should be disguised in a prominent feature. Gate design should minimize flow and stress lines, which may impair adhesion.

1.11 SUMMARY OF PLASTICS

Plastics, and particularly thermoplastics, have undergone a tremendous development effort during the 1980s in an attempt to raise their high-temperature performance characteristics, as shown in Table 1.20.[68] Most of these developments are directed toward specialty or niche applications such as automobile and aerospace

TABLE 1.19   Characteristics of Major Plated Plastics[65]

	ABS	Poly-propylene	Poly-sulfone	Poly-arylether	Modified PPO
Flow	AA	AA	BA	BA	A
Heat distortion under load	A	BA	AA	AA	A
Plateability	AA	A	BA	AA	BA
Thermal cycling	BA	A	AA	AA	AA
Warpage	A	BA	A	A	A
Mold definition	AA	BA	A	A	A
Coefficient of expansion	A	A	A	A	A
Water absorption	BA	AA	A	BA	AA
Material cost	A	AA	BA	BA	BA
Finishing cost	AA	BA	BA	AA	BA
Peel strength	A	AA	AA	BA	BA

Polymers are rated according to relative desirability of various characteristics: AA = above average; A = average; BA = below average.

TABLE 1.20 Properties of Representative High-Temperature Thermoplastics[68]

Polymer	Common designation	Morpho-logy*	Glass transition, °F	Tensile strength, ksi	Tensile modulus, ksi	Elongation, %	Fracture toughness, G_{IC}, in·lb/in²	Notched Izod, ft·lb/in	Major suppliers
Polyimide	N†	A	700	16.0	580	6	—	—	E. I. du Pont
Polyimide	LARC-TPI	A	507	17.3	540	4.8	—	1.0	Rogers, Mitsui Toatsu
Polyimide	K-III†	A	484	14.8	546	14	11	—	E. I. du Pont
Polyetherimide	PEI	A	423–518	15.2	430	60	19	1.0	GE Plastics
Polyamide-imide	PAI	A	527	9.2–13.0	400–667	1.4–30	19.4	2.7	Amoco, Dow Chemical
Polyarylimide	J-2†	A	320	15.0	460	25	—	—	E. I. du Pont
Polyimidesulfone	PISO₂	A	523	9.1	719	1.3	8	—	Hoechst-Celanese
Polysulfone	PSF	A	374	10.2	360	>50	14	1.2	Amoco, BASF
Polyarylsulfone	PASF	A	428	10.4	310	60	20	1.2	Amoco
Polyarylene sulfide	PAS	SC	419	14.5	470	7.3	—	0.8	Phillips 66
Polyphenylene sulfide	PPS	SC	194	12.0	630	5	—	3.0	Phillips 66, Hoechst-Celanese, Mobay, GE Plastics
Polyether sulfone	PES, HTA‡	A	446–500	12.2	380	>40	11	1.6	ICI, BASF
Polyetherketone	PEK	SC	329	16.0	580	—	—	1.52	ICI, BASF
Polyetherketoneketone	PEKK	SC	311	—	—	—	—	—	E. I. du Pont
Polyetheretherketone	PEEK	SC	289	14.5	450	>40	>23	1.6	ICI
Poly(EKEKK)	PEKEKK	SC	343	—	—	—	—	—	BASF
Polyarylene ketone	PAK, HTX‡	SC	509	12.7	360	13	—	—	ICI, BASF
Liquid crystal polymer	LCP, SRP‡	C	662	20.0	2400	4.9	6.9	2.4	Dartco (Amoco), Hoechst-Celanese, ICI

* A—amorphous; SC—semicrystalline; C—crystalline.
† Trade name of E. I. du Pont.
‡ Trade name of ICI.

1.73

components, and the high cost of these materials will ensure large-scale usage of the more moderately priced commodity-type thermoplastics and thermosets.

Another niche area for plastics is in applications where resistance to ionizing radiation exposure is required.[69-71] An excellent summary of the radiation resistance of a variety of materials is presented in Fig. 1.35.[72]

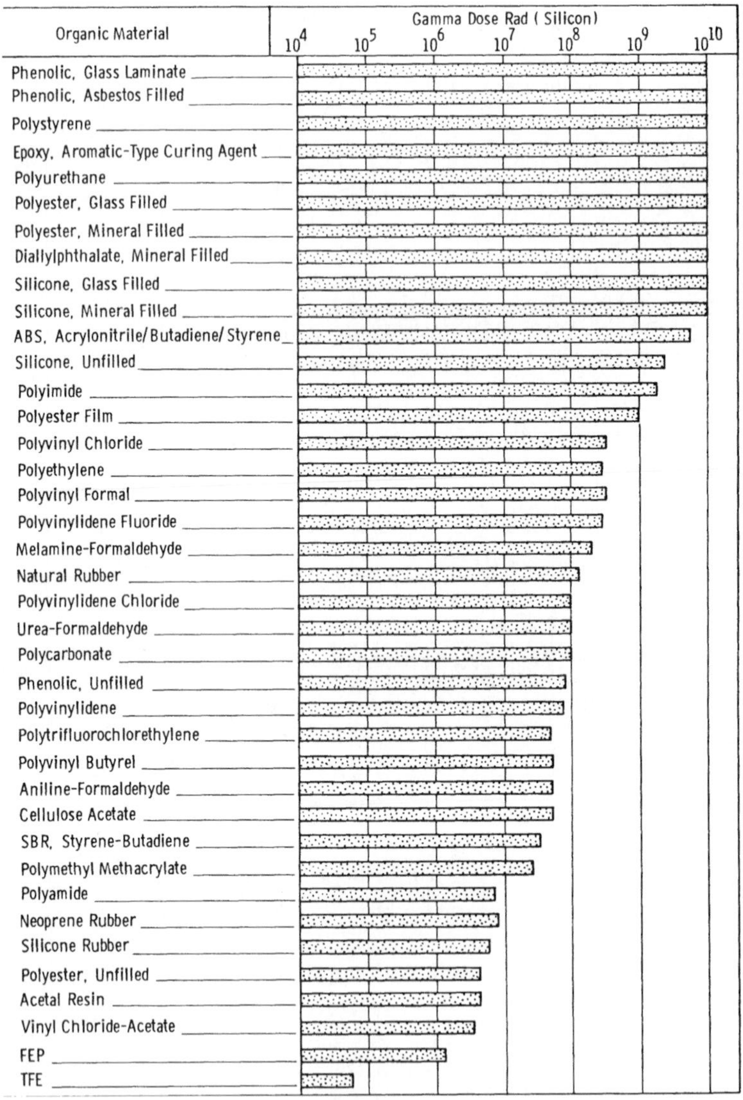

FIGURE 1.35 Relative radiation resistance of organic materials.[72]

1.12 THE NATURE OF ELASTOMERS

ASTM D1566-90[73] defines elastomers as "macromolecular material that returns rapidly to approximately the initial dimensions and shape after substantial deformation by a weak stress and release of the stress." It also defines a rubber as: "material that is capable of recovering from large deformations quickly and forcibly, and can be, or already is, modified to a state in which it is essentially insoluble (but can swell) in solvent, such as benzene, methyl ethyl ketone, and ethanol toluene azeotrope."

A rubber in its modified state, free of diluents, retracts within 1 min to less than 1.5 times its original length after being stretched at room temperature (20 to 27°C) to twice its length and held for 1 min before release. More specifically, an elastomer is a rubberlike material that can be or already is modified to a state exhibiting little plastic flow and quick and nearly complete recovery from an extending force. Such material before modification is called, in most instances, a raw or crude rubber or a basic high polymer and by appropriate processes may be converted into a finished product.

When the basic high polymer is converted (without the addition of plasticizers or other diluents) by appropriate means to an essentially nonplastic state, it must meet the following requirements when tested at room temperature (60 to 90°F; 15 to 32°C):

1. It is capable of being stretched 100 percent.
2. After being stretched 100 percent, held for 5 min, and then released, it is capable of retracting to within 10 percent of its original length within 5 min after release.[73]

The rubber definition with its swelling test certainly limits it to only the natural latex-tree source, whereas the elastomer definition is more in line with modern new synthetics.

1.12.1 Polymer Basis of Elastomers

The important discoveries leading to present polymer science can possibly be credited as beginning with Staudinger's work in the later 1920s. The 1930s saw work by Corothers and Flory on polycondensates such as esters and polyamides (nylon).[2] Also, the 1930s saw the important start of the addition polymers, resulting in high molecular weights of vinyl and diene polymers. The 1940s showed rapid advance of the addition polymers: copolymerization, control of molecular weights for plasticity, emulsion polymerization, development of physics in polymers, solution properties of polymers, intrinsic viscosities, effect of cis or trans isomerism, and mechanical behavior of polymers. In the 1950s chemists learned the detailed structure of polymers, and by means of special catalysts (Ziegler) developed stereospecific polymers and finally a true synthetic natural rubber.

Many carbon-hydrogen chain elements, it is found, tend to gather and behave in basic groups. The hydrogen in a basic group could be substituted by some other element or group to form a new group. These basic groups could react with each other in a head-to-tail fashion to make large molecules and larger macromolecules. Also, a group can react with a different group to form a new larger group, which can then react with itself in a head-to-tail fashion to produce the desired new macromolecule. These basic groups are called monomers; and when they are head to tail or polymerize into long chains, they then form polymers. The reaction between different basic groups to form a new basic recurring group is called copolymerization. These copolymers can also recur in a head-to-tail fashion to form long-chain molecules. The basic monomers are usually gases or very light liquids; and, as polymerization continues, the molecular weight and viscosity both increase until solidification and formation of the gum or solid products results. The degree of polymerization can be controlled so that the end product has the desired processing properties for the rubber industry but lacks the final properties for engineering use. The long chains in this state have a great deal of mobility between them, which also varies greatly with slight changes in temperature. Some such materials are usable as thermoplastic materials; but generally, to obtain the properties of elasticity, this mobility must be arrested. These long chains can react further with themselves or with other chemicals. The type of reaction desired for elastomers is not a continuation of chain length at this point, but a tying or linking of the chains together at certain points to form a network and thus reduce plastic mobility. As the chains lie close to each other, especially when the reactive centers are near each other as double bonds between carbon atoms, they can be made to join to each other directly or through some element such as sulfur, which will hold the two chains together. These joiners can be loosely called linking agents, and more loosely, vulcanization agents; and again the cross-linked material in the rubber industry can now be called vulcanized material.

One of the basic monomers is gaseous ethylene, which is polymerized to the well-known polymer, polyethylene. Polyethylene is the basis for many other monomers such as styrene, acrylonitrile, isobutylene, vinyl alcohol, and vinyl chloride; which in turn can be polymerized to form polystyrene, polyacrylonitrile, polyisobutylene, polyvinyl alcohol, and polyvinyl chloride. Another basic monomer is butadiene, which is the basis for other monomers such as isoprene and chloroprene. The reactions of these two groups of monomers yield, for example, the familiar copolymers of butadiene-styrene (GR-S), butadiene-acrylonitrile (buna-N), and isoprene with isobutylene (butyl), which constitutes a basic elastomer series.

The mechanical properties of solid insulators are often more important than their electrical properties. Most organic materials are good insulators if they can be formed into continuous films that exclude moisture. The general principles relating mechanical properties of polymers to structures have been known for many years. Rubbers, plastics, and fibers, for example, are not intrinsically different materials. Their differences are a matter of degree rather than kind. If the forces of attraction between the molecular chains are small and the chains do not fit readily into a geometric pattern, lattice, or network, the normal thermal motion of the atoms tends to cause the chains to assume a random, more or less coiled arrangement. These conditions lead to a rubberlike character. In practical rubbers, a few cross-links are added to prevent slippage of the molecular chains and permanent deformation under tension. With such polymers, when the stress is released, the normal thermal motion of the atoms causes them to return to a random coiled arrangement. If the forces between the chains are strong and the chains fit easily into a regular geometric pattern, the material is a typical fiber. In cases where the forces are moderate and tendency to form a regular lattice is also moderate, the result is a typical plastic. Some polymers are made and used as three different materials: rubber, plastic, and fiber. Polyethylene, for example, is used as a substitute for natural rubber in wire covering, as a plastic in low-loss standoff insulators and insulating films, and as a fiber in acid-resistant filter cloths where high fiber strength is not as important as chemical resistance.

The rubber industry as such was long occupied in processing gumlike natural rubber. The user expected a material from this industry to be elastic in nature. Therefore, the chemical plants, in their designing of polymers for the rubber industry, kept two requirements in mind: substitutes for natural rubber with its elasticity, and also polymers that could be processed with the conventional equipment. Plastics, as such, behave quite differently from rubber in processing, and this fact led to the establishment of a new industry. Therefore, the incorporation of plastic materials into the rubber inventory was not accomplished. This separation of the two industries has resulted in separate technical groups and societies.

Elastomers, when compared with other engineering materials, are characterized by large deformability; lack of rigidity; large energy-storage capacity; nonlinear stress-strain curves; high hysteresis; large variations with stiffness, temperature, and rate of loading; and compressibility of the same order of magnitude as most liquids. Certain of the elastomeric materials possess additional useful characteristics to a relative degree, such as corrosive chemical resistance, oil resistance, ozone resistance, temperature resistance, and resistance to other environmental conditions.

1.12.2 Commercially Available Elastomers

The nomenclature for elastomers, common names, ASTM designations, relative cost, and general characteristics are summarized in Table 1.21.[74] A listing of representative trade names and suppliers is given in Ref. 75. Further discussion of these commercially available elastomers is presented hereafter, and more detail is given in Babbit.[76]

Natural Rubber (NR).   As the name implies, this is a naturally occurring product mostly derived from the Hevea brasiliensis tree under cultivation in large plantations around the Malay peninsula area. There are, however, other sources, such as the wild rubbers of the same tree growing in Central America; guayule rubber, coming from shrubs grown mostly in Mexico; and balata. Balata is a resinous material of cis-trans isomerism and cannot be tapped like the Hevea tree sap. The balata tree is actually cut down and boiled to extract balata, which cures to a hard and tough product used as golf-ball covers.

Although not broadly recognized, there are many grades of natural rubber that must be selected for specific uses. When natural rubber is produced, the tree is tapped and the free-flowing latex is coagulated with an acid and coagulant, then milled or creped to reduce moisture, and smoked or air-dried for further drying. Smoking of rubber also adds mildew protection by the introduction of creosote. Most natural rubber is from the Hevea tree, but the grading that affects properties is based generally on the cleanliness of the latex before coagulation.

Another rubber is plantation leaf gutta-percha. As the name implies, this material is produced from the leaves of trees grown in bush formation or on plantations. These leaves are plucked, and the rubber is boiled out as with the balata. Gutta-percha has been used successfully for submarine-cable insulation for more than 40 years, a lasting tribute to the pioneers who made this selection. The dielectric constant of gutta-percha was reported to be 2.6, while reports during the same period for Hevea rubber were 3.90 to 4.31.

Needless to say, natural rubber, properly selected, processed, and compounded, produces very high quality rubber properties and was the original basis of electrical insulation. It is still a good basis in comparison with the new synthetic elastomers. It possesses high tensile strength, resilience, and tear, wear, and cut-through resis-

TABLE 1.21 Elastomer Compound Selection and Service Guide[74]

Base Polymer / Common Name		Natural Rubber	Synthetic Rubber	Butadiene	GRS	Butyl	E P Rubber
Chemical Name		Polyisoprene	Synthetic Polyisoprene	Polybutadiene	Styrene Butadiene	Isobutylene Isoprene	Ethylene Propylene
SAE J200 – ASTM D-2000 Classification		AA	AA	AA	AA, BA	AA, BA	CA
ASTM D-735, SAE J-14: MIL-R-3065 (MIL-STD-417)		R(N)	R(S)	R(S)	R(S)	R(S)	R(S)
ASTM Designation (D 1418)		NR	IR	BR	SBR	IIR	EPDM, EPM
Coefficient of Thermal Expansion – 10^{-4} Per °F (Gum)		37	37	37.5	37	32	32
Relative Cost SBR = 1.00		1.14	1.00	1.15	1.00	1.25	1.00
Solubility Parameter (Hildebrands) (Note 2)		$9.5^{\pm3.0}$	$9.5^{\pm3.0}$	$9.5^{\pm2.5}$	$9.5^{\pm2.5}$	$8.5^{\pm2.0}$	$8.5^{\pm2.0}$
Hydrogen Bonding (Gordys) (Note 2)		<19.0	<19.0	<18.0	<18.0	<9.5	<15.0
Dipole Moment (Debyes) (Note 2)		<4.5	<4.5	<3.0	<3.0	<2.5	<2.7
Weight of Base Polymer	Lb./cu. inch	0.033	0.033	0.032	0.034	0.033	0.031
	Specific Gravity	0.92	0.91	0.91	0.94	0.92	0.86
Durometer Range Available		30-90	40-80	40-90	40-80	40-90	30-90
Tensile Strength, PSI – Reinforced (Max)		4500	4500	3000	3500	3000	3000
Elongation, % – Reinforced (Max)		650	650	650	600	850	600
Compression Set		A	A	B	B	C-B	B-A
Resilience		A	A	A	B	C	B
Permeability Coef.-Nitrogen, 10^{-8} CM³. SEC⁻¹. ATM⁻¹		6.12	6.12	20.0	4.8	0.25	6.4
Electrical Resistivity (Polymer)		A	A	A	A	A	A
Creep, Drift or Strain Relaxation		A	B	B	A	C	C-B
Mechanical Properties	Impact Strength	A	A	B	A	B	B
	Abrasion Resistance	A	B	A	A	C	B
	Tear Resistance	A	A	B	C	B	C
	Cut Growth	A	A	C	B	A	B
(Hot) Tensile Strength % Decrease	212° F	−32	−44	−45	−44	−57	−49
	350° F	−84	−	−	−72	−87	−78
(Hot) Elongation % Change	212° F	+17	+6	−7	−25	−10	−21
	350° F	−33	−	−	−47	−32	−48
Strain Relaxation at 212° F		B	B	B	B	C	C-B
Heat Aging at 212° F		B-C	B-C	C	B	B-A	A
Flame Resistance		D	D	D	D	D	D
Low Temperature	Stiffening – °F	−20 to −50	−20 to −50	−30 to −60	0 to −50	−10 to −40	−20 to −50
	Brittle Point – °F	−80	−80	−100	−80	−80	−90
Weather-Sunlight Aging		D	NR	D	D	A	A
Oxidation		B	B	B	C	A	A
Ozone Cracking		NR	NR	NR	NR	A	A
Radiation		B	C-B	D	B	B	B
Water		A	A	A	B-A	A	A
Steam		B	B	B	B	C	B-A
Alkali Dilute/Concentrated		A/C-B	C-B/C-B	C-B/C-B	C-B/C-B	A/A	A/A
Acid Dilute/Concentrated		A/C-B	C-B/C-B	C-B/C-B	C-B/C-B	A/A	A/A
Ketones, Oxygenated Solvents (Note 3)		B	B	B	B	A	B-A
Chlorinated Hydrocarbons, Degreasers		NR	NR	NR	NR	NR	NR
Aliphatic Hydrocarbons, Kerosene, etc.		NR	NR	NR	NR	NR	NR
Aromatic Hydrocarbons, Benzol, Toluol, etc.		NR	NR	NR	NR	NR	NR
LP Gases, Fuel Oils		NR	NR	NR	NR	NR	NR
Alcohols (Note 3)		B-A	B	B	B	B-A	B-A
Brake Fluid, Non-Petroleum Base		B-A	B	B	B-A	B	B-A
Synthetic Lubricants – Diester		NR	NR	NR	NR	NR	NR
Animal and Vegetable Oils		D-B	D-B	D-B	D-B	B-A	B
Hydraulic Fluids	Petroleum Base	NR	NR	NR	NR	NR	NR
	Water Glycol	B-A	B-A	B-A	B	B-A	A
	Silicate Ester	B-A	B-A	B-A	B-A	B-A	B-A
	Phosphate Ester	B	B	B	B	A(250° Max)	A(300°+)
Lubricating Oils	High Aniline (190°+)	NR	NR	NR	NR	NR	NR
	Low Aniline Point	NR	NR	NR	NR	NR	NR
Refrigerants	Ammonia	B	B	B	B	B	B
	Fluorinated	R-12, 13, 22	R-12, 13, 22	R-12, 13, 22	R-12, 13, 22	R-12, 13, 22	R-12, 13, 22
	Methyl Chloride	D	D	D	D	C	D
Refrigerant + Oil	Fluorinated	NR	NR	NR	NR	NR	NR
Taste		C-B	C-B	C-B	C-B	C-B	B
Odor		B-A	B	B	B	B	B
Non-Staining		A	A	B	D-B	B	B
Bonding to Rigid Materials		A	A	A	A	C-A	C-B

A - Excellent   B - Good   C - Fair   D - Use with Caution   NR - Not Recommended

Hypalon	Neoprene	Nitrile (High)	Nitrile (Med High)	Acrylic	Polysulfide	Fluorocarbon	Fluorosilicone	Silicone	Urethane	Polyalkylene Oxide Polymer
Chloro-sulfonated Polyethylene	Chloroprene	Butadiene Acrylonitrile	Butadiene Acrylonitrile	Polyacrylate	ST	Fluorinated Hydrocarbon	Fluoro-Vinyl Silane	Polysiloxane	Polyester/Polyether Urethane	Epichlorohydrin
CE	BC, BE	BF, BG, BK	BF, BG	DF, DH	AK	HK	FK	FC, FE, GE	BG	DK, DJ
SC	SC	SB, SA	SB	TB	SA	TB	TA, TB	TA	SB	SA
CSM	CR	NBR	NBR	ACM, ANM	PTR	FPM	FVS	PSi, PVSi, Si, VSi	AU, EU	CO, ECO
27	34	39	39	10	42	88	45	45	27	36
1.30	1.25	1.40	1.40	3.50	2.50	45.00	50.00	12.00	4.00-10.00	3.00
$9.9^{-2.5}$	$10.0^{-2.7}$	$11.0^{-4.0}$	$11.0^{-4.0}$	$11.5^{-3.5}$	$12.0^{-3.0}$	$9.5^{-2.5}$	$9.7^{-2.7}$	$10.5^{-2.5}$	$10.5^{-2.5}$	$11.0^{-2.0}$
<19.0	<18.9	<19.0	<19.0	<20.5	<18.7	$13.0^{-5.3}$	$9.8^{-8.3}$	<17.9	<20.0	<13.0
<3.5	<4.5	<4.5	<4.5	<4.5	<4.5	<4.4	<4.5	<3.0	<3.5	<3.8
0.043	0.045	0.036	0.036	0.039	0.049	0.067	0.051	0.040 (1.1)	0.045	0.049 (0.046)
1.18	1.25	1.00	1.00	1.09	1.35	1.86	1.4	1.1-1.6	1.25	1.36/1.27
45-100	30-95	40-95	40-95	40-90	40-85	60-90	40-80	30-90	35-100	40-90
4000	4000	4000	4000	2500	1500	3000	1500	1500	5000	2500
500	600	650	650	450	450	300	400	900	750	350
C-B	B	B	B	B	D	B-A	C-B	B-A	D	A-D
C	A	B	B	B	C	C	C	D-A	C-A	C-A
0.7 to 0.9	0.89	0.18	0.31	0.88	0.66	0.20	165	200.0	0.95/16	.17/.66
B	C	D-C	D-C	C	C	B	A	A	B	B
C	B	B	B	C	D	B	B	C-A	C-A	B
B	B	C	C	D	D	B	D	D-C	B-A	C-A
A	A	A	A	C-B	D	B	D	C-B	A	C-B
B	B	B	B	D-C	D	B	D	C-B	A	C-A
B	B	B	B	C-B	D	B	D	C-B	A	B
-57	-50	-50	-55	-67	-26	-72	-56	-12	-56	-45
-82	-74	-75	-76	-82	-44	-87	-60	-41	-83	-67
-49	-35	-30	-35	-45	-34	-33	+9	-29	-32	-45
-69	-45	-57	-60	-61	-38	-63	-17	-48	-54	-60
C	B	B	B	C	D	B-A	B-A	A	D	C-B
B-A	B-A	B	B	B	C-B	A	A	A	B	B-A
B-A	B-A	D	D	D	D	A	A	A	D	B-D
-30 to -50	+10 to -50	+30 to -20	0 to -40	+35 to +10	-10 to -45	+10 to -10	-70	-60 to -180	-10 to -30	-15 to -40
-70	-85	-40	-85	-20	-68	-60	-85	-90 to -180	-60 to -200	-10 to -50
A	B	D	D	A	B	A	A	A	A	B
A	A	B	B	A	B	A	A	A	B	B
A	A	C	C	B	A	A	A	A	A	A
B	B	B	B	B	C	C-B	C-B	C-B	B	-
B	B	A	B	D	B	A	A	A	C-B	B
B	B	C-B	C-B	NR	D	B	C-B	C-B	D	B-C
A/A	A/A	B/B	B/B	C/C	B/B	A	A/B	A/A	C/D	B-D
A/A	A/A	B/B	B/B	C/C	C/NR	B/C	A/B	B/C	C/D	B/C
B	C	D	D	D	A	NR	D	B-C	D	C
D	D	C-B	C	B	C-A	A	B-A	NR	C-B	A
C	C	A	A	A	A	A	A	D-C	B	B-A
B	B	B-A	B	C-B	C-B	A	B-A	NR	C	B-A
B	B	A	A	B	A	A	A	C	C-B	A
A	A	C-B	C-B	D	B	C-A	C-B	C-B	B	C-B
C	C	NR	NR	D	D	C	A	A	NR	D
D	D	B-A	D	D-C	C-B	A	A	NR	D	B
B	B	B	B	B	A	A	A	A	A	A
C-B	D-C	B-A	C-B	A	A	A	A	NR	B	A
B	B	C	B	C-B	A	A	A	A	C-B	B
B	B	B	B	B	B	A	A	NR	NR	B
C	C	D	D	D	B-A	B-A (Note 4)	NR	B	NR	NR
A	B	B	A	A	A	A	A	C	A	A
B	A	A	A	A	A	A	A	B	B	A
B	A	B	B	C	NR	NR	A	A	D	NR
R-11, 12, 13, 22	R-11, 12, 13, 21, 22	R-11, 12, 13	R-11, 12, 13	R-11, 12, 13, 22	R-11, 12, 13, 22	R-11, 12, 13	R-11, 12	NR	R-12	R-12, 22/12
NR	NR	NR	NR	NR	B	B	R-11, 12	NR	NR	NR
R-11, 12, 22	R-11, 12, 22	R-11, 12	R-11, 22	R-11, 12, 13, 22	R-11, 12, 13, 22	R-11, 12	R-11, 12	NR	R-12, +13/NR	B-A
C-B	C-B	C-B	C-B	C-B	D-C	C-B	B	B	B	B
B	C-B	B	B	C-B	D	B	B	B	B	B
A	B-A	D-C	C-B	B	D-B	C-B	A	A	B	A
A	B-A	B-A	B-A	C	C-B	C-B	C-B	C-B	B-A	C-A

Notes: 1. Physical properties shown on this table apply to general class of base polymers.
2. A. Beerbower - Esso R & E Co.   3. Molecular weight under 30.   4. NR in Skydrol 500.

tance. It has good electrical-insulation properties, but by special compounding can be made conductive to a specific resistance as low as 2000 $\Omega \cdot$ cm. It has very low compression set and good resistance to cold flow, and it can be compounded to function as low as $-70°F$. Resilience is the main superior quality over the synthetics. On the negative side, it is not as good for sunlight, oxygen, and ozone resistance as some of the synthetics, but special compounding offsets this to a degree. Its aging properties are not as good as those of the synthetics, but synthetics will harden with time, while natural rubber will show a reversal and start softening. Also, natural rubber is not satisfactory in resistance to oils from petroleum, vegetable, or animal origin. On the other hand, it has good resistance to acids and bases with the exception of oxidizing agents. Natural rubber sees wide usage in tires and mechanical goods.[77]

Isoprene Rubber (IR). This is the synthetic-rubber equivalent of natural rubber. As a synthetic, isoprene offers the rubber industry the purity of all synthetics and the freedom from the worry about world supply experienced during World War II. Synthesizing to controlled molecular weights gives the industry uniform processing qualities and a bonus on the side—odorlessness.

Styrene-Butadiene Copolymer (SBR). There are many manufacturers of SBR, but generally it is known as GR-S or buna-S. This was used for the synthetic-rubber tire that was developed because of the natural-rubber shortage during World War II; and since the government undertook the development, it became known as the Government Rubber Styrene type (GR-S). There are now many variations of SBR, generally basic in the ratio of butadiene to styrene, temperature of polymerization, and type of chemicals used during polymerization. This last aspect is important in its initial choice for electrical insulation, since the type of salts and moisture content affect the quality of insulation.

The original purpose of GR-S, of course, was to substitute for the then hard-to-obtain imported natural rubber. In this role it became and still is a workhorse of the industry, even though it does not match the superior physical properties of natural rubber. It lacks in tensile strength, elongation, resilience, hot tear, and hysteresis; but it more than makes up for this in low cost, cleanliness, slightly better heat aging, and slightly better wear than natural rubber for passenger tires. Generally speaking, natural rubber was not used with its ultimate properties but in a cheapened and compounded-down product to meet the desired engineering needs. In this respect, SBR is certainly no longer a substitute, but the prime choice.

Neoprene (CR). Neoprene was actually the first commercial synthetic rubber. This elastomer is also a major industry workhorse. Discovered in the laboratories of Notre Dame University and developed by E. I. du Pont de Nemours and Company, it was first used in the 1930s as an oil-resisting rubber. Today it is classified as a moderately oil-resisting rubber with very good weather- and ozone-resisting properties and other properties closer to those of natural rubber than SBR. Because of its crystallizing nature, it has inherent high tensile strength, elongation, and wear properties at pure gum (not extended or hardened) levels. In this respect, it compares with natural rubber, whereas SBR needs reinforcing materials for good tensile strength and thus has low pure gum properties. Electrically, neoprene ranks below natural rubber or SBR because of its polar chlorine group. It can be compounded for temperatures as low as $-67°F$ but crystallizes rapidly, and it is not as good in this respect as either natural rubber or SBR. It has excellent flame resistance and is, in fact, self-extinguishing. Because of this property, it is a must in coal-mining operations and other areas where fire is a potential hazard. Also, it has good resistance to oxidative chemicals. A comprehensive review of halogen-containing elastomers is given in Bhowmick and Stephens.[78]

Nitrile Rubber (NBR).   NBR is another of the most widely used elastomers. With its nitrile group, it is above neoprene in polarity and thus poorer in electrical insulation. Its main advantage is that it is considerably more resistant to oils, fuel, and solvents than neoprene. Its other properties fall in line with those of the butadiene group, and thus it closely matches the physical properties of SBR except that it has much better heat resistance. Also, considering cost (slightly above neoprene), it has overall excellent properties along with broader resistance to chemicals than any other polymer. It has a gain in oil resistance over that of neoprene, but at a sacrifice to weathering and ozone resistance. Hydrogenated versions of nitrile have been produced recently[79] and are seeing considerable usage in automotive under-the-hood applications.

Butyl Rubber (IIR).   Butyl rubber's outstanding physical properties are low air permeability (about one-fifth that of natural rubber) and high energy-absorbing qualities. It has excellent weathering and ozone resistance, excellent flexing properties, excellent heat resistance, good flexibility at low temperature, tear resistance about that of natural rubber, tensile strength about in the range of SBR, and very good insulation properties. It has very poor resistance to petroleum oils and gasoline, but excellent resistance to corrosive chemicals, dilute mineral acids, vegetable oils, phosphate ester oils, acetone, ethylene, glycol, and water. It is also very nonpolar.

EPM Copolymer and EPDM Terpolymer.   Both the EPM copolymer and EPDM terpolymer, which were commercialized in the early 1960s, have seen tremendous growth in market share due to inherent oxygen and heat resistance and a good balance of other properties. They also have the ability to accept high loading of reinforcing agents, fillers, and plasticizers. Babbit[76] gives a more in-depth review of these materials.

Hypalon (CSM).   A very close match to neoprene is Hypalon, and in many instances, substitutions can be made with hardly any difference in properties. However, Hypalon does offer some exceptional added properties over neoprene, such as improved heat and ozone resistance, improved electrical properties, improved color stability, and improved chemical resistance. Light-colored neoprene will darken with age and Hypalon will not; so if neoprene properties are desired along with color resistance, Hypalon is the choice. Its very good electrical properties plus added heat resistance, color stability, and oil resistance are the reasons why it is preferred for spark-plug boots over the costlier silicone rubber and for dust covers on high-voltage ground cables for television picture tubes.

Acrylic Rubber (NBR).   Oils for hot applications are often fortified with sulfur-bearing chemicals. This sulfur, in a system such as automotive transmissions, will continue to react and harden a nitrile-base rubber to a point of excessive hardening. Acrylic rubber is not affected by hot-sulfur-modified oils; and for this reason, it is used in applications of this sort. Where test temperatures are normally run at 250°F for nitrile base, 302°F is used for the acrylic rubber in conjunction with transmission oils. Acrylics, therefore, can be said to have better oil, heat, and ozone resistance than nitrile, and good sunlight resistance; but they are not as good as nitrile for low-temperature work. They are usually used in hot-oil applications.

Polysulfide Rubber (PTR). Polysulfide is especially resistant to hydrocarbon solvents, aliphatic liquids, or blends of aliphatic with aromatic. The common alcohols, ketones, and esters used in paints, varnishes, and inks have little effect on it. It is also resistant to some chlorinated solvents, but preliminary tests should be made before it is used for this purpose. Nitrile-base rubbers will swell considerably in such solvents, and polysulfide has a special use in applications using such solvents. Compared with nitrile, it has poor tensile strength, pungent odor, poor rebound, high creep under strain, and poor abrasion resistance. Polysulfide has most of its applications in solvent-carrying hose, printer's rolls, and newspaper blankets; and because of its excellent weathering properties, it is used heavily for calking purposes.

Silicones (FSI, PSI, VSI, PVSI, SI). Silicones are an elastomer family of their own, since their polymer backbone consists of silicone and oxygen atoms rather than the carbon structure of all other elastomers. Silicon is in the same chemical group as carbon; and since it is a more stable element, chemists predicted more stable compounds from it if it could be substituted for carbon in the chain. Its outstanding property is its heat resistance (500°F+). It is also very flexible at below −100°F, and it has very good electrical properties. The outstanding property of silicone elastomers for insulation purposes is that their decomposition product on extreme heat is still an insulating silicon dioxide, whereas the decomposition product for organic compounds is conductive carbon black, which can sublime and thus leave nothing for insulation. The first user of silicone during World War II took advantage of this property in purposely undersizing motors where weight was a factor and letting them run hot. Silicone has resistance to oxidation, ozone, and weathering corona; and special modifications by inclusion of halogen groups can make oil-resistant fluorosilicones.

Silicones generally do not have high tensile strength, but their overall properties of good compression set, improved tear resistance, and stability over a wide temperature range certainly make them suitable for many engineering uses. Silicones are the most heat-resisting elastomers available today and the most flexible at low temperature. The basic structure is modified with groups such as vinyl or fluoride, which enhance properties such as tear and oil resistance, so that there is now a family of silicone rubbers covering a wide range of physical and environmental needs.

Urethane (AU, EU). Urethanes cross-link as well as undergo chain extension to produce a wide variety of compounds. They are available as castables, or liquids, and as solids, or gums, to be used on conventional rubber-processing equipment; and as thermoplastic resins to be processed similarly to polyvinyl chloride. They can be cured to tough elastic solid rubbers with outstanding load-bearing properties; and as thermoplastic resins they can be processed similarly to polyvinyl chloride. They can be cured to tough elastic solid rubbers with outstanding resilience, abrasion resistance, and tensile strength. The outstanding properties are exceptionally high wear and two to three times the tensile strength of natural rubber. Urethanes exhibit very good resistance to oils, solvents, oxidation, and ozone. They have poor resistance to hot water and are not recommended for temperatures above 175°F; they also are quite stiff at low temperatures. They are useful where load-bearing and wear properties are desirable, and can be used on insulation as a protective coating over an underlay.

Fluoroelastomers (FPM). Again the halogen groups indicate stability, and these are stable. The fluorinated rubbers are exceptionally good for high-temperature service, but they are below silicones in this respect. They resist most of the lubricants, fuels, and hydraulic fluids encountered in aircraft and missiles; a wide variety of chemicals, especially the corrosive variety; and also, most chlorinated solvents. They have good physical properties, somewhere near those of SBR at the higher hardness levels. FPM is valuable in automotive use for its extreme heat and oil resistance and is on a much higher level in this respect than the acrylic elastomers. It has weathering properties superior to those of neoprene, but is very expensive.

Butadiene (BR). A stereospecific controlled structure like isoprene rubber, its outstanding properties are excellent resilience and hysteresis (almost equivalent to those of natural rubber) and superior abrasion resistance compared with SBR. Butadiene is most similar to SBR and finds wide use as an admixture with SBR in tire treads to improve wear qualities. It is somewhat difficult to process, and for this reason it is hardly ever used in amounts larger than 75 percent of the total polymer in a compound.

Epichlorohydrin Rubber (CO, ECO). Epichlorohydrin rubbers are recognized as having excellent resistance to swelling in oils and fuels, in addition to good chemical resistance to acids, bases, and water. They also have very good aging properties, including ozone resistance. The high chlorine content imparts good to fair flame retardance. Epichlorohydrin rubber possesses a blend of the good properties of neoprene and nitrile.

1.12.3 Physical Properties

The proper selection and application of elastomers are difficult for design engineers in many instances, because engineering terms in conventional usage have different meanings when applied to rubber properties. Elastomers are organic materials and react in a completely different manner from metals. For this reason some of the more common definitions are given here.[80] Another excellent source for rubber engineering terms is Lindley.[81]

Tensile Strength. In rubber, tensile strength refers to the force per unit of original cross section on elongating to rupture. As such, it is not really an important property in itself, since rubber is rarely used in tensile applications. However, it is an indication of other qualities that correlate with tensile strength. Properties that improve with tensile strength are wear and tear resistance, resilience, cut resistance, stress relaxation, creep, flex fatigue, and in some polymers such as neoprene better ozone resistance. In insulation, it should indicate better cut and abrasion resistance and, generally, a tougher and more durable shield.

Elongation. This is the maximum extension of a rubber at the moment of rupture. As a rule of thumb, a rubber with less than 100 percent elongation will usually break if doubled over on itself.

Hardness. Hardness is an index of the resistance of rubber to deformation and is measured by pressing a ball or blunt point into the surface of the rubber. The most commonly used instrument is the durometer, made by the Shore Instrument Company. There are several Shore instruments, namely, A, B, C, D, and 0, designed to give different readings, from soft sponge up through ebonite-type materials. Shore A durometer readings are the most common, and these are the readings appearing in most specifications. On the Shore A scale, 0 would be soft and 100 hard; as a comparison, a rubber band would be about 35 and rubber tire tread 70. Hard rubber (ebonite) is much more solid and is read on the Shore D scale; as a comparison, a hard-rubber pipe stem may read about 60 Shore D and a bowling ball about 90 Shore D. Most products will fall between 40 and 90. (Unless otherwise specified, Shore A is assumed.)

Modulus. In rubber, modulus refers to the force per unit of original cross section to a specific extension. Most modulus readings are taken at 300 percent elongation, but lower extensions can be used. It is a ratio of the stress of rubber to the tensile strain but differs from that of metal, as it is not a Young's modulus stress-strain-type curve. Stress-strain values are extremely low for slight extensions but increase logarithmically with increased extension. Rubber has the useful property, for some products, of being extensible to 10 times its original length. The modulus can vary for the same hardness and does affect the stiffness of insulation.

Hysteresis. This term denotes energy loss per loading cycle. This mechanical loss of energy is converted into heating of the rubber product and could reach destructive temperatures.

Heat Buildup. This term is used to express the temperature rise in a rubber product resulting from hysteresis. However, it also can mean use of high frequencies on rubber where the power factor is too high.

Permanent Set. The deformation of a rubber that remains after a given period of stress followed by release of stress is called permanent set.

Stress Relaxation. This is the loss in stress that remains after the rubber has been held at constant strain over a period of time.

Creep. In rubber, creep refers to change in strain when the stress is held constant.

Compression Set. In rubber, this is the permanent creep that remains after the rubber has been held at either constant strain or stress and in compression for a given time. Constant strain is most generally used and is reported as a percentage of the permanent creep divided by the amount of original strain. A strain of 25 percent is most common.

Abrasion Resistance. One of the remarkable properties of rubber and one most difficult to measure is abrasion resistance. The term refers to the resistance of a rubber composition to wear. It is usually measured by the loss of material when a rubber part is brought into contact with a moving abrasive surface. It is specified as percent of volume loss of sample as compared with a standard rubber composition, and it is almost impossible to correlate these relative values to life expectancy. Also, since many formulations contain wax-type substances which exude to the surface, test results can be erroneous because of the lubricating effect of the waxes.

Flex Fatigue.   This is the result of rubber fracturing after being subjected to fluctuating stresses.

Impact Resistance.   This is the resistance of a rubber to abrading or cutting when hit by a sharp object.

Tear Resistance.   This is a measure of the stress needed to continue rupturing a sheet of rubber, usually after an initiating cut.

Flame Resistance.   This is the relative flammability of a rubber. Some rubbers will burn profusely when ignited, whereas others are self-extinguishing when the igniting source is removed.

Low-Temperature Properties.   These properties indicate a stiffening range and brittle point of rubber. Stiffening range is the more useful of these two, but is difficult to measure. Brittle point has little meaning unless the deforming force and rate are known. Time is also an important consideration, since some characteristics change as the temperature is lowered and held: hardness, stress-strain rate, and modulus, for example. With many materials crystallization occurs, at which time the rubber is brittle and will fracture easily. There are many tests to measure cold-temperature properties; but for a general comparison it is simplest to bend the specimens manually at the test temperature for a difference in "feel" of stiffness or actual breakage. All tests must be made in the cold box and all precautions taken so that the rubber is not warmed during the test. The specimens (0.075 standard ASTM slabs) warm very quickly and will produce false results unless extreme care is taken in the testing.

Heat Resistance.   No rubber is completely heat-resistant; time and temperature have their aging effects. Heat resistance is measured usually as change in tensile strength, elongation, and durometer readings from the original values, usually after a 72-h period.

Aging.   Elastomeric properties can be destroyed only by further chain growth and linkage, which would result in a hard, rigid material, or a chain rupture, which would result in a plastic or resinous mass. The deteriorative agents mostly considered in this category are sunlight, heat, oxygen, stress with atmospheric ozone, atmospheric moisture, and atmospheric nitrous oxide. Chain growth or cross-linkage will usually decrease elongation and increase hardness and tensile strength, whereas chain rupture will have the opposite effect. Some elastomers will continue to harden and some to soften, and some will show an initial hardening followed by softening. All are irreversible responses.

Radiation.   Deteriorating effects of radiation are similar and complementary to those of aging. It has been fairly well demonstrated that damage is dependent only on dosage, or the amount of radiation energy absorbed, irrespective of the form of radiation within a factor of 2. Dosage rate is an important factor only where significant degradation from other agents, such as oxygen, has access to the system. Surprisingly enough, the least heat-resistant of the elastomers displays the most radiation resistance. Figure 1.35 shows radiation-resistance data for some of the more commonly used elastomers,[72] and most of the testing procedures for rubbers can be found in an ASTM book of standards.[82] Thermoplastic elastomers are treated in detail in Chap. 5.

REFERENCES

1. C. A. Harper, *Handbook of Plastics and Elastomers,* McGraw-Hill, New York, 1975.
2. P. J. Flory, *Principles of Polymer Chemistry,* Cornell Univ. Press, Ithaca, NY, 1953.
3. "The Plastic Industry and the Energy Shortage," *Celanese Plastics Co. Bull.,* 1974.
4. H. F. Mark, *Giant Molecules,* Time Inc., New York, 1966.
5. J. E. Hauck, "Engineers' Guide to Plastics," *Mater. Eng.,* Feb. 1967.
6. "Standard Tests on Plastics," *Celanese Plastics Co. Bull.,* 1970.
7. "Plastics," ASTM Stds., sec. 8, vol. 08.01-08.04, 1990.
8. C. A. Harper, *Handbook of Materials and Processes for Electronics,* McGraw-Hill, New York, 1972.
9. P. Grafton, "Commonly Used Plastics," Boonton Molding Co., Boonton, NJ, 1973.
10. R. D. Deanin and S. B. Driscoll, "Cost per Unit Property of Plastics," *Mod. Plast.,* Apr. 1973.
11. Staff Rep., "What Are Fire Officials Squawking About?," *Mod. Plast.,* Aug. 1973.
12. R. H. Wehrenberg, "Flammability of Plastics: The Heat Is on for Large-Scale Tests," *Mater. Eng.,* pp. 35–40, Feb. 1979.
13. R. C. Masek, "Flammability Ratings. What They Mean and Don't Mean," *Insul. Circuits,* pp. 25–28, Nov. 1978.
14. A. Tewarson, "Fire Behavior of Polymeric Materials," *IEEE Elec. Insul. Mag.,* vol. 6, pp. 20–23, May/June 1990.
15. Staff Rep., "Flammability Report," *Mod. Plast.,* Nov. 1972.
16. ASTM E906, "Standard Test Method for Heat and Visible Smoke Release Rate for Materials and Products," Am. Soc. for Testing and Materials, Philadelphia, Pa., 1983.
17. "Improved Flammability Standards for Materials Used in the Interiors of Transport Category Airplane Cabins," *Federal Resister,* pt. II, vol. 50, no. 73, Apr. 16, 1985; "Final Rule; Findings Concerning Comments," *ibid.,* vol. 53, no. 165, Aug. 25, 1988.
18. J. T. Howarth et al., "Flame-Retardant Additives," *Plast. World,* Mar. 1973.
19. British Plastics Federation, *Flame Retardants '90,* Elsevier Appl. Sci., London, 1990.
20. L. K. English, "Intumescent Coatings: First Line of Defense against Fines," *Mater. Eng.,* pp. 39–43, Feb. 1986.
21. "Fire Protection Coatings, Work by Forming Foam Like Chars," *Prod. Eng.,* pp. 36–37, Feb. 1976.
22. R. B. Seymour, "Fillers for Molding Compounds," *Mod. Plast.,* MPE Suppl., 1966–1967.
23. "Plaskon Plastics and Resins," Tech. Bull., Allied Chemical Corp., Plastics Div. (regularly updated).
24. "Dapon Diallyl Phthalate Resin," Tech. Bull., FMC Corp. (regularly updated).
25. J. Chottiner, "Dimensional Stability of Thermosetting Plastics," *Mater. Eng.,* Feb. 1962.
26. S. Sherman, "Epoxy Resins," *Mod. Plast.,* MPE Suppl., 1976.
27. C. A. May and Y. Tanaka, *Epoxy Resins,* Marcel Dekker, New York, 1973.
28. F. Petruccelli, "Aminos," *Mod. Plast.,* MPE Suppl., 1974.
29. G. B. Sunderland and A. Nufer, "Aminos," *Mach. Des.,* Plastic Reference Issue, June 1966.
30. R. W. Martin, *The Chemistry of Phenolic Resins,* Wiley, New York, 1956.
31. T. S. Carswell, *Phenoplastics—Their Structure, Properties and Chemical Technology,* Interscience, New York, 1947.
32. "Ricon High Vinyl 1,2 Liquid Polybutadiene," Product Bull., CCS-110, Advanced Resins, Inc., Grand Junction, CO, Feb. 1980.

33. T. J. Czarnomski, "Unsaturated Polyesters," *Mod. Plast.,* MPE Suppl., 1974.

34. G. J. Kookootsedes, "Silicones," *Mod. Plast.,* MPE Suppl., 1974.

35. "Dow Corning Molding Compounds," Tech. Bull., Dow Corning Corp. (regularly updated).

36. *Mod. Plast.,* MPE Suppl., 1990.

37. "Marbon ABS Plastics," Tech. Data Bull., Borg-Warner Corp. (regularly updated).

38. J. E. Hauck, "Long-Term Performance of Plastics," *Mater. Eng.,* Nov. 1965.

39. G. A. Patten, "Heat Resistance of Thermoplastics," *Mater. Eng.,* July 1962.

40. J. E. Hauck, "Alloy Plastic to Improve Properties," *Mater. Eng.,* July 1967.

41. H. E. Barkén and A. E. Javitz, "Plastics Molding Materials for Structural and Mechanical Applications," *Elec. Manuf.,* May 1962.

42. "Delrin Acetal," Design Handbook, E. I. du Pont de Nemours and Co. (regularly updated).

43. Tech. Data Bull. on AF Delrin, E. I. du Pont de Nemours and Co. (regularly updated).

44. Staff Report, "Contemporary Thermoplastic Materials Properties," *Plast. World,* Sept. 17, 1973.

45. "Lucite Acrylic Resins," Design Handbook, Tech. Bull., E. I. du Pont de Nemours and Co. (regularly updated).

46. G. P. Koo et al., "Engineering Properties of a New Polytetrafluoroethylene," *SPE J.,* Sept. 1965.

47. "Mechanical Design Data for Teflon Resins," Tech. Booklet, E. I. du Pont de Nemours and Co. (regularly updated).

48. P. West, "Wide Space Ranging Applications Developing for Piezoelectric Polymers," *Adv. Mater.,* vol. 8, May 26, 1986.

49. "Surlyn Ionomers," Tech. Bull., E. I. du Pont de Nemours and Co. (regularly updated).

50. "Zytel Nylon Resins," Tech. Bull., E. I. du Pont de Nemours and Co. (regularly updated).

51. "Parylene Conformal Coatings," Tech. Bull., Union Carbide Corp. (regularly updated).

52. W. B. Isaacson, "Polyarylsulfone," *Mod. Plast.,* MPE Suppl., 1974.

53. "Astrel 360 Plastic," Tech. Bull., 3M Corp. (regularly updated).

54. A. G. Bager, "Bayer Polycarbonate Films," Tech. Bull., Mobay Chemical Co. (regularly updated).

55. "Lexan Polycarbonate Resin," Tech. Booklet, General Electric Co.

56. "Noryl and PPO Resins," Tech. Bull., General Electric Co.

57. "Celanex," Tech. Data Bull., Celanese Plastic Co.

58. V. J. Leslie, "Polyethersulfone," *Mod. Plast.,* MPE Suppl., 1974.

59. J. S. Eickert, "Phenylene Oxide-Based Resin," *Mod. Plast.,* MPE Suppl., 1974.

60. R. V. Jones, "Polyphenylene Sulfide," *Mod. Plast.,* MPE Suppl., 1974.

61. L. R. Guenin, "Polystyrene," *Mod. Plast.,* MPE Suppl., 1974.

62. R. K. Walton, "Polysulfone," *Mod. Plast.,* MPE Suppl., 1974.

63. "Film Properties Charts," *Mod. Plast.,* 1974.

64. T. P. Kinsella et al., "Painting Plastics," *Plast. World,* Apr. 1973.

65. F. A. Leary, "Characteristics of Platable Plastics," *Plast. Des. Process.,* June 1971.

66. "Polyflon Plated TFE," Tech. Bull., Polyflon Corp.

67. Staff Rep., "Proper Part Design for Profitable Plating," *Plast. World,* Feb. 1971.

68. L. K. English, "Raising the Ceiling of High-Temperature Thermoplastics," *Mater. Eng.,* pp. 67, 69, 70, May 1988.

69. C. R. Ruffing, "Radiation Effects on Electrical/Electronic Materials," *Insul. Circuits,* pp. 23–26, May 1971.

70. J. W. Hitchon, "Radiation Stability of Plastics and Rubbers," Materials Development Div., AERE, Harwell Memo 325, Jan. 1983.

71. H. Schonbacher, "How Plastics Perform under Nuclear Radiation," *Mod. Plast.,* pp. 64, 67, 68, Dec. 1985.

72. L. K. English, "How High-Energy Radiation Affects Polymers," *Mater. Eng.,* pp. 41–44, May 1986.

73. ASTM D1566, Am. Soc. for Testing and Materials, Philadelphia, PA, 1990.

74. D. L. Hertz, "Compound Selection and Service Guide," Seals Eastern, Red Bank, NJ.

75. *Rubber World Mag., Blue Book,* pp. 383–497, 1988.

76. R. O. Babbit, *The Vanderbilt Rubber Handbook,* R. T. Vanderbilt Co., 1978.

77. D. A. Meyer and J. A. Welch, "Design and Engineering with Natural Rubber," presented at the 109th Mtg. of the Rubber Div. of the Am. Chem. Soc., Minneapolis, MN, Apr. 27–30, 1976.

78. A. K. Bhowmick and H. L. Stephens, "Halogen Containing Elastomers," in *Handbook of Elastomers,* Marcel Dekker, New York, 1988, pp. 443–483.

79. J. R. Dunn, G. C. Blackshaw, and J. Timar, "Compression Set and Sealing Force Retention of HNBR," *Elastomers,* pp. 20–25, Dec. 1986.

80. "Elastomers: Tailor Made Design Materials," *Des. News,* Mar. 31, 1965.

81. P. B. Lindley, "Engineering Design with Natural Rubber," Malaysian Rubber Producers Research Assoc., 1978.

82. *Annual Book of ASTM Standards,* Am. Soc. for Testing and Materials, Philadelphia, PA, vol. 09.01, 1990.

CHAPTER 2
LAMINATES AND REINFORCED PLASTIC MATERIALS AND PROCESSES

Ronald N. Sampson
Technology Seminars, Inc.
Pittsburgh, Pennsylvania

2.1 LAMINATES

2.1.1 Definition

Laminates are plastic materials formed by bonding together reinforcing sheets or webs with a polymer or plastic. Heat and pressure are usually required. A common laminate is plywood. The reinforcing sheets may be fabric, mats, or paper, and the binder polymer may be either a thermoset or thermoplastic polymer. Thermosetting resins can be combined with reinforcing fibers directly in a molding press by pouring a liquid resin onto a fiber mat or fabric, or by combining the resin and reinforcement in a treater. In the treater process, the thermosetting resins are often dissolved in a solvent to aid the wetting and saturation of the reinforcement web. This process is shown in Figs. 2.1 and 2.2. The reinforcing web is pulled from the delivery roll through a resin-impregnating bath or tank where the viscosity and temperature of the resin are carefully controlled. The saturated web then enters the heating section of the treater. When the product requires a cosmetic appearance, the treater is arranged vertically. When the appearance is not critical, the treater is horizontal and the impregnated web is supported through the horizontal treater on a conveyor (sticks) as shown in Fig. 2.2. Horizontal treaters can reach speeds of 600 ft/min, and vertical treaters run at speeds from 10 to 100 ft/min. In the treater, the impregnated web is heated to boil away the solvent and to begin the polymerization of the thermosetting resin. This step is known as "B staging" or "prepreging." It is desirable to cure the resin enough so that the web is not tacky or dry, and is handleable but yet is uncured enough to permit an additional cure at the lamination press. As the sheets exit the treater, they are sheared, stacked, and stored in temperature- and humidity-controlled rooms.

The prepreg (preimpregnated material) is then delivered to the lamination press. Here the required number of prepreg sheets are stacked between polished steel

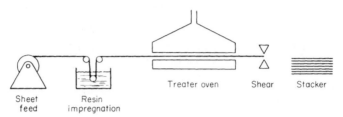

FIGURE 2.1   Horizontal treater.

plates. The plates are further stacked together to form a pack that is then placed into the laminating press's platens. Each pack can consist of ten laminates, and the press may have 24 platens, making a press load of up to 240 laminates per cycle. The press is closed and the laminates are consolidated and polymerized (cured) at temperatures from 250 to 400°F, pressures from 200 to 3000 lb/in², and times from 3 to 120 min. Molded laminates are then trimmed to final sizes. Some materials must be postcured by heating them in ovens.

Thermoplastic laminates are made by combining polymer solutions or films on either a treater or roll combiners.

2.1.2   Design Factors

The presence of reinforcing fibers lends unusual properties to laminates and reinforced plastics.

FIGURE 2.2   Decorative laminate treater.

Physical Properties. The physical properties of plastic materials are substantially improved by combining them with reinforcing fibers. Tensile strengths, flexural strengths, and compressive strengths are higher in laminates than the neat polymers. In particular, the strength per unit weight (specific strength) permits reinforced materials to outperform metals as discussed in Chap. 3. Reinforcements often greatly improve impact strengths, sometimes by a factor of 10. Reinforced plastics, unlike steel, do not demonstrate a clear yield point. Figure 2.3 shows stress-strain curves for several materials, including laminates. Reinforced plastics can be used very close to their ultimate strength without fear of excessive stretching.

Design safety factors should be observed for reinforced plastics. Based on the standardized properties of laminates, rods, and tubes, a minimum safety factor of 4 for mechanical strength and 6 for electrical strength is recommended. If repeated impact loads are expected, a safety factor of 10 is advisable.

Reinforced plastics—and particularly laminates—are anisotropic, i.e., their properties differ depending on the direction of measurement. For example, a laminate made of a fabric has physical properties controlled by the weave of the fabric and the number and density of the threads in the warp and woof directions. Both of these values are different from the values in the z or thickness direction. This is clearly evident in the shear properties. Thermal expansion and thermal conductivity properties are also anisotropic. In the x, y, and z planes of a laminate, the thermal expansion values differ as shown in Table 2.1. Even the reinforcing fibers themselves may have anisotropic properties as is also shown in Table 2.1.

Electrical Properties. The dielectric constant ϵ' of laminates and reinforced plastics is closely related to the nature and amount of reinforcing fiber present. Like

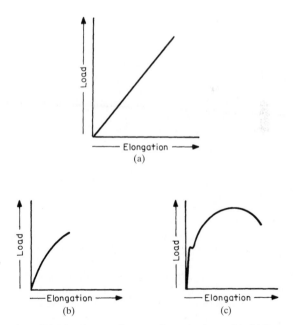

FIGURE 2.3 Stress-strain curves for various materials. (*a*) Reinforced plastics; (*b*) wood and most metals; (*c*) steel.

TABLE 2.1 Thermal Coefficient of Expansion of Epoxy Glass Fabric Laminate and of Kevlar Fiber

Material	Direction	Coefficient of thermal expansion, $10^{-6}/°C$
Epoxy	x	16.0
	y	12.7
	z	200
Kevlar fiber	x	-2
	z	60

many other electrical properties, the dielectric constant follows a "rule of mixtures," which states that the properties of mixtures is dependent on the relative volume fractions of the components. Therefore, as the percent of the reinforcement approaches 100 percent, the dielectric constant approaches the dielectric constant of the reinforcement.

Also the dielectric strength of laminates (and many other polymer structures) is highly dependent on the thickness of the sample. The dielectric strength decreases with increasing thickness. The phenomenon is largely statistical, i.e., the thicker the material between two electrodes, the higher is the probability that a defect or discontinuity will occur.

The dissipation factor (tan δ) does not follow the rule of mixtures as closely as the ϵ'. Instead the resin phase determines the dissipation factor except when the reinforcing fibers are also organic in nature.

The dielectric strength of laminates is highly dependent on the direction of the electric field stress. Samples tested across the thickness of a laminate have higher dielectric strengths than samples tested end-to-end along the fiber length. The differences can be substantial, and dielectric strength values obtained from manufacturers should state the electrode orientation. Because of these electric property relationships, laminates with a high resin content have better electrical properties but poorer physical properties than those with lower resin content.

Other unique properties are related to the nature of the fiber reinforcement. For example, the ability to punch holes in laminates rather than drill holes is related to the type and amount of fiber. Some laminates, i.e., those with fibers of glass, quartz, or Kevlar, do not punch well; others like cellulose do punch well.

Tubes and Rods.   Many of the thermosetting materials that are laminated into flat sheets can also be made into tubes and rods. Round tubes are made by rolling impregnated prepreg sheets onto mandrels between heated rollers. The resin is then cured or polymerized by oven baking in the case of rolled tubes or by pressing in heated molds for molded tubing.

Molded rods are made of plies of prepreg and are molded into shape in cylindrical molds. Rods can also be machined from flat sheets with properties similar to those of the sheets. Both rods and tubes are highly anisotropic.

Stability.   Both thermosetting resins and thermoplastic polymers are used as binders in laminates, but thermosetting materials are more common. The thermosetting materials are superior in stability, thermal resistance, and electrical properties.

The dimensional stability of laminates is an important property. Because the laminate is formed of dissimilar materials under high temperatures and pressures, substantial stresses can be molded into the laminate. These stresses can be relieved by humidity, strain, or temperature and result in warping or twisting of the laminate. Troublesome in most applications, such dimensional instability is of major importance in copper-clad printed wiring boards.

The thickness of laminates is limited by the reactivity of the resin matrix. Laminates are made in presses with heat and pressure. The limiting factors on thickness are the thermal conductivity of the polymerizing resin and the thermal conductivity of the cured laminate. Heat must be conducted into the laminate from the press platens in order to begin the polymerization process. The resins, as they polymerize, produce substantial reactive heat that then must be conducted out of the laminate. Hence it is difficult to make very thick laminates. In general, laminates are available from 0.002 to about 2 in. Some special grades can be made from 4 to 10 in.

2.1.3 Reinforcing Fibers

Many reinforcing fibers are used in laminates and reinforced plastics. The type of reinforcement that is used depends on cost, the properties required, and the nature of the resin system. Although fiber glass is the most common reinforcement, many others are used. Table 2.2 lists common reinforcements and the special properties they contribute to reinforced plastics. Detailed information on fibers can be found in Ref. 1.

Glass Fibers.   Glass fibers are formed continuously from a melt in special fiber-forming furnaces. Six formulations are made. The most common is E glass. This glass resists moisture and results in products with excellent electrical properties. Table 2.3 lists the chemical composition of E glass and other types of glass. C glass is designed for use where optimum chemical resistance is required. D glass has very good electrical properties—particularly the dielectric constant—and is used in electronic applications. S glass is used for its high strength and stiffness, while R glass is a lower cost fiber than S glass. Table 2.4 shows the physical and electrical properties of these fibers.

TABLE 2.2   Properties of Reinforced Plastic and of Reinforcing Fiber

Reinforcing fiber	Mechan-ical strength	Electrical prop-erties	Impact resis-tance	Chemical resis-tance	Machin-ing and punching	Heat resis-tance	Moisture resis-tance	Abrasion resis-tance	Low cost	Stiffness
Glass strands	X		X	X		X	X		X	X
Glass fabric	X	X	X	X		X	X			X
Glass mat			X	X		X	X		X	X
Asbestos		X	X			X				
Paper		X			X				X	
Cotton/linen	X	X	X		X				X	
Nylon		X	X	X				X		
Short inorganic fibers	X	X							X	
Organic fibers	X	X	X	X	X			X		
Ribbons		X					X			
Metals	X		X			X				X
Polyethylene	X		X		X					X
Aramid	X	X	X			X		X		X
Boron	X		X			X				X
Carbon/graphite	X		X			X				X
Ceramic	X		X						X	

TABLE 2.3 Composition of Glass Fibers

Ingredient	Glass type*					
	A	C	D	E	R	S and S2
SiO_2	72	65	74	52–56	60	65
CaO	10	14	0.5	16–25	9	—
Al_2O_3	0.6	4	0.3	12–16	25	25
MgO	2.5	3	—	0–5	6	10
B_2O_3	—	6	22	5–10	—	—
TiO_2	—	—	—	0–1.5	—	—
Na_2O	14.2	8	1.0	0–2	—	—
K_2O	—	—	1.5		—	—
Fe_2O_3	—	0.2	—	0–0.8	—	—
So_3	0.7	0.1	—	—	—	—
F_2	—	—	—	0–1.0	—	—

* Values represent weight percent of total.

TABLE 2.4 Properties of Glass Fibers

	Glass type					
	A	C	D	E	R	S and S2
	Physical properties					
Specific gravity	2.50	2.49	2.16	2.52–2.61	2.55	2.49
tensile strength, lb/in²	350,000	400,000	350,000	500,000	640,000	665,000
Tensile elastic modulus	9,800,000	10,000,000	7,500,000	10,500,000	12,475,000	12,600,000
Elongation at 72°F, %	—	—	—	3–4	—	5.4
Poisson's ratio	—	—	—	0.22	—	—
	Thermal properties					
Softening point, °F	1300	1380	1420	1540–1555	1481	1778
Coefficient of thermal expansion—in/ in/°F × 10^{-7}	90	40	17	28–33	74	13–17
Thermal conductivity (k) BTU in/hr/ ft²/°F at 72°F	—	—	—	7.2	6.9	—
Specific heat at 72°F BTU/lb/°F	—	—	—	0.197	—	—
	Optical properties					
Index of refraction	1.512	1.541	1.47	1.56	—	1.523
	Electrical properties					
Dielectric constant, 72°F, 10^6 Hz	6.90	6.24	3.56	6.1–6.7	6.2	5.34
Loss tangent, 72°F, 10^6 Hz	0.0085	0.0052	0.0005	0.001	0.0015	0.002
	Features					
	Soda-lime glass with limited water re-sistance	Chemical glass	Lowest dielectric constant	Electrical grade	High strength plus mod-ulus	Highest physical properties
	Uses					
	Surface mat	Chopped roving surface mat yarns	Yarn	Roving fabrics yarn	Yarn	Roving yarn

Glass filaments are made in a variety of diameters as shown in Table 2.5. The filaments of small diameters are not usually used to reinforce plastics, but those from G diameter to about T are commonly used. The filaments are formed into a strand with 204, 400, 800, 1000, 2000, 3000, or 4000 filaments to a strand. A sizing agent is applied to the filaments to bond them into the strand and to give them environmental and abrasive protection. Subsequently, coupling agents are added to the finished products to enhance adhesion of the resin matrix to the glass fibers. Silanes, chrome complexes, and polymers are used as coupling agents. The strands are then used to manufacture the various types of glass reinforcements shown in Table 2.6. Glass fabrics, glass mats, and chopped strands are the most common reinforcements in reinforced plastics.

Chopped Strands.   Strands of glass (usually E glass) are mechanically cut into lengths from 0.25 to 2 in and are used in reinforced molding compounds. The larger lengths are used with thermoset resins in bulk molding compounds, premix, putty compounds, rope compounds, and injection molding materials. Thermoplastic polymers use short fibers or even milled fibers.

Glass Mats.   Glass strands are cut and dropped onto a moving belt where a polymer binder is applied to hold the mat together. Mats vary in weight from 0.75 to 3 oz/ft^2 and widths of up to 10 ft.

Continuous strand mat is made when the strands are not cut but deposited continuously in a swirling pattern onto a moving belt where a binder has been applied.

TABLE 2.5   Glass Fiber Diameters

Filament	Diameter range, × 10^5 in	Nominal diameter × 10^5 in	μm
B		10.3	2.6
B	10.0–14.9	12.5	3.2
B		14.7	3.7
BC	13.0–18.0		
C	15.0–19.9	17.5	4.4
D	20.0–24.9	22.5	5.7
DE	23.0–27.9	25.0	6.6
E	25.0–29.9	27.5	7.0
F	30.0–34.9	32.5	8.3
G	35.0–39.9	36	9.1
G	35.0–39.0	37.5	9.5
H	40.0–44.9	42.5	10.8
J	45.0–49.9	47.5	12.1
K	50.0–54.9	52.5	13.3
L	55.0–59.0	57.5	14.6
M	60.0–64.9	62.5	15.9
N	65.0–69.9	67.5	17.1
P	70.0–74.9	72.5	18.4
Q	75.0–79.9	77.5	19.7
R	80.0–84.9	82.5	21.0
S	85.0–89.9	87.5	22.2
T	90.0–94.9	92.5	23.5
U	95.0–99.9	97.5	24.8
V	100.0–104.9	102.5	26.0
W	105.0–109.9	107.5	27.3

TABLE 2.6 Glass Fiber Reinforcement Forms[1]

Form of fiber glass reinforcement	Definition and description	Range of grades available	General types of sizing applied	General usage in RP/C (and secondary uses if any)
Twisted yarns	Single-end fiberglass strands twisted on standard textile tube-drive machinery.	B to K fiber, S or Z twist, 0.25–10.0 twists/in, many fiber and yardage variations.	Starch.	Into single and plied yarns for weaving; many other industrial uses (and decorative uses).
Plied yarns	Twisted yarns plied with reverse twist on standard textile ply frames.	B to K fiber, up to $^4/_{18}$ ply—many fiber and yardage variations.	Starch.	Weaving industrial fabrics and tapes in many different cloth styles; also heavy cordage.
Fabrics	Yarns woven into a multiplicity of cloth styles with various thicknesses and strength orientations.	D to K fibers, 2.5–40 oz/yd^2 in weight.	Starch size removed and compatible finish applied after weaving.	Wet lay-up for open molding, prepreg, high pressure lamination, and also some press molding.
Chopped strands	Filament bundles (strands) bonded by sizings, subsequently cured and cut or chopped into short lengths. Also the reverse, i.e., chopped and cured.	G to M fiber, $^1/_8$–$^1/_2$ in or longer lengths, various yardages.	LSB and RTP.	Compounding, compression, transfer and injection press molding.
Roving	Gathered bundle of one or more continuous strands wound in parallel and in an untwisted manner into a cylindrical package.	G or T fiber used. Roving yields 1800 to 28 yd/lb and packages 15- to 450-lb weight (size of up to 24 × 24 in).	HSB, LSB, and RTP.	Used in all phases of RP/C. Some (HSB) are used in continuous form, e.g., filament winding, and others (LSB) are chopped, as in sheet molding compound for compression molding.
Woven roving	Coarse fabric, bidirectional reinforcement, mostly plain weave, but some twill. Uni- and multidirectional nonwoven rovings are also produced.	K to T fiber, fabric weights 10–48 oz/yd^2.	HSB.	Mostly wet lay-up, but some press molding.
Chopped strand mats	Strands from forming packages chopped and collected in a random pattern with additional binder applied and cured; some "needled" mat produced with no extra binder required.	G to K fiber, weights 0.75–6.0 oz/ft^2.	HSB and LSB.	Both wet lay-up and press molding.

2.8

Form of fiber glass reinforcement	Definition and description	Range of grades available	General types of sizing applied	General usage in RP/C (and secondary uses if any)
Mats, continuous strand (swirl)	Strands converted directly into mat form without cutting with additional binder applied and cured, or needled.	Nominally M to R fiber diameter, weights 0.75–4.5 oz/ft².	LSB.	Compression and press molding, resin transfer molding, and pultrusion.
Mat-woven roving combinations	Chopped strand mat and woven roving combined into a drapable reinforcement by addition of binder or by stitching.	30–62 oz/yd².	HSB.	Wet lay-up to save time in handling.
Three-dimensional reinforcements	Woven, knitted, stitched, or braided strands or yarns in bulky, continuous shapes.	—	HSB.	Molding, pultrusion.
Milled fibers	Fibers reduced by mechanical attrition to short lengths in powder or nodule form.	Screened from $\frac{1}{32}$–$\frac{1}{4}$ in. Actual lengths range 0.001–$\frac{1}{4}$ in. Several grades.	None, HSB, and RTP.	Casting, potting, injection molding, reinforced reaction injection molding (RRIM).
Related forms:				
Glass beads	Small solid or hollow spheres of glass.	Range 1–53 μm diameter, bulk densities; hollow = 0.15–0.38 g/cm³, solid = 1.55 g/cm³.	Usually treated with cross-linking additives.	Used as filler, flow aid, or weight reduction medium in casting, lamination, and press molding.
Glass flake	Thin glass platelets of controlled thickness and size.	0.0001 in and up thick.	None, or treated with doupling agent.	Used as a barrier in, or to enhance abrasion resistance of, linear resins; coatings used for corrosion-resistance applications. Also used in RRIM for increased dimensional stability.

The physical properties of continuous strand mats are better than those of cut mats, but the material is less homogeneous. Continuous strand mats vary in weight from 0.75 to 4.5 oz/ft² and widths of up to 6 ft.

Woven roving is a mat fabric made by weaving multiple strands collected into a roving into a coarse fabric. The physical properties of woven roving are intermediate between mats and fabrics. These constructions are used in low-pressure lamination and in pultrusion. Various thicknesses are made in widths of up to 10 ft.

Glass Fabrics. Many different fabrics are made for reinforced plastics. E glass is used for most of them, and filament laminates of D, G, H, and K are common. Glass filaments are combined into strands, and the strands are plied into yarns. The yarns are woven into fabrics on looms. The machine direction of the loom is called the warp, while the cross direction is the fill (also called woof or weft). The number of yarns can be varied in both warp and fill to control the weight, thickness, appearance, and strength of the fabric. When each fill yarn is laced alternately over and then under the warp, the type of fabric (called the weave) is known as plain. Modifying this one-on-one arrangement by having yarns cross two or more adjacent yarns or staggering the crossing along the warp results in other weaves called basket, satin, eight-harness satin, leno, and mock leno, all of which are used by the reinforced plastic industry.

Table 2.7 lists the fabrics used in electrical-grade laminates; Table 2.8 lists those used in mechanical laminates; Table 2.9 lists those used in laminates for marine applications; and Table 2.10 lists scrim fabrics used as facing layers for smooth, cosmetic applications.

Asbestos. Asbestos fibers (crysotile) are used in laminates to impart flame resistance and thermal stability. Used extensively in the past, they are listed in NEMA Standard LI-1 as Grade AA, but concerns about asbestosis now limit their use and they are being phased out of the laminate industry.

Paper. Paper—usually kraft—is widely used as a reinforcement. Saturated with phenolics, it is made into a common printed wiring board. Combined with melamine,

TABLE 2.7 Electrical-Grade Glass Fabrics[1]

Fabric style no.	Count Yarns/in Yarns/5 cm	Warp yarn	Yarn count TEX	Fill yarn	Yarn count TEX	Weave	Mass oz/yd² g/m²	Thickness in mm	Breaking strength lb/in N/(5 cm)
104	60 × 52	ECD	900 1/0	ECD	1800 1/0	Plain	0.58	0.0012	40 × 15
	118 × 102	EC5	5.5 1 × 0	EC5	2.75 1 × 0	Plain	19.7	0.030	350 × 131
108	60 × 47	ECD	900 1/2	ECD	900 1/2	Plain	1.43	0.0020	70 × 40
	118 × 93	EC5	5.5 1 × 2	EC5	5.5 1 × 2	Plain	48.5	0.051	613 × 350
112	40 × 39	ECD	450 1/2	ECD	450 1/2	Plain	2.10	0.0032	90 × 80
	79 × 77	EC5	11 1 × 2	EC5	11 1 × 2	Plain	71.2	0.081	788 × 701
116	60 × 58	ECD	450 1/2	ECD	450 1/2	Plain	3.16	0.0040	125 × 120
	118 × 114	EC5	11 1 × 2	EC5	11 1 × 2	Plain	107.0	0.102	1095 × 1051
2112	40 × 39	ECD	225 1/0	ECD	225 1/0	Plain	2.10	0.0034	82 × 80
	79 × 77	EC5	22 1 × 0	EC5	22 1 × 0	Plain	71.6	0.076	720 × 700
2116	60 × 58	ECD	225 1/0	ECD	225 1/0	Plain	3.16	0.0040	125 × 120
	118 × 114	EC5	22 1 × 0	EC5	22 1 × 0	Plain	107.0	0.102	1095 × 1050
7628	44 × 32	ECG	75 1/0	ECG	75 1/0	Plain	6.00	0.0068	250 × 200
	87 × 63	EC9	66 1 × 0	EC9	66 1 × 0	Plain	203.0	0.173	2189 × 1751
7642	44 × 20	ECG	75 1/0	ETG	37 1/0	Plain	6.87	0.0110	250 × 120
	87 × 39	EC9	66 1 × 0	ET9	134 1 × 0	Plain	232.0	0.279	2190 × 1050
1080	60 × 47	ECD	450 1/0	ECD	450 1/0	Plain	1.44	0.002	70 × 40
	118 × 93	EC5	11 1 × 0	EC5	11 1 × 0	Plain	48.8	0.051	610 × 350

TABLE 2.8   Mechanical-Grade Glass Fabrics[1]

Fabric style no.	Count Yarns/in Yarns/5 cm	Warp yarn Yarn count TEX			Fill yarn Yarn count TEX			Weave	Mass oz/yd² g/m²	Thickness in mm	Breaking strength lb/in N/(5 cm)
1581	57 × 54	ECG	150	1/2	ECG	150	1/2	8-H Satin	8.90	0.0090	350 × 340
	112 × 106	EC9	33	1 × 2	EC9	33	1 × 2	8-H Satin	302.0	0.228	3065 × 2977
1582	60 × 56	ECG	150	1/3	ECG	150	1/3	8-H Satin	13.90	0.0140	490 × 450
	118 × 110	EC9	33	1 × 3	EC9	33	1 × 3	8-H Satin	471.0	0.355	4291 × 3940
1583	54 × 48	ECG	150	2/2	ECG	150	2/2	8-H Satin	16.10	0.0160	650 × 590
	106 × 94	EC9	33	2 × 2	EC9	33	2 × 2	8-H Satin	545.0	0.406	5692 × 5166
1584	44 × 35	ECG	150	4/2	ECG	150	4/2	8-H Satin	26.00	0.0260	950 × 800
	87 × 69	EC9	33	4 × 2	EC9	33	4 × 2	8-H Satin	880.0	0.670	8318 × 7005
3706	12 × 6	ECG	37	1/0	ECG	37	1/2	Leno	3.70	0.0086	140 × 120
	24 × 12	EC9	134	1 × 0	EC9	134	1 × 2	Leno	125.0	0.218	125 × 1050
7781	57 × 54	ECDE	75	1/0	ECDE	75	1/0	8-H Satin	8.95	0.0090	350 × 340
	112 × 106	EC6	66	1 × 0	EC6	66	1 × 0	8-H Satin	304.0	0.228	3065 × 2977
7626	34 × 32	ECG	75	1/0	ECG	75	1/0	Plain	5.40	0.0066	225 × 200
	67 × 63	EC9	66	1 × 0	EC9	66	1 × 0	Plain	183.0	0.168	1970 × 1751
181	57 × 54	ECD	225	1/3	ECD	225	1/3	Satin	8.9	0.009	350 × 340
	22 × 21	EC5	22	1 × 3	EC5	22	1 × 3	Satin	302	0.229	3065 × 2971

TABLE 2.9   Marine-Grade Glass Fabrics[1]

Fabric style no.	Count Yarns/in Yarns/5 cm	Warp yarn Yarn count TEX			Fill yarn Yarn count TEX			Weave	Mass oz/yd² g/m²	Thickness in mm	Breaking strength lb/in N/(5 cm)
1800	16 × 14	ECK	18	1/0	ECK	18	1/0	Plain	9.60	0.0130	450 × 350
	31 × 28	EC13	275	1 × 0	EC13	275	1 × 0	Plain	326.0	0.330	3940 × 3065
2532	16 × 14	ECH	25	1/0	ECH	25	1/0	Plain	7.25	0.0100	300 × 280
	31 × 28	EC10	200	1 × 0	EC10	200	1 × 0	Plain	246.0	0.254	2627 × 2452
7500	16 × 14	ECG	75	2/2	ECG	75	2/2	Plain	9.66	0.0140	450 × 410
	32 × 28	EC9	66	2 × 2	EC9	66	2 × 2	Plain	327.0	0.356	3940 × 3590
7533	18 × 18	ECG	75	1/2	ECG	75	1/2	Plain	5.80	0.0080	250 × 220
	35 × 35	EC9	66	1 × 2	EC9	66	1 × 2	Plain	197.0	0.203	2189 × 1926
7544	28 × 14	ECG	75	2/2	ECG	75	2/4	Basket	18.0	0.0220	750 × 750
	55 × 28	EC9	66	2 × 2	EC9	66	2 × 4	Basket	610.0	0.559	6567 × 6567
7587	40 × 21	ECG	75	2/2	ECG	75	2/2	Mock Leno	20.5	0.0300	750 × 450
	79 × 41	EC9	66	2 × 2	EC9	66	2 × 2	Mock Leno	695.0	0.761	6567 × 3940

TABLE 2.10   Scrim Glass Fabrics[1]

Fabric style no.	Count Yarns/in Yarns/5 cm	Warp yarn Yarn count TEX			Fill yarn Yarn count TEX			Weave	Mass oz/yd² g/m²	Thickness in mm	Breaking strength lb/in N/(5 cm)
1610	32 × 28	ECG	150	1/0	ECG	150	1/0	Plain	2.41	0.0040	115 × 100
	63 × 55	EC9	33	1 × 0	EC9	33	1 × 0	Plain	84.1	0.102	1010 × 965
1650	20 × 10	ECG	150	1/0	ECG	75	1/0	Plain	1.60	0.0040	80 × 70
	39 × 20	EC9	33	1 × 0	EC9	66	1 × 0	Plain	54.3	0.102	700 × 615

it becomes decorative high-pressure laminates used in counter tops, furniture, and wall panels. Alpha cellulose (a highly refined cellulose), cotton linters, or wood pulp are combined with resins—mostly thermosetting ones—to make compression-, transfer-, and injection-molding compounds. Paper reinforcement is inexpensive and easy to machine, drill, and punch, and it contributes good electrical properties. It is sensitive to moisture and cannot withstand high temperatures.

Cotton or Linen.   Fabrics of cotton or linen are used as reinforcements in several grades of laminates—usually impregnated with a phenolic resin. Laminates with these fabrics have better water resistance than paper-based laminates. They machine well and have good physical properties—particularly their impact strength and abrasion resistance. Their electrical properties are poor. They are used for gears and pulleys. In chopped or macerated form, they can be used in molding compounds.

Nylon.   Nylon fibers are used with phenolic resins to produce tough laminates with good electrical properties. They are listed as Grade N-1 in NEMA Standard LI-1. Creep or cold flow often limits their application.

Short Inorganic Fibers.   Several short inorganic fibers are used to reinforce molding compounds. With a high aspect ratio, these fibers are usually used in combination with other mineral fillers in both thermosetting- and thermoplastic-molding materials. Fiber contents can reach 40 percent. The presence of the fiber improves the physical properties and stiffness while impact strengths are slightly improved. The biggest improvement is usually in the heat deflection temperature (HDT). For example, the HDT of nylon 6/12 can be increased from 217°F to 322°F by adding 30 percent by weight of a short fiber. Less desirable effects can be increased viscosities, decreased flow, and embrittlement. Short inorganic fibers include Dawsonite, Fiberkal, Franklin fiber, magnesium oxysulfate, milled glass, phosphate fiber, processed mineral fiber, Tismo, and Wollastonite.

Polyethylene.   Polyethylene can be made into a high-strength, high-modulus fiber trade-named Spectra by Allied-Signal Inc.[17] The polyethylene is an ultrahigh molecular weight polymer up to 5 million, compared to conventional polyethylene with a molecular weight of about 200,000. Spectra is produced in two fibers, 900 and 1000, with the 1000 grade being higher strength and modulus as shown in Table 2.11. The density of Spectra is the lowest of all fibers, which makes the use of aerospace laminates especially attractive in applications like wing tips and helicopter seats. However, laminates must not be exposed to temperatures over 250°F. Spectra has very attractive electrical properties with a low dielectric constant and a loss tangent of 0.0002. Table 2.12 shows the dielectric constant of epoxy-based Spectra fabrics compared to other reinforcements. Table 2.13 shows fiber electrical properties. These properties are useful in radomes. Spectra laminates are virtually transparent to radar and they are used as periscope housings. The performance in water and the resistance to chemicals is shown in Table 2.14. Like aramid, Spectra is used in ballistic applications like helmets and aircraft panels. Sporting equipment based on excellent impact properties include canoes, water skis, tennis racquets, and golf club shafts. The fibers are often combined with glass, graphite, or aramid. More discussion is in Chap. 3.

Ribbons or Flakes.   These unlikely reinforcing forms have a small but important role in reinforcing materials. Mica is a complex potassium aluminum silicate that can be prepared in flake form. It is used broadly as a reinforcing filler, especially in connection with gel coat systems used in polyester hand lay-up processes where it pro-

TABLE 2.11 Polyethylene Fiber Tensile Properties

	Spectra*		Aramid	
	900	1000	LM	HM
Tenacity, psi ($\times 10^3$)	373	435	406	406
Modulus, psi ($\times 10^6$)	17.4	24.8	9.0	18.0
Elongation, %	3.5	2.7	3.6	2.8

* Tradename of Allied Signal.

TABLE 2.12 Electrical Properties of Composites

	Dielectric constant		
System	Fiber	Resin	Composite
Spectra*/epoxy	2.2	2.8	2.42
Quartz/epoxy	3.78	2.8	3.35
Aramid/epoxy	3.85	2.8	3.39
E glass/epoxy	6.31	2.8	4.56

* Tradename of Allied Signal.

TABLE 2.13 Fiber Electrical Properties

Material	Dielectric Constant	Loss Tangent
Spectra*	2.0–2.3	0.0002–0.0004
E glass	4.5–6.0	0.0060
Aramid	3.85	0.0100
Quartz	3.78	0.0001–0.0002

* Tradename of Allied Signal.

TABLE 2.14 Chemical Resistance of Spectra* and Aramid

	% Retention of fiber tensile strength			
	Spectra[†]		Aramid	
Agent	6 months	2 years	6 months	2 years
Sea water	100	100	100	98
Hydraulic fluid	100	100	100	87
Kerosene	100	100	100	97
Gasoline	100	100	93	†
Toluene	100	96	72	†
Glacial acetic acid	100	100	82	†
1M hydrochloric acid	100	100	40	†
5M sodium hydroxide	100	100	42	†
Ammonium hydroxide (29%)	100	100	70	†
Perchloroethylene	100	100	75	†
10% detergent solution	100	100	91	†
Clorox	91	73	0	0

* Tradename of Allied Signal.
† Samples not tested due to physical deterioration.

vides a stiff, strong impact-resistant surface with good water resistance. Its biggest use is in swimming pools, tubs, shower enclosures, and boat hulls. Mica is also used to strengthen epoxy castings (see Chap. 4) and is used with polyurethanes in automotive parts. Thermoplastics like PET and PBT use mica flakes to control warping of large parts; mica is also used in polyethylene and polypropylene to enhance their physical properties and stiffness. Glass flakes made by mechanically fracturing ribbons are similarly used.

Glass ribbons 0.0005-in thick can be handled like filaments. Wrapped on a mandrel, they form pipes with excellent chemical-resistant properties. Highly notch-sensitive, they are difficult to machine or drill.

Metals. Nickel, stainless steel, and silver can be made in fiber form. In addition, conventional glass fibers can be aluminized. These are used with various plastics to make EMI-compatible housings, thermally conductive heat sinks, and electrically conductive heaters.

Aramid. Aramid is a generic name for aromatic polyamides (nylon). The reinforcement is available as a paper (Nomex from DuPont and TP Technora from Teijin America) and as a fiber (Kevlar from DuPont). The fiber is used in applications in much the same way as glass fiber: It is used in filament wounds, it is pultruded, and it is woven into fabrics for reinforced plastics and laminates. Thin (2 mil, 1 oz per sq yd) to thick (30 mil, 16 oz per sq yd) fabrics are made. Properties of Kevlar fibers are shown in Table 2.15. Properties of various laminates are shown in Table 2.16. The use of aramid fabrics in printed circuit boards is increasing, and more data can be found in this chapter in the section on copper-clad laminates. Aramid paper is used in circuit boards[18] to improve crack resistance and to impart a smooth surface, but a high coefficient of thermal expansion, poor resin adhesion, and difficult drilling and machining characteristics limits the paper. Properties of aramid fiber used in paper are given in Table 2.17.

Aramid fabrics are used in a wide variety of aircraft and aerospace components where the low density of aramid is critical: Boat hulls are stronger and lighter and ride better because of the vibration-dampening qualities of the fiber. The light weight is especially useful in laminated canoes and kayaks. Other sport applications include downhill skis, tennis racquets, and golf shafts.

Boron, Carbon, Graphite, or Ceramic Fiber. These fibers are characterized by exceptionally high moduli, tensile strength, thermal resistance, and cost. They are used in advanced composites and are described in detail in Chap. 3.

TABLE 2.15 Properties of Aramid Fibers

Property	Type				
	29	49	149	68	129
Tensile strength, 10^3 lb/in^2	525	525	500	525	610
Tensile modulus, 10^6 lb/in^2	12	18	25	16	16
Yarns					
No. size	5	7	3	3	3
Denier range	200–3000	55–2840	380–1420	1420–2840	840–1420
Rovings					
No. size	2	6	1	1	—
Denier range	9000–15,000	4320–22,720	7100	7100	

TABLE 2.16   Properties of Aramid Laminates[22]

Reinforcement	Resin system	T_g (°C)	Dielectric constant at 1 MHz	Z-axis CTE 25°C to 260°C (ppm/°C)
Woven E glass	Standard FR-4 epoxy	130	4.5	190
Woven E glass	Multifunctional epoxy	145	4.5	175
Woven E glass	High-temperature epoxy	170	4.5	135
Woven E glass	High-performance epoxy	210	4.0	105
Woven E glass	BT/epoxy	175	4.2	130
Woven E glass	Cyanate ester	250	3.6	65
Woven E glass	Polyimide	275	4.2	50
Woven S glass	Standard FR-4 epoxy	130	4.1	180
Woven S glass	Cyanate ester	240	3.7	65
Woven D glass	Standard FR-4 epoxy	130	3.9	180
Woven D glass	Cyanate ester	240	3.4	62
Woven quartz	Polyimide	275	4.2	50
Woven aramid	Standard FR-4 epoxy	125	3.8	200
Woven aramid	High-temperature epoxy	165	4.0	160
Woven aramid	Polyimide	260	3.6	65
Paper aramid	High-temperature epoxy	165	4.0	210
E glass filament	High-performance epoxy	210	4.0	90

TABLE 2.17   Properties of Aramid Paper Fiber

Density, g/cm³	1.39
Tensile strength, 10^3 lb/in²	600
Tensile modulus, 10^6 lb/in²	13
Elongation, %	4.5
Coefficient of thermal expansion /°C	6
Water absorption, %	2
Specific heat, cal/g °C	0.26

2.1.4   Laminating Resins—Thermoset

Phenolic Resins.   Phenolic resins are among the oldest plastics. They are available in a large number of variations, depending on the nature of the reactants, the ratios used, and the catalysts, plasticizers, lubricants, fillers, and pigments employed. These laminates are excellent general-purpose materials. They have a wide range of mechanical and electrical properties, and by selecting the correct reinforcement material, products with many properties can be obtained. Phenolics are used widely for both high- and low-temperature service.

Phenolics are often used because their cost is lower than those of laminates formed with most other resins. They have excellent water resistance and are frequently chosen for marine applications. They can be used in gears, wheels, and pulleys because of their wear and abrasion resistance and in printed circuits and terminal blocks because of their stability under a variety of environmental conditions. Two disadvantages limit their use. Because of the nature of the resin, the laminates are available only in dark colors, usually brown or black. In addition, they have rather poor resistance to electric arcs. High filler loading can improve the low-power arc resistance as measured by ASTM D 495, but the presence of moisture, dirt, or high voltages usually results in complete arcing breakdown. In addition, although their water resistance is good, i.e., their physical properties are retained, their water absorption is high, reaching 14 percent in some paper-based grades of laminate.

Melamine Polymers. Melamine-formaldehyde polymers are widely used in the home in the form of high-quality molded dinnerware. This application takes advantage of their heat resistance and water-white color. Melamine-surfaced decorative laminates are also used in the home and are described in Sec. 2.1.4. Melamine resin can be combined with a variety of reinforcing fibers, but the best properties are obtained when glass cloth is used as the reinforcement material. Table 2.18 shows that its electrical properties are maximized when the filler is glass. Melamines retain their properties at low temperatures. Table 2.19 shows the physical properties of melamine with several reinforcements at temperatures as low as –65°F. The arc and track resistance of melamines is very good and is competitive with the best epoxies and alkyds. This property is used in high-power arcing applications like circuit breakers, particularly on military ships. Melamines are often used where resistance to caustics is required. Other uses include switch gear, terminal blocks, circuit-breaker parts, slot wedges, and bus bar supports. Melamine laminates have poor dimensional stability, particularly when the part is exposed to alternating cycles of high and low humidity.

Epoxy Resins. Epoxy resins consist of a large family with varying polymers, hardening agents, catalysts, fillers, and special compounds. They are particularly useful in laminates because they can be B-staged or made into prepreg, i.e., the polymerization of the polymer can be controlled and stopped before it is complete (see Sec. 2.1.1). Epoxies are resistant to chemicals, are strong, and have superior adhesion and good electrical properties. They are very useful in electronic applications as circuit boards, conformal coatings, and adhesives. Epoxies are used in cryogenic applications where resistance to resin crazing is desired. Figure 2.4 shows the tensile properties of laminates to –423°F. In general, their electric properties worsen with temperature, limiting the electrical application of epoxies to about 250°F. They are also used in marine applications because of their resistance to water. With the correct choice of hardener, laminates can be made with good retention of physical properties up to 400°F. Laminates are used in electrical and electronic applications. Figures 2.5, 2.6, and 2.7 show volume resistivity, dielectric constant, and dissipation factor with temperature and frequency. Epoxy laminates

TABLE 2.18 Electrical Properties of Melamine Laminates as a Function of the Reinforcement[3]

Property	Reinforcement	
	Glass	Cellulose
Insulation resistance, ohms...............	10^9	10^7
Arc track resistance, volts (D 2303).......	1,500	1,000
Dielectric strength, volts/mil............	300	250

TABLE 2.19 Mechanical Properties of Melamine at Varying Temperatures[4]

Property	Tempera-ture, °F	Melamine† glass	Electrical melamine-formalde-hyde§	α-Melamine-formalde-hyde¶
Tensile strength, psi × 10⁻³	77	32.5	5.4	7.8
	10	32.9	6.7	6.9
	−40	38.1	5.7	6.9
	−65	37.2	5.6	6.7
Modulus of elas-ticity, psi × 10⁻⁶	77	2.130	1.060	1.270
	10	2.290	1.610	1.640
	−40	1.430	1.540	1.730
	−65	1.580	1.390	1.880
Elongation at break, %	77	2.15	0.54	0.62
	10	2.17	0.50	0.44
	−40	2.36	0.40	0.39
	−65	2.75	0.38	0.37
Work to produce failure, ft-lb/in.³	77	31.0	1.32	2.08
	10	36.7	1.40	1.33
	−40	44.3	1.02	0.86
	−65	52.2	0.97	1.13
Proportional limit, psi × 10⁻³	77	15.8	33.2	25.9
	10	7.1	10.5	9.5
	−40	5.4	5.7	5.6
	−65	7.8	6.9	6.7
Izod impact strength, ft-lb/ in. notch (D 256)	77	11.12	0.33	0.31
	10	12.64	0.37	0.28
	−40	13.43	0.32	0.28
	−65	14.68	0.28	0.29

* Laminated melamine, glass fabric base.
† Melamine-formaldehyde, cellulose filler, electrical grade.
‡ Melamine-formaldehyde, α-cellulose filler.

designated FR-4 by NEMA are the workhorse of the printed circuit board indus-try. The resistance to water and ease of handling of FR-4 are excellent for boats and marine applications, and the toughness of FR-4 is used in large power genera-tors and transformers.

Polyesters. Commercial polyesters usually consist of an unsaturated ester polymer dissolved in a monomer such as styrene. The resins are usually combined with glass fibers since many cellulosic reinforcements inhibit their cure. Table 2.20 shows the effect of reinforcement media on several properties. One of the chief virtues of these resins is the excellent chemical resistance of the laminates, and, hence, they are used for corrosion-resistant tanks, exhaust ducts, scrubbers, and plating equipment. Note that special grades of polyester are used for chemical service. These materials resist acids and bases for long times at 200°F. They are attacked by most chlorinated sol-vents. Table 2.21 shows the effect of a variety of chemicals on a chemically resistant polyester. The chemical resistance of these laminates also can be influenced by the

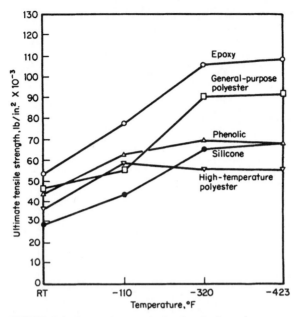

FIGURE 2.4   Cryogenic strength of various laminates.[5]

type of glass reinforcement, the fabrication method, and the amount of stress in the product. A closely related polymer family, the vinyl esters, has even more superior chemical resistance and some processing advantages.

Polyester laminates are widely used in the switchgear industry. They can be made highly arc-, track-, and flame-resistant. The polyester resins themselves possess good arc and track resistance, which is further enhanced by the addition of mineral fillers, particularly aluminum oxide trihydrate. Flame resistance is obtained by modifying the basic resin, usually by the incorporation of chlorine, bromine, or phosphorus. Polyesters are moderate in cost. They are used in applications requiring dimensional stability, such as furniture, and water resistance, such as boats and bathtubs. They are used in automotive applications, such as grills, hoods, fenders, and interior parts. Resins stabilized to sunlight are widely used in corrugated sheet in screens, patio roofs, and dividers. Low-profile polyesters are resins to which any of several thermoplastic polymers have been added. This addition decreases the resin shrinkage and, thereby, produces exceptionally smooth surfaces. These resins are used in automotive, truck, and recreational vehicle body parts. Surfacing resins are polyesters that contain polymers or waxes to eliminate air inhibition and thixotropic agents for viscosity control. They are used mostly in hand lay-up applications such as boat hulls.

In the manufacture of large decorative components like hot tubs, bathtubs, shower enclosures, and vanity tops, vacuum-formed acrylic sheets are used as both a decorative surface and a forming mold. Glass polyester is molded to the reverse side of the acrylic. The ability to fabricate polyesters at ambient conditions of temperature and pressure results in very large moldings like wall panels, barge covers, railcar covers, boats, and water treatment domes. The ability of the laminates to diffuse light results in greenhouse panels, skylights, trailer roofs, and awnings.

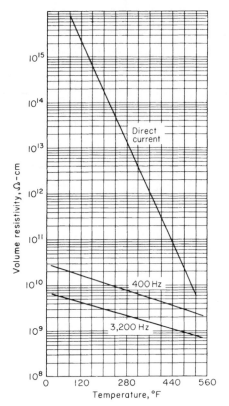

FIGURE 2.5 Volume resistivity versus temperature at 400 and 3200 Hz and direct current for a high-temperature epoxy.[6]

Silicone Resins.   The silicone resins were the first of the true high-temperature plastics. Silicone polymers are similar to organic polymers except that certain carbon atoms are replaced by silicon and oxygen atoms. Silicone polymers are available in the form of liquids, gels, elastomers, and brittle solids. The latter polymers, when combined with glass fabrics, constitute the silicone laminates used mostly in the electronics and electrical industries. These materials are useful from cryogenic temperatures to about 500°F. The cryogenic properties of the silicone laminates are given in Table 2.22, where their physical properties are compared with those of other laminating resins. The dielectric properties of silicones are particularly useful. The dissipation factor and dielectric constant are low at room temperature and stay relatively constant to 300°F. Figures 2.8–2.10 compare glass-cloth-reinforced silicones with other laminates. Two silicones are shown, one designated for low-pressure laminating and the other for high-pressure laminating, and they are compared with other laminates. Notice the flat slope of the silicones. The curves for the epoxies characteristically show an inflection point near the glass transition temperature, and their losses get progressively worse. The response of dissipation factor to frequency is shown in Fig. 2.10. Thermal aging does little to influence these properties. Table 2.23

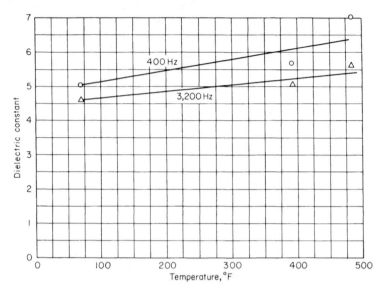

FIGURE 2.6 Dielectric constant versus temperature at 400 and 3200 Hz for a high-temperature epoxy.[6]

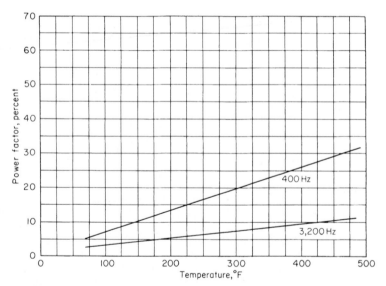

FIGURE 2.7 Dissipation factor versus temperature at 400 and 3200 Hz for a high-temperature epoxy.[6]

TABLE 2.20 Relationship of Properties to Reinforcement for Polyester Laminates

Reinforcement material	Moisture absorption, %	Ultimate tensile strength, psi × 10⁻³				Ultimate flexural strength, psi × 10⁻³			
		At room temperature		At −65°F (frozen)		At room temperature		At −65°F (frozen)	
		Dry	Wet	Dry	Wet	Dry	Wet	Dry	Wet
Glass cloth......	0.25	52.9	48.7	64.7	62.8	64.8	47.7	72.9	69.4
Glass mat.......	0.28	13.0	11.8	16.2	13.3	31.1	23.1	31.9	29.4
Cotton cloth....	2.6	5.5	6.0	5.9	7.2	13.2	....	12.8	

shows the effect of aging at 572°F on the dielectric constant, dissipation factor, surface resistivity, and electric strength of a silicone laminate ⅛-in thick formed from 116 glass cloth. Because of the presence of the silicon atom, silicone laminates have good arc and track resistance. Table 2.24 shows typical tracking properties. The physical properties of silicones are not greatly influenced by aging, but compared with laminates that are based on other resins, the flexural and tensile strengths of silicones are not unusually high. Figures 2.11 and 2.12 show the influence of heat aging at 250°C on flexural and compressive strength. Silicone laminates are used as radomes, structures in electronics, heaters, rocket components, slot wedges, ablation shields, coil forms, and terminal boards.

Polyimide Resins. Polyimide (PI) can be prepared as either thermoplastic or thermoset resins. Since the melt and processing temperatures of thermoplastic resins are so close to their decomposition temperatures, some of the thermoplastics are called pseudothermoplastics. Although the thermoplastics can be combined with fiber reinforcements to make laminates, most laminated applications rely on thermosetting polyimides. The thermosetting resins are difficult to process. They are dispersed into highly polar solvents with moderately high boiling points. At this point, the imides are in a polyamic acid precursor condition. Fabrics are impregnated in conventional treaters, and the solvents are removed to form a prepreg. Presses for polyimides must reach about 500°F where the resin flows and fuses, and the imidization reaction occurs, forming the polyimide. Some resins can be autoclaved at 200 lb/in², but curing temperatures are high. Autoclaved parts must be postcured to ensure imidization and to develop all the properties. Laminates have flexural strengths approaching 50,000 lb/in² with a flexural modulus of 3×10^6 lb/in². The laminates have low creep and can be used continuously from cryogenic temperatures to 450°F and for short times up to 900°F. They resist burning without chemical modification and have a low dissipation factor (0.005) and dielectric constant (3.9). Table 2.25 shows the properties of a reinforced polyimide compared to a thermoplastic polyimide. Polyimides are very sensitive to alkaline chemicals and will dissolve in concentrated, hot sodium hydroxide. They are also somewhat moisture sensitive, gaining 1 percent in weight after 1000 h at 50 percent RH and 72°F. They are used as printed wiring boards and as structural components in aircraft, copiers, and automobiles.

TABLE 2.21 Chemical Resistance of a Polyester Laminate

Acids	Concentration, %	Max temp, °F
Acetic	10	220
Acetic	25	200
Acetic	50	160
Acetic	75	160
Acrylic	100	Amb
Benzene sulfonic	All	220
Benzoic	All	220
Boric	All	220
Butyric	50	220
Chloroacetic	25	200
Chloroacetic	50	140
Chromic	5	200
Chromic	10	140
Chromic	30	NR
Citric	All	220
Fatty acids	All	220
Fluoboric	All	220
Fluosilicic*	25	Amb
Formic	All	Amb
Gluconic	50	Amb
Hydrobromic	48	160
Hydrochloric	10	220
Hydrochloric	20	180
Hydrochloric	37	160
Hydrochlorous	20	160
Hydrocyanic	10	160
Hydrofluoric*	10	180
Hydrofluoric*	20	NR
Hypochlorous	20	160
Lactic	All	220
Maleic	All	220
Nitric	5	160
Nitric	60	Amb
Nitrous oxide fumes	...	140
Oleic	All	220
Oleum	All	NR
Oxalic	Sat'd	220
Perchloric	30	Amb
Phosphoric	80	220
Stearic	All	220
Sulfuric	25	220
Sulfuric	50	220
Sulfuric	70	160
Superphosphoric 115% P_2O_5	...	200
Tannic	All	220
Tartaric	All	220
Trichloroacetic	50	220

Alcohols	Concentration, %	Max temp, °F
Amyl	All	210
Benzyl	All	Amb
Butyl	All	Amb
Ethyl	All	Amb
Methyl	All	Amb

Alkalies	Concentration, %	Max temp, °F
Ammonium hydroxide	5	180
(Aqueous ammonia)	10	140
Ammonium hydroxide	20	140
(Aqueous ammonia)	29	100
Calcium hydroxide	25	160
Potassium hydroxide	25	160
Sodium hydroxide	5	200
Sodium hydroxide	10	160
Sodium hydroxide	25	160
Sodium hydroxide	50	Amb

Bleaches	Concentration, %	Max temp, °F
Calcium hypochlorite	20	200
Chlorine dioxide	15	220
Chlorine water	Sat'd	160
Chlorite	...	210
Hydrogen peroxide	30	140
Hydrosulfite	...	220
Peroxide	...	220
Sodium hypochlorite	5¼	†
Sodium hypochlorite	15	†
Textone	...	220

Gases	Concentration, %	Max temp, °F
Bromine	...	NR
Carbon dioxide	...	220
Carbon monoxide	...	220
Chlorine, dry	...	220
Chlorine, wet	...	220
Sulfur dioxide, dry	...	220
Sulfur dioxide, wet	...	220
Sulfur trioxide	...	220

Organic Materials	Concentration, %	Max temp, °F
Acrylonitrile	100	NR
Diethyl ether	100	NR
Diethyl ketone	100	NR
Dimethyl formamide	100	NR
Dimethyl phthalate	100	Amb
Dimethyl sulfoxide	100	NR
Dioxyl phthalate	100	Amb
Diphenyl ether	100	120
Ethyl acetate	100	NR
Ethylene chlorohydrin	100	220
Ethylene glycol	All	220
Formaldehyde solution	All	†
Furfural	5	160
Furfural	20	Amb
Gasoline	100	Amb

TABLE 2.21 Chemical Resistance of a Polyester Laminate (*Continued*)

Organic Materials (Continued)	Concentration, %	Max temp, °F	Salts (Continued)	Concentration, %	Max temp, °F
Isopropyl palmitate	100	220	Barium sulfide	All	140
n-Heptane	100	Amb	Calcium chlorate	All	220
Kerosene	100	Amb	Calcium chloride	All	220
Orange oil	100	100	Calcium sulfate	All	220
Phenol	10	NR	Copper chloride	All	220
Phthalic anhydride	100	220	Copper cyanide	All	220
Pyridine	100	NR	Copper sulfate	All	220
Sour crude oil	100	220	Ferric chloride	All	220
Toluene diisocyanate	100	Amb	Ferric nitrate	All	220
Triethanolamine	100	Amb	Ferric sulfate	All	220
			Ferrous chloride	All	220
			Ferrous nitrate	All	220
Solvents			Ferrous sulfate	All	220
			Lead acetate	All	220
Acetone	10	180	Magnesium carbonate	All	140
Acetone	100	NR	Magnesium chloride	All	220
Benzaldehyde	100	NR	Magnesium sulfate	All	220
Benzene	100	NR	Mercuric chloride	All	220
Carbon disulfide	100	NR	Mercurous chloride	All	220
Carbon tetrachloride	100	Amb	Nickel chloride	All	220
Chlorobenzene	100	NR	Nickel nitrate	All	220
Chloroform	100	NR	Nickel sulfate	All	220
Dichlorobenzene	100	NR	Potassium aluminum		
Ethylene chloride	100	NR	sulfate	All	220
Ethylene dichloride	100	NR	Potassium bicarbonate	All	160
Ethyl ether	100	NR	Potassium carbonate	10	140
Methylene chloride	100	NR	Potassium carbonate	25	100
Methyl ethyl ketone	100	NR	Potassium carbonate	50	Amb
Monochlorobenzene	100	NR	Potassium chloride	All	220
Naphtha	100	Amb	Potassium dichromate	All	220
Nitrobenzene	100	NR	Potassium ferricyanide	All	220
Tetrachloroethylene	100	Amb	Potassium ferrocyanide	All	220
Toluene	100	NR	Potassium nitrate	All	220
Trichloroethylene	100	NR	Potassium permanganate	All	220
Trichloromonofluoro-			Potassium persulfate	All	220
methane	100	Amb	Potassium sulfate	All	220
Water, distilled	...	210	Silver nitrate	All	220
Xylene	100	NR	Sodium acetate	All	220
			Sodium bicarbonate	10	160
			Sodium bisulfate	All	220
Salts			Sodium carbonate	10	180
			Sodium carbonate	32	160
Aluminum chloride	All	220	Sodium cyanide	All	220
Aluminum potassium			Sodium chloride	All	220
sulfate	All	220	Sodium chlorite	10	Amb
Aluminum sulfate	All	220	Sodium ferricyanide	All	220
Ammonium bicarbonate	10	160	Sodium nitrate	All	220
Ammonium bicarbonate	50	160	Sodium nitrite	All	220
Ammonium carbonate	50	Amb	Sodium silicate	All	220
Ammonium chloride	All	220	Sodium sulfate	All	220
Ammonium nitrate	All	220	Sodium sulfide	All	220
Ammonium persulfate	All	180	Sodium sulfite	All	220
Ammonium sulfate	20	220	Stannic chloride	All	220
Aniline sulfate	All	220	Stannous chloride	All	220
Antimony trichloride	All	220	Trisodium phosphate	25	Amb
Barium carbonate	All	220	Zinc chloride	All	220
Barium chloride	All	220	Zinc sulfate	All	220

TABLE 2.21 Chemical Resistance of a Polyester Laminate (*Continued*)

Plating Solutions	Concentration, %	Max temp, °F	Other	Concentration, %	Max temp, °F
Cadmium cyanide.......	...	200	Aluminum chloro-		
Chrome*...............	...	160	hydroxide...........	50	Amb
Gold..................	...	200	Glycerine..............	100	220
Lead..................	...	200	Linseed oil.............	100	220
Nickel................	...	200	Sorbitol solutions........	All	120
Silver................	...	200	Succinonitrile, aqueous...	...	Amb
Tin fluoborate*.........	...	200	Sulfonated detergents....	100	160
Zinc fluoborate*........	...	200	Urea-ammonium nitrate		
			fertilizer mixture......	...	100
			8-8-8 fertilizer...........	...	100

NR = not recommended. Amb = ambient temperature. Sat'd = saturated.
* 10 to 20 mils of synthetic surfacing mat such as Dynel or Orlon should be used to reinforce surface in contact with chemical.
† Satisfactory up to maximum stable temperature for product.

The use of polyimide-printed wiring boards has been greatly increasing as the electronics industry requires higher temperatures and lower losses. Table 2.26 shows properties of a typical polyimide copper-clad laminate. They are also used in high-temperature electrical machines such as transit motors.

Some polyimides have been modified with fluorine-containing monomers, resulting in improved thermal capability. These laminates have an improved dielectric constant of 2.4 and glass transition temperature of greater than 423°F and can be used at 700°F.

Polyimides are also reacted with amides to form polyamide-imides. These resins are thermoplastic and have been used in special laminates.

Bismaleimides (BMI) are formed from methylene dianiline and maleic anhydride; they are laminating resins that prepreg well and can be produced on conventional treating and pressing equipment. They are being used as printed wiring boards and in some aerospace applications. Table 2.27 lists the properties of several BMI resins. The bismaleimides can also be reacted with triazine to form bismaleimide-triazine resins (BT) and further reacted with epoxies to make BT-epoxy copolymers. These resins have glass transition temperatures of 190°C compared with 120°C for conventional epoxies. Figure 2.13 shows the dielectric loss factor (dielectric constant multiplied by the dissipation factor) for a typical BT laminate.

TABLE 2.22 Strength of Laminates Made with Style 181 Glass Fabric

Resin	Tensile strength, $lb/in^2 \times 10^{-3}$				Flexural strength, $lb/in^2 \times 10^{-3}$				Compressive strength, $lb/in^2 \times 10^{-3}$			
	72°F	−110°F	−320°F	−424°F	72°F	−110°F	−320°F	−424°F	72°F	−110°F	−320°F	−424°F
Epoxy	61	98	115	118	94	129	168	172	72	107	138	135
Phenolic	52	66	66	65	77	100	110	110	70	88	100	90
Polyester	41	61	57	55	63	75	70	65	28	42	48	44
Silicone	30	47	70	76	38	46	67	65	21	39	43	46
Phenolic silicone	34	55	63	61	69	102	105	110	46	60	60	56

TABLE 2.23 Effect of Accelerated Heat Aging on Silicone-Glass Laminates

Hours aging at 572°F	Condition A	Condition D 48/50
Dielectric constant at 1 MHz		
0	3.78	3.82
24	3.72	3.76
48	3.77	3.80
96	3.75	3.79
192	3.71	3.90
Dissipation factor at 1 MHz		
0	0.0009	0.0027
24	0.0013	0.0024
48	0.0012	0.0030
96	0.0012	0.0033
192	0.0012	0.0230
Surface resistivity, ohms		
0	3.4×10^{13}	8.5×10^{13}
24	4.2×10^{14}	4.6×10^{13}
48	7.9×10^{14}	2.6×10^{14}
96	3.3×10^{14}	1.5×10^{14}
192	3.6×10^{14}	5.5×10^{13}
Perpendicular electric strength in air, volts/mil		
0	>328	>337
24	>379	>361
48	>375	349
96	351	315
192	106	100

TABLE 2.24 Arc and Track Resistance of Silicones

Test*	Method	Value
IEC	Critical tracking index	320 V
ASTM D495	Arc test	>240 s
ASTM D2132	Dust-fog test	1 h
ASTM D2303	Inclined plane	1.5 kV
ASTM D2302	Differential wet track	
	Discharge power	1.3 W
	Time to track	8 s

* IEC = International Electrotechnical Commission and ASTM = American Society for Testing and Materials.

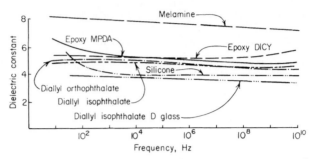

FIGURE 2.8 Dielectric constant of various laminates versus frequency.[7]

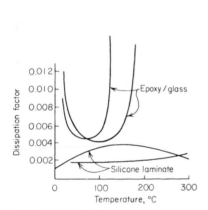

FIGURE 2.9 Dissipation factor versus temperature for silicone and epoxy laminates at 60 Hz.[7]

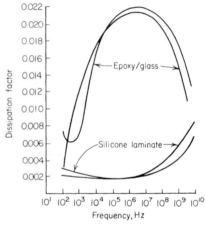

FIGURE 2.10 Dissipation factor versus frequency for silicone and epoxy laminates at 60 Hz.[7]

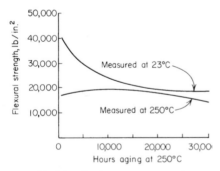

FIGURE 2.11 Effect of long-term aging on flexural properties of silicone laminates.[8]

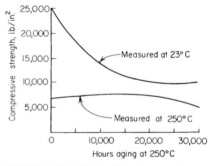

FIGURE 2.12 Effect of long-term aging on compressive properties of silicone laminates.[8]

TABLE 2.25 Properties of Polyimides[9]

Property	Thermoplastic PI	50% glass-filled thermoset PI
Heat-deflection temperature at 1.82 MPa, °C	315	350
Glass transition temperature (T_g), °C	300	310
Maximum resistance to continuous heat, °C	300	325
Coefficient of linear expansion, cm/cm/°C $\times 10^{-5}$	5.0	1.3
Tensile strength, MPa	96	44
Percent elongation	3	1
Flexural strength, MPa	172	145
Compressive strength, MPa	241	234
Notched Izod impact strength, J/m	100	300
Hardness, Rockwell	E60	M118
Specific gravity	1.4	1.6

TABLE 2.26 Properties of Polyimide Copper-Clad Laminate

Property	Unit	Typical value
Peel strength		
As received	lb/in	6.0
After thermal stress	lb/in	6.0
Volume resistivity		
As received	megohm-cm	3×10^6
At elevated temperature	megohm-cm	3×10^6
Surface resistivity		
As received	megohm	7×10^7
At elevated temperature	megohm	5×10^6
Dimensional stability		
After stress	in/in	0.0003
After bake	in/in	0.0003
Permittivity @ 1 MHz		4.2
Loss tangent @ 1 MHz		0.022
Arc resistance	sec	126
Flammability	UL 94	V-1
Water absorption	%	1.8
Electric strength	volts/mil	920
Glass transition	°C	250

Typical properties of a 0.006″ laminate, clad 1/1, at 55% resin content. Properties of other thicknesses may vary. Except as noted, IPC TM-650 test methods were used.

2.1.5 Laminating Resins—Thermoplastic

Thermoplastic polymers offer many advantages to the laminator when compared to traditional thermoset materials. Laminates made with thermoplastic binders have better impact strength, improved crack resistance, reuse of scrap, better formability and damping, and simpler part fabrication in that no chemical modifications like catalysts

TABLE 2.27   Properties of Bismaleimides[9]

Property	BMI	BMI/60% glass	BMI/60% carbon	BMI/70% carbon
Tensile strength, MPa	97	—	1725	570
Tensile modulus, GPa	4.1	—	148	16
Flexural strength, MPa	210	480	2000	725
Coefficient of expansion, $10^{-5}/°C$	3.1	—	—	—
Maximum use temperature, °C	260	225	215	175

or solvents are needed at the molding facility. However, thermoplastics have proven very difficult to combine with reinforcements, making the processing of most polymers complex and expensive. The difficulties are caused by two factors: viscosity and the stiff nature of the prepreg. The melt viscosity of most thermoplastic polymers is extremely high, even at high temperatures. The viscosity of some polymers is much like chewing gum. In order to flow the polymer into a reinforcement such as glass fabric, the melt viscosity needs to be of the order of 1000 poise, and few polymers are this low. To avoid this, some polymers are dissolved in solvents, but the solvents required tend to be expensive and difficult to use without complex recovery and environmental control systems. When this viscosity problem can be overcome, the resulting prepreg is usually stiff, boardlike, tack-free, and hard to use except in the most simple systems.

To overcome these limits, thermoplastic prepregers have derived four methods of preparing prepreg.

1. A few polymers do have rather low melt viscosities. These polymers, like some polysulfones, can be impregnated on a treater similar to that shown in Figure 2.1. Other polymers are dissolved in a solvent and treated similarly. However, the prepregs are stiff with no tack.

2. Films of a polymer are brought into contact with webs of reinforcement like glass fabric or mat, and, under heat and pressure, the films are forced to impregnate the fabric to form a prepreg. Some manufacturers use continuous belt-press laminators. Polypropylene is favored, and the prepreg can be stamped or hot formed into automotive parts where appearance is not critical.

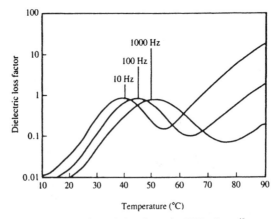

FIGURE 2.13   Dielectric loss factor for BT laminate.[10]

3. Combinations of reinforcing fibers and thermoplastic fibers are made where the thermoplastic becomes the binder in finished parts. Polymer yarns can be plied with glass or graphite yarns and pultruded or woven into fabrics or mats. Prepregs have good drape and handling. In some cases slit polymer films are used instead of yarns.

4. New developments use fine thermoplastic powders. The polymer powder is mixed with water to form a slurry that is coated onto reinforcing fibers or webs. When the water is evaporated, the powder sticks to the reinforcement. Prepregs are drapable and pliant, and they can be combined in stacks by adding water to cause some tack. The tendency for the powder to fall off is solved by some prepregers by encapsulating the powdered fiber in another polymer sheath. Prepregs based on powders are expensive. They are limited in polymer choice and used mostly on advanced composites. More discussion on composites is in Chap. 3.

Thermoplastic polymers used in prepreg are polysulfone, polyphenylene sulfide, polyetheretherketone, polyamide imide, polybutylene terephthalate, nylon 6, and polypropylene. Table 2.28 lists the properties of a polypropylene–glass-mat laminate.

2.1.6 NEMA Industrial Laminates

The electrical and electronic industries use large quantities of laminate materials for primary and secondary insulations. Over the years the industry has organized and maintained standards on the manufacture, testing, and performance of laminated thermosetting products in the form of sheets, rods, and tubes. These standards are described in "Industrial Laminated Thermosetting Products,"[12] which describes sheets, round tubes and rods of reinforced plastic materials. Reinforcements include paper, fabrics, and glass mats. Resins include phenolic, melamine, silicone, epoxy, and polyester resins. There are two categories in the specification. Copper-clad laminates are covered with copper foil for use in printed wiring boards, while unclad laminates do not have a foil. Unclad laminates are used in many industrial applications, but copper-clad laminates are unique to the electronic printed circuit board industry. The following sections describe the specific NEMA grades.

TABLE 2.28 Glass-Reinforced Polypropylene Stampable Sheet (30 Percent Glass Mat)

Property	Value
Tensile strength (lb/in^2)	8000
Elongation (%)	2
Tensile modulus (lb/in^2)	900,000
Compressive strength (lb/in^2)	6500
Flexural strength (lb/in^2)	10,500
Izod impact (ft-lb/in)	3.6
Heat deflection temp (°F) at	
264 lb/in^2	250
66 lb/in^2	305
Water absorption (% in 24 h)	0.05

Unclad Laminates

Paper-Base Grades with Phenolic Resin

- *Grade X:* These laminates are used for mechanical applications and are affected by humidity. They are not usually used in electrical applications as an insulating material. The impact strength is low. Rolled tubes machine easily and have low dissipation factors. Rods are not available.

- *Grade XP:* This grade is intended for hot punching at 212°F or higher, but laminates less than 0.062 in may be punched at room temperature. Grade XP has lower physical properties than Grade X but is somewhat better in resistance to humidity. Rods and tubes are not made from this material.

- *Grade XPC:* These laminates can be punched or cut at room temperature. They have lower flexural strength than Grade XP and are less rigid. Rods and tubes are not made from this material.

- *Grade XX:* These laminates can be used as electrical insulations. They do not punch well but are easily machined. Rolled tubes can be threaded, but they are not as strong as Grade X. Moisture resistance is somewhat better than Grade X. Molded tubes resist humidity better than the rolled tubes, but seams have lower electric strength. Rods are available.

- *Grade XXP:* This is a punchable type of Grade XX. Laminates must be punched hot (>212°F). The electrical properties are better than Grade XX. Rods and tubes are not available.

- *Grade XXX:* This is an excellent electrical laminate, and it resists humidity and creep. Rolled tubes have electrical properties superior to other grades. Molded tubes and rods are similar in properties to the laminate.

- *Grade XXXP:* This is a punchable grade of XXX. Hot punching (>212°F) is required. Electrical properties are better than Grade XXX. The material can be used when high resistivities are needed at high humidity. Rods and tubes are not available.

- *Grade XXXPC:* This grade is punchable at room temperature and has properties similar to Grade XXXP. Rods and tubes are not available.

- *Grade ES-1:* This grade and the other ES grades are used as name-plate stock. The surface layer is black or other colors and the center core is white melamine paper. Engraving exposes the white section.

- *Grade ES-2:* This material has a white surface and a black core made from phenolic impregnated paper.

- *Grade ES-3:* This grade is similar to ES-1, with various colored surfaces and a black core.

Fabric-Base Grades with Phenolic Resin Binder

- *Grade C:* Laminates are molded from cotton fabrics weighing more than 4 oz per square yard. The material is used for pulleys, gears and other applications requiring impact resistance, but marginal electrical properties limit its use in electrical devices. Both rods and rolled tubes are available.

- *Grade CE:* This grade is similar to Grade C in construction, but a different resin system provides improved electrical properties and humidity resistance. Electrical characteristics are similar to Grade XX. However, its use at voltages greater than 600 is not intended. Molded tubes are available, but the seams are weak electrically. The properties of rods are the same as laminates.

- *Grade L:* This mechanical grade can be machined with better edges and surfaces than the other cotton-based grades. The material is not intended for electrical applications.

- *Grade LE:* These laminates can be used in electrical applications and are stronger than grade XX. However, like Grade CE, they should not be used as primary electrical insulations, and voltages are limited to less than 600. Rolled and molded tubes and rods are available.

Asbestos-Base Grade

- *Grade AA:* Laminates of this grade are more heat resistant than the preceding grades. The material is not used in electrical devices, and the laminates expand in humid environments.

Glass-Base Grades

- *Grade G-3 (glass fabric with phenolic resin binder):* These laminates have good mechanical properties and are used in nonelectrical applications. The impact resistance is excellent as is the resistance to heat. Rods and rolled tubes are available.

- *Grade G-5 (glass fabric with melamine resin binder):* This grade is an excellent electrical insulation and widely used in marine electrical devices, particularly where arc resistance is needed. The material is hard and abrasion resistant and does not burn readily. Electrical applications are limited to conditions below 120°F. Rods and rolled tubes are available.

- *Grade G-7 (glass fabric with silicone resin binder):* This material has excellent heat resistance and arc resistance. It resists burning and has good electrical properties. Its physical properties are excellent.

- *Grade G-9 (glass fabric with moisture-resistant melamine resin binder):* This material retains its electrical properties better than class G-5 under wet conditions. Other characteristics are similar to G-5. Rods and tubes are available.

- *Grade G-10 (glass fabric with epoxy resin binder):* Laminates of G-10 have very high mechanical strengths at room temperature. Electrical properties are also very good. Rods and rolled tubes are available.

- *Grade G-11 (glass fabric with heat-resistant epoxy resin binder):* This material is similar to Grade G-10 but can be used at higher temperatures. Rods and rolled tubes are available.

Nylon-Cloth Grade with Phenolic Resin Binder

- *Grade N-1:* Based on nylon fabrics, these laminates are used for electrical insulation at humid conditions where their impact resistance is good. They creep readily, particularly at elevated temperatures.

Flame-Resistant Grades

- *Grade FR-1:* This grade has a phenolic resin binder and paper reinforcement. It is similar to Grade XP but burns less readily. Rods and tubes are not available.

- *Grade FR-2:* This grade is similar to Grade XXXPC but is fire retardant.

- *Grade FR-3:* These laminates are reinforced with paper and impregnated with an epoxy resin. The fire resistance equals or exceeds V-1, and electrical properties are good. The flexural strength exceeds that of Grade XXXPC.

- *Grade FR-4:* This grade is the most common grade used in printed wiring applications. It is made with epoxy resin and glass fabric reinforcement and has properties similar to G-10 but with a flame resistance of V-0. Rods and rolled tubes are available.

- *Grade FR-5:* This grade is made with epoxy resin and glass fabric and is similar to Grade G-11 but with a flame resistance of V-0. Rods and tubes are available.

Composite-Base Laminates

- *Grade CEM-1:* These materials consist of a glass fabric surface and a cellulose paper core impregnated with a flame-resistant epoxy resin. Designed for punchability and ease of drilling and machining, they are used in printed wiring applications.

- *Grade CEM-3:* This grade is made with a glass fabric surface and a glass mat core impregnated with an epoxy resin. It can be punched and machined. Its properties are similar to those of FR-4.

The properties of NEMA-type laminates are shown Table 2.29. The available thicknesses of laminates are a function of the type of resin binder and the thermal conductivity of the reinforcement. Cotton-based laminates can be molded up to 10 in thick while some glass-based grades are limited to 4 in. Key physical properties are highly influenced by the reinforcing form. Paper-based laminates have a higher water absorption than do glass-fabric-based grades. Phenolic-resin-based grades have higher water absorption than do epoxy-resin-based grades. Izod impact strengths are highest for those laminates based on glass fibers, and the same is true for flexural strengths. The electric strength of laminates is strongly affected by humidity. The electric strength paper-based grades decreases dramatically as humidity increases—sometimes by a factor of 8. That of the glass fiber grades also decreases, but only slightly. Grades G-5, G-7, and G-9 are used in arcing applications, and they all exceed 180 in ASTM D495 arc resistance test. In most applications the dimensional stability of laminates is important. Laminates may contain considerable internal strain that can result in deformations called warp and twist.

Tubes and rods are manufactured differently than laminates (see Sec. 2.1.2). This manufacturing difference imposes restraints on their dimensions and tolerances. Tubes are wound up to an outside diameter of 50 in. The seams of rods and tubes may be sources of electrical and mechanical weakness.

Electrical and Mechanical NEMA Grades of Polyester Glass. NEMA maintains standards on six compositions of glass mat with polyester resins. GPO-1 laminates are suitable for general purpose use. GPO-2 laminates are specified when flame resistance is needed. GPO-3 laminates are resistant to carbon tracking and are flame-resistant. Grades GPO-1P, GPO-2P, and GPO-3P are corresponding grades designed for punching with a small loss in electrical and mechanical properties. These laminates have good electrical properties and excellent arc and tracking resistance. Because of these properties, they are used in electrical power applications in switches, circuit breakers, switch gear, motors, and transformers. Properties are covered in Ref. 12.

Copper-Clad Laminates. Copper-clad laminates are those laminates with copper bonded to one or both sides for use as printed wiring boards. Not all NEMA-grade laminates are used as copper clad and not all laminates used in the electronics industry are NEMA-grade materials. However, more than 90 percent of all copper-clad laminates used in the industry are NEMA-grade laminates and more than 80 percent are grade FR-4. The copper foil used is 99.5 percent pure and may contain some silver. Generally electrodeposited foil is used rather than rolled copper, and the foil is usually treated to improve the adhesion of the laminating resin to the foil surface. The NEMA grades used as copper-clad laminates are XXXP, XXXPC, FR-1, FR-2, FR-3, FR-4, FR-5, G-10, G-11, CEM-1 and CEM-2. Some of these grades are described in MIL-P-13949. Table 2.30 gives typical properties of copper-clad laminates.

Electronic designers continue to require ever increasing performance[19] from printed wiring boards by desiring better agreement in thermal expansion between boards and ceramic chips, by increasing the speed of circuit signals, by making line traces and holes smaller, by needing higher glass transition temperatures (T_g), by using higher voltages, and by demanding smoother board surfaces. While FR-4 continues to dominate copper-clad usage, the laminate industry is responding to these needs with a variety of high-performance laminates.

Signal speed in circuits is influenced by the dielectric constant of the circuit board according to the equation:

$$\text{Signal propagation delay} = 1.017\sqrt{\epsilon'} \tag{2.1}$$

Speeds are also related to frequency, conductor length, board thickness, and conductor spacing.

Another concern is to better match the coefficients of thermal expansion (CTE) of the circuit board (CTE FR-4 = 15/°C) and of the ceramic chip package (CTE Alumina = 6.3 in/in/°C). Large mismatches of CTE cause solder cracking of surface mount leads.

The CTE of a mixture of materials in a laminate is related to the CTE of each component (C), the Young's moduli (M), and the volume fractions (V). By the Shapery equation[20]:

$$\text{CTE laminate} = C_1 M_1 V_1 + \frac{C_n M_n V_n}{M_1 V_1} + M_n V_n \tag{2.2}$$

Other concerns are to decrease the weight, increase thermal conductivity, improve electrical properties, and increase the T_g of the circuit board.

FR-4 is difunctional with a T_g of 120°C. By adding a multifunctional epoxy, the T_g is increased to 150°C. High-temperature FR-4 is made by using mostly multifunctional epoxies to get a T_g of 190°C. Further improvement is found in high-performance epoxies, where other copolymers are added to multifunctional epoxies to raise the T_g to greater than 190°C. One such modification is obtained by reacting an epoxy with bismaleimide and triazine to form BT/epoxy.

Cyanate esters are being used in circuit boards because of their low dielectric constant of 2.91 and a high T_g of 290°C. Prepregs and laminates are made with conventional technology, and they are tougher and more moisture resistant than FR-4. Laminates are also used in radomes, antennas, and aerospace structures.[21]

Polyimides are combined with glass, Kevlar, and quartz because of their high temperature capability and good electrical properties. Kevlar-based laminates have a dielectric constant of 3.6. Properties for a glass/polyimide laminate are given in Table 2.26 and compared with other laminates in Table 2.31.

Epoxy resins can be alloyed with polyphenylene oxide to form an interpenetrating network. Laminates, trade-named Getek by General Electric, are processed exactly like FR-4 but have a T_g of 210°C and a lower CTE. Table 2.32 gives data for Getek.

A different polymer is polybutadiene, the dielectric constant of which is only 2.2. The polymer can be made into a prepreg and handled like FR-4. The resin has a low specific gravity of 0.98 g/ml, and it highly resists moisture. It can be polymerized with epoxies.

Bayer has recently announced a polyurethane alternative to epoxies for circuit board applications. Trade-named Blendur the laminate has a T_g of 572°F. It is inherently flame retardant.

Polyester resins are the base for copper-clad laminates produced by Glasteel Industrial Laminates. Designated CRN-5 and CRM-7, they are made of glass fabric,

TABLE 2.29 Properties of NEMA Laminates[13]

Properties	NEMA grade								
	X	XX	XXX	XP	XPC	XXP	XXXP-C	C	CE
Tensile strength, 10^3 lb/in^2									
Lengthwise	20	16	15	12	10.5	11	12.4	10	8
Crosswise	16	13	12	9	8.5	8.5	9.5	8	7
Compressive strength, 10^3 lb/in^2									
Flatwise	36	34	32	25	22	25	25	37	39
Edgewise	19	23	25.5					23.5	24.5
Flexural strength, 10^3 lb/in^2 ($\frac{1}{8}$ in-thick)									
Lengthwise	25	15	13.5	14	16	14	12	17	16.5
Crosswise	22	14	11.8	12	10	12	10.5	16	14
Modulus of elasticity in flexure, 10^6 lb/in^2									
Lengthwise	1.8	1.4	1.3	1.2	1	0.9	1	1	0.9
Crosswise	1.3	1.1	1	0.9	0.8	0.7	0.7	0.9	0.8
Shear strength, 10^3 lb/in^2	12	11	10	8		11	11	12	11
Izod impact, ft-lb/in of notch									
Flatwise	4	1.3	1					3.2	2.3
Edgewise	0.5	0.35	0.35					1.7	1.4
Rockwell hardness, M scale	110	105	110	95	75	100	95	103	103
Specific gravity	1.36	1.34	1.32	1.33	1.34	1.32	1.3	1.36	1.36
Coeff. of thermal expansion, cm/cm/°C $\times 10^{-5}$	2	2	2	2	2	2	2	2	2
Water absorption, %/24 h									
$\frac{1}{16}$-in	6	2	1.4	3.6	5.5	1.8	0.75	4.4	2.3
$\frac{1}{8}$-in	3.3	1.3	0.95	2.2	3	1.1	0.55	2.5	1.6
$\frac{1}{2}$-in	1.1	0.55	0.45					1.2	0.75
Dielectric strength, V/mil; perpendicular to laminations; short time test									
$\frac{1}{16}$-in	700	700	650	650	600	700	650	200	500
$\frac{1}{8}$-in	500	500	470	470	425	500	470	150	360
Dissipation factor, 1 megacycle									
Condition A	0.06	0.045	0.038	0.06		0.04	0.038	0.1	0.1
Dielectric constant, 1 megacycle									
Condition A	6	5.5	5.3	6		5	4.6		5.3
Insulation resistance, MΩ									
Condition: 96 h/90% relative humidity, 95°F		60	1000			500	50,000		
Maximum constant operating temperature, °F	285	285	285	285		285	275	265	265
Temp. indices, NEMA, °C									
Electrical	130	130–140	130–140	130	130	130–140	125	85–115	85–115
Mechanical	130	130–140	130–140	130	130	130–140	125	85–125	85–125
Flame resistance classification, UL 94	94HB	94HB	94HB	94HB	94HB, V-0	94HB	94HB	94HB	94HB
Bond strength, lb	700	800	950					1800	1800
Thickness, in									
Minimum	0.010	0.010	0.015	0.010	0.020	0.015	0.031	0.031	0.03
Maximum	2	2	2	0.250	0.250	0.250	0.250	10	2
Standard colors	Natural (variable tan) and black	Natural (variable tan) and black	Natural (variable tan) and black	Natural (variable tan) and black	Natural (variable tan) and black	Natural (variable tan) and black	Natural (variable tan)	Natural (variable tan) and black	Natural (variable tan) and black
Standard finishes	Semigloss	Glossy, semigloss	Glossy, semigloss	Semigloss	Semigloss	Semigloss	Semigloss	Semigloss	Semigloss

	NEMA grade												
L	LE	N-1	G-3	G-5	G-7	G-9	G-10	G-11	GPO-1	GPO-2	GPO-3	GPO-1P	GPO-2P
13 9	12 8.5	8.5 8	23 20	37 30	23 18.5	37 30	40 35	40 35	9	9	9	8	8
35 23.5	37 25		50 17.5	70 25	45 14	70 25	60 35	60 35	30 20	30 20	30 20	20	20
16 14	16 14	10 9.5	18 16	55 45	20 18	55 35	55 45	55 45	18 18	18 18	18 18	16 16	16 16
1 0.85	1 0.85	0.6 0.5	1.5 1.2	1.7 1.5	1.4 1.2	2.5 2	2.7 2.2	2.8 2.3	1.2 1.0	1.2 1.0	1.2 1.0	1.2 1.0	1.2 1.0
	11.5	14			17	20	19	19	14	14	14	10	10
1.35 1.10	1.8 1	4 2.0			8.5 5.5	12 5.5	7 5.5	7 5.5	8	8	8	5	5
105	105	105	100	120	100	120	111	112	100	100	100	100	100
1.35	1.33	1.15	1.65	1.90	1.68	1.9	1.80	1.80	1.5–1.9	1.5–1.9	1.7–1.95	1.5–1.9	1.5–1.9
2	2	8	2	2	1	1	0.9	0.9	2	2	2	2	2
2.5 1.6 0.9	1.95 1.3 0.7	0.6 0.4 0.35	2.7 2.0 1.5	2.7 2.0 1.5	0.55 0.35 0.20	0.8 0.7 0.4	0.25 0.15 0.1	0.25 0.15 0.1	1.0 0.7 0.35	0.8 0.6 0.25	0.6 0.5 0.25	1.0 0.7	0.8 0.6
200	500 360	600 450	700 600	350 260	400 350	400 350	700 550	700 550	300	300	300	300	300
	0.055	0.038			0.003	0.017	0.025	0.025	0.03			0.03	
	5.8	3.9			4.2	7.2	5.2	5.2	4.3			4.3	
	30	50,000		100	2500	10,000	200,000	200,000					
265	265				465	285	285	300					
85–115 85–125	85–115 85–125	140 170		140	170 220	140	130 140	140–170 160–180		130 160	120 140		
94HB	94HB	94HB		94V-0	94V-0	94V-0	94HB	94HB		94V-0			
1600	1600	1000	850	1570	850	1700	2000	1600	850	850	850		
0.010 2	0.015 2	0.010 1			0.010 2	0.010 3.5	0.010 1	0.010 1	0.031 2	0.031 2	0.031 2	0.062 0.188	0.062 0.188
Natural (variable tan) and black	Natural (variable tan) and black	Natural (variable tan)			Natural (white to tan)	Grayish brown	Natural (green to tan)	Natural (green to tan)	Natural (tan)				
Semigloss	Semigloss	Semigloss			Semigloss	Semigloss	Semigloss	Semigloss					

TABLE 2.29 Properties of NEMA Laminates[13] (*Continued*)

Properties	GPO-3P	FR-2	FR-3	FR-4	FR-5	CEM-1	CEM-3
Tensile strength, 10^3 lb/in^2							
Lengthwise	8	12.5	14	40	40		
Crosswise		9.5	12	35	35		
Compressive strength, 10^3 lb/in^2							
Flatwise	20	25	29	60	60		
Edgewise				35	35		
Flexural strength, 10^3 lb/in^2 ($\frac{1}{8}$ in-thick)							
Lengthwise	16	12	20	55	55	35	40
Crosswise	16	10.5	16	45	45	28	32
Modulus of elasticity in flexure, 10^6 lb/in^2							
Lengthwise	1.2	1.1	1.5	2.7	2.7		2.3
Crosswise	1.0	0.9	1	2.2	2.2		2.1
Shear strength, 10^3 lb/in^2	10		11	19	19	15	14
Izod impact, ft-lb/in of notch							
Flatwise	5			7	7	1.6	1.6
Edgewise				5.5	5.5	1.2	1.4
Rockwell hardness, M scale	100	97	100	110	110		
Specific gravity	1.7–1.95	1.33	1.36	1.85	1.85		1.75
Coeff. of thermal expansion, cm/cm/°C × 10^{-5}	2	2	2	1	0.9		
Water absorption, %/24 h							
$\frac{1}{16}$-in	0.6	0.75	0.65	0.25	0.25	0.3	0.25
$\frac{1}{8}$-in	0.5	0.55	0.5	0.15	0.15		
$\frac{1}{2}$-in				0.1	0.1		
Dielectric strength, V/mil; perpendicular to laminations; short time test							
$\frac{1}{16}$-in		650	600	700	700	400	
$\frac{1}{8}$-in		470	475	550	550		
Dissipation factor, 1 megacycle							
Condition A		0.038	0.035	0.025	0.025	0.035	0.025
Dielectric constant, 1 megacycle							
Condition A		4.6	4.6	5.2	5.2	4.6	5.2
Insulation resistance, MΩ Condition: 96 h/90% relative humidity, 95°F		20,000	100,000	200,000	200,000	100,000	200,000
Maximum constant operating temperature, °F			265	285	300		265
Temp. indices, NEMA, °C							
Electrical		75–105	90–110	130	140–170	130	130
Mechanical		75–105	90–110	140	160–180	140	140
Flame resistance classification, UL 94		94V-0	94V-0	94V-0	94V-0	94V-0	94V-0
Bond strength, lb			950	2000	1600		
Thickness, in							
Minimum	0.062	0.031	0.031	0.002	0.002	0.031	0.031
Maximum	0.188	0.250	0.250	1	1	0.093	0.093
Standard colors		Natural	Natural (gray to tan)	Natural (green to tan)	Natural (green to tan)	Natural (variable tan)	Natural (variable green)
Standard finishes		Semigloss	Semigloss	Semigloss	Semigloss	Semigloss and dull	Semigloss and dull

TABLE 2.30 Typical Properties of Copper-Clad Laminates

	Type of laminate						
Property*	Paper, phenolic	Paper, fire-retardant phenolic	Paper, epoxy	Glass fabric, epoxy	Glass fabric, fire-retardant epoxy	Glass fabric-paper, epoxy	Glass fabric-glass mat, epoxy
Flexural strength, lb/in^2	12,000	12,500	14,000	55,000	65,000	55,000	62,000
Peel strength, lb/in (1 oz cu)	6	6	8	10	10	10	10
Water absorption, % 24 hr	0.75	0.75	0.65	0.10	0.10	0.17	0.10
Volume resistivity, $\Omega \cdot$cm	10^{11}	10^{11}	10^{12}	10^{14}	10^{14}	10^{13}	10^{14}
Surface resistivity, Ω	10^6	10^6	10^7	10^{12}	10^{12}	10^{12}	10^{12}
Dielectric constant	4.6	4.8	4.6	5.2	5.5	4.5	4.8
Dissipation factor	0.040	0.038	0.035	0.020	0.022	0.025	0.021
Electric strength, kV parallel to lam.	20	10	40	50	50	60	60
Arc resistance, s	20	20	50	120	120	90	115
Flammability	Burns	V-1	V-1	Burns	V-0	V-0	V-0

* Based on 0.059-in laminates tested by IPC-TM-650.

glass paper, and glass mat to provide lower-cost printed circuits. These laminates are compared with others in Table 2.33.

Polytetrafluoroethylene (PTFE) is used as a circuit board material in microwave applications. With conductors on the surface, these boards are called stripline. PTFE is characterized by a low dielectric constant and low dissipation factors—among the lowest of all insulating materials. Because they are soft, they can be cut and shaped with ease. Their electrical performance at very high frequencies is excellent. Table 2.34 shows the properties of glass-reinforced PTFE. Its dielectric properties are compared with other printed wiring board materials in Tables 2.35 and 2.36, and Table 2.37 shows the physical properties of glass-reinforced PTFE. Soft to begin with, PTFE grows even softer with increasing temperatures, requiring special handling during soldering and heating processes. The softness also contributes to creep (cold flow) under stress. The coefficient of thermal expansion of PTFE is high and is not similar to those of the other printed circuit materials, causing deformation and stress during thermal cycles and cracking of plated-through holes.

2.1.7 Decorative Laminates

Decorative laminates are high-pressure laminates with a paper base, similar in many respects to industrial laminates except for the special attractive surfaces. These laminates have a core of sheets of phenolic-resin-impregnated kraft paper. On top of this is placed a special grade of paper with a decorative pattern printed on the surface and impregnated with a clear melamine resin. The decorative pattern may be wood grain, solid color, or any other design. On top of this is placed another sheet of paper called an overlay that is impregnated with a melamine with an index of refraction that is similar to that of the cellulose in the paper. This overlay protects the decorative sheet and provides the unique abrasion and stain resistance of these laminates. The stack then goes into large, multiple-opening presses. Here the sheets are bonded together at a temperature of 300°F and pressures of up to 1500 lb/in^2. The sheets are trimmed, and the back is sanded to achieve tolerances and provide for adhesive bonding.

TABLE 2.31   Properties of Printed Wiring Board Materials

Material	Dielectric constant (ϵ')	Dissipation factor	Coefficient of thermal expansion, in/in/ °C × 10⁻⁶	Glass transition temperature (T_g), °C
Glass epoxy—NEMA FR-4	4.7	0.021	15	120
Glass epoxy-tetrafunctional	4.4	0.023	15	137
Glass epoxy—multifunctional	4.3	0.023	15	155
Polyimide glass	3.4	0.022	12	250
Bis-benzocyclobutene	2.7	0.0008	13	350
Fluorinated polyimide	2.9	0.0004	12	260
Polyimide Kevlar	3.6	0.008	5	230
Polyimide quartz	4.0	0.005	7	250
Epoxy Kevlar	3.6	0.017	6	125
Teflon glass	2.2	0.0025	8	—
Cyanate ester-triazine	3.8	0.006	9	247
BT-Epoxy	4.1	0.018	15	175

TABLE 2.32   Properties of Epoxy/Polyphenylene Oxide Laminate

Dielectric constant, 1 MHz	4.3
Dissipation factor, 1 MHz	0.015
Dielectric strength, V/mil	750
Volume resistivity, Mohm·cm	10
Surface resistivity, Mohm	10
Arc resistance, sec	60
Glass transition temperature, °C	210
CTE z axis, /°C	4
CTE x,y axis, /°C	13
Water absorption, %	0.3
Flammability, UL 94	V-O

The top surface of the decorative laminate may be any of several finishes with varying gloss. The gloss finish has a mirrorlike surface with a gloss-meter reading of 80 to 100. Furniture finish with a reading of 36 to 55 is often used as table tops. Satin finish (15 to 34) is popular. Velvet finish (7 to 14) is a popular finish on walls and partitions. Low-glare finish with a gloss of 4 to 10 is especially attractive on wood-paneled walls, and oil rub finish (2 to 5) can be treated with furniture polish or oil and results in a fine craftsmanlike wood-grain surface.

Decorative laminates, available in several grades, can be used for different applications. Table 2.38 shows the grades available and describes typical applications. Postforming grades are capable of being bent to produce curved surfaces. They must be rapidly heated to 315 to 325°F and pressed to shape in a form or mold with moderate pressure. The resistance of these laminates to chemical and household preparations is excellent. The chemical resistance is shown in Table 2.39.

These laminates are usually bonded to a substrate such as plywood, chipboard, or composition board. It is recommended that a balancing sheet be used when these laminates are bonded. This sheet consists of several layers of phenolic-impregnated paper similar to the core in the decorative laminate. When it is bonded between the laminate and the substrate, the balance sheet prevents moisture absorption and minimizes warpage of the structure. Adhesives that are used include urea formaldehydes, and phenolic, casein, epoxy, and polyvinyl acetate.

Decorative laminates are not structural materials and should be treated as veneers except when they are bonded to structural cores. Although the face wears well in abrasive environments, the materials are handled as brittle, rigid, thin structures. The grain direction and dimensional stability of the laminates are similar to those of wood. With varying humidity, the width of the laminate changes twice as much as the length; therefore, in critical applications, the machine or grain direction should be established.

Decorative laminates are standardized in NEMA publications LD3, *High Pressure Decorative Laminates,* and LD3.1, *Performance, Applications, Fabrication and Installation of High Pressure Decorative Laminates.*

2.2 REINFORCED PLASTICS

2.2.1 Definitions

Reinforced plastics are those plastics or polymers to which fibers that increase the physical properties—especially the resistance to impact and the heat deflection temperature—have been added. The fiber diameter is usually 10 mm or less, and the fibers are longer than 100 mm. The fiber-length-to-diameter ratio is greater than 10. Glass fibers are most common, but others (notably carbon, graphite, aramid, and boron fibers) are also used. In a reinforced plastic, the resin matrix is the continuous phase, and the fiber reinforcement is the discontinuous phase. The function of the resin is to bond the fibers together to provide shape and form and to transfer stresses on the structure from the resin to the fiber. Only high-strength fibers with high moduli are used, i.e., E glass fibers having a tensile strength of over 500,000 lb/in^2 and a modulus of over 10.5×10^6 lb/in^2.

Different kinds of mineral fillers are added to plastics for many reasons, but this does not necessarily make the fillers reinforcements. True fiber reinforcements dramatically increase the structure's tensile, compressive, flexural, and impact strengths and cause great stiffness. Because of this increased stiffness, those plastics that are prized for their great flexibility are not usually used with stiff fiber reinforcement.

2.2.2 Thermoset-Reinforced Plastics

Virtually all thermosetting polymers can be reinforced with fibers. Glass fibers are usually longer than ¼ in. Properties of thermoset-molded materials are shown in App. C. Several materials reinforced with graphite or carbon fibers—especially alkyds and epoxies—are being used to provide electromagnetic interference protection (EMI). Thermoset-reinforced molding materials of phenolic, alkyd, or epoxy are used extensively in the electronics industry for connectors, terminals, sockets, and housings.

Polyester resins are particularly useful in reinforced plastics. The resins can be cured at room temperature or at elevated temperatures and at low molding pressures. They have good electrical and physical properties and can be molded in contact with many surfaces, making inexpensive and highly versatile molds. One fabrication method used with polyesters is pultrusion and is also used with epoxies and vinyl esters (see Sec. 2.3.7 and Chap. 3).

Polyesters are used extensively in manufacturing very large components such as swimming pools, truck parts, saunas, boat hulls, bath and shower enclosures, and building components. The fiber glass reinforcements in these applications are mat, woven roving, and fabric. Molds are of plastic, wood, plaster, or metal. The polyester

TABLE 2.33 Polyester Copper-Clad Laminates

Grade	Composition	General properties	Applications	Alternates	Comments
XXXPC	Phenolic/kraft paper	Excellent punching at room temperature Low flex strength Poor wet electrical properties	Automotive	FR-2	No domestic suppliers available
FR-1	Phenolic/kraft paper	Punchable at room temperature Extremely poor wet electrical properties Lesser grade than XXXPC/FR-2	Video game controls Calculators	FR-1	As a clad laminate Not flame-retardant
FR-2	Phenolic/kraft paper	Brittle but punchable with heat Similar physical/electrical properties as XXXPC	Automotive Consumer electronics	FR-2	No domestic sources available
FR-3	Epoxy/kraft paper	Extremely brittle but punchable with heat Good wet electricals for paper laminate	Automotive Video game controls Calculators	CEM-1 CRM-7	No domestic sources available Popular in far east and Europe
FR-6 (MC2)	Polyester/glass mat	Outstanding electrical properties	Telephone sets	CEM-1 CEM-3	No domestic sources available Popular in far east and Europe
*CRM-5 (MC3)	Polyester/glass fabric surface Polyester/glass paper core	Punchable at room temperature Higher flex strength than paper products	Consumer electronics	CEM-1	Excellent high frequency properties in a punchable grade Suitable for low drift applications
		Outstanding electrical properties Punchable at room temperature Double flex strength of FR-6	Automotive Telecommunications Consumer electronics Automotive Telecommunications	CRM-5 CRM-7 FR-4 CEM-3	Used where outstanding electrical properties are required Product can be drilled/punched

	Construction	Properties	Applications	Comparable	Comments
CRM-7	Polyester/glass paper surface Polyester/glass mat core	Same physical/electrical properties as FR-6 Smoother surface finish	Consumer electronics Automotive Telecommunications	CEM-1 CRM-5	Smoother surface suitable For fine line applications
CEM-1	Epoxy/glass fabric surface Epoxy/wood paper core	Punchable at room temperature Good electricals but less than polyester Good flex and impact strength	Consumer electronics	FR-2 CRM-5	Good all-around laminate for punching Performance varies greatly between suppliers
CEM-3	Epoxy/glass fabric surface Epoxy/glass paper core	Punchable but harder than CEM-1 Good electricals Suitable for pth applications	Computer peripherals Keyboards	CRM-5 FR-4	One domestic source and difficult to obtain. Popular in Far East for two-sided applications
FR-4	Epoxy/glass fabric	High flex and impact strength Excellent electrical properties Excellent for pth applications	Computer applications Telecommunications Military	CEM-3 CRM-5 FR-5	Most popular laminate Material for drilled pth applications
G-10	Epoxy/glass fabric Not flame-retardant	High flex and impact strength Excellent electrical properties Excellent dimensional stability	Structural applications	FR-4	Used as an unclad laminate for structural parts Not flame-retardant
FR-5	Modified epoxy/glass fabric	Improved hot flex strength over FR-4 Excellent electricals	Military applications	Polyimide	Original FR-5 materials have disappeared Properties vary greatly between suppliers

*Trademark of Glasteel Industrial Laminates.

TABLE 2.34 Properties of Glass-Filled Polytetrafluoroethylene[11]

Property	Test method*	Condition	Unit†	Direction	Typical value
Dielectric constant, ϵ'	IPC-650, TM 2.5.5.5	10GHz/23°C	—	z	2.94 ± 0.04
Thermal coefficient of ϵ'	IPC-650, TM 2.5.5.5	10GHz/0–100°C	ppm/K	z	≈ 0
Dissipation factor, tan δ	IPC-650, TM 2.5.5.5	10GHz/23°C	—	z	0.0012
Volume resistivity	ASTM D257	A	MΩ	z	10^6
Surface resistivity	ASTM D257	A	MΩ	z	10^7
Tensile modulus	ASTM D638	23°C	MPa ($10^3 \times$ lb/in²)	x,y	828 (120)
Ultimate stress			MPa ($10^3 \times$ lb/in²)	x,y	6.9 (1.0)
Ultimate strain			%	x,y	7.3
Compressive modulus	Estimated		MPa ($10^3 \times$ lb/in²)	z	2300 (360)
Water absorption	IPC-TM-650,2.6.2.1	D23/24	%	—	0.1
	ASTM D570	D48/50	%	—	0.13 max.
Specific gravity	ASTM D792	23°C	—	—	2.1
Specific heat	Calculated		J/g/K (BTU/lb/°F)	—	0.93 (0.22)
Thermal conductivity	Rogers TR2721	100°C	W/m/K (BTU in/ft²/h/°F)	—	0.44 (3.0)
Coefficient of thermal expansion	ASTM D3386	10K/min	ppm	z	24
				x,y	16

* IPC = Institute for Interconnecting and Packaging Electronic Circuits; and ASTM = American Society for Testing and Materials.
† SI units given first, with other frequently used units in parentheses.

TABLE 2.35   Dielectric Constant and Dissipation Factor of Materials at 1 MHz[11]

Material	ϵ_r	Tan δ	$(\epsilon_r^{1/2}$ tan $\delta) \times 1000$
PTFE/glass	2.2	0.0008	1.19
RO2800*	2.8	0.0014	2.34
Polyimide/quartz	3.4	0.005	9.22
Polyimide/Kevlar	3.6	0.008	15.2
Polyimide/glass	4.5	0.010	21.2
Epoxy/glass	4.8	0.022	48.2
Alumina	10	0.0001	0.32

* PTFE/glass with ceramic filler.

TABLE 2.36   Dielectric Constant and Dissipation Factor of Materials at Microwave Frequencies[11]

Material	2.6 GHz*		3.0 GHz*		3.5 GHz*		4.0 GHz*		10.0 GHz†	
	ϵ_r	tan δ	ϵ_r	tan δ	ϵ_r	tan δ	ϵ_r	tan δ	ϵ_r	tan δ
PTFE/glass	2.16	0.0009	2.17	0.0011	2.18	0.0011	2.18	0.0010	2.20	0.0009
RO2800‡	2.80	0.0012	2.81	0.0015	2.82	0.0014	2.84	0.0013	2.84	0.0014
Polyimide/glass	—	—	4.47	0.0083	4.52	0.0089	4.58	0.0094	—	—
Epoxy/glass	4.30	.0105	4.32	0.0135	4.35	0.0143	4.41	0.0148	—	—

* Waveguide cavity perturbation method per ASTM D 2520.
† Stripline resonator method per ASTM D 3380.
‡ PTFE/glass with ceramic filler.

TABLE 2.37   Thermal and Mechanical Properties of Glass-Reinforced PTFE and Other Materials[11]

Material	CTE-xy, ppm/°C	CTE-z, ppm/°C	Thermal conductivity, W/m/K	Modulus of elasticity, 1000 × lb/in^2	Density, lb/in^3
PTFE/glass	24	261	0.26	0.14	0.079
RO2800*	16–19	24	0.44	0.06	0.072
Polyimide/quartz	6–8	34	0.13	4.0	0.07
Polyimide/KEVLAR	3.4–6.7	83	0.12	4.0	0.06
Polyimide/glass	11.7–14.2	60	0.35	2.8	0.066
Epoxy/glass	12.8–16	189	0.18	2.5	0.065
Alumina	6.5	6.5	16.8	37.0	0.13
Copper	16.9	16.9	394	17.0	0.324

* PTFE/glass with ceramic filler.

parts can be prepared with a smooth surface (gel coat) to improve their appearance. The gel coats are filled extensively with mineral fillers to impart a smooth appearance and to make the resin sufficiently thixotropic to stay on vertical mold surfaces. In these large parts, the polyester is catalyzed to cure at room or ambient temperature. Special resins and catalysts are needed to avoid inhibition of the curing process by oxygen. Special molds can produce highly decorative effects, such as masonry effects or woodlike appearances, on products such as rocky water falls, brightly colored playground equipment, and brick walls.

Epoxy materials are used in filament winding. This process is described in Chap. 3. Epoxy resins are used for their low shrinkage and good physical properties.

TABLE 2.38 Types and Applications of Decorative Laminates

Grade	Application	Data
$\frac{1}{16}$-in. general-purpose	Most widely used. For vertical and horizontal use requiring tough, long-wearing properties such as counters, tables, sinks, furniture, bars, and doors	For bonding to other structures. Supplied oversize. Should be bonded to plywood or Novoply
0.050-in. post-forming	Used where inside or outside radii are required, such as sink and counter tops, and restaurant tables	Should be bonded to a solid base. Formable with inside radii of $\frac{3}{16}$ in. and outside radii of $\frac{3}{4}$ in.
$\frac{1}{32}$-in. vertical	For furniture, cabinets, and walls or other vertical applications. Also where small radii postformability is needed	Similar to postforming grade. Minimum outside radius is $\frac{3}{8}$ in.
0.050-in. fire-resistant	Underwriters' label. Used on buses, ships, and public buildings	On incombustible core has flame spread of 15, fuel contributed -10, smoke developed -5
$\frac{1}{16}$-in. cigarette-proof	Will withstand lighted cigars and cigarettes. Used on tables and bars	Similar to general-purpose but contains an aluminum foil for heat dissipation. Meets MIL-T-17171B type II. Withstands 550°F for 10 min without blister

TABLE 2.39 Chemical Resistance of Decorative Laminates

Chemicals not affecting laminates:

Gasoline	Olive oil
Naphtha	Ammonia solution (10%)
Water	Citric acid
Alcohols	Coffee
Amyl acetate	Mustard
Acetone	Sodium bisulfite
Carbon tetrachloride	Wax and crayons
Moth spray	Urine
Fly spray	Shoe polish
Soaps	Trisodium phosphate
Detergents	

Chemicals that may stain laminates—stain can be removed by buffing with mild abrasive cleanser:

Tea	Ink
Beet juice	Iodine solution
Vinegar	Mercurochrome solution
Bluing	Phenols (Lysol)
Dyes	

Chemicals that may damage laminates:

Hypochlorite bleaches	Potassium permanganate
Hydrogen peroxide	Berry juices
Mineral acids	Silver nitrate
Lye solutions	Gentian violet
Sodium bisulfate	Silver protein (Argyrol)

Filament-wound parts are used in aerospace applications and in cherry-picker booms and tubing.

Polyurethane resins are often used in the molding process called reactive injection molding (RIM), which is also called resin transfer molding (see Chap. 9). Polyurethanes are selected for their low viscosity and rapid polymerization. Applications include large automotive parts and building components.

2.2.3 Thermoplastic-Reinforced Plastics

Many thermoplastic materials can be reinforced with fibers. Glass fibers with lengths less than ¼ in are used. Commonly "milled" fibers (fibers less than ¼ in) are selected. The fibers improve the physical properties—in particular the heat deflection temperature. Properties of thermoplastic-reinforced plastics are shown in App. C. Some thermoplastics are being reinforced with graphite fibers to give electromagnetic interference protection. Aramid fibers with thermoplastics results in excellent wear and abrasion resistance. Although fibers can be used with any thermoplastic, the following materials are most important:

Nylon materials are combined with glass fibers to control brittleness. Tensile strengths are improved by a factor of three, and the heat-deflection temperature increases from 150 to 500°F.

Polycarbonate compounds are made at 10, 20, 30, and 40 percent glass fiber loading. The physical properties are greatly improved.

Thermoplastic polyester compounds show a two-fold increase in tensile strength and a four-fold improvement in flexural strength with the addition of 40 percent by weight of glass fibers.

Other polymers greatly benefiting from the addition of glass fibers are polyphenylene sulfide, polypropylene, and polyether sulfone.

Reinforced thermoplastics are injection molded and some materials are extruded. Some suppliers are offering reinforced thermoplastic-pultruded products.

2.3 MANUFACTURING PROCESSES FOR REINFORCED PLASTICS

2.3.1 Thermoset Molding

Thermoset-reinforced molding compounds are molded in conventional compression, transfer, and injection molding presses. Descriptions of the processes and equipment are given in Chap. 9.

Mechanical properties of reinforced plastic molding compounds are greatly influenced by the condition of the reinforcing fiber and the quality of the molding process. The length of the fiber is important. The longer the fiber length in the product, the higher the mechanical properties will be. Another important parameter is the ratio of resin to glass. This must be optimized for each product and molding operation, but mechanical properties increase with increasing fiber content. Molding pressures influence the properties of the product. Properties improve as pressure increases because of better wetting of the fiber and decreased voids. A pressure limit is reached for each product past which further improvement is not gained.

Several types of reinforced plastic molding materials are used in thermoset molding. Bulk molding compounds (BMC) are materials made by combining a resin and chopped fibers in a Sigma Blade mixer and mixing until the fibers are well wetted and the material has the consistency of modeling clay. Fibers are usually E glass ⅛ to ½ inch long. Most of the BMC are based on polyesters, but vinyl esters and some epoxies are used. The material is often extruded into logs or rope for ease of handling at the mold. The strength of BMC is lower than others because the fibers are degraded in the mixer, and fibers orient somewhat in the mold, usually not in optimum alignment. The glass content can be varied between 10 and 20 percent by weight. BMC is molded in presses at 300 to 400°F and 500 psi. Many electrical parts requiring modest strength are made from BMC. Other applications include microwave dishes, table tops, and steam iron skirts.

Sheet molding compounds (SMC) are made by compounding a polyester resin with fillers, pigments, catalysts, mold release agents, and special thickeners that react with the polymer to greatly increase the viscosity. The resin mixture is spread onto a moving nylon film. The resin passes under cutters that chop roving into fibers from ½ to 2 inches long. A second film is placed on top, sandwiching the compound inside. This passes through rollers that help the resin to wet the fibers, and then the material is rolled up to mature for 24 to 72 hours. As the compound matures, it thickens and reaches a nontacky final consistency of leather. The nylon films are removed before the compound is molded. Resin content is 65 to 75 percent by weight. Molding conditions are 400°F and 2000 psi for 1 to 2 minutes. Because the fibers have not been greatly damaged in the preparation of the compound, the mechanical properties are much better than BMC, although fiber orientation in the mold is still a problem. Molded parts have good surfaces, so much SMC is used in automotive applications. Surface appearances are further improved by adding paraffins, polyethylene, acrylic, or other polymers to the polyester. This controls the shrinkage of the polymer, and such compounds are called low-profile or low-shrinkage SMC.

The process by which SMC is produced has been modified by some suppliers to produce thick molding compound (TMC). In this process the resin and fiber are intimately mixed on counter-rotating rollers before being deposited on the moving nylon film. Thickening chemicals may or may not be added. The compound has a high cross section. SMC usually has a weight of about 20 oz/ft^2 and TMC a weight of about 8 lb/ft^2.

Because polyesters are easy to mold at low pressures and modest temperatures, many laminators prepare flat-sheet laminates and even more complex shapes like trays by placing glass mat in the female section of a mold, pouring formulated liquid resin onto the reinforcement and closing the mold. However, as draws become deeper in boxlike shapes, flat-mat reinforcement can not bend to the shapes needed, and the laminator must use preforms. Preforms are made by spraying cut fibers and a dilute binder resin onto a perforated metal screen roughly the shape of the product. A vacuum holds the fibers in place. The screen is put into an oven, and the binder is cured. Preforms are removed from the screen and inserted into the compression mold with liquid resin. In the preform process, glass length is controlled from ½ to 3 inches, and glass content in the molded product can vary from 15 to 50 percent.

Polyester architectural panels, skylights, trailer panels, and corrugated sheets are made in a continuous process not unlike SMC. Glass mat, chopped fibers, and fabrics are laid on a carrier film like polyethylene or nylon, and resin is spread on. A top film is added, and rollers work the resin into the glass to wet it. Corrugations are formed either by molds or rollers as the laminate enters a tunnel oven to cure.

2.3.2 Thermoplastic Molding

Thermoplastic-reinforced molding compounds are molded in conventional injection machines and extruders. Descriptions of the processes and equipment are given in Chap. 9.

Thermoplastic laminates are made by the methods described in Sec. 2.1.5. Some thermoplastic laminates (mostly laminates of polypropylene and glass) are made on continuous double-belt presses. In this process, rolls of glass mat are combined with films of polypropylene at the nip end of the belt press. Continuous stainless steel belts top and bottom compress the material, as it passes first into a heated zone to melt the polypropylene and consolidate it into the fibers, and second it passes into a cold zone to freeze the polymer. This product is usually not used in this form but sold to fabricators for stamped or thermally formed parts. Thermosetting decorative trim has also been made on these continuous laminators.

Stamped thermoplastic-glass laminates are made in special compression presses with rapid closing features. Laminates are blanked and heated to about 520°F and then inserted into a cold mold. The press closes rapidly at 1400 in/min to contact with the blank and applies pressure at 60 in/min to a maximum pressure of 3000 psi. Cycle times are about 45 sec. The thermoplastics used in this process are polypropylene, polybutylene terephthalate, polycarbonate, and polyphenylene sulfide.

2.3.3 Hand Lay-Up

The simplest method of making large reinforced plastic parts is the hand lay-up or contact lay-up method. It is frequently employed by boat and swimming pool manufacturers as shown in Fig. 2.14. Hand lay-up can be used on either male or female molds. The molds can be wood, metal, plastic, or plaster. When wood or plaster is used, the pores must be sealed with varnishes or lacquers. After the sealing compound is dry, the mold must be coated or covered with a release agent such as floor wax or a film-forming polymer like polyvinyl alcohol (PVA). In many applications a gel coat is applied to the mold surface. This coating is a thick, pigmented resin

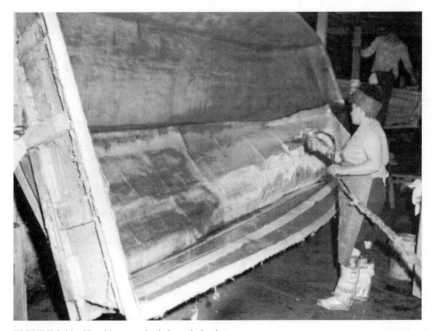

FIGURE 2.14   Hand lay-up of reinforced plastics.

that contributes a smooth, colorful outer surface to the molding. Gel coats are usually 20 to 40 mils thick and are permitted to cure to a tacky surface. At this point the reinforcing web (glass cloth, mat, or woven roving) is placed on the mold, and a low-viscosity resin, generally a room-temperature-curing polyester or epoxy, is painted or sprayed onto the reinforcement. The air from the reinforcement is removed carefully with a paint roller or a rubber squeegee. Additional layers of reinforcement and resin are added until the correct thickness is obtained. The exposed topmost layer is covered with polyethylene, cellophane, or polyester film to make a smooth surface. Some polyesters are slightly inhibited by air, and thus film overlays help to complete the cure. When the resin has hardened, the part is removed and sometimes given a postbake. Although this process is cheap and easy to perform, the moldings are not of the highest quality and may not completely have the properties expected. The laminates produced by this method contain more resin than those produced by other methods, and more voids may be present. Little control can be exercised over the uniformity of the wall thickness. Typical cure times for polyesters are about 10 hours. Epoxies are used in special applications. They can be cured in as little as 3 hours at 180°F. Hand lay-up is very labor intensive but is an excellent molding process when large sizes and few units are needed.

A special case of hand lay-up is the spray-up process (Fig. 2.15). This is accomplished with a special type of spray gun. Molds similar to those used in hand lay-up are used. The gun chops glass fiber into predetermined lengths, mixes them with resin, and deposits them onto the mold surface. The mold surface is often pretreated with a mineral-filled resin that has been partially cured. This surface is called a gel coat. When enough glass has been deposited, an impregnated glass mat has been formed. From this point on, the process is identical to the hand lay-up process. The process is attractive because it makes use of glass fiber in its least expensive form. Parts with complicated shapes can be made, and size is not critical. Skilled operators are needed if control over the wall thickness is required.

2.3.4 Vacuum Bag Molding

Vacuum bag molding is one of the most popular methods of making prototype or short-run parts. Molds similar to those just discussed are used. Either wet resins or prepregs can be employed. Figure 2.16 shows the molding process. Over the mold surface is placed a release sheet such as PVA or release-treated polyester film. The reinforcement is placed onto the release film and impregnated by brushing or spraying if prepregs are not being used. Next, bleeders are installed along the top of the mold. These are made of burlap, muslin, Osnaburg cloth, or felt, and are intended to provide a space from which the air in the mold can be evacuated. The impregnated reinforcement is sometimes covered with a perforated release film: PVA, polyethylene, Teflon, or treated polyester film. Special release fabrics are used also. PVA can be used up to about 150°F and is tougher and more tear-resistant than cellophane. PVA is highly sensitive to humidity, being water-soluble. Polyethylene is used where temperatures permit (room temperature only). Polyvinyl chloride (PVC) bags are useful up to 150°F and are less expensive than PVA. Certain PVC polymers inhibit polyester resins and prevent a tack-free cure. When high curing temperatures or steam-autoclave curing are required, the bag should be made of neoprene rubber or some similar elastomer. Neoprene and some other rubbers also inhibit polyester resins, and if polyesters are to be cured, the bag should have a film such as Mylar or Kapton between the lay-up and the bag.

FIGURE 2.15   Spray-up of reinforced plastic part.

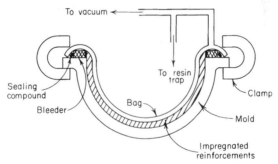

FIGURE 2.16   Vacuum bag molding.

The edge of the bag must be well sealed to prevent loss of vacuum and introduction of air into the lay-up. A common method is to use zinc chromate paste, which is supplied in ribbons or beads. The paste is pressed into intimate contact with the mold, and the bag is pressed into the paste. This material holds a vacuum well and is easily removed from the assembly after molding. Other rubber gaskets can be used along with hold-down or clamp rings, either to replace or to supplement the paste. Getting a good seal is critical to the success of the vacuum bag process, and care exercised at this step will avoid much difficulty. Vacuum is then applied, bringing air pressure to bear on all the surfaces. The use of a resin trap in the vacuum line is essential. The trap plays the same role that an oil separator plays in an air line. As the resin softens and flows during the molding, it will enter the vacuum line. If it is not removed, it will decrease the vacuum, clog and ruin the lines, and, if it is permitted to enter the pump, will cause pump failure. Mechanical working of the bag with a roller or squeegee aids in air removal when wet resins are used.

If prepregs are used, the whole assembly is placed in an oven to cure. If prepregs requiring temperatures greater than 300°F are employed, all the release sheets should be made of Mylar or Teflon. A typical cure schedule consists of increasing the temperature to 250°F in 30 min, 300°F in 15 min, and to 350°F in 20 min. When the part is cured, the temperature should be reduced to about 150°F before the vacuum is released.

A modification of the vacuum bag process is pressure bag molding. A lid is placed over the mold and clamped down to contain about 40 psi. Steam is sometimes used to accelerate cure of the resin. This modification can be used only on female molds.

Advantages of the vacuum bag process are:

1. The molds are relatively inexpensive.
2. The process is faster than hand lay-up.
3. Prepregs can be used.
4. The product is not as rich in resin as those produced by the hand lay-up process, and its properties are better.
5. The process does not require highly skilled personnel.

Disadvantages of the process are:

1. Only the surface adjacent to the mold surface is smooth. The other surface may be duller and show folds and marks from the bag.
2. The part still contains more resin that those made by high-pressure techniques, and hence its properties are not optimal.
3. Certain shapes with complicated surfaces or sharp corners are difficult to form.
4. Dry spots and voids are fairly common.

2.3.5 Autoclave Molding

Autoclave molding is similar to the preceding methods except that the pressure plate is removed and the entire mold and the lay-up are placed in an autoclave. Steam or hot pressurized air up to 100 lb/in² is admitted, supplying both pressure and heat. If steam is used, the covering bag must not have holes or leaks because steam will enter the lay-up and ruin the molding. The chief advantages of this method are faster cycles and improved properties. See Chap. 3 for additional information.

2.3.6 Reaction Injection Molding

This process for manufacturing reinforced plastics has, with small variations, several closely related processes: RTM (resin transfer molding), SRIM (structural reaction injection molding), and RRIM (reinforced reaction injection molding).

RIM refers to pumping a resin into a closed mold that has been preloaded with reinforcements. The reinforcements may have several forms. If the reinforced plastic part is simple, the reinforcement can be precut glass mat that is placed by hand in the mold. If a deeper draw is needed a fabricated glass preform is manufactured as described in Sec. 2.3.1. A third common process is to use a continuous strand glass mat prepared with a special binder. This mat can be precut and then shaped thermally on a mold equiped with clamps and heaters. This produces a three-dimensional fiber preform ready for molding. When the mold has been loaded with the preform, it is clamped shut, and liquid catalyzed resin is heated and pumped under modest pressure into the mold, saturating the preform and pushing out the air through vents. Cycle times vary from 1 to 20 minutes depending on size and complexity of the part. Weights vary from a few ounces to several hundred pounds.

Molds can be made of reinforced plastic for short runs of product or steel for higher volume runs. Polyesters are often used, but polyurethanes, vinyl esters, and epoxies can be molded. The advantages of RIM are:

- Two finished surfaces are molded.
- Gel coats can be used.
- Little postmold finishing is required.
- Tooling costs are low to modest.
- Volatiles are minimum and easily controlled.
- Inserts and cores can be molded.
- The process is energy-efficient.
- Complex parts are possible, such as fenders, pickup truck beds, and tub and shower units.

2.3.7 Pultrusion

Pultrusion is a continuous method for making reinforced plastics where reinforcing rovings are pulled through a resin bath and into a die where the composite is cured. Since the process is continuous, there is no theoretical limit on the length of the product. Figure 2.17 shows the process schematically, and Fig. 2.18 shows the delivery and impregnation sections of a typical pultrusion facility.

The reinforcing fibers that are used determine the physical properties of a pultruded product. Because of its costs and physical properties, fiber glass is usually used—particularly E glass. With fiber glass, the product has highly directional properties. The tensile and flexural strengths in the fiber direction are extremely high, exceeding 100,000 lb/in^2, while the transverse properties are much lower. Transverse properties can be enhanced somewhat by incorporating glass mat, glass cloth, or chopped fibers, but with an attendant reduction in longitudinal property. For this reason, the direction of prime stress must be carefully considered in the design of a pultruded part. The physical properties of a 60 percent by weight fiber glass-polyester pultruded part are shown in Table 2.40.

Polyesters, epoxies, acrylics, vinyl esters, and silicones have all been used in pultrusion, but 90 percent of the products use vinyl esters and polyesters because of their favorable costs, desirable properties, and easy handling characteristics. Liquid systems with low viscosity (about 2000 cP) are used to ensure rapid saturation of the

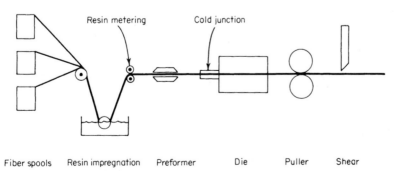

FIGURE 2.17 Pultrusion process.

FIGURE 2.18   Pultrusion equipment.

fiber bundles. The resin determines the weather resistance, thermal resistance, burning characteristics, moisture sensitivity, chemical resistance, and most of the electrical properties. The curing rate of the resin in the heated die must be carefully controlled. Most of these resins are exothermic (give off heat while curing), and too rapid curing results in voids and cracks, while too slow curing produces a poorly cured product with poor shear and compression strength.

Dies used in pultrusion are usually quite simple and inexpensive. Steel dies up to 6 ft long are often used with production rates of 5 ft/min. Dies are heated electrically or with radio frequency (rf) generators of up to 10 kW at 100 MHz. When rf heating is used, the dies are made of Teflon.

Typical uses of pultruded shapes are fishing rods, ladders, I beams, electrical pole-line hardware, tool handles, and electric-motor wedges. In applications where pul-

TABLE 2.40   Properties of Pultruded Parts

Property	Test method	Value
Flexural strength	ASTM D 790-63	100,000 lb/in.²
Tensile strength	ASTM D 638-64T	120,000 lb/in.²
Compressive strength	ASTM D 695-63T	40,000 lb/in.²
Impact strength	ASTM D 256-56	40 ft-lb/in. notch
Electric strength:		
Perpendicular	ASTM D 149-64	450 V/mil
Parallel	ASTM D 14-64	60 kV
Arc resistance	ASTM D 495-61	130 s
Water absorption	ASTM D 570-63	0.1%

TABLE 2.41 Pultruded Parts

Size	Shaping die and equipment pulling capacity influence size limitations
Shape	Straight, constant cross sections; some curved sections possible
Reinforcements	Fiberglass Carbon fiber Aramid fiber
Resin Systems	Polyester Vinyl ester Epoxy Silicones
Fiberglass Contents	Roving, 40–80% by weight Mat, 30–50% by weight Woven roving, 40–60% by weight
Mechanical Strengths	Medium to high, primarily unidirectional, approaching isotropic
Labor Intensity	Low to medium
Mold Cost	Low to medium
Production Rate	Shape and thickness related

trusions are used such as cables, special termination devices must be used to avoid interrupting the fiber continuity and to achieve the best properties. A summary of pultrusion product characteristics is shown in Table 2.41[14]. Pultrusion design guidelines are given in Fig. 2.19.

Pulforming is a variation of pultrusion. The process uses continuous roving, a resin impregnation bath, and forming dies, but the dies are two-piece and recirculate. They clamp onto the roving and cure it as it moves. This forms tapered cross sections like hammer handles.

2.3.8 Filament Winding

Filament winding refers to the process of wrapping resin-impregnated continuous fibers around a mandrel, forming a surface of revolution. When enough layers have been wound, the part is cured and the mandrel is removed. The continuous filaments are generally tapes or rovings of fiber glass, but other reinforcements are used also. Epoxies are popular impregnating resins, although others are used. Typical properties of glass-epoxy filament-wound parts are shown in Table 2.42. Of special notice are the high tensile and modulus values. Filament-wound parts contain a high percentage of reinforcing fiber, and the properties of filament-wound parts approach the properties of the fiber. The role of the resin is to bind adjacent fibers together and to transmit stress from fiber to fiber. The resin content is determined by the amount of resin on the glass and the winding tension.

Minimum inside radius, inches		1/32" Roving Shapes 1/16" Mat Shapes	Corrugated sections		Yes, Longitudinal
Molded-in holes	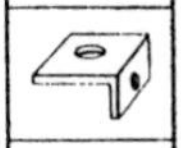	No	Metal inserts	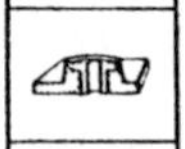	No
Trimmed in mold	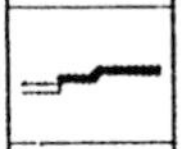	Yes	Bosses	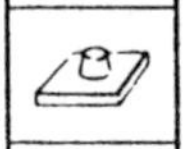	No
Core pull & slides	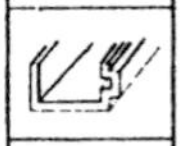	Yes	Ribs		Yes, Longitudinal
Undercuts	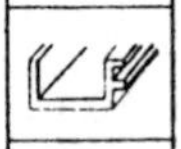	Yes	Molded-in labels		Yes, but not recessed
Minimum recommended draft, in./deg.	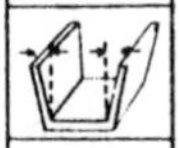	No Limitation	Raised numbers		No
Minimum practical thickness, inches		Roving – .040" Mat. – .060"	Finished surfaces (reproduces mold surface)		2
Maximum practical thickness, inches	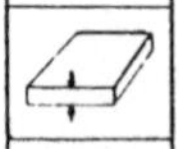	Roving – 3 0" Mat. – 1.0"	Hollow sections	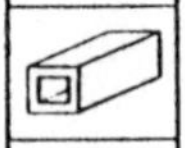	Yes, Longitudinal
Normal thickness variation, inches	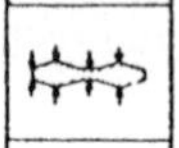	As Required	Wire inserts	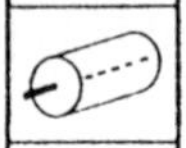	Yes, Longitudinal
Maximum thickness buildup		As Required	Embossed Surface		No

FIGURE 2.19 Pultrusion design guides.

When designing filament-wound parts, the fibers should be oriented in the direction of stress. If, because of the geometry of the pair, it is impossible to do so, a balanced structure is used wherein all the fibers regardless of direction have equal stress. Careful design consideration must be given to holes and flanges to avoid interrupting the continuous nature of the fibers.

Many other variables must be controlled to optimize the structure. These include the winding angles, relative location of the layers of differing winding angles, winding tension, width of the roving bond, type of fiber, and number of ends of fiber in the roving. The type of mandrel used contributes to the quality of the product. Most mandrels are machined and polished steel with a slight taper (about ½°). Other mandrels can be made of plaster, salt, low-melting alloys, and plastic. When closed-end structures are wound, water-softening plasters are often used.

Winding machines are of two types—helical and polar. Helical winders consist of a rotating mandrel and a reciprocating roving delivery system. Both motions are controlled to fix the winding pattern and fiber angle. It takes two layers to cover the mandrel with fiber-containing crossovers, except for those structures with a wind angle of 90° (called circumferential or hoop wraps). Helical winding produces open-ended structures such as cylinders.

To produce closed-end structures like bottles, a polar winder is used. In this machine, the mandrel is generally stationary, and the delivery arm that revolves around it inclines to the proper angle. After each rotation, the mandrel is indexed one fiber band width, resulting in a two-layer application without fiber crossovers. Hoop wraps can also be added to the center portion of the structure.

TABLE 2.42   Properties of Filament-Wound Products

Property	Value
Specific gravity	2.0
Density, lb/in.^{3}	0.072
Thermal conductivity, Btu/(hr)(ft^2)(°F/in.)	2.2
Thermal expansion, per °F $\times$ 10^{-6}	7.0
Specific heat, Btu/(lb)(°F)	0.227
Maximum use temperature, °F	<400
Hoop tension, psi $\times$ 10^{-3}:	
Unidirectional windings	230
Helical windings	135
Compressive strength, psi $\times$ 10^{-3}	70
Flexural strength, psi $\times$ 10^{-3}	100
Bearing strength, psi $\times$ 10^{-3}	35
Shear strength, psi $\times$ 10^{-3}:	
Interlaminar	6
Cross	18
Modulus of elasticity, tension, psi $\times$ 10^{-6}	6.0
Modulus of rigidity, torsion, psi $\times$ 10^{-6}	2
Dielectric strength, step-by-step, volts/mil	400

The resin used can be applied at the time of the winding process (wet winding), or prepreg rovings can be used. Such prepreg rovings are available from C glass, E glass, S glass, boron, and graphite in 12, 20, 30, and 60 ends. The resin constitutes 17 to 23 percent by weight of the roving. In most cases, the materials are shipped frozen and must be kept refrigerated. Available resins include epoxy, phenolic, diallyl phthalate, polyester, and polyimide resins.

Filament-wound structures are used in rocket-motor cases, pipelines, cherry-picker booms, lightning arresters, chemical-storage tanks, sporting equipment, shotgun barrels, and circuit-breaker parts. More information on filament winding can be found in Chap. 3. Many filament winding processes use tape-laying machines that precisely introduce glass cloth or mat tapes at the required locations as described in Chap. 3.

2.3. Reinforced Plastics—Summary

The design of reinforced plastics is a complicated process that requires the placement of reinforcing fibers in the positions where they will contribute significantly to resisting the stresses imposed on the structure. Ref. 15 is a fine source for design considerations. Table 2.43 gives some design rules for reinforced plastics configurations.

2.4 STANDARDS

Many standards have been developed for reinforced plastics, laminates, and composites. In the United States, the most active bodies* have been:

American National Standards Institute (ANSI)
11 West 42nd Street
New York, NY 10036

Institute for Interconnecting and Packaging Electronic Circuits (IPC)
7380 North Lincoln Avenue
Lincolnwood, IL 60646

* See also App. E

TABLE 2.43 Design Rules for Reinforced Plastics

Design rules	Spray-up, hand lay-up (contact)	Matched-metal-die molding preform mat	Sheet molding compound	Cold molding
Min inside radius, in.	$\frac{1}{4}$	$\frac{1}{8}$	$\frac{1}{16}$	$\frac{1}{4}$
Molded-in holes..	Large	Yes, parallel or perpendicular to ram action	Yes	No
Trimmed-in mold.	No	Yes	Yes	No
Undercuts.......	Yes	No	Yes	No
Min draft recommended $\frac{1}{4}$- to 6-in. depth	0°	1–3°	1–3°	2°
6-in. + depth		3° + or as required	3° + or as required	3°
Min practical thickness, in.	0.060	0.030	0.050	0.080
Max practical thickness, in.	0.500+	0.250	1	0.500
Normal thickness variation, in.	±0.020	±0.008	±0.005	±0.010
Max thickness buildup	As desired	2:1 max	As desired	2:1
Corrugated sections	Yes	Yes	Yes	Yes
Metal inserts....	Yes	Yes	Yes	NR
Surfacing mat....	Yes	Yes	No	Yes
Max size part to date, ft²	3,000	200	50	50
Limiting size factor	Mold size	Press dimension	Press capacity	
Metal edge stiffness	Yes	Yes	No	No
Bosses..........	Yes	NR	Underside only	NR
Ribs............	Yes	NR	As required	No
Molded-in labels	Yes	Yes	Yes	Yes
Raised numbers..	Yes	Yes	Yes	Yes
Gel-coat surface..	Yes	Yes	No	Yes
Shape limitations.	None	Moldable	Moldable	Moldable
Translucency....	Yes	Yes	No	Yes
Finished surfaces.	One	Two	Two	Two
Strength orientation	Random	Random	Random	Random
Typical glass loading, % by wt	20–30	25–40	15–35	20–35

American Society for Testing and
Materials (ASTM)
1916 Race Street
Philadelphia, PA 19103

National Electrical Manufacturers
Association (NEMA)
2101 L Street, N.W.
Washington, DC 20037

Federal Government
General Services Administration
(various local addresses)

Underwriters Laboratories (UL)
333 Pfingsten Road
Northbrook, IL 60062

Federal specifications are defined in Department of Defense's *Index of Specifications and Standards (DODISS), D7.14:981* and *Index of Federal Specifications, Standards and Commercial Item Descriptions, GS2.8/2:981.* These are available from the U.S. Government Printing Office.

For many years most test specifications for reinforced plastics were covered by *Federal Test Method Standard No. 406, Plastics: Methods of Testing.* This is gradually being abandoned as the Department of Defense replaces these with ASTM methods.

Reinforced plastics are also covered in many countries other than the United States. The most active of these foreign standards bodies* are:

British Standards Institute

Canadian Standards Association

French Association for
Standardization (Association
Française de Normalisation)

European Association of Aerospace
Manufacturers (Association
Européenne des Constructeurs de
Materiel Aerospatial)

German Institute for Standards
(Deutsches Institut für
Normunge.V.)

Japanese Standards Association

Other countries maintain standards and are listed in Ref. 16. Many of these standards can be obtained from the American National Standards Institute. A complete listing is also given in App. E of this handbook.

The following is a list of sponsoring organizations and specification titles describing reinforced plastics, composites, and laminates.

2.4.1 Laminates

Clad Laminates

- ASTM

D 1867	Copper Clad Thermosetting Laminates for Printed Wiring
D 5109	Test Methods for Copper-Clad Thermosetting Laminates for Printed Wiring Boards

- U.S. government

MIL-P-13949	Plastic Sheet, Laminated, Metal Clad
MIL-P-55110	Printed Wiring Boards

- IPC

IPC-L-108B	Specification for Thin Metal Clad Base Materials for Multilayer Printed Boards

---

* See also App. E

IPC-L-109B	Glass Cloth, Resin Impregnated (B Stage) for High-Temperature Multilayer Printed Boards
IPC-L-112A	Standard for Foil Clad Composite Laminate
IPC-L-115B	Specification for Rigid Metal Clad Base Materials for Printed Boards
IPC-L-125A	Specification for Plastic Substrates, Clad or Unclad for High Speed/High Frequency Interconnections
IPC-EG-140	Specification for Finished Fabric Woven From E Glass for Printed Circuit Boards
IPC-SG-141	Specification for Finished Fabric Woven From S Glass for Printed Circuit Boards
IPC-A-142	Specification for Finished Fabric Woven From Aramid for Printed Circuit Boards
IPC-QF-143	Specification for Finished Fabric Woven From Quartz for Printed Circuit Boards
IPC-RF-245	Performance Specification for Rigid-Flex Printed Board
IPC-D-275	Design Standard for Rigid Printed Boards and Rigid Printed Board Assemblies
IPC-RB-276	Qualification and Performance Specification for Rigid Printed Boards
IPC-D-300G	Printed Board Dimensions and Tolerances
IPC-HF-318A	Microwave End Product Board Inspection and Test
IPC-D-322	Guidelines for Selecting Printed Wiring Board Sizes Using Standard Panel Sizes
IPC-D-325	Documentation Requirements for Printed Boards
IPC-BP-421	General Specification for Rigid Printed Board Backplanes with Press-Fit Contacts
IPC-HM-860	Specifications for Multilayer Hybrid Circuits
IPC-PE-740	Technology Manual—Troubleshooting Guide for Printed Board Manufacture and Assembly
IPC-ML-960	Qualification and Performance Specification for Mass Lamination Panels for Multilayer Printed Boards

- NEMA

| Publication No. LI 1 | Industrial Laminated Thermosetting Products |

- UL

| 746E | Polymeric Materials—Industrial Laminates, Filament Wound Tubing, Vulcanized Fibre and Materials Used in Printed Wiring Board |
| UL 796 | Printed Wiring Boards |

Decorative Laminates

- U.S. government

| MIL-P-17171 | Plastic, Laminate, Decorative, High Pressure |

- NEMA

Publication No. LD 3-1985	High-Pressure Decorative Laminates
Publication No. LD 3.1-1985	Performance, Application, Fabrication, and Installation of High-Pressure Decorative Laminates

Polyester-Glass-Reinforced Sheet

- ASTM

D 1532	Polyester Glass Mat Sheet Laminate

- U.S. government

L-P-383	Plastic Material, Polyester Resin, Glass Fiber Base, Low Pressure Laminate
MIL-R-7575	Resin, Polyester, Low Pressure Laminating
MIL-P-21607	Resins, Polyester, Low Pressure Laminating, Fire Retardant
MIL-P-24364	Plastic Sheet, Laminated, Thermosetting Electrical Insulating Sheet, Glass Mat
MIL-P-24364/1	Plastic Sheet, Laminated, Thermosetting Electrical Insulating Sheet, Glass Mat Grade GPO-N1 (Classes 130, 155, and 180)
MIL-P-24364/2	Plastic Sheet, Laminated, Thermosetting Electrical Insulating Sheet, Polyester Glass Mat Grade GPO-N2 (Class 130)
MIL-P-24364/3	Plastic Sheet, Laminated, Thermosetting Electrical Insulating Sheet, Polyester Glass Mat, Grade GPO-N3
MIL-R-25042	Resin, Polyester, High Temperature Resistant, Low Pressure Laminating
MIL-P-25395	Plastic Material, Heat Resistant, Low Pressure Laminated Glass Fiber Base, Polyester Resin
MIL-P-43038	Plastic Molding Material, Polyester, Low Pressure Laminating
MIL-R-46068	Resin, Polyester, Bisphenol—A Type
MIL-P-46169	Plastic, Sheet Molding Compound, Polyester, Glass Fiber Reinforced

- NEMA

Publication No. LI 1	Industrial Laminated Thermosetting Products

Rod, Tubing, Filament Winding, and Pultrusion

- ASTM

D 348	Testing Rigid Tubes Used for Electrical Insulation
D 349	Testing Laminated Round Rods Used for Electrical Insulation
D 668	Measuring Dimensions of Rigid Rods and Tubes Used for Electrical Insulation
D 1180	Test for Bursting Strength of Round Rigid Plastic Tubing

| D 3647 | Practice for Classifying Reinforced Plastic Pultruded Shapes According to Composition |
| D 3917 | Specification for Dimensional Tolerance of Thermosetting Glass Reinforced Plastic Pultruded Parts |

- U.S. government

L-P-509	Plastic Sheet, Rod and Tube Laminated, Thermoset
MIL-P-79	Plastic Rods and Tube, Thermosetting, Laminated
MIL-P-5431	Plastic Phenolic Graphited, Sheets, Rods, Tubes, and Shapes
MIL-T-10652	Tubing, Plastic Glass Fabric Base, Low Pressure Laminated
MIL-C-47257	Compound, Epoxy, Filament Winding
MIL-P-82540	Plastic Material, Polyester Resin, Glass Fiber Base, Filament Wound Tube
MIL-B-83369	Boron Filament Reinforcement, Continuous, Epoxy Resin Impregnated
MIL-R-87120	Rod, Pultruded, Graphite Fiber Reinforced, Processing

- NEMA

| Publication No. LI 1 | Industrial Laminated Thermosetting Products |
| Publication No. TC 14-1984 (R1986) | Filament—Wound Reinforced Thermosetting Resin Conduit and Fittings |

- UL

| 746 E | Polymeric Materials—Industrial Laminates, Filament Wound Tubing, Vulcanized Fibre and Materials Used in Printed Wiring Board |

Unclad Laminates

- ASTM

D 229	Testing Rigid Sheet and Plate Materials Used for Electrical Insulation
D 257	Tests for D-C Resistance or Conductance of Insulating Materials
D 495	Test for High Voltage, Low Current, Dry Arc Resistance of Solid Electrical Insulation
D 618	Conditioning Plastics and Electrical Insulating Materials for Testing
D 709	Standard Specification for Laminated Thermosetting Materials
D 785	Test for Rockwell Hardness of Plastics and Electrical Insulating Materials
D 883	Definition of Terms Relating to Plastics
D 2304	Thermal Evaluation of Rigid Electrical Insulating Materials
D 3841	Specification for Glass Fiber Reinforced Polyester Plastic Panels

D 4357	Specification for Plastic Laminates Made from Woven Roving and Woven Yarn Glass Fabrics

- U.S. government

L-P-509	Plastic Sheet, Rod and Tube, Laminated Thermoset
MIL-P-997	Plastic Material, Laminated, Thermosetting, Electrical Insulations, Sheets, Glass Cloth, Silicone Resin
MIL-R-9299	Phenolic, Laminating
MIL-R-9300	Resin, Epoxy, Low Pressure Laminating
MIL-P-9400	Plastic, Laminate and Sandwich Construction Parts, Aircraft Structural, Process Specification Required
MIL-P-15035	Plastic Sheet, Laminated, Thermosetting, Cotton Fabric Base, Phenolic Resin
MIL-P-15037	Plastic Sheet, Laminated, Thermosetting, Glass Cloth, Melamine Base
MIL-P-15047	Plastic Material, Laminated, Thermosetting, Nylon Fabric Base, Phenolic Resin
MIL-P-17549	Plastic Laminates, Fibrous Glass Reinforced, Marine, Structural
MIL-P-18177	Plastic Sheet, Laminated, Thermosetting, Paper Base, Epoxy Resin
MIL-P-18324	Plastic Material, Laminated Phenolic for Bearings
MIL-P-19161	Plastic Sheet, Laminated, Glass Cloth, Polytetrafluoroethylene Resin
MIL-P-22324	Plastic Sheet, Laminated, Thermosetting, Paper Base, Epoxy Resin
MIL-I-24204	Insulation, Electrical, High Temperature, Bonded, Synthetic Fiber Paper
MIL-P-24364	Plastic Sheet, Laminated, Thermosetting Sheet, Glass Mat
MIL-S-25392	Sandwich Construction, Plastic Resin, Glass Fabric Base, Laminated Facings and Polyurethane Foamed-In-Place Core, for Aircraft Structural Applications
MIL-P-25421	Plastic Material, Glass Fabric Base, Epoxy Resin, Low Pressure Laminated
MIL-R-25506	Resin Solution, Silicone, Low Pressure, Laminating
MIL-P-25515	Plastic Material, Phenolic Resin, Glass Fiber Base, Laminated
MIL-P-25518	Plastic Material, Silicone Resin, Glass Fiber Base, Low Pressure Laminated
MIL-P-25525	Plastic Material, Phenolic Resin, Glass Fiber Base, Laminated
MIL-P-25770	Plastic Materials, Asbestos Base, Phenolic Resin, Low or High Pressure Laminate
MIL-L-62474	Laminate, Aramid Fabric Reinforced, Plastic
MIL-L-64154	Laminate, Fiberglass, Fabric Reinforced, Phenolic

MIL-HDBK-17	Polymer Matrix Composites: Volume 1: Guidelines
MIL-HDBK-727	Design Guidance for Producibility
MIL-HDBK-731	Nondestructive Testing Methods of Composite Materials—Thermography
MIL-HDBK-732	Nondestructive Testing Methods of Composite Materials—Acoustic Emission
MIL-HDBK-733	Nondestructive Testing Methods of Composite Materials—Radiography
MIL-HDBK-787	Nondestructive Testing Methods of Composite Materials—Ultrasonics

- IPC

| IPC-L-125A | Specification for Plastic Substrates, Clad or Unclad, for High Speed/High Frequency Interconnections |

- NEMA

| Publication No. LI 1 | Industrial Laminated Thermosetting Products |
| Publication No. LI 6 | Relative Temperature Indices of Industrial Thermosetting Laminates |

- UL

| UL 746D | Polymeric Materials—Fabricated Parts |
| 746E | Polymeric Materials—Industrial Laminates, Filament Wound Tubing, Vulcanized Fibre and Materials Used in Printed Wiring Board |

2.4.2 Prepreg

- IPC

| IPC-L-109B | Specification for Resin Preimpregnated Fabric (Prepreg) for Multilayer Printed Boards |

- NEMA

| Publication No. LI 1 | Industrial Laminated Thermosetting Products |

- U.S. government

MIL-P-13949/11	Plastic Sheet, Laminated, Materials, GE Base Material, Glass Cloth, Resin Preimpregnated
MIL-P-13949/12	Plastic Sheet, Laminated, Materials, GF Base Material, Glass Cloth, Resin Preimpregnated
MIL-P-13949/13	Plastic Sheet, Laminated, Materials, GI Base Material, Glass Cloth, Resin Preimpregnated
MIL-P-46187	Prepreg, Unidirectional Tape, Carbon-Graphite Fiber Polyimide (PMR-15) Resin Impregnated, 316°C
MIL-L-46197	Laminate, S-2 Glass Fabric Reinforced, Polyester Resin Preimpregnated
MIL-M-46861	Molding Material, Glass Mat, Epoxy Coated

MIL-M-46862 Molding Material, Glass, Epoxy Impregnated

MIL-G-83410 Graphite Fiber, Resin Impregnated Tape and Sheet, for Hand Lay-Up

MIL-S-83474 Shims, Molded, Filled Resin Compound and Sheet Prepreg

- IPC

IPC-L-109B Specifications for Resin Preimpregnated Fabric (Prepreg) for Multilayer Printed Boards

2.4.3 Reinforced Plastics

Armor

- U.S. government

MIL-A-46165 Armor, Woven Glass Roving Fabrics

MIL-A-46166 Plastic Laminates, Glass Reinforced, for Use in Armor Composites

Thermoplastic-Reinforced Plastics

- ASTM

D 3935 Specification for Polycarbonate Unfilled and Reinforced Materials

D 4067 Specification for Reinforced and Filled Polyphenylene Sulfide Injection Molding and Extrusion Materials

- U.S. government

L-P-395 Plastic Molding Material, Nylon, Glass Fiber Reinforced

MIL-P-21347 Plastic Molding Material, Polystyrene, Glass Fiber Reinforced

MIL-P-46109 Plastic Molding Material, Polypropylene, Glass Fiber Reinforced

MIL-P-46115 Plastic Molding and Extrusion Material, Polyphenylene Oxide, Modified, Glass Fiber Reinforced

MIL-P-46137 Plastic Molding and Extrusion Material, Acetal, Glass Fiber Reinforced

MIL-P-46161 Plastic Molding Material, Polyterephthalate, Thermoplastic, Glass Fiber Reinforced

MIL-P-46174 Plastic Molding Material, Polyphenylene Sulfide, Glass Fiber Reinforced

MIL-P-46180 Plastic Molding Material, Polyamide, Glass Fiber Reinforced

MIL-P-81390 Plastic Molding Material, Polycarbonate, Glass Fiber Reinforced

Thermosetting-Reinforced Plastics

- ASTM

D 700 Specification for Phenolic Molding Compounds

D 704 Specification for Melamine Formaldehyde Molding Compounds

D 705	Specification for Urea Molding Compounds
D 1201	Specification for Thermosetting Polyester Molding Compounds
D 1636	Specification for Allyl Molding Compounds
D 1763	Specification for Epoxy Resins
D 3013	Specification for Epoxy Molding Compounds
D 4617	Specification for Phenolic Compounds

- U.S. government

MIL-M-14*	Molded Plastics and Molded Plastic Parts
TT-R-371	Resin, Phenol-Formaldehyde, Para-Phenyl
MIL-P-43043	Plastic Molding Material, Premix, Polyester, Glass Fiber Reinforced
MIL-M-46069	Molding Plastic, Glass/Epoxy Premix
MIL-C-46866	Compound, Epoxy Molding and Application of
MIL-M-46891	Molding Compound, Phenolic Resin, Asbestos Reinforced
MIL-P-46892	Plastic Molding Material, Epoxy, Glass Fiber Reinforced
MIL-C-47221	Compound, Molding, Epoxy Resin
MIL-C-47224	Compound, Molding, Transfer, Epoxy Resin, Single Component
MIL-M-81255	Plastic Molding Material, Asbestos, Phenolic
MIL-P-82650	Plastic Molding Material, Glass, Phenolic

REFERENCES

1. J. V. Milewski in H. S. Katz (ed.), *Handbook of Reinforcements for Plastics*, Van Nostrand Reinhold, NY, 1987.
2. W. Brenner et al., *High Temperature Plastics*, p. 60, Van Nostrand Reinhold, NY, 1962.
3. W. A. Laurie, "New Concepts for Thermosets," *Reg. Tech. Conf. Plastics in Electrical Insulation*, Society of Plastics Engineers, 1964.
4. H. A. Tish, "Mechanical Properties of Rigid Plastics at Low Temperature," *Modern Plastics* **33(11)** 1955.
5. A. H. Landrock, "Properties of Plastics and Related Materials at Cryogenic Temperatures," *Plastics Rept. No. 20*, Picatinny Arsenal, 1965.
6. P. E. Kueser et al., "Electric Conductor and Electrical Insulation Materials, Topical Report," *NASA Rept. CR-54092*, 1964.
7. H. Raech and J. M. Kreinik, "Prepreg Materials for High Performance Dielectric Applications," *20th Annual Technical Conference, Reinforced Plastics Division*, Society of Plastics Industry, 1965.
8. W. F. Herberg and E. C. Elliott, "Reinforced Silicone Plastics Retain Physical and Electrical Properties with Little Regard to Temperature or Frequency," *Plast. Des. Processing* **6(1)** 1966.

---

* This specification includes alkyd, allyl, melamine, phenolic, and polyester molded plastic and molded plastic parts.

9. R. B. Seymour, "Reinforced Plastics Properties and Applications," ASM International, Materials Park, OH, 1991.

10. J. T. Gotro and B. K. Appelt, "Characterization of a Bismaleimide Triazine Resin for Multilayer Printed Wiring Boards," *IBM Res. Develop.* **32(5)** Sept. 1988.

11. D. J. Arthur, "Electrical and Mechanical Characteristics of Low Dielectric Constant Printed Wiring Boards," *Pub. 0179-027-1.0,* Rogers Corp, Chandler, AZ.

12. "Industrial Laminated Thermosetting Products," *Std. Pub. No. LI 1,* National Electrical Manufacturers Association, 1989.

13. *Modern Plastics Encyclopedia, Modern Plastics,* McGraw-Hill Inc., NY, 1990.

14. R. W. Meyer, *Handbook of Pultrusion Technology,* Chapman and Hall, NY, 1985.

15. K. Ashbee, *Fundamental Principles of Fiber Reinforced Composites,* Technomic Publishing Co., Lancaster, PA, 1989.

16. F. T. Traceski, *Specifications and Standards for Plastics and Composites,* ASM International, 1990.

17. D. S. Cordova and D. S. Donnelly, *Spectra Extended Chain Polyethylene Fibers,* Allied-Signal Technologies, Petersburg, VA.

18. "TL-01 Laminate Board," Bulletin No. L 100.0, Teijin America Inc., New York, NY.

19. T. Senese, "Trends in Electronic Substrate Technology," *Elec. Packaging & Production* 30(9), Sept. 1990.

20. C. Guiles, "Mechanical & Electrical Properties of Laminates for High Performance Printed Wiring Boards," *Proceedings 5th International SAMPE Electronics Conference,* June 1992.

21. V. McConnel, "Tough Promises from Cyanate Esters," *Adv. Composites,* June 1992.

22. M. M. Zurenda, "Bare Materials: Meeting the Challenge of High Density Packaging," *Elec. Packaging and Production,* 33(5), May 1993.

CHAPTER 3
ADVANCED COMPOSITE MATERIALS AND PROCESSES

S. T. Peters
Process Research, Materials Consultant
Mountain View, California

3.1 INTRODUCTION

Modern structural composites, frequently referred to as advanced composites, are a blend of two or more components, one of which is made up of stiff, long fibers; and the other, for polymeric composites, a resinous binder or *matrix* that holds the fibers in place. The fibers are strong and stiff relative to the matrix and are generally orthotropic (having different properties in two different directions). The fiber, for advanced structural composites, is long, with length-to-diameter ratios of over 100. The fiber's strength and stiffness are much greater, perhaps several times more than the matrix material. When the fiber and the matrix are joined to form a composite, they both retain their individual identities and both directly influence the composite's final properties. The resulting composite is composed of layers (laminae) of the fibers and matrix stacked to achieve the desired properties in one or more directions.

Modern composite materials evolved from the simplest mixtures of two or more materials to gain a property that was not there before. The Bible mentions the combining of straw with mud to make bricks. The three key historical steps leading to modern composites were the commercial availability of fiberglass filaments in 1935, the development of strong aramid and carbon fibers in the late 1960s and early 1970s, and the promulgation of analytical methods for structures made from these fibers.[1–3]

Large-scale use of fiberglass-reinforced products began in 1949 at almost 4,275,000 kg (10 million lbs) and rose almost tenfold in 1979 to 3,600,000,000 kg (8 billion lbs). The gains in consumption were primarily due to the needs for nonconductive electrical components, noncorroding and noncorrosive storage containers and transfer lines, and sporting goods. Although no industry component could offset the effect of the reduced U.S. Defense budget cuts on the composite industry, the annual growth rate for advanced composites in the U.S. has been predicted to be 7.1 percent and 6.4 percent for the rest of the world. These growth figures are primarily driven by the expected needs of the automotive industry.[4] The technologies for matrices and for fabrication of useful structures with fiberglass reinforcements were directly useful for the stronger fibers that were commercialized in the two decades after 1970. Along with the new fibers, new matrices were developed, and new fabrication techniques were commercialized. These developments were possible in part because military aircraft designers were quick to realize that these materials could increase the speed, maneuverability, or range of an aircraft by lowering weight of substructures. The first military aircraft to use composites was the F-111, in its boron/epoxy-reinforced horizontal stabilizers. Applications for boron and other fibers have continued to increase steadily in aircraft up to the Bell-Boeing V-22 (Fig. 3.1), which is in production with approximately 60% of its weight composed of advanced composites.

Designers of aircraft structures were quick to capitalize on the high strength-to-weight or modulus-to-weight ratios of composites. Table 3.1 shows that there are other advantages. These advantages translate not only into aircraft but also into everyday activities such as longer drives with a graphite-shafted golf club (because more of the mass is concentrated at the clubhead) or less fatigue and pain because a graphite composite tennis racquet has inherent damping. Generally, the advantages shown in Table 3.1 accrue for any fiber/composite combination, and the disadvantages are more obvious with some. These advantages have now resulted in composite applications far outside the aircraft industry with many more reasons for use as shown in Table 3.2. Proper design and material selection can circumvent many of the disadvantages.

FIGURE 3.1 The Bell-Boeing V-22.

TABLE 3.1   Reasons for Using Composites

Reason for use	Material selected	Application/driver
Lighter, stiffer, stronger	Boron, all carbon/graphites, some aramid	Military aircraft, better performance commercial aircraft, operating costs
Controlled or zero thermal expansion	Very high modulus carbon/ graphite	Spacecraft with high positional accuracy requirements for optical sensors
Environmental resistance	Fiberglass, vinyl esters, Bisphenol-A fumarates, chlorendic resins	Tanks and piping, corrosion resistance to industrial chemicals, crude oil, gasoline at elevated temperatures
Lower inertia, faster startups, less deflection	High-strength carbon/ graphite, epoxy	Industrial rolls, for paper, films
Lightweight, damage tolerance	High-strength carbon/ graphite, fiberglass, (hybrids), epoxy	CNG tanks for "green" cars, trucks and buses to reduce environmental pollution
More reproducible complex surfaces	High-strength or high-modulus carbon/graphite/ epoxy	High-speed aircraft. Metal skins cannot be formed accurately
Less pain and fatigue	Carbon/graphite/epoxy	Tennis, squash, and racquetball racquets. Metallic racquets are no longer available
Reduces logging in "old growth" forests	Aramid, carbon/graphite	Laminated "new" growth wooden support beams with high modulus fibers incorporated
Reduces need for intermediate support and resists constant 100% humidity atmosphere	High-strength carbon/ graphite-epoxy	Cooling tower driveshafts
Tailorability of bending and twisting response	Carbon/graphite-epoxy	Golf shafts, fishing rods
Transparency to radiation	Carbon/graphite-epoxy	X-ray tables
Crashworthiness	Carbon/graphite-epoxy	Racing cars
Higher natural frequency, lighter	Carbon/graphite-epoxy	Automotive and industrial driveshafts
Water resistance	Fiberglass (woven fabric), polyester, or isopolyester	Commercial boats
Ease of field application	Carbon/graphite, fiberglass-epoxy, tape, and fabric	Freeway support structure repair after earthquake

TABLE 3.2 Advantages/Disadvantages of Advanced Composites

Advantages	Disadvantages
Weight reduction (High strength or stiffness-to-weight ratio)	Cost of raw materials and fabrication
Tailorable properties Can tailor strength or stiffness to be in the load direction	Transverse properties may be weak
Redundant load paths (fiber to fiber)	Matrix is weak, low toughness
Longer life (no corrosion)	Matrix subject to environmental degradation
Lower manufacturing costs because of less part count	Difficult to attach
Inherent damping	Analyses for physical properties and mechanical properties are difficult, and analysis for damping efficiency has not reached a consensus
Increased (or decreased) thermal or electrical conductivity	Nondestructive testing is tedious
Better fatigue life	Acceptable methods for evaluation of residual properties have not reached a consensus

3.2 MATERIAL SYSTEMS

An advanced composite laminate can be tailored so that the directional dependence of strength and stiffness matches that of the loading environment. To do that, layers of unidirectional material called laminae are oriented to satisfy the loading requirements. These laminae contain fibers and a matrix. Because of the use of directional laminae, the tensile, flexural, and torsional shear properties of a structure can be disassociated from each other to some extent, and a golf shaft, for example, can be changed in torsional stiffness without changing the flexural or tensile stiffness. This allows for almost infinite variations in the shafts to accommodate individual needs. It also allows for altering the stiffness of a forward-swept aircraft wing to respond to the incoming loads (Fig. 3.2). This is not an option with isotropic metal. Fibers can be of the same material within a lamina or several fibers mixed (hybrid). The common commercially available fibers are as follows:

- Fiberglass
- Graphite
- Aramid
- Polyethylene
- Boron
- Silicon carbide
- Silicon nitride, silica, alumina, alumina silica

Fiberglass is a product of silica sand, limestone, boric acid, and other ingredients that are dry-mixed, melted (at approximately 1260°C), and then drawn into fibers.

FIGURE 3.2   NASA-Grumman forward-swept aircraft wing.

The advantages of fiberglass are its high tensile strength and strain to failure (Table 3.3), but heat and fire resistance, chemical resistance, moisture resistance, and thermal and electrical properties are also cited as reasons for its use. It is by far the most widely used fiber, primarily because of its low cost, but its mechanical properties are not comparable with other structural fibers.

Graphite fibers have demonstrated the widest variety of strengths and moduli and have the greatest number of suppliers (Table 3.4). The fibers begin as an organic fiber, rayon, polyacrylonitrile, or pitch called the precursor. The precursor is then stretched, oxidized, carbonized, and graphitized. There are many ways to produce these fibers,[6] but the relative amount of exposure at temperatures from 2500–3000°C results in greater or less graphitization of the fiber. Higher degrees of

TABLE 3.3   Glass Fibers in Order of Ascending Modulus Normalized to 100 Percent Fiber Volume (Vendor Data)

Type	Nominal tensile modulus GPa (psi × 10^6) strand	Nominal tensile strength MPa (psi × 10^6) strand	Ultimate strain (%)	Fiber density kg/m^3 (lb/in^3)	Typical suppliers
E	72.5 (10.5)	3447 (500)	4.8	2600 (0.093)	Pittsburgh Plate Glass, Manville Co., Owens Corning Fiberglass
R	86.2 (12.5)	4400 (638)	5.1	2530 (0.089)	Vetrotex St. Gobain, Certainteed
Te	84.3 (12.2)	4660 (675)	5.5	2530 (0.089)	Nittobo
S-2, S	88 (12.6)	4600 (665)	5.2	2490 (0.090)	Owens Corning

From Ref. 5.

TABLE 3.4 Carbon and Graphite Fibers in Order of Ascending Modulus Normalized to 100 Percent Fiber Volume (Vendor Data)

Class of fiber	Nominal tensile modulus GPa (psi × 10⁶) strand	Nominal tensile strength MPa (psi × 10³) strand	Ultimate strain (%)	Fiber density kg/m³ (lb/in³)	Suppliers/ typical products
High tensile strength	227 (33)	3996 (580)	1.60	1750 (0.063)	Amoco, T-300; Hercules, AS-4
High strain	234 (34)	4100 (594)	1.95	1790 (0.064)	Courtaulds Grafil, 33-600
Intermediate modulus	275 (40)	5133 (745)	1.75	1740 (0.062)	Hercules, IM-6; Amoco, T-40; Courtaulds Grafil, 42-500
Very high strength	289 (42)	7027 (1020)	1.82	1820 (0.066)	Toray, T-1000
High modulus	358	2482 (360)	0.70	1810 (0.065)	Amoco, T-50; Celanese, G-50
High modulus (pitch)	379 (55)	2068 (300)	0.50	2000 (0.072)	Amoco, P-55
Ultrahigh modulus (pan)	517 (75)	1816 (270)	0.36	1960 (0.070)	Celanese, GY-70
Ultrahigh modulus (pitch)	517 (75)	2068 (300)	0.40	2000 (0.072)	Amoco, P-75
Extremely high modulus (pitch)	689 (100)	2240 (325)	0.31	2150 (0.077)	Amoco, P-100

From Ref. 5.

graphitization usually result in a stiffer fiber (higher modulus) with greater electrical and thermal conductivities. Pitch fibers above 689 GPa (10^8 psi) tensile modulus have thermal conductivity greater than copper and have been used in spacecraft for thermal control applications. Figures 3.3 and 3.4[7] show the effect of processing temperature on tensile strength, tensile modulus, and thermal and electrical conductivity.

The organic fiber Kevlar 49,® an aramid, essentially revolutionized pressure-vessel technology because of its great tensile strength and consistency coupled with low density, resulting in much more weight-effective designs for rocket motors. Aramid composites are still widely used for pressure vessels but have been largely supplanted by the very high strength graphite fibers. Aramid composites have relatively poor shear and compression properties; careful design is required for their use in structural applications that involve bending. The polyethylene fibers have these property drawbacks as well but also suffer from low melting temperatures, which

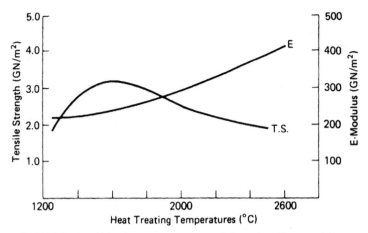

FIGURE 3.3 Graphitization temperature vs. modulus and tensile strength for carbon graphite fibers (Ref. 6).

limits their use to composites that cure or operate below 149°C (300°F). Moreover, they are susceptible to degradation by ultraviolet light exposure. Both of these types of fibers have wide usage in personal protective armor. In spite of the drawbacks, both of these fibers are enjoying strong worldwide growth. Table 3.5 shows their properties.

Boron fibers, the first fibers to be used on production aircraft, are produced as individual monofilaments upon a tungsten or carbon substrate by pyrolylic reduction of boron trichloride (BCl_3) in a sealed glass chamber (Fig. 3.5). The relatively large cross-section fiber is used today primarily in composites that are processed at temperatures that would attack graphite fibers. The graphite core is protected by the unreactive boron. Boron fibers have been used extensively in metal matrix

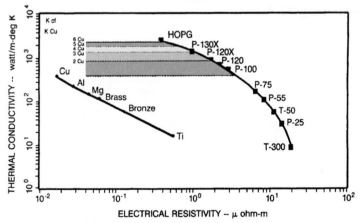

FIGURE 3.4 Electrical and thermal conductivity of carbon fibers and metals (Ref. 7).

TABLE 3.5   Organic Fibers in Order of Ascending Modulus Normalized to 100 Percent Fiber Volume (Vendor Data)

Type	Nominal tensile modulus GPa (psi × 10⁶) strand	Nominal tensile strength MPa (psi × 10³) strand	Ultimate strain (%)	Fiber density kg/m³ (lb/in³)	Suppliers/ product
Aramid (medium modulus)	62 (9.0) 80 (11.6) 70 (10.1)	3617 (525) 3150 (457) 3000 (440)	4.0 3.3 4.4	1440 (0.052) 1440 1390	DuPont, Kevlar 29; Enka, Twaron; Teijin, Technora
Oriented polyethylene	117 (17)	2585 (375)	3.5	968 (0.035)	Allied Fibers, Spectra 900
Aramid (intermediate modulus)	121 (18) 121 (18)	3792 (550) 3150 (457)	2.9 2.0	1440 (0.052) 1450	DuPont, Kevlar 49; Enka, Twaron HM
Oriented polyethylene	172 (25)	3274 (471)	2.7	968 (0.035)	Allied Fibers, Spectra 1000
Aramid (high modulus)	186 (27)	3445 (500)	1.8	1440 (0.052)	DuPont, Kevlar 149

From Ref. 5.

composites and in composites requiring a high confidence in compressive properties. The other fibers shown (Table 3.6) have varying uses, and most are still in development. Silicon carbide continuous fiber is produced in a CVD (chemical vapor deposition) process similar to boron and has many mechanical properties identical to boron. The other fibers show promise in metal matrix composites as high-temperature polymeric ablative reinforcements, in ceramic-ceramic composites, or in microwave-transparent structures (radomes or microwave printed wiring boards).

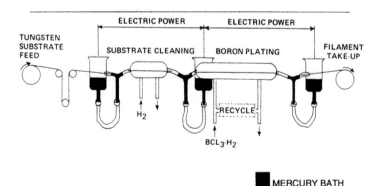

FIGURE 3.5   Production of boron fiber (from Ref. 8).

TABLE 3.6 Boron and Ceramic Fibers, Normalized to 100 Percent Fiber Volume (Vendor Data)

Type	Nominal tensile modulus GPa (psi × 10⁶)	Nominal tensile strength MPa (psi × 10³)	Fiber density kg/m³ (lb/in³)	Suppliers
Boron	400 (58)	3520 (510)	2.55–3.30 (0.093)	Textron, Huber, Nippon, Tokai
Silicon carbide	425 (62)	611	3.56 (0.125)	Textron, Dow Corning, Nippon Carbon
Silicon nitride	300	2500	2.5	Tonen
Silica	66		1.80–2.50	Enka, Huber
Alumina	345 (50)	1380 (200)	3.71 (0.134)	DuPont, FP
Alumina boria silica	(27)	300	2.71 (0.098)	3MS, Nexel 312

3.2.1 Matrix Systems

If the continuous fibers are combined with a suitable matrix, unidirectional composite properties such as those shown in Table 3.7, result.

The functions and requirements of the matrix are to:

- Keep the fibers in place in the structure
- Help to distribute or transfer load
- Protect the filaments, both in the structure and before and during fabrication
- Control the electrical and chemical properties of the composite
- Carry interlaminar shear

In terms of structural needs, the matrix should:

- Minimize moisture absorption
- Wet and bond to fiber
- Flow to penetrate the fiber bundles completely and eliminate voids during the compacting/curing process
- Be elastic to transfer load to fibers
- Have strength at elevated temperature (depending on application)
- Have low temperature capability (depending on application)
- Have excellent chemical resistance (depending on application)
- Have low shrinkage
- Have a low coefficient of thermal expansion
- Have reasonable strength, modulus, and elongation (elongation should be greater than fiber)
- Be easily processable into the final composite shape
- Have dimensional stability or be able to maintain its shape

TABLE 3.7 Properties of Typical Unidirectional Graphite/Epoxy Composites (Fiber Volume Fraction, $V_f = 0.60$) Elastic Constants, Strengths, Strains, and Physical Properties

	High strength	High modulus
Elastic constants, GPa (psi $\times 10^6$)		
Longitudinal modulus, E_L	145 (21)	220 (32)
Transverse modulus, E_T	10 (1.5)	6.9 (1.0)
Shear modulus, G_{LT}	4.8 (0.7)	4.8 (0.7)
Poisson's ratio (dimensionless), ν_{LT}	0.25	0.25
Strength properties, MPa (10^3 psi)		
Longitudinal tension, F_L^{tu}	1240 (180)	760 (110)
Transverse tension, F_T^{tu}	41 (6)	28 (4)
Longitudinal compression, F_L^{cu}	1240 (180)	690 (100)
Transverse compression, F_T^{cu}	170 (25)	170 (25)
Inplane shear, F_{LT}^{su}	80 (12)	70 (10)
Interlaminar shear, F^{Lsu}	90 (13)	70 (10)
Ultimate strains, %		
Longitudinal tension, ϵ_L^{tu}	0.9	0.3
Transverse tension, ϵ_T^{tu}	0.4	0.4
Longitudinal compression, ϵ_L^{cu}	0.9	0.3
Transverse compression, ϵ_T^{cu}	1.6	2.8
Inplane shear	2.0	—
Physical properties		
Specific gravity	1.6	1.7
Density (lb/in^3)	0.056	0.058
Longitudinal CTE, 10^{-6} in/in°F (10^{-6} m/m°C)	−0.2	−0.3
Transverse CTE, 10^{-6} m/m°C (10^{-6} in/in°F)	32 (18)	32 (18)

From Ref. 9.

There are many matrix choices available, and each type has an impact on the processing technique, physical and mechanical properties, and environmental resistance of the finished composite. The first key choice is between thermoplastic and thermoset composite matrix. Thermoplastic matrices (polymers that can be repeatedly softened by increase in temperature and hardened by decrease in temperature) have been developed to increase hot-wet use temperature (see Sec. 3.5.6) and the fracture toughness of composites. Although thermoplastic composites are still not in general usage, their properties are well documented because of sponsorship of development programs by the U.S. Air Force. Table 3.8 shows the relative advantages and disadvantages of both thermoplastics and thermoset matrices. Thermoplastic composites have relatively high potential advantage because their large-scale use has yet to be realized. Some special considerations must be made for thermoplastics; they are:

- Because high temperatures (up to 300°C) are required for processing, special autoclaves, processes, ovens, and bagging materials may be needed.
- The fiber finishes used for thermosetting resins are not compatible with thermoplastic matrices, necessitating alternative treatment.
- Thermoplastic composites can have greater or much less solvent resistance than a thermoset material. If the stressed matrix of the composite is not resistant to the solvent, the attack and destruction of the composite may be nearly instantaneous.

TABLE 3.8 Composite Matrix Trade-Offs

Property	Thermoset	Thermoplastic	Notes
Resin cost	Low to medium-high, based on resin requirements	Low to high. Premium thermoplastic prepregs are more than thermoset prepregs	Will decrease for thermoplastics as volume increases
Formulation	Complex	Simple	
Melt viscosity	Very low	High	High melt viscosity interferes with fiber impregnation
Fiber impregnation	Easy	Difficult	
Prepreg tack/drape	Good	None	Simplified by commingled fibers
Prepreg stability	Poor	Good	
Composite voids	Good (low)	Good to excellent	
Processing cycles	Long	Short to long (long processing degrades polymer)	
Fabrication costs	High for aerospace, low for pipes and tanks with glass fibers	Low (potentially); some shapes still cannot be processed economically	
Composite mechanical properties	Fair to good	Good	
Interlaminar fracture toughness	Low	High	
Resistance to fluids/solvents	Good	Poor to excellent (choose matrix well)	Thermoplastics stress craze
Damage tolerance	Poor to excellent	Fair to good	
Resistance to creep	Good	Not known	
Data base	Very large	Small	
Crystallinity problems	None	Possible	Crystallinity affects solvent resistance
Other		Thermoplastics can be reformed to make an interference joint	

A thermosetting matrix, in contrast to the thermoplastic resins, will set at some temperature (room temperature or above) and cannot be reshaped by subsequent heating. In general, thermosetting polymers contain two or more ingredients: a resinous matrix and a curing agent. Solidification of the composite matrix starts either when the resin and curing agent are mixed or when the matrix is heated to cause a reaction between the resin and its curing agent.

The common thermoset matrices for composites include the following:

- Polyester and vinylesters
- Epoxy
- Bismaleimide
- Polyimide
- Cyanate ester and phenolic triazine

Each of the resin systems has some drawbacks that must be accounted for in design and manufacturing plans. Polyester matrices have been in use for the longest period, and they are used in the widest variety and greatest amount of structures. Polyesters are macromolecules that are the result of condensation polymerization between dibasic acids or anhydrides and dihydric alcohols (glycols). The usable polymers may contain up to 50 percent by weight of unsaturated monomers and solvents such as styrene. Polyesters cure via a catalyst (usually a peroxide), resulting in an exothermic reaction. This reaction can be initiated at room temperature. The catalyst reaction results in a decreased sensitivity to the amount of curative or the extent of mixing. This decreased sensitivity disposes the polymer to small-scale, low-technology operations such as boat building. The resulting polymer is nonpolar and very water resistant. Isopolyester resins, which have been described as the most water-resistant polymer of the polyester group, were chosen as the matrix material for the 17 United States Navy minehunters being built by Intermarine USA in Savannah, Georgia (Fig. 3.6). Because of the rapid uncontrolled cure, there is up to 9 percent cure shrinkage that limits their use with the high-strength or high-modulus fibers. Table 3.9[10] shows the properties of several clear castings.

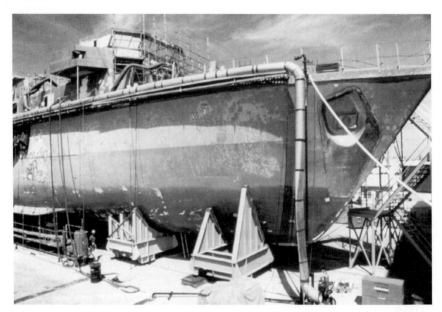

FIGURE 3.6   Glass fiber, isopolyester minehunter being built for the U.S. Navy. (*Courtesy of Amoco Chemical Co.*)

TABLE 3.9 Neat Resin Casting Properties of Polyester-Related Matrices

Material	Barcol hardness	Tensile strength		Tensile modulus		Elongation	Flexural strength		Flexural modulus		Compressive strength		Heat deflection temperature	
		MPa	ksi	10^{-2} Pa	10^{-5} psi	%	MPa	ksi	10^{-2} Pa	10^{-5} psi	MPa	ksi	°C	°F
Orthophthalic	—	55	8	34.5	5.0	2.1	80	12	34.5	5.0	—	—	80	175
Isophthalic	40	75	11	33.8	4.9	3.3	130	19	35.9	5.2	120	17	90	195
BPA fumarate	34	40	6	28.3	4.1	1.4	110	16	33.8	4.9	100	15	130	265
Chlorendic	40	20	3	33.8	4.9	—	120	17	39.3	5.7	100	15	140	285
Vinylester	35	80	12	35.9	5.2	4.0	140	20	37.2	5.4	—	—	100	212

From Ref. 10.

   The most widely used matrices for advanced composites have been the epoxy resins. These resins cost more than polyesters and do not have the high-temperature capability of the bisimalimides or polyimides; but because of the advantages shown in Table 3.10 they are widely used. The epoxy group

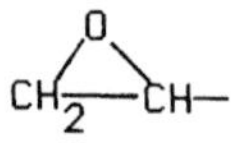

$$CH_2 \overset{O}{-} CH-$$

is the feature that causes the thermosetting reaction and distinguishes epoxy resins from all other matrices. The basic epoxy in use is diglycidyl ether of bisphenol A (DGEBA), a white crystalline solid that looks chemically like:

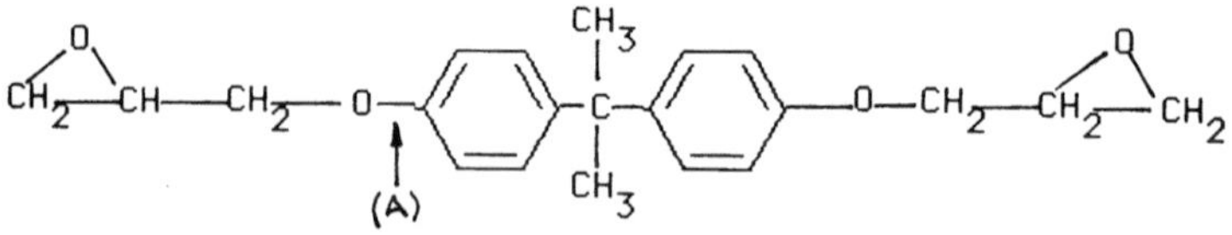

The impure forms of this molecule, if it is liquid and has an epoxy equivalent weight of 175–195 g/eq, could be Shell's Epon 828 or Dow's DER 383. These are widely used liquid epoxy resins. By inserting or subtracting chemical submolecules at (A) such as:

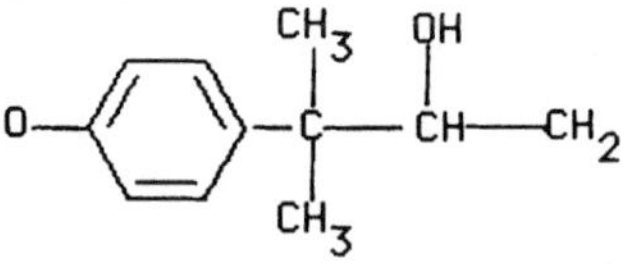

the resin manufacturer can arrive at a wide variety of viscosities, molecular weights, and resultant physical properties. The epoxy resin that has been most frequently used as a base for graphite-fiber, high-performance composites is the solid resin tetraglycidylmethylenedianiline (TGMDA, MY 720, Ciba-Geigy):

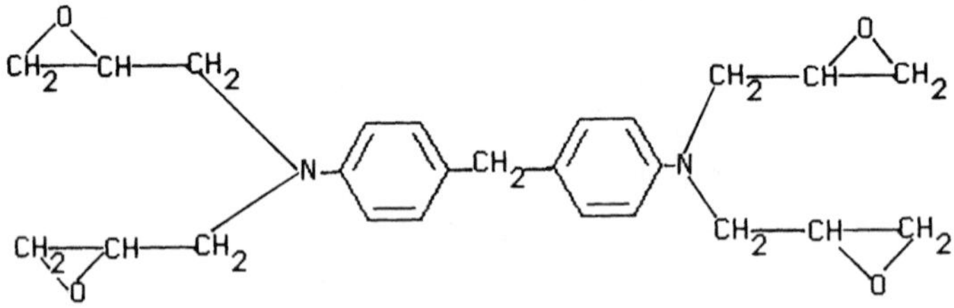

   This solid resin has exceptional stiffness, resulting in increased compressive strength that is widely used in aircraft structural laminates. Virtually all prepreg producers in the United States use a TGMDA resin.[11] The selection of a curing

TABLE 3.10 Epoxy Resin Selection Factors

Advantages
- Adhesion to fibers and to resin
- No by-products formed during cure
- Low shrinkage during cure
- Solvent and chemical resistance
- High or low strength and flexibility
- Resistance to creep and fatigue
- Good electrical properties
- Solid or liquid resins in uncured state
- Wide range of curative options

Disadvantages
- Resins and curatives somewhat toxic in uncured form
- Absorb moisture
    Heat distortion point lowered by moisture absorption
    Change in dimensions and physical properties due to
        moisture absorption
- Limited to about 200°C upper temperature use (dry)
- Difficult to combine toughness and high temperature resistance
- High thermal coefficient of expansion
- High degree of smoke liberation in a fire
- May be sensitive to ultraviolet light degradation
- Slow curing

---

agent or system is usually not an option to the user of the composite unless the process selected for manufacture involves "wet" resins, i.e., hand lay-up, filament winding, resin-transfer molding, or pultrusion. Cure technique selection and the formulation of the system are performed by the prepregger, who also marries the resin and the fiber and advances the cure of the mixture so that it has reasonable handling characteristics and shelf stability. The prepregger has three general choices for an epoxy curing agent: aromatic amines, anhydrides, and catalytic methods of curing the resin component. Table 3.11 shows the effect of these curatives on neat resin properties. Prepreggers aiming their product line towards aircraft uses have attempted to maximize high-temperature, hot-wet properties along with high shear and compressive strengths by utilization of the elevated-temperature-curing aromatic amine curing agent DDS (4,4′ Diaminodiphenyl sulfone), usually with a catalyst, and the high-temperature, stiff epoxy matrix resin TGMDA (tetraglycidylmethylenedianiline). The user can help to optimize the properties of the resin/fiber system by properly specifying all environmental limits and defining the fabrication parameters and strength requirements. Thus, some property like hot-wet strength retention might be traded off for ease of processing if the composite will never experience an environmental temperature greater than 20°C.

There are two resin systems in common use for higher temperatures: bismaleimides and polyimides. New designs for aircraft demand a 177°C (350°F) operating temperature not met by the other common structural resin systems. The primary bismaleimide (BMI) in use is based on the reaction product from methylene dianiline (MDA) and maleic anhydride: bis (4 maleimidophenyl) methane (MDA BMI):

Variations of this polymer with compounded additives to improve impregnation, tack, or drape are now marketed by Rhone-Poulenc, Inc., Technochemie (Shell Chemical Co.), BTL Specialty Resins, and Ciba-Geigy. These resins are then prepregged with a suitable reinforcement to result in the high-temperature mechanical properties shown in Table 3.12. The BMIs are notable because of their easy epoxylike processing that results in a high-temperature laminate. However, the BMI prepreg raw material costs are still very high.

Two newer resin systems have been developed and have found applications in widely diverse areas. The cyanate ester resins, marketed by Ciba-Geigy, have shown

TABLE 3.11 Typical Properties of DGEBA Cured with Various Hardeners

Property hardener	Aliphatic amine	BF³MEA	Anhydride	Aromatic amine
Cure schedule	2 h @ 120°C (250°F)	16 h @ 93°C (200°F) and 2 h @ 150°C (300°F)	4 h @ 150°C (300°F)	2 h @ 150°C (300°F)
Tensile strength at room temperature, MPa (ksi)	42.7 (6.2)	58.6 (8.5)	82.7 (12.0)	80 (11.6)
Tensile modulus at room temperature, GPa (10^6 psi)	2.3 (0.500)	—	3.1 (0.450)	2.7 (0.395)
Elongation, %	1.35	2.3	—	5.7
Flexural strength at room temperature, MPa (ksi)	103 (15.0)	112 (16.3)	131 (19.0)	131 (19.0)
Flexural modulus at room temperature, GPa (10^6 psi)	2.48 (0.360)	—	2.90 (0.420)	2.93 (0.425)
Heat deflection temperature, °C (°F)	91 (195)	128 (260)	109 (230)	175 (350)
Izod impact strength, kJ/m² (ft-lbf/n²)	0.46 (0.22)	0.57 (0.27)	1.5 (0.7)	—

From Ref. 12.

TABLE 3.12 Approximation of Mechanical Properties of BMI Composites

Property	Unreinforced homopolymer	Glass-reinforced homopolymer	Carbon-reinforced homopolymer	Carbon-reinforced homopolymer
Reinforcement, vol%	0	60	60	70
Service temperature, °C (°F)	260 (500)	177–232 (350–450)	177–232 (350–450)	149–204 (300–400)
Flexural strength, MPA (ksi)				
At room temp.	210 (30)	480 (70)	2000 (290)	725 (105)
At 230°C (450°F)	105 (15)	290 (42)	1340 (194)	—
Flexural modulus, GPa (10^6 psi)				
At room temp.	4.8 (0.7)	17.2 (2.5)	126 (18.4)	71 (10.3)
At 230°C (450°F)	3.4 (0.5)	15.1 (2.2)	57 (8.2)	—
Interlaminar shear strength, MPa (ksi)				
At room temp.	—	—	117 (17)	—
At 230°C (450°F)	—	—	59 (8.6)	—
Tensile strength, MPa (ksi)				
At room temp.	97 (14)	—	1725 (250)	570 (83)
At 230°C (450°F)	76 (11)	—	—	—
Tensile modulus, GPa (10^6 psi)				
At room temp.	4.1 (0.6)	—	148 (21.5)	15.9 (2.3)
At 230°C (450°F)	2.8 (0.4)	—	—	—

From Ref. 13.

superior dielectric properties and much lower moisture absorption than any other structural resin for composites. The dielectric properties have enabled their use as adhesives in multilayer microwave printed circuit boards, and the low moisture absorbance has made them the resin of universal choice for structurally stable spacecraft components. The physical properties of cyanate ester resins are contrasted to those of a representative BMI resin in Table 3.13.[15,16]

The PT resins also have superior elevated-temperature properties, along with excellent properties at cryogenic temperatures. Their resistance to proton radiation under cryogenic conditions was a prime cause for their choice for use in the superconducting supercollider, subsequently canceled by the U.S. Congress. The PT resins are available in several viscosities, ranging from a viscous liquid to powder, facilitating their use in applications that use liquid resins such as filament winding or resin-transfer molding.

Polyimides are the highest temperature polymer in general advanced composite use with a long-term upper temperature limit of 232°C (450°F) or 316°C (600°F). Two general types are condensation polyimides that release water during the curing reaction and addition-type polyimides with somewhat easier process requirements. Table 3.14[14] shows several commercial types.

TABLE 3.13 Mechanical and Physical Properties of Cyanate Ester and BMI Resin Castings

Property	Arocy B (Cyanate)	Primaset PT	Matrimid 5292 (BMI)
Tensile strength, Mpa (ksi)	88 (12.7)	41.3 (6,000)	82 (11.9)
Tensile elongation, %	3.2	2.0	2.3
Flexure strength, MPa (ksi)	174 (25.2)	110 (16,000)	167 (24.2)
Young's modulus, GPa (ksi)	Flexure	Flexure	Tensile
25°C	3.17 (.46)	4.0 (.59)	4.28 (.62)
149°C	—		2.42 (.35)
163°C	2.55 (.37)		—
204°C	—		2.00 (.29)
G_{ic}, J/m^2 (in-lb/in^2)	140 (.80)		170 (.97)
HDT, °C dry (wet)	254 (197)		273 (217)
T_g, °C by DMA (TMA)	289 (257)	up to 399, (750)	295 (273)
CTE by TMA ppm/°C (40 to 200°C)	64	50 (28)	63

Source: Refs. 15 and 16.

TABLE 3.14 Commercial Polyimides Used for Structural Composites

	Upper temperature capability	
	°C	°F
Condensation		
Monsanto Skybond 700, 703	316	600
DuPont NR-150B2 (Avimid N)	316	600
LARC TPI	300	572
Avimid K-III	225	432
Ultem	200	400
Addition		
PMR-15 (Reverse Diels-Adler nadic end-capped)	316	600
LARC 160 (Reverse Diels-Adler nadic end-capped)	316	600
Thermid 600 (Acetylene end-capped)	288	550
BMIs (Bismaleimides, maleimide end-capped)	232	450

From Ref. 14.

There are several problems that consistently arise with thermoset matrices and prepregs that do not apply to thermoplastic composite starting materials. Because of these problems, if raw material and processing costs were comparable for the two matrices, the choice would always be thermoplastic composites without regard to the other advantages resulting in the composite. These problems lead to a great increase in quality control efforts that may result in the bulk of final composite structure costs. They are:

Frequent Variation from Batch to Batch
- Effects of small amounts of impurities
- Effects of small changes in chemistry
- Change in matrix component vendor or manufacturing location

Causes of Void Generation

- Premature gelation
- Premature pressure application
- Effects on interlaminar shear and flexural modulus because of water absorption

Change in Processing Characteristics

- Absorbed water in prepreg
- Length of time under refrigeration
- Length of time out before cure
- Loss of solvent in wet systems

There are other resins in general commercial and aerospace use that are not treated here because they are not in wide use with the modern fibers. Thus, consult Refs. 17, 18, and 19 for more details in the uses and properties of phenolics, amino plastics, and allyls.

The following general notes are more or less applicable to all thermoset matrices:

- The higher the service temperature limitation, the less strain to failure.
- The greater the service temperature, the more difficult the processing that may be due to:
  1. Volatiles in matrix
  2. Higher melt viscosity
  3. Longer heating curing cycles
- The greater the service temperature or the greater the curing temperature, the greater the chance for development of color in the matrix.
- Higher service temperatures and higher curing temperatures may result in better flame resistance (although this is not evident for epoxies with curing temperatures between 250°F and 350°F).

3.2.2 Fiber Matrix Systems

The end user sees a composite structure. Someone else, probably a prepregger, combined the fiber and the resin system, and someone else caused the cure and compaction to result in a laminated structure. A schematic of the steps is shown in Fig. 3.7. In many cases, the end user of the structure has fabricated the composite from prepreg. The three types of continuous fibers, roving, tape, and woven fabric available as prepregs (roving, tape, and woven fabric) give the end user many options in terms of design and manufacture of a composite structure. Although the use of dry fibers and impregnation at the work (i.e., filament winding, pultrusion, or hand layup) is very advantageous in terms of costs, there are many advantages to the use of prepregs as shown in Table 3.15, particularly for the manufacture of modern composites. In general, fabricators skilled in manufacturing from prepreg will not care to use wet processes.

The prepreg process for thermoset matrices can be accomplished by feeding the fiber continuous tape, woven fabric, or roving through a resin-rich solvent solution and then removing the solvent by hot tower drying. The excess resin is removed via a doctor blade or metering rolls, and then the product is staged to the cold-stable prepreg form (B stage). The newer technique, the hot melt procedure for prepregs, is gradually replacing the solvent method because of environmental concerns. A film

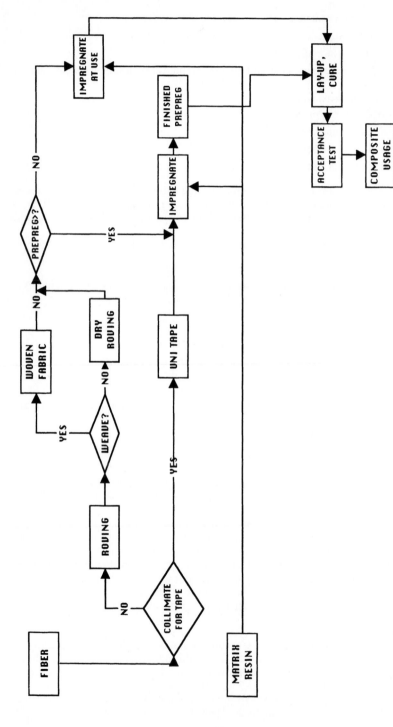

FIGURE 3.7 The manufacturing steps in composite structure fabrication.

TABLE 3.15   Advantages of Prepregs over Wet Impregnation

- Prepregs reduce the handling damage to dry fibers.
- Prepregs improve laminate properties by better dispersion of short fibers.
- Prepregs allow the use of hard-to-mix or proprietary resin systems.
- Prepregs allow more consistency because there is a chance for inspection before use.
- Heat curing provides more time for the proper laydown of fibers and for the resin to move and degas before cure.
- Increased curing pressure reduces voids and improves fiber wetting.
- Most prepregs have been optimized as individual systems to improve processing.

---

of resin that has been cast hot onto release paper is fed, along with the reinforcement, through a series of heaters and rollers to force the resin into the reinforcement. Two layers of resin are commonly used so that a resin film is on both sides of the reinforcement; one of the release papers is removed, and the prepreg is then trimmed, rolled, and frozen. The two types of prepregging techniques, solvent and film, are shown in Figs. 3.8 and 3.9.[20] The solvent technique has been largely replaced for advanced fibers because of environmental pollution concerns and a need to exert better control over the amount of resin on the fiber.

Unidirectional Ply Properties.   The manufacturer of the prepreg reports an areal weight for the prepreg and a resin percentage, by weight. Since fiber volume is used to relate the properties of the manufactured composites, the following equations can be used to convert between weight fraction and fiber volume.

$$W_f = \frac{w_f}{w_c} = \frac{\rho_f V_f}{\rho_c V_f} = \frac{\rho_f}{\rho_c} V_f \tag{3.1}$$

$$V_f = \frac{\rho_c}{\rho_f} W_f = 1 - V_m \tag{3.2}$$

where W_f = weight fraction of fiber
   w_f = weight of fiber
   w_c = weight of composite
   ρ_f = density of fiber
   ρ_c = density of composite
   V_f = volume fraction of fiber
   V_m = volume fraction of matrix

A percentage fiber that is easily achievable and repeatable in a composite and convenient for reporting mechanical and physical properties for several fibers is 60 percent. The properties of unidirectional fiber laminates are shown in Tables 3.7, 3.16, 3.17, and 3.18. These values are for individual lamina or for a unidirectional composite, and they represent the theoretical maximum (for that fiber volume) for longitudinal in-plane properties. Transverse, shear, and compression properties will show maximums at different fiber volumes and for different fibers, depending on how the matrix and fiber interact. These properties are not reflected in strand data. These values may also be used to calculate the properties of a laminate that has fibers oriented in several directions. Using the techniques shown in 3.2.5, the methods of description for ply orientation must be introduced.

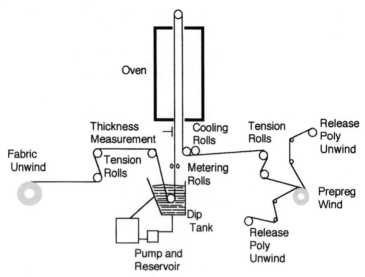

FIGURE 3.8   Schematic of the typical solution prepregging process.

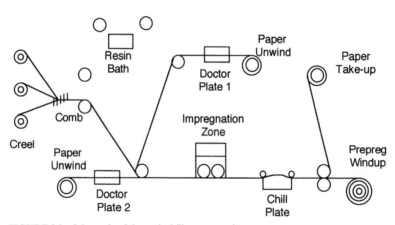

FIGURE 3.9   Schematic of the typical film prepregging process.

3.2.3   Ply Orientations, Symmetry, and Balance

Ply Orientations.   One of the advantages of using a modern composite is the potential to orient the fibers to respond the load requirements. This means that the composite designer must show the material, the fiber orientations in each ply, and how the plies are arranged (ply stackup). A shorthand code for ply fiber orientations has been adapted for use in layouts and studies.

TABLE 3.16 Properties of Unidirectional Glass/Epoxy Composites (Fiber Volume Fraction, $V_f = 0.60$) Elastic Constants, Strengths, Strains, and Physical Properties

	E glass	S glass
Elastic constants, GPa (10^6 psi)		
Longitudinal modulus, E_L	45 (6.5)	55 (8.0)
Transverse modulus, E_T	12 (1.8)	16 (2.3)
Shear modulus, G_{LT}	5.5 (0.8)	7.6 (1.1)
Poisson's ratio (dimensionless), ν_{LT}	0.19	0.28
Strength properties, MPa (10^3 psi)		
Longitudinal tension, F_L^{tu}	1020 (150)	1620 (230)
Transverse tension, F_T^{tu}	40 (7)	40 (7)
Longitudinal compression, F_L^{cu}	620 (90)	690 (100)
Transverse compression, F_T^{cu}	140 (20)	140 (20)
Inplane shear, F_{LT}^{su}	60 (9)	60 (9)
Interlaminar shear, F^{Lsu}	60 (9)	80 (12)
Ultimate strains, %		
Longitudinal tension, ϵ_L^{tu}	2.3	2.9
Transverse tension, ϵ_T^{tu}	0.4	0.3
Longitudinal compression, ϵ_L^{cu}	1.4	1.3
Transverse compression, ϵ_T^{cu}	1.1	1.9
Inplane shear	—	3.2
Physical properties		
Specific gravity	2.1	2.0
Density (lb/in^3)	0.075	0.72
Longitudinal CTE, 10^{-6} m/m°C (10^{-6} in/in°F)	3.7 (6.6)	3.5 (6.3)
Transverse CTE, 10^{-6} m/m°C (10^{-6} in/in°F)	30 (17)	32 (18)

From Ref. 9.

Figure 3.10 illustrates the ply lay-up descriptions. Each ply (lamina) is shown by a number representing the direction of the fibers in degrees, with respect to a reference (x) axis. Zero-degree fibers of both tape and fabric are normally aligned with the largest axial load (axis). See Fig. 3.10a.

Individual adjacent plies are separated by a slash in the code if their angles are different (Fig. 3.10b).

The plies are listed in sequence, from one laminate face to the other, starting with the ply first on the tool and indicated by the code arrow with brackets indicating the beginning and end of the code.

Adjacent plies of the same angle of orientation are shown by a numerical subscript (Fig. 3.10c).

When tape plies are oriented at angles equal in magnitude but opposite in sign, (+) and (−) are used. Each (+) or (−) sign represents one ply. A numerical subscript is used only when there are repeating angles of the same sign. Positive and negative angles should be consistent with the coordinate system chosen. An orientation shown as positive in one right-handed coordinate system may be negative in another. If the y and z axis directions are reversed, the ± 45 plies are reversed (Fig. 3.10d).

TABLE 3.17 Properties of Unidirectional Aramid/Epoxy Composite Kevlar 49 (Fiber Volume Fraction, $V_f = 0.60$) Elastic Constants, Strengths, Strains, and Physical Properties

Elastic constants, GPa (10^6 psi)	
Longitudinal modulus, E_L	76 (11)
Transverse modulus, E_T	5.5 (0.8)
Shear modulus, G_{LT}	2.1 (0.3)
Poisson's ratio (dimensionless), ν_{LT}	0.34
Strength properties, MPa (10^3 psi)	
Longitudinal tension, F_L^{tu}	1380 (200)
Transverse tension, F_T^{tu}	30 (4.3)
Longitudinal compression, F_L^{cu}	280 (40)
Transverse compression, F_T^{cu}	140 (20)
Inplane shear, F_{LT}^{su}	60 (9)
Interlaminar shear, F^{Lsu}	60 (9)
Ultimate strains, %	
Longitudinal tension, ϵ_L^{tu}	1.8
Transverse tension, ϵ_T^{tu}	0.5
Longitudinal compression, ϵ_L^{cu}	2.0
Transverse compression, ϵ_T^{cu}	2.5
Inplane shear	—
Physical properties	
Specific gravity	1.4
Density (lb/in³)	0.050
Longitudinal CTE, 10^{-6} m/m°C	−4
(10^{-6} in/in°F)	−2.2
Transverse CTE, 10^{-6} m/m°C (10^{-6} in/in°F)	70
	40

From Ref. 9.

Symmetric laminates with an even number of plies are listed in sequence, starting at one face and stopping at the midpoint. A subscript S following the bracket indicates only one half of the code is shown (Fig. 3.10e).

Symmetric laminates with an odd number of plies are coded as a symmetric laminate except that the center ply, listed last, is overlined to indicate that half of it lies on either side of the plane of symmetry (Figs. 3.10f–3.10h).

Symmetry. The geometric midplane is the reference surface for determining if a laminate is symmetrical. In general, to reduce out-of-plane strains, coupled bending and stretching of the laminate, and complexity of analysis, symmetric laminates should be used. However, some composite structures (e.g., filament-wound pressure vessels) can achieve geometric symmetry so that symmetry through a single laminate wall is not necessary if it constrains manufacture. To construct a midplane symmetric laminate, for each layer above the midplane there must exist an identical layer (same thickness, material properties, and angular orientation) below the midplane (see Fig. 3.10e).

Balance. All laminates should be balanced to achieve inplane orthotropic behavior. To achieve balance, for every layer centered at some positive angle $+\theta$ there

TABLE 3.18 Properties of Typical Unidirectional Boron/
Epoxy Composite Boron (Fiber Volume Fraction, $V_f = 0.50$)
Elastic Constants and Strengths

Elastic constants, GPa (10^6 psi)	
Longitudinal modulus, E_L	207 (30)
Transverse modulus, E_T	19 (2.7)
Shear modulus, G_{LT}	4.8 (0.7)
Poisson's ratio (dimensionless), ν_{LT}	0.21
Strength properties, MPa (10^3 psi)	
Longitudinal tension, F_L^{tu}	1320 (192)
Transverse tension, F_T^{tu}	72 (10.4)
Longitudinal compression, F_L^{cu}	2430 (350)
Transverse compression, F_T^{cu}	276 (40)
Inplane shear, F_{LT}^{su}	105 (15)
Interlaminar shear, F^{Lsu}	90 (13)
Ultimate strains, %	
Longitudinal tension, ϵ_L^{tu}	0.6
Transverse tension, ϵ_T^{tu}	0.4
Longitudinal compression, ϵ_L^{cu}	—
Transverse compression, ϵ_T^{cu}	—
Inplane shear	—
Physical properties	
Specific gravity	2.0
Density (lb/in^3)	0.072
Longitudinal CTE, 10^{-6} m/m°C (10^{-6} in/in°F)	4.1 (2.3)
Transverse CTE, 10^{-6} m/m°C (10^{-6} in/in°F)	19 (11)

From Ref. 9.

must exist an identical layer oriented at $-\theta$ with the same thickness and material properties. If the laminate contains only 0° and/or 90° layers, it satisfies the requirements for balance. Laminates may be midplane symmetric but not balanced and vice versa. Fig. 3.10e is symmetric and balanced, whereas Fig. 3.10g is balanced but unsymmetric.

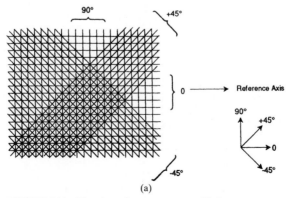

FIGURE 3.10 Ply orientations symmetry and balance.

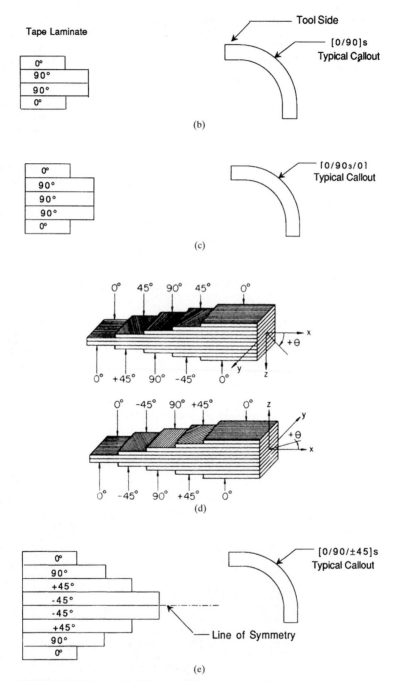

FIGURE 3.10 *(Continued)* Ply orientations symmetry and balance.

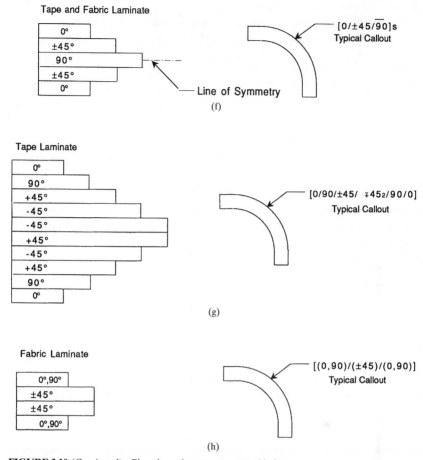

FIGURE 3.10 *(Continued)* Ply orientations symmetry and balance.

3.2.4 Quasi-Isotropic Laminate

The goal of composite design is to achieve the lightest, most efficient structure by aligning most of the fibers in the direction of the load. Many times there is a need, however, to produce a composite that has some isotropic properties similar to metal because of multiple or undefined load paths. A quasi-isotropic laminate layup accomplishes this for the x and y planes only; the z plane through the laminate thickness is quite different and lower. Most laminates produced for aircraft applications have been, with few exceptions, quasi-isotropic. One exception was the X-29 (Fig. 3.2). As designers become more confident and have access to a greater database with fiber-based structures, more applications will evolve. For a quasi-isotropic laminate, the following are requirements:

- It must have three layers or more.
- Individual layers must have identical stiffness matrices and thicknesses.
- The layers must be oriented at equal angles. For example, if the total number of layers is n, the angle between two adjacent layers should be $360°/n$. If a laminate is constructed from identical sets of three or more layers each, the condition on orientation must be satisfied by the layers in each set, for example: $[0°/\pm 60°]_s$ or $[0°/\pm 45°/90]_s$.[3]

3.2.5 Easy Methods of Analysis

There are a number of methods in common use for the analysis of composite laminates. The use of micromechanics, i.e., the application of the properties of the constituents to arrive at the properties of the composite ply, can be used to:

1. Arrive at "back of the envelope" values to determine if a composite is feasible
2. Arrive at values for insertion into computer programs for laminate analysis or finite element analysis
3. Check on the results of computer analysis

The rule of mixtures holds for composites. The micromechanics formula to arrive at the Young's modulus for a given composite is:

$$E_c = V_f E_f + V_m E_m$$

and
$$V_f + V_m = 1 \tag{3.3}$$

$$= V_f E_f + E_m(1 - V_f)$$

where E_c = Composite or ply Young's modulus in tension for fibers oriented in direction of applied load
V = Volume fraction of fiber (f) or matrix (m)
E = Young's modulus of fiber (f) or matrix (m)

But, since the fiber has much higher Young's modulus than the matrix (Table 3.7 vs. Table 3.9), the second part of the equation can be ignored.

$$E_f \gg E_m$$
$$E_c \approx E_f V_f \tag{3.4}$$

This is the basic rule of mixture and represents the highest Young's modulus composite, where all fibers are aligned in the direction of load. The minimum Young's modulus for a reasonable design (other than a preponderance of fibers being orientated transverse to the load direction) is the quasi-isotropic composite and can be approximated by:

$$E_c \approx \frac{3}{8} E_f V_f \tag{3.5}$$

(The quasi-isotropic modulus E of a composite laminate is $3/8E_{11} + 5/8E_{22}$, where E_{11} is the modulus of the lamina in the fiber direction and E_{22} is the transverse modulus of the lamina.[23] The transverse modulus for polymeric-based composites is a small fraction of the longitudinal modulus (see E_T in Table 3.7) and can be ignored for preliminary estimates, resulting in a slightly lower-than-theoretical value for E_c for a quasi-isotropic laminate. This approximate value for quasi-isotropic modulus represents the lower bound of composite modulus. It is useful for comparisons of composite properties to those of metals and to establish if a composite is appropriate for a particular application.)

The following formulas also can be used to obtain important data for unidirectional composites:

$$\text{Density} = \rho_c = V_f \rho_f + V_m \rho_m \tag{3.6}$$

$$\text{Poisson's ratio} = \nu_{12} = \nu_f V_f + \nu_m V_m \tag{3.7}$$

$$\text{Transverse Young's modulus} = E_2 = \frac{E_{2m}(1 + \xi \eta_2 V_f)}{1 - \eta_2 V_f} \tag{3.8}$$

and values for η_2 and ξ can be seen in Ref. 22 and Ref. 3. The matrix is isotropic.

3.2.6 Carpet Plots

The analysis of a multilayered composite, if attempted by hand calculations, is not trivial. Fortunately, there are a significant number of computer programs to perform the matrix multiplications and the transformations.[10, 37] However, the use of carpet plots is still in practice in U.S. industry, and these plots are useful for preliminary analysis. The carpet plot graphically shows the range of properties available with a specific laminate configuration. For example, if the design options include $[\pm 0/90]_s$ laminates, a separate carpet plot for each value of θ would show properties attainable by varying percentage of $\pm \theta$ plies versus 90-degree plies. A sequence of these charts would display attainable properties over a range of θ values. The computer programs described in Ref. 10 can be programmed to produce such charts for arbitrary laminates.

Figure 3.11 shows a sample carpet plot[24] of extensional modulus of elasticity E_x for Kevlar 49/epoxy with [0/±45/90] construction. As expected, the chart shows $E_x = 76$ GPa (11×10^6 psi) with all 0° plies, and $E_x = 5.5$ GPa (0.8×10^6 psi) with all 90° plies. With all 45° plies, an axial modulus is only slightly higher—8 GPa (1.1×10^6 psi)—than the all-90° value predicted for this material. A quasi-isotropic laminate (Sec. 3.2.4) with 25 percent 0°, 50 percent ±45°, and 25 percent 90° produces an intermediate value of $E_x = 29$ GPa (4.2×10^6 psi).

3.3 COMPOSITE FABRICATION TECHNIQUES

The goals of the composite manufacturing process are to:

- achieve a consistent product by controlling fiber thickness, fiber volume, and fiber directions.
- minimize voids.
- reduce internal residual stresses.
- process in the least costly manner.

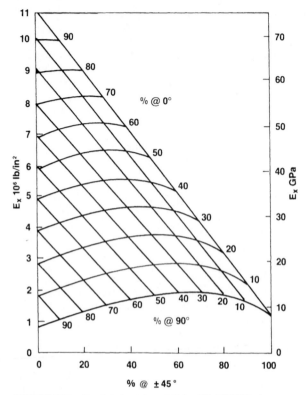

FIGURE 3.11   Predicted axial moduli for [0/±45/90] Kevlar epoxy laminates (from Ref. 24).

The procedures to reach these goals involve iterative processes to select the three key components:

- Composite material and its configuration
- Tooling
- Process

Once material selection has been completed, the first step leading to the acceptable composite structure is the selection of tooling, which is intimately tied to process and material. For all curing techniques, the tool must be:

- Strong and stiff enough to resist the pressure exerted during cure
- Dimensionally stable through repeated heating and cooling cycles
- Light enough to respond reasonably quickly to the changes in cure cycle temperature and to be moved in the shop
- Leakproof so that the vacuum and pressure cycles are consistent

Table 3.19 shows some thermal, physical, and mechanical properties for different tool materials.

TABLE 3.19 Typical Tooling Materials for Advanced Composites

Material	Density kg/m³	Young's modulus in tension E, GPa	Temperature limitation °C	Coeffcient of thermal expansion m/m°C × 10⁻⁶
Silicone rubber	1605		260	81–360
Aluminum alloy	2714	68.9	204	22.5
Steel	7833		*	12.5
Electroless nickel	—	—	*	12.6
Cast iron	7474	165	*	10.8
Fiberglass	1950	20	177†	11.7–13.1
Carbon fiber epoxy (T-300)	1577	66	177†	2.8–3.6
Cast ceramic	3266	10	*	0.81
Monolithic graphite	1522	13	*	2.7–3
Low expansion nickel alloys	8137	144	*	1.4–1.7
Carbon fiber epoxy pitch 55	1720	227	177†	<1

* Above composite.
† Limited by resin matrix.

The tool face is commonly the surface imparted to the outside of the composite and must be smooth, particularly for aerodynamic surfaces. The other surface frequently may be of lower finish quality and is imparted by the disposable or reusable vacuum bag. This surface can be improved by the use of a supplemental metal tool known as a caul plate. (Press curing, resin transfer molding, injection molding, and pultrusion require a fully closed or two-sided mold.) Figure 3.12 shows the basic components of the tooling for vacuum bag or autoclave-processed components. Table 3.20 shows the function of each part of the system. Tooling options have been augmented by the introduction of elastomeric tooling, wherein the thermal expansion of an elastomer provides some or all of the pressure curing cure, or a rubber blanket is used as a reusable vacuum bag. The volumetric expansion of an elastomer can be used to fill a cavity between the uncured composite and an outer mold. The use of elastomeric tooling can provide the means for fabricating complex boxlike structures such as integrally stiffened skins with a cocured substructure in a single curing operation.[26]

Tooling and the configuration of the reinforcement have a great influence on the curing process selected and vice versa. The probable reinforcement configuration that facilitates the completion of the finished composite is shown in Table 3.21. The choice between unidirectional tape and woven fabric has been frequently made on the basis of the greater strength and modulus attainable with the tape particularly in applications which compression strength is important. There are other factors that should be included in the trade, as shown on Table 3.22.

3.3.1 Lay-up Technique

Lay-up techniques along with composite cure control have received the greatest attention for processing. In efforts to reduce labor costs of composite fabrication to which lay-up has traditionally been the largest contributor, mechanically assisted, controlled tape laying and automated integrated manufacturing systems have been developed. Table 3.23 shows some of the considerations for choosing a lay-up tech-

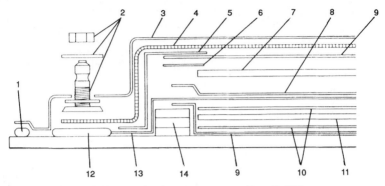

FIGURE 3.12 Typical vacuum bag lay-up components (from Ref. 25).

TABLE 3.20 Functions of Vacuum Bag Components

Component of process*	Functions
1. Bag sealant	Temporarily bonds vacuum bag to tool
2. Vacuum fitting and hardware	Exhausts air, provides convenient connection to vacuum pump
3. Bagging film	Encloses part, allows for vacuum and pressure
4. Open-weave breather mat	Allows air or vacuum transfer to all of part
5. Polyester tape (wide)	Holds other components of bag in place
6. Polyester tape (narrow)	Holds components in place
7. Caul sheet	Imparts desired contour and surface finish to composite
8. Perforated release film	Allows flow of resin or air without adhesion
9. Nonperforated release film	Prevents adhesion of laminate resin to tool surface
10. Peel ply	Imparts a bondable surface to cured laminate
11. Laminate	
12. 1581-style glass breather manifold	Allows transfer of air or vacuum
13. 1581-style glass bleeder ply	Soaks up excess resin
14. Stacked silicone-edge dam	Forces excess resin to flow vertically, increasing fluid pressure

* Numbers refer to Fig. 3.12.

TABLE 3.21 Common Reinforcement Configuration for Manufacturing Process

Reinforcement configuration	Prepreg tape	Prepreg or (dry) tow	Prepreg or (dry) woven or nonwoven fabric	Other woven preforms, chopped fibers
Hand lay-up	X		X, (X)	X
Automatic tape lay-down	X			
Filament winding		X, (X)	X, (X)	X, (X)
Resin transfer molding		(X)	(X)	X
Pultrusion		(X)		X
Fiber placement		X		

TABLE 3.22   Fabric vs. Tape Reinforcement

Tape advantages	Tape disadvantages
Best modulus and strength efficiency	Poor drape on complex shapes
High fiber volume achievable	Cured composite more difficult to machine
Low scrap rate	Lower impact resistance
No discontinuities	Multiple plies required for balance and symmetry
Automated lay-up possible	Higher labor costs for hand lay-up
Available in thin plies	
Lowest cost prepreg form	
Less tendency to trap volatiles	
Fabric advantages	Fabric disadvantages
Better drape for complex shapes	Fiber discontinuities (splices)
Single ply is balanced and may be	Less strength and modulus efficient
essentially symmetric	Lower fiber volume than tape
Can be laid up without resin	More costly than tape
Plies stay in line better during cure	Greater scrap rates
Cured parts easier to machine	Warp and fill properties differ
Better impact resistance	Fabric distortion can cause part warping
Many forms available	

nique. In addition to any cost savings by the use of automated techniques for long production runs, there are two key quality assurance factors that promote the automated techniques. They are the greatly reduced chance that release paper or film could be retained, which would destroy shear and compressive strength if undetected, and the reduced probability of the addition or loss of an angle ply, which would cause warping due to the laminate's lack of symmetry and balance.

The goals in the curing-compacting process are the same for all techniques. Namely:

- Composite with minimal voids and porosity. This is to assure the preservation of matrix-dominated properties and matrix-influenced properties such as compression.

- Full cure of the composite. This results in the most environmentally stable composite as well as maximizing matrix-dominated properties.

- Reproducible, consistent fiber volumes. This assures consistent composite mechanical response.

When these goals are not fully achieved, there are production slowdowns with excessive rework and repair, and there is extensive nondestructive technique testing in addition to the production of expensive scrap composite.

There have obviously been many studies to determine the critical parameters in the process and for autoclave curing of thin, composite aircraft components. There are now recommendations dealing with prepreg properties such as tack and drape, lay-up environment (moisture), laminate thickness, ply orientation and drop-offs, and bleeder amount and placement.[29]

All curing techniques use heat and pressure to cause the matrix to flow and wet out all the fibers before the matrix solidifies. Generally, the percent matrix weight is higher before cure initiation; the matrix flows out of the laminate and takes the excess resin with the potential voids. The matrix exhibits a low-viscosity phase during the cure and then advances in viscosity rapidly after a gel period at an interme-

TABLE 3.23 Considerations in Composite Lay-Up Technique

	Manual	Flat tape	Contoured tape
Orientation accuracy	Least accurate	Automatic Dependent on tape	Somewhat dependent on tape accuracy and computer program
Ply count	Dependent on operator, count mylars	Dependent on operator	Program records
Release film retention	Up to operator	Automatic	Automatic removal
Labor costs	High	86% Improvement	Additional improvement
Machine costs	N/A	Some costs	Approximately $1 million or greater
Production rate	Low (1.5 lb/hr)	10 lb/hr	Approximately same as flat tape
Machine "up" time	N/A	Not a consideration	Complex program and machine make this a consideration
Varying tape widths	Not a concern	Easily changed	Difficulty in changing
Tape lengths	Longer tapes more difficult	Longer is more economical	Longer tape is more economical
Cutting waste	Scrap on cutting	Less scrap	Least scrap due to back and forth lay-down
Compaction pressure	No pressure	Less voids	Least voids
Programming	N/A	N/A	Necessary

diate temperature (Fig. 3.13). Because of the many changes going on in the matrix, determining the actual point at which cure occurs is very difficult and cure techniques (by almost all methods) have evolved into a stepped cure with gradual application of heat to avoid formation of voids. An arbitrary 1 percent void limit has been adopted for most autoclaved composites; filament wound and pultruded composites will have higher void volumes depending upon the application.

Each technique for compaction and cure is unique in terms of types of structures for which it can be used and also in terms of the forms of starting materials. For instance, a unidirectional prepreg is optimum for autoclave cure, but because of potential fiber washout, a fabric prepreg may be more appropriate for a press-cured laminate. The advantages and disadvantages of fabric vs. unidirectional tape fiber forms for composite manufacture are shown in Table 3.22.

An autoclave is essentially a closed pressurized oven; the most common epoxy laminates are cured at an upper temperature of 177°C (350°F) and 6 MPa (100 psig). Figure 3.14 shows a typical cure schedule. Autoclaves are still the primary tool in advanced composite processing and have been built up to 16 m (55 feet) long at 6.1 m (20 feet) diameter. Since autoclaves are so expensive to build and operate, many other methods of curing compacting composites have been promoted. The two newest and most attractive methods are fiber placement and resin-transfer molding, discussed below.

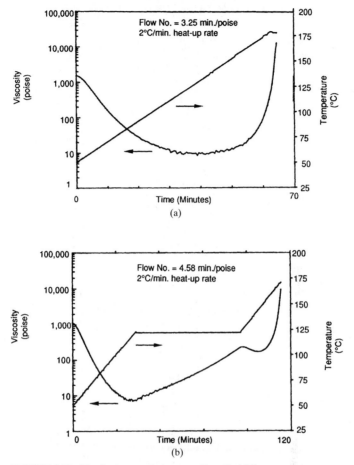

FIGURE 3.13 Matrix changes during cure (from Ref. 27).

Each resin fiber combination has at least one optimum cure technique, depending on the proposed environment. The prepreg manufacturer supplies a time-temperature cycle that may have been used to manufacture the test specimens that were used to generate preliminary mechanical and physical properties. Frequently it has been necessary to modify the preliminary cure cycle because of part configuration (thickness) or because of production economics.[32] There are several developments that help to inject science into the otherwise hit-or-miss and time-consuming procedure of cure cycle optimization. They are:

- Development of dielectric sensors and signal processing to gain a knowledge of the resin condition (viscosity) at any point during the cure process
- Use of thermocouple data, transducer outputs, and dielectric information to interactively control the curing process within an autoclave

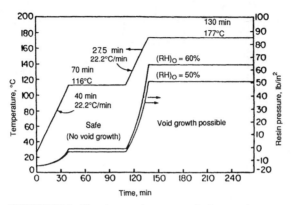

FIGURE 3.14 Time-temperature cure cycle for autoclave cure to prevent moisture-induced void growth (from Ref. 28).

• Development of computer programs supported by test results to predict the physical changes that will occur during cure

• Analytical techniques such as rheometric dynamic scanning and differential scanning calorimetry that give information on changes in viscosity and the amount of heat absorbed or liberated during cure

3.3.2 Resin Transfer Molding

Previous discussions have centered on moving resin out of the laminate to reduce voids. Resin transfer involves the placement of dry fiber reinforcement into a closed mold and then injecting a catalyzed resin into the mold to encapsulate the reinforcement and form a composite. The impetus for the use of this process comes from the large cost reductions that can be realized in raw materials and lay-up. The process can utilize low injection pressures, i.e., 551 KPa (80 psig); therefore, the tooling can be lower cost plastic rather than metal. The process is most appropriately used for nonaerospace composites but has been extended to many advanced applications. RTM manufacturing considerations are shown in Table 3.24. The advantages of RTM include the possibility of producing very large (Fig. 3.15) and complex shapes efficiently and inexpensively, reducing production times with the ability to include inserts in the composite. For large structures such as boats, the SCRIMP process uses a vacuum bag instead of the two plates of a typical mold. Advantages quoted for this technique are: one-sided mold, better control of and higher fiber volume, and lower porosity in the composite. It has been used for structures that need higher quality than can be obtained by other RTM processes. Table 3.25 shows the range of applications for the RTM technique. RTM is also a way of preparing a composite structure from a knitted preform. Knitting and braiding and sewn tridimensionally reinforced preforms offer complex shapes that are not attainable by other techniques, the techniques can possibly lower costs due to reduction of labor, and the product may gain increased impact resistance due to the multiple, interlocked directions of fiber.

TABLE 3.24 RTM Manufacturing Considerations

Materials	Tooling	Technology
Fiber type	Mold material	Resin viscosity
Preform complexity	Mold surface finish	Flow modelling
Preform cost	Resin pump type	Gating and vent design
Inserts	Integral or oven heating	Composite strength, stiffness, fiber volume, transverse properties
	Clamping method	Vacuum assist or not
	Tool durability	

TABLE 3.25 RTM Composite End Uses

Composite use	Part
Industrial	Solar collectors
	Electrostatic precipitator plates
	Fan blades
	Business machine cabinetry
	Water tanks
Recreational	Canoe paddles
	Large yachts
	Television antennas
	Snowmobile bodies
Construction	Seating
	Baths and showers
	Roofing
Aerospace	Airplane wing ribs
	Cockpit hatch covers
	Speedbrakes
	Escape doors
Automotive	Crash members
	Leaf springs
	Car bodies
	Bus shelters

3.3.3 Filament Winding

Filament winding is a process by which continuous reinforcements in the form of rovings or tows (gathered strands of fiber) are wound over a rotating mandrel. The mandrel can be cylindrical, round, or any other shape as long as it does not have reentrant curvature. Special machines (Fig. 3.16) traversing a wind eye at speeds synchronized with the mandrel rotation, control the winding angle of the reinforcement and the fiber lay-down rate. The reinforcement may be wrapped in adjacent bands or in repeating bands that are stepped the width of the band and that eventually cover the mandrel surface. Local reinforcement can be added to the structure using circumferential windings, local helical bands, or by the use of woven or unidirectional cloth. The wrap angle can be varied from low-angle helical to high-angle circumferential or "hoop," which allows winding from about 4°–90° relative to the mandrel axis; newer machines can "place" fiber at 0°.

FIGURE 3.15 RTM (SCRIMP) process for injecting a boat hull. (© *Billy Black, courtesy of Seemann Composites.*)

There are advantages and disadvantages to filament winding when compared to other methods. The most obvious advantages, summarized in Table 3.26, are cost savings, both capital and recurring labor, and the ability to build a structure that is larger than autoclave capacity. The disadvantages of filament winding can be worked around. Fabricators of large rocket motors have used plaster mandrels that can be stripped, reduced in size, and passed out through the relatively small port. Reverse curvature can be formed into a positive curvature by the addition of oriented fibers or mats, or if the curvature is necessary to the design, such as on an airfoil, it can be accomplished by removing the uncured structure from the mandrel and using alternate means of compaction to form the composite. Newer filament winding machines have the capacity to change the wind angle at any point over the part surface. This gives the option of actually winding the fiber into a reverse curvature by selecting the wind angle that will follow a hyperboloidal path into a smooth recess without bridging or slipping. In Peters et al. to Westinghouse, "Method of forming a joint between a tubular composite and a metal ring," U.S. Patent 4,701,231, this technique has been used to wind in an in situ metallic end ring for a composite-to-metal joint, eliminating the need for further bolting or pinning and providing a measure of fail-safe operation. The fiber path can be altered by pins or sawtooth to avoid slipping or bridging. Mandrels are expensive, but they are less expensive than two-sided molds, and their reusability contributes to the cost-effectiveness of filament winding. The poor external surface can be smoothed somewhat by proper selection of resin and fiber, use of surfacing mat or filled smoothing compounds at some weight penalty, or by compaction and cure in a female die mold using vacuum bag or autoclave pressure. Figure 3.17 shows the Beech Starship graphite epoxy fuselage that was filament-wound and then expanded into tooling during cure to form a smooth outer skin.

Thermoset resins have generally been used as the binders for the reinforcements. These resins can be applied to the dry roving at the time of winding (wet winding) or

FIGURE 3.16   Filament winding, helical bands. (*Courtesy Westinghouse Marine Division.*)

may be applied previously and gelled to a B stage as prepreg. The fiber can be impregnated and rerolled without B staging and used promptly or refrigerated. Prepreg and wet rerolled materials are useful because of the opportunity they afford to perform quality control checks early.

The cure of the filament wound composite is generally conducted at elevated temperatures without the addition of any process for composite compaction.

The filament winding process, like pultrusion, can employ wet resin systems to result in potentially lower-cost composite structures.

FIGURE 3.17   Filament-wound aircraft fuse-lage with smooth skin. (*Courtesy of Fibertek Div. of Alcoa.*)

TABLE 3.26 Comparison of Filament Winding with Other Fiber Deposition, Compacting, Curing Process

---

Advantages:
  1. Highly repetitive and accurate in fiber placement (from part to part and from layer to layer).
  2. Can use continuous fibers over the whole component area (without joints); can orient fibers easily in load direction. This simplifies the fabrication of aircraft fuselages and reduces the joints.
  3. Avoids capital expense and the recurring expense for inert gas of autoclave.
  4. Large and thick-walled structures can be built, larger than any autoclave.
  5. Mandrel costs can be lower than other tooling costs. There is only one tool, the male mandrel, which sets the inside diameter and the inner surface finish. The outer surface is uncontrolled and may be rough.
  6. Lower cost for large numbers of components since there is less labor than many other processes.
  7. Costs are relatively low for material since fiber and resin can be used in their lowest-cost form rather than as prepreg and no preforming is necessary. (Preforming is necessary for RTM and may be a significant recurring expense.)

Disadvantages:
  1. Shape of the component must permit mandrel removal. Long tubular shapes will generally have a taper. Different mandrel materials because of differing thermal expansion and differing laminate lay-up percentages of hoops versus helical plies will demonstrate varying amounts of difficulty in removal of the part from the mandrel.
  2. Cannot generally wind reverse curvature. To wind a reverse curvature, wind the exact shape on a positive dummy mandrel insert and then remove the insert and place the fiber.
  3. Cannot change fiber path easily (in one lamina). It can be done by use of pins or slip of the tow. Fiber placement is the only fabrication method capable of "steering" the fiber.
  4. Needs a mandrel, which sometimes can be complex or expensive. Usually the mandrel is less costly than the dies or molds for forming methods other than pultrusion or RTM.
  5. Generally poor external surface that may hamper aerodynamics. A better outside surface can be obtained by: use of outer clamshell molds; external hoop plies or thinner tows on last ply; or shrink tape or porous TFE-Glass tape overwrap.

The resin system in a filament wound composite serves the same function as it does in composite structures fabricated by other means (see Sec. 3.2.3).

Handling guidelines for a wet resin system that are unique to wet filament winding are:

- Viscosity should be $2P_a \cdot s$ or lower.
- Pot life should be as long as possible (preferably over 6 hours).
- Toxicity should be low.

The other approach is to use a wet-rerolled system, essentially a wet resin that is applied to the fiber beforehand and then kept in the freezer until use. Table 3.27 shows a comparison of the three methods.

Fibers are used in tow (graphite) or (roving) glass and Kevlar. The two terms define a gathered parallel bunch of fibers with essentially no twist. The fibers that are made in individual processes (e.g., boron) have not been used extensively in filament-winding applications due to their stiffness.

Fibers. The most widely used fiber for filament winding and the fiber that has had the longest period for development is fiberglass, which has been marketed in several

TABLE 3.27   Comparison of Prepreg vs. Wet and Wet-Rerolled Filament Winding Materials

	Prepreg	Wet winding	Wet-rerolled winding
Cleanliness	Best	Worst	Almost equal to prepreg, is away from winder
Fiber availability	Poor. Not all fibers are available; many necessitate special order	Best. Any fiber that system will handle	Best—all fibers
Control of resin content	Best. Constant speed and viscosity	Poor. Speed of mandrel varies, viscosity of resin may vary	Better. Process is away from winding and is faster; little viscosity change
Quality assurance	Highest. Can be done far ahead	Worst. Imposes quality control procedures onto factory floor and can lead to errors	Good. Can be done ahead
Ability to use complex resin systems	Yes. Hot melts available	Very difficult. Requires complex impregnators to remove solvents or liquify hot melts	Difficult. Still requires complex impregnators
Large data base resin systems	Yes	Commercial resins generally not available as liquids; the wet systems with large data bases may be proprietary	Same as wet winding
Graphite fibers encapsulated (to prevent electronics shortcircuits)	Yes	No	Graphite fibers not released at winder
Storage	Must be refrigerated and storage records maintained	Easy mix at winder; dry fibers have long shelf life	Must be stored like prepreg but shorter storage life; records must be kept
Fiber damage	Depends on impregnator; fiber is handled twice	May require special equipment; less damage potential because of less handling	All handling of fiber is under control of user
Cost	Highest	Lowest	Slightly above wet but also requires capital investment for impregnation equipment
Large roving package	Depends on impregnator	Whatever is available dry from fiber manufacturers	Whatever is available dry from fiber manufacturers
Room-temperature cure	Not possible	Possible	Possible
Simple resin formulation	Possible	Necessary	Necessary
Winding speed	Can be highest. Resin throw from fiber is minimized	Lowest speed	Intermediate. Resin can be staged to lower resin throw
Stability on nongeodesic path	Highest possible	Lowest. Wet resin may cause slippage	Intermediate. Resin can be staged to increase tack

From Ref. 5.

grades in the United States for over 40 years. During that period of time many different glasses were developed, including leaded glass and beryllium high-modulus glass, all of which are no longer produced or are in limited supply. The various types of glass that continue to be useful for composite structures were shown in Table 3.2. A glass fiber, Hollex, has been recently reintroduced into the market by Owens Corning. The fiber is a hollow S-2 fiber that follows the rules of mixtures, resulting in better mechanical properties than E glass, with up to 30 percent weight savings. It is proposed for weight-sensitive structures such as aircraft sandwich composites and radomes.[30]

The table shows the common description for the glass and the nominal tensile strength and tensile modulus of strands and composite. The composite data is theoretical value that reflects the ideal, or maximum strength and stiffness. A tensile test will generally not reflect these theoretical values for strength. Burst tests of subscale pressure vessels can utilize over 80 percent of the fiber's strength, depending on the configuration of the vessel and the resin system used; other tests using necked-down or straight-sided specimens will result in lower percentages of the fiber's ultimate strength. The maximum number of fibers per strand is usually important information for filament winding since it influences handling ease and per-ply thickness. The fiber density is included so that the rule of mixtures equations involving fiber volume and resin volume can be used to evaluate void volume and theoretical mechanical values. The usefulness of glass continues for filament winding because of its low cost, dimensional stability, its moderate strength and modulus, and its ease of handling. Compressive strength of glass-reinforced composites is relatively high and has led to their selection in underwater deep diving applications. The electrical properties of glass-reinforced composites have resulted in their use in radomes, printed circuit boards, and many other products that require high dielectric strength.

Aramid Organic Fibers. Table 3.4 shows the properties of several organic reinforcing fibers. Aramid fibers were introduced by DuPont in 1972. The aramids, initially useful because of their strength- and modulus-to-weight ratios (called specific strength and specific modulus) have also shown great consistency with a low coefficient of variation, permitting high design allowables. The specific tensile strength of Kevlar® was, at its introduction, the highest of any fiber. Graphite fibers, because of advances in processing, have the highest values now. (Specific strength and modulus based on fiber values are simply tensile strength or tensile modulus/density and are a good measure of structural efficiency of a fiber for airborne applications.) The values for tensile moduli may be near to those developed in a composite structure, but the tensile strength values for fibers may be quite different because of factors such as translation efficiency, possibility of flaws, processing damage, or incorrect fiber orientations. The values must be reduced for quasi-isotropic laminates; good composite designs seek to take advantage of the higher unidirectional properties while still retaining enough fibers in alternate directions to react unplanned or off-axis loads. The aramids have relatively poor shear and compression properties in a composite; components, such as pressure vessels that avoid these stresses, are most efficient.

Polyethylene Organic Fibers. High-strength polyethylene fibers, announced in 1984, have very high specific strength and modulus, due in part to the fiber's low density. Like the aramids, this fiber demonstrates poor compression and shear strength in composites but has a high degree of energy absorption, making it suitable for ballistic protection. Unlike the aramids, the polyethylene fiber has a low temperature tolerance (essentially zero strength at 150°C), so it cannot be combined with any resin system that cures at 177°C or greater.

Other Organic Fibers. A new fiber, PBO, poly(p-phenylene-benzobisoxoazole), developed by Dow Chemical Co. and tested by Brunswick Composites under Air

FIGURE 3.18 Completed PBO spherical test vessel. (*Courtesy of Lincoln [formerly Brunswick] Composites.*)

Force funding shows great promise due to its high strength (5.5 GPa, 798 ksi) and low density (1560 kg/m^3, 0.056 lbs/in^3). Pressure vessels fabricated with the fiber and a proprietary Brunswick Composites resin system LRF-0092 (Fig. 3.18) demonstrated a performance factor 30 percent better than test vessels fabricated with the highest-performing graphite fibers.[31] The development continues at Toyabo.

Carbon/Graphite Fibers. The largest variety of available strengths and moduli can be obtained with graphite fibers. Graphite fibers have improved dramatically in terms of modulus, tensile strength, and strain to failure since their introduction in the late 1960s. Improvements in surface finish have made handling for filament winding easier. As tensile modulus is increased by heat treatment, the tensile strength generally decreases. The intermediate modulus fibers recently introduced were the only exception. The amount of graphitization increases with increasing modulus, which results in increases in thermal and electrical conductivity. Unfortunately, the cost of the fiber also dramatically increases, primarily because there is less demand for the high-modulus fibers, and large-scale production economics have not yet been realized. All fibers except ultrahigh modulus pitch and extremely high modulus pitch have been reported to have been filament-wound. Special precautions must be taken during filament winding when the modulus of the fiber is greater than 50 million; large bend radius rollers and special impregnators may be necessary.

Netting Analysis. An analysis technique has traditionally been used for filament-wound structures. It is called netting analysis. Netting analysis is a simplified procedure for predicting stresses in a fiber-reinforced composite by neglecting the contribution of the resin system. The procedure has intuitive appeal, and it gained acceptance early in the development of composite structures. The technique applies static equilibrium principles with no consideration of strain compatibility. Even with such theoretical flaws, netting analysis continues to survive, and it is sometimes useful for preliminary sizing and for predicting failure in simple structures such as pressure vessels. Netting analysis must always be followed and confirmed by computer analysis based on micro- and macro-mechanics as shown in Ref. 22.

Cylindrical Pressure Vessels. As an example application, consider a filament-wound cylinder of radius R pressurized with an internal pressure P_i. If the vessel is wound with only helical ($\pm\theta$) fibers, with an allowable fiber stress $\sigma_{f\theta}$ determine the helical fiber thickness ($t_{f\theta}$) and the wind angle θ.

The forces in the axial direction are summed as:

$$N_m = \sigma_{f\theta}\, t_{f\theta}\cos^2\theta = \frac{P_i R}{2} = \frac{P_i D}{4} \tag{3.9}$$

Solving Eq. (3.9), the helical fiber thickness required to contain internal pressure is:

$$t_{f\theta} = \frac{P_i R}{2\sigma_{f\theta}\cos^2\theta} \tag{3.10}$$

Summing forces in the circumferential (h) direction produces:

$$N_h = \sigma_{f\theta}\, t_{f\theta}\sin^2\theta = P_i R \tag{3.11}$$

Using $t_{f\theta}$ from Eq. (3.10) in (3.11) shows that $\tan^2\theta = 2$, or $\theta = \pm54.7°$. This is the wind angle required for a pressurized cylinder with only helical windings with a 2:1 hoop-stress-to-axial-stress ratio.

Summing forces in the axial direction produces the helical fiber thickness required to contain the internal pressure, given by Eq. (3.11). Using this value for $t_{f\theta}$ when summing forces in the hoop direction produces

$$t_{f90} = \frac{P_i R}{2\sigma_{f90}}(2 - \tan^2\theta) \tag{3.12}$$

where σ_{f90} is the fiber stress in the hoop fibers and the second term accounts for the portion of the hoop load carried by the helical fibers.

In the winding process, N spools are used to form a winding band of width W. Each spool has a cross-sectional area CSA, the values of which are listed in Table 3.28 for several rovings.

CSA can also be calculated from the following:

$$\text{CSA} = \frac{1/\text{Density}}{\text{Yield}} \tag{3.13}$$

TABLE 3.28 Cross-Sectional Area of Typical Rovings

Fiber type	CSA (mm²)	CSA (in²)
E-Glass type 30 (675 yield)	0.289	44.73×10^{-5}
S-2 Glass (20 end)	0.268	41.52×10^{-5}
Kevlar 49 (4 end)	0.352	54.55×10^{-5}
Thornel 300 12000 filament tow	0.466	72.20×10^{-5}
Type AS4 12000 filament tow	0.486	75.40×10^{-5}
Celion 12k 12000 filament tow	0.465	72.00×10^{-5}

From Ref. 5.

The theoretical fiber thickness of each ply as wound is:

$$t_{fp} = \frac{N(\text{CSA})}{W} \tag{3.14}$$

The fiber stress at failure in a composite laminate is less than that for a unidirectional coupon test or a strand tensile test. This reduction is called the translation efficiency; typical values are 70 to 80 percent of the strand value. For example, the ultimate stress for Kevlar 49 fibers is 3800 MPa (550 ksi) in a strand tensile test (Table 3.4). A high value of fiber stresses for Kevlar 49 in pressure-vessel applications is 2750 MPa (400 ksi), a translation efficiency of 73 percent. Translation efficiency is affected by fiber damage during processing into a composite, voids, complex stress states, and the like, which are present in the laminate and not in a strand tensile specimen.

In a cylindrical pressure vessel, translation efficiencies are usually lower for helical fibers than for hoop fibers. This is because the helical patterns include more overlaps and crossovers as well as discontinuities at the dome cylinder junction and in the polar boss region. For these reasons, the helical hoop fiber stress ratio used in design is usually between 75 and 100 percent.

In winding, the number of plies is selected so that the ply thickness in Eq. (3.14) divides equally into the total fiber thickness required, per Eqs. (3.10) and (3.12). Band widths are selected in conjunction with the number of spools so that a band thickness (including resin) of about .10 to .76 mm (0.004 to 0.30 inches) is produced. This provides a band width that is neither too thick nor too thin. For helical windings, the band width W_θ is selected to cover the cylinder with no overlap with a whole number of bands wound at the angle θ. The helical ply thickness (including resin) is then:

$$t_{C\theta} = \frac{N_\theta(\text{CSA})}{W_\theta V_{f\theta}} \tag{3.15}$$

where $V_{f\theta}$ is the fiber volume for the helical windings. For the hoop windings, the band width W_{90} is selected so that the hoop windings do not deviate appreciably from the 90° wind angle. The hoop ply thickness (including resin) is then:

$$t_{C90} = \frac{N_{90}(\text{CSA})}{W_{90} V_{f90}} \tag{3.16}$$

where V_{f90} is the fiber volume fraction for the hoop plies. Table 3.29 provides typical ranges of fiber volume fractions attained with different fiber systems for helical and hoop plies. Actual fiber volume fraction depends on several process and geometric considerations, including resin viscosity, mandrel diameter, winding tension, wind angle, processing time, B-stage temperature, and external pressure during cure. The fiber volume fraction is determined by using in-processing thickness measurements. Good process control produces repeatable dimensions.

TABLE 3.29 Typical Fiber Volume Fractions

Fiber	Helical	Hoop
Glass	0.55–0.60	0.65–0.70
Kevlar	0.55–0.60	0.65–0.70
Graphite	0.50–0.55	0.60–0.65

From Ref. 5.

Example Calculation.   The above equations are used to provide preliminary thickness for the standard ASTM D 2528 pressure vessel designed for a 45 MPa (6600 psi) burst pressure and an inside diameter of 146 mm (5.75 in). If the winding machine is set up for a 5.8-mm (0.23-inch) hoop band width and 5.1-mm (0.2-inch) helical band width, the specified helical wind angle θ in the cylinder is ~12°. The vessel is to be wound with Kevlar 49 four-end aero grade roving (Table 3.4). Design allowables are chosen as 2.93 GPa (425 ksi) hoop fiber stress and 2.21 GPa (320 ksi) helical fiber stress. This uses translation efficiencies of 77 percent for the hoops and 58 percent in the helicals, with a 75 percent ratio of helical-to-hoop fiber stress. (Translation efficiencies are empirically determined from burst bottle testing.)
The required helical fiber thickness from Eq. (3.10) is:

$$t_{f\theta} = \frac{P_i R}{2\sigma_{f\theta}\cos^2\theta} = \frac{45 \times (5 \times 10^6) \times 73}{2 \times (2.21 \times 10^9) \times \cos^2 12} = 0.79 \text{ mm } (0.031 \text{ inch})$$

Assuming three helical layers (six helical plies), the calculated number of helical spools from Eq. (3.16) is:

$$N_\theta = \frac{t_{f\theta}W_\theta}{\text{CSA}} = \frac{(0.79/6)5.1}{0.352} = 1.9$$

The design would use $N_\theta = 2$, the closest whole number. It may be desirable to repeat the calculations with a slightly modified band width in order that the calculated number of spools is closer to a whole number.
The required hoop fiber thickness from Eq. (3.16) is:

$$t_{f90} = \frac{P_i R}{2\sigma_{f90}}(2 - \tan^2\theta) = \frac{45 \times (5 \times 10^6) \times 73 \times (2 - \tan^2 12)}{2 \times (2.93 \times 10^9)} = 1.11 \text{ mm } (0.0436 \text{ in})$$

Assuming nine hoop plies, the number of hoop spools from Eq. (3.16) is:

$$N_{90} = \frac{t_{f90}W_{90}}{\text{CSA}} = \frac{(1.11/9)5.8}{0.352} = 2$$

Assuming a fiber volume fraction of 0.6 for the helical and 0.65 for the hoops (Table 3.26), the total composite thickness from Eqs. (3.15) and (3.16) is:

$$t_c = \frac{0.79}{0.6} + \frac{1.11}{0.65} = 3.0 \text{ min } (0.12 \text{ inch})$$

The winding parameters and amount of fiber to be used are found as follows:

1. Determine number of circuits per helical layer (Fig. 3.19).

$$Z = \text{number of circuits to close} = \frac{\pi D \cos\theta}{W_\theta} \qquad (3.17)$$

where D = diameter of cylinder
θ = wind angle

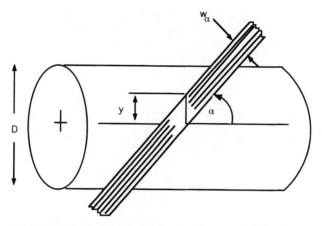

FIGURE 3.19 Band width relation for filament winding a circular cylinder, where $W = y \cos_\alpha$ and $Z = (\pi D \cos\alpha)/W_\alpha$.

2. Calculate weight of fiber used in one complete helical pattern.

$$\frac{\text{weight}}{\text{pattern}} = \frac{(2)ZN_\theta[(ML/\cos\theta) + TL]}{\text{fiber yield}} \qquad (3.18)$$

where
$\quad ML$ = mandrel length + tag end
$\quad TL$ = turnaround length (domes)
fiber yield = one tow length per weight

3. Repeat calculations for hoop fibers. It is not desirable to modify any parameters except for band width when changing from helical to hoops. Test to see that parameters are consistent by solving for hoop fiber volume. If the fiber volume for hoop layers is higher than for helicals, the parameters are consistent.

4. Calculate weight of fiber.

$$\frac{\text{weight}}{\text{pass}} = \frac{N_{90}\pi D}{\text{fiber yield}} \left(\frac{\text{winding length}}{\text{band width}} \right) \qquad (3.19)$$

Total the hoop and helical layers and add an estimate for scrap.

3.3.4 Fiber Placement

Fiber placement, developed by Hercules Aerospace Co., is a cross between filament winding and automatic tape lay-down, retaining many of the advantages of both. The natural outgrowth of adding multiple axes of control to filament winding machines results in control of the fiber lay-down so that nonaxisymmetric surfaces can be wound. This involves the addition of a modified tape lay-down head to the filament winding machine and much more. The Cincinnati-Milacron machine additions include in-process compaction, individual tow cut/start capabilities, a resin tack control system, differential tow pay-out, low tension on fiber, and enhanced off-line programming. The machines are capable of winding the shapes shown in Fig. 3.20 and

FIGURE 3.20   Versatility of shapes fabricated by fiber placement. (*Courtesy of Hercules Aerospace Co.*)

can change fiber paths such as shown on Fig. 3.21. Table 3.30 shows the advantages quoted for the technique. The disadvantages are, presently, the cost of the machines (very high when compared to some filament winding machines), the dependence upon computers and electronics rather than mechanical means of directing fiber laydown, the necessity of a long production run to justify the capital and setup costs, and the cost and complexity of mandrels.

3.3.5   Pultrusion Braiding and Weaving

Pultrusion.   Pultrusion is an automated process for the manufacture of constant volume/shape profiles from composite materials. The composite reinforcement is continuously pulled through a heated die and shaped (Fig. 3.22) and cured simultaneously. If the cross-sectional shape is conducive to the process, it is the fastest and most economical method of composite production. Straight and cured configurations can be fabricated with square, round, hat-shaped, angled I or T-shaped cross sections from vinylester, polyester, or epoxy matrices with E and S glass, Kevlar and graphite reinforcements. Some of the available cross-sectional shapes and limitations are shown in Table 3.31. The curing is effected by combinations of dielectric preheating and microwave or induction (with conductive reinforcements like carbon graphite) while the shape traverses the die. The resin systems, predominantly polyester, can be wet or prepreg, but the cure rates will be much more rapid than those for filament winding. The process lends itself to long lengths of components with needs for reinforcement in the 0° direction. Many uses have been found for the process; pultrusion is the primary fabrication technique for reinforced plastic booms, ladder components, light poles, and conduits. The primary reasons for the use

RADIUS: 25" 20" 15" 10" 5"

FIBER STEERING WITH DIFFERENTIAL TOW PAYOUT

FIGURE 3.21 Special fiber path options available with fiber placement. (*Courtesy of Cincinnati Milacron, Inc.*)

of the technique are design considerations driven by commercial uses, such as cost, weight, electrical properties, and environmental resistance, but the process has also been used to produce components from advanced composite materials. An application that uses the cost-effective technique for high-volume production is the pultruded, graphite-reinforced driveshaft, which replaces two metal shafts, universal joints, and hangers and results in a quieter product because of inherent composite damping (Fig. 3.23). The process has some limitations:

- The part must be of constant cross section over its length; it cannot be tapered.
- Transverse strength will be somewhat lower than for other manufacturing methods.
- Curved shapes require special machines.
- Thick sections are difficult because of exotherm with rapidly curing resins.
- Cure shrinkage may cause dimensional and mechanical problems.
- There will generally be a cut edge, and this must be cut carefully to reduce delamination and then sealed to prevent moisture or other attack to fiber or interface.
- Joints are more difficult because of fiber orientation and lack of section changes that would allow geometric locking.

TABLE 3.30   Fiber Placement Processing Advantages

Flexibility	Compaction	Material usage
Full range of fiber orientations	Continuous in-process debulking	Wide use of advanced thermoset material systems near prepreg tape equivalent
Nongeodesic path generation constant ply thickness over complex shapes localized reinforcements	Complex surface fabrication Concave/convex Nonaxisymmetric and axisymmetric shapes	
Continuous fibers over three-dimensional shapes (without joints)		
Large structures		

Adapted from Ref. 33.

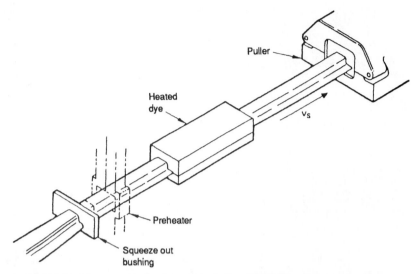

FIGURE 3.22 Elements of pultrusion schematic (from Ref. 34).

FIGURE 3.23 Graphite-reinforced pultruded driveshaft for GMT 400 trucks. (*Courtesy Spicer division of Dana.*)

TABLE 3.31 Pultrusion Design Guidelines

Minimum inside radius		Roving, 0.79 mm mat, 1.6 mm	Corrugated sections		Yes, longitudinal
Molded-in holes		No	Metal inserts		No
Trimmed in mold		Yes	Bosses		No
Core pull and slides		Yes	Ribs		Yes, longitudinal
Undercuts		Yes	Molded-in labels		Yes, but not recessed
Minimum recommended draft		No limitation	Raised numbers		No
Minimum practical thickness		Roving, 1.0 mm Mat, 1.5 mm	Finished surfaces (reproduces mold surface)		Two
Maximum practical thickness		Roving, 75 mm Mat, 25 mm	Hollow sections		Yes, longitudinal
Normal thickness variation		See ASTM D 3917[22]	Wire inserts		Yes, longitudinal
Maximum thickness buildup		As required	Embossed surface		No

Source: From J. D. Martin and J. E. Sumerak in Reinhart,[35] p. 542.

Pultrusion has also been combined with filament winding to achieve high transverse properties with advanced composite starting materials as shown in Fig. 3.24.

Braiding, Weaving, and Other Preform Techniques.   Braiding, weaving, knitting, and stitching (with low-modulus fibers) represent methods of forming a shape, generally referred to as preforming, with the composite fibers before impregnation. The shape may be the final product or some intermediate form such as a woven fabric. New techniques allow prepregs to be used, and the introduction of three-dimensional braids has extended braiding to airborne structural components with high fracture-toughness requirements and high damage tolerance. The braiding process is continuous and is amenable to round or rectangular shapes or smooth curved surfaces and can transition easily from one shape to another. Resin systems are gen-

FIGURE 3.25 Continuous braiding machine. (*Courtesy B.C.M.E., Ltd.*)

FIGURE 3.24 Filament-wound/pultruded graphite epoxy bridge beam. (*Courtesy Sci. Div. of Harsco.*)

erally epoxies or polyester, and the fiber options are similar to those for filament winding; the single stiff fibers such as boron cannot endure the tight bend radii.

In the process, the mandrel is fed through the center of the machine at a controlled rate, and fibers from a large number of rolls are deposited at multiple angles, usually 45° and 90° (Fig. 3.25). Zero-degree plies can be laid from tape or placed by a triaxial braider.

Braiding, in some applications, has turned out to be almost half the cost of filament winding because of labor savings in assembly and simplification of design.[36] Fiber volume of a braided composite will generally be lower than for other methods.

The other fabric preforming techniques are weaving, knitting, and the nonstructural stitching of unidirectional tapes. The weaving and knitting are compared to braiding in Table 3.32. Stitching simply uses a nonstructural thread, such as nylon or Dacron, to hold dry tapes at selected fiber angles. Preforming in this manner results in a higher-cost raw material but saves labor costs. The stitched preform has known, stable fiber orientations similar to woven fabric without the crossovers.

3.4 OVERVIEW OF MECHANICS OF COMPOSITE MATERIALS

The axes in Fig. 3.26 are special and are called the ply axes or material axes. The 1-axis is in the direction of the fibers and is called the longitudinal axis or the fiber axis. The longitudinal axis is typically the highest stiffness and strength of any direction. Any direction perpendicular to the fibers (in the 2, 3 plane) is called a trans-

TABLE 3.32 Comparison of Fabric Formation Techniques

	Braiding	Weaving	Knitting
Basic direction of yarn introduction	One (machine direction)	Two (0°/90°, warp and fill)	One (0° or 90°, warp or fill)
Basic formation technique	Intertwining (position displacement)	Interlacing (by selective insertion of 90° yarns into 0° yarn system)	Interlooping (by drawing loops of yarns over previous loops)

From Ref. 36.

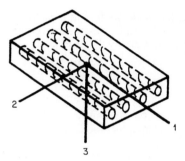

FIGURE 3.26 The unidirectional ply.

verse direction. Sometimes, to simplify analysis and test requirements, ply properties are assumed to be the same in any transverse direction. This is the transverse isotropy assumption; it is approximately satisfied for most unidirectional composite plies. These properties are typically modified by transformation relative to the laminate axis, where these may not be the same as the ply axes. These calculations involve matrix manipulations. For the simplest ply stresses along the ply axes, the strains produced by the stresses are:

$$\left\{ \begin{matrix} \varepsilon_1 \\ \varepsilon_2 \\ \gamma_{12} \end{matrix} \right\} = \begin{bmatrix} \dfrac{1}{E} & -\dfrac{\gamma_{21}}{E_2} & 0 \\ -\dfrac{\gamma_{12}}{E_1} & \dfrac{1}{E_2} & 0 \\ 0 & 0 & \dfrac{1}{G_{12}} \end{bmatrix} \left\{ \begin{matrix} \sigma_1 \\ \sigma_2 \\ \tau_{12} \end{matrix} \right\} \tag{3.19}$$

where ε = strain
υ = Poisson's ratio
E = Young's modulus
G = shear modulus
σ = tensile stress
τ = shear stress
γ = shear strain

In a multidirectional laminate there can be as many as 21 stiffness constants. Strength predictions are equally as complicated for two reasons: because of directional differences, i.e., compression is not always equal to tension, and because of the complexity of the several failure theories. As the complexity of the matrix calculations increases, it becomes evident that errorless mathematical manipulations are impossible without the aid of computers. Fortunately there are a number of software packages that can accomplish these manipulations.[37-39] The output of the computer, however, is only as good as the input information. Users must view the data in handbooks and computer material property files as preliminary and should verify the necessary constants and failure properties of the composite materials and processes by subscale and full-scale tests. Since there are no standard, accepted design allowables for composite materials, most organizations that extensively use the materials have had to develop their own data to supply A- or B-basis allowables. Because of the possible numbers of permutations of resins, fibers, and curing techniques, it will be some time before standardized strength and moduli values can be published as is done now for most metallic structural materials.

3.5 DESIGN OF COMPOSITES

The design process for composites involves both laminate design and component design and must also include considerations of manufacturing process and eventual environmental exposure. These are some steps that can simplify the process.

3.5.1 Laminate Design Recommendations

1. Take advantage of the orthotropic nature of the fiber composite ply.
   - To carry inplane tensile or compressive loads, align the fibers in the directions of these loads.
   - For inplane shear loads, align most fibers at $\pm45°$ to these shear loads.
   - For combined normal and shear in-plane loading, provide multiple or intermediate ply angles for a combined load capability.
2. Intersperse the ply orientations.
   - If a design requires a laminate with 16 plies at $\pm45°$, 16 plies at $0°$, and 16 plies at $90°$, use the interspersed design $[90_2/\pm45_2/0_2]_{4S}$ rather than $[90_8/\pm45_8/0_8]_S$. Concentrating plies at nearly the same angle ($0°$ and $90°$ in the above example) provides the opportunity for large matrix cracks to form. These produce lower laminate allowables, probably because large cracks are more injurious to the fibers and more readily form delaminations than the finer cracks occurring in interspersed laminates.
   - If a design requires all $0°$ plies, some $90°$ plies (and perhaps some off-angle plies) should be interspersed in the laminate to provide some biaxial strength and stability and to accommodate unplanned loads. This improves handling characteristics and serves to prevent large matrix cracks from forming.
   - Locally reinforce with fabric or mat in areas of concentrated loading. (This technique is used to locally reinforce pressure vessel domes.)
   - Use fabric, particularly fiberglass or Kevlar, as a surface ply to restrict surface (handling) damage.
   - Insure that the laminate has sufficient fiber orientations to avoid dependence on the matrix for stability. A minimum coverage of 6 to 10 percent of total thickness in 0, ±45, and 90 directions is recommended.
3. Select the lay-up to avoid mismatch of properties of the laminate with those of the adjoining structures, or provide a shear/separator ply.

- Poisson's ratio. If the transverse strain of a laminate greatly differs from that of adjoining structure, large interlaminar stresses are produced under load.

- Coefficient of thermal expansion. Temperature change can produce large interlaminar stresses if coefficient of thermal expansion of the laminate differs greatly from that of adjoining structure.

- The ply layer adjacent to most bonded joints should not be perpendicular to the direction of loading. Thicken the composite in the joint area, soften the composite by adding fiberglass or angle plies, and select the highest strain-capability adhesive.

4. Use multiple ply angles. Typical composite laminates are constructed from multiple unidirectional or fabric layers that are positioned at angular orientations in a specified stacking sequence. From many choices, experience suggests a rather narrow range of practical construction from which the final laminate configuration is usually selected. The multiple layers are usually oriented in at least two different angles, and possibly three or four; $\pm\theta°$, $0°/\pm\theta°$, or $0°/\pm\theta°/90°$ cover most applications, with θ between 30 and 60°. Unidirectional laminates are rarely used except when the basic composite material is only mildly orthotropic (e.g., certain metal matrix applications) or when the load path is absolutely known or carefully oriented parallel to the reinforcement (e.g., stiffener caps).

An important observation concerning practical laminate construction is the concept of midplane symmetry with respect to stacking sequence (i.e., uncoupled response) and balanced construction with regard to the inplane angular orientations (i.e., orthotropic behavior as opposed to anisotropic). If a laminate does not exhibit midplane symmetry, the stretching and bending behavior can be highly coupled. The severity of this coupling is inversely proportional to the number of layers within the laminate. The fewer the layers, the worse the problem. This heterogeneous nature should be avoided. In general, seek a symmetric laminate to satisfy the design. When symmetric and balanced laminates are used, bending, twisting, and warping effects are reduced.

When unsymmetrical laminates are used, such as for the faces of a sandwich structure or sides of a cylinder, the entire laminate is generally made symmetric because of geometry. If an unsymmetric laminate must be used alone, it is very prudent practice to avoid those of less than eight layers.

5. Use midplane symmetry. The usual reference surface for determining if a laminate is symmetrical is the geometric midplane. A midplane symmetric laminate is highly desirable and to be preferred. Extremely few cases exist where an unsymmetrical stacking sequence should even be considered.

To construct a midplane symmetric laminate, for each layer above the midplane there must exist an identical layer the same distance below the midplane, with a repetition of thickness, material properties, and angular orientation. Thus, the lamination stacking sequence will possess a mirror image about the geometric midsurface. As Figs. 3.10e and 3.10f show, both even- and odd-number-ply laminates can satisfy these criteria. Note that the use of a woven fabric at identical distances from the midplane may not confer symmetry. One of the layers will have to be flipped because the fabric may not be identical on both surfaces.

6. Observe effects of stacking sequence. Once the orientations of a laminate are specified, the stacking sequence will control the flexural rigidities of a laminated plate. Thus, the stability, vibration, and static bending behavior are controlled by the relative dispersion and thickness distribution of these oriented layers.

Interlaminar stresses (normal and shear) are influenced by the laminate stacking sequence. For example, a $(\pm45°_2/0°_2)_s$ laminate generates compressive normal stresses at the laminate boundaries, which are ten times as high as the compressive stresses produced by a $(45°/0°/\pm45°/0°/-45°)_s$ laminate when each is loaded in compression. Tensile free-edge stresses would result if these laminates were subjected to axial tensile loads. However, if the 0° layers were stacked on the outside, tensile (delamination) stresses would be induced at the edges as a result of compressive loads. The interlaminar normal stresses can be minimized, particularly for fatigue applications, by optimizing the stacking sequence in reference to the direction of load. The interlaminar stress problem is normally referred to as the free-edge effect. Such stresses are usually dominant at the free edges of laminates. Their magnitude can be significant, especially for fatigue conditions such as thermal cycling. The major free-edge-effect width is over a very narrow region (approximately equal to the laminate thickness). Furthermore, the magnitudes of such stresses are proportional to the thicknesses of nondispersed oriented layers. Hence, avoid stacks of layers at the same orientation; alternate or disperse the layers.

If plies must be terminated in the laminate, provide for equal steps of 0.10 or greater and cover the step area with an additional ply (preferably fabric) to avoid interlaminar shear failure to provide some load redistribution and to avoid ply edge peeling.

Adjacent layers, if possible, should be positioned with a maximum orientation differential of 60°. Potential macrocracking of layers from induced thermal stress during cool-down from curing is sufficient reason to avoid this problem. As the adjacent orientation differential increases, so do thermal stresses (e.g., $\pm45°$ or $0°/90°$ laminates are representative of the worst-case situations).

Macrocracking of layers within a laminate is an irreversible process. Although it may not affect static strengths, it can lead to reduced fatigue life. Generally, the frequency of macrocracking increases with thermal cycling and can quickly lead to reduced laminate stiffnesses, changes in Poisson's ratio, and coefficients of thermal expansion. In addition, layer cracking may interact with free-edge stresses, which can trigger delamination. Fabric layers (as opposed to unidirectional) are very resistant to cracking.

3.5.2 Component Design Recommendations

The following recommendation relates to material and processes interactions with environmental concerns and cost-effective fabrication: Electrode potential-corrosive galvanic cells can be produced whenever two materials of different electrode potential are in electrical contact in the presence of an electrolyte. This is an important consideration for graphite/epoxy laminates that are electrically conductive. Many metals are anodic with respect to graphite/epoxy and are corroded when they are part of a galvanic cell with graphite/epoxy. An electrolyte, or a fluid capable of transferring electrons, is very difficult to avoid in practice. In designs with graphite/epoxy laminates:

- Avoid magnesium and magnesium alloys because of their high potential relative to graphite/epoxy.
- Carefully investigate the potential of other metals and provide special intersurface protection such as a fiberglass ply or other nonconductor, as required.

3.5.3 Component Fabrication Recommendations

- Consider fabrication requirements during preliminary design. Avoid selecting a design that might be technically elegant but could not be built consistently or reliably. The design of a composite structure requires early concurrent engineering to insure that the composite can be manufactured.

- Select the simplest design to manufacture. If there is a hard choice between ease of analysis and ease of manufacture, choose the latter.

- Design for a composite structure rather than using composite as a substitute for a metal.

- The designer/composites technologist should perform preliminary analysis. There are many tools available for preliminary composite sizing and layout, including excellent computer programs for use on microcomputers.[37-39] Use these computer programs and closed form analyses early, and defer the relatively high costs of finite element analysis.

- Reduce the part count. Consolidate as many substructures together as possible, cocure adhesive bonded assemblies, and build in stiffeners. This is the primary method of reducing costs.

- Use the greatest tow-size, thickest tape, or fiber as practical. Larger tow sizes will generally be much less expensive on a weight basis and will tend to decrease fabrication time but may have a negative effect on physical properties and surface finish.

- Use the most appropriate fabrication method for the part. Note the constraints associated with each technique, e.g., filament winding results in nonsmooth outer surface; pultrusion does not normally result in a quasi-isotropic laminate. For large structures, the relative cost of filament winding is less than one fourth that of hand lay-up and less than half that of the best tape laying machine.[40]

- Use the least expensive form of composite raw materials. Use a wet resin and dry fiber, if appropriate to the process. This method reduces the cost of filament winding. In Ref. 40 the cost for the large MX launch canister wet filament wound by Hercules was $45/lb—less than the cost of just the prepreg tape material for many composite parts.

- Consider the use of fabric if appropriate to the structure. Although raw material costs are higher and some physical properties are reduced, woven fabric shortens lay-up times, avoids tape spread-out problems at abrupt contour changes, and remains in place during cure (no distortion or fiber washout).

- Use composite and metals together to obtain a synergistic part. Avoid making the component of either metal or composite only when the judicious use of both will make a better, cheaper part.

- Make in situ joints if possible. A metal-to-composite in situ joint made concurrently with the composite structure saves costs in terms of machining, surface preparation quality assurance, and adhesive application. Generally the technique will require the use of a film adhesive.

- Select verification methods carefully. For coupon testing, avoid extensive tests for matrix-dominated properties when loads are almost totally reacted by fiber and vice versa. Make witness panels as a part of the component if possible. Coupon testing does not remove the need for full-scale component testing and vice versa.

- Water absorption will change the mechanical and physical properties of a composite laminate. Matrix-dominated shear, compression, and bending moduli and strength may be reduced and coefficient of expansion may be increased. Account for these changes in design.
- Graphite composites have poor impact resistance. If impact is a consideration, add glass plies.

3.5.4 Damage Tolerance

One of the primary hurdles for the widespread use of composites, in addition to cost, has been concerns for the damage tolerance. Composite materials show no yield behavior, defects are hard to find, and their effects are not always predictable.

Almost all applications for advanced composites, except one-time-use structures like rocket motors, involve repeated cyclic application of stresses. Defects can be introduced by the manufacturing processes or by unplanned occurrences such as impacts. The effect of these defects must be evaluated because their growth during cyclic stresses could cause delaminations or other failures of the structure. Thus, the structure must be damage-tolerant to ensure that it will not fail catastrophically during its operating life. Three suggestions for safety of advanced composite structures are:

- Defects or damage that are undetectable should not affect the life of the structure.
- Detectable damage should be able to be sustained by the composite structure for a period of time until it is found.
- Damage during a mission or flight should be sustained to as great an extent as possible to complete the mission.

The operational stresses in the composite structure can be kept low by increasing thicknesses, but that approach would negate the obvious weight advantage of composites and would also make the product more expensive. Also, some stresses such as interlaminar or transverse will not be reduced but may be actually increased by additional plies. Increased attention to processing controls can reduce or eliminate many defects such as voids, delaminations, or ply buckles, but not all (e.g., ply drops and free edges). Costs escalate when nonstandard controls are exercised in production. The advantage of low-maintenance costs of composites over product lifetime is enhanced because of good fracture-toughness properties of the composite.

The fracture toughness of the laminate material can be measured by a number of methods, some of which are shown in Fig. 3.27. The results will vary with different lay-up orientations. Other methods may be similar to the test specimens shown in ASTM-E-399 and may involve impact. Presently, the most accepted technique for evaluating the fracture toughness of a laminate or structure is a compression test after impact. Compression is a valid indicator; without matrix support the fibers will buckle, while the fibers do not need as much matrix support for tensile loading.

Fracture toughness is dependent on matrix and fiber properties, fiber orientations, and stacking sequence. The emphasis on thermoplastic composites by the U.S. Air Force has been directed towards increasing the fracture toughness of composites.

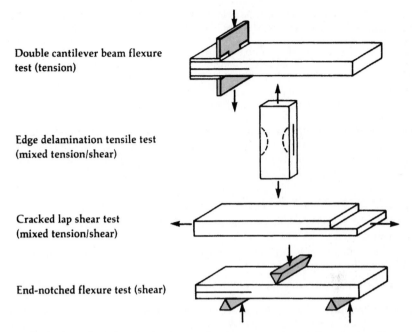

Double cantilever beam flexure
test (tension)

Edge delamination tensile test
(mixed tension/shear)

Cracked lap shear test
(mixed tension/shear)

End-notched flexure test (shear)

FIGURE 3.27 Typical methods of measuring fracture toughness of composites (from Ref. 41).

3.5.5 Composite Repairs and Adhesive Bonding

There will be damage to composite structures, particularly in aircraft, that may not be visible but still of concern. The damage to a composite structure may be more than that incurred by a comparable metal structure under an identical impact due to the lack of strain capability of the composite. The metal may yield to show the impact site; the composite may delaminate within and not reveal the damaged area. Inplane tensile strength may not always be compromised by impact damage to the matrix, but a damaged matrix cannot stabilize fibers under compressive bending or shear stresses, and compression is usually the critical loading mode in aircraft structures. A modest impact to a composite can result in severe undetected internal damage in the form of delamination.

The objectives of most repair/maintenance programs are as follows:

1. Investigate and quantify the extent of damage.
2. Determine the comprehensiveness and location of the repair effort.
   a. Repair to full properties (with all flaws detected and corrected), or,
   b. Repair to acceptable properties that will allow the structure to be usable to some high, predetermined percentage of full-scale operation, or,
   c. Repair to some emergency level that will allow use at a low percentage of operational effectiveness (i.e., to fly a damaged aircraft to depot maintenance base).
3. Select the repair configuration.

4. Define the materials and processes to be used, i.e., adhesives and cure cycle limitations.

Most repairs to aircraft composites involve damaged and possibly wet honeycomb in addition to the composite. A nomograph for determining acceptable conditions for these repairs is shown in Fig. 3.28.[43] The first priority is preparing the composite for repair, which may involve drilling holes for a bolt-on patch or reducing the moisture to a level that will allow heating the composite and the adhesive for cure.

Repair can involve the simple injection of a wet resin through drilled holes into the delaminations and then adding blind fasteners to defer the onset of local buckling, or the use of wet or dry prepregs in conjunction with adhesives to form a plug or flush on bonded-on patch. A summary of the available techniques for repair of solid and honeycomb laminates is shown in Table 3.33. Several of the most important points for consideration, summarized in Ref. 42, are:

1. Match the modulus of the original material and the fiber direction as much as possible.

2. Repair techniques are sufficiently complex that a high degree of technician skill and training is necessary.

3. For aircraft, nonmetallic repairs must incorporate lightning-strike protection.

Adhesive Bonding. There are overwhelming reasons to adhesive-bond composites to themselves and to metals, and, conversely, there are substantial reasons for using mechanical methods such as bolts or rivets (see Table 3.34).

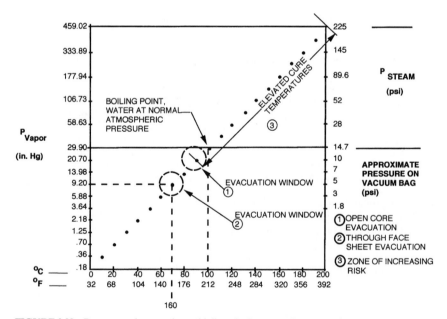

FIGURE 3.28 Pressure and evacuation guidelines for honeycomb core repair.

TABLE 3.33 Summary of Available Composite Repair Techniques

Method	Advantages	Disadvantages	Ease of Repair	Structural integrity
Bolted patch	No surface treatment; no refrigeration, heating blankets, or vacuum bags required	Bolt holes weaken structure; bolts can pull out	Fast	Low
Bonded patch	Flat or curved surface; field repair	Not suitable for high temperatures or critical parts	Fast, but depends on cure cycle of adhesive	Low to medium
Flush aerodynamic	Restores full design strength; high-temp capability	Time-consuming; usually limited to depot; requires refrigeration	Time-consuming	High
Resin injection	Quick; may be combined with an external patch	May cause plies to separate further	Fast	Low
Honeycomb, fill in with body filler	Fast; restores aerodynamic shape	Limited to minimal damage	Fast	Low
Honeycomb, remove damage, replace with synthetic foam	Restores aerodynamic shape and full compressive strength	Some loss of impact strength, gain in weight	Relatively quick	High
Honeycomb, remove damage, replace with another piece of honeycomb	Restores full strength with nominal weight gain	Time-consuming; requires spare honeycomb	More difficult	High

From Ref. 42.

The studies on adhesives, surface preparation, test specimen preparation, and design of bonded joints reported for the PABST Program[45] gave much more credibility to the concept of a bonded aircraft and provided reliable methods of transferring loads between composites and metals or other composites.

There are a number of decisions that must be made before bonding to a composite is attempted. These are:

1. Surface preparation technique.
    a. Composite. Peel ply, manual abrasion, or, if possible, cocure.
    b. Metal.
        (1) Aluminum. Phosphoric anodize is accepted, preferred airframe method.
        (2) Titanium. Several types of etches available.
        (3) Steel. Time between surface preparation and bonding is concern.
2. Type of adhesive.
    a. Film adhesive. Reproducible chemistry; early, frequent quality assurance (because resin is premixed by the manufacturer); requires special storage and generally cannot accommodate varying bond line thicknesses.

TABLE 3.34 Reasons For and Against Adhesive Bonding

For	Against
Higher strength-to-weight ratio	Sometimes difficult surface preparation techniques cannot be verified 100% effective
Manufacturing cost is lower	Changes in formulation
Better distribution of stresses	May require heat and pressure
Electrically isolated components	Must track shelf life and out-time
Minimized strength reduction of composite	Adhesives change values with temperature
Reduced maintenance costs	May be attacked by solvents or cleaners
Corrosion of metal adherend is reduced (no drilled holes)	Common statement: "I won't ride in a glued-together airplane"
Better sonic fatigue resistance	

    b. Paste adhesive. Longer shelf life; minimal changes in storage; accommodates varying bond line thicknesses but quality assurance is performed after the structure is bonded.
3. Cure temperature of adhesive. Maximum cure temperature of adhesive should be below the composite cure temperature unless they are cocured. The cure temperature or upper use temperature of the adhesive may dictate the maximum environmental-exposure temperature of the component. There are several cure temperature ranges:
    a. Room temperature to 225°F. Generally for paste adhesives for noncritical structures.
    b. 225°F to 285°F. For nonaircraft critical structures. Cannot be used for aircraft generally because moisture absorption may lower HDT (Heat Deflection Temperature) below environmental operating temperature.
    c. 350°F and above. For aircraft structural bonding. Higher cure temperature may mean more strain discontinuity between adhesive and both mating surfaces (residual stresses in bond).
4. Joint design. The primary desired method of load transfer through an adhesive bonded joint is by shear. This means that the design must avoid peel, cleavage, and normal tensile stresses. One practical application of a method avoiding other than shear stresses is the use of a rivet bonded construction (Fig. 3.29). The direction of the fibers in the outer ply of the composite (against the adhesive) should not be 90° to the expected load path.

3.5.6 Environmental Effects

The primary environmental concern for all polymeric matrix composites is the effect of moisture intrusion into the raw materials as well as the finished components. Moisture in the raw materials causes voids and delaminations of the composite, and moisture in the finished composite structure can result in the effects shown in Table 3.35. Moisture enters the composite through cracks or voids in the matrix and diffuses through the resin or through the fiber-to-matrix interface (if open to atmosphere). Except for fiberglass and Kevlar, the advanced composite fibers do not

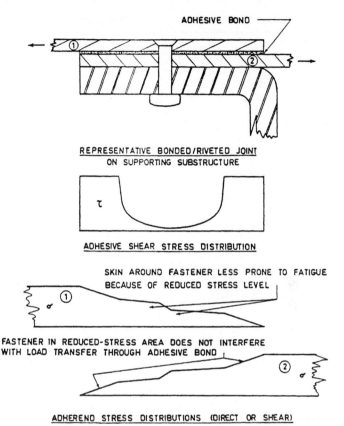

FIGURE 3.29   Load transfer in rivet-bonded construction (from Ref. 45).

transport water, and none of the fibers except for fiberglass are substantially affected by water intrusion. Fig. 3.30 shows typical degradation techniques for fiberglass and Kevlar.[46] Most of the effects of moisture are reversible, and the affected property will be restored when the composite is dried. It may take a long time to dry since the moisture must diffuse out.

Many users of composites require a systematic evaluation of the reaction of each composite (fiber and resin), to a broad list of environments. Many of these environments are listed in test documents such as MIL-STD-810. The evaluation can result in the approval of a composite structure because it is the same or very close to one already tested (similarity), because it is theoretically unresponsive (analysis), or because the actual structure or a series of coupons has been tested by the composite manufacturer (test). It is in the best interest of both the fabricator and the user when the resolution of the composites' response to an environment can be obtained at minimum cost, i.e., verification by similarity. Some of these environments, the predicted effects, and references are shown in Table 3.36 (adapted from Ref. 5, pp. 8-1 to 8-13).

TABLE 3.35   Possible Effects of Absorbed Moisture
on Polymeric Composites

- Plasticization of epoxy matrix
  Reduction of glass transition temperature;
  reduction of usable range
- Change in dimensions due to matrix swelling
- Enhanced creep and stress relaxation
  Increased ductility
- Change in coefficient of expansion
- Reduction in ultimate strength and stiffness
  properties in matrix-dominated properties
  Transverse tension
  Inplane shear
  Interlaminar shear
  Longitudinal compression
  Fatigue properties (change may be beneficial)
- Fiber-dominated properties are generally not affected
- Change in microwave transmissibility properties

3.6 COMPOSITE TESTING

To insure consistent, reproducible components, three levels of testing are employed:
incoming materials testing, in-process testing, and control and final structure verification.

3.6.1 Incoming Materials Testing

Incoming materials testing seeks to verify the conformance of the raw materials to
specifications and to insure processibility. The levels of knowledge of composite raw
materials do not approach those for metals, which can be bought to several consensus specifications and will appear generally identical from any manufacturer.
Although there are fewer suppliers for composite raw materials, the numbers of permutations of resins, fibers, and manufacturers prevents the kind of standardization
necessary to be able to buy composite raw materials as if they were alloys. ASTM
(American Society for Testing and Materials), SAE/AMS/NOMETCOM (Society
of Automotive Engineers, Aeronautical Materials Standards/Nonmetallic Materials
Committee), and SACMA (Suppliers of Advanced Composite Materials Association) attempt to standardize the raw materials and their test methods by publication
of specifications. However, these standards have not reached the level of use to
allow complete dependence upon them without supplier-user interaction and user
testing.

The fabricators of composites will rely on specifications for control of fiber, resin,
and/or the prepreg as shown in Table 3.37. Many prepreg resin and fiber vendors will
certify only to their own specifications, which may differ from those shown; users
should consult the vendors to determine what certification limits exist before committing to specification control. The purpose of incoming testing is to achieve a consistently reliable product that can be verified to meet the user's requirements with
his processing techniques; thus, the testing should reflect, if possible, the individuality of the processing method.

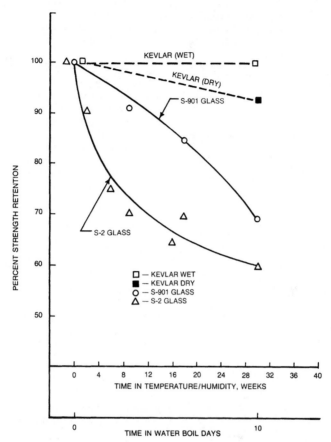

FIGURE 3.30 Composite response to humidity (from Ref. 46).

As part of raw materials verification, composite design effort, and final product verification, mechanical testing of composite test specimens will be performed. The testing of composite materials offers unique challenges because of the special characteristics of composites. Factors not considered important in metals testing are very important in testing composites. For example, composites are anisotropic, with properties that depend on the direction in which they are measured. Strain and load rate must be carefully monitored; specimen conditioning (drying, storage, and the like) can importantly affect results; and even humidity conditions at the time of specimen fabrication or test can significantly affect the material. Fiber content and void content, which can also vary with manufacturing conditions, have important effects on material properties. All personnel involved in the generation and use of test results, from fabrication of the material to final data interpretation, must be familiar with these factors and their influence on test results. Refs. 5 and 47 provide reference monographs on composites testing. Refs. 48 and 49 provide excellent descriptions of the factors that influence test results.

TABLE 3.36 Environmental Effects on Composite Structures

Environment	Effects (matrix or fiber)	Comments
Moisture-related		
Rain	Matrix softens and swells	Rain may cause water intrusion into composite and joints
Humidity	Matrix softens and swells	Effect can be aggravated by high, >1%, void content
Salt Fog	Matrix softens and swells	Corrosion of metal attached to graphite epoxy composite can be increased
Deep submergence	Matrix	Interface may be affected; low void content is requirement for compressive inputs
Rain and sand erosion	Both	Protect surfaces with paint or elastomer; composite response is significantly different from metals
Galvanic corrosion	Neither	Graphite composites can cause corrosion of metals; composite is unaffected
Radiation		
Solar radiation (earth)	Matrix	Unprotected aramid or polyethylene is affected
Ultraviolet	Matrix	Unprotected aramid or polyethylene is affected
Nonionizing space	Both	Composites can be destroyed; most metals are unaffected
Solar (LEO)	Both	Composites can be destroyed; most metals are unaffected
Temperature		
High	Matrix	Fibers (except polyethylene) are generally more resistant than matrix
Low	Matrix	
Extreme (ablation)	Both	
Thermal cycling (shock)	Matrix	
Miscellaneous environments		
Solvents, fuels	Matrix	Determine for each matrix, stressed and unstressed; effects may be aggravated by voids
Vacuum	Matrix	
Fungus	Matrix	Each material must be evaluated; may affect organic fibers
Fatigue	Both	Effects may be diminished by voids

Adapted from Ref. 5.

3.6.2 In-Process Testing

In-process testing seeks to ensure the repeatability of the process so as to reach a consistent fiber volume, resin content, and fiber placement. Since the processing from the start of the cure is a continuous part of the batch processing, there can be no test sample testing. The control is essentially interrogation of the curing process, modification of the process to accommodate the batch-to-batch inconsistencies of the input materials, and correlation of the effects of changes in input materials and processes to arrive at limits on the process variables and inputs. There have been frequent problems of resin exotherm and cure temperature overshoot with large or thick structures that could ruin a component. New autoclaves have been provided with interactive controls and "smart" operating software to accommodate changes in cure dynamics. An interactive "smart" curing machine can respond to the curing temperature overshoot automatically and rapidly. Table 3.38 shows techniques for in-process monitoring. Since composite structures can be large and complex to cost effectively replace metal assemblies, the labor input value of these structures can be considerable. The benefit of testing the incoming materials and of verifying the processing is obvious. Ideally, with feedback control and complete control over the incoming materials, there would be lessened need for final product inspection.

TABLE 3.37   Test Methods for Composite Raw Materials

Component	Test	Test method	Reference
Resin	Moisture content	Karl Fisher titration*	ARP* 1610
	Comparison identity	Chromatography (several types)*	ARP 1610
	Chlorine (epoxies)	Hydrolyzable chlorine*	ARP 1610
	Comparison	Color*	ASTM D-1544
	Comparison	Density	ASTM D-1475
	Comparison	Refractometric	User-defined
	Comparison	Infrared identity*	ARP 1610
	Comparison	Viscosity	ASTM D-445, 1545
	Reactivity	Weight per epoxide	ASTM D-1652
Fiber	Strand mechanicals	Strand tensile strength and modulus	AMS 3892
	Yield	Weight/length	AMS 3892
	Sizing	Extraction	ARP 1610, AMS 3901, AMS 3892
	Fuzz		User-defined
	Ribbonization		User-defined
	Packaging		AMS 3901, AMS 3894
	Density		ASTM D-3800
Prepreg	Fiber mechanicals	Tow test or unidirectional tensile	ASTM D-2343, D-3379, D-4018, D-3039
	Resin content	Extraction	ASTM C-613, D-1652, D-3529
	Resin identity	Infrared identity	ARP 1610
	Curative identity, %	Extraction, infrared	ARP 1610
	Nonvolatiles		ASTM D-3530
	Resin flow		ASTM D-3531
	Gel time		ASTM D-3532
	Tack		AMS 3894
	Visual imperfection	Optical	AMS 3894
	Areal weight	Weight	ASTM D-3529, D-3539

ARP: Aerospace Recommended Practice, SAE, Society of Automotive Engineers, 400 Commercial Drive, Warrendale, PA 15096.
AMS: Aerospace Materials Specification, SAE.
ASTM: American Society for Testing & Materials, 1916 Race Street, Philadelphia, PA 19103.
* Also applicable to prepreg.

TABLE 3.38 In-Process Monitoring of Advanced Composites

Parameter controlled (measured)	Desired location control*	Method	Notes
Temperature	In laminate, top bottom center on tool, air temp	Thermocouple	Includes rate of heating and cooling[†]
Vacuum	Composite surface	Gage (transducer)	[†]
Pressure	Composite surface in composite vessel	Gage (transducer), springback go/ no-go gage	[†]
Cure rate	In composite	Dielectric analysis optical (fluoresence),[‡] FTIR spectra[‡]	[†]
Viscosity	In composite	Dielectric analysis ultrasonic,[‡] acoustic waveguides[‡]	[†]

See Ref. 47 for further details.
* Frequency and number of monitored factors will decrease as production confidence is gained.
[†] Information part of adaptive control system.
[‡] Not in general use yet.

3.6.3 Final Product Verification

Nearly all final composite structure verification includes mechanical tests of tag end or coprocessed coupons and nondestructive tests on the structural laminate. Mechanical testing is also used to verify the materials and processes prior to committing to the lay-up and cure of a larger composite structure. Table 3.39 shows the key data and test methods that are needed to support design or production of a composite structure.

It can be seen that there are several tests in current use to determine some properties. Many of the tests are still evolving and will eventually become standardized through ASTM or other consensus organizations. Also, this list does not exhaust the available test methods and nonstandard tests abound for composites fabricated by alternate techniques (e.g., filament winding, Ref. 46).

Nondestructive test techniques are preferred to verify the final structure because coprocessed or tag-end test specimens suffer from edge effects and may not mirror the fiber angles, resin, or fiber content. The component may also have discrepancies such as fiber kinking and wash-out that would not be reflected in the test specimens. Confidence is gained in a structure because of careful control of incoming materials, controlled processing, adherence of tag-end or coprocessed test specimens to required test values, and the meeting of NDI standards. The simplest NDI techniques that involved coin tapping and visual observance have evolved into those shown in Table 3.40 that can be used individually or, in some cases, concurrently.

TABLE 3.39 Advanced Composite Test Methods in Current Use

Test	Test Method(s)
Tensile stress and modulus axial and transverse	ASTM D-3039
Compression	ASTM D-695 (modified for high-modulus composites); Celanese,* ASTM D-3410; IITRI, ASTM D-3410; sandwich beam, ASTM D-3410
Shear	Iosipescu, ASTM D-3518 Rail shear, ASTM D-4255 Short beam, ASTM D-2344
Flexure	ASTM D-790
Fracture toughness	ASTM E-399 NASA 1092* End-notched flexure mode II Cracked lap shear (Mixed modes I and II) Edge delamination NASA 1092
Impact	Instrumented drop-weight impact Tensile impact Compression after impact (Boeing BSS-7260)
Fatigue	ASTM D-3479, D-671
Coefficient of thermal expansion	User-defined
Coefficient of moisture expansion	User-defined
Single fiber tension	ASTM D-3379
Single fiber tensile creep	†
Open-hole tension, compression	Boeing BMS 8-276
Thermal conductivity	User-defined
Bolt bearing	ASTM D-953

Adapted from Ref. 49.
* Refers to the original design agency, or the agency who prepared the reference specification or to the commercial company whose internal specification was used throughout the industry.
† No general consensus specification exists.

3.7 SAFETY ISSUES WITH COMPOSITE MATERIALS

Composite materials, particularly in the raw material form, present toxic hazards to the users and should be handled carefully. Essentially there is a need for management inputs and controls on the following aspects of composite fabrication:

- Material handling
- Training (to gain awareness of hazards and proper use of toxic materials)
- Isolation of some operations
- Use of personal protective equipment (if necessary)

TABLE 3.40 NDE Test Methods for Composites

	Radiography	Computer tomography	Ultrasonics	Acoustic emission	Acoustic ultrasonics	Thermography	Optical holography
Principal characteristic detected	Differential absorption of penetrating radiation	Conventional x-ray technology with computer digital processing	Changes in acoustic impedance caused by defects	Defects in part stressed generate stress waves	Uses pulsed ultrasound stress wave stimulation	Mapping of temperature distribution over the test area	3D imaging of a diffusely reflecting object
Advantages	Film provides record of inspection, extensive data base	Pinpoint defect location; image display is computer controlled	Can penetrate thick materials, can be automated	Remote and continuous surveillance	Portable, quantitative, automated, graphic imaging	Rapid, remote measurement, need not contact part, quantitative	No special surface preparation or coating required
Limitations	Expensive, depth of defect not indicated, radiation safety	Very expensive, thin wall structure might give problems	Water immersion or couplant needed	Requires application of stress for defect detection	Surface, contact surface geometry critical	Poor resolution for thick specimens	Vibration-free environment required heavy base needed

Adapted from Ref. 52.

- Personal hygiene
- Significance of warnings and labels
- Housekeeping
- Dispensing and storage
- Emergency instructions

The typical materials that are encountered are shown in Table 3.41. This table is not exhaustive, and the user should consult the Material Safety Data Sheets (MSDS) and more in-depth references.[53] The liquid resins and catalysts are the highest noted hazards because there are so many avenues of attack to our bodies and because there are many different effects. Once the composite structure is fabricated, the fibers and resins are rendered essentially innocuous, and machining, drilling, and the like pose minimal hazards. Machining dusts should be contained and properly disposed of to reduce the nuisance.

TABLE 3.41   Commonly Encountered Hazards with Advanced Composite Materials

Hazard	Form	Reported Hazards
Epoxy resins	Liquid or as prepreg	Dermatitis. Some may be potential skin carcinogens
Epoxy curing agents Aromatic amines	Liquids, but primarily in prepreg	Liver, kidney damage, jaundice, hazards associated with carcinogens, anemia
Aliphatic amines	Liquids	Severe bases and irritants, visual disturbances
Polyaminoamides	Liquids	Somewhat less irritating, may cause sensitization
Amide		Slight irritant
Anhydride	Liquid, solid, prepreg	Severe eye irritants, strong skin irritants
Phenolic and amino resins	Usually as dry prepreg	Free phenol also released on cure. Formaldehyde is a strong eye, skin, and respiratory irritant
Bismaleimide resins	Liquid or paste, generally as prepreg	Skin irritant and sensitization, not fully characterized yet
Thermoplastic resins	Solids or prepreg	Exercise care with molten materials and provide good ventilation. Consult MSDS
Reinforcing materials Graphite fibers	Airborne dusts	Skin irritation. Prevent exposure and protect skin
Aramid fibers		Airborne dusts
Fiberglass		Mechanical irritation to skin, eyes, nose, and throat

REFERENCES

1. Jones, Robert M., *Mechanics of Composite Materials,* Scripta Book Co., Washington, DC, 1975.
2. Tsai, Steven W. and H. Thomas Hahn, *Introduction to Composite Materials,* Technomic Publishing Co., Westport, CT, 1980.
3. Agarwal, Bhagwan D. and Lawrence J. Broutman, *Analysis and Performance of Fiber Composites,* Wiley Interscience, New York, 1980.
4. Kline and Co., "The Changing Advanced Polymer Composite Market," 1993.
5. Peters, S. T., W. D. Humphrey, and R. F. Foral, *Filament Winding: Composite Structure Fabrication,* SAMPE Publishers, Covina, CA, 1991.
6. Delmonte, John, *Technology of Carbon and Graphite Fiber Composites,* Van Nostrand Reinhold Co., New York, 1981, p. 59.
7. Bacon, Roger and Miles Towne, Amoco performance products, personal communication.
8. Schwartz, Mel M., *Composite Materials Handbook,* McGraw-Hill, New York, 1983, p. 2.65.
9. Foral, Ralph F. and Stanley T. Peters, *Composite Structures and Technology Seminar Notes,* 1989.
10. Dudgeon, Charles D., in *Engineered Materials Handbook,* vol. 1, Theodore Reinhart, Tech. Chairman, ASM International, 1987, p. 91.
11. Mones, Eleno T. and Roger J. Morgan, "FTIR Studies of the Chemical Structure of High Performance Composite Matrices," UCRL Report 85789, Lawrence Livermore National Laboratory, University of California, 1981.
12. Puglisi, Joseph S. and Mohammad A. Chaudhari, "Engineering Plastics" in J. N. Epel et al. (eds.), *Engineered Materials Handbook,* vol. 2, ASM International, 1988, p. 241.
13. Harvey, James A. in Ref. 10, p. 256.
14. Scola, D. A. in Ref. 12, p. 78.
15. Shimp, David A., in S. T. Peters (ed.), *Handbook of Composites,* Chapman & Hall, in press.
16. Couch, Brian P. and Lawrence A. McAllister, in *35 SAMPE Symposium,* April, 1990, pp. 2298–2310.
17. Harrington, Herman J. in Ref. 12, p. 242.
18. Fisher, James J. in Ref. 12, p. 230.
19. Giler, Michael R. in Ref. 12, p. 227.
20. Mayorga, G. D. in S. M. Lee (ed.), *International Encyclopedia of Composites,* vol. 4, VCH Publishers, New York, 1991.
21. Shimp, David A. in Ref. 12, p. 232.
22. Tsai, Steven W., *Composites Design,* 4th ed., Think Composites, Dayton, OH, pp. 10-2–10-6.
23. Tsai, Steven W. and Nicholas J. Pagano, in S. W. Tsai, J. C. Halpin, and Nicholas J. Pagano (eds.), *Composite Materials Workshop,* Technomic Publishing Co., Lancaster, PA, 1978, p. 249.
24. Dupont Data Manual for Kevlar 49 Aramid, May 1986, p. 95.
25. Sanders, Robert E. and Saad Taha, in Ref. 10, p. 703.
26. Foston, Marvin and R. C. Adams in Ref. 10, pp. 591–601.
27. ICI Fiberite Data Sheet, March 15, 1989, pp. V, b-6.
28. Dave, R., J. L. Kardos, and M. P. Dudokovic in *Proceedings of the American Society for Composites,* First Technical Conference, Oct., 1986, Charles F. Browning, Chairman, Technomic Publishing Co., Lancaster, PA, p. 151.

29. Campbell, Flake C. et al., *J. Adv. Materials,* July 1995, pp. 18–33.

30. Hartman, D. R. and S. Schuster, *SAMPE Proceedings,* vol. 38, May 1993, p. 1529.

31. Humphrey, W. Donald and Murali Vedula, *SAMPE Proceedings,* vol. 40, May 1995, p. 1478.

32. Loos, Alfred C. and George S. Springer, *J. Comp. Mater.,* March 17, 1983, pp. 135–169.

33. Enders, Mark L. and Paul C. Hopkins, *SAMPE International Symposium and Exhibition,* vol. 36, April 1981, pp. 778–790.

34. Goldsworthy, W. Brandt, in Stuart M. Lee (ed.), *Reference Book for Composites Technology,* vol. 2, Technomic Publishing Co., Lancaster, PA, 1989, p. 180.

35. Martin, Jeffrey D. and Joseph E. Sumerak in Ref. 10, p. 542.

36. Ko, Frank K. in Ref. 10, p. 519.

37. Genlam, General Purpose Laminate Program, Think Composites, Dayton, OH.

38. Brown, Richard T. in Ref. 10, pp. 268–274.

39. Kibler, John J. in Ref. 10, pp. 275–279.

40. Freeman, W. T. and B. A. Stein, *Aerospace America,* Oct. 1985, pp. 44–49.

41. Klein, Allen J., *Advanced Materials and Processes,* Oct. 1985, pp. 43–46.

42. Seidl, A. L. "Repair of Composite Structures on Commercial Aircraft," 15th Annual Advanced Composites Workshop, Northern California chapter of SAMPE, Jan. 27, 1989.

43. Seidl, A. I. in Ref. 15.

44. Klein, Allen J., *Advanced Composites Magazine,* July/August 1987, p. 51.

45. Potter, D. L., *Primary Adhesively Bonded Structure Technology (PABST) Design Handbook for Adhesive Bonding,* Douglas Aircraft Co., McDonnell Douglas Corporation, Long Beach, CA, Jan. 1979.

46. Peters, S. T. and W. D. Humphrey in Ref. 10, pp. 512–514.

47. Kranbuehl, David E., in Stuart M. Lee (ed.), *International Encyclopedia of Composites,* vol. 1, pp. 531–543, VCH Publishers, New York, 1990.

48. Whitney, J. M., I. M. Daniel, and R. B. Pipes, *Experimental Mechanics of Fiber Reinforced Composite Materials,* SESA Monograph No. 4, The Society for Experimental Stress Analysis, Brookfield Center, CT, 1982.

49. Carlsson, L. A. and R. B. Pipes, *Experimental Characterization of Advanced Composite Materials,* Prentice-Hall, Englewood Cliffs, NJ, 1987.

50. *Annual Book of ASTM Standards,* American Society for Testing and Materials, Philadelphia, PA.

51. Adams, Donald F., "Test Methods for Composite Materials," Seminar Notes, Technomic Publishing Co., Lancaster, PA.

52. Munjal, A., *SAMPE Quarterly,* Jan. 1986.

53. Safe Handling of Advanced Composite Materials Components: Health Association, Arlington, VA, April 1989.

CHAPTER 4
LIQUID AND LOW-PRESSURE RESIN SYSTEMS

Leonard S. Buchoff
Elastomeric Technologies, Inc.
Hatboro, Pennsylvania

4.1 INTRODUCTION

The materials covered in this chapter are resins that are liquids at room temperature or solids that melt at temperatures up to 200°C. The chapter does not cover materials that are liquids by virtue of being dissolved in solvents like solvent-based adhesives, or being suspended in water, such as rubber. Their common attribute is conversion to a solid by the action of heat or a curing agent. Most of the materials are cross-linked during the curing, resulting in an infusible, insoluble thermoset. Some, such as cast acrylic, may be converted to a thermoplastic. In a third group that is typified by vinyl plastisols, the conversion from liquid to solid is accomplished by a physical process, with no chemical change. In general, if solvents are used with these resins, they play a minor role in the conversion of the formulation, or function in ways other than a solvent. They may enter into the reaction, as styrene does in polyester.

Liquid resins are useful in the fields of encapsulation, art, adhesives, laminates, foams, coatings, construction, resurfacing, centrifugal casting, embedments, tooling, and model making. In recent years there has been a rapid growth of reinforced thermosets, particularly in the automotive field. This growth has triggered many new compounds tailored specifically to be mixed and injected into closed molds, which contain the reinforcement. Other chapters in this handbook deal with adhesives, foams, laminates, coatings, and electrical insulation. In this chapter, properties and guides to the selection of materials and compounds are presented.

4.1.1 Advantages of Liquid Resins

Liquid resins are important because of their versatility and ease of application. Compounding is often accomplished by dispersing curing agents, fillers, and other ingredients at room temperature with a simple propeller mixer. Fillers are wetted better by the liquid resins than by molten polymer, and the fillers can be chemically bound. A variety of properties is readily available by changing formulations. Stiffness, color, strength, and electrical properties can be tailored to specific needs by changing proportions and types of ingredients.

A range of cure temperatures and times is available. Most classes of liquid resins can be formulated to cure at room temperature, eliminating the need for ovens. Equipment in general can be very simple and inexpensive for small prototype production, but highly automated liquid dispensers, mold handlers, reinforcement placers, and peripheral equipment are used for high-volume production. Often disposable containers can be used for hand mixing to prepare material for repairs or use in remote areas, such as in mixing field repairs on electrical equipment, in resurfacing highways and concrete floors, or in installing sealants or tiles.

Liquid resins also allow a wide range of mold constructions. Inexpensive, one-shot paper molds and wooden molds can be used to construct entire walls and boats. Plaster of paris, plastic, sheet metal, foam, and many other materials have been used in mold construction for castings.

There is no theoretical limitation to the size of castings that can be produced. Insulators weighing several tons have been cast from epoxies, and nylon parts much too large to be molded have been cast from caprolactams. Liquid resins can also be poured around small electrical components, flowers, or other delicate objects without altering their shape or changing their positions. Proper compounding can produce castings with negligible shrinkage and optical clarity. Castings can be produced with a minimum of internal stress. They will therefore be stronger, more dimensionally stable, and less subject to stress cracking than parts made by other processes.

Part and mold design as well as the placement of reinforcement are calculated with sophisticated computer software. Computer-aided design and engineering have dramatically changed the way parts are designed and manufactured. Most of the trials and errors have been wrung out of the processes.

4.1.2 Selection of the Resin

Many of the materials covered in this chapter are used in electrical and electronic parts. A comparison of the properties of potting resins is given in Tables 4.1 and 4.2. Table 4.3 identifies resin characteristics that influence the choice of materials in electrical applications. Table 4.4 is included to aid in selecting materials for nonelectrical applications, as the properties requirements are not necessarily the same as in electrical applications.

4.1.3 Safety and Environmental Considerations

In the past few years there has been a great deal of concern and governmental activity to eliminate or curtail the use of many chemicals. The health hazards associated with exposure to several resins, curing agents, solvents, and additives have greatly influenced the selection of materials used and the precautions taken in their use. Specific considerations are given for most materials covered. More detailed recommendations are presented in Sec. 4.2.

4.1.4 Processing Liquid Resins

Electrical components and circuits are embedded in liquid resins by casting, potting, liquid injection molding, encapsulation, or impregnation. Fabricated parts are pro-

TABLE 4.1 Comparison of Properties of Liquid Resins

Material	Cure shrinkage	Adhesion	Thermal shock	Electrical properties	Mechanical properties	Handling properties	Cost
Epoxy:							
Room temp. cure.........	Low	Good	Fair*	Fair	Fair	Good	Moderate
High temp........	Low	Good	Fair*	Good	Good	Fair to good	Moderate
Flexible.........	Low	Excellent	Good	Fair	Fair	Good	Moderate
Polyesters:							
Rigid........	High	Fair	Fair*	Fair to good	Good	Fair	Low to moderate
Flexible.........	Moderate	Fair	Fair	Fair	Fair	Fair	Low to moderate
Silicones:							
Rubbers.........	Very low	Poor	Excellent	Good	Poor	Good	High
Rigid........	High	Poor	Poor	Excellent	Poor	Fair	High
Polyurethanes:							
Solid.........	Low	Excellent	Good	Fair	Good	Fair	Moderate
Foams........	Variable	Good	Good	Fair	Good for density	Fair	Moderate
Butyl LM	Low	Good	Excellent	Good	Poor	Fair	Low to moderate
Butadienes........	Moderate	Fair	Poor*	Excellent	Fair	Fair	Moderate

* Depends on filler.

TABLE 4.2 Properties of Various Liquid Resins[1]

Property	ASTM	Rigids (silica-filled)			Flexibilized epoxies			Liquid elastomers				Rigid polyurethane
		Styrene-polyester	Epoxy	Silicone	Epoxy-polysulfide (50-50)	Epoxy-polyamide (50-50)	Epoxy-polyurethane (50-50, diamine cure)	Polysulfide	Silicone	Polyurethane (diamine cure)	Plastisol	Prepolymer
					Electrical Properties							
Dielectric strength, V/mil*	D 149	425	425	350	350	430	640	340	400	350	200	
Dielectric constant:												
60 Hz	D 150	3.7	3.8	3.7	5.6	3.2	.....	7.2	3	8.2	5.6	1.05
10^3 Hz	D 150	3.7	3.6	3.6	5.4	3.2	4.9	7.2	3	7.3	4.9	1.06
10^6 Hz	D 150	3.6	3.4	3.6	4.8	3.1	.....	7.2	3	.....	3.6	1.04
Dissipation factor:												
60 Hz	D 150	0.01	0.02	0.008	0.02	0.01	.....	0.01	0.005	0.08	0.12	0.004
10^3 Hz	D 150	0.02	0.02	0.004	0.02	0.01	0.04	0.01	0.004	0.09	0.1	0.003
10^6 Hz	D 150	0.02	0.03	0.01	0.06	0.02	.....	0.02	0.003	.....	0.12	0.003
Surface resistivity, Ω/square:												
Dry	D 257	10^{15}	10^{15}	10^{15}	10^{12}	10^{14}	.....	10^{11}	10^{13}	.....	10^{10}	$>10^{12}$
After 96 h at 95°F, 90% RH	D 257	10^{11}	10^{13}	10^{15}	10^{10}	.....	.....	10^{10}	10^{12}	.....	10^{10}	$>10^{12}$
Volume resistivity (dry), Ω-cm	D 257	$>10^{16}$	$>10^{15}$	10^{15}	10^{12}	10^{14}	10^{14}	10^{11}	10^{14}	10^{12}	10^{10}	$>10^{14}$
					Mechanical Properties							
Tensile strength, lb/in.²	D 638, D 412	10,000	9,000	4,000 ·	1,800	4,600	6,000	800	275	4,000	2,400	110
Elongation, %	D 412	.....	.....	.....	30	.....	10	450	200	450	300	.....
Compression strength, lb/in.²	D 695, D 575	25,000	16,000	13,000	.....	7,000	.....	.....	.....	.....	.....	85

4.4

	ASTM											
Flexural modulus of rupture, lb/in.²	D 790	10,000	12,000	8,000	.....	8,300	.....	.....	.....	.....	.....	.....
Izod impact strength (notched), ft-lb.	D 256	0.3	0.4	0.3	.....	.....	.....	.....	.....	.....	.....	.....
Shore hardness	D 676, D 1484	.....	.....	.....	40D	80D	.....	45A	35A	90A	80A	.....
Penetration at 77°F, mils	D 5	.....	.....	.....	.....	.....	.....	1	2	4.5	.....	.....
Shrinkage on cooling or curing, % by vol.		6	3.5	8	3	3	6	.....	.....	.....	.....	.....
Physical Properties												
Specific gravity	D 71, D 792	1.6	1.6	1.8	1.2	1.0	1.2	1.2	1.1	1.1	1.2	0.1†
Min cold flow, °F.		.....	.....	.....	.....	.....	.....	.....	.....	.....	.....	.....
Softening or drip point, °F.	D 36	.....	.....	.....	.....	.....	.....	.....	.....	.....	.....	.....
Heat-distortion temp, °F.	D 648	230	165	300	.....	.....	100	.....	.....	.....	.....	.....
Max continuous service temp, °F.		250	250	480	225	175	250	200	350	210	150	165
Coefficient of thermal expansion per °F × 10⁻⁶	D 696	26	22	44	44	44	110	.....	128	110	.....	19
Thermal conductivity, Btu/(ft²)(h)(°F/ft).		0.193	0.29	0.23	.....	.....	0.77	.....	0.13	.....	0.09	0.02
Cost, $/lb.		0.25	0.50	2.75	0.90	0.80	1.10	0.85	4.25	1.30	0.30	1.75

* Short-time test on ⅛-in specimen.
† 6 lb/ft³.

TABLE 4.3   Characteristics Influencing Choice of Resins in Electrical Applications

Resin	Cure and handling characteristics	Final part properties
Epoxies........	Low shrinkage, compatible with a wide variety of modifiers, very long storage stability, moderate viscosity, cure under adverse conditions	Excellent adhesion, high strength, available clear, resistant to solvents and strong bases, sacrifice of properties for high flexibility
Polyesters......	Moderate to high shrinkage, cure cycle variable over wide range, very low viscosity possible, limited compatibility, low cost, long pot life, easily modified, limited shelf life, strong odor with styrene	Fair adhesion, good electrical properties, water-white, range of flexibilities
Polyurethanes...	Free isocyanate is toxic, must be kept water-free, low cure shrinkage, solid curing agent for best properties	Wide range of hardness, excellent wear, tear, and chemical resistance, fair electricals, excellent adhesion, reverts in humidity
Silicones .... (flexible)	Some are badly cure-inhibited, some have uncertain cure times, low cure shrinkage, room temp cure, long shelf life, adjustable cure times, expensive	Properties constant with temperature, excellent electrical properties, available from soft gels to strong elastomers, good release properties
Silicones........ (rigid)	High cure shrinkage, expensive	Brittle, high-temperature stability, electrical properties excellent, low tensile and impact strength
Polybutadienes..	High viscosity, reacted with isocyanates, epoxies, or vinyl monomers, moderate cure shrinkage	Excellent electrical properties and low water absorption, lower strength than other materials
Polysulfides.....	Disagreeable odor, no cure exotherm, high viscosity	Good flexibility and adhesion, excellent resistance to solvents and oxidation, poor physical properties
Depolymerized rubber	High viscosity, low cure shrinkage, low cost, variable cure times and temperature, one-part material available	Low strength, flexible, low vapor transmission, good electrical properties
Allylic resins....	High viscosity for low-cure-shrinkage materials, high cost	Excellent electrical properties, resistant to water and chemicals

duced from liquid resins by casting, rotational molding, centrifugal casting, pultrusion, liquid injection molding, reaction injection molding, resin transfer molding, and variations of these techniques. Entirely new processes are being used that depend on ultraviolet, laser, or other high-energy sources to cure the resins. Features of these processes are presented in the following subsections.

Casting.   Prepared liquid-resin compound is poured into a stationary mold and cured. If objects are to be embedded, they are prepositioned in the mold or introduced after the resin is poured.

  Advantages.   Minimum equipment is required, delicate parts can be embedded; a wide choice of optical effects is possible; there is no limit to part size; and there exists a wide choice of mold constructions.

  Disadvantages.   Bubbles and voids are problems. High-viscosity formulations are difficult to handle.

TABLE 4.4 Characteristics of Liquid Resins for Nonelectrical Applications

Resin	Handling characteristics	Final part properties	Typical applications
Cast acrylics	Low to high viscosity, long cure times, bubbles a problem, special equipment necessary for large parts	Optical clarity, excellent weathering, resistance to chemicals and solvents	In glazing, furniture, embedments, impregnation
Cast nylon	Complex casting procedure, very large parts possible	Strong, abrasion-resistant, wear-resistant, resistance to chemicals and solvents, good lubricity	Gears, bushings, wear plates, stock shapes, bearings
Phenolics	Acid catalyst used, water given off in cure	High density unfilled, brittle, high temperature, brilliance	Billiard balls, beads
Vinyl plastisol	Inexpensive, range of viscosities, fast set, cure in place, one-part system, good shelf life	Same as molded vinyls, flame-resistant, range of hardness	Sealing gaskets, hollow toys, foamed carpet backing
Epoxies	Cures under adverse conditions, over a wide temp range, compatible with many modifiers, higher filler loadings	High strength, good wear, chemical and abrasion resistance, excellent adhesion	Tooling, fixtures, road and bridge repairs, chemical-resistant coating, laminates, and adhesives
Polyesters	Inexpensive, low viscosity, good pot life, fast cures, high exotherm, high cure shrinkage, some cure inhibition possible, wets fibers easily	Moderate strength, range of flexibilities, water-white available, good chemical resistance, easily made fire retardant	Art objects, laminates for boats, chemical piping, tanks, aircraft, and building panels
Polyurethanes	Free isocyanate is toxic, must be kept water-free, low cure shrinkage, cast hot	Wide range of hardnesses, excellent wear, tear, and chemical resistance, very strong	Press pads, truck wheels, impellers, shoe heels and soles
Flexible soft silicone	High-strength materials, easily inhibited, low cure shrinkage, adjustable cure time, no exotherm	Good release properties, flexible and useful over wide temperature range, resistant to many chemicals, good tear resistance	Casting molds for plastics and metals, high-temperature seals
Allylic resins	Long pot life, low vapor pressure monomers, high cure temperature, low viscosity for monomers	Excellent clarity, abrasion-resistant, color stability, resistant to solvents and acids	Safety lenses, face shields, casting impregnation, as monomer in polyester
Depolymerized rubber	High viscosity, low cost, adjustable cure time	Low strength, low vapor transmission, resistant to reversion in high humidity	Roofing coating, sealant, in reservoir liners
Polysulfide	High viscosity, characteristic odor, low cure exotherm	Good flexibility and adhesion, excellent resistance to solvents and oxidation, poor physical properties	Sealants, leather impregnation
Epoxy vinyl esters	Low viscosity, high cure shrinkage, wets fibers easily, fast cures	Excellent corrosion resistance, high impact resistance, excellent electrical insulation properties	Absorption towers, process vessels, storage tanks, piping, hood scrubbers, ducts and exhaust stacks
Cyanate esters	Low viscosity at room temperature or heated, rapid fiber wetting, high temperature cure	High operating temperature, good water resistance, excellent adhesive properties and low dielectric constant	Structural fiber-reinforced products, high temperature film adhesive, pultrusion and filament winding

Material Used.   Epoxies, polyesters, acrylics, polyurethanes, silicones.

Equipment.   Equipment ranges from simple molds and hand mixers to elaborate metering and mixing machines and automated lines.

Applications.   Electrical embedding; casting art statuary; embedding biological specimens; building facades; sinks, bathtubs, toilets; road and bridge repairs.

Rotational Molding.   The catalyzed resin is poured into a mold, and the mold is rotated biaxially in a heated chamber. The resin is distributed solely by gravitational force. Centrifugal force is not a factor.

Advantages.   Large hollow parts can be produced; molds are inexpensive; parts are stress-free.

Disadvantages.   Heating and cooling cycles of heavy molds are time-consuming. Room-temperature-setting materials would reduce this problem.

Material Used.   Plastisols at present; epoxies, silicones, and polyurethanes in the future.

Equipment.   Heavy machinery is necessary for handling large molds. Carousel machines having multiple molds are used to speed production. Heating is done with hot air, hot salt spray, or heating oil.

Applications.   Toys, novelty items, tank bodies.

Centrifugal Casting.   The catalyzed resin is introduced into a rapidly rotating mold, where it forms a layer on the mold surfaces and hardens. Vacuum is sometimes used to aid in bubble removal.

Advantages.   Reinforcing elements can be accurately placed; high viscosity, rapidly curing material can be used; dense bubble-free parts; precise wall thickness is possible; finished inside diameter.

Disadvantages.   Fillers separate rapidly; limited to toroidal shapes.

Material Used.   Polyurethane, plastisols, other liquid elastomers, epoxy.

Equipment.   High-speed rotating machinery; vacuum system sometimes used.

Applications.   Elastomer sheets, reinforced pipe, rubber track.

Liquid Injection Molding.   Catalyzed resin is metered into closed molds, curing rapidly, and the finished parts are ejected automatically.

Advantages.   Liquids can be metered accurately; low injection pressure; rapidly curing materials can be automatically mixed immediately before entering the molds; flash is easily avoided.

Disadvantages.   Low-viscosity resin necessary; expensive equipment; precision molds needed.

Material Used.   Polyurethanes, epoxies, silicones, polyesters, vinyl plastisols.

Equipment.   Injection presses; automatic mixing and dispensing machine; multiple molds.

Applications.   Polyurethane shoe soles and heels; transparent caps for light-emitting diodes; large structural polyester parts.

Pultrusion.   Reinforced-plastics profiles are produced in continuous lengths. Fibrous glass in the form of roving, mat, or cloth is impregnated with catalyzed resin, drawn through a heated die, and the cured shapes are cut to appropriate lengths.

Advantages.   High-strength products; continuous lengths available; potentially inexpensive.

Disadvantages.   Expensive equipment; complex specialized dies; limited speed.

Material Used.   Polyester and epoxy; low viscosity, internally lubricated.

Equipment. Impregnation tank, heated die, RF generator or oven, puller, diamond-faced cutoff saw.

Applications. Fishing rods, ladders, electrical-pole-line hardware, standard structural shapes.

Resin Transfer Molding. Liquid reactive system is metered into a closed mold that contains reinforcement. The part is removed after the resin has cured.

Advantages. Low tooling cost; large parts such as truck sections can be made; reinforcement is placed where needed.

Disadvantages. Long cycles, typically 30 min.

Material Used. Slow-curing low-viscosity epoxy and polyester.

Equipment. Low-cost molds; metering and mixing equipment; reinforcement cutting and forming equipment.

Applications. High-temperature aerospace parts, automotive and truck sections.[2]

Reaction Injection Molding. Reactive liquid is injected into a closed mold containing reinforcement. After rapid curing the part is removed.

Advantages. Mold cycles as short as 30 s; more economical than thermoplastic injection molding; high-strength structural parts are possible.

Disadvantages. Rapid curing, low-viscosity resin required; molds more expensive than those used in resin transfer molding; size limitation of parts.

Resins Used. Polyurethanes, polyureas, polydicyclopentadiene, acrylamate, caprolactam.

Reinforcements Used. Glass, polyester, carbon, aramid fiber.

Equipment. Molds; metering and mixing machines.

Applications. Automobile body panels; window glass frame; furniture.[3,4]

Rapid Prototyping. Prototypes are produced by curing photosensitive compounds with ultraviolet radiation. A beam of ultraviolet light controlled by a computer sweeps out a pattern that represents one layer of the part wanted. The liquid resin polymerizes to a solid where the ultraviolet light strikes it. The process is repeated until the entire part is formed. Acrylic functional compounds are frequently used.[5,6]

4.2 EPOXIES

In the past 40 years, epoxy resins have gained wide acceptance in such industrial fields as adhesives, coatings, castings, potting, building construction, chemical-resistant equipment, and boats. The properties that have made the epoxies popular in so many fields are their versatility, excellent adhesion, low cure shrinkage, good electrical properties, compatibility with a great number of materials, resistance to chemicals and weathering, dependability, and ability to cure under adverse conditions.

Epoxies can be compounded to produce a wide range of handling, curing, and final-part properties by choice of the basic resins, curing agents, fillers, and modifiers. Each of these topics is discussed separately in this section. As the curing agent becomes an integral part of the cured compound, its choice is a controlling influence on the curing and final properties of the mixture. Fillers and modifiers are used to tailor the liquid viscosity and cured properties to the application.

4.2.1 Basic Types of Epoxy Resins

Epi-Bis Epoxies. Most of the epoxies used today are a liquid reaction product of epichlorohydrin and bisphenol A, often called epi-bis epoxies. Table 4.5 lists the trade names of some of the major epi-bis resins and the companies that produce them. Others in this group range from resins of increasing viscosity to solids melting up to 175°C. Generally the higher the melting point, the less curing agent is needed. The cured properties of all these resins are similar, but the toughness increases as the melting point increases. Although most of the epi-bis resins are light amber, transparent colorless epoxies are available for optical embedments.

The electronic industry demands epoxy resins with minimum ionic contamination, particularly sodium and chlorine. Most manufacturers supply epi-bis resins with fewer than 100 ppm ionics, some less than 1 ppm of chlorine and sodium.[7,8] Table 4.6 lists the properties of typical cured epi-bis resin with several types of curing agents.

FIGURE 4.1 High voltage bushing cast from a cycloaliphatic epoxy. (*Courtesy of Union Carbide Corporation.*)

Cycloaliphatic Epoxies. These resins are characterized by the saturated ring present in the structure. They are free of the hydrolyzable chlorine, sometimes present in the epi-bis epoxies, which adversely affects some electronic devices. The cycloaliphatics have superior arc-track resistance, good electrical properties under adverse conditions, good weathering properties, high heat-deflection temperatures, and good color retention. Some members of this group are low in viscosity and serve as reactive diluents in laminate structures.

While most cycloaliphatics are cured only with anhydrides, others do react with amines. Table 4.7 shows how the properties of these materials can be modified by blending a flexible and a rigid resin. Massive high-voltage insulators weighing 5000 lb have been cast with cycloaliphatics.[11,12]

Extensive testing has shown that bushings cast from cycloaliphatic resins resist high-voltage breakdown and tracking better than bushings cast from other epoxies. Fig. 4.1 shows typical high-voltage cycloaliphatic epoxy bushings.

TABLE 4.5 Trade Names and Suppliers of Epoxy Resins

Resin characteristic	Shell (Epon)	Dow (D.E.R.)	Hi-Tek (Epi-Rez)	Ciba-Geigy (Araldite)	Reichhold (Epotuf)
Standard	828	331	510	6010	37–140
Low viscosity, unmodified	825	332	508	6004	37–139
Flexible	871	732	50,821	508	37–151
Low viscosity, butyl glycidyl ether modified	815	334	5071	506	37–130
Low viscosity, phenyl glycidyl ether modified	820	336			37–135

TABLE 4.6 Properties of Epi-Bis Resin Cured with Various Hardeners[9]

Hardener	phr*	Tensile strength, lb/in.² at 25°C	Tensile modulus, lb/in.² ×10⁻⁶	Tensile elongation, %	Dielectric constant, 60 Hz at 25°C	50°C	100°C	Dielectric strength, S/T, V/mil	Dissipation factor, 60 Hz at 25°C	50°C	100°C	150°C	Volume resistivity, MΩ-cm	Arc resistance, s (ASTM D 495)
Aliphatic amines:														
Diethyl amino propylamine	7	8,500	……	5–6	……	……	……	……	……	……	……	……	……	58
Aminoethyl piperazine	20	10,000	……	……	……	……	……	……	……	……	……	……	……	62
Aromatic amines:														
Mixture of aromatic amines	23	12,600	0.4	6–5	4.4	……	4.6	……	0.007	……	0.002	……	$>1 \times 10^{10}$	80
p,p'-Methylene dianiline	27	9,500	0.5	4–5	4.4	……	4.7	420	0.007	……	0.003	……	$>1 \times 10^{10}$	83
m-Phenylene diamine	14.5	13,000	……	4	……	……	……	483	……	……	……	……	……	78
Diaminodiphenyl sulfone	30	7,000	0.47	……	4.22	……	4.71	410	0.004	……	0.068	……	$>2 \times 10^{8}$	65
Anhydrides:														
Methyl tetrahydrophthalic anhydride	84	12,200	0.44	6–4	3.0	3–0	3.0	377	0.006	0.005	0.005	0.09	$>2 \times 10^{9}$	110
Phthalic anhydride	78	8,500	0.5	……	3.5	……	3.8	390	0.004	……	0.006	……	$>5 \times 10^{9}$	95
"Nadic" methyl anhydride	80	11,400	0.40	5–6	3.3	3–4	3.4	443	0.003	0.002	0.004	……	$>4.7 \times 10^{10}$	110
BF₃MEA	3	5,900	0.4	1–8	3.6	……	4.1	480	0.004	……	0.049	……	13×10^{9}	120

* Parts hardener per hundred parts resin.

TABLE 4.7   Cast-Resin Data on Blends of Cycloaliphatic Epoxy Resins[10]

Resin ERL-4221,* parts.........	100	75	50	25	0
Resin ERR-4090,* parts.........	0	25	50	75	100
Hardener, hexahydrophthalic (HHPA), phr†	100	83	65	50	34
Catalyst (BDMA), phr†	1	1	1	1	1
Cure, h/°C	2/120	2/120	2/120	2/120	2/120
Postcure, h/°C	4/160	4/160	4/160	4/160	4/160
Pot life, h/°C	>8/25	>8/25	>8/25	>8/25	>8/25
HDT (ASTM D 648), °C	190	155	100	30	−25
Flexural strength (D 790), lb/in.²...	14,000	17,000	13,500	5,000	Too soft
Compressive strength (D 695), lb/in.².....................	20,000	23,000	20,900	Too soft	Too soft
Compressive yield (D 695), lb/in.².....................	18,800	17,000	12,300	Too soft	Too soft
Tensile strength (D 638), lb/in.²...	8,000–10,000	10,500	8,000	4,000	500
Tensile elongation (D 638), %.....	2	6	27	70	115
Dielectric constant (D 150), 60 Hz:					
25°C	2.8	2.7	2.9	3.7	5.6
50°C	3.0	3.1	3.4	4.5	6.0
100°C	2.7	2.8	3.2	4.9	Too high
150°C	2.4	2.6	3.3	4.6	Too high
Dissipation factor:					
25°C	0.008	0.009	0.010	0.020	0.090
100°C	0.007	0.008	0.030	0.30	Too high
150°C	0.003	0.010	0.080	0.80	
Volume resistivity (D 257), Ω-cm	1×10^{13}	1×10^{12}	1×10^{11}	1×10^{8}	1×10^{6}
Arc resistance (D 495), s	>150‡	>150	>150	>150	>150

* Union Carbide Corp.
† Parts per 100 resin.
‡ Systems started to burn at 120 s. All tests stopped at 150 s.

Cycloaliphatic epoxide systems show advantages over glycidyl ether systems in their ability to produce high-strength bonds on poorly cleaned or even oily metal surfaces. Cycloaliphatic epoxies are often reacted with polyols, which act as plasticizers.[13]

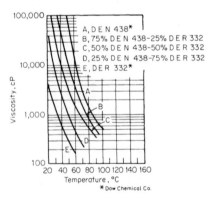

FIGURE 4.2   Viscosity of blends of a novolac epoxy resin with an epi-bis resin as a function of temperature.[7]

Novolac Epoxies.   Phenolic or cresol novolacs are reacted with epichlorohydrin to produce these resins. The novolacs are high-viscosity liquids or semisolids and are often mixed with other epoxies to improve handling properties. Fig. 4.2 shows the viscosities of blends of a novolac resin with an epi-bis resin versus temperature. They cure more rapidly than epi-bis epoxies and have higher exotherms. The cured novolacs have higher heat-deflection temperatures than the epi-bis resins (Table 4.8). The novolacs have excellent resistance to solvents and chemicals. This property is compared in Table 4.9 with that of an epi-bis resin.

TABLE 4.8   Heat-Deflection Temperatures* of Blends of Novolac Epoxy and Epi-Bis Resins[16]

Hardener	D. E. N. 438†	75/25	50/50	25/75	D. E. R. 332‡
TETA..........	§	§	133	126	127
MPDA..........	202	192	180	...	165
MDA...........	205	193	190	186	168
5% BF₃MEA....	235	...	...	204	160
HET...........	225	213	205	203	196

 * Heat-distortion temperature, °C (stoichiometric amount of curing agent—except BF₃ MEA—cured 15 h at 180°C).
 † Novolac resin, Dow Chemical Co.
 ‡ Epi-bis resin, Dow Chemical Co.
 § The mixture reacts too quickly to permit proper mixing by hand.

TABLE 4.9   Comparison of Chemical Resistance for Cured Epoxy Novolac and Epi-Bis Resin[16]

Chemical	Weight gain,* %	
	D.E.N. 438†	D.E.R.331‡
Acetone.......................................	1.9	12.4
Ethyl alcohol.................................	1.0	1.5
Ethylene dichloride...........................	2.6	6.5
Distilled water...............................	1.6	1.5
Glacial acetic acid...........................	0.3	1.0
30% sulfuric acid.............................	1.9	2.1
3% sulfuric acid..............................	1.6	1.2
10% sodium hydroxide..........................	1.4	1.2
1% sodium hydroxide...........................	1.6	1.3
10% ammonium hydroxide........................	1.1	1.3

 One-year immersion at 25°C; cured with methylene dianiline; gelled 16 h at 25°C, postcured 4½ h at 166°C; D.E.N. 438 cured additional 3½ h at 204°C.
 * Sample size ½ by ½ by 1 in.
 † Novolac epoxy resin, Dow Chemical Co.
 ‡ Epi-bis resin, Dow Chemical Co.

Other Epoxy Resins.   Resorcinol diglycidyl ether has a viscosity of 400 Cp and one of the highest epoxy concentrations available in a casting resin. This resin exhibits low cure shrinkage, rapid curing, good resistance to moisture and chemicals, and good adhesive qualities. It is often used to reduce the viscosity of epi-bis compounds.

Epoxy resins based on bisphenol F have some important advantages over the bis A epoxies. The structure of bisphenol F is that of bisphenol A without the methyl groups on the carbon between the two phenols. Compared to bis A epoxies, Bis F epoxies have a lower room temperature viscosity (3,900 cps), a crystallization time twice as long, better resistance to sulfuric acid, acetone and methanol, but twice the price.[14] The cured epoxy bis F is generally tougher than bis A cured with the same curing agent, but has a lower Tg.

Dow's Tactix* 556 is an epoxy resin with a dicyclopentadiene backbone and a novolac-like structure. It produces cured products with low moisture absorption and good property retention in high moisture 150°C service.[15]

 * Trademark, Dow Chemical Co.

Chlorinated or brominated bisphenol A is used to produce flame-resistant epoxies. As the halogen content increases, the viscosity generally increases. Table 4.10 compares the properties of systems prepared from a chlorinated epi-bis resin and from a standard epi-bis resin.

Epoxy resins have been synthesized that contain high weight percents of fluorine. They can be cured with silicone amines to yield products with very low moisture pickup, 0.25 percent at 20°C. These materials are extremely stable on high temperature and outdoor weathering.[18]

Highly aromatic and multifunctional epoxies are available that have glass-transition temperatures of up to 310°C.[19] N,N,N',N'-tetraglycidyl 4',4'-diaminodiphenyl (TGMDA) has been the principal resin used by the aerospace industry for the past 25 years.

Resins that make epoxy formulations flexible are straight-chain compounds terminated in epoxy groups. Other polyhydric compounds such as glycerin are used as starting points for epoxy resins. These generally have low viscosities, high reactivity, and poorer physical, electrical, and chemical resistance properties than epi-bis resins.

Epon* resin 58000-series products contain large amounts of elastomer adducts and are typically used in combination with conventional epoxy resins to produce toughened products.[20]

---

* Trademark, Shell Chemical Co.

TABLE 4.10   Comparison of Chlorinated and Nonchlorinated Epi-Bis Resin[17]

Property	Resin			
	A	*B*	*C*	*D*
Composition (parts by weight):				
Epi-Rez 5161*.	100	. . .	100	
Epi-Rez 510†.	. . .	100	. . .	100
m-Phenylenediamine.	8.8	14.5		
70/30 HHPA/HET.	. . .	. . .	64	116
DMP-10.	. . .	. . .	0.5	0.5
Gel time at 250°F (min).	20	10	30	30
Physical properties:				
Ultimate flexural strength, lb/in.².	16,750	21,000	13,850	18,950
Izod impact (ft-lb/in. notch).	0.30	0.32	0.27	0.26
Hardness:				
Rockwell M.	111	110	106	112
Shore D.	80	80	80	80
Heat-distortion temp, °C.	122	145	133	126
% water absorption, 24 h.	0.06	0.17	0.06	0.09
% weight loss, 168 h at 350°F.	5.76‡	1.11	1.45	1.22
% weight loss, 168 h at 300°F.	1.01	0.51	0.22	0.27
Flame extinction§, s:				
Vertical.	1	Keeps burning	2	Keeps burning
Horizontal.	1	50	2	50

Each composition was cured at 250°F for 2 h, followed by a postcure of 3 h at 350°F.
* Chlorinated epi-bis resin, Rhone-Poulenc.
† Epi-bis resin, Rhone-Poulenc.
‡ Severe decomposition.
§ Description of test: sample 6 by ½ by ⅛ in. Test sample (¼ in) is placed in the hottest part of a Meker-burner flame for 10 s. The flame is removed, and the time (in seconds) is measured for the sample to stop burning.

4.2.2 Curing Agents

The curing agent used in a formulation plays a major role in determining the handling properties, curing schedule, and final part properties. The effects of curing agents are being discussed with regard to epi-bis resins, but the data will be generally applicable to most other types of epoxies.

The major classes of epoxy curing agents are aliphatic amines, aromatic amines, acids and anhydrides, catalysts, latent curing agents, and flexibilizing curing agents. Mixtures of curing agents are used to modify the curing schedule of the compound and cured properties.

Aliphatic Amines. Aliphatic amines such as diethylenetriamine (DETA) are widely used for curing epoxies at room temperature in small masses or in films. These materials are highly exothermic so that a large mixture may overheat during cure. Disadvantages of these materials include a short pot life and a tendency to pick up moisture from the air, which inhibits the cure. DETA and its adducts are used in potting, adhesives, patching, coatings, and wet-lay-up laminates.

Slower curing agents (for example, dimethylaminopropylamine and diethylaminopropylamine) are preferred in applications such as the manufacture of castings, molding, and other parts using large resin masses. They often require heating for complete cure. The aliphatic amines produce cured epoxies that are useful up to 80 to 100°C and have reasonably good electrical, physical, and chemical-resistant qualities up to 70°C. Table 4.11 lists typical amine catalysts in the order of their reactivity.

Aromatic Amines. Aromatic amines such as metaphenylenediamine (MPDA) and methylene dianiline (MDA) are solid, moderate-temperature epoxy curing agents. They are often used in liquid eutectic mixtures because of the inconvenience of dissolving the solids in hot resin. Epoxy compositions containing aromatic amines are typically cured for 2 h at 80°C, followed by 2 h at 150°C. Their resistance to chemicals and their electrical properties are good, and their physical properties are among the highest of any epoxy system. The upper use temperature is 135°C. These systems

TABLE 4.11 Approximate Order of Activity of Typical Amine Catalysts[21] (Arranged in Decreasing Order of Activity)

Type of amine	Source
DMP-30 (tri-dimethyl-aminomethyl-phenol)	Rohm & Haas Co.
Triethylenetetramine	Carbide & Carbon Chemical Co.
Diethylenetriamine	Carbide & Carbon Chemical Co.
Dimethylaminopropylamine	Carbide & Carbon Chemical Co.
DMP-10 (dimethyl-aminomethyl-phenol)	Rohm & Haas Co.
Diethylaminopropylamine	Carbide & Carbon Chemical Co.
Benzyldimethylamine	Rohm & Haas Co.
Piperidine	Matheson Co.
Diethylamine	Carbide & Carbon Chemical Co.
Dimethylaminopropionitrile	Matheson Co.
Shell catalyst D(2-ethyl hexoic acid salt of tri-dimethyl-aminomethyl phenol)	Shell Chemical Company
m-Phenylenediamine (Shell catalyst CL)	E. I. du Pont de Nemours & Co., National Aniline Division of Allied Chemical & Dye Corp.

will B-stage overnight, that is, the mixture will partially cure to a solid, which is soluble and can be melted and cured on further heating. Molding compounds can be produced this way. The aromatic amines are used in potting compositions, laminates, and chemical resistant castings.

In recent years MDA has been found to be a carcinogen, and its use has been sharply limited. A number of substitutes with properties similar to those of MDA and MPDA have been proposed. These include diamino ethyl benzene, O-phenylene diamine and its mixtures,[22] and 1,2 diaminocyclohexane.[23]

Polycycloaliphatic polyamines and cycloaliphatic amine cured epoxies exhibit the high strength, Tg, and modulus characteristic of aromatic diamine-cured systems combined with the ductility and toughness characteristic of cycloaliphatic diamine-cured systems.[24]

Acids and Anhydrides.   Epoxies cured with acids and anhydrides are generally useful at higher temperatures than those cured with other agents. Very large castings are produced with these anhydrides because the exotherms are generally low. Most cycloaliphatic epoxies are cured with them. The chief disadvantages of the anhydrides are the necessity to melt the materials that are room-temperature solids and the long high-temperature cures. Tertiary amines are often used to shorten the cure cycles.

Phthalic anhydride (PA) is an inexpensive room-temperature solid. The epoxy must be kept hot to prevent the PA from coming out of solution. At the high temperature needed to cure this material, the PA sublimes out of the mix, creating a production inconvenience and safety hazard. Very large unfilled castings can be produced that are tough and readily machinable and possess good electrical properties. Methyl nadic anhydride (MNA) is used extensively in laminating and casting because it is a liquid at room temperature, and the cured systems have excellent electrical properties and very low weight loss at high temperature. The weight loss of epi-bis epoxies cured with various hardeners is shown in Fig. 4.3.

Chlorendic or HET anhydride imparts flame resistance to epoxies and produces materials with excellent high-temperature electrical properties. After an initial very low weight loss at high temperature, HET anhydride-cured epoxies abruptly decompose in a relatively short time (Fig. 4.4).

Pyromellitic dianhydride (PMDA) dissolves readily in hot epoxies. The mixture cures rapidly at 175°C, yielding a brittle, moderate-strength product. The high-temperature properties of this material are exceptional, particularly in adhesives.

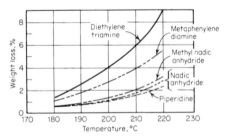

FIGURE 4.3 1,000-h weight loss of an epi-bis resin cured with various hardeners.[26]

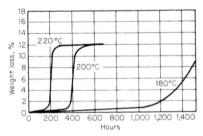

FIGURE 4.4 Weight loss at various temperatures for a HET anhydride-cured epoxy resin.[26]

Trimellitic anhydride esterified with glycol are used in powder coatings. Azelaic and polysabacic anhydrides are used in fluidized beds and transfer molding compounds. Tg's of 200°C and 2–4-minute cure times are possible.[25]

Catalysts. Tertiary amines such as benzyldimethylamine and tri-dimethyl aminomethyl phenol (Rohm and Haas's DMP-30) are sometimes used as sole curing agents, but generally they are used to accelerate the curing of anhydride-epoxy mixtures. Piperidine has a high odor level and produces epoxy castings that have outstanding high-temperature stability, even though the heat-deflection temperature is only 75°C. Boron trifluoride complexes can also be classified as latent agents. They have adverse effects on wet electrical characteristics.

Latent Curing Agents. Latent curing agents such as dicyandiamide are used to produce one-part systems that have long room-temperature shelf lives but cure rapidly at elevated temperatures. They are used in potting compounds, laminates, and adhesives. A new class of latent curing agent is being used that is insoluble in the epoxy at room temperature but goes into solution at moderate temperature to effect a rapid cure.

Flexibilizing Curing Agents. Flexibilizing curing agents are generally long aliphatic chains containing amino groups. Ancamides (Pacific Anchor) and the Versamids (Henkel) are members of this group. They cure slowly at room temperature to produce tough, water-resistant products. Adhesion is excellent, and the ratio of curing agent to resin is not critical. Emery's trimer acid is an example of another class of flexibilizer. It is viscous and insoluble at room temperature and has a long cure cycle. The electrical properties are better than those of other flexibilized systems, and the cured epoxies are usable over a wide temperature range. Aliphatic polyanhydrides are efficient flexibilizers and thermal shock improvers. Polyadipic polyanhydride (PADA), polyazelaic polyanhydride (PAPA), and polysebacic polyanhydride (PSPA) are three examples. Accelerators are used to decrease cure time.[27]

Polyurethane amine curing agents produce a high degree of flexibility and toughness.[28]

4.2.3 Epoxy Modifiers

Reactive Diluents. Reactive diluents such as butyl glycidyl ether reduce the viscosity of the epoxy mixture, as shown in Fig. 4.5. There is relatively little change in the cured properties when these diluents are used in concentrations of up to 5 parts per hundred resin (phr). Above this concentration, there is a deterioration of electrical, mechanical, and chemical-resistance properties.

Reactive Flexibilizers. Reactive flexibilizers such as polysulfides, urethanes, polybutadienes, and polyesters are copolymerized with epoxies to increase flexibility, improve adhesion, modify electrical properties, and improve resistance to chemicals. Table 4.2 gives properties of modified epoxies. The dielectric constant and the dissipation factor of a phthalic-anhydride-cured epoxy polybutadiene as a function of temperature and frequency are covered in Sec. 4.8.2 (see Figs. 4.31 and 4.32).

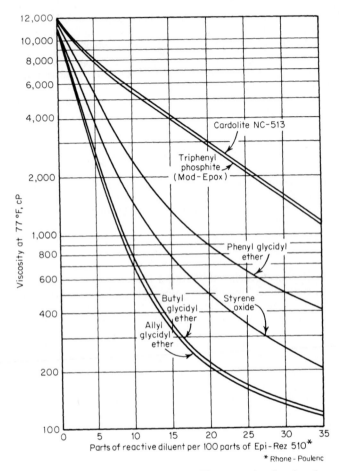

FIGURE 4.5   Effect of various reactive diluents on the viscosity of an epoxy resin.[29]

Nonreactive Extenders.   Nonreactive extenders such as thermoplastics, asphaltums, and waxes are used to modify properties and reduce the cost of the compounds. Polyarylene ether sulfone increases the fracture toughness fivefold with only a slight drop in Tg.[19] These find application in areas of high-volume usage, such as paving and construction.

New encapsulants are under development based on simultaneous interpenetrating networks, for example, epoxy and poly(n-butyl acrylate) for improved resistance to crack growth.

Inclusion of 15 percent preformed thermoplastic rubber particles give consistent cured epoxy products.[20]

4.2.4 Fillers

Most commercial epoxy compounds contain one or more fillers in the form of powders, fibers, or flakes. Fillers are used to modify viscosity, increase pot life, reduce exotherm, reduce curing shrinkage, modify density, improve heat resistance, modify thermal conductivity (usually to increase), reduce coefficient of thermal expansion, increase strength, improve machinability, increase hardness and wear resistance, modify electrical properties, increase chemical and solvent resistance, modify friction characteristics, improve thermal shock resistance, improve adhesion, and impart color. Generally the fillers should be low in cost; reproducible in composition, particle size, and shape; easy to disperse in the compound; and low in density; and they should not increase the viscosity of the mixture excessively. The filler should stay in suspension or, at worst, be able to be resuspended with a minimum of stirring. Table 4.12 lists some of the more commonly used fillers and their characteristics. The effects of several fillers on the properties of epoxy and polyester resins are given in Table 4.13. Figs. 4.6–4.11 show the effects of filler concentration on exotherm, cure shrinkage, coefficient of thermal expansion, thermal conductivity, arc resistance, and viscosity. Fig. 4.12 illustrates the surface resistivity versus relative humidity of epoxy castings containing various fillers.

TABLE 4.12 Characteristics of Some Commonly Used Fillers

Filler	Characteristics
Silica	Inexpensive, hard, abrasive, lightweight, easily mixed in, good electrical properties, resistant to chemicals and weathering, difficult to machine, high loading possible
Calcium carbonate	Inexpensive, lightweight, improves machinability, high loadings possible with minimum viscosity increase, poor water and acid resistance, easily mixed in, poor electrical properties
Clay	Inexpensive, used as extender; some grades do not mix well
Aluminum oxide	Very hard, abrasive, castings are abrasion-resistant and must be ground to size, increases thermal conductivity, produces translucent compounds; some types used to help keep other fillers in suspension
Calcium silicate	Fibrous material increases impact strength
Glass spheres	Available in a variety of graded sizes, easily mixed in, good packing density, reduces thermal expansion
Hollow spheres	Glass, phenolic, thermoplastic available, reduces density, reduces thermal conductivity and dielectric constant
Fibers	Glass, asbestos, Dacron,* cotton, nylon, increases impact strength, high viscosity
Metal powders and particles	Heavy, settles rapidly, easy to mix in Aluminum—powder and pellets, increased thermal conductivity, easy to machine, castings are malleable Silver—flakes and powder, silver-coated copper less expensive, high electrical and thermal conductivity Copper, bronze, brass—flakes and powder, increased thermal and electrical conductivity, pigment Stainless steel—flakes, weather resistance, moisture barrier
Finely divided silica	Thickens compounds to reduce sagging
Pigments	To produce color and opacity, most easily incorporated from dispersions in epoxy resins or plasticizers

* Trademark of E. I. du Pont de Nemours & Company, Inc.

TABLE 4.13 Effects of Several Fillers on Properties of Typical Epoxy and Polyester Resins[30]

Property	Epoxy resins					Polyester resins			
	Unfilled	Calcium carbonate	Mica	Glass	Alumina	Mica	Glass	Calcium carbonate	Unfilled
Coefficient of linear expansion	72	57	43	.....	70.5	65	70	84	97
Thermal conductivity, W/(in.²)(°C)(in.)	0.008	0.014	0.012	0.012	0.0109	0.0074	0.0073	0.0069	0.0075
Water absorption, mg	24	20	22	24	60	50	40	50	50
Specific gravity	1.2	1.58	1.41	1.52	1.87	1.38	1.55	1.5	1.23
Compressive strength, lb/in.²	15,900	7,540	5,700	.....	26,600	19,600	25,000	18,500	18,000
Tensile strength, lb/in.²	9,700	6,000	5,650	.....	3,700	5,000	5,000	4,250	7,000
Dielectric strength, V/mil	320	370	420	370	270	400	380	280	250–300
Brinell hardness (500-kg load, 10-mm ball)	.....	.....	.....	.....	76	49	49	65	49

Power factor, tan δ, MHz:									
1..........	0.029	0.026	0.035	0.026	0.0194	0.0213	0.0165	0.0173	0.018
10.........	0.029	0.026	0.034	0.026	0.0188	0.0204	0.0171	0.0171	0.031 (at 1 kHz)
20.........	0.028	0.026	0.032	0.026	0.0183	0.0197	0.0170	0.0170	0.045 (at 50 Hz)
50.........	0.026	0.025	0.030	0.025	0.0176	0.0185	0.0165	0.0167	0.016
100.........	0.020	0.023	0.026	0.023	0.0177	0.0176	0.0160	0.0164	
Dielectric constant, MHz:									
1..........	3.9	4.45	4.05	4.05	5.05	3.62	4.33	4.75	3.13
10.........	3.75	4.2	3.9	3.9	4.92	3.53	4.4	4.63	3.18 (at 1 kHz)
20.........	3.7	4.2	3.85	3.85	4.9	3.5	4.35	4.60	3.91 (at 50 Hz)
50.........	3.65	4.15	3.75	3.8	4.85	3.47	4.29	4.55	4.3
100.........	3.6	4.03	3.7	3.75	4.8	3.44	4.25	4.51	
Remarks.........	......	Best general character-istics	Best for high dielectric strengths	......	Highest thermal conductivity	Lowest coeffi-cient of linear expansion			

* ×10⁻⁶ per °C; measured at 60–80°C for epoxy resins and at 60°C for polyester resins.

4.21

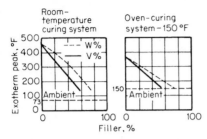

FIGURE 4.6 Effect of filler concentration on exotherm of an epoxy resin.[31]

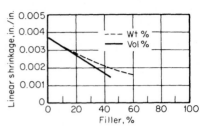

FIGURE 4.7 Effect of filler concentration on shrinkage of an epoxy resin.[31]

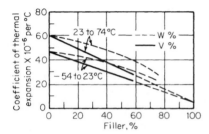

FIGURE 4.8 Effect of filler concentration on coefficient of thermal expansion of an epoxy resin.[13]

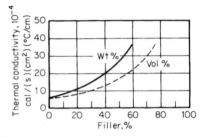

FIGURE 4.9 Effect of filler concentration on thermal conductivity of an epoxy resin.[13]

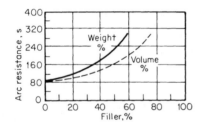

FIGURE 4.10 Effect of filler concentration on arc resistance of an epoxy resin.[13]

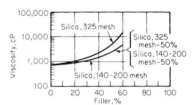

FIGURE 4.11 Effect of fillers on viscosity of an epoxy resin.[13]

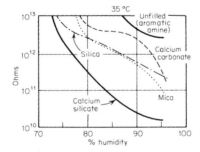

FIGURE 4.12 Surface resistivity of an epoxy resin containing various fillers as a function of percent humidity.[32]

4.2.5 Applications

Epoxy resins have found wide use in such fields as adhesives, potting, art reproductions, construction, jigs and fixtures, and laminates. Coatings consume 50 percent or more of the epoxies produced. Undersea pipes and tanks of epoxy composites are joined with epoxy cement;[33] large contoured jet nozzles have been cast.[34] Synthetic foam forms the hull of deep-submergence vessels. Binder for terrazzo flooring, heavy-duty floor retopping, missile casings, chemical-resistant paint, and laminates for the process industry are only a few of the applications of the versatile epoxies.

Reactive transfer molding compounds are used to make propeller blades, industrial fan blades, support beams, skateboards, and snow skis. Ultraviolet-radiation-curable epoxy coatings are used as protective coatings for printed wiring boards.

4.2.6 Safety Considerations

Aliphatic amines are skin sensitizers, and some can cause respiratory difficulties. Piperidine's toxicity is the subject of governmental restrictions. Monoepoxides used for viscosity reduction are often skin sensitizers. Toxicological problems have been observed with some traditional resins such as vinylcyclohexane dioxide, and these have been discontinued.

Good Practice Procedures.   Cover benches and seats in the mixing area with disposable paper. Keep tools clean. Discard drums and containers; do not reuse. Be sure that there is sufficient air ventilation with the fumes directed away from you, particularly when using volatile amines. Wash before eating, before relief periods, and after work. If epoxies spill on your clothing, change to clean clothing and wash well. Apply protective cream before working with resins.

Any employee who develops dermatitis or redness of the skin while handling resins or other materials should report this immediately to his or her supervisor, who should refer such employees to a physician.[35]

4.3 POLYESTERS

Polyesters are becoming increasingly popular because of their low cost, ease of use, and versatility. They find applications in laminates, castings, art objects, industrial construction, insulation, embedments, and molding.

4.3.1 Basic Types of Polyesters

Compounds, Coatings, and Adhesives.   Other advantages of polyesters are their high impact resistance, good weathering resistance, transparency, and good surface effects. Disadvantages include poor adhesion, high cure shrinkage, and inhibition of cure by air and some fillers. There are three required components in a polyester formulation: (1) an unsaturated polyester, prepared from organic acids and polyols, (2) a monomer, and (3) a curing system. There are usually also fillers, reinforcement, and other modifiers.

Saturated and unsaturated organic acids such as maleic and phthalic acid are reacted with hydroxyl-containing chemicals such as ethylene glycol to produce unsaturated polyesters that are very thick or solid at room temperature. The general-purpose polyester, when blended with a monomer (see Sec. 4.3.2) and cured, produces rigid, rapidly curing, transparent castings, the properties of which are shown in Table 4.14. A second type produces a flexible material with the properties given in Table 4.15. This type of polyester is tougher and slower curing, and it produces lower exotherms and less cure shrinkage. Flexible polyesters absorb more water and are more easily scratched but show more abrasion resistance than the rigid type. The two types can be blended to produce intermediate properties, as shown in Table 4.16.

Polyesters that are inherently fire-resistant are produced from chlorinated anhydrides such as chlorendic anhydride or tetrabromophthalic anhydride. Their properties are similar to those of other polyesters except for a higher density.

A group of polyesters are available that are used where resistance to chemicals is important. Some of these incorporate bisphenol A in their structures; others contain isophthalic anhydride or neopentylglycol. They also have improved impact resistance, flexural strength, corrosion resistance, and crack resistance over the general-purpose polyesters.[36,37]

4.3.2 Monomers

Styrene is the most widely used monomer because of its low cost, high solvency, low viscosity, reactivity, and the desirable cured properties it imparts to the polyester.

TABLE 4.14   Properties of Typical Polyester Producing Rigid Castings

Product specifications at 25°C	
Flash point, Seta closed cup, °F	89
Shelf life, minimum, months	3
Specific gravity	1.10–1.20
Weight per gallon, lb	9.15–10.0
% styrene monomer	31–35
Viscosity, Brookfield model LVF, #3 spindle at 60 r/min, cP	650–850
Gel time	
150–190°F, min	4–7
190°F to peak exotherm, min	1–3
Peak exotherm, °F	385–425
Color	Amber clear
Typical physical properties (clear casting)	
Barcol hardness (ASTM D 2583)	47
Heat-deflection temperature, °C (°F) (ASTM D 648)	87 (189)
Tensile strength, lb/in^2 (ASTM D 638)	8000
Tensile modulus, 10^5 lb/in^2 (ASTM D 638)	5.12
Flexural strength, lb/in^2 (ASTM D 790)	13,500
Flexural modulus, 10^5 lb/in^2 (ASTM D 790)	6.0
Compressive strength (ASTM D 695)	22,000
Tensile elongation, % at break (ASTM D 638)	1.5
Dielectric constant (ASTM D 150)	
At 60 Hz	2.97
At 1 MHz	2.87
Power factor	
At 1 kHz	0.005
At 1 MHz	0.017
Loss tangent at 1 MHz	0.017

TABLE 4.15 Properties of Typical Polyester Producing Flexible Material

Product specifications at 25°C	
Flash point, Seta closed cup, °F	89
Shelf life, minimum, months	3
Specific gravity	1.13–1.25
Weight per gallon, lb	9.4–10.4
% styrene monomer	18–22
Viscosity, Brookfield model LVF, #3 spindle at 60 r/min, cP	1100–1400
Gel time	
SPI 150–190 °F, min	6–8
190°F to peak exotherm, min	4.5–6.0
Peak temperature, °C (°F)	102–116 (215–240)
Color	Amber clear
Typical physical properties (clear casting)	
Tensile strength, lb/in^2 (ASTM D 638)	50
Tensile elongation, % (ASTM D 638)	10
Flexural strength, lb/in^2 (ASTM D 790)	Yields
Hardness, Shore D (ASTM D 2240)	15

Abrasion and wear properties increase in most polyester laminates when the styrene content is raised from 21 to 31 percent. Fig. 4.13 shows cure time and gel time versus styrene content. The effect of the styrene content on the mechanical properties is shown in Table 4.17. The effect on wear and hardness characteristics is illustrated in Figs. 4.14 and 4.15. The choice of monomer is predicated on both the curing characteristics and the properties of the resulting polyester. Fig. 4.16 presents exotherm curves for the various monomers. The important properties of the monomers are given in the following.

Alpha Methyl Styrene.   Used with styrene at a concentration of 2 to 4 percent; reduces exotherm without affecting the cure time excessively; used in large castings; higher concentrations decrease hardness and tensile strength.

Methyl Methacrylate.   Used in combination with styrene; increases durability, color retention, resistance to fiber erosion; improves weathering resistance; proper concentration will change index of refraction of the polyester to match that of glass and thus

TABLE 4.16 Properties of Blend of Rigid and Flexible Polyesters

Flexible polyester*	30%	20%	10%	5%	—
Rigid polyester†	70%	80%	90%	95%	100%
Tensile strength, lb/in^2 (ASTM D 638)	5200	8100	7300	6800	6500
Tensile elongation, % (ASTM D 638)	10.0	4.8	1.7	1.3	1.0
Flexural strength, lb/in^2 (ASTM D 790)	8100	13,200	15,600	14,600	13,500
Flexural modulus, 10^5 lb/in^2 (ASTM D 790)	2.40	3.75	5.60	5.80	6.00
Barcol hardness (ASTM D 2583)	0–5	20–25	30–35	35–40	40–45
Heat-deflection temperature, °C (°F) (ASTM D 648)	51 (124)	57.5 (136)	63 (145)	85 (185)	88 (190)

* Polylite 31-820, Reichhold Chemicals, Inc.
† Polylite 31-000, Reichhold Chemicals, Inc.

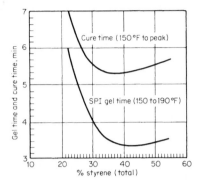

FIGURE 4.13 Gel time and cure time of polyester resin as a function of styrene content.[38]

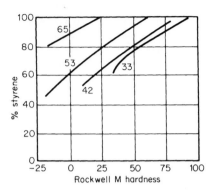

FIGURE 4.14 Styrene content vs. Rockwell M hardness of triethylene maleate polyester castings.[23]

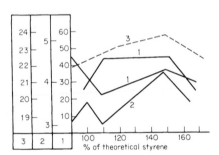

FIGURE 4.15 Wear characteristics of a typical polyester resin as a function of styrene content.[24] Curve 1, Taber abrasion, milligrams loss per 1,000 cycles. Curve 2, diamond threshold tear. Curve 3, Taber scratch factor.

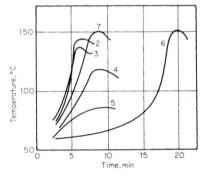

FIGURE 4.16 Exotherm curves for polyesters containing different monomers.[41] (1) Styrene. (2) Vinyl acetate. (3) Acrylonitrile. (4) Methyl methacrylate. (5) Diallyl phthalate. (6) Triallyl cyanurate.

produce a nearly transparent laminate; lower boiling point, higher cost, and greater cure shrinkage than styrene.

Vinyl Toluene. Compared with styrene, it has shorter cure time, higher boiling point, lower cure shrinkage, and higher viscosity.

Diallyl Phthalate. Low volatility gives prepregs long shelf life, lower cure shrinkage, lower exotherms, reduced odor, increased flexural strength and Izod impact resistance, and higher cost.

Triallyl Cyanurate. Greatly improved high-temperature stability; greater cure shrinkage; more brittle.

Divinyl Benzene. Used with styrene. Castings are harder, more heat resistant, brittle.

Chlorostyrene. Increased chemical resistance.

TABLE 4.17 Effect of Styrene Concentration on Propylene Glycol-Maleate-Phthalate Polyester Resins

Mole ratio maleate/ phthalate	Styrene, %	Peak exotherm, °F	Flexural strength, lb/in.²	Flexural modulus lb/in.² × 10⁵	Tensile strength, lb/in.²	Tensile modulus lb/in.² × 10⁵	HDT °F	Elongation, %	Water absorption, % after 24 h at 77°F
40/60........	20	323	20,700	6.6	8,200	8.4	147	1.2	0.17
40/60........	30	347	16,100	5.9	7,900	7.8	158	1.31	0.21
40/60........	40	349	14,100	5.9	9,100	6.8	172	1.73	0.17
40/60........	50	340	15,700	5.3	9,500	7.5	176	1.85	0.17
50/50........	20	340	20,000	6.5	8,100	8.5	158	1.3	0.19
50/50........	30	380	19,000	5.8	8,300	8.1	194	1.32	0.23
50/50........	40	392	17,700	6.3	9,200	7.9	201	1.7	0.21
50/50........	50	396	15,500	5.2	8,000	7.2	199	1.7	0.20
60/40........	20	356	19,500	6.5	8,000	8.7	169	....	0.23
60/40........	30	400	17,200	5.7	8,600	7.9	219	1.38	0.25
60/40........	40	407	17,800	6.0	7,200	7.7	226	1.46	0.25
60/40........	50	404	17,700	5.1	6,600	7.0	225	1.23	0.28

4.27

4.3.3 Curing Systems

The curing system does not greatly change the properties of the final polyester product, but the storage life, mold or press time, postcure (if any), and maximum temperature developed are controlled. The reactivity of the polyester used as well as the size and shape of the product affect the choice of systems. This versatility of the cure schedule, from very rapid room temperature cure to extended temperature cure, is one of polyester's most attractive attributes. The catalyst is generally a peroxide such as benzoyl peroxide or methyl ethyl ketone peroxide. Promoters used singly or in combination and retarders complete the cure system.

4.3.4 Fillers

The fillers used with polyesters are much the same as those used with epoxies. They are used to reduce shrinkage, modify processing properties, increase hardness, increase thermal conductivity, decrease thermal expansion, and increase wear and chemical resistance. Greater care must be used in picking the correct filler with the polyester, as some fillers will reduce the curing rate or even inhibit the cure completely. Calcium carbonate is the most widely used filler for polyesters because of its low cost and its compatibility and the high loadings possible.

4.3.5 Applications

Polyesters are ideal for do-it-yourselfers as well as in large industrial operations, because they are easy to formulate, mix, and cure, with a minimum of elaborate equipment. Polyesters are most widely used in glass cloth laminates and with chopped glass fibers in molding compositions because they wet the glass well, bonding to it and producing strong, attractive, long-lasting composites. For accurate reproduction of mold details, an additive has been developed that produces zero cure shrinkage in polyester-glass bulk.[42]

Polyester-fiberglass laminates are used in boats, automobile bodies, translucent building panels, bathtub and shower units, chemical storage tanks and piping, chemical-resistant tank cars, fishing rods, and snowmobiles. Bowling balls, synthetic marble, and furniture parts are cast from polyesters. Water-extended polyester (WEP) has been used widely to make plaques, lamp bases, and furniture. In this process 50 to 80 percent water is incorporated in the mix before casting. This results in a lower cost, lighter weight, easier mixing, more controllable curing, and the ability to fabricate finished castings with conventional woodworking tools.[43,44] Hobbyists and artists use polyesters extensively to embed objects and produce works of art, such as those shown in Figs. 4.17 and 4.18.

Polymer concrete, that is, polyester styrene with coarse aggregate, is used to rehabilitate portland cement on bridge decks and other critical areas.[45]

4.3.6 Safety Considerations

Polyesters are not a serious allergy or high-toxicity problem. Styrene must be kept below levels mandated by regulatory agencies, generally 50 to 100 ppm. This has led to increased use of less volatile monomers such as vinyl toluene and p-methylstyrene.

Reduced styrene emission resins generally depend on the formation of a film on the surface of the resin. The film formers have limited solubility and form a barrier layer that inhibits styrene evaporation. Interlaminar bond strengths can suffer.

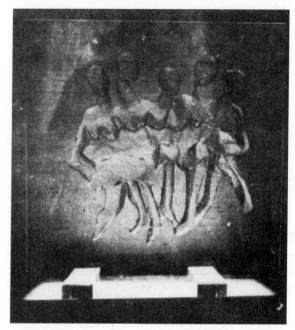

FIGURE 4.17 "The Group"—polyester resin cast in an RTV silicone mold attached to an acrylic sheet. (*Courtesy of Thelma R. Newman.*)

FIGURE 4.18 "The Crowd"—polyester resin cast in an RTV silicone mold attached to an acrylic sheet. (*Courtesy of Thelma R. Newman.*)

4.4 POLYURETHANES

Polyurethanes can be formulated to produce a range of materials from elastomers as soft as a Shore A of 5 to tough solids with a Shore D of 90.

Fig. 4.19 shows the range of hardness used in various applications. Polyurethanes can have extremely high abrasion resistance and tear strength, excellent shock absorption, resistance to a broad spectrum of solvents, good electrical properties, and excellent resistance to oxygen aging. They have limited life in high-humidity and high-temperature applications. Fig. 4.20 presents the effect of humidity aging on the hardness of polyurethanes derived from various polyols. The liquid components must be kept dry during storage and fabrication.

The polyurethanes are reaction products of an isocyanate, a polyol, and a curing agent. Because of the hazards involved in handling free isocyanate, prepolymers of the isocyanate and the polyol are generally used in the casting.

4.4.1 Types of Polyurethanes

The polyurethanes are classified according to the polyol used. The polyols are generally polyethers and polyesters. Polyether-based urethanes are more resistant to hydrolysis and have higher resilience, good energy-absorption characteristics, good hysteresis characteristics, and good all-around chemical resistance. A polyester-

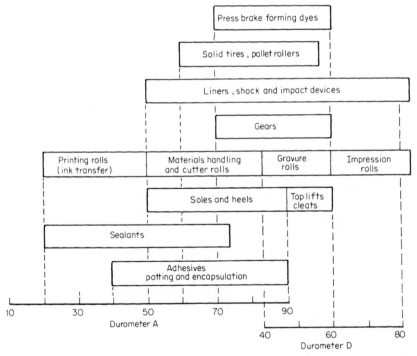

FIGURE 4.19 Applications of urethane elastomers and their relation to elastomer hardness.[30]

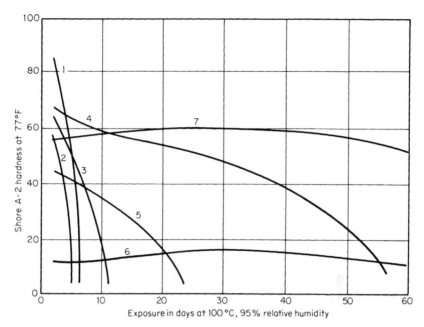

FIGURE 4.20   Hardness of polyurethanes as a function of humidity exposure.[47] (1) Polyester-urethane. (2) Polyesterurethane. (3) Polycaprolactone. (4) Polyetherurethane. (5) Polyetherure-thane. (6) Polybutadiene polyol urethane. (7) Polybutadiene polyol urethane.

based material will generally be stiffer and will have higher compression and tensile moduli, higher tear strength and cut resistance, higher operating temperature, lower compression set, optimum abrasion resistance, and good oil and fuel resistance.[48] Table 4.18 summarizes the characteristics of the two materials. Table 4.19 lists the properties of a family of polyether-based polyurethanes. Resistance to temperature exposure is illustrated in Fig. 4.21.

Castor-oil-based potting compounds are convenient to use. They have improved hydrolytic stability and poorer physical properties compared to polyether- or polyester-based urethanes. CasChem markets a family of polyurethane systems based on ricinoleate polyols derived from castor oil.[50] The properties of a member of this family are given in Table 4.20. Polybutadiene-based polyurethanes have outstanding resistance to humidity, but they have higher viscosities and only moderate physical properties.

Aliphatic prepolymers, isophorone diisocyanate-based products, are light-stable and meet stringent optical clarity requirements. Some 1,4-butanediol-cured polyurethanes have been FDA approved for use in applications that will come in contact with dry, aqueous, and fatty foods.[51]

4.4.2   Curing Agents

The choice of curing agent influences the curing characteristics and final properties. Diamines are the best general-purpose curing agents; the highest physical properties are produced with MOCA [4,4′-methylene-bis(2 chloroaniline)]. As most useful diamines are room-temperature solids, they must be melted and mixed

TABLE 4.18 Typical Properties of Urethane Rubbers (High-Performance Castables)[48]

Properties	Ester-based			Ether-based		
	20–50	60–75	80–90	50–80	75–80	80–90
Hardness ranges:						
Shore A	20–50	60–75	......	......	......	80–90
Shore D	......	......	80–90	50–80	75–80	......
Tensile strength, lb/in.²	80–3,500	4,500–6,500	6,500	4,800–10,000	8,000–9,500	4,400–5,000
Elongation, %	525–650	460–615	550–660	250–450	270–300	450–575
100% modulus, lb/in.²	25–150	265–410	530–1,000	1,100–5,500	4,100–5,000	700–1,050
300% modulus, lb/in.²	45–300	560–795	1,000–1,900	2,500–4,700	......	1,600–1,700
Tear strength, pli*						
Die C	30–128	168–370	420–650	450–2,000	......	550–675
Split	10–34	30–200	250–400	310–1,100	115–125	85–120
Compression set, % (22 h at 158°F)	6–1.7	1.4–25	25–38	0.8–33	......	28–35
Bashore rebound, % (78°F)	19–21	8–24	27–32	23–35	48–50	50–55

* Pounds per lineal inch.

TABLE 4.19 Properties of Family of Polyether-Based Polyurethane Resins[49]

	Adiprene Product No.								
	L-100	L-100	L-42	L-83	L-100	L-167	L-200	L-213	L-315
Compound:									
Adiprene*	100	100	100	100	100	100	100	100	100
MOCA†			8.8	10.3	12.5	19.5	23.2	25	26
1,4-Butanediol	3.2								
Trimethylolpropane	0.9								
Methylene dianiline		9.6							
Dioctyl phthalate		50							
Mix temp, °F	212	150	212	212	212	158	185	175	170
Cure, h at °F	16/212	1/212	3/212	3/212	3/212	2/212	2/100	2/212	2/212
Physical properties:									
Hardness, Durometer A	60	75	80	83	90	95			
Hardness, Durometer D					40	48	58	75	75
100% modulus, lb/in.²	275	550	400	700	1,100	1,800	3,000	3,800	4,300
300% modulus, lb/in.²	500	800	625	1,200	2,100	3,400	7,800		
Tensile strength, lb/in.²	2,100	2,850	3,000	4,400	4,500	5,000	8,200	7,500	11,000
Elongation at break, %	500	625	800	575	450	400	315	250	270
Tear strength, Graves, lb/in.	150			400	500	600			725
Tear strength, split, lb/in.	12	70	70	85	75	150	135	145	115
Abrasion resistance, NBS index, %	25	600	110	160	175	285	370	350	435
Compression set, method B, 22 h at 158°F	9	27	45	35	27	45	40		
Compression set, method A, 22 h at 158°F, 1350 lb/in.²					9	9			9
Resilience—Yerzley, %	72	70	70					50	
Resilience—Bashore, %	60	60	60		65	39		50	48
Specific gravity	1.06	1.06	1.08	1.05	1.10	1.14	1.14	1.19	1.19

* Du Pont Co.
† Du Pont's registered trademark for its curing agent 4,4'-methylene-bis-(2-chloroaniline).

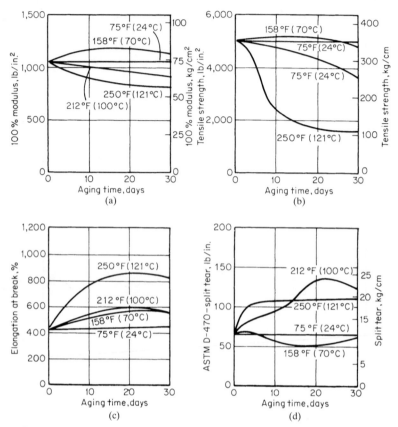

FIGURE 4.21 Physical properties as a function of time at elevated temperature of a polyether polyurethane.[49] (a) 100 percent modulus vs. aging time at elevated temperature. (b) Tensile strength vs. aging time. (c) Elongation at break vs. aging time. (d) Split tear vs. aging time.

hot, reducing pot life and complicating the casting procedure. The other major class of curing agents, the polyols, are more convenient to use, but the products have lower physical properties. The properties of these curing agents are summarized in Table 4.21.

There has been a concerted drive to find non-MOCA curing agents for polyurethanes because MOCA has been judged a carcinogen. OSHA set the acceptable exposure level for MOCA at 0.02 ppm as an 8-h time-weighted average; no short-term limits have been suggested.[52] Several alternative amines are available, including trimethylene glycol di-p-aminobenzoate (Polacure from Polaroid), 4,4-methylene-bis(orthochloroaniline) (Cyanacure from American Cyanamid Company), and 4,4'-methylene-bis(3-chloro-2,6-diethyl aniline) (Lonacure from Lonza Chemical).[53]

4.4.3 Applications

Most polyurethanes make use of the good abrasion resistance and low coefficient of friction. Polyurethanes are used in roller coatings and press pads, where they last ten

TABLE 4.20 Properties of Castor Oil Polyurethane System

Handling (300 g mass at 25°C)	
All operations should be conducted at room temperature in well ventilated areas.	
Mix viscosity, cP	600
Gel time	
Catalyzed, min	60
Uncatalyzed, min	140
Exotherm, Δ°C	45
Curing	
Cure over night at room temperature (25°C) followed by an oven cure at 80°C for 6 h. This system will also fully cure without baking in approximately 2 weeks at room temperature.	
Physical specifications at 25°C	
Specific gravity	1.07
Hardness, Shore A	60
Tensile strength, lb/in^2	600
Elongation, %	200
Tear strength, lb/lin. in	129
Coefficient of linear expansion, in/in · °C	20.6×10^{-5}
T_f,°C*	−28
Shear modulus, lb/in^2	
At 0°C	5400
At −10°C	30000
At −20°C	37000
At −30°C	53000
At −40°C	64000
Compression set, %†	
At 100°C	0.35
At 121°C	2.1

* Temperature at which modulus of rigidity equals 45,000 lb/in^2.
† Method B, 1 h at elevated temperature.

times longer than the rubber they replace. Other applications are gaskets, encapsulation, casting molds, timing belts, bumpers, wear strips, liners of equipment-handling abrasive materials, heels, and soles. The outstanding wear resistance is shown in the truck wheel pictured in Fig. 4.22. Figure 4.23 compares a polyurethane impeller with a steel impeller that has had the same usage.

A very large and increasing market for polyurethanes is in reaction injection molding, described earlier in this chapter. Several companies have developed compounds that cure rapidly in the mold to produce high-impact car panels and other parts. Table 4.22 lists typical properties of a family of elastomeric polyurethanes for reaction injection molding.[54] In RRIM (Reinforced Reaction Injection Molding) glass reinforcement in the form of short milled fibers are present in the polyol stream. In SRIM (Structural RIM), a preformed glass mat is placed in the mold before the resin introduction. RRIM molding pressures are 1450–2900 psi. In one large application, car window gaskets are molded directly onto the window.[55] In another application polyurethane containing internal mold release is impregnated into a combination of chopped fiberglass and fiberglass fabric to produce a low-cost bumper. Demold time is 75 seconds.[56] Polyurethanes are also used in flexible adhesives and wear-resistant coatings for bowling balls and pins, gymnasium floors, and golf balls.

TABLE 4.21   Properties of Polyurethane Resins Cured with Diamines and Polyols[49]

Property	Diamine cures	Polyol cures
Resilience	Medium	Medium to high
Reactivity	Medium to high	Low to medium
Modulus	Medium to high	Low
Tensile strength	Medium to high	Low to medium
Ultimate elongation	Medium to high	High
Tear strength	High	Low to medium
Abrasion resistance:		
Sliding	High to very high	Low to medium
Impact	Medium to high	High to very high
Compression set	Medium	Low
Hardness	High	Low to medium

FIGURE 4.22   Truck wheel cast from a polyurethane elastomer. (*Courtesy of E. I. du Pont de Nemours & Company, Inc.*)

The largest use of polyurethanes is in rigid and flexible foams. Rigid foams blown with halocarbons have the lowest thermal conductivity of any commercially available insulation. They are used in refrigerators, picnic boxes, and building construction. Flexible foam is used in furniture, clothing, packaging, and shock and vibration mounts. Bayer Corp. markets hydrochlorofluorocarbons (HCFCs) as replacements for ozone-destroying CFC-11.[57]

4.4.4   Hydrolytic Stability

Water affects polyurethanes in two ways: temporary plasticization and permanent degradation. Moisture plasticization results in a slight reduction in hardness and tensile strength. The original properties are restored when the absorbed water is removed.

Hydrolytic (chemical) degradation causes a permanent reduction in the cured physical and electrical properties. The functional groups present in the chain are hydrolyzed, resulting in both chain breaking and loss of cross-linking. Disastrous failures of polyurethane potting compounds in humid atmospheres have been reported.

In an attempt to accelerate the evaluation of these resin systems, a number of aging tests have been developed. These include (1) exposure to an environment of 95 percent relative humidity at 100°C, (2) exposure to an environment of 15 lb/in^2 steam at 110°C, and (3) continuous immersion in boiling water. A minimum Shore A hardness of 30 after a 28-day exposure to 95 percent relative humidity at 100°C is a widely accepted criterion for satisfactory performance. Fig. 4.20 shows the change in hardness

FIGURE 4.23 Cast polyurethane impeller (*right*) compared with a steel impeller with equivalent service. (*Courtesy of E. I. du Pont de Nemours & Company, Inc.*)

of polyurethanes derived from various polyols when they are exposed to 95 percent relative humidity at 100°C. Ricinoleate-based polyurethane resins have stabilities between those of the polyether polyurethanes and the polybutadiene polyurethanes.

4.5 SILICONES

Silicone synthetic polymers are partly organic and partly inorganic in nature. They have a combination of properties that make them the unique choice in many applications. Silicone elastomers remain flexible as low as −80°C and stable at temperatures as high as 300°C for extended periods of time. They are virtually unchanged after long weathering, and their excellent electrical properties remain constant with temperature and frequency. They are strongly hydrophobic, but they readily transmit water vapor and other gases. Almost all castable materials readily release from

TABLE 4.22 Properties of a Family of Elastomeric Polyurethanes

Property*	MP-3000	110-7	110-25	110-50	110-80	110-80GR II†
Density, lb/ft³	62	62	65	65	65	71
Hardness, Shore D	20	35	50	61	65	71
Flexural modulus, lb/in²						
At −30°C	8800	22,000	64,000	120,000	140,000	400,000
At 22°C	3000	7000	25,000	53,000	75,000	200,000
At 65°C	2500	6000	18,000	35,000	45,000	144,000
Flexural modulus ratio, −30°C/65°C	3.52	3.7	3.55	3.43	3.1	2.9
Tensile strength, lb/in²	1600	2000	3500	3800	3500	4700
Tensile modulus at 100% elongation, lb/in²	575	1000	1400	2425	—	—
Elongation, %	450	300	340	280	110	20
Tear strength (die C), lb/lin. in	190	300	490	600	500	470
Heat sag, 4 in Overhang, at 1 h at 121°C, in	—	0.3	0.4	0.2	0.08	0.10
Notched Izod impact, ft · lb/in	—	—	—	11.0	4.9	2.0

* These items are provided as general information only. They are approximate values but are not part of the product specification.
† Flow direction parallel.

silicone surfaces without pretreatment. Silicone rubber's use has been limited by its high cost, low strength, high thermal expansion, cure inhibition of some compounds, and occasionally poor cure of some types of silicones. The principal companies supplying silicone resins are General Electric, Dow Corning Corporation, Stauffer Chemical Company, Union Carbide Corporation, Wacker, and Shin Etsu.

4.5.1 Types of Silicones and Their Applications

RTV Silicone Elastomers. The room-temperature vulcanizing (RTV) silicones are available in a wide range of viscosities. Their cure time and temperature are readily adjustable by varying the type and amount of catalysts used. Most of the newer materials cure well in deep sections. The electrical properties are excellent, and the physical properties are adequate for electrical encapsulation. The properties of a series of RTVs are listed in Table 4.23; chemical resistance is shown in Table 4.24. The low modulus over the −65 to 200°C range places a minimum of stress on the electrical components during thermal cycling and shock. These RTVs are also used as molds for casting art objects, furniture parts, and low melting metals, as well as for formed-in-place gaskets.

Flexible Resins. Flexible resins, or vinyl silicones, are two-part systems used in much the same applications as the silicone elastomers. The encapsulating grades include Dow Corning's Sylgards and General Electric's RTV 615. These materials are water-white or pigmented. The properties of a series of these materials are given in Table 4.25. The excellent resistance to heat aging is shown in Fig. 4.24. The clarity of these materials allows easy identification and replacement of electrical parts. They are used in aircraft canopies as clear, flexible buffers between the rigid plastic outer faces. These flexible resins are stronger than the other RTVs and can be cured very rapidly in thick sections at elevated temperatures. Their chief disadvantage is their cure inhibition by a variety of materials, including sulphur, sulphur-containing compounds, amines and certain nitrogen containing compounds, organic acids, organotin RTV silicone rubber catalysts, residual catalysts on molds, acid residues from cured silicone rubbers, plasticized PVC, amine-cured epoxies, polysulphides, adhesive tapes, solder flux, some natural and synthetic rubbers, and contaminated ovens.[58] This problem can be overcome by coating the inhibiting surface with a barrier coating before casting.

Another group of related flexible resins is widely used in mold making because of its high tear strength, low cure shrinkage, and good release properties. These materials are cure-inhibited by the same materials as the encapsulating flexible resins. Properties are listed in Table 4.26 for one family of these compounds. The properties of a family of condensation cured moldmaking silicone rubbers is given in Table 4.27. These materials are not inhibited by the materials that affect the addition cured silicones.

One-Part Silicones. These materials are used primarily as adhesives and sealants. They cure by reacting with moisture in air, liberating such materials as acetic acid and methanol. The acetic-acid-liberating compounds must be used with caution near copper or other materials that might be attacked. This cure mechanism makes it imperative that at least one of the adherents be permeable to water and that relatively thin layers be formed. The bond is excellent to most materials, including metals, glass, ceramics, plastics, and silicone rubbers. The one-part silicones are available in a variety of colors, flame resistance, solvent resistance, and viscosity. They resist

TABLE 4.23 Properties of Family of RTV Silicones

Typical properties	RTV 11	RTV 12	RTV 21	RTV 31	RTV 41	RTV 60	RTV 88
Key feature	General purpose	Soft general purpose	General purpose	High temperature	General purpose	High temperature flowable	High temperature paste
Color	White	Clear	Pink	Red	White	Red	Red
Cure conditions	RT	RT	RT	RT	RT	RT	RT
Mix ratio (base to curing agent), by weight	200:1	20:1	200:1	200:1	200:1	200:1	200:1
Silicone content, %	100	99	100	100	100	100	100
Solvent type	—	Min. Spirits	—	—	—	—	—
Viscosity at 25°C (77°F), cP	12,000	1500	30,000	20,000	40,000	40,000	700,000
Specific gravity	1.18	1.00	1.31	1.45	1.31	1.47	1.48
Hardness, Shore A	45	18	45	55	45	55	55
Tensile strength, lb/in² (kg/cm²)	350 (25)	100 (7)	400 (28)	500 (35)	350 (25)	750 (53)	850 (60)
Elongation, %	150	200	180	100	150	120	110
Tear resistance (die B), lb/in (kg/cm)	20 (3.5)	10 (2.0)	25 (4.5)	30 (5.0)	25 (4.5)	40 (7.0)	40 (7.0)
Dielectric strength, 75 mil at 25°C, 77°F, V/mil (kV/mm)	500 (19.7)	400 (15.7)	500 (19.7)	500 (19.7)	500 (19.7)	500 (19.7)	500 (19.7)
Dielectric constant (1000 Hz)	3.4	3.0	3.4	3.6	3.4	3.7	3.7
Dissipation factor (1000 Hz)	0.02	0.001	0.01	0.02	0.02	0.02	0.02
Volume resistivity, $\Omega \cdot cm$	9×10^{14}	1×10^{13}	9.0×10^{14}	2.9×10^{15}	9.0×10^{14}	1.3×10^{14}	1×10^{14}
Useful temperature range, °F (°C)	−65 to 400 (−54 to 204)	−75 to 400 (−60 to 204)	−65 to 400 (−54 to 204)	−65 to 500 (−54 to 260)	−65 to 400 (−54 to 204)	−65 to 500 (−54 to 260)	−65 to 500 (−54 to 260)

Source: General Electric Company.

TABLE 4.24 Typical Chemical Resistance of RTV Silicone Rubber[59]

Test fluid	Tensile-strength change, %	Volume change, %
MIL-L-7808 (70 h/300°F)	−40	+20
Silicone oil SF-96(100) (70 h/300°F)	−55	+30
ASTM No. 3 (70 h/300°F)	−50	+30
Skydrol 500 (70 h/212°F)	−65	+85
JP-4 (70 h/80°F)	−15	+80
5% NaCl in distilled water* (70 h/80°F)	0	+1

* Cured 144 h/80°F.

TABLE 4.25 Properties of Family of Silicone Resins

	170	170 fast cure	182	184	186
Physical and chemical					
Physical nature (as cured)	Flexible rubber	Flexible rubber	Rubberlike	Rubberlike	Rubber
Color	Black	Black	Clear	Clear	Translucent
Viscosity at 77°F (25°C), poises	25	25	39	39	450
Specific gravity at 77°F (25°C) (ASTM D 702)	1.33	1.33	1.05	1.05	1.12
Shelf life, months	12	12	12	12	6
Pot life at 77°F (25°C) (ASTM D 702), h	¼	—	8	2	2
Cure time/temperature, h/°C	8/25	4.5 min/25	4/65	24/25	24/25
Refractive index at 77°F (25°C) (ASTM D 1218)	—	—	1.43	1.41	1.43
Radiation resistance, Mrad	200	—	200	200	200
Useful temperature range, °C	−55–225	−55–225	−55–200	−55–200	−55–200
Thermal conductivity at 77–212°F (25–100°C), cal/cm · s · °C)	7.5×10^{-4}	8.0×10^{-4}	3.5×10^{-4}	3.5×10^{-4}	4.1×10^{-4}
Thermal shock, at −67–266°F (−55–130°C), 10 cycles (MIL-I-16923G)	Pass	—	Pass	Pass	Pass
Self-extinguishing (UL 94)	Yes	Yes	Yes	Yes	Yes
Volume coefficient of thermal expansion at 77–302°F (25–150°C), cm³/cm³ · °C	8.0×10^{-4}	—	9.6×10^{-4}	9.6×10^{-4}	9.0×10^{-4}
Specific heat at 77°F (25°C), cal/g · °C	0.35	—	0.34	0.34	0.34
Corrosion resistance (MIL-S-23586D)	Good/pass	—	Good/pass	Good/pass	Good/pass
Mechanical					
Tensile strength, lb/in² (ASTM D 412)	350	380	700	700	700
Elongation, % (ASTM D 412)	150	150	100	100	420
Hardness, Shore A (ASTM D 2240)	40	48	40	35	32
Tear strength, lb/in (ASTM D 412)	25	13.5	15	15	90
Shrinkage after 3 days at 77°F (25°C), %	Nil	Nil	Nil	Nil	Nil
Deep section cure, in thickness	Yes	Yes	Yes	Yes	Yes
Electrical					
Arc resistance, s (ASTM D 495)	120	—	115	115	119
Dielectric constant (ASTM D 150)					
At 100 Hz	3.15	3.12	2.66	2.66	2.79
At 100 kHz	3.10	3.02	2.65	2.65	2.78
Dissipation factor (ASTM D 150)					
At 100 Hz	0.008	0.015	0.0009	0.0009	0.0009
At 100 kHz	0.002	0.004	0.001	0.001	0.001
Dielectric strength, V/mil (ASTM D 149)	450	—	450	450	450
Volume resistivity, Ω · cm (ASTM D 257)	1.0×10^{15}	3.16×10^{14}	2×10^{15}	2×10^{15}	3.0×10^{15}

Source: Dow Corning Corporation.

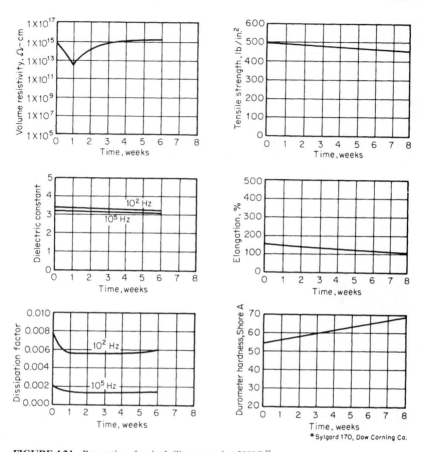

FIGURE 4.24 Properties of a vinyl silicone aged at 200°C.[37]

ozone, oxidation, moisture, and harsh chemicals, and have high dielectric strength. They are used in formed-in-place gaskets, high-vacuum feedthrough seals, construction and swimming-pool sealants, plastic piping seals, on automobile windshields, and as printed-circuit-board coatings. Dow Corning 790 Sealant was used to fill large and small cracks in Mount Rushmore National Memorial in 1991.[62]

Liquid Silicone Rubber. Liquid silicone rubbers (LSRs) are two-part injection-molded compounds that can be cured rapidly. They can cure in 20 s at 175°C, and new grades cure at 95°C. They have low compression set, low durometer, outstanding adhesion, and are biocompatible. Properties of a typical LSR are given in Table 4.28. Fig. 4.25 shows a group of products molded with LSRs. The LSRs have been coinjection-molded with thermoplastic to produce vent flaps and door-locking systems.[63] LSRs cost 50 percent more than moldable silicone rubber, and the initial cost of a mold can be three times higher (very high precision is required to prevent flash formation), but the cure time is four to five times faster with the LSRs.[64]

TABLE 4.26 Typical Properties*

	Addition cure materials						
	Silastic® silicone rubber						
	E	J	K	L	M	R	T
As supplied							
Specific gravity	1.12	1.28	1.29	1.28	1.29	1.28	1.13
As catalyzed							
Appearance	White	Green	Green	Green	Teal Blue	Purple	Translucent
Viscosity, poise	880	880	990	980	860	450	552
As-cured physical properties†							
Durometer hardness,							
Shore A, points	37	57	55	38	59	47	38
Tensile strength, psi	850	800	600	550	640	600	830
Elongation, percent	360	250	320	350	240	350	350
Tear strength, die B, ppi	110	90	85	55	85	80	95
Useful temperature range, percent	←			−55 to 200°C (−67 to 392°F)			→
Linear shrink, percent							
After 24 hrs @ 77°F (25°C)	Nil‡	Nil‡	Nil‡	Nil‡	Nil‡	Nil‡	Nil‡
After 7 days @ 77°F (25°C)	0.1	0.1	0.1	0.1	0.1	0.1	0.1

* These values are not intended for use in preparing specifications. Silastic is a trademark of Dow Corning Company.
† Based on sample thickness of 125 mils, cured 24 hours at room temperature.
‡ Shrinkage not measurable after curing 24 hours at room temperature.

4.5.2 Electrical Applications

High-purity silicone compounds are widely used to protect integrated circuit chips as thin coatings that act as a buffer to the epoxy-molded package or as a "glob top" that acts as sole protection. The intermediate coatings contain less than 2 ppm of sodium and potassium and provide protection against alpha particles. The clear materials serve as coatings for light-emitting diodes and as light pipes in optocoupler devices.[65] Elastoplastic conformal coatings for printed wiring boards are harder and more wear-resistant than the silicone rubbers.[66] Dow Corning has introduced new materials that cure in seconds under UV light.[69]

4.6 CAST PHENOLICS

Unlike most of the materials covered in this chapter, the phenolics cure by a condensation mechanism, generating water in the process. This water is finely dispersed in the casting, producing a brilliance found in few other plastics. This and the high density of the unfilled castings have made this material desirable in beads, knife handles, paperweights, and billiard balls. These products are either cast directly to final shape or machined from cast rods and sheets. The phenolics are not widely used in electrical applications because of the corrosive effects of the acid catalyst and the

TABLE 4.27 Typical Properties*

	Condensation cure materials										
	Dow Corning high strength moldmaking silicone rubber								Dow Corning Silicone rubber		
	HS II			HS III			HS IV		3110	3112	3120
As supplied											
Specific gravity	1.21			1.16			1.16		1.17	1.35	1.47
Catalyst used	10:1	20:1	10:1	10:1	10:1	10:1	10:1	10:1	10:1	10:1	10:1
	Clear	Clear	Colored	Clear	Colored	Clay	Clear	Colored	1 Catalyst	1 Catalyst	1 Catalyst
As catalyzed											
Appearance	White	White	Pink	White	Pink	Green	White	Pink	White	White	Red
Viscosity, poise	240	240	240	160	160	180	150	150	130	280	250
As-cured physical properties[†]											
Durometer hardness, Shore A, points	20	20	16	10	10	14	5	5	45	60	60
Tensile strength, psi	600	600	600	500	500	425	435	435	320	600	800
Elongation, percent	500	500	470	500	500	400	730	730	230	120	120
Tear strength, die B, ppi	140	140	135	130	130	125	110	110	20	30	40
Upper temperature limit, °C	200	200	200	200	200	200	200	200	200	200	300
Linear shrink, percent											
After 24 h @ 77°F (25°C)	.20	.20	.26	.23	.20	.01	0.14	0.14	—	—	—
After 7 days @ 77°F (25°C)	.30	.30	.48	.48	.48	.21	0.15	0.15	—	—	—

* These values are not intended for use in preparing specifications.
† Based on sample thickness of 125 mils, cured 24 hours at room temperature.

TABLE 4.28 Properties of a Typical LSR*

Color	9280-30 Colorless, translucent	9280-40 Colorless, translucent	9280-50 Colorless, translucent	9280-60 Colorless, translucent	9280-70 Colorless, translucent	94-595-HC Colorless, translucent	94-599-HC Colorless, translucent	590 Off-white	591 Off-white
Extrusion rate, grams per minute (90 psi ⅛″ Orifice)	60	100	60	45	100	135	60	500	530
Viscosity, Pa·s	4500 (0.9 sec^{-1}) 450 (10 sec^{-1})	2000 (0.9 sec^{-1}) 290 (10 sec^{-1})	3100 (0.9 sec^{-1}) 455 (10 sec^{-1})	4600 (0.9 sec^{-1}) 500 (10 sec^{-1})	5500 (0.9 sec^{-1}) 475 (10 sec^{-1})	2800 (0.9 sec^{-1}) 260 (10 sec^{-1})	8700 (0.9 sec^{-1}) 570 (10 sec^{-1})	470 (0.9 sec^{-1}) 75 (10 sec^{-1})	225 (0.9 sec^{-1}) 70 (10 sec^{-1})
Specific gravity	1.14	1.13	1.14	1.14	1.15	1.12	1.15	1.23	1.53
Physical properties, as molded (5 min, 150°C)									
Durometer, Shore A, points	29	38	48	59	68	40	47	37	52
Tensile strength, MPa (psi)	9.7 (1400)	9.0 (1300)	9.5 (1380)	9.4 (1370)	8.8 (1270)	9.7 (1400)	10.0 (1450)	7.0 (1010)	3.5 (510)
Elongation, percent	940	725	620	440	400	580	575	545	340
Modulus, 100%, MPa (psi)	0.5 (70)	0.8 (120)	1.7 (240)	3.1 (450)	4.0 (580)	1.0 (150)	1.4 (200)	0.8 (115)	1.9 (280)
Tear strength, die B, kN/m (ppi)	35.0 (200)	37.7 (215)	52.5 (300)	59.3 (310)	49.0 (280)	32.4 (185)	32.4 (185)	15.8 (90)	12.3 (70)
Compression set, percent (22 h, 177°C)	40	26	25	40	60	50	23	80	65
Bashore resilience	40	55	58	61	58	60	53	54	50
Circle shrink, percent	2.68	2.58	2.52	2.42	2.65	2.40	2.65	2.05	1.77
Physical properties, post cured (4 h, 200°C)									
Durometer, Shore A, points	36	41	51	62	73	44	53	42	59
Tensile strength, MPa (psi)	9.0 (1300)	8.8 (1280)	9.1 (1320)	9.4 (1360)	7.7 (1120)	8.9 (1290)	10.1 (1460)	6.9 (1000)	4.8 (700)

Property	700	545	415	360	255	460	450	375	180
Elongation, percent	700	545	415	360	255	460	450	375	180
Modulus, 100%, MPa (psi)	1.0 (140)	1.4 (200)	2.6 (370)	3.7 (530)	5.0 (725)	1.4 (200)	1.9 (280)	1.3 (195)	3.4 (500)
Tear strength, die B, kN/m (ppi)	35.0 (200)	38.5 (220)	49.0 (280)	45.5 (260)	13.1 (75)	32.4 (185)	29.8 (170)	12.3 (70)	12.3 (70)
Compression set, percent (22 h, 177°C)	12	6	8	14	26	14	20	11	8
Bashore resilience	44	55	57	59	58	58	50	56	52
Circle shrink, percent	3.49	3.35	3.19	3.10	3.28	3.24	3.47	2.89	2.39
Electrical properties, as molded									
Dielectric strength, kV/mm (volts/mil)	18.6 (473)	18.5 (470)	17.7 (452)	17.7 (451)	17.8 (453)	19.9 (507)	18.1 (460)	18.6 (474)	18.0 (459)
Dielectric constant at 100 Hz	2.92	2.98	2.94	3.00	2.91	2.87	2.91	2.88	3.23
Dielectric constant at 10,000 Hz	2.90	2.95	2.92	2.98	2.88	2.86	2.89	2.87	3.16
Volume resistivity (ohm·cm)	2.7×10^{14}	3.8×10^{14}	3.6×10^{14}	5.2×10^{14}	6.9×10^{14}	6.1×10^{14}	5.5×10^{14}	2.6×10^{14}	4.8×10^{14}
Dissipation factor at 100 Hz	0.0040	0.0033	0.0030	0.0028	0.0008	0.0016	0.0047	0.0028	0.0152
Dissipation factor at 10,000 Hz	0.0020	0.0020	0.0017	0.0016	<0.0001	0.0011	0.0022	0.0013	0.0049
After 70 h immersion in 100°C (212°F) water									
Dielectric strength, kV/mm (volts/mil)	17.3 (440)	18.0 (457)	17.1 (436)	17.6 (448)	17.6 (447)	18.8 (478)	17.7 (449)	18.3 (466)	18.3 (464)
Dielectric constant at 100 Hz	3.04	3.01	2.99	3.04	2.94	2.93	2.96	2.96	3.35
Dielectric constant at 10,000 Hz	3.02	2.99	2.98	3.02	2.93	2.92	2.95	2.93	3.23
Volume resistivity (ohm·cm)	9.1×10^{14}	6.6×10^{14}	6.5×10^{14}	5.6×10^{14}	1.0×10^{15}	7.6×10^{14}	7.0×10^{14}	7.7×10^{14}	5.0×10^{14}
Dissipation factor at 100 Hz	0.0013	0.0011	0.0012	0.0010	0.0018	0.0007	0.0011	0.0004	0.0180
Dissipation factor at 10,000 Hz	0.0008	0.0004	0.0007	0.0008	0.0007	0.0005	0.0007	0.0003	0.0061

* Properties obtained on 1.9 mm-thick (.075 inch) slabs.

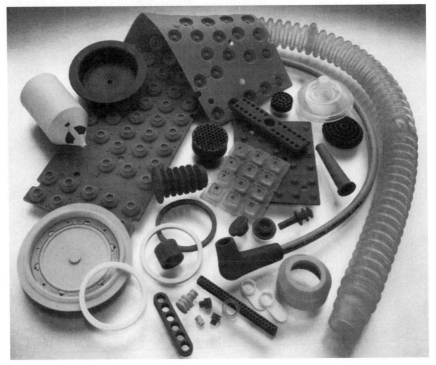

FIGURE 4.25   Parts molded from liquid silicone rubber. (*Courtesy of Dow Corning Corporation.*)

water vapor given off in curing. Phenolic dip coatings have been used successfully for protecting electronic components. Properties of phenolic casting resins are listed in Table 4.29.

4.7 ALLYLIC RESINS

Allylic resins have been used in optical parts because of their excellent clarity, hardness, abrasion resistance, and color stability. Their electrical properties are outstanding, but the casting difficulties and moderate mechanical properties have limited their use in potting. The allylics have never competed with acrylics for glazing because of their higher price and poorer weathering properties. Low vapor pressure of the monomers and long pot life have led to their use in prepregs and as impregnants.

4.7.1 Types of Allylics

Diallyl phthalate and diallyl isophthalate are the allylics produced in largest quantity. Diethylene glycol bis(allyl carbonate), used largely in optical castings, is marketed by PPG Industries as CR-39. Triallyl cyanurate (TAC) and other allylics are used as monomers in polyesters. Diallyl chlorendate (DAC) is used in flame-resistant compositions.

TABLE 4.29   Properties of Cured Cast-Phenolic Resins[68]

Property	Test method	Test values
Inherent properties:		
Specific gravity	D 792	1.30–1.32
Specific volume, in.³/lb	D 792	21.3–20.9
Refractive index, n_D	D 542	1.58–1.66
Cured properties:		
Tensile strength, lb/in.²	D 638–D 651	6,000–9,000
Elongation, %	D 638	1.5–2.0
Modulus of elasticity in tension, 10^5 lb/in.²	D 638	4–5
Compressive strength, lb/in.²	D 695	12,000–15,000
Flexural strength, lb/in.²	D 790	11,00–17,000
Impact strength, ft-lb/in. notch		
(½- × ½-in. notched bar, Izod test)	D 256	0.25–0.40
Hardness, Rockwell	D 785	M93–M120
Thermal conductivity,		
10^4 cal/(s)(cm²)(°C)(cm)	C 177	3–5
Specific heat, cal/(°C)(g)	.....	0.3–0.4
Thermal expansion, 10^{-5} per °C	D 696	6–8
Resistance to heat (continuous), °F	.....	160
Heat-distortion temperature, °F	D 648	165–175
Volume resistivity (50% RH and 23°C),		
Ω-cm	D 257	10^{12}–10^{13}
Dielectric strength (⅛-in. thickness), V/mil:		
Short time	D 149	350–400
Step-by-step	D 149	250–300
Dielectric constant:		
60 Hz	D 150	5.6–7.5
10^5 Hz	D 150	5.5–6.0
10^6 Hz	D 150	4.0–5.5
Dissipation (power) factor:		
60 Hz	D 150	0.10–0.15
10^3 Hz	D 150	0.01–0.05
10^6 Hz	D 150	0.04–0.05
Arc resistance, s	D 495	200–250
Water absorption (24 h, ⅛-in. thickness), %	D 570	0.3–0.4
Burning rate	D 635	Very low
Effect of sunlight	.....	Colors may fade
Effect of weak acids	D 543	None to slight, depending on acid
Effect of strong acids	D 543	Decomposed by oxidizing acids; reducing and organic acids, none to slight effect
Effect of weak alkalies	D 543	Slight to marked, depending on alkalinity
Effect of strong alkalies	D 543	Decomposes
Effect of organic solvents	D 543	Attacked by some
Machining qualities	.....	Excellent
Clarity	.....	Transparent, translucent, opaque

4.7.2   Processing Characteristics

The vapor pressure of all the allyl ester monomers is very low, as shown in Table 4.30. The low odor level is also a plus. The peroxide-catalyzed materials have a pot life extending to over a year and cure rapidly at temperatures above 150°C. If castings are made directly from monomers, the cure shrinkage is about 12 percent. Incorporation of prepolymer in the casting composition reduces the shrinkage considerably but raises the viscosity, precluding electrical potting.

TABLE 4.30 Properties of Allyl Ester Monomers[72]

Property	Diallyl phthalate	Diallyl iso-phthalate	Diallyl maleate	Diallyl chloren-date	Diallyl adipate	Diallyl diglycollate	Triallyl cyanurate	Diethylene glycol bis-(allyl carbonate)
Density, at 20°C	1.12	1.12	1.076	1.47	1.025	1.113	1.113	1.143
Molecular wt.	246.35	246.35	196	462.76	226.14	214.11	249.26	274.3
Boiling pt, °C, 4 mm Hg	160	181	111	......	137	135	162	160
Freezing pt, °C	−70	−3	−47	29.5	−33	......	27	−4
Flash pt, °C	166	340	122	210	150	146	>80	177
Viscosity at 20°C, cP	12	16.9	4.5	4.0	4.12	7.80	12	9
Vapor pressure at 20°C, mm Hg	27	......	......	......	......	......	......	......
Surface tension at 20°C, dynes/cm	34.4	35.4	33	......	32.10	34.37	......	35
Solubility in gasoline, %	24	100	23	......	100	4.8	80	......
Thermal expansion, in./(in.)(°C)	0.00076							

4.48

4.7.3 Properties

The properties of the allylic resins are tabulated in Tables 4.31–4.34. Their low tensile strength and impact resistance preclude the use of DAP and DAIP in unfilled castings where physical abuse is likely. In most applications of CR-39, it is used in competition with glass or polymethyl methacrylate (PMMA), where its abrasion resistance (Fig. 4.26), high heat distortion, or impact resistance is needed. CR-39 is widely used in casting lenses. A similar product, CR-307, contains monomer additives to make the polymer softer and acceptable for imbibing photochromatic chemicals into the surface layers.[70]

HIRI casting resin has a higher refractive index (1.5302) than CR-39 and a lower cure shrinkage. The higher refractive index allows the cast lens to be thinner and lighter than those made with CR-39.[71]

Electrical properties are excellent. The variations of dissipation factor, dielectric constant, and dielectric strength with temperature and frequency are given in Figs. 4.27–4.30. The surface and volume resistivities remain high after prolonged exposure to high humidity. Resistance to solvents and acids is excellent, and resistance to alkalies is good.

TABLE 4.31   Properties of Two Cured Allyl Esters[59]

Property by ASTM procedures	Diallyl phthalate	Diallyl isophthalate
Dielectric constant:		
25°C and 60 Hz	3.6	3.5
25°C and 10⁶ Hz	3.4	3.2
Dissipation factor:		
25°C and 60 Hz	0.010	0.008
25°C and 10⁶ Hz	0.011	0.009
Volume resistivity, Ω-cm at 25°C	1.8×10^{16}	3.9×10^{17}
Volume resistivity, Ω-cm at 25°C (wet)*	1.0×10^{14}	
Surface resistivity, Ω at 25°C	9.7×10^{15}	8.4×10^{12}
Surface resistivity, Ω at 25°C (wet)*	4.0×10^{13}	
Dielectric strength, V/mil at 25°C†	450	422
Arc resistance, s	118	123–128
Moisture absorption, %, 24 h at 25°C	0.09	0.1
Tensile strength, lb/in.²	3,000–4,000	4,000–4,500
Specific gravity	1.270	1.264
Heat-distortion temp., °C at 264 lb/in.²	155 (310°F)	238‡ (460°F)
Heat-distortion temp., °C at 546 lb/in.²	125 (257°F)	184–211 (364–412°F)
Chemical resistance, % gain in wt.		
After 1 month immersion at 25°C in:		
Water	0.9	0.8
Acetone	1.3	−0.03
1% NaOH	0.7	0.7
10% NaOH	0.5	0.6
3% H₂SO₄	0.8	0.7
30% H₂SO₄	0.4	0.4

* Tested in humidity chamber after 30 days at 70°C (158°F) and 100% relative humidity.
† Step by step.
‡ No deflection.

TABLE 4.32 Properties of Cured Diethylene
Glycol bis(Allyl Carbonate)[72]

Property	Value
Specific gravity, 25°C	1.32
Hardness, Rockwell M	100
Heat distortion, 10 mil, 264 lb/in.², °C	65
Izod impact, ft-lb/in.	0.3
Compressive strength, lb/in.²	22,500
Flexural strength, lb/in.²	9.000
Flexural modulus, lb/in.² × 10^{-6}	0.30
Tensile strength, lb/in.²	6,000
Volume resistivity, Ω-cm	4.1×10^{14}
Surface resistivity, Ω	3.4×10^{11}
Dielectric strength, V/mil	560
Dissipation factor, 1 MHz, %	5.6
Dielectric constant, 1 MHz	3.5
Arc resistance, s	249
Water absorption, 24 h, %	0.2

TABLE 4.33 Properties of Cast Alkyd–Triallyl
Cyanurate Resin[72]

Property	Value
Specific gravity	1,346
Flexural strength, lb/in.²	7,200
Flexural modulus, lb/in.² × 10^{-6}	0.75
Heat distortion, °C	270
Refractive index	1.550
Dielectric constant:	
60 Hz	3.97
1 MHz	3.60
9,375 MHz	2.85
Dissipation factor, 60 Hz	0.016
Water absorption, % 24 h	0.39

TABLE 4.34 Properties of Cast Diallyl Orthophthalate–Diallyl
Chlorendate Copolymer (80% DAP, 20% DAC)[72]

Property	Value
Hardness, Rockwell M	118
Heat distortion, °F	324
Impact, Izod notched, ft-lb/in.	0.27
Dissipation factor, %:	
60 Hz	1.0
1 kHz	1.0
1 MHz	1.3
Dielectric constant:	
60 Hz	3.6
1 kHz	3.5
1 MHz	3.5
Volume resistivity, Ω-cm	8.9×10^{15}
Burning rate:*	
ASTM 635-56T	Self-extinguishing
ASTM 747-49, in./min	0.15

Note: 10% antimony oxide contained in this resin mixture.

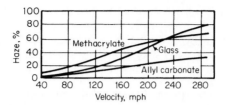

FIGURE 4.26 Effect of falling emery on haze of transparent optics.[73]

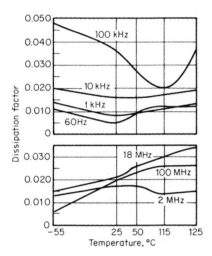

FIGURE 4.27 Dissipation factor of cast diallyl phthalate as a function of frequency and temperature.[74]

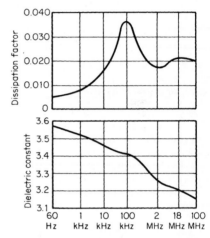

FIGURE 4.28 Dissipation factor and dielectric constant of cast diallyl phthalate as a function of frequency.[74]

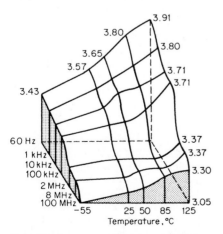

FIGURE 4.29 Dielectric constant of cast diallyl phthalate as a function of temperature and frequency.[74]

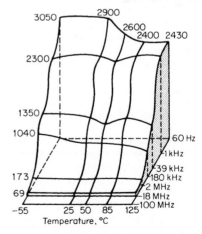

FIGURE 4.30 Dielectric strength of cast diallyl phthalate as a function of temperature and frequency.[74]

4.7.4 Applications

Diallyl phthalate and diallyl isophthalate are seldom used as cast homopolymers except for small electrical parts. CR-39 cast into sheet and other shapes has been used in glazing, safety-goggle lenses, face shields, sunglasses, tape take-up reels, and transparent plotting boards.

DAP is used for impregnation of ferrous and nonferrous castings because of its low viscosity, excellent sealant characteristics, low resin bleed-out, and ease of cleanup. Wood impregnated with DAP has reduced water absorption and increased impact, compressive, and shear strengths. Oppressive odors of wet lay-ups of styrene polyesters were eliminated by adding 3 percent DAP.[75]

DAP and DAIP-glass laminates have high-temperature electrical properties superior to those of most other structural laminates. The cure cycles are shorter, and little or no postcure is necessary to provide usable strength up to 250°C.

4.8 POLYBUTADIENES

The polybutadienes have a hydrocarbon structure that is responsible for the excellent resistance to chemicals and solvents, electrical properties that are good over a range of frequencies, temperatures, and humidity, and resistance to high temperature. The basic materials are monopolymers or copolymers that react through terminal hydroxyl groups, terminal carboxyl groups, vinyl groups, or a combination of these. These materials have limited use as casting compounds because of their high viscosity.

4.8.1 Hydroxyl-Terminated Resins

Atochem North America, Inc., part of the Elf Aquitaine group, markets Poly BD, a family of hydroxyl-terminated butadiene homopolymers and copolymers with styrene or acrylonitrile. These materials are generally reacted with isocyanates to produce polyurethanes that have excellent resistance to boiling water. Typical properties are shown in Table 4.35. Fig. 4.20 compares the water resistance of these polyurethanes with those derived from other polyols.

4.8.2 Modified Resins

Monopolymers and copolymers with acrylonitrile are the Hycar series, produced by the B. F. Goodrich Chemical Company. The incorporation of acrylonitrile increases the viscosity and imparts oil resistance, adhesion, and compatibility with epoxy resins. Carboxy-terminated butadiene acrylonitrile copolymer (CTBN) is generally cured with epoxies using standard epoxy curing agents. CTBN enhances impact strength, thermal shock resistance, peel strength, low-temperature shear strength, and crack resistance of epoxy compositions. The epoxy content can be varied to produce materials ranging from elastomers to toughened rigid resins, as shown in Table

TABLE 4.35 Properties of Polybutadiene Polyurethanes[47]

Property	Soft	Medium		Hard		
	Gumstock oil-extended	Gumstock	Mixed polyols, one-step	Urea urethanes	Urea urethanes two-step	Mixed polyols
Hardness, Shore A	32	52	80	83	95	70 Shore D
Tensile strength, lb/in.²	240	510	2,200	1,500	3,130	6,500
% elongation	520	420	315	230	475	30
Modulus, lb/in.², 100%	100	190	765	930	475	
Tear strength, pli	32	42	175	235	1,410	

Other Typical Values:

Property	One-shot system	Prepolymer
Shrinkage, in./in.	. . .	0.012
Thermal conductivity, Btu/(ft²)(h)(in./°F)	1.95	1.50
Water absorption, %	. . .	0.38 (48 h)

4.53

TABLE 4.36 Properties of Blends of Carboxyl-Terminated Polybutadienes and Epi-Bis Epoxies[76]

Property	A	B	C	D	E	F	G	H
Hycar* CTBN	100	100	100	100	50	25	12.5	
DER-331†	12.5	25	50	100	100	100	100	100
Triethylene tetramine	0.75	1.5	3	6	6	6	6	6
Cured at room temp:								
Work life	>14 days	>6 days	>2 h	>2 h	2 h	2 h	70 min	45 min
Cure time		2 weeks	1 week	24 h	4 h	3 h	2½ h	2½ h
Cured 16 h at 105°C:								
Tensile strength, lb/in.²	133	555	1,060	2,250	>3,000	>3,000	>3,000	>3,000
Elongation at break, %	230	175	75	50	>10	>10	>10	<10
Hardness:								
Duro A	45	83	83	63	60	83	83	88
Duro D								
Flexural strength, lb/in.²				3,193	4,940	9,060	12,675	
Heat-distortion temp, ASTM D 648, °C:								
Temp				44	52	68	69	73
180° bend	OK	OK	OK	OK	Fail	Fail	Fail	Fail
Pan cure hardness, inst./10 in.:								
Duro A	47/41	61/55	88/85	55/47	60/55	75/73	76/74	82/81
Duro D								

* Registered trademark of B. F. Goodrich Chemical Co.
† A bisphenol A type of epoxy resin supplied by the Dow Chemical Company.

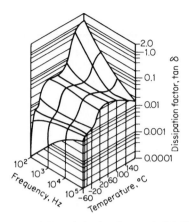

FIGURE 4.31 Dissipation factor of phthalic-anhydride-cured epoxy polybutadiene as a function of temperature and frequency.[77]

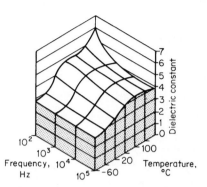

FIGURE 4.32 Dielectric constant of phthalic-anhydride-cured epoxy polybutadiene as a function of temperature and frequency.[77]

4.36. Amine-terminated butadiene-acrylonitrile copolymers (ATBN) are also used to modify epoxy resins but are formulated on the amine hardener side of the mix. Some commercial and potential applications include electrical potting compounds (with transformer oil), sealants, and moisture-block compounds for telephone cables. The dissipation factor and the dielectric constant of phthalic-anhydride-cured epoxy butadienes are given in Figs. 4.31 and 4.32. An epoxy-terminated nitrile elastomer diluted with styrene serves as a modifier for polyester in bulk molding compounds and sheet mold compound.

Butadienes are available with methacrylate groups that can react with unsaturated polyester and vinyl esters (epoxy methacrylates) to produce tough polymers.

4.8.3 Resins Containing Pendant Vinyl Groups

Ricons, produced by Advanced Resins, Inc., are a series of 1,4 butadienes with 15 to 95 percent 1,2-pendant vinyl groups. The cured materials have a dielectric constant of 2.4, a moisture pickup of 0.02 to 0.05 percent, and are almost unaffected by 200-h exposure at 260°C. The Ricons are often mixed with vinyl toluene monomer to lower their viscosity and modify cured properties. Dicumyl peroxide is the preferred catalyst. Low-temperature cure yields weak, cheesy materials. Temperatures above 150°C are necessary to produce usable materials. Disadvantages of these materials are high cure shrinkage (6 to 9 percent) and low strength. They are used in electrical potting, impregnation, oil-well cables, radomes, and laminates.

Liquid styrene butadienes (LSBs) with high vinyl content have good corrosion resistance, a dielectric strength of 800 to 900 V/mil in ⅛-in slabs, and can withstand 220°C for 6000 h with good retention of electrical properties. Its high viscosity makes this material difficult to use in encapsulation. Table 4.37 presents typical properties of an all-hydrocarbon monomer formulation using LSBs.

TABLE 4.37   Properties of All-Hydrocarbon Monomer Formulation Using LSBs

	A	B	C	D	E	F
Material, parts by weight						
Ricon 100 (LSB)	80.0	40.0	70.0	60.0	76.0	76.0
Vinyl toluene	20.0	—	30.0	40.0	20.0	20.0
Styrene	—	57.0	—	—	—	—
Divinyl benzene	—	3.0	—	—	—	—
Vinyl silane (A-172)	—	—	—	1.5	0.7	0.5
TMPTM	—	—	—	—	4.0	4.0
Lupersol 101	1.0	—	1.0	2.0	3.0	2.0
Dicumyl peroxide	1.0	—	—	—	—	—
Polyethylene powder	—	—	10.0	—	—	—
Curnene hydroperoxide	—	2.0	—	—	—	—
Cobalt napthenate	—	0.1	—	—	—	—
Ground silica	—	—	—	100	—	200
Cure conditions	2.5 h at 122°C	8 h at 80°C	2 h at 110°C	2 h at 94°C	1 h at 138°C	1 h at 135°C
	1 h at 138°C	8 h at 140°C	2 h at 180°C	1 h at 130°C	2 h at 162°C	1 h at 162°C
	2 h at 162°C			1 h at 150°C		
Physical						
Shrinkage, in/in	0.06	—	0.04	0.016	0.06	—
Flexural strength, lb/in^2	4800–6500	920	2000–3000	8000–10,500	5000	7700
Hardness, Shore	90(D)	100(A)	100(A)	80–90 (D)	85–93 (D)	85 (D)
Electrical						
Dielectric constant						
At 60 Hz	2.5	2.7	2.6	2.8	2.8	4.0
At 1 kHz	2.6	2.7	2.8	2.5	2.8	—
At 1 MHz	2.7	3.0	2.6	3.2	2.7	3.5
Dissipation factor						
At 60 Hz	0.003	0.003	0.003	0.008	0.003	0.008
At 1 kHz	0.002	0.005	0.002	0.005	0.003	—
At 1 MHz	0.002	0.006	0.002	0.006	0.003	0.008
Dielectric strength, V/mil	470	—	470	445	800	472
Surface resistivity, 10^{14} Ω	3.5	3.8	1.0	2.5	1.0	—
Volume resistivity, 10^{15} Ω · cm	2.6	2.7	2.2	2.5	1.4	—

4.9   DEPOLYMERIZED RUBBER

Hardman, Inc. produces three liquid polymers that can be vulcanized to rubbers. All three can be cured with *p*-quinone dioxime and lead dioxide at room temperature and at elevated temperature with sulfur. Ultraviolet initiators can be used.

4.9.1   Depolymerized Virgin Butyl Rubber

Trademarked Kalene, this material is similar to butyl LM rubber, which had been marketed by Exxon. The properties are similar to those of conventional butyl rubber such as the lowest vapor-transmission rate of any elastomer, resistance to degradation in high humidity, high-temperature environments (very little change after 120°C steam for 1000 h), excellent electrical properties (volume resistivity of 5.5 × 10^{15}, and a dielectric constant of 3.1), resistance to soil bacteria, excellent weathering, and resistance to chemicals and oxidation. Because of the high viscosity of the rubber, oil or solvent must be added to produce a pourable compound. This material is useful in applications where it can be applied under pressure and protected from physical abuse. Use of Kalene includes roofing coating, reservoir liners, aquarium sealants, and conformal coatings.

4.9.2 Depolymerized Polyisoprene

Cis-1,4-Polyisoprenes are produced by depolymerizing polyisoprene or natural rubber. The former material is used as a reactive plasticizer for adhesive tape, abrasive and friction products such as grinding wheels and automobile brake linings, wire and cable sealants, and in hot melts. The natural rubber product serves as a base for cold molding compounds for arts and crafts, asphalt modifier, as potting, or in molds.

4.10 POLYSULFIDE RUBBER

Morton Thiokol, Inc. markets a series of liquid polysulfides that can be oxidized to rubbers. These rubbers have excellent resistance to most solvents, good water and ozone resistance, very low specific permeability to many highly volatile solvents and gases, objectionable odor, poor physical properties, and a service-temperature range of −55 to 150°C. A typical formulation and the resulting properties are given in Table 4.38.

TABLE 4.38 Typical Physical Properties of Polysulfide Converted by p-Quinone Dioxime–Diphenylguanidine and Trinitrobenzene Cures[79]

	GMF-DPG cure	TNB cure
Compounding Recipes, Parts:		
Thiokol LP-3*.	100	100
Titanox A†.	50	50
2,4,6-Trinitrobenzene.	. . .	4
p-Quinone dioxime.	7	
Diphenylguanidine.	3	3
Sulfur.	. . .	1
Vulcanizate properties (cured 16 h at 158°F):		
Shore A sheet hardness.	40	40
300% modulus, lb/in.².	290	180
Tensile strength, lb/in.².	350	250
Elongation, %.	400	600
Compression set, %‡ (2 h at 158°F, 25% comp.).	40	40
Low-temperature torsional flexibility,§ absolute modulus:		
G_{rt}, lb/in.²/100° twist.	120	103
$G_{5,000}$, °F.	−56	−56
$G_{10,000}$.	−60	−60
$G_{20,000}$.	−61	−63

* Morton-Thiokol Corp.
† Titanium pigments incorporated.
‡ ASTM method D 395-49T, method B.
§ Determined according to ASTM method D 1043-49T; G_{rt} denotes the torsional modulus at room temperature, whereas the values for G_{5000}, $G_{10,000}$, and $G_{20,000}$ denote the temperature at which the torsional moduli were 5000, 10,000, and 20,000 lb/in², respectively.

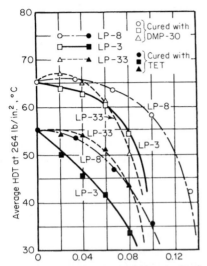

Polysulfide ploymer, parts per 100 parts of resin

FIGURE 4.33 Effect of flexibilizer (polysulfide polymer) content on the heat-distortion temperature of an epi-bis resin.[80]

Polysulfides have been used extensively as flexibilizing coreactants with epoxy resins. The effect of the concentration of the polysulfide on the heat-distortion temperature of epoxy formulations is shown in Fig. 4.33; properties of one mixture are included in Table 4.2.

A series of epoxy-terminated polysulfide polymers is available that has all the attributes of the polysulfide polymer without the mercaptan odor. It is used at 25 to 50 parts per hundred resin of epoxy to yield a material that retains lap shear strength during water immersion better than other toughened epoxies.

Applications of polysulfide rubbers include impregnation of leather to impart water and solvent resistance, protective coatings for metals, binder in gaskets, sealing of aircraft gasoline tanks, deck caulking, patching hose, printers' rollers, and adhesives. In the electrical industry, flexible-cable connections are sealed with polysulfide compounds.

4.11 ETHYLENE-PROPYLENE LIQUID POLYMERS

Trilene from Uniroyal Chemical Company, Inc., is a family of ethylene propylene copolymers and ethylene-propylene nonconjugated diene terpolymers.[81] The terpolymers have pendant unsaturation. These materials are randomly polymerized to produce liquid elastomers with stable, saturated hydrocarbon backbones. Viscosity is 100,000 Cp at 95°C. The Trilenes can be cured at room temperature with p-quinone dioxime systems or two-component peroxide systems. They can also be cured at elevated temperature with sulfur. The properties of a group of electrical encapsulation compounds are shown in Table 4.39. The dielectric constant of the cured materials is 2.1 to 2.2 and the volume resistivity is $10^{16}\,\Omega$-cm. They have very low moisture permeability, low strength, and they are ultraviolet-resistant. Applications include automobile and construction sealants, waterproofing membranes, and electrical encapsulants. From a safety standpoint, these materials have an extremely low level of dermal, oral, and inhalation toxicity and low eye irritation.

4.12 CAST NYLON

There is no theoretical limitation to the size of nylon parts that can be cast. Castings weighing as much as 1500 lb have been produced. The process lends itself to the production of large, complex shapes where the reduced cost of the mold required at atmospheric pressures offsets the curing time required.[82] The nylon 6 is polymerized in

TABLE 4.39 Typical Properties of Hydrocarbon Electrical Encapsulation Compounds

	A	B	C	D
Trilene 65	100	40	30	20
Paraffinic oil (Sunpar 150)	—	60	70	80
Sartomer SR-350	5	20	20	20
Viscosity before cure, at 25°C, cP	3×10^6	64,000	22,000	4000
Cured tensile strength,* lb/in²	420	210	90	30
Elongation, %	130	180	110	165
Volume resistivity, $\Omega \cdot$ cm	4×10^{15}	2×10^{15}	0.4×10^{15}	0.4×10^{15}

* Cured with 5 parts per 100 resin t-butyl perbenzoate for 10 min at 305°F.

place from caprolactam anionically. Castings have a 15 percent increase in tensile strength and a 17 percent increase in flexural modulus over the injection molded product. Typical properties of cast nylon are shown in Table 4.40. Complex castings can be made having a minimum of internal stress, resulting in better dimensional stability and greater strength. The incorporation of internal lubricants, impact modifiers, and glass reinforcement has expanded the applications areas for cast nylon. Special lubricants are uniformly dispersed in the base monomer, yielding a compound with high pressure velocity (PV) capabilities and an extremely low coefficient of friction. They are also longer wearing and self-lubricating. Molybdenum disulfide can be added for increased permanent lubricity, higher crystallinity, and better wear resistance.

TABLE 4.40 Properties of Cast Nylon[84]

Property	ASTM No.	Value
Specific gravity	D 792	1.15–1.17
Tensile strength, lb/in.²	D 638	11,000–14,000
Elongation, %	D 638	10–60
Modulus of elasticity, lb/in.²	D 638	350,000–450,000
Compressive strength, lb/in.²	D 695	
At 0.1% offset	.....	9,000
At 1.0% offset	.....	12,000
Shear strength, lb/in.²	D 732	10,500–11,500
Hardness (Rockwell)	D 785	112–120
Tensile impact, ft-lb/in.²	.....	80–130
Deformation under load, %, 122°F, 2,000 lb/in.²	D 621	0.5–1.0
Stiffness, lb/in.²	D 747	200,000–400,000
Heat-distortion temp:		
66 lb/in.², °F	D 648	425
264 lb/in.², °F	D 648	200–425
Flammability	D 635	Self-extinguishing
Coefficient of linear thermal expansion, in./(in.)(°F)	D 696	5.0×10^{-5}
Dielectric strength (short-time 0.040 in. thick), V/mil	D 149	500–600
Dielectric constant:		
60 Hz	D 150	3.7
10³ Hz	D 150	3.7
10⁶ Hz	D 150	3.1
Power factor:		
60 Hz	D 150	0.02
10³ Hz	D 150	0.02
10⁶ Hz	D 150	0.02
Water absorption, 24 h, %	D 570	0.6–1.2
Water absorption, saturation, %	D 570	5.5–6.5

Cast-nylon gears are used widely because of their resistance to abrasion and galling, their light weight, and their superior bearing properties. Nylon's resistance to strong alkalis and most solvents makes it useful as tank linings. Other applications include gears, elevator buckets, crane bearings, wear plates, ball-valve seats, bushings, roll covers, and slide bars such as those shown in Fig. 4.34.

A copolymer of caprolactam and laurylactam is not as hard as nylon 6 but is more resilient. It is used in valve seats in the gas and oil industries. Blends of urethane-filled cast nylon are used in specialized applications requiring a high degree of impact resistance. The complexities and controls necessary in casting nylon and the patent situation limit the practitioners to a few specialized companies.

A family of elastomer modified nylon 6 copolymers has been formulated to be processed by reaction injection molding. The caprolactam, which is solid at room temperature, is heated to above 69°C to produce 100 cps viscosity. This material can be introduced into glass fiber mats, producing parts with very low thermal coefficient of expansion. Minimum cure time is 120 s. The scrap can be reground and extruded in the normal way.[83]

FIGURE 4.34 Cast-nylon slide bar. (*Courtesy of Polymer Corporation.*)

4.13 ACRYLICS

Castings of acrylic esters have excellent optical clarity; outstanding weather resistance; resistance to discoloring; resistance to acids, alkalis, and oils; and excellent electrical properties, especially arc resistance. The physical properties of three commonly used acrylic-ester monomers are given in Table 4.41. Methyl methacrylate is the most widely used monomer of the family. Properties of cured methyl methacrylate are given in Table 4.42. Of the polymethyl methacrylate (PMMA) produced, 45 percent is sold as cast sheets.[87] It is used as glazing where breakage is a problem. As a thermoplastic, it can be formed into a variety of shapes such as aircraft canopies, furniture, and decorative forms. Tubes, rods, and decorative embedments are also produced. A 5000-lb window was cast for a rain forest exhibit.[88]

Ethyl methacrylate is sometimes used in place of the methyl ester because its higher boiling point reduces the possibility of bubbles during curing. Butyl methacrylate produces a permanently soft polymer that is easily sliced, allowing embedded specimens to be cut into thin sections. Polymethyl α-chloroacrylate sheet has a higher heat-distortion temperature, hardness, and scratch resistance than polymethyl methacrylate, but it is much more expensive. Methacrylate polymers modified with α-methyl styrene have exceptional stability to ultraviolet light and a high heat distortion point.[89]

TABLE 4.41 Physical Constants of Three Commonly Used Acrylic-Ester Monomers[85]

Properties	Methyl acrylate	Ethyl acrylate	Methyl methacrylate
Molecular weight	86	100	100
Boiling point, °C	79.6–80.3	99.3–99.7	100.6–101.1
Freezing point, °C	< -75	< -75	-48.2
Refractive index (N_d at 25°C)	1.401	1.404	1.412
Specific gravity, 25/25°C	0.952	0.919	0.940
Flash point, °F:			
Closed cup, Tagliabue	26.6	48.2	50.0
Open cup, Cleveland	60	85	85
Viscosity, cSt at 25°C	0.503	0.596	0.569
Vapor pressure, mm Hg at °C:			
2	...	21	18
10	46	27	24
20	72	40	35
30	112	61	53
40	177	93	81
50	270	142	124
60	395	210	189
70	565	300	279
80	765	425	397
90	...	582	547
Solubility in water at 30°C, % by wt	5.2	1.82	1.50
Heat of polymerization, kcal/mole	19–20	...	13
Density, lb/gal	7.95	7.65	7.8

TABLE 4.42 Typical Properties of Clear Transparent Acrylic Sheeting[86]

Property	Continuously cast	Conventional cell-cast	Extruded
Mechanical:			
Tensile strength, lb/in.^{2}	10,600	10,800	9,400
Tensile modulus ($\times 10^{-6}$), lb/in.^{2}	0.47	0.47	0.45
Tensile elongation, %	5.0	5.5	4.0
Flexural strength, lb/in.^{2}	16,500	16,900	14,900
Hardness, Barcol	50	50	47
Hardness, Rockwell M	98	98	89
Optical:			
Haze, %	1	1	3
Luminous transmittance, %	92	92	91
Light reflectance, %	96	97	91
Thermal:			
Heat-distortion temp, °F	204–209	204–217	162–183
Miscellaneous:			
Specific gravity	1.19	1.19	1.19
Water absorption (wt. gain), %	0.2	0.2	0.3

Acrylics are slowly cured with peroxide catalysts under very rigidly controlled conditions. Because of the skill and experience needed to produce high-quality castings, a few specialized companies produce most of the finished products.

Other uses for acrylics include dentures, fillings, and incorporation in polyesters. Wear-resistant flooring can be made by impregnating the wood with methyl methacrylate monomer and polymerizing it with radiation from a cobalt-60 source.[90]

In recent years there has been a proliferation of linear and cross-linked acrylics used in high-precision castings of kitchen and bath fixtures. One commercial product consists of 65 percent quartz powder and 30 percent PMMA. The remaining ingredients are held highly confidential by the suppliers.

A large number of modified resins have been developed for specific applications. One series, Modar, is used in resin transfer molding, pultrusion, filament winding, hand lay-up, spray-up, foam casting, low profile casting, cold pressing, and bulk molding compound (BMC). Thermoset reaction injection molding acrylics with fast mold times were introduced in 1984.[91]

Another family of acrylics consists of a series of materials with various bead sizes and concentrations of residual benzoyl peroxide. These are mixed with methyl methacrylate monomer and cured to final form. A cross-linking agent such as ethylene glycol dimethylacrylate is added to increase the resistance to crazing and improve surface hardness.[92]

4.13.1 Safety Considerations

PMMA is generally regarded as being biologically inert. The monomer is only moderately toxic when absorbed into the body, but for a few individuals it can be a powerful sensitizer of the skin. The effect of the vapors are not cumulative; concentrations over 100 ppm may be irritating. The monomer is highly flammable; it has a low flash point and should be used in a flame-free, well-ventilated area. A ventilation system should be installed where PMMA will be machined, ground, or processed at high temperature. Periodic cleaning is necessary to remove dust that might present an explosion or fire hazard. High-velocity filing should be avoided to minimize the possibility of a spark causing a dust explosion.

4.14 VINYL PLASTISOLS

A plastisol is a liquid dispersion of a fine-particle-size polyvinyl chloride in a plasticizer. Plastisols are converted to flexible shapes by heating to 175 to 190°C. This solidification is produced by a purely physical change, the PVC absorbing the plasticizer. The property range shown in Table 4.43 is produced by varying the amounts and types of resin and plasticizer used. If a more rigid part is needed, part or all of the plasticizer can be replaced by a cross-linking reactive monomer such as diallyl orthophthalate. A typical formulation for a rigidsol is given in Table 4.44. The properties of this material are listed in Table 4.45.

TABLE 4.43 Property Ranges of Plastisol Systems[93]

Viscosity	As required from 1,000 to more than 1 million cP
Tensile strength	From 500 to 2,700 lb/in.^{2}
Elongation	From 150 to 600%
Hardness	From 10 on the Shore A$_2$ durometer scale to 65 on the Shore D durometer scale
Heat resistance	Continuous exposure at 200°F for 2,000 h and intermittent exposure at 450°F for 2 h or less
Chemical resistance	Excellent for most acids, alkalies, detergents, oils, and solvents
Flexibility	Good to temperatures as low as −65°F
Flame resistance	Self-extinguishing
Dielectric strength	Min of 400 V/mil at room temp in 3-mil thicknesses; 800 V/mil in 10-mil thicknesses
Volume-resistivity range	At 25°C, $14.5 \times 10^8 - 2.44 \times 10^{16}$ Ω-cm

TABLE 4.44   Allylic Rigidsol Formulation[72]

Ingredient	Parts by wt.	
	For thick section parts	For thin section parts
Blacar 1716 (Cary Chemical)...................	70	70
Geon 202 (B. F. Goodrich Co.)................	30	30
Diallyl orthophthalate (FMC Corp.)............	60	60
Dyphos (National Lead).......................	10	5
Benzoyl peroxide............................	0.1	0.2
Dicumyl peroxide (Hercules Powder Co.)........	0.2	1.0
Polyethylene glycol-400 monolaurate............	1.0	1.0
Thermolite-13 (M & T Chemical Co.)............	...	2.0

TABLE 4.45   Properties of Allylic Vinyl Rigidsols[72]

Property	Value
Hardness, Shore D.............................	84
Tensile strength, lb/in.^{2}.........................	5,800
Elongation, %................................	7
Flexural strength, lb/in.^{2}........................	9,200
Heat-distortion temp, at 264 lb/in.^{2}, °F............	126
Volatility, % loss..............................	0.2
Kerosene extraction, % loss......................	0.2

Plastisols are used where the extra expense of the finely divided PVC particles used is justified by the production conveniences of the material. Manufacturing processes used include dip coating, slush molding, and fabric coating. Advantages are the ability to produce hollow parts, inexpensive molds, in-place forming, and larger-part capability. Applications include beer-cap gasketing, coated glass bottles, coated battery holders, armrests, strippable coatings, prosthetic devices, crown seals,[94] die-coated yarn, storage-tank linings,[93] footballs, bottles, dolls' heads,[93] molds,[95] and mechanically frothed carpet backing.[96]

4.15 CYCLIC THERMOPLASTICS

The General Electric Company has developed a series of experimental low-molecular-weight cyclic thermoplastics. When reaction-molded polycarbonate pellets are first heated to about 200°C, they become a low-viscosity liquid. Further heating between 250 and 300°C with anionic catalysts produces a 50,000- to 100,000-molecular-weight plastic. It can be used for pultrusion and in making parts that are too large for conventional injection molding.[97] Polyarylate, polyetherketone, polyethersulfone, and polyetherimide have been processed in the same way. The melt readily wets glass fibers. These materials have been used in resin transfer molding, filament winding, and casting.[98] The cyclic oligomer carbonates can be cross-linked with epoxies and other resins to produce a higher molecular weight with enhanced properties.[99]

GE and Ford are working cooperatively with NIST to develop entire automobile bodies of cyclic thermoplastic-glass-reinforced RIM. The goal is better solvent resistance and lower tooling costs.[103]

4.16   EXPANDING MONOMER

Epolin, Inc. is marketing a line of materials that form polymers that expand on curing. The line consists of spiro orthocarbonates such as bismethylene spiroorthocarbonate or those containing norbornane with melting points above 250°C. They can be disbursed in epoxies and other materials to produce two-phase polymerizing systems. The monomer expands as it goes into solution and the final product expands. This results from ring opening that increases the size of the polymer, counteracting the usual cure shrinkage.

4.17   VINYL ESTERS

Vinyl esters, also called epoxy vinyl esters, are made by reacting an epoxy with acrylic or methacrylic acid. The resulting monomer contains a vinyl (unsaturated double bond) group on both ends. This vinyl group is reacted with a monomer such as styrene to cure by chain extension and cross linking. Curing is usually accomplished with a peroxide and an accelerator in a manner similar to polyester curing. Vinyl esters have been made from epoxies such as bis A, brominated bis A, novolac, and elastomer modified materials.

Vinyl esters are used to fabricate a wide range of corrosion-resistant fiber-reinforced plastic (FRP) products using all conventional fabricating techniques. The properties that make vinyl esters useful in FRP applications include low viscosity (250–400 cps at 25°C), high impact resistance, outstanding resistance to corrosion by many different chemicals at room and elevated temperatures, good adhesion to glass fiber, and excellent high-temperature aging.

The dielectric constant ranges from 3.27 to 3.54, and the coefficient of thermal expansion varies from 31–41 ppm/°C. The high cure shrinkage (6.9–9.9 percent) limits the unfilled uses of vinyl esters. Table 4.46 presents the properties of a family of these materials.

Vinyl ester resins are being used in industrial equipment and structures such as absorption towers, process vessels, storage tanks, piping, hood scrubbers, ducts and exhaust stacks.[104]

4.18   CYANATE ESTERS

Cyanate esters are a family or aryl dicyanate monomers and their prepolymer resins that contain the reactive cyanate ($—O—C\equiv N$) functional group. When heated, this cyanate functionality undergoes an exothermic cyclotrimerization reaction to form triazine ring connecting units, resulting in gelation and formation of thermoset polycyanurate plastics.

Cyanate ester monomers range from low-viscosity liquids to meltable solids. They cure by the catalytic action of solutions of zinc, cobalt, or copper carboxylates or acetylacetonate chelates in nonylphenol.

One of the series, AroCy® L-10 from Ciba-Geigy has a room-temperature viscosity of 140 cps. The homopolymer will develop useful properties in 3 hours at 177°C (205°C Tg, 93% conversion); but a postcure at 225–250°C is required for full cure (240–250°C Tg). Cyanate esters also function as epoxy converters. Incorporation of 50–65 percent epoxy resin will provide essentially complete conversion at 177°C. Table 4.47 presents the cured properties of AroCy L-10 alone and in mixtures with other cyanate esters and epoxies.

TABLE 4.46 Comparison of Typical Room-Temperature Properties* of ⅛-inch Clear Castings Made with Derakane Resins[†]

Property	Derakane 411 resin	Derakane 441-400 resin	Derakane 470-36 resin	Derakane 8084 resin	Derakane 510C-350 resin	Derakane 510A resin	Derakane 510N resin
Tensile strength, psi	11–12,000	12–13,000	10–11,000	10–11,000	11–12,000	10–11,000	10–11,000
Tensile modulus, psi × 10^5	4.9	5.2	5.1	4.6	5.0	5.0	5.0
Elongation, %	5.0–8.0	7.0–8.0	3.0	10.0	5.0	4.0	3.0–4.0
Flexural strength, psi	16–18,000	22–24,000	18–20,000	16–18,000	16–18,000	16–18,000	18–20,000
Flexural modulus, psi × 10^5	4.5	5.1	5.5	4.4	5.6	5.2	5.3
Heat distortion temp., °F	210–220	240–250	295–305	170–180	220–230	220–230	245–255
Barcol hardness	35	35	40	30	40	40	40

* Typical properties; not to be construed as specifications.
[†] The data given are indicated to highlight the property differences existing between the various families of Derakane resins. Specific values listed for a particular resin represent typical properties for other members of that resin family.

Amorphous thermoplastic resins such as polysulfone, polyethersulfone, polyetherimide, and thermoplastic polyimide can be dissolved in L-10 up to a concentration of 25 percent. Phase separation occurs at gelation temperatures in the range of 121–177°C, providing up to fivefold increases in fracture toughness with little sacrifice in Tg.

Cyanate esters are used in the manufacturing structural composites such as filament winding, resin transfer molding, and pultrusion.[105]

TABLE 4.47 Cured State Properties[a]

	A	B	C	D	E	F	G
Composition (weight)							
AroCy L-10[b]	100	100	50	50	50	75	35
AroCy M-50[b]	—	—	50	—	—	—	—
AroCy B-30[b]	—	—	—	50	—	—	—
AroCy F-40[h,c]	—	—	—	—	50	—	—
Epoxy resin A[d]	—	—	—	—	—	25	—
Epoxy resin B[e]	—	—	—	—	—	—	65
Nonylphenol	6	2	2	2	2	2	2
Copper naphthenate, 8% Cu	0.25	—	—	—	—	—	—
Zinc naphthenate, 8% Zn	—	0.15	0.15	0.15	0.12	—	—
Copper acetylacetonate, 24% Cu	—	—	—	—	—	0.05	0.05
Cure							
Gel temperature, °C	105	105	121	121	121	121	121
Minutes to gel	55	15	230	5	2	135	165
Hours at 177°C	3	1	1	1	1	2	2
Hours at 210°C	—	1	1	1	1	2	2
Hours at 250°C	—	2	2	2	2	—	—
% conversion	93	98	NT[f]	NT[f]	NT[f]	>95	>95
Properties							
Tg (DMA), °C							
Dry	205	259	251	262	270	238	220
After 208-hour water boil	187	225	232	209	223	195	207
Flexure strength, ksi	27.2	23.5	20.3	22.8	22.1	24.8	11.7
Flexure modulus, msi	0.44	0.40	0.44	0.40	0.43	0.44	0.40
Flexure strain, %	8.1	7.7	5.0	6.9	5.6	6.9	3.1
% Water absorption							
64 hours at 100°C	1.47	1.67	1.62	1.88	1.26	1.46	1.28
208 hours at 100°C	1.65	1.80	1.73	2.03	1.38	1.65	1.33
Dielectric constant, 1 MHz							
Dry	2.99	2.98	2.91	2.96	2.92	3.06	3.18
48-hour water boil	3.33	3.37	3.31	3.36	3.18	3.37	3.53
% Weight loss at 235°C							
24 hours	0.38	0.51	0.65	0.51	0.38	16.8	1.17
300 hours	1.09	1.13	1.56	1.30	1.47	Out	5.77
Flammability, UL-94	Burns	Burns	V-1	Burns	V-0	V-0	Burns
Density, g/cm³	1.222	1.228	1.170	1.205	1.331	1.333	1.201

[a] Properties determined on ⅛-inch thick castings.
[b] Ciba-Geigy.
[c] Available commercially as AroCy F-40S solution, 75% solids in MEK.
[d] Diglycidyl ether of tetrabromobisphenol A, 48% Br.
[e] Diglycidyl ether of bisphenol A, WPE 183.
[f] NT = not tested.

4.19 PHOTOPOLYMERS

There has been a great deal of development and commercialization of photopolymers in the recent past in the fields of adhesives, decorative coatings, printed wiring board solder masks, laminates, and pultrusion. UV radiation is used most often, although visible light and electronic beam radiation are used in specific applications. Acrylic monomers are the most widely used starting materials. Acrylated urethanes and epoxies have gained popularity, and pure epoxies are being used in applications requiring the highest degree of chemical inertness.

A series of materials has been developed by DuPont for rapid prototyping with photopolymers. The various members of the series are described as flexible and opaque, tough and transparent, and rigid and high-accuracy. They are priced at $135/kg. Layer thickness ranges from 0.003 to 0.015 in. Excess resin is removed by rinsing in solvent such as isopropyl alcohol. Heat and/or ultraviolet light exposure is used to postcure the parts.[106] In the past few years there has been a dramatic switch from acrylics to epoxies. Software that controls the process has advanced rapidly. Accuracy and finish of the parts have improved.[107]

REFERENCES

1. C. V. Lundberg, "A Guide to Potting and Encapsulation Materials." *Mater. Eng.*, May 1960.
2. J. K. Rogers, "RTM and SRIM—Ready for Mainstream Markets?," *Mod. Plast.*, Nov. 1990.
3. E. Galli, "RIM Applications Increase with New Resin Systems," *Plast. Des. Forum*, Jan./Feb. 1991.
4. "Structural RIM Systems; An Industry Review," Tech. Bull., Polyurethane Div., Mobay Corp.
5. "Rapid Prototyping Shapes up as Low-Cost Modeling Alternative," *Mod. Plast.*, Aug. 1990.
6. "Will Rapid Prototyping Be Part of Your Future?," *Plast. Des. Forum*, Jan./Feb. 1991.
7. "Quatrex Electronic Grade Resins," Tech. Bull., Dow Chemical Co.
8. "Shell Resins for Encapsulation and Adhesives," Tech. Bull., Shell Chemical Co.
9. "Bakelite Liquid Epoxy Resins and Hardeners," Tech. Bull., Union Carbide Corp.
10. "Bakelite Cycloaliphatic Epoxides," Tech. Bull., Union Carbide Corp.
11. F. Thomsen, "Centrifugal Casting," *Kunstoffe*, Oct. 1969.
12. J. Delmonte, "Epoxies for Massive Castings," *Plast. Des. Process.*, Jan. 1969.
13. "Cycloaliphatic Epoxide Systems," Tech. Bull., Union Carbide Corp.
14. "D.E.R. 354 Bisphenol-F Epoxy Resin," Tech. Bull., Dow Chemical Co.
15. "Tactix Performance Polymers for Advanced Composites and Adhesives," Tech. Bull., Dow Chemical Co.
16. "Dow Epoxy Novolac Resins," Tech. Bull., Dow Chemical Co.
17. "Epi-Rex 5161—Flame Resistant Resin," Tech. Bull., Celanese Resins Div., Celanese Coatings Co.
18. J. R. Griffith, "Epoxy Resins Containing Fluorine," *Chemtech*, May 1982.
19. A. J. Klein, "Epoxy Update," *Adv. Composites*, Sept./Oct.
20. "Resin Systems for Structural Applications," Tech. Bull., Shell Chemical Co.

21. "Thiokol Liquid Polymer/Epoxy Casting Compounds," Tech. Bull., Thiokol Chemical Corp.
22. D. Sudkaher et al., "Curing of Epoxy Resins I: Aromatic Diamines."
23. "Specialty Intermediates," Tech. Bull., Milliken and Co.
24. "Epoxy Curing Agent for Industrial Composites," Tech. Bull., Air Products and Chemicals, Inc.
25. Anhydrides and Chemicals—private communication.
26. L. S. Buchoff and W. R. Sherwin, "Properties of High Temperature Epoxy Systems," *Rubber Plast. Age,* Sept. 1962.
27. M. S. Rhodes, "Polyanhydride Flexibilizing Hardeners for Epoxy Resins," *Insulation/Circuits,* Dec. 1977.
28. "Toughened Epoxy Resin Systems for Structural Applications," Tech. Bull., Shell Chemical Co.
29. "Epoxy Resins for Plastics," Tech. Bull., Celanese Resins Div., Celanese Coatings Co.
30. Drummer, "Embedded and Printed Circuits," *Electrotechnol.,* May 1953.
31. J. F. Formo and R. E. Isliefson, "Producing Special Properties in Plastics with Fillers," in *Reg. Tech. Conf., Soc. of Plastics Engineers,* Fort Wayne, Ind., May 1959.
32. L. S. Buchoff, "Effect of Humidity on Surface Resistance of Filled Epoxy Resins," in *Annual National Tech. Conf., Soc. of Plastics Engineers,* Chicago, 111, Jan. 1960.
33. "Marina Goes to Epoxy: 5900 Others Follow," *Mod. Plast.,* June 1971.
34. H. Glick and Y. Tuan, "Epoxy Casting of Large Contoured Nozzles," *SPE J.,* Sept. 1969.
35. Tech. Bull. SC: 106-86, Shell Chemical Co.
36. "In Gel Coats, the Iso's Have It," *Mod. Plast.,* Nov. 1970.
37. J. Sommer, "Isophthalic Polyesters, New Tool for Industry," *Austral. Chem. Eng.,* Sept. 1968.
38. Tech. Bull. 5-1, Inmont Corp.
39. T. M. Church and C. Berenson, *Ind. Eng. Chem.,* 1955.
40. C. B. Sias in *15th Annual Conf., Reinforced Plastics Div., Soc. of the Plastics Industry,* 1960.
41. E. Benkhe, *Kunststoffe Rundsch.,* 1957.
42. "Latest Development in Low Profile Polyesters: Zero Shrinkage," *Mod. Plast.,* Mar. 1970.
43. "New Water Filler Polyesters," *Plast. World,* Nov. 1969.
44. R. H. Leitheiser et al., "Water Extended Polyester Resin (WEP), New Low Cost Molding Material," in *27th Annual Tech. Conf., Soc. of Plastics Engineers,* May 1964.
45. *Mod. Plast.,* Nov. 1988.
46. "Plastics," Reference Issue, *Mach. Des.,* Dec. 1968.
47. "Poly bd, the Best of Two Worlds," Arco Tech. Bull., Arco Chemical Co., Div. of Atlantic Richfield Co.
48. "Processing Versatility Comes to Urethane Elastomers and Plastics," *Mater. Eng.,* Feb. 1972.
49. "Adiprene," Tech. Bull., Elastomer Chemicals Dept., du Pont de Nemours & Co.
50. "System 97," Tech. Bull., CasChem, Inc.
51. "Polyurethane Specialty Products Product Reference Guide," Tech. Bull., Air Products and Chemicals, Inc.
52. "OSHA Regulates MOCA Exposure Levels," *Plast. Technol.,* May 1989.
53. R. D. Cody and R. Polansly, "PU Prepolymer Curative Increases Cast Elastomer Hardness and Resilience," *Elastomerics,* Dec. 1989.
54. "Bayflex and Baydur RIM Polyurethane Systems," Tech. Bull., Mobay Corp.

55. "PUR-RIM Technology Suppliers Concentrate on Reinforced Applications," *Modern Plastics*, June 1995.

56. K. L. Parks et al. "Advancement in Low-cost Bumper Systems," *Modern Plastics*, May 1995.

57. Research and Development column, "Three Prongs," *Plastics Engineering*, May 1995.

58. "How to Process Sylgard® Brand Elastomers," Tech. Bull., Dow Corning Corp.

59. Directory/Encyclopedia Issue, *Insulation/Circuits*, June/July 1972.

60. "Information about Sylgard Elastomers," Tech. Bull., Dow Corning Corp.

61. "Silicone Moldmaking Materials from Dow Corning," Tech. Bull. Form No. 10-168E-94, Dow Corning Corp.

62. "Dow Corning Silicone Construction Sealants," Tech. Bull., Dow Corning Corp.

63. P. A. Toensmeier, "Another Option in Part Design: Liquid Injection Molding Silicones," *Mod. Plast.*, June 1989.

64. "Sealants & Encapsulants," Tech. Bull., General Electric Corp.

65. "HiPec Semiconductor Protective Materials," Tech. Bull., Dow Corning Corp.

66. "Materials for High Technology Applications," Tech Bull., Dow Corning Corp., 1989.

67. "STI—Silastic Liquid Silicone Rubber Product Selector Guide," Tech. Bull., Dow Corning Corp.

68. "Plastics Property Chart," *Mod. Plast.*, MPE Suppl., 1970–1971.

69. Tech. Bull., Form. No. 10-31-90, Dow Corning Corp.

70. "CR-307 Monomer," Tech. Bull., PPG Industries.

71. "HIRI™ Casting Resin," Tech. Bull., PPG Industries.

72. H. Raech, Jr., *Allylic Resins and Monomers*, Van Nostrand Reinhold, New York, 1965.

73. "CR-39 Allyl Diglycol Carbonate in Cast Sheets and Forms," Tech. Bull., PPG Industries.

74. J. J. Chapman and L. J. Frisco, "The Electrical Properties of Diallyl Phthalate Resin (Dapon)," *FMC Final Rep.*, Jan. 1, 1956 to Jan. 31, 1957.

75. "DAP Suppresses Styrene Odors in Polyester Aircraft Parts," *Plast. World*, Dec. 1971.

76. "Hycar Reactive Liquid Polymers," Tech. Bull., B. F. Goodrich Chemical Co.

77. C. G. Fitzgerald et al., "Epoxy-Butadiene Resins," *SPE J*, Jan. 1957.

78. R. E. Drake, "1,2-Polybutadienes—High Performance Resins for the Electrical Industry," paper 362, Tech. Bull., Advanced Resins Inc.

79. "Polymer LP-3," Tech. Bull., Morton-Thiokol Chemical Corp.

80. A. J. Breslau, "Heat Distortion Properties of Polysulfide Polymer-Modified Epoxy Resins," *Plast. Technol.*, Apr. 1957.

81. " 'Trilene' as an Electronics Encapsulant," *Tech. Bull.*, Uniroyal Chemical Co.

82. H. Hemmel et al., "Activated, Anionic Polymerization of Lactams and Its Utilization in Pressureless Casting," *Kunststoffe*, July 1969.

83. C. Robertson, "Thermoplastic RIM: An Introduction to NYRIM® Nylon Block Copolymer Technology," *Structural Plastics Technical Conference*, 1991.

84. "Monocast and Nylatron GSM Nylon," Tech. Bull., The Polymer Corp.

85. M. B. Horn, *Acrylic Resins*, Reinhold, New York.

86. "Now—Cast Acrylic in Rolls of Continuous Sheet," *Plast. Des. Process.*, Nov. 1965.

87. "New Material, New Technology Produce Close-Tolerance Cast Acrylic Parts," *Mod. Plast.*, Oct. 1985.

88. "Acrylic Panel Provides 'Window' on the Tropics," *Plast. Eng.*, Jan. 1991.

89. "Methyl Methacrylate," *Chem. Purchasing*, Sept. 1971.

90. A. E. Witt and J. E. Morrissey, "Economics of Making Irradiated Wood-Plastic Products," *Mod. Plast.*, Jan. 1972.

91. " 'Modar' Modified Acrylic Resin," Tech. Bull., ICI Chemicals and Polymers Ltd.

92. " 'Diakon' Reactive Polymers," Tech. Bull., ICI Chemicals and Polymers Ltd.

93. "New Plastisol Markets Emerge," *Plast. World*, Apr. 1969.

94. J. W. Hull, "Progress in Vinyl Dispersions," *Plast. Technol.*, Oct. 1968.

95. "Flexible Mold Materials for Casting," *Plast. Des. Process.*, Nov. 1971.

96. E. T. Simoneau, "Silicone Surfactants in Mechanically Frothed Vinyl Plastisol Foams," *Rubber World*, Aug. 1970.

97. "Experimental Polycarbonate Pultrudes like a Thermoset," *Mod. Plast.*, Sept. 1988.

98. " 'Cyclics' Technology May Make TP Composites Common," *Mod. Plast.*, Nov. 1989.

99. N. R. Rosenquist and L. P. Fontana, "Formation of Crosslinked Networks via Polymerization of Oligomeric Cyclic Carbonates in the Presence of Polyfunctional Co-Reactants," *Polymer Preprints*, vol. 30, no. 2, 1989.

100. N. C. Slone, *SPE J.*, 1960.

101. "Electronic Materials Selector Guide," Tech. Bull., GE Silicones.

102. "Materials for High Technologies Applications," Tech. Bull., Dow Corning Corp.

103. General Electric Corporation—private communication.

104. "Derakane Epoxy Vinyl Ester Resins," Tech. Bull., Dow Plastics.

105. "Matrix Resins, Explore the Possibilities," Tech. Bull., Ciba-Geigy Corp.

106. Brochure 51 ver. 03.05.94, Tech. Bull., DuPont.

107. 3D Systems—private communication.

CHAPTER 5
THERMOPLASTIC ELASTOMERS

Charles P. Rader
Marketing Technical Service Principal
Advanced Elastomer Systems, L.P.
Akron, Ohio

5.1 INTRODUCTION

A thermoplastic elastomer (TPE) is a rubbery material with properties and functional performance very similar to those of a conventional thermoset rubber, yet it can be fabricated in the molten state as a thermoplastic. ASTM D 1566[1] defines TPEs as "a diverse family of rubberlike materials, that unlike conventional vulcanized rubbers, can be processed and recycled like thermoplastic materials." The great majority of TPEs meet the standard ASTM definition of a rubber since (1) they recover quickly and forcibly from large deformations, (2) they can be elongated by more than 100 percent, (3) their tension set is less than 50 percent, and (4) they are insoluble in boiling organic solvents.

Much controversy has arisen over whether a TPE is a rubber or a plastic. In reality, it is both. In terms of properties and functional use, a TPE is a rubber; in terms of processing and fabrication, it is a thermoplastic. Above its melting point (T_m), a TPE is fluid and can be molded or extruded with the same equipment and methods as commonly employed for thermoplastics. Below its T_m, a TPE functions as a flexible, elastic rubber. Unlike a thermoset rubber, a TPE can be recycled repeatedly to recover scrap from processing and, in most cases, to allow material reuse at the end of the useful life of a TPE article.

During the last decade, the term *thermoplastic elastomer* has become the accepted generic description of these materials and has superseded earlier designations such as *elastoplastic*.[2] Literature references to specific classes or types of TPEs may include proprietary trade names and narrower designations of specific types of material. Generic terms such as "blend," "thermoplastic vulcanizate," and "block copolymer" refer to TPEs with distinctly different elastomeric performance to be described in detail in a following section.

The scope and purpose of this chapter is to describe this emerging technical and commercial discipline, which has shown rapid growth during the 1970s, 1980s, and

1990s. Fig. 5.1 indicates how TPEs bridge the range of materials available to the rubber and plastics industries, with greater overlap with thermoset rubbers. The rapid growth of this new technology is described, along with the types of TPEs available and how they compare with thermoset rubbers. Processing of TPEs is often a critical feature of their commercial usefulness and is discussed in detail. A discussion of TPE applications and markets will serve to illustrate how this technology connects and integrates with the established rubber and plastics industries. Finally, the distinct advantages of TPEs over thermoset rubbers in regard to recovery and recycle are described.

The published literature of TPEs has risen exponentially as this technology has grown toward adulthood during the decades of the 1980s and 1990s. TPEs have been the subject of numerous books,[2-6] review articles,[7-14] and technical and commercial symposia.[15-31] The Rubber Division of the American Chemical Society has formed a permanent topical interest group devoted exclusively to TPE technology,[32] and the Society of Plastics Engineers has generated a special interest group devoted to TPEs, with focus on their fabrication into useful articles. This chapter provides leading references for an introduction to TPE literature.

5.2   TPEs IN THE RUBBER AND PLASTICS INDUSTRIES

The birth of TPEs occurred in the 1950s with the development of thermoplastic polyurethanes (TPUs).[33] These were the first true commercial TPEs, evolving from the discovery of polyurethanes in the 1930s.[34] Heavily plasticized polyvinyl chloride (PVC) was commercialized by B. F. Goodrich Company just before World War II and has many desirable rubberlike properties but does not meet the accepted definition of a rubber.[1]

In the early 1960s, Shell Development Company introduced another class of TPEs, the styrene-diene block copolymers, to the industrial market.[35] These were followed in the 1970s by copolyester (COP) TPEs from duPont de Nemours & Company[36] and thermoplastic polyolefin-elastomer blends, known as TPOs, from Uniroyal.[37] Blends of PVC and nitrile rubber (NBR) with some elastomeric properties and thermoplastic processability were also developed during this period,[37] but these have had more significance in Japan than in North America and Europe. Elastomeric alloys (EAs, now called thermoplastic vulcanizates) were commercialized in 1981 by Monsanto Chemical Company,[38] based on compositions of thermoplastic and crosslinked rubber. TPEs with the highest performance and highest cost, the elastomeric polyamides (PEBAs), were introduced in the early 1980s. Materials sup-

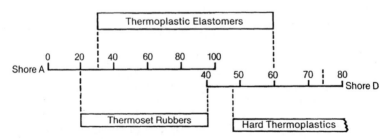

FIGURE 5.1   TPEs bridge the hardness ranges of rubbers and plastics.

pliers continue to develop and market new TPEs of all types under a variety of trade names to meet specific market needs.

The growth of TPEs has been truly phenomenal, with an annual worldwide growth rate of 8 to 9 percent/year between 1970 and 1990. Fig. 5.2 depicts this growth, showing TPE usage growing from 100,000 metric tons (MT) in 1970 to 650,000 MT in 1990 and 950,000 MT in 1995. Numerous predictions have been made of future TPE usage,[39] with annual growth rates of 5.5 to 10 percent forecast for the next decade. Fig. 5.2 forecasts as most probable a growth rate of 7.5 percent per year between 1995 and 2000 and 6.5 percent between 2000 and 2010. While the percent growth forecast for TPEs is debatable, certain trends can clearly be gleaned from this forecast:

1. TPE growth will continue to be both brisk and greater than that of either the rubber or plastics industry.

2. The percentage growth rate will continue to decrease, with the absolute magnitude of the growth not decreasing until well after 2000.

3. TPEs in 1994 captured approximately 15 percent of the nontire segment of the rubber products market. This percentage should double before 2010.

4. Within the next decade, TPEs will not penetrate significantly the tire segment of rubber usage, which consumes slightly more than one half of the rubber produced. This penetration must await the development of polymer systems presently not known.

The history of TPEs is given in condensed form in Table 5.1. These materials are presently in their early adulthood. This growth is not expected to level off until the second decade of the twenty-first century.

About 65 percent of the growth in TPEs will come from replacement of thermoset rubber compounds in existing parts.[40] New applications requiring rubber performance will account for nearly 25 percent, and the final 10 percent of TPE growth will come from displacement of some soft thermoplastics, where the higher performance of TPEs would be a benefit to the user. This growth will be primarily at the expense of thermoset rubbers. TPEs will directly replace them in some applications and deny them new business in many others.

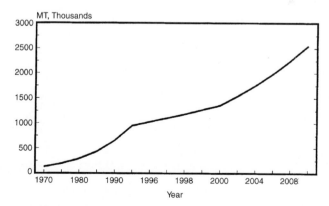

FIGURE 5.2 Worldwide growth curve for TPEs, actual to 1995, projected for growth rate of 7.5 percent from 1995 to 2000 and 6.5 percent from 2000 to 2010.

TABLE 5.1 History of
Thermoplastic Elastomers

Decade	Stage
1950s	Birth
1960s	Infancy
1970s	Childhood
1980s	Adolescence
1990s	Adulthood
After 2000	Maturity

Past communication between the plastics and rubber industries has been limited at best (Fig. 5.3), although it has been better in Europe than in North America. The plastics person knows how to fabricate an article from a TPE and understands the economic advantages of thermoplastics processing. Yet the rubber person is the one with the necessary knowledge and skills for marketing TPE articles. The growth of TPE usage has forced these two giant industries to communicate far more closely than they have ever had to before.

This lack of communication between rubber and plastics people has been a pronounced impediment to the fabricators, users, and suppliers of TPEs. Yet it has provided thermoplastics processors with a major opportunity, since they can now use their existing equipment and techniques to fabricate TPE articles and participate in a market to which they previously had no access. The entry of the plastics processor into the rubber products market has forced more thermoset rubber processors to invest in thermoplastics fabrication equipment to avoid the loss of market share.

TPEs have been described as having one foot in the plastics industry and the other in the rubber industry (Fig. 5.4).[11] With projected annual TPE business growth rates of more than 7 percent, it is little wonder that both of these mature industries have embraced this young and rapidly expanding discipline. The emerging field of TPEs will continue to build bridges and increase communication between these two historically separate and independent industries to generate permanent change in

Rubber People   Plastics People

FIGURE 5.3 Communication between rubber and plastics people has traditionally been quite poor.

each. The traditional barriers between rubber and plastics are rapidly disappearing because of the continuum of materials now available between soft rubbers and hard, rigid thermoplastics.

5.3 COMPARISON OF TPEs
AND THERMOSET RUBBERS

To understand how TPEs have succeeded in replacing thermoset rubber in a wide range of parts, the advantages and disadvantages of TPEs must be examined in comparison with thermoset rubbers. Among the practical advantages offered by TPEs over thermoset rubbers are the following:

1. Processing is simpler and requires fewer steps. Fig. 5.5 contrasts the simple thermoplastics processing, used to make TPE parts, with the multistep process required for conventional thermoset rubber parts. Each processing step adds an incremental cost to the finished part and, in the case of thermoset rubbers, may generate significant amounts of scrap. The speed, efficiency, and economy of thermoplastics processing can lower the final cost of a rubber part markedly.

2. Processing time cycles are much shorter for TPEs. Molding cycles need only be long enough for the TPE part to cool sufficiently to hold its shape when ejected from the mold. These times are on the order of seconds, compared to several minutes for thermoset rubber parts, which must be held in the mold while the chemical cross-linking (vulcanization) reaction takes place.

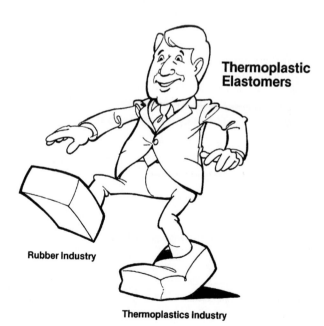

Thermoplastic Elastomers

Rubber Industry

Thermoplastics Industry

FIGURE 5.4 TPEs have grown up with one foot in the rubber industry and the other in the plastics industry.

3. TPEs require little or no compounding with other materials. They are available fully compounded and ready for a wide range of uses. Their compositional consistency is higher than that of thermoset rubbers, which must be mixed with curatives, stabilizers, processing aids, and specialty additives such as flame retardants. Fewer weighing and mixing steps means tighter quality control for TPEs, resulting in more uniform performance and handling than with thermoset rubbers. The cost of quality-control testing can therefore be lower with TPEs and the quality of the fabricated article actually higher.

4. TPE part production consumes less energy, a result of the fewer and simpler processing steps. Like other thermoplastics, TPEs are melted, the parts are shaped from the melt, and then they are cooled to freeze them into the desired shape. Thermoset rubbers require mixing energy to prepare compounded stocks with the necessary additives and crosslinking agents. These are then shaped (Fig. 5.5) into the desired part and held in a heated mold or curing chamber for several minutes.

5. TPE scrap (regrind) may be recycled. Such scrap is generated, for example, in runners and sprues from injection molding and during startup and shutdown of any processing unit. Thermoset rubber scrap is often discarded, representing an added cost to the process and a load on the environment. The amount of scrap relative to usable parts becomes increasingly significant as the part size decreases. Most TPEs will tolerate several regrind-recycle steps without significant change in properties (Fig. 5.6).

6. TPE parts can readily be recycled after they have given a normal, useful lifetime of service. It has recently been demonstrated that recycled material from used TPE automobile parts can be remolded to give properties essentially undiminished from the virgin material.[41] The recycle of thermoset rubber parts is far less practical than that of TPE ones, with the recoverable value much less.

7. Thermoplastics processing allows tighter control on part dimensions than thermoset rubber processing. Dimensional tolerances can be more precise by a factor of 2 to 3. Thus, fabricated parts are more uniform in dimensions and overall quality.

8. A TPE is typically lower in specific gravity than a comparable thermoset rubber containing carbon black or inorganic fillers. This differential can range from 15 to 50 percent. These materials are purchased on a weight basis but used on a volume

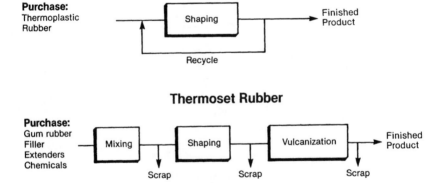

FIGURE 5.5 TPE processing consists of a single step, compared to three or more for thermoset rubbers.

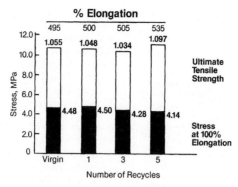

FIGURE 5.6 Retention of physical properties of TPE after multiple processing recycles.

basis, so more parts can be produced from a given weight of TPE than from the same weight of thermoset rubber.

9. TPEs permit thermoplastic fabrication methods not feasible for thermoset rubbers. These methods include blow molding, coextrusion with rigid thermoplastics, thermoforming, heat welding, and film blowing. These techniques greatly improve the efficiency and cost of rubber part fabrication through the use of TPEs. In numerous areas, the use of one or more of these novel techniques has reduced the cost per fabricated article by 50 percent or more. These techniques thus enable newer, more costly (on a weight basis), higher-technology TPE materials to be cost-competitive with conventional thermoset rubbers.

10. Automated, high-speed fabrication methods can be used with TPEs, mainly due to the tighter dimensional tolerances to which they can be molded or extruded. This makes possible a massive reduction in fabrication and assembly costs.

It should not be surprising that there are offsetting disadvantages to the use of TPEs compared to thermoset rubbers:

1. To thermoset rubber processors, TPEs belong to a new technology requiring unfamiliar processing equipment and techniques. Thermoplastics processors are familiar with this technology and have the necessary equipment, though they are generally not familiar with the markets for rubber articles. The capital investment for thermoplastics equipment is often a major hurdle for a thermoset rubber processor to participate in the market for TPE parts.

2. Many TPEs must be dried before processing. While this is a familiar step to thermoplastics processors, it is not necessary for thermoset rubbers. Drying equipment is usually not available in a rubber shop.

3. TPEs melt at a specific elevated temperature, above which a part will not maintain its structural integrity. Crosslinked thermoset rubbers do not display such melting behavior and are limited in upper service temperature only by chemical degradation such as oxidation. Thus a TPE part would not be suitable for brief use (a few seconds to a few minutes) at a temperature above its melting point. This would likely pose no problem for a thermoset rubber, since it has no melting point.

4. TPEs require moderately high production volume for good processing economics. Thermoplastics tooling costs are generally higher than those for thermoset

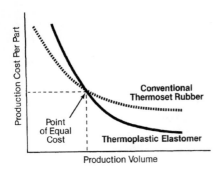

FIGURE 5.7 Sensitivity of cost per part to number of parts produced for thermoplastic elastomers and thermoset rubbers.

rubber parts, some of which are compression molded in volumes of only a few hundred per year. Fig. 5.7 shows schematically the production cost (material plus fabrication) versus part volume (number of parts per year) for the same part made from a TPE and from a thermoset rubber.

Each use of a TPE requires that these advantages and disadvantages be weighed against each other to determine the suitability of the TPE. A compounded rubber stock is often less costly on a weight basis than a competitive TPE. Lower processing costs and specific gravity can more than compensate for the material cost difference. The needed equipment investment and production volumes must be weighed against the fabrication savings and material cost differences. In many cases, the advantages have been found to outweigh the disadvantages, as attested by the phenomenal growth of TPE uses over the past thirty years.

5.4 CHEMISTRY, MORPHOLOGY, AND PROPERTIES OF TPEs

Virtually all TPEs consist of at least two polymeric phases: a hard thermoplastic phase and a soft elastomeric phase. The properties of the resulting TPE will be derived from the properties of each of the two phases and their mutual interaction. The two phases may result from simply mixing two different polymers, as in a blend of a hard thermoplastic such as polypropylene (PP) with a soft elastomer such as ethylene-propylene terpolymer (EPDM rubber), to give a thermoplastic elastomeric olefin (TEO). Dynamic vulcanization (under conditions of high shear and temperature) of the elastomer phase of such a blend gives rise to a thermoplastic vulcanizate (TPV), with properties much closer to those of a conventional thermoset rubber. The two phases of a TPE may also be present as alternating hard and soft segments along a common polymer backbone. This is the case for block copolymers, the basis for many commercially important TPEs. Table 5.2 compares the performance characteristics of six different generic classes of TPEs.

The performance characteristics of a TPE depend on the T_m of the hard thermoplastic phase and the glass-transition temperature (T_g) of the soft elastomeric phase. The useful temperature range for a TPE is between T_m and T_g. Within this range, the TPE displays its desirable elastomeric properties. At temperatures above T_m the hard thermoplastic phase melts, and the TPE becomes fluid and can be processed by usual thermoplastics techniques. Below T_g the TPE becomes brittle and loses all of its useful elastomeric characteristics.

Passage of a TPE through T_g and through T_m is reversible and can take place many times, since both are physical (and not chemical) transformations. On the other hand, the passage of a crosslinked thermoset rubber through T_g is reversible, but that through T_m does not exist due to the thermoset nature of the material. Thus

TABLE 5.2 Key Properties of Generic Classes of TPEs*

Property	Styrenic	TEO	TPV	Copolyester	Polyurethane	Polyamide
Specific gravity[†]	0.90–1.20	0.89–1.00	0.94–1.00	1.10–1.40	1.10–1.30	1.00–1.20
Shore hardness	20A–60D	60A–65D	35A–50D	35D–72D	60A–55D	60A–65D
Low temperature limit, °C	−70	−60	−60	−65	−50	−40
High temperature limit, °C (continuous)	100	100	135	125	120	170
Compression set resistance, at 100°C	P	P	G/E	F	F/G	F/G
Resistance to aqueous fluids	G/E	G/E	G/E	P/G	F/G	F/G
Resistance to hydrocarbon fluids	P	P	F/E	G/E	F/E	G/E

* P = Poor; F = Fair; G = Good; E = Excellent.
[†] Does not include grades containing a special flame-retardant package, which generally raises the specific gravity 20 to 30 percent.

on heating a crosslinked thermoset rubber to progressively higher temperatures, nothing will happen until the onset of chemical attack from either the environment or internal decomposition. This chemical attack is irreversible since it entails the destruction of the crosslinks between the elastomer chains and, to a lesser extent, the chain backbones themselves.

The simplest type of TPE is the TEO polymer blend, a simple mixture of a thermoplastic polymer with a compatible elastomeric polymer. Each of the polymeric components has its own phase, but the interaction between the phases is quite weak, if it exists at all. The properties of the blend are approximately those predicted directly from the properties of the components, and a knowledge of the T_m and T_g for the hard and soft phases, respectively. The hard phase must be continuous, or at least cocontinuous, in order for the blend to be thermoplastic and melt at T_m. Commercial blends include those of EPDM rubber with PP or polyethylene (PE) and a few blends of NBR with PVC.[37]

A TPV is a TPE produced by dynamic vulcanization, the process of intimate melt mixing of a rubbery polymer and a thermoplastic to vulcanize the rubbery polymer and generate a TPE with properties closer to those of a thermoset rubber than are those of a comparable unvulcanized composition. As with the TEOs, the hard thermoplastic phase must be continuous or cocontinuous (Fig. 5.8) for the TPV to be thermoplastic and melt at T_m. These TPEs consist of finely divided particles of highly crosslinked rubber in a continuous matrix of thermoplastic. Major thermoplastic-elastomer interaction results from the great interfacial area between the two phases. The crosslinking of the elastomer phase prevents reaggregation of the fine elastomer particles when the high shear of mixing is removed. Commercial TPVs include compositions with a polyolefin, often PP, as the hard phase and crosslinked EPDM, NBR, natural rubber (NR), butyl rubber (IIR) or ethylene vinyl acetate (EVA) as the soft phase.

Block copolymer TPEs contain both hard and soft segments along a common polymer chain (Fig. 5.9). At temperatures below the effective T_m of the hard phase, the hard segments will aggregate into rigid domains, making the TPE a solid. At these temperatures, the soft segments are present as amorphous rubbery domains that impart at least some elastomeric nature to the TPE. In the useful service temperature region of a block copolymer TPE, the hard phase will restrict the motion of

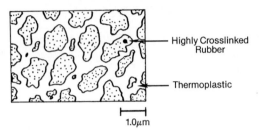

1.0μm

FIGURE 5.8   Morphology of TPV.

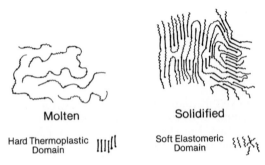

Molten   Solidified

Hard Thermoplastic   ⅢⅢ   Soft Elastomeric   ⟨⟨⟨⟩⟨
Domain   Domain

FIGURE 5.9   Morphology of block copolymer TPEs.

the soft phase segments in much the same way as the crosslinks of a thermoset rubber restrict the motion of the elastomer chains. TPE block copolymers include polymers of styrene and dienes (the styrenics), COPs, polyurethanes, and PEBAs.

In contrast to TPEs, thermoset rubbers derive their useful elastic characteristics from network formation via chemical crosslinks between the elastomer chains.[42] This crosslinking is a thermally irreversible process that must be carried out after a functional part has been shaped, usually while the part is held under pressure in a hot mold. Useful thermoset rubbers are based on a wide variety of polymer types, from NR to synthetic products, based on polymerization or copolymerization of dienes and other monomers. The ability to choose a polymer type and mix it with a chemical crosslinking agent and other additives chosen to suit the intended use of the rubber, leads to an endless list of thermoset rubber compounds available to the parts producer. Fig. 5.10 shows a cost-performance comparison for generic classes of TPEs and also for several thermoset rubbers. To a first approximation, the properties and performance of a given TPE class will be comparable to those of the thermoset rubber at the same position on the cost-performance chart. Thus the styrenics are rational candidates to replace NR and styrene-butadiene rubber (SBR), and the TPVs logical replacements for EPDM and polychloroprene.

Whereas a TPE will melt at a specific temperature (150 to 250°C), these same temperatures will cause no immediate change in a thermoset rubber. Over much longer periods of time (days, months, years) such temperatures can bring about the chemical degradation (primarily oxidation) of both TPEs and thermoset rubbers. At increasingly higher temperatures (over 250°C), both TPEs and thermoset rubbers are progressively thermally degraded by pyrolysis and oxidation.

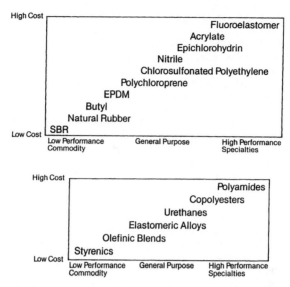

FIGURE 5.10 Relative cost and performance of thermoset rubbers and TPEs.

5.5 GENERIC CLASSES OF TPEs

5.5.1 Rationale for Classification

Commercial TPEs may be classified rationally on the basis of their chemistry and morphology. Virtually all of them have two polymeric phases—a soft rubbery one and a hard thermoplastic one—and thus fall into one of three broad categories:

1. Block copolymers with alternating soft and hard segments
2. Blends of a rubbery polymer and thermoplastic
3. Rubbery polymer-thermoplastic compositions in which the former is highly vulcanized and finely dispersed in the latter

The block copolymers include the styrenics, COPs, TPUs, and PEBAs. The blend TPEs have commonly been known as TPOs, a designation that is progressively yielding to the more descriptive term of TEOs. The TPEs with a highly vulcanized rubbery polymer intimately dispersed in a thermoplastic are known as TPVs. The TPVs are the closest approach of a TPE to the properties and performance of a conventional thermoset rubber vulcanizate.

5.5.2 Styrenic Block Copolymers

Commonly referred to as styrenics, these TPEs[43,44] are copolymers having the S-D-S structure, where S is a hard segment of polymerized styrene or styrene derivative and D is a soft central segment of polymerized diene or hydrogenated diene units. Polybutadiene (B), polyisoprene (I), and polyethylenebutylene (EB) are the most commonly used soft diene segments (D). Fig. 5.11 shows the chemical structures of typical styrenic TPEs.

As indicated in Fig. 5.9, the polystyrenic segments, with molecular weights of 5000 to 8000, form aggregates on cooling with many of the properties of polystyrene itself. The melting point of such TPEs based on styrene is thus close to that for polystyrene (100 to 110°C). This ordered arrangement of the styrene and diene components gives elastomeric properties that differ greatly from the properties of ordinary SBR, a random copolymer of the same monomers.

The block structure of these TPEs thus gives rise to a morphology that generates properties clearly superior to those of a random styrene-diene polymer. Figure 5.12 gives the tensile stress-strain curves of an S-B-S block copolymer TPE and a vulcanized SBR (the random copolymer). The rigid polystyrene blocks serve to reinforce (or anchor) the much softer diene blocks in much the same way as the crosslinks and added carbon black of an SBR reinforce the thermoset elastomer chains. In each case, the polymer system is rendered harder, tougher, more resistant to deformation (higher modulus), and more suitable for use as a rubber. For block copolymers with constant polystyrene content, the tensile modulus and break strength do not depend on the molecular weight, provided the polystyrene blocks are sufficiently large to form well-defined domains.

The characteristics of these TPEs thus depend on the relative proportions of the polymerized styrene and diene units as well as the chemical nature of the monomers. At low styrene levels, the TPEs will be soft and rubbery with relatively low tensile properties. With increasing styrene content, the TPE progressively becomes like a harder rubber, a leatherlike material, and finally a glossy, hard material similar to an impact-modified polystyrene. The residual olefinic double bond in block polymers derived from B and I renders these TPEs susceptible to oxidative degradation. Removal of this unsaturation by hydrogenation, as in the case of S-EB-S materials, makes them much more resistant to oxidation and more suited to higher-temperature applications.

Chemical modification of the monomer in the hard blocks of the polymer can increase the tensile properties and raise the upper temperature for satisfactory service. Thus styrenic TPEs derived from α-methylstyrene have both higher tensile strengths and service temperatures than those from styrene. Unfortunately, α-methylstyrene is much more difficult to polymerize than styrene, and its block copolymers are thus not commercial.

Another requirement for good properties of a styrenic TPE is that hard segments be at both ends of the polymer. Thus the properties of a copolymer with an S-D-S

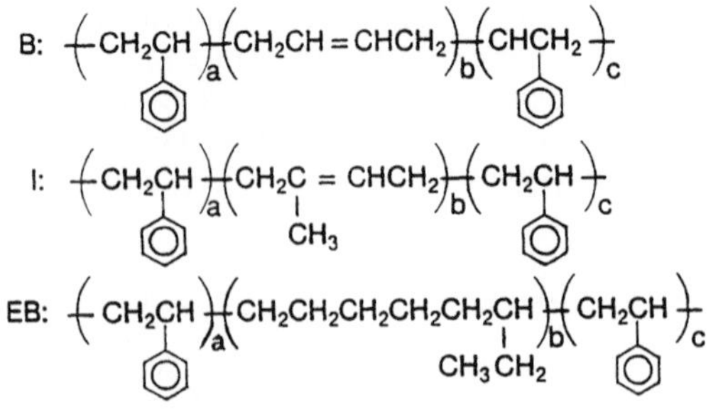

FIGURE 5.11   Structures of three common styrenic block copolymer TPEs; a and c = 50 to 80, b = 20 to 100.

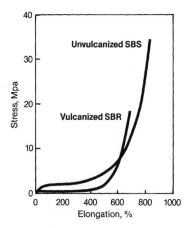

FIGURE 5.12 Comparison of tensile stress-strain curves of vulcanized SBR thermoset to that of unvulcanized S-B-S styrenic TPE.

structure are markedly superior to those of D-S-D or D-S structures. The best performance is obtained when each end of a soft diene segment is attached to polystyrene segments that are hard and relatively immobile.

The upper limit of the useful service temperature range of these TPEs (Table 5.2) is determined by the T_m of the polystyrene segments (95 to 110°C). The lower limit of this range is fixed by the T_g (about –70°C) of the soft diene blocks. As in all block copolymers, the two phases are chemically bonded together, and the hard segments are therefore very efficient in restricting the motion of the soft diene segments. This gives the TPE its high resistance to deformation (modulus) and its elasticity. The modulus of the TPE will vary directly with the number of diene blocks per unit volume and inversely with the length of these blocks.

Commercially available styrenic TPEs range in hardness from 20 Shore A to 60 Shore D. In the soft range below 60 Shore A, there are more styrenic TPEs available than any other generic class. Styrenics have a broad service temperature range, from about –70 to about 100°C, but the upper limit is sufficiently low to restrict their use to lower-temperature applications. They are very resistant to water and other polar fluids but lose much of their effectiveness if exposed to oils, fuels, and other nonpolar organic solvents. Their compression and tension set (resistance to plastic deformation under stress) is good (that is, low) at ambient temperatures but becomes progressively poorer with increasing temperature. Above 70°C their compression set is poor.

The common styrenic TPEs with B or I segments are among the lowest-cost, lowest-performance classes of TPEs (Figure 5.10). Their performance, however, is adequate for a wide variety of nondemanding rubber applications. Styrenic TPEs with EB soft blocks have greater resistance toward oxidative attack since the EB block is saturated. Thus they are suitable for higher temperatures than styrenics with butadiene or isoprene as soft blocks. The styrenics therefore have the largest usage of any generic class of TPEs, about 500,000 MT per year worldwide. Although new styrenic TPEs continue to be developed, these materials are growing at a significantly slower rate than other generic classes of TPEs.

The huge market for styrenic TPEs includes applications as fully compounded materials which can function as replacements for thermoset rubber, leading to uses in such articles as shoe soles, sporting goods, and a variety of other applications not requiring service above 70°C, hydrocarbon fluid resistance, or exceptionally good physical properties. The styrenics are the generic class of TPEs most widely used as compounding ingredients in other useful compositions. They are major constituents in adhesives, resins, sealants, caulking, motor vehicle lubricants, and thermoset automotive body parts. Asphalt is markedly upgraded in performance to a much tougher, more durable material by addition of a small amount of styrenic TPE. These materials are also useful impact modifiers for brittle plastics such as PP and, of course, polystyrene.

5.5.3 Copolyesters

Polymers with the -A-B-A-B- structure, where A and B are alternating hard and soft polymeric segments connected by ester linkages, are known as COPs.[45,46] Fig. 5.13 shows the structure of a commercial COP. These block copolymers differ from simple polyester thermoplastics, which are typically hard crystalline polymers of organic dibasic acids and diols. As shown in Fig. 5.10 and Table 5.2, COPs have an excellent combination of properties and are priced accordingly higher than TEOs or TPVs.

The morphology of COPs is that shown in Fig. 5.9. These materials perform as TPEs if the structures of A and B are chosen to give rubbery properties to the copolymer over a useful temperature range. The glass transition point of the soft segment defines T_g for the COP and should be low enough to prevent brittleness at the lowest temperature to which the working material will be exposed. The structure of this segment provides the flexibility and mobility needed for elastomeric performance. The melting point of the hard segment should be high enough to allow the material to maintain a fabricated shape at the highest temperature experienced, but low enough to allow processing on standard thermoplastics equipment. The necessary characteristics for a soft segment are provided by polyether linkages in the example shown in Fig. 5.13.

COPs have a material cost higher than that of most thermoset rubber compounds. This higher cost can be more than offset by the high strength and modulus of the COP, which permits thinner parts and markedly lower part weights. The efficiency of thermoplastics processing can combine with the lower part weight to give pronounced cost savings relative to a thermoset rubber. Thus COP TPEs have been quite successful in replacing thermoset rubber in the fabrication of numerous articles.

Traditionally, rubber people have tended to solve problems through compositional change of the material (that is, compounding or chemistry). In contrast, plastics people have solved problems through design changes (that is, engineering) to enable optimal use of a given material. Each of these approaches has been quite effective for many years. COPs—and TPEs in general—have made major strides forward in combining both of these approaches (materials technology and part design) to give improved properties and performance at lower bottom-line cost than that achievable with a thermoset rubber.[46,47]

The hardness of most commercial COP TPEs is greater than 30 Shore D (Table 5.3), near the upper end of the thermoset rubber range. Table 5.3 gives the mechanical properties of a series of COPs with hardness ranging from 40 to 63 Shore D. As hardness increases, these COPs show progressively more thermoplastic behavior and less rubbery performance. This is demonstrated in Fig. 5.14, which gives stress-strain curves for COPs of 40, 55, and 63 Shore D hardness. For these three examples, stress is proportional to strain only at low strain values, with plastic yielding occurring above 10 or 20 percent elongation. The practical performance range for use of COPs is thus limited to this low-strain region, but proper part design makes this degree of deformation adequate for many applications.

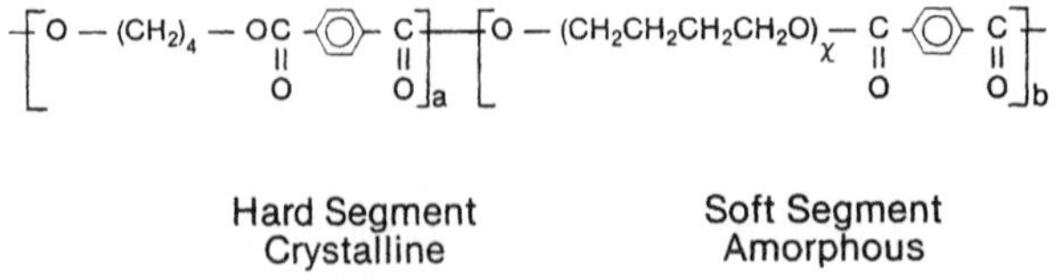

Hard Segment   Soft Segment
Crystalline   Amorphous

FIGURE 5.13   Structure of a commercial COP TPE; $a = 16$ to $40, x = 10$ to 50, and $b = 16$ to 40.

TABLE 5.3   Properties of Copolyester TPEs

Property	40D	55D	63D
Hardness, Shore D (ASTM D 2240)	40	50	63
Specific gravity (ASTM D 792)	1.16	1.20	1.22
Tensile strength, MPa (ASTM D 638)	28	40	41
Ultimate elongation, pct (ASTM D 638)	540	500	430
Flexural modulus, MPa (ASTM D 790)	55	200	300
Tear strength, kN/m (ASTM D 1004)	100	160	175
Izod impact, J/cm (ASTM D 256)	No break	No break	0.3

COPs have a useful service temperature range from –40 to 150°C. The lower limit is set by the soft segment T_g and the upper limit by oxidative attack of the polymer chains. Retention of physical properties measured at elevated temperatures (70 to 150°C) is quite good. They have very good resistance to a wide range of fluids—aqueous salt solutions, polar organics, and hydrocarbons. The ester linkages in the polymer backbone render them susceptible to hydrolysis in both acids and bases. Thus COPs cannot be recommended for service in concentrated acids and bases, especially at elevated temperatures.

The impact resistance of COPs (Table 5.3) is excellent, a result of their elastomeric nature. They are also resilient, with low hysteresis and heat buildup for uses requiring rapid, repeated flexing. In their elastic, low-strain region, COPs have very good resistance to flex fatigue and to tensile and compressive creep.

COPs are excellent materials for the fabrication of harder rubber articles, if their cost can be tolerated. On a weight basis, their cost is 50 to 150 percent higher than that of the TPVs and TEOs. The rubber parts user must compare this difference to the material properties and performance needed for a specific application. Some workers consider the COPs as elastomeric engineered thermoplastics, even though they are generally considered as a generic class of TPEs.[46]

5.5.4   Thermoplastic Polyurethanes

The first commercial TPEs were the TPUs, which have the same block copolymer morphology as the styrenics and COPs (Fig. 5.9).[33,48,49] Their general structure is -A-B-A-B-, where A represents a hard crystalline block derived by chain extension

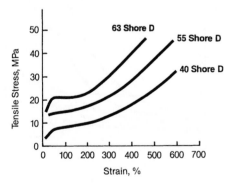

FIGURE 5.14 Tensile stress-strain curves for three different hardness COP TPEs.

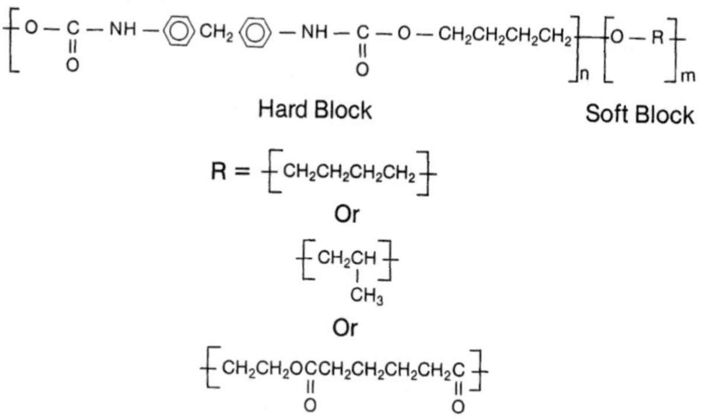

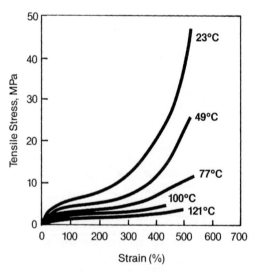

FIGURE 5.15   General structure of a TPU; n = 30 to 120, m = 8 to 50.

of a diisocyanate with a glycol. The soft block is represented by B and can be derived from either a polyester or a polyether. Fig. 5.15 gives typical TPU structures, both polyester and polyether types.

The urethane linkages in the hard blocks are capable of a high degree of inter- and intramolecular hydrogen bonding. Such bonding increases the crystallinity of the hard phase and can influence the mechanical properties—hardness, modulus, tear strength—of the TPU.

As with other block copolymers, the nature of the soft segments determines the elastic behavior and low-temperature performance. TPUs based on polyester soft blocks have excellent resistance to nonpolar fluids and high tear strength and abrasion resistance. Those based on polyether soft blocks have excellent resilience (low heat buildup, or hysteresis), thermal stability, and hydrolytic stability.

FIGURE 5.16   Tensile stress-strain curves of 86 Shore A hardness TPU at temperatures from 23 to 121°C.

TPUs can be made much softer than COPs, down to 60 Shore A (Table 5.2). Recent reports show that TPUs with hardness as low as 50 Shore A can be prepared by the use of hard blocks derived from aliphatic (rather than aromatic) diisocyanates.[50] Fig. 5.16 shows the variation of TPU tensile properties with temperature. As hardness decreases, properties such as modulus, tensile strength, and fluid resistance also decrease. These properties are primarily determined by the ratio of hard to soft phase, the length and length distribution of the segments, the crystallinity of the hard segments, and the morphology of the copolymer. The tensile properties of a TPU can often be improved by postcuring, a heat treatment following a molding step.

TPUs are noted for their outstanding abrasion resistance and low coefficient of friction on other surfaces. TPUs have a specific gravity comparable to that of a carbon-black-filled rubber, and they do not enjoy the density advantage over thermoset rubber compounds that is inherent in other TPEs.

TPUs deteriorate slowly but noticeably between 130 and 170°C by both morphological changes and chemical degradation. Melting of the hard phase causes morphological changes and is reversible, while oxidative degradation is slow and irreversible. Both processes become progressively more rapid with increasing temperature. Polyether soft blocks give TPUs with greater resistance to thermal and oxidative attack than TPUs based on polyester blocks.

TPUs are polar materials and are therefore resistant to nonpolar organic fluids such as oils, fuels, and greases, but they are readily attacked and even dissolved by polar organic fluids such as dimethylformamide and dimethylsulfoxide.[51] TPUs behave like COPs toward water and aqueous solutions, being resistant to these media except at very high or low pH. Polyether TPUs are more resistant to such hydrolytic degradation than polyester TPUs.

TPUs have been commercial longer (since 1958) than any other class of TPEs.[33,34] Their premium cost can be justified for those applications requiring a high level of abrasion resistance and toughness or a low coefficient of friction. These applications include caster wheels, shoe soles, automotive fascia, and heavy-duty hose and tubing.

5.5.5 Elastomeric Polyamides

The newest and highest-performance class of TPEs are block copolymeric PEBAs.[52,53] Amide linkages connect the hard and soft segments of these TPEs, and the soft seg-

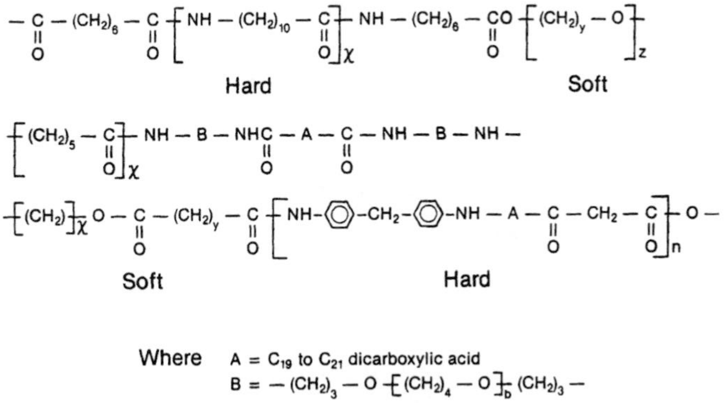

Where $A = C_{19}$ to C_{21} dicarboxylic acid
        $B = -(CH_2)_3 - O -[(CH_2)_4 - O]_b (CH_2)_3 -$

FIGURE 5.17   Structures of three PEBA TPEs.

ment may have a polyester or polyether structure similar to the COPs. The morphology of PEBAs is that of typical block copolymers, as shown in Fig. 5.9. The structures for three commercial PEBAs are given in Fig. 5.17. The amide linkages connecting the hard and soft blocks of PEBAs are more resistant to chemical attack than either an ester or a urethane bond. For this reason, PEBAs typically have higher temperature and chemical resistance than TPUs or COPs, and their cost is greater.

The structure of the hard and soft blocks also contributes to the performance characteristics of PEBAs. The soft segments may consist of polyester, polyether, or polyetherester chains. Polyether chains give better low-temperature properties and resistance to hydrolysis, while polyester chains in the soft segment give better fluid resistance and resistance to oxidation at elevated temperatures. As in other block copolymer TPEs, the nature of the hard segments determines the melting point of the PEBAs and their performance at elevated temperatures.

PEBAs cover a wide hardness range, from a high of 65 Shore D down to 60 Shore A and can therefore be softer than COPs but not as soft as some TPUs. These elastomeric PEBAs have useful tensile properties at ambient temperatures and excellent retention of these properties at higher temperatures. Fig. 5.18 shows that a 90 Shore A PEBA retains more than 50 percent of its tensile strength and modulus at 100°C. Annealing a PEBA above the melting point of the hard phase can result in significant increases in tensile strength, modulus, and ultimate elongation. PEBAs are second only to TPUs in abrasion resistance, and show excellent fatigue resistance and tear strength.

With service temperatures ranging from –40 up to 170°C, PEBAs give the highest performance of any generic class of TPE. They are also the most expensive. Polyester PEBAs give excellent properties retention after aging at 175°C for five days. The hardness, modulus, and fluid resistance all increase with the proportion of hard segments in a PEBA. These TPEs have good resistance toward hydrocarbon fuels, oils, and greases. Their good resistance to water and aqueous solutions decreases as the temperature is raised. Polyester PEBAs are sensitive to hydrolysis in humid air at higher temperatures, whereas polyether PEBAs are more resistant to moisture.

Processing temperatures (220 to 290°C) for PEBAs are higher than those for other TPEs because of their higher melting points. As with the TPUs, PEBAs are

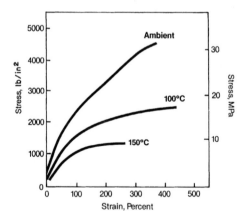

FIGURE 5.18 Tensile stress-strain curves of a 90 Shore A PEBA at ambient and elevated temperatures.

very hygroscopic and thorough drying is required before processing. Recommended drying conditions are 4 to 6 hours at 80 to 110°C for injection molding and extrusion.

Their physical properties, chemical stability, and resistance to temperatures above 135°C make PEBAs competitors for silicone rubber and fluoroelastomers. Applications for PEBAs include hose, tubing, gaskets, and protective covers for use in high-temperature, aggressive environments.

5.5.6 Thermoplastic Elastomeric Olefins

TEOs are simple blends of a rubbery polymer (such as EPDM or NBR) with a thermoplastic (such as PP or PVC).[37,54-56] Each polymer will have its own phase, and the rubber phase will have little or no crosslinking (that is, vulcanization). In earlier literature, TEOs were called TPOs, a term that has fallen progressively into disuse. The polymer present in larger amount will usually be the continuous phase, with the thermoplastic being favored, due to its lower viscosity. The discontinuous phase should have a small particle size for the best properties of the TEO. Reasonable compatibility of the two polymers requires that their solubility parameters and polarities be similar.[57]

EPDM rubber and PP are the constituents of the most common TEOs. Blends of NBR and PVC are also significant but less common in Europe and North America than in Japan.[37] TEOs are similar to thermoset rubbers since they can be compounded with a variety of the same additives and fillers to meet specific market needs. These additives include carbon black, plasticizers, antidegradants, and fillers, all of which tend to concentrate in the soft rubber phase of the TEO. The properties of the TEO can sometimes be altered by lightly crosslinking the rubber phase.[58,59]

An EPDM/PP TEO will have a T_m near that of the hard PP phase, and a T_g close to that for the soft EPDM (plus additives) phase. These TEOs thus melt in the range of 150 to 165°C and can be processed above these temperatures. They show excellent low-temperature performance with brittle points often below –60°C. The T_m clearly determines the upper theoretical service temperature limit of these TEOs. The maximum long-term service temperature is usually 25 to 50°C below the T_m and is determined primarily by the resistance of the polymers to oxidative attack.

EPDM/PP TEOs compete directly with styrenic TPEs as low-cost, low-specific-gravity (0.9 to 1.0) materials with fair to good mechanical performance and environmental resistance. They range in hardness (Table 5.2) from 60 Shore A up to 65 Shore D, with the harder products being more commonly found in commercial applications. The harder TEOs are essentially impact-modified thermoplastics and not true rubbers. The softer TEOs are rubbery at room temperature, but these characteristics are rapidly lost at elevated temperatures. EPDM/PP TEOs are therefore generally useful only below 70 to 80°C.

At ambient temperatures (0 to 40°C), TEOs are quite rubberlike in properties such as modulus, tear strength, and resistance to set. As the temperature is raised, however, these properties decrease quite sharply, much more than those of a thermoset rubber. This is a result of the low level of crosslinking of the elastomer phase and the low level of interaction between this phase and the continuous thermoplastic phase. This low level of crosslinking also renders a TEO highly vulnerable to fluids with a similar solubility parameter (or polarity). Thus EPDM/PP TEOs have very poor resistance to hydrocarbon fluids such as the alkanes, alkenes, or alkyl-substituted benzenes, especially at elevated temperatures.

The absence of unsaturation in the polymer backbones of both PP and EPDM makes these polymers and the TEOs derived from them very resistant to degradation by oxidation or ozone attack. The nonpolar nature of EPDM/PP TEOs makes them highly resistant to water, aqueous solutions, and other polar fluids such as alcohols and glycols, but they swell extensively with loss of properties when exposed to halocarbons and hydrocarbons such as oils and fuels. The TEOs derived from NBR and PVC blends are much more resistant to these aggressive fluids, with the exception of the halocarbons. EPDM/PP TEOs have good electrical properties (Table 5.4) such as resistivity, dielectric strength, and power factor, allowing their use as primary electrical insulation where temperature and fluid resistance are not critical.

TEOs are one of the lower-performance, lower-cost classes of TPEs (Fig. 5.10).[55] Their performance and properties are generally inferior to those of a thermoset rubber. Yet they can be suitable for uses where (1) the maximum service temperature is modest (below 80°C), (2) fluid resistance is not needed, and (3) a high level of creep and set can be tolerated. Thus TEOs are the closest generic class to a commodity TPE and are marketed more on the basis of cost rather than performance, competing directly with the lower-cost general-purpose rubbers (NR, SBR, and the like). TEOs are also the closest approach of a TPE to the traditional practice of rubber compounding and mixing. They can be prepared with the same techniques and equipment as used for thermoset rubber, the principal difference being the need for a higher processing temperature (above the T_m of the thermoplastic). The amounts of elastomer, thermoplastic, plasticizer, and other ingredients can be varied to achieve specific desired properties in much the same manner as used by rubber compounders for more than a century.

EPDM/PP TEOs were first commercialized in 1972 and have grown to a worldwide usage of about 240,000 MT. They are used mainly in external automotive and electrical applications up to 80°C. Automotive uses include exterior trim such as bumpers, fascia, and nonsealing moldings, while under-the-hood uses are limited because of temperature and fluid-resistance requirements in the engine compartment. As Fig. 5.10 indicates, the TEOs are lower in both cost and performance than the TPVs, which are similar to TEOs in compositional chemistry.

5.5.7 Thermoplastic Vulcanizates

TPVs differ from TEOs in that the rubber phase (Fig. 5.8) is highly vulcanized (crosslinked). This phase of a TEO has little or no crosslinking. As a result, the properties and performance of a TPV are much closer to those of a conventional thermoset rubber. Key properties for distinguishing between a TPV and a TEO for a

TABLE 5.4 Electrical Properties of Styrenic, TEO, and TPV TPEs

Electrical property	Styrenics	TPOs	TPVs
Dielectric strength, volts/mil (ASTM D 149, 40 mil thickness)	660	650	960
Dielectric constant (ASTM D 150, 60 Hz)	2.4	2.3	2.4
Volume resistivity, Ω-cm (ASTM D 257)	4×10^{16}	2×10^{16}	2×10^{16}
Surface resistivity, Ω (ASTM D 257)	6×10^{16}	$>1 \times 10^{16}$	2×10^{16}
Power factor (ASTM D 150, 60 Hz)	0.0008	0.0004	0.0009

given elastomer-thermoplastic system are (1) resistance to plastic deformation (that is, low compression and tensile set), (2) greater resistance to fluids, (3) good retention of properties at elevated temperature relative to those at ambient temperature, and (4) low creep and stress relaxation.[60,61]

In earlier literature these TPEs have been called elastomeric alloys.[60] Most workers in the TPE field have come to prefer the term TPV over EA,[38] since the former conveys more clearly and discretely the specific nature of these materials.

TPVs are prepared by a process of dynamic vulcanization,[60,62,63] defined as "the process of intimate melt mixing of a rubbery polymer and a thermoplastic to vulcanize (or crosslink) the rubbery polymer and thus generate a TPE with properties closer to those of a thermoset rubber than those of a comparable unvulcanized composition." Upon melt mixing of the thermoplastic and rubbery polymers under high shear, the less viscous thermoplastic will tend to become the continuous phase with the more viscous rubber dispersed in it.[64] The dispersed rubber particles will then vulcanize, forming a three-dimensional polymer network within each particle, and become trapped since they cannot recombine into larger aggregates.

The most common polymer system in TPVs is PP/EPDM rubber, however, a number of other polymer systems have been used commercially. These include PP/NBR,[65] PP/butyl and PP/halobutyl,[66] PP/NR,[67] and PP/EVA/EPDM.[68]

In a pioneering paper, Coran, Patel, and Williams explored the dynamic vulcanization of nine different thermoplastics with eleven different elastomers.[69] Decades of research will be needed to map out fully the commercial possibilities arising from these 99 thermoplastic/elastomer combinations. From thermoplastics and elastomers with comparable solubility parameters (such as EPDM or NR with PP), TPVs can be prepared directly. From polymers with differing solubility parameters (that is, PP and NBR), one must resort to compatibilization techniques.[70]

The morphology of a TPV is best understood as a dispersion of very small, highly crosslinked elastomer particles in a continuous phase of hard thermoplastic (Fig. 5.8).[71] The size of the elastomer phase particles is one key to the performance of the TPV. As the size of these particles decreases, the ultimate tensile properties of the TPV increase. Fig. 5.19 shows that this increase is not linear, and particle sizes approaching 1 μm give properties that are surprisingly good, almost reaching those of a corresponding thermoset rubber and vastly exceeding those of a TEO from the same polymers. The TPVs are the closest approach of a TPE to true rubberlike performance, such as that offered by a conventional thermoset rubber.

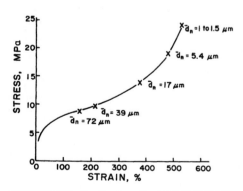

FIGURE 5.19 Tensile strength of EPDM/PP TPVs as a function of dispersed EPDM particle size.

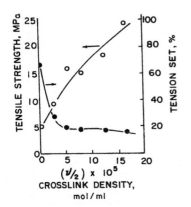

FIGURE 5.20 Effect of TPV crosslink density on tensile strength and tension set.

The second key to TPV performance is the degree of crosslinking of the soft elastomeric phase. It has been demonstrated that highly crosslinking the elastomeric phase gives properties that are surprisingly good and allows the TPV to capture many of the applications previously devoted to a thermoset rubber.[63] A high degree of crosslinking of the elastomeric phase is a characteristic that distinguishes a TPV from a TEO. In an EPDM/PP TPV, it is highly desirable for the EPDM phase to be fully crosslinked. Fig. 5.20 shows the effect of crosslink density (as measured by conventional Flory-Rehner swelling techniques) on the tensile strength and tension set of an EPDM/PP TPV.[72] The tensile strength increases progressively with crosslink density. With crosslinking, the tension set decreases precipitously from that of the blend (zero crosslink density, a TEO) and then levels off at a respectably low level comparable to that of a conventional thermoset rubber. Thus, a high level of crosslinking of the dispersed elastomer phase is necessary for the rubberlike performance of a TPV.

Table 5.5 compares an EPDM/PP composition with a crosslinked EPDM phase (a TPV) to an uncrosslinked EPDM/PP blend (a TEO). The improvements in the TPV in Table 5.5 compared to the TEO include higher tensile strength, higher modulus, lower compression and tension set, and greatly improved oil resistance. In addition to those listed, additional property enhancements in the TPV include improved fatigue resistance and greater retention of physical properties at elevated temperatures.

Fig. 5.21 compares the fatigue resistance of an EPDM/PP TPV to that of several thermoset rubbers,[73] showing the TPV to have outstanding resistance. Thermoset rubber compounds with good fatigue resistance are generally poor in compression and tension set because the two tend to be mutually exclusive. TPVs, however, have excellent fatigue resistance and a set resistance comparable to that of thermoset rubbers, an unusual combination of properties.

The anisotropy (or directionality) of TPV properties has been found much closer to that of thermoplastics than to thermoset rubbers. Thus the parameter of direction must be considered when properties are measured. Generally, the ultimate tensile strength

TABLE 5.5 Comparison of Crosslinked and Uncrosslinked EPDM/PP Compositions*

Property	Crosslinked	Uncrosslinked
Hardness, Shore A	84	81
Ultimate tensile strength, MPa	13.1	4.0
Stress at 100% elongation, MPa	5.0	2.8
Ultimate elongation, %	430	630
Tension set, %	14	52
Compression set, %	31	78
Swell in ASTM #3 oil, %	52	162

* Parts by weight: EPDM polymer, 91; polypropylene, 55; carbon black, 37; oil, 36.

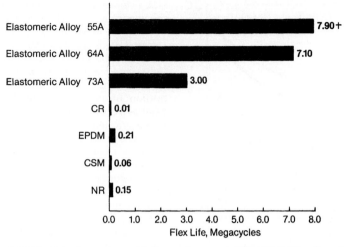

FIGURE 5.21   Comparison of fatigue resistance of EPDM/PP TPV to that of thermoset vulcanizates. Monsanto fatigue-to-fail test, 100 percent elongation.

(UTS) and ultimate elongation are significantly greater (15 to 50 percent) in the direction perpendicular to the flow of molten TPV in the molding or extrusion process than in the direction parallel to the flow (Figure 5.22). This anisotropy is an artifact (Table 5.6) of the process history of the TPV article, being greater for injection molding than for extrusion (which employs a lower shear rate). It increases progressively with the shear rate at which the TPV is fabricated. TPV anisotropy can be reduced by subsequent heating and annealing of a fabricated part.

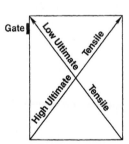

FIGURE 5.22   Anisotropy of injection-molded EPDM/PP TPV plaque.

The tensile stress-strain curves of a material distinguish between degrees of thermoplastic and elastomeric behavior. Fig. 5.23 shows typical tensile stress-

TABLE 5.6   Variation of Tensile Properties with Direction for TPVs

Property	64 Shore A	80 Shore A	40 Shore D
Tensile strength, MPa			
Strong direction	7.1	11.0	17.9
Weak direction	4.8	8.8	15.0
Stress at 100% elongation, MPa			
Strong direction	3.2	4.6	8.0
Weak direction	4.1	6.1	10.2
Ultimate elongation, %			
Strong direction	502	500	610
Weak direction	317	340	470

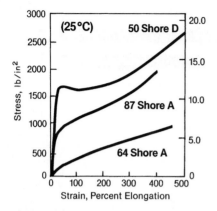

FIGURE 5.23 Tensile stress-strain curves for three different hardness EPDM/PP TPVs.

strain curves for three EPDM/PP TPVs, ranging from a soft 64 Shore A to a hard 50 Shore D material. The curve for the softest TPV is similar to that for a conventional thermoset rubber, deviating only moderately from linearity. The medium hardness TPV has a higher proportion of PP in its composition and exhibits a distinct "knee" but, on balance, is clearly a rubber. The hardest TPV actually has a yield point and has a tension set too large[1] to qualify as a rubber.

The tensile behavior of a TPV of a given hardness over a range of temperatures is shown in Fig. 5.24. At very low temperatures, the TPV becomes more like a rigid thermoplastic, and the stress-strain curve develops a yield point. The other notable feature of this family of curves is that the properties at elevated temperatures are significantly high fractions of the room-temperature values. For example, at 100°C, the tensile strength and modulus are still 40 to 50 percent of the 25°C values. At 100°C, a simple blend (an EPDM/PP TEO) would retain only 10 to 20 percent of its room-temperature properties.

A commonly reported property of a material is the UTS, or the stress developed when the material is strained to failure. For TPEs, the UTS is often significantly lower than that for a thermoset rubber of the same hardness. This is of little practi-

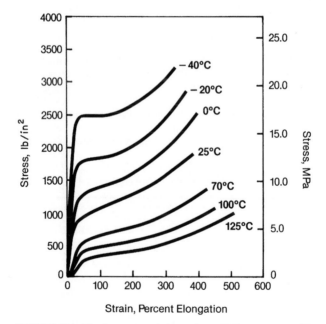

FIGURE 5.24 Tensile stress-strain curves for an 87 Shore A EPDM/PP TPV at different temperatures.

cal importance because very few applications require a rubbery material to operate near its tensile breaking point. For more than a century, UTS has been used as a criterion for ascertaining the degree of vulcanization of a thermoset rubber. With TPVs, and TPEs in general, the use of UTS in delineating material quality and predicting product performance can be both specious and misleading. In many rubber articles, a TPE has replaced a thermoset rubber of much greater tensile strength and performed excellently.

TPVs (Table 5.7) have mechanical properties such as modulus (compression or tensile), tear strength, abrasion resistance and compression set resistance which make them suited for a broad range of rubber applications. As discussed, their resistance to fatigue is superior to that of a thermoset rubber of comparable hardness.[73] The properties and performance of these materials are approximately near the middle of the spectrum of TPEs (Fig. 5.10).

The useful long-term (weeks, years) service temperature range for TPVs is between the T_g of the soft elastomer phase and the temperature at which oxidative degradation of the TPV becomes significant. Because there is no unsaturation in the polymer backbone of EPDM rubber, TPVs based on EPDM will have a higher service temperature limit (approaching T_m) than those based on unsaturated elastomers such as NBR and NR. For EPDM/PP TPVs this range is –60 to 135°C; for NBR/PP TPVs it is –40 to 125°C; and it is –40 to 100°C for NR/PP TPVs. The lower temperature of each of these ranges is determined by the brittle point of the elastomer. The upper temperature is based on properties retention after continuous aging in hot air for 1000 hours.

EPDM PP TPVs are affected very little by water and other polar fluids such as aqueous solutions, acids, and bases. Just as with thermoset rubbers, nonpolar fluids such as oils or fuels cause varying degrees of swelling, fluid absorption, and loss of properties by the TPV. Resistance to such fluids is consistent with the fluid resistance

TABLE 5.7 Properties of TPV TPEs

Property	EPDM/ PP 73A	EPDM/ PP 87A	NBR/ PP 70A	NBR/ PP 40D	NR/PP 70A	NR/PP 40D
Hardness, Shore (ASTM D 2240, 5-s delay)	73A	87A	70A	40D	70A	40D
Specific gravity (ASTM D 297)	0.98	0.96	1.00	0.97	1.04	1.01
Tensile strength, MPa (ASTM D 412)	8.3	15.9	6.2	16.6	7.6	14.3
Ultimate elongation, % (ASTM D 412)	410	530	265	420	380	540
Stress at 100% elongation, MPa (ASTM D 412)	3.2	6.9	3.3	9.0	3.7	8.2
Tear strength, kN/m (ASTM D 624)	28	49	32	76	29	73
Brittle point, °C (ASTM D 746)	<–60	–61	–40	–28	–50	–40
Tension set, % (ASTM D 412)	14	33	10	37	16	39
Compression set, 22 h at 100°C, % (ASTM D 395B, 25% compression)	33	52	28	46	32	60
Weight change, ASTM #3 oil, 166 h at 100°C, % (ASTM D 471)	65	42	0	5	97	26

of the elastomeric component of the TPV. Thus an NBR/PP TPV is more oil-resistant than an EPDM/PP TPV, which, in turn, is more resistant than an NR/PP TPV. However, the swelling and property loss for a TPV is generally less than that observed for the same elastomer in a thermoset compound. An NBR/PP TPV is as oil-resistant as an NBR compound; an EPDM/PP TPV has oil resistance superior to an EPDM thermoset rubber and is equivalent to neoprene. An NR/PP TPV usually has better oil resistance than an NR compound.

The TPVs—EPDM/PP, NBR/PP, and NR/PP—have high resistance to attack by ozone, known to attack elastomers at olefinic carbon-carbon double bonds in the polymer backbone. This result is not surprising for the EPDM TPV, since this elastomer has no primary chain unsaturation. The unsaturation in the backbone of NBR and NR renders this result with these TPVs to be quite surprising, especially for NR/PP. These findings can readily be understood if PP is the continuous phase and the elastomer the dispersed phase. This results in the molded or extruded NBR/PP or NR/PP part having a thin skin of PP, a polymer with no olefinic unsaturation. Thus the ozone must first penetrate the chemically resistant PP skin before attacking the olefinic unsaturation.

The highly rubberlike properties of the TPVs have enabled them to perform as engineered thermoplastic rubbers. In numerous application areas they have directly replaced premium-performance thermoset rubber compounds to a degree unparalleled by any other class of TPEs. Prominent among these uses are demanding automotive applications (Figs. 5.25 and 5.26), electrical insulation and connectors (Table 5.4 and Fig. 5.27), compression seals (Fig. 5.28), appliance parts, medical devices (Fig. 5.29), and food and beverage contact applications (Fig. 5.30).

5.5.8 Other TPEs

An emerging group of TPEs are those based on reactor technology involving metallocene catalysts.[74] These catalysts effect the copolymerization of ethylene and alpha olefins to give a polymer with a narrow molecular weight distribution, controlled level of long-chain branching, and homogeneous comonomer distribution. The poly-

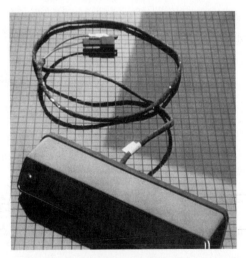

FIGURE 5.25   TPV gasket for brake light mounted in rear window of motor vehicle.

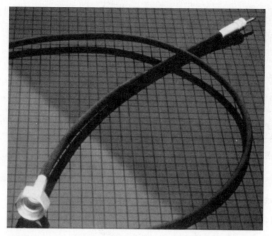

FIGURE 5.26 Protective TPE cover for automotive brake cable.

FIGURE 5.27 Electrical cable insulated with an EPDM/PP TPV.

mer structure is similar to that of a block copolymer. Produced in a single reactor, these TPEs have a major cost advantage over TEOs and TPVs. They likely will compete most directly with the TEOs. It is highly uncertain whether or not their properties and performance can be brought up to the level of the TPVs. These reactor materials show promise of becoming the lowest cost group of TPEs. Much of this cost advantage will be lost if further compounding is needed for adequate performance.

Blends of PVC and NBR have found some use in areas where the service temperature is at or near ambient and oil resistance is needed,[75] such as the hose and electrical wire and cable markets. Normally custom compounded for a specific use, their applications have been somewhat limited except in Japan. These TPEs may be classified as TEOs (Sec. 5.5.6).

Another group of TPEs in the TEO category are NR/PP and NR/PE blends, now finding some limited uses.[76] Other TPEs have been derived from fluoroelastomers (for high temperature use),[77] acrylate copolymers,[78] and silicones.[79]

5.6 PROCESSING OF TPEs

5.6.1 Processing Economics

The rapid growth of the TPE market in the past two decades has resulted largely from the processing advantages these materials offer over thermoset rubbers. All of

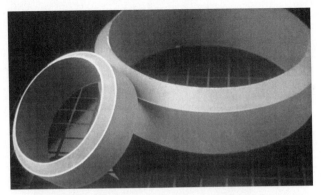

FIGURE 5.28   Compression seal injection molded from TPE.

FIGURE 5.29 Peristaltic pump tubing, a mechanically demanding application, exploits the properties of some of the newer TPEs for the transport of medical and other fluids.

FIGURE 5.30 Injection-molded dispensing valve for seltzer bottle.

the advantages enumerated in Sec. 5.3 (with the exception of specific gravity) are related to the use of fast, low-cost thermoplastics processing, compared to the slower, labor-intensive techniques used for thermoset rubber.

The cost per unit weight of a TPE is almost always greater than that for a competitive compounded thermoset rubber stock. However, the savings in part-fabrication cost by thermoplastics processing (plus the savings from the lower specific gravity of TPEs) can often far outweigh the differences in material cost of a TPE compared to a thermoset.

To properly assess the cost issue for a TPE versus a competitive thermoset rubber, it is critical to focus on the total, bottom-line cost (material plus fabrication) of the rubber part. Table 5.8 gives such an analysis for an actual TPV use, a windshield wiper reservoir seal (Fig. 5.31). The part had previously been assembled from a brass screen and a compression-molded neoprene article. It was replaced by a part injection molded in one step from a TPV to give a 30 percent weight reduction and a cost decrease of more than one-half, with no loss of functional performance. The bulk of the cost reduction arises from the ease, efficiency, and economy of thermoplastics processing and improvements in design.

A second cost example is given in Table 5.9 for an architectural glazing

TABLE 5.8 Cost Comparison for Windshield Wiper Reservoir Seal

Material	TPV, 73 Shore A hardness	Neoprene thermoset
Part weight, g	3.2	4.7
Material cost, $U.S./kg	3.96	2.20
Fabrication process	Injection molding, one step	Compression molding, subsequent assembly with brass screen
Cost per part, $U.S.	0.13	0.30

FIGURE 5.31 Windshield wiper reservoir seal—left, assembled from thermoset rubber part and brass screen; right, injection-molded from TPV.

seal (Fig. 5.32). The use of the engineered TPV generates a 10 percent cost savings over neoprene and a 60 percent savings over silicone rubber.

An important factor in fabricated-part economics is the size or wall thickness, especially for injection molding. Thermoset rubber injection or compression molding cycles are limited by relatively long cure times, during which the material must be held under pressure in a hot mold while crosslinking (vulcanization) takes place. These times are commonly on the order of 3 to 30 minutes. Thermoplastic molding cycle times are limited by the time needed to fill the mold and the (much greater) time required to cool the parts sufficiently to maintain shape when ejected from the mold. These times are typically on the order of a few seconds (10 to 40) for parts weighing up to 50 to 100 g.

Fig. 5.33 shows that as the part size becomes smaller, the greater material cost of a TPE compared to a thermoset rubber is outweighed by the relatively low molding cost of the TPE and its lower specific gravity, making the total TPE part cost less than that of the same part in thermoset rubber. As the part size increases, the thermoplastic molding cycle time increases because of the required time to freeze a larger block of material sufficiently to prevent distortion when removed from the mold. At part weights in the range of 250 to 300 g, this longer cooling time obviates much of the processing advantage of a TPE, causing the total part cost to become about the same as that for the same part molded from a thermoset rubber. For even larger parts, little or no processing cost advantage is observed for a TPE, and the lower material cost of a thermoset rubber dominates the total part cost.

This rule of thumb must be applied judiciously, however, because the critical factor affecting cooling time for a TPE part is the cross-sectional thickness, not the part

TABLE 5.9 Cost Comparison for Extruded Architectural Glazing Seal

Material	TPV, 73 Shore A hardness	Neoprene thermoset	Silicone thermoset
Part weight, g/m	14	18	17
Material cost, $U.S./m	0.039	0.046	0.098
Total cost, $U.S./m	0.102	0.112	0.246

FIGURE 5.32 Architectural glazing seal extruded from TPE.

weight. It is possible to have a very large, thin part such as a diaphragm that has a fast cooling time and is still more economical to mold in a TPE than in a thermoset rubber. Each case must be considered in detail, including part geometry as it affects fill and especially cooling times.

5.6.2 Rheology

TPEs are highly nonnewtonian fluids in the molten state. Their apparent viscosity depends strongly on the shear rate of the process to which they are subjected, with the observed viscosity decreasing sharply as the shear rate is increased (Fig. 5.34). This characteristic behavior means that molding operations should be carried out at high shear rates so the resulting low viscosity allows quick filling of the mold. It also means parts may be removed rapidly from the mold (a low shear process) without distortion, even though the interior of the part is still molten.

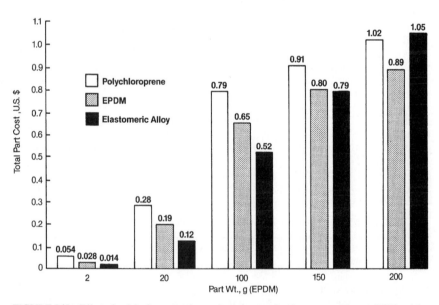

FIGURE 5.33   Effect of article size on total manufacturing cost for thermoset rubber and TPE articles.

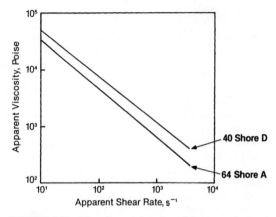

FIGURE 5.34 Viscosity as a function of shear rate for hard and soft TPEs.

TPE viscosity is far less sensitive to temperature than to shear rate (Fig. 5.35). It should be noted that the variation of viscosity in Fig. 5.34 is log-log, whereas that in Fig. 5.35 is direct. The implication for processing operations is that input of mechanical energy is far more effective in changing the flow of the melt than is input of thermal energy. In an injection molding process, for example, raising the shear rate by using a higher injection pressure will have a more noticeable impact on fill time than a temperature increase.

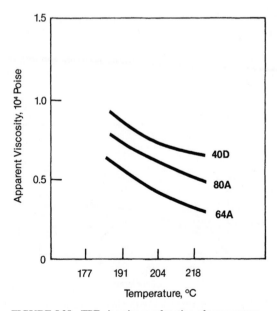

FIGURE 5.35 TPE viscosity as a function of temperature.

5.6.3 Need for Drying TPEs

For optimum performance, most TPEs benefit from drying before processing. Many TPEs can absorb sufficient moisture in a short time to cause difficulties such as poor surface appearance (or splay) on injection molded parts and rough surfaces and porosity in extrusions. Fig. 5.36 shows the moisture pickup for two TPVs under ambient plant conditions. A moisture pickup of 0.10 percent or greater can give highly significant processing problems, even in the case of nonpolar hydrocarbon TPEs. An exposure of less than one day can result in a moisture content that is outside the acceptable processing window.

While some TPEs have been developed to specifically avoid the need for drying, most should be dried immediately before use at a temperature of 70 to 100°C for 2 to 6 hours, depending on the specific TPE. Because of its high surface area, regrind material for recycling to the process should be dried an extra 1 to 2 hours. It is advisable to combine recycled regrind with virgin TPE in a constant ratio, with the regrind fraction as low as practical. Desiccant dryers are common in thermoplastics processing plants and should be used with TPEs.

5.6.4 Extrusion

Simple extrusion (Fig. 5.37) can be used to manufacture a variety of TPE shapes such as tubing, hose, sheet, and complex profiles. Coextrusion of a TPE with a thermoplastic or another TPE is possible if the materials are compatible and have melting points in the same range. Coextrusion is a cost-effective way to take advantage of the features of a soft TPE as a sealing surface while using a rigid polyolefin or hard TPE member for support. Crosshead extrusion is a widely used technique for applying a TPE insulating or jacketing layer on an electrical wire or cable, and for applying a TPE cover over a reinforced hose.

A thermoplastic extruder with a screw length-to-diameter (L/D) ratio of at least 20:1, and preferably 24:1 to 32:1, is recommended for most TPEs.[80] The extruder should be capable of operating in the melt temperature range of 170 to 250°C. A variety of screw designs—polyolefin type, flighted barrier, pin mixing, and Maddock mixing—have been used successfully. Screw cooling is generally not used. Typical

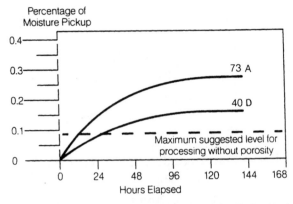

FIGURE 5.36 Moisture pickup as a function of time for hard and soft TPVs.

FIGURE 5.37   Extrusion of TPE tubing.

polyolefin-type metering screws with compression ratios in the range of 2:1 to 4:1 are normally used, although other thermoplastics screws have been used with success. Screen packs of 20-40-60 mesh are used to provide a clean melt stream and even melt flow. Finer mesh screens can be used depending on the viscosity of the particular TPE and its sensitivity to the higher temperatures caused by the finer screen.

Extruders commonly used for thermoset rubbers have been used for TPEs; however, their use is almost universally not recommended. Thermoset rubbers are normally extruded in the 120 to 140°C range, whereas thermoplastics require 170 to 250°C. Thus a rubber extruder must usually have auxiliary heaters, with a concomitant loss in temperature control. Further, a rubber extruder commonly has an L/D ratio of 14:1 to 16:1, too short for adequate homogenization of the TPE melt prior to entering the extruder die. Thus the use of a thermoplastics extruder will give a more readily controlled process with a broad process window, in contrast to one difficult to control with a narrow process window.

The polymer melt temperature in the extrusion process should be about 30 to 70°C above the melting point of the TPE. For most TPEs, this corresponds to a melt temperature of 170 to 250°C.

Thermoplastic materials generally exhibit a die swell on exiting from an extrusion die. TPEs tend to show a die swell significantly lower than that of typical thermoplastics. This swell must be accounted for in designing dies and adjusting extrusion conditions to achieve an accurately sized profile. The die swell for TPEs increases with increasing hardness, decreasing temperature, and increasing shear rate (such as higher screw revolutions per minute). Fig. 5.38 shows the die swell for TPV TPEs at extrusion shear rates.

TPE rubber sheet is produced by extrusion processing because the calendering process normally used with thermoset rubbers does not achieve the shear rates

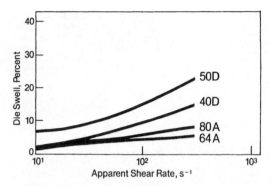

FIGURE 5.38   Die swell as a function of shear rate for
TPEs of different hardness, 204°C.

needed to produce a uniformly heated melt. Sheet extrusion of TPEs can produce
thicknesses from about 0.2 to 4 mm and widths of 2 m or more. Very thin TPE sheeting can be produced by the blown film process used for many thermoplastics.[81] This
technique allows hard TPE sheet to be made down to a thickness as low as 0.05 mm,
and even soft TPEs can yield sheets 0.2 mm thick.

The TPE extrusion process is much simpler and lower in cost than that for thermoset rubbers because TPEs do not require a final vulcanization step. A TPE extrudate need only be cooled (commonly in a circulating water trough) and it is ready
for use. This contrasts with a thermoset rubber extrudate which must be heated,
either in batch processes with autoclaves or continuously with complex equipment
to maintain the necessary pressure during the curing step. Thus the extrusion of a
TPE totally avoids the need for vulcanization equipment, with a massive savings in
capital investment and ultimate article cost.

5.6.5   Injection Molding

Injection molding (Fig. 5.39) is the most widely used process for fabricating TPE
parts,[80] because it exploits so fully the processing advantages of TPEs over thermoset
rubbers. Shorter molding cycles and elimination of scrap by regrinding quite often outweigh the generally higher material cost of a TPE compared to a thermoset. Sprues
and runners from injection molding can be recycled, and the use of hot-runner molding can eliminate them entirely, as well as avoiding a labor-consuming trimming step.[82]

TPEs can be molded in the same type of equipment used for injection molding
thermoplastics such as PP and PE.[83] Molding machines should be sized to give clamp
pressures in the range of 40 to 70 MPa over the entire shot area, including runners
and sprues, and barrel capacities of about four shots.

Good part definition and integrity are obtained by adequate mold packing, which
gives strong weld lines and minimizes shrinkage. This is achieved by mold designs
with a balanced layout—equal-pressure drops and equal-length melt flow paths to
each cavity. Runners should be as short as feasible and fully round in order to minimize their surface-to-volume ratio. Small gates are recommended to provide high
shear rates for uniformity of the material as it enters the cavity. Vents should be
about 0.03 mm deep and located in the mold as far as possible from the gate, and
wherever knit lines are found.

FIGURE 5.39 Injection molding of a TPE.

TPEs are quite fluid at the shear rates normally employed in thermoplastic injection molding (above 500 sec^{-1}), and machine configurations and operating conditions should be selected accordingly. High injection pressure should be used for rapid mold filling, 4 seconds or less, followed by a short hold period of 1 to 10 seconds to allow gate freeze-off. The cooling cycle (typically 8 to 50 seconds) in the mold depends on part size and thickness, and need only be long enough to solidify the outer skin of the part sufficiently to allow ejection from the mold without distortion. Mold release agents are normally not needed with TPEs, which separate quite readily from the mold surface.

Operating conditions should be selected to give melt temperatures in the range of 20 to 50°C above the melting point of the TPE in order to allow adequate mold packing and minimize shrinkage. Good mold packing will give shrinkage of about 1.5 to 4.0 percent for most TPEs. This shrinkage can be controlled to give dimensional tolerances that are one half to one third of those obtained for thermoset rubber parts.

5.6.6 Blow Molding

TPEs can be blow molded (Fig. 5.40) to produce hollow rubber shapes in the same manner and equipment used for thermoplastics, either via injection or extrusion blow molding techniques.[84,85] Extrusion blow molding is the simpler process in which a hollow molten parison is extruded vertically downward into a mold cavity. As the mold is clamped around the parison, blowing takes place, forcing the molten TPE against the water-cooled mold. The part is then cooled to give it sufficient structural integrity for removal from the mold, normally by gravity. Injection blow molding is similar except that very close control of the parison dimensions is obtained by injection molding the parison. While still hot, the molded parison is transferred into a blow mold for final part shaping. The need for two sets of tooling makes this process more costly but allows for precise wall-thickness control, as well as molded-in fittings, screw threads, and so on. Injection blow molding is very capital-intensive and thus requires large production volumes for favorable economics.

Blow molding cannot be used to produce hollow rubber parts from conventional thermoset rubbers. These must be injection-molded over a manually removable collapsible core to form thin-walled rubber shapes such as boots, bellows, and covers. Blow molding is uniquely suited to TPEs as a material for hollow, thin-walled rubber articles.

Compared to thermoplastics, the blow ratios (cavity diameter to parison diameter) of TPEs may be limited and the parison diameter should be as large as possible relative to the mold cavity size. Molten TPE parisons maintain their shapes well prior to blowing because of their high viscosity at low shear rates. Melt temperatures are in the same range as for extrusion and injection molding. Cycle times are normally in the 10-to-240-second range and can be minimized by rapid circulation of water in the mold.

5.6.7 Other Fabrication Methods

Another thermoplastics processing technique that is suited to TPEs but not to thermoset rubbers is thermoforming (Fig. 5.41). A sheet of TPE is heated to 10 to 40°C above its softening point, then pressure or vacuum is used to stretch the softened sheet over or into a mold.[86] Draw ratios up to 3:1 (thickness of original sheet to final part thickness) can be attained with hard TPEs in the 40 to 50 Shore D range. The maximum draw ratio decreases with decreasing hardness. For suitable parts, this process can be a very fast, low-cost production technique. Limitations include the need to avoid undercuts and sharp corners on the mold and adequate taper angles to allow easy removal of the part from the mold.

Welding (Fig. 5.42) is a fast and simple method for bonding a TPE to itself or to another compatible thermoplastic material. This technique is especially useful in forming corners on rectangular glazing seals and gaskets. The surfaces to be joined must be heated above the melting point and held together under slight pressure until the joint cools and solidifies. Surface heating can be by direct contact with hot air or a hot surface, radiation heating, or vibration (as in spin welding). Miter cutting equipment and heat welding fixtures are available, by which complex extruded profiles with very thin lips and ribs may be readily heat-welded. Robot welders are available that can heat weld-lap joints of TPE sheets many feet in length. Bond strength can be as much as 70 to 80 percent of the tensile strength of the sheet, if the welding is done properly.

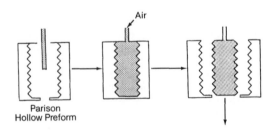

FIGURE 5.40   Schematic depiction of blow molding.

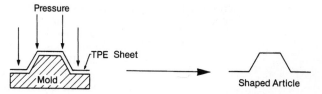

FIGURE 5.41 Schematic depiction of thermoforming.

FIGURE 5.42 Schematic depiction of heat welding.

FIGURE 5.43 Foamed TPE tubing produced by extrusion. A thin skin of solid TPE covers the cellular interior.

TPEs can readily be foamed via extrusion to lower their macroscopic hardness and density (Fig. 5.43). Closed-cell foamed tubing, sheeting, and profiles with a thin, solid skin are now routinely fabricated by extruding TPEs with a chemical and/or physical blowing agent. The bulk specific gravity can be reduced from 1.0 to 0.7 with chemical foaming agents such as azodicarbonamide. Physical blowing agents such as fluorocarbons may be used to lower the specific gravity down to approximately 0.3.[87] Much interest has been focused on a recent invention employing water (steam) as a TPE blowing agent, thus obviating the environmental problems resulting from the use of fluorocarbons.[88]

5.7 APPLICATIONS AND MARKETS FOR TPEs

TPEs find uses in virtually all the applications where thermoset rubbers are used, with the major exception of pneumatic vehicle tires, which account for slightly more than one half of all worldwide rubber consumption. There are no current TPE materials which meet automotive tire performance and assembly requirements.[89] There are, however, a number of TPEs that are being used for solid, nonpneumatic tires, and caster wheels for carts, lawn equipment, and toys (Fig. 5.44).

FIGURE 5.44 Caster wheel injection-molded onto PP hub by insert molding of TPE tread. Compatibility of PP and TPE generates a weld bond.

The replacement of latex-dipped rubber articles (such as condoms and surgical gloves) has been considered an application area most difficult for TPEs to penetrate. Today, condoms derived from styrenic TPEs are a commercial reality.[90] These devices do not have several of the disadvantages—immune reactions and poor ozone resistance—of those derived from NR latex.

Nontire automotive uses where the service requirements are not too demanding (temperature below 70°C and little or no fuel and oil resistance required) have provided numerous markets for both the styrenic and TEO materials.[91] These uses include air dams, weatherstripping, rub strips, bumpers, fascia, dashboard trim, plugs, and grommets, to mention a few.

The higher-performance COPs, TPUs, and TPVs are used in those areas where the service-temperature range, mechanical abuse, and fluid resistance demand a higher level of performance. Such applications include seals and gaskets, convoluted grease-filled boots for steering and front-wheel drive, and assemblies and covers for safety air bags (Fig. 5.45). The average automobile now emerging from a North American assembly line contains from 8 to 12 kg of different TPEs.

Growth of TPE usage in nonautomotive applications has also been rapid. The mechanical rubber goods (MRG) TPE market includes uses in building construction, appliances, tools, and business machines, to name a few.[92] Specific parts are as diverse

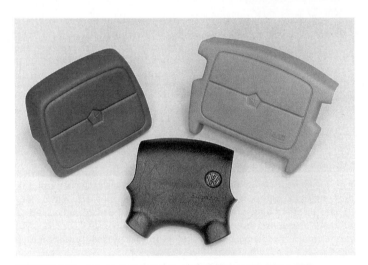

FIGURE 5.45 Automotive air bag covers injection-molded from TPE.

as dishwasher sump boots, architectural window glazing (Fig. 5.32), flashlight housings, hand-tool gaskets, typewriter and printer rollers, and household plumbing seals. TPEs are extensively used in wire and cable insulation (Fig. 5.27). Low dielectric constant, high electrical resistance, high dielectric strength and low power factor make hydrogenated styrenic block copolymers and EPDM/PP TPVs especially useful in electrical applications. Flame-retardant TPEs are available for electrical uses requiring rubber that will not support combustion.

Food and potable water contact articles are increasingly inviting applications for TPEs, as are medical and health-care rubber parts.[93] Many of the toxicological concerns associated with the cure systems and additives of thermoset rubbers do not apply to TPEs. A growing number of uses is being found for TPEs in food-processing equipment, beverage-dispenser pumps, food-container seals (Fig. 5.30), peristaltic pump tubing, syringe stoppers, catheters, and hospital tubing and sheeting. TPEs may be sterilized by steam, ethylene oxide, or high-energy radiation with no significant loss in properties or functional performance.

The ongoing market need for softer TPEs has been addressed by the styrenic and TPV materials. Styrenics are now available down to the 20 to 30 Shore A hardness range and TPVs down to 35 Shore A. The need for TPEs with air permeability competitive with thermoset butyl rubber has now been met by PP/butyl-halobutyl TPVs.[94] The marketplace continues to demand TPEs for service at higher temperatures and with fluid resistance at these temperatures. The high cost of the silicones, fluoroelastomers, and polyphosphazines will continue to fuel this demand for many years to come.

5.8 THE RECOVERY AND RECYCLE OF TPEs

The thermoplastic processability of TPEs makes an excellent fit with today's trend to recycle the materials from articles that have given a normal lifetime of service. At the forefront of this recovery and recycle effort is the reuse of polymeric materials such as rubber, plastics, and paper.[95] The recycle of these materials arises directly from the need to reduce our solid waste, which must be disposed of some way, and to conserve our natural resources.

It is well established that thermoplastics are far more suitable for recycling than either thermoset plastics or thermoset rubbers.[96] The proper recovery of value from spent pneumatic tires is currently a problem of the first magnitude.[97] This problem arises primarily from the thermoset nature of these articles.

TPEs, like thermoplastics, can readily be recycled by simple remelting and reshaping the material in virtually the same manner that process scrap (regrind) from the fabrication of virgin material can be reused. As with process scrap, it is necessary that the material not be contaminated prior to or during reprocessing. To be suitable for recycling, a used TPE part must not have undergone significant chemical changes, either through environmental attack (such as oxidation or chemical attack) or fluid contamination (migration of oils or greases into it). Even with some contamination the TPE can be recycled, but for applications requiring lower performance.

The refabrication of a TPE material requires only a reversible physical change. Thus, in theory, it should be possible to recycle a TPE numerous times (Fig. 5.6) as long as the material remains chemically unchanged. On the other hand, the recycle of a thermoset rubber article requires an irreversible chemical change, that of devulcanization, the cleavage of the crosslinks between the polymer chains. As a result, the recovery of value from thermoset rubber articles is quite difficult, being limited

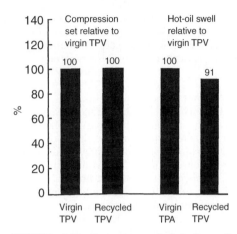

FIGURE 5.46 Comparison of hot-oil swell (ASTM #3) and compression set of virgin TPV and identical material recycled from used steering gear boots.

primarily to shredding (to generate padding or filler) or to incineration (to generate energy). The recycle of TPEs, like thermoplastics, is limited primarily by logistics—the mundane tasks of collection, sorting, segregation, and cleaning.[95,96] The technology for their recycle is well worked out and known. To the contrary, the recycle of conventional rubber articles is limited by lack of technology, a result of their thermoset nature.

The feasibility of recycling used TPE articles has recently been reported for TPV rack and pinion steering gear boots.[41] These boots had been in service for more than five years in contact with lubricating grease in this demanding under-the-hood automotive application. These boots were removed from the steering assembly, cleaned with steam, ground into small particles, dried, and remolded. The properties of the remolded TPV were essentially the same as those of the virgin TPV. Fig. 5.46 compares the oil swell and compression set (two key indicators of TPV performance) of the recycled and virgin materials.

5.9 ACKNOWLEDGMENTS

The author is grateful to Jennifer L. Digiantonio for her capable, cheerful, competent assistance in preparing this chapter. He also wishes to thank Advanced Elastomer Systems, L.P. for permission to prepare and publish it.

REFERENCES

1. ASTM D 1566, "Standard Terminology Relating to Rubber," American Society for Testing and Materials, Philadelphia, PA, vol. 9.01, 1995.

2. B. M. Walker (ed.), *Handbook of Thermoplastic Elastomers,* Van Nostrand Reinhold, New York, 1979.

3. A. Whelan and K. L. Lee (eds.), *Developments in Rubber Technology—3: Thermoplastic Rubbers,* Applied Science Publishers, London, 1982.

4. N. R. Legge, G. Holden, and H. E. Schroeder (eds.), *Thermoplastic Elastomers—A Comprehensive Review,* Hanser Publications, Munich, Germany, 1987.

5. B. M. Walker and C. P. Rader (eds.), *Handbook of Thermoplastic Elastomers,* 2d ed., Van Nostrand Reinhold, New York, 1988.

6. S. K. De and A. K. Bhowmick, *Thermoplastic Elastomers from Rubber-Plastic Blends,* Ellis Horwood Ltd., Chichester, England, 1990.

7. R. L. Arnold and C. P. Rader, "Thermoplastic Elastomers," in *Handbook of Plastics, Elastomers, and Composites,* 2d ed., C. Harper (ed.), McGraw-Hill, New York, 1992.

8. M. T. Payne and C. P. Rader, chap. 14 in *Elastomer Technology Handbook,* N. P. Cheremisinoff (ed.), CRC Press, Boca Raton, FL, 1993.

9. C. P. Rader, chap. 15 in *Know Your Plastics,* 2d ed., Plastics Industry Association, Melbourne, Australia, 1992.

10. J. C. West and S. L. Cooper, *Nippon Gomu Kyokaishi,* 45, 1984.

11. C. P. Rader, "Thermoplastic Elastomers," *RAPRA Review Reports,* R. Meredith (ed.), vol. 1, no. 3, Pergamon Press, London, 1987.

12. A. D. Thorn, "Thermoplastic Elastomers: A Review of Current Information," Rubber and Plastics Research Association of Great Britain, Shawbury, United Kingdom, 1980.

13. C. P. Rader and J. Stemper, *Progress in Rubber and Plastics Technology,* 50, 1990.

14. Laboratoire de Recherche et de Controle du Caoutchouc, "Etude Bibliographique des Caoutchouc Thermoplastiques," *Rapport Technique,* 122, Montrogue, France, 1980.

15. N. R. Legge, S. Davison, H. E. De Le Mare, G. Holden, and M. K. Martin, "Block Polymers and Related Materials," R. W. Tess and G. G. Poehlein (eds.), *ACS Symposium Series 285,* American Chemical Society, Washington, D.C., 1985.

16. American Chemical Society, *Thermoplastic Elastomers Symposium, 121st National Meeting of the Rubber Division,* Philadelphia, PA, 1982.

17. American Chemical Society, *Thermoplastic Elastomers Symposium, 127th National Meeting of the Rubber Division,* Los Angeles, CA, 1985.

18. American Chemical Society, *Thermoplastic Elastomers Symposium, 134th National Meeting of the Rubber Division,* Cincinnati, OH, 1988.

19. Society of Plastics Engineers, *Thermoplastic Elastomers and Engineering Properties and Structure Symposium, ANTEC '88, 46th Annual Technical Conference,* Atlanta, GA, 1988.

20. "Thermoplastic Elastomers—Threat or Opportunity?," *Symposium, European Rubber Journal and RAPRA Technology Ltd.,* London, 1988.

21. *Thermoplastic Elastomers—II. Processing for Performance Symposium (European Rubber Journal and RAPRA Technology Ltd.),* London, Apr. 7, 1989.

22. Schotland Business Research, *TPE '88, First International Conference on TPE Markets and Technology,* Orlando, FL, Mar. 9–11, 1988.

23. American Chemical Society, *Thermoplastic Elastomers Symposium, 138th National Meeting of the Rubber Division,* Washington, D.C., 1990.

24. American Chemical Society, *Thermoplastic Elastomers Symposium, 142nd National Meeting of the Rubber Division,* Nashville, TN, 1992.

25. American Chemical Society, *Thermoplastic Elastomers Symposium, 146th National Meeting of the Rubber Division,* Pittsburgh, PA, 1994.

26. Schotland Business Research, *TPE '89, Second International Conference on TPE Markets and Products,* Orlando, FL, Mar. 16–18, 1989.

27. Schotland Business Research, *TPE '90, Third International Conference on TPE Markets and Products,* Detroit, MI, Mar. 28–30, 1990.

28. Schotland Business Research, *TPE '94, Fourth International Conference on TPE Markets and Products,* Orlando, FL, Feb. 13–15, 1991.

29. *TPE Europe '91,* Schotland Business Research, Luxembourg, Oct., 1991.

30. Society of Plastics Engineers, *TPE RETEC '93,* Cincinnati, OH, Oct. 19–20, 1993.

31. Society of Plastics Engineers, *TPE RETEC '95,* Wilmington, DE, Sept. 26–27, 1995.

32. *Rubber and Plastics News,* Oct. 31, 1988.

33. C. S. Schollenberger, H. S. Scott, and G. R. Moore, *Rubber World,* 549, 1958.

34. O. Bayer, H. Rinke, L. Siefken, L. Ortner, and H. Schild, German Patent 728,981, 1937.

35. N. R. Legge, *Chemtech,* 630, 1983.

36. M. Brown and W. K. Witsiepe, *Rubber Age,* 35, 1972.

37. J. R. Wolfe, chap. 6 in *Thermoplastic Elastomers—A Comprehensive Review,* N. R. Legge, G. Holden, and H. E. Schroeder (eds.), Hanser Publications, Munich, Germany, 1987.

38. G. E. O'Connor and M. A. Fath, *Rubber World,* 26, 1982.

39. R. Eller, in "Modern Plastics Encyclopedia," McGraw-Hill, New York, NY, 1993, p. 91; L. White, *European Rubber Journal,* May, 1995, p. 24; R. J. School, chap. 15 in *Elastomer Technology Handbook,* N. P. Cheremisinoff (ed.), CRC Press, Boca Raton, FL, 1993; C. P. Rader, "Thermoplastic Elastomers: Bridging the Gap Between Rubber and Plastics," Decision Resources, Burlington, MA, 1991; *Rubber and Plastics News,* July 27, 1992, p. 6; R. J. School, Paper 11 in *146th National Meeting of the Rubber Division,* American Chemical Society, Pittsburgh, PA, 1994.

40. C. P. Rader in *First International Conference on Thermoplastic Elastomer Markets and Technology,* Schotland Business Research, Orlando, FL, 1988.

41. M. Alderson and M. T. Payne, *Rubber World,* May, 1993, p. 22.

42. A. Y. Coran, chap. 7 in *Science and Technology of Rubber,* F. R. Eirich (ed.), Academic Press, Inc., New York, 1978.

43. G. Holden and N. R. Legge, chap. 3 in *Thermoplastic Elastomers—A Comprehensive Review,* N. R. Legge, G. Holden, and H. E. Schroeder (eds.), Hanser Publications, Munich, Germany, 1987.

44. W. M. Halper and G. Holden, chap. 2 in *Handbook of Thermoplastic Elastomers,* 2d ed., B. M. Walker and C. P. Rader (eds.), Van Nostrand Reinhold, New York, NY, 1988.

45. R. K. Adams and G. K. Hoeschele, chap. 8 in *Thermoplastic Elastomers—A Comprehensive Review,* N. R. Legge, G. Holden, and H. E. Schroeder (eds.), Hanser Publications, Munich, Germany, 1987.

46. T. W. Sheridan, chap. 6 in *Handbook of Thermoplastic Elastomers,* 2d ed., B. M. Walker and C. P. Rader (eds.), Van Nostrand Reinhold, New York, 1988.

47. M. W. Bednarik, B. D. Wilson, R. E. Lietz, and C. P. Rader in *1990 SAE International Congress and Exposition,* Detroit, 1990.

48. E. C. Ma, chap. 7 in *Handbook of Thermoplastic Elastomers,* 2d ed., B. M. Walker and C. P. Rader (eds.), Van Nostrand Reinhold, New York, 1988.

49. W. Meckel, M. Goyert, and W. Wieder, chap. 2 in *Thermoplastic Elastomers—A Comprehensive Review,* N. R. Legge, G. Holden, and H. E. Schroeder (eds.), Hanser Publications, Munich, Germany, 1987.

50. A. T. Chen, C. P. Smith, J. M. O'Connor, and R. R. Wells, Paper 25 in *Fall Meeting of the Rubber Division,* American Chemical Society, Pittsburgh, PA, Oct. 11–14, 1994.

51. C. Hepburn, "Polyurethane Elastomers," Applied Science Publishers, London, England, p. 355, 1982.

52. W. J. Farrissey and T. M. Shah, chap. 8 in *Handbook of Thermoplastic Elastomers,* 2d ed., B. M. Walker and C. P. Rader (eds.), Van Nostrand Reinhold, New York, 1988.

53. *Rubber and Plastics News,* Jan. 22, 1990, p. 26.

54. R. A. Ranalli, chap. 2 in *Developments in Rubber Technology—3: Thermoplastic Elastomers,* A. Whelan and K. S. Lee (eds.), Applied Science Publishers, New York, 1982.

55. C. D. Shedd, chap. 3 in *Handbook of Thermoplastic Elastomers,* 2d ed., B. M. Walker and C. P. Rader (eds.), Van Nostrand Reinhold, New York, 1988.

56. S. Danesi and E. Garagnani, *Kautschuk und Gummi Kunststoffe,* 195, 1984.

57. A. F. M. Barton, *Handbook of Solubility Parameters and Other Cohesion Parameters,* CRC Press, Boca Raton, FL, 1983, p. 1.

58. A. M. Gessler and W. H. Haslett, Jr., U.S. Patent 3,037,954, June 5, 1962.

59. W. K. Fischer, U.S. Patent 3,758,643, Sept. 11, 1973; U.S. Patent 3,835,201, Sept. 10, 1974; U.S. Patent 3,862,106, Jan. 21, 1975.

60. C. P. Rader, chap. 4 in *Handbook of Thermoplastic Elastomers,* 2d ed. B. M. Walker and C. P. Rader (eds.), Van Nostrand Reinhold, New York, 1988.

61. T. Burton, J. L. Delanaye, and C. P. Rader, *Rubber and Plastics News,* Dec. 12, 1988.

62. A. Y. Coran, chap. 7 in "Thermoplastic Elastomers—A Comprehensive Review," N. R. Legge, G. Holden, and H. E. Schroeder (eds.), Hanser Publications, Munich, Germany, 1987.

63. S. Abdou-Sabet and M. A. Fath, U.S. Patent 4,311,628, Jan. 19, 1982; A. Y. Coran and R. P. Patel, U.S. Patent 4,104,210, Aug. 1, 1978; U.S. Patent 4,130,524, Dec. 19, 1978; A. Y. Coran, B. Das, and R. P. Patel, U.S. Patent 4,130,535, Dec. 19, 1978.

64. A. Y. Coran and R. P. Patel, Paper 41 in *Fall Meeting of the Rubber Division,* American Chemical Society, Nashville, TN, Nov. 3–6, 1992; J. M. A. Alvarez, E. F. Miron, and J. M. T. Lopez, U.S. Patent 5,198,496, Mar. 30, 1993; R. Anderlink and H. G. Fritz, *Kautschuk und Gummi Kunststoffe,* July, 1992, p. 527.

65. S. Abdou-Sabet, Y. L. Wang, and E. F. Chu, *Rubber and Plastics News,* Nov. 4, 1985.

66. R. C. Puydak, D. R. Hazelton, B. A. Graham, and N. R. Dharmarajan, U.S. Patent 5,100,947, Mar. 31, 1992; A. Y. Coran and R. P. Patel, U.S. Patent 4,130,534, Dec. 19, 1978.

67. P. E. F. Cudby, J. Patel, and A. J. Tinker in *Fall Meeting of the Rubber Division,* American Chemical Society, Nashville, TN, Nov. 3–6, 1992; R. S. George and R. Joseph, *Kautschuk und Gummi Kunststoffe,* Nov., 1994, p. 816.

68. D. R. Hazelton, R. C. Puydak, and D. A. Booth, U.S. Patent 5,086,121, Feb. 4, 1992.

69. A. Y. Coran, R. P. Patel, and D. Williams, *Rubber Chemistry and Technology,* vol. 55, p. 116, 1982; ibid. 58, p. 1014, 1985.

70. A. Y. Coran and R. P. Patel, *Rubber Chem. Technol.,* vol. 56, p. 1045, 1983.

71. S. Abdou-Sabet and R. P. Patel, *Rubber Chem. Technol.,* vol. 64, p. 769, 1991.

72. A. Y. Coran and R. P. Patel, *Rubber Chem. Technol.,* vol. 53, p. 141, 1980.

73. C. P. Rader and K. E. Kear, *Rubber and Plastics News,* May 6, 1986.

74. *Plastics Engineering,* June, 1995, p. 21; B. Vernyi, *Plastics News,* June 26, 1995, p. 1; J. H. Schut, *Plastics World,* May, 1995, p. 47; *Modern Plastics,* Oct., 1994, p. 23.

75. M. Stockdale, Paper 36 in *Fall Meeting of the Rubber Division,* American Chemical Society, Washington, D.C., Oct. 9–12, 1990.

76. S. Al-Malaika and E. J. Amir, *Polymer Degradation and Stability,* 347, 1986; B. Kuriakose and S. K. De, *International Journal of Polymeric Materials,* 101, 1986; B. Kuriakose and S. K. De, *Polymer Engineering and Science,* 630, 1985.

77. Daikin Kogyo Co. Ltd., *Japanese Chemical Week,* Oct. 21, 1982; H. Kamiya and M. Saito, U.S. Patent 5,354,811, Oct. 11, 1994.

78. P. A. Mancinelli and S. O. Norris, *Adhesives Age,* Sept., 1985; P. Nicholas, U.S. Patent 5,244,978, Sept. 14, 1993.

79. Dow Corning, Ltd., *Plastics and Rubber Weekly,* Nov. 27, 1982.

80. C. P. Rader and J. R. Richwine, *Rubber and Plastics News,* Feb. 11, 1985.

81. P. D. Gage, Paper 56 in *Fall Meeting of the Rubber Division,* American Chemical Society, Washington, D.C., Oct. 9–12, 1990.

82. D. V. Rosato and D. V. Rosato, *Injection Molding Handbook,* Van Nostrand Reinhold Co., New York, 1986.

83. I. I. Rubin, *Injection Molding Theory and Practice,* John Wiley & Sons, Inc., New York, 1972.

84. J. G. D'Auteuil, D. E. Peterson, and C. P. Rader, *Journal of Elastomers and Plastics,* vol. 265, 1989.

85. J. G. D'Auteuil, *Journal of Polymer Engineering,* Dec., 1990.

86. E. Van Issum, D. E. Peterson, and B. K. Weider, Presented at ANTEC '86, 43rd Annual Technical Conference, Society of Plastics Engineers, Boston, MA, 1986.

87. D. E. Peterson, R. L. Arnold, and G. L. Dumbauld, *Rubber World,* Dec., 1992, p. 19.

88. G. L. Dumbauld, U.S. Patent 5,070,111, Dec. 3, 1991.

89. C. P. Rader, Paper 27 in *Fall Meeting of the Rubber Division,* American Chemical Society, Nashville, TN, Nov. 3–6, 1992.

90. N. R. Legge and R. C. Mowbray, *Rubber World,* Oct., 1994, p. 43; R.G. Wheeler, U.S. Patent 5,360,590, Nov. 1, 1994; S. Walters, *Rubber and Plastics News,* Feb. 14, 1994, p. 1.

91. P. C. Killgoar, Jr., in "Handbook of Thermoplastic Elastomers, Second Edition," B. M. Walker and C. P. Rader (eds.), Van Nostrand Reinhold, New York, NY, 1988, Chapter 10; P. C. Killgoar, Jr., Presented at 134th National Meeting of the Rubber Division, American Chemical Society, Cincinnati, 1988; M. A. Wright, N. R. Hamblin, and C. P. Rader, Chemtech, July, 1988, p. 354.

92. J. H. Muhs, chap. 12 in *Handbook of Thermoplastic Elastomers,* 2d ed., B. M. Walker and C. P. Rader (eds.), Van Nostrand Reinhold, New York, 1988.

93. J. L. Williams, chap. 14 in *Handbook of Thermoplastic Elastomers,* 2d ed., B. M. Walker and C. P. Rader (eds.), Van Nostrand Reinhold, New York, 1988.

94. T. Ouhadi and D. S. T. Wang in *Scotland Business Research Conference,* Luxembourg, Germany, Oct. 22–23, 1991.

95. C. P. Rader, S. Baldwin, D. D. Cornell, G. Sadler, and R. Stockel (eds.), *Recycling of Plastics, Rubber and Paper—A Pragmatic Perspective,* ACS Symposium Series, American Chemical Society, Washington, D.C., 1995.

96. M. M. Russo and C. P. Rader in *Thermoplastic Elastomers Symposium, Society of Plastics Engineers, RETEC,* Wilmington, DE, Sept. 26–27, 1995; C. P. Rader and R. C. Wegelin in *Northeast Regional Rubber and Plastics Exposition, Society of Plastics Engineers,* Mahwah, NJ, Sept. 22, 1994; E. P. Purgley, C. P. Rader, and E. A. Gonzalez, *Rubber and Plastics News,* Aug. 31, 1992, p. 15.

97. J. R. Serumgard and A. L. Eastman in *Recycling of Plastics, Rubber and Paper—A Pragmatic Perspective,* C. P. Rader, S. Baldwin, D. D. Cornell, G. Sadler, and R. Stockel (eds.), ACS Symposium Series, American Chemical Society, Washington, D.C., 1995.

CHAPTER 6
PROTECTIVE AND DECORATIVE COATINGS

Carl P. Izzo
Industry Consultant
Export, Pennsylvania

6.1 INTRODUCTION

The history of protective and decorative organic coatings is almost as old as the history of humankind. From cave dwellers decorating their walls with earth pigments ground in egg whites to factory workers protecting products with E-coat primers and urethane acrylic enamels, these coatings are still composed of film-forming vehicles, pigments, solvents, and additives.[1] Significant changes have been made in the vehicles, which are the film-forming resinous portions of the coating.[2] Since the introduction of phenolic synthetic resin vehicles in the 1900s, coatings have been designed to increase production and meet performance requirements at lower costs. These developments were highlighted by the introduction of nitrocellulose lacquers for the automotive and furniture industries, followed by the alkyds, epoxies, vinyls, polyesters, acrylics, a host of other resins, and finally the polyurethanes. The first fifty years of the 20th century were the decades of discovery.

In the 1960s, the decade of technology, just as coatings were becoming highly developed, another variable, environmental impact, was added to the equation. Los Angeles County officials, who found that volatile organic compound (VOC) emissions were a major source of air pollution, enacted Rule 66 to control the emission of solvents that cause photochemical smog. These solvents, which are used in coatings for viscosity and flow control, evaporate during application and cure and are emitted to the atmosphere. To comply with Rule 66, the paint industry reformulated its coatings using exempt solvents, which presumably did not produce smog. California's Rule 66 was followed by other local air-quality standards and finally by the establishment of the U.S. Environmental Protection Agency (EPA), the charter of which, under the law, is to improve air quality by reducing solvent emissions.

In the 1970s, the decade of conservation, the energy crisis resulted in shortages and price increases for solvents and synthetic-coating-resin raw materials. Also affected was the distribution of natural gas, the primary fuel for curing ovens, which caused shortages and price increases. In response to those pressures, the coatings industry developed low-temperature curing coatings in an effort to reduce gas consumption.

The 1980s, the decade of restriction, saw an end to the energy crisis and the beginning of more restrictive air-quality standards. However, energy costs remained high. The importance of transfer efficiency, the percentage of an applied coating that actually coats the product, was recognized by industry and the EPA. This led to the development and use of coatings application equipment and coatings methods having higher transfer efficiencies. The benefits of using higher-transfer-efficiency coating methods are threefold—reduced coating material usage, lower solvent emissions, and lower costs.

The 1990s are considered the decade of compliance. Resin and coating suppliers have developed compliance coatings—waterborne, high solids, and powder. Equipment suppliers have developed devices to apply these new coatings. Today the coatings and equipment suppliers' investments in research and development are paying dividends. Coating materials and application equipment have improved. Primers are applied by electrocoating. One-coat finishes are replacing two coats in many cases. High-solids and waterborne liquid coatings are replacing conventional solvent-thinned coatings. Powder coatings usage has increased dramatically. Radiation-cured coatings and vapor-cured coatings are finding more applications. Coatings and solvent usage as well as coating-application costs are reduced. Air-quality standards are being met.

Coatings today are considered engineering materials. Their performance characteristics must not only match service requirements, they must also meet governmental regulations. In the past, the selection of a coating depended mainly on the service requirements and application method. Now, more than ever before, worker safety, environmental impact, and economics must be considered.

Coatings are applied to most industrial products by spraying. Figure 6.1 shows a typical industrial spray booth. In 1890 Joseph Binks invented the cold-water paint-spraying machine, the first airless sprayer, which was used to apply whitewash to barns and other building interiors. In 1924 Thomas DeVilbiss used a modified medical atomizer, the first air-atomizing sprayer, to apply a nitrocellulose lacquer on the Oakland automobile. Since then, these tools have remained virtually unchanged, and, until the enactment of the air-quality standards, they were used to apply coatings at 25 to 50 percent volume solids at transfer efficiencies of 30 to 50 percent. Using this equipment, the remainder of the nonvolatile material—the overspray—coated the floor and walls of spray booths and became hazardous or nonhazardous waste, while the solvents—the VOCs—evaporated from the coating during application and cure to become pollutants. Now finishes are applied by highly transfer-efficient application equipment.

Even the best coatings will not perform their function if they are not applied on properly prepared substrates. For this reason surfaces must first be cleaned by removing oily soils, corrosion products, and particulates, and then pretreated before the application of any protective and decorative coating.

After coatings are applied, they form films and cure. Curing mechanisms can be as simple as solvent evaporation or as complicated as free-radical polymerization. Basically, coatings can be classified as baking or air drying, which usually means room-temperature curing. The curing method and times are important in coating selection, because they must be considered for production schedules and equipment.

The purposes of this chapter are threefold: (1) to stress the importance of environmental compliance in coating operations; (2) to acquaint the reader with surface preparation, coating materials, application equipment, and curing methods; and (3) to aid in the selection of coating materials.

FIGURE 6.1 Spray application is the most widely used method for applying coatings to industrial products. (*Courtesy of American Iron and Steel Institute.*)

6.2 ECOLOGY

In the past, changes in coating materials and coating application lines were discussed only when lower prices, novel products, new coating lines, or new plants were considered. Today, with rising material costs, rising energy costs, and more restrictive governmental regulations, they are the subject of frequent discussions. Coating material and solvent costs, which are tied to the price of crude oil, have risen since the 1970s, as has the cost of natural gas, which is the most used fuel for coating bake ovens. The EPA has imposed restrictive air-quality standards. The Occupational Safety and Health Act (OSHA) and the Toxic Substances Control Act (TOSCA) regulate the environment in the workplace and limit workers' contact with hazardous materials. These factors have increased coating costs and the awareness of product finishers. To meet the challenge, they must investigate and use alternative coating materials and processes for compliance and cost-effectiveness.

Initial attempts to control air pollution in the late 1940s resulted in smoke control laws to reduce airborne particulates. The increased use of the automobile and industrial expansion during that period caused a condition called photochemical smog (smog created by the reaction of chemicals in the atmosphere to sunlight) in major cities throughout the United States. Los Angeles County officials recognized that automobile exhaust and VOC emissions were major sources of smog, and they enacted an air-pollution regulation called Rule 66. Rule 66 forbade the use of specific solvents that produced photochemical smog and published a list of exempt solvents for use in coatings. Further study by the EPA has shown that, if given enough time, even the Rule-66-exempt solvents will produce photochemical smog in the atmosphere.

The Clean Air Act of 1970 and its 1990 amendments, formulated by the EPA, established national air-quality standards that regulate the amount of solvents emitted. The EPA divided the 50 states into 250 air-quality regions, each of which is responsible for the implementation of the national air-quality standards. It is important to recognize that many of the local standards are more stringent than the national. For this reason, specific coatings that comply with the air-quality standards of one district may not comply with another's. Waterborne, high-solids, powder, electrophoretic, and radiation-cured coatings will comply. The use of precoated metal can eliminate all the compliance problems.

Not only because the EPA mandates the reduction of VOC emissions, but also because of economic advantages, spray painting, which is the most used application method, must be done more efficiently. The increased efficiency will reduce the amount of expensive coatings and solvents used, thereby reducing production costs.

6.3 SURFACE PREPARATION

The most important step in any coating operation is surface preparation, which includes cleaning and pretreatment. For coatings to adhere, surfaces must be free from oily soils, corrosion products, and loose particulates. New wood surfaces are often coated without cleaning. Old wood and coated wood must be cleaned to remove oily soils and loose, flaky coatings. Plastics are cleaned by using chemicals to remove mold release. Metals are cleaned by media blasting, sanding, brushing, and by solvents or aqueous chemicals. The choice of a cleaning method depends on the substrate and the size and shape of the object.

After cleaning, pretreatments are applied to enhance coating adhesion and, in the case of metals, corrosion resistance. Some wood surfaces require no pretreatment, while others require priming of knots and filling of nail holes. Cementitious and masonry substrates are pretreated, using acids, to remove loosely adhering contaminants and to passivate the surfaces. Some plastic surfaces may be paintable after cleaning to remove mold release and other contaminants, while others may require priming, spark discharge, or chemical etching to ensure coating adhesion. Metals, the most common industrial substrates, are generally pretreated using phosphates, chromates, and oxides to passivate their surfaces and provide corrosion resistance. Since most of the industrial substrates coated are metals, their cleaning and pretreatment are described in more detail in the next sections. As with the choice of a cleaning method, the choice of a pretreatment method depends on the size and shape of the object.

6.3.1 Metal Surface Preparation

Oily soils must be removed before any other surface preparation is attempted. Otherwise these soils may be spread over the surface. These soils can also contaminate abrasive cleaning media and tools. Oily soils can be removed faster using liquid cleaners that impinge on the surface or in agitated immersion baths. It is often necessary to heat liquid cleaners to facilitate soil removal.

Abrasive Cleaning.   After removal of the oily soils, surfaces are abrasive-cleaned to remove rust and corrosion by media blasting, hand or power sanding, and hand or power brushing. Media blasting consists of propelling by gases under pressure, materials such as sand, metallic shot, nut shells, plastic pellets, and dry-ice crystals so that they impinge on the surfaces to be cleaned. High-pressure water-jet cleaning is similar to media blasting.

Alkaline Cleaning.   To remove oily soils, aqueous solutions of alkaline phosphates, borates, and hydroxides are applied to metals by immersion or spray. After cleaning, the surfaces are rinsed with clear water to remove the alkali. These materials are not effective for removing rust and corrosion.

Detergent Cleaning.   Aqueous solutions of detergents are used to remove oily soils in much the same way as alkaline cleaners. Then they are rinsed with cold water to flush away the soils.

Emulsion Cleaning.   Heavy oily soils and greases are removed by aqueous emulsions of organic solvents such as mineral spirits and kerosene. After the emulsified solvent has dissolved the oily soils, they are flushed away using a hot-water rinse. The remaining oily residue must be removed using clean solvent, alkaline, or detergent cleaners.

Solvent Cleaning.   Immersion, hand wiping, and spraying using organic solvents are effective methods for removing oily soils. Since these soils will contaminate solvents and wipers, it is important to change them frequently. Otherwise, oily residues will remain on substrates. Safe handling practices must be followed because of the hazardous nature of most organic solvents.

Steam Cleaning.   Detergent and alkaline cleaners applied using steam cleaners is a well known degreasing method. The impingement of the steam and the action of the chemicals will dissolve and flush away heavy greases and waxes. Hot-water spray cleaning using chemicals is nearly as effective as steam cleaning.

Vapor Degreasing.   Vapor degreasing has been a very popular cleaning method for removing oily soils. Boiling solvent condenses on the cool surface of the product and flushes away oily soils, but does not remove particulates. Since this process uses chlorinated solvents, which are under regulatory scrutiny by governmental agencies, its popularity is declining.

6.3.2 Metal Surface Pretreatment

Cleaning will remove oily soils but will generally not remove rust and corrosion from substrates to be coated. Abrasive cleaning will remove corrosion, and for this reason it is also considered a pretreatment, because the impingement of blasting media and the action of abrasive pads and brushes roughen the substrate and therefore enhance adhesion. The other pretreatments use aqueous chemical solutions, which are applied by immersion or spray techniques. Pretreatments for metallic substrates used on industrial products are discussed in this section. Because they provide corrosion protection to ferrous and nonferrous metals, chromates are used in pretreatment stages and as conversion coatings. They are being replaced by nonchromate chemicals.

Aluminum.   Aluminum is cleaned to remove oily soils and corrosion products by solvents and chemical solutions. Cleaned aluminum is pretreated using chromate conversion coating and anodizing. Phosphoric-acid-activated vinyl wash primers, which are also considered pretreatments, must be applied directly to metal and not over other pretreatments.

Copper. Copper is cleaned by solvents and chemicals and then abraded to remove corrosion. Bright dipping in acids will also remove corrosion. Cleaned surfaces are often pretreated using chromates and vinyl wash primers.

Galvanized Steel. Galvanized steel must be cleaned to remove the oil or wax that is applied at the mill to prevent white corrosion. After cleaning, the surfaces are pretreated using chromates and phosphates. Vinyl wash primer pretreatments can also be applied on galvanized steel surfaces having no other pretreatments.

Steel. Steel surfaces are cleaned to remove oily soils and, if necessary, pickled in acid to remove rust. Clean steel is generally phosphate-pretreated to provide corrosion resistance. Other pretreatments for steel are chromates and wash primers.

Stainless Steel. Owing to its corrosion resistance, stainless steel is usually not coated. Otherwise the substrate must be cleaned to remove oily soils and then abraded to roughen the surface. Wash primers will enhance adhesion.

Titanium. Cleaned titanium is pretreated like stainless steel.

Zinc and Cadmium. Zinc and cadmium substrates are pretreated like galvanized steel.

6.4 COATING SELECTION

To aid in their selection, coatings will be classified by use in finish systems, physical state, and resin type. Finish systems can be one-coat or multicoat that use primers, intermediate coats, and top coats. Primers provide adhesion and corrosion protection to substrates. Top coats provide weather, chemical, and physical resistance, and generally determine the performance characteristics of finish systems. Performance properties for coatings formulated with the most commonly used resins are shown in Table 6.1.

In coating selection, intended service conditions must be considered. To illustrate this point, consider the differences between service conditions for toy boats

TABLE 6.1 Performance Properties* of Common Coating Resins[3]

Resin type	Humidity resistance	Corrosion resistance	Exterior durability	Chemical resistance	Mar resistance
Acrylic	E	E	E	G	E
Alkyd	F	F	P	G	G
Epoxy	E	E	G	E	E
Polyester	E	G	G	G	G
Polyurethane	E	G	E	G	E
Vinyl	E	G	G	G	G

* E—excellent; G—good; F—fair; P—poor.

and for battleships. Table 6.2 shows the use of coating finish systems in various service conditions.

It is not the intention of this chapter to instruct the reader in the chemistry of organic coating but rather to aid in selecting coatings for specific applications. Therefore the coating resin's raw materials feed stock and polymerization reactions will not be discussed. On the other hand, generic resin types, curing, physical states, and application methods are discussed.

A resin's physical state can help determine the application equipment required. Solid materials can be applied by powder coating methods. Table 6.3 lists resins applied as powder coatings. Liquids can be applied by most of the other methods, which are discussed later. Many of the coating resins exist in several physical states. Table 6.4 lists the physical states of common coating resins.

It is important to realize that in selecting coatings, tables of performance properties of generic resins must be used only as guides, because coatings of one generic type, such as acrylic, epoxy, or polyurethane, are often modified using one or more of the other generic types. Notable examples are acrylic alkyds, acrylic urethanes, acrylic melamines, epoxy esters, epoxy polyamides, silicone alkyds, silicone epoxies,

TABLE 6.2   Typical Industrial Finish Systems[3]

Service conditions	Primer	One-coat enamel	Intermediate coat	Top coat
Interior				
Light duty		×		
Heavy duty	×			×
Exterior				
Light duty		×		
Heavy duty	×			×
Extreme duty	×		×	×

TABLE 6.3   Plastics Used in Powder Coatings

	Fluidizing conditions			Fluidized-bed powder		
		Cure or fusion				
Resin	Preheat temperature, °F	Temperature, °F	Time, min	Maximum operating temperature, °F	Adhesion	Weather resistance
Epoxy	250–450	250–450	1–60	200–400	Excellent	Good
Vinyl	450–550	400–600	1–3	225	Poor	Good
Cellulose acetate butyrate	500–600	400–550	1–3	225	Poor	Good
Nylon	550–800	650–700	1	300	Poor	Fair
Polyethylene	500–600	400–600	1–5	225	Fair	Good
Polypropylene	500–700	400–600	1–3	260	Poor	Good
Penton	500–650	450–600	1–10	350	Poor	Good
Teflon	800–1000	800–900	1–3	500	Poor	Good

TABLE 6.4 Physical States of Common Coating Resins[3]

Resin type	Conv. solvent	Waterborne	High solids	Powder coating	100% solution liquid	Two-component liquid
Acrylic	x	x	x	x		
Alkyd	x	x	x			
Epoxy	x	x	x	x	x	x
Polyester	x		x	x	x	x
Polyurethane	x	x	x	x	x	x
Vinyl	x	x	x	x	x	

silicone polyesters, vinyl acrylics, and vinyl alkyds. While predicting specific coating performance properties of neat resins is simple, predicting the properties of modified resins is difficult, if not impossible. Parameters causing these difficulties are resin modification percentages and modifying methods such as simple blending or copolymerization. The performance of a 30 percent copolymerized silicone alkyd is not necessarily the same as one which was modified by blending. These modifications can change the performance properties subtly or dramatically.[3]

There are nearly 1000 coatings manufacturers in the United States, each having various formulations that could number in the hundreds. Further compounding the coating selection difficulty is the well known practice of some coating manufacturers that adds small amounts of a more expensive, better performing resin to a less expensive, poorer performing resin and calls the product by the name of the former. An unsuspecting person, whose choice of such a coating is based on properties of the generic resin, can be greatly disappointed. Instead, selections must be made on the basis of performance data for specific coatings or finish systems. Performance data are generated by the paint and manufacturing industries during the conduct of standard paint evaluation tests. Test methods for coating material evaluation are listed in Table 6.5.

Table 6.6 shows the electrical and physical properties, environmental resistance, flammability, repairability, film formation method, application method, and typical uses of coatings for most of the available resins. This information can be used to guide the selection of coatings. However, it is important to remember the aforementioned warnings.

6.5 COATING MATERIALS

Since it is the resin in the coating's vehicle that determines its performance properties, coatings can be classified by their resin types. The most widely used resins for manufacturing modern coatings are acrylics, alkyds, epoxies, polyesters, polyurethanes, and vinyls.[3] In this section the resins used in coatings are described.

6.5.1 Common Coating Resins

Acrylics. Acrylics are noted for color and gloss retention in outdoor exposure. Acrylics are supplied as solvent-containing, high-solids, waterborne, and powder

TABLE 6.5 Specific Test Methods for Coatings

Test	ASTM	Fed. STD. 141a, method	MIL-STD-202, method	Fed. STD. 406, method	Others
Abrasion	D 968	6191 (Falling Sand) 6192 (Taber)	....	1091	Fed. Std. 601, 14111
Adhesion	D 2197	6301.1 (Tape Test, Wet) 6302.1 (Microknife) 6303.1 (Scratch Adhesion) 6304.1 (Knife Test)	....	1111	Fed. Std. 601, 8031
Arc resistance	D 495	....	303	4011	
Dielectric constant	D 150	....	301	4021	Fed. Std. 101, 303
Dielectric strength (breakdown voltage)	D 149 D 115	....	....	4031	Fed. Std. 601, 13311
Dissipation factor	D 150	....	....	4021	
Drying time	D 1640 D 115	4061.1			
Electrical insulation resistance	D 229 D 257	....	302	4041	MIL-W-81044, 4.7.5.2
Exposure (exterior)	D 1014	6160 (On Metals) 6161.1 (Outdoor Rack)			
Flash point	D 56, D 92 D 1310 (Tag Open Cup)	4291 (Tag Closed Cup) 4294 (Cleveland Open Cup)	....	....	Fed. Std. 810, 509
Flexibility		6221 (Mandrel) 6222 (Conical Mandrel)	....	1031	Fed. Std. 601, 11041
Fungus resistance	D 1924	....	....	....	MIL-E-5272, 4.8 MIL-STD-810, 508.1 MIL-T-5422, 4.8
Hardness	D 1474	6211 (Print Hardness) 6212 (Indentation)			
Heat resistance	D 115 D 1932	6051			
Humidity	D 2247	6071 (100% RH) 6201 (Continuous Condensation)	103 106A	....	MIL-E-5272, Proc. 1 Fed. Std. 810, 507
Impact resistance		6226 (G.E. Impact)	1074		
Moisture-vapor permeability	E 96 D 1653	6171	7032		
Nonvolatile content		4044			
Salt spray (fog)	B 117	6061	101C	6071	MIL-STD-810, 509.1 MIL-E-5272, 4.6 Fed. Std. 151, 811.1 Fed. Std. 810, 509
Temperature-altitude			....	....	MIL-E-5272, 4.14 MIL-T-5422, 4.1 MIL-STD-810, 504.1
Thermal conductivity	D 1674 (Cenco Fitch) C 177 (Guarded Hot Plate)	....	....	....	MIL-I-16923, 4.6.9
Thermal shock		....	107	....	MIL-E-5272, 4.3 MIL-STD-810, 503.1
Thickness (dry film)	D 1005 D 1186	6181 (Magnetic Gage) 6183 (Mechanical Gage)	....	2111, 2121, 2131, 2141, 2151	Fed. Std. 151, 520, 521.1
Viscosity	D 1545 D 562 D 1200 D 88	4271 (Gardner Tubes) 4281 (Krebs-Stormer) 4282 (Ford Cup) 4285 (Saybolt) 4287 (Brookfield)			
Weathering (accelerated)	D 822	6151 (Open Arc) 6152 (Enclosed Arc)	....	6024	

A more complete compilation of test methods is found in J. J. Licari, *Plastic Coatings for Electronics*, McGraw-Hill Book Company, New York, 1970.

The major collection of complete test methods for coatings is *Physical and Chemical Examination of Paints, Varnishes, Lacquers, and Colors*, by Gardner and Sward, Gardner Laboratory, Bethesda, MD. This has gone through many editions.

TABLE 6.6 Properties of Coatings by Polymer Type

Coating type	Electrical properties				Maximum continuous service temperature, °F	Physical characteristics			
	Volume resistivity, ohm-cm (ASTM D 257)	Dielectric strength, volts/mil	Dielectric constant	Dissipation factor		Adhesion to metals	Flexibility	Approximate Sward hardness (higher number is harder)	Abrasion resistance
Acrylic	10^{14}–10^{15}	450–550	2.7–3.5	0.02–0.06	180	Good	Good	12–24	Fair
Alkyd	10^{14}	300–350	4.5–5.0	0.003–0.06	200 250 T.S.	Excellent	Fair to good Low temperature—poor	3–13 (air dry) 10–24 (bake)	Fair
Cellulosic (nitrate butyrate)		250–400	3.2–6.2		180	Good	Good Low temperature—poor	10–15	
Chlorinated polyether (Penton*)	10^{15}	400	3.0	0.01	250	Excellent	Good		
Epoxy-amine cure	10^{14} at 30°C 10^{10} at 105°C	400–550	3.5–5.0	0.02–0.03 at 30°C	350	Excellent	Fair to good Low temperature—poor	26–36	Good to excellent
Epoxy-anhydride, Dicy		650–730	3.4–3.8	0.01–0.03	400	Excellent	Good to excellent Low temperature—poor	20	Good to excellent
Epoxy-polyamide	10^{14} at 30°C 10^{10} at 105°C	400–500	2.5–3.0	0.008–0.02	350	Excellent	Good to excellent Low temperature—poor	20	Fair to good
Epoxy-phenolic	10^{12}–10^{13}	300–450			400	Excellent	Good Low temperature—fair		Good to excellent
Fluorocarbon TFE	10^{18}	430	2.0–2.1	0.0002	500	Can be excellent; primers required.	Excellent		
FEP CTFE	10^{18} 10^{18}	480 500–600	2.1 2.3–2.8	0.0003–0.0007 0.003–0.004	400 400	Can be excellent; primers required	Excellent		
Parylene (polyxylylenes)	10^{14}–10^{17}	700	2.6–3.1	0.0002–0.02	240°F (air) 510°F (inert atm.)	Good	Good		

Material	Resistivity	Dielectric strength	Dielectric constant	Dissipation factor	Temp.	Adhesion	Flexibility	Elongation	
Phenolics	10^9–10^{12}	100–300	4–8	0.005–0.5	350	Excellent	Poor to good Low temperature—poor	30–38	Fair
Phenolic-oil varnish					250	Excellent	Good Low temperature—fair		Poor to fair
Phenoxy	10^{13}–10^{14}	500	3.7–4.0	0.001	180	Excellent	Excellent		
Polyamide (nylon)	10^{13}–10^{15}	400–500	2.8–3.6	0.01–0.1	225–250	Excellent			
Polyester	10^{12}–10^{14}	500	3.3–8.1	0.008–0.04	200	Good on rough surfaces; poor to polished metals.	Fair to excellent	25–30	Good
Chlorosulfonated (polyethylene (Hypalon))†		400 3,000 (10 mil)	6–10	0.03–0.07	250	Good	Elastomeric	Less than 10	
Polyimide	10^{16}–10^{18}	500–700	3.4–3.8	0.003	500	Good	Fair to excellent		Good
Polystyrene	10^{10}–10^{19}	450–500 3,800 (1 mil)	2.4–2.6	0.0001–0.0005	140–180		Poor to fair		
Polyurethane	10^{12}–10^{13}	550	6.8 (1 kHz) 4.4 (1 MHz)	0.02–0.08	250	Often poor to metals. (Excellent to most non-metals.)	Good to excellent. Low temperature—poor	10–17 (castor oil) 50–60 (polyester)	
Silicone	10^{14}–10^{16}	300–800	3.0–4.2	0.001–0.008	500	Varies, but usually needs primer for good adhesion.	Excellent Low temperature—excellent	12–16	Fair to excellent
Vinyl chloride (poly-)	10^{11}–10^{15}	400	3–9	0.04–0.14	150	Excellent, if so formulated.	Excellent Low temperature—fair to good	5–10	
Vinyl chloride (plastisol, organisol)	10^{10}–10^{16}		2.3–9	0.10–0.15	150	Requires adhesive primer.	Excellent Low temperature—fair to good	3–6	
Vinyl fluoride	10^{12}–10^{14}	260 1,200 (8 mil)	6.4–8.4	0.05–0.15	300	Excellent, if fused on surface.	Excellent Low temperature—excellent		
Vinyl formal (Formvar‡)	10^{13}–10^{15}	850–1,000	3.7	0.007–0.2	200	Excellent			

* Trademark of Hercules Powder Co., Inc., Wilmington, DE.
† Trademark of E. I. du Pont de Nemours & Co., Wilmington, DE.
‡ Trademark of Monsanto Co., St. Louis, MO.

6.11

TABLE 6.6 Properties of Coatings by Polymer Type (*Continued*)

Coating type	Resistance to environmental effects						Film formation		Application method	Typical uses
	Chemical and solvent resistance	Moisture and humidity resistance	Weatherability	Resistance to microorganisms	Flammability	Repairability	Method of cure	Cure schedule		
Acrylic		Good	Excellent resistance to UV and weather	Good	Medium	Remove with solvent.	Solvent evaporation	Air dry or low-temperature bake	Spray, brush, dip	Coatings for circuit boards quick dry protection for markings and color coding.
Alkyd	Solvents—poor Alkalies—poor Dilute acids—poor to fair	Poor	Good to excellent	Poor	Medium	Poor	Oxidation or heat	Air dry or baking types	Most common methods	Painting of metal parts and hardware.
Cellulosic (nitrate butyrate)	Solvents—good Alkalies—good Acids—good	Fair		Poor to good	High	Remove with solvents.	Solvent evaporation	Air dry or low-temperature bake	Spray, dip	Lacquers for decoration and protection. Hot-melt coatings.
Chlorinate polyether (Penton*)		Good			Low		Powder or dispersion fuses	High temperature fusion	Spray, dip, fluid bed	Chemically resistant coatings.
Epoxy-amine cure	Solvents—good to excellent Alkalies—good Dilute acids—fair	Good	Pigmented—fair; clear—poor (chalks)	Good	Medium	No	Cured by catalyst reaction	Air dry to medium bake	Spray, dip, fluid bed	Coatings for circuit boards. Corrosion-protective coatings for metals.
Epoxy-anhydride, Dicy	Solvents—good Alkalies—good Dilute acids—	Good		Good	Medium	No	Cured by chemical reaction	High bakes 300 to 400°F	Spray, dip, fluid bed, impreg.	High-bake, high-temperature-resistant dielectric and corrosion coatings.
Epoxy-polyamide	Solvents—fair Alkalies—good Dilute acids—poor	Good		Good	Medium	No	Cured by coreactant	Air dry or medium bake	Spray, dip	Coatings for circuit boards. Filleting coating.
Epoxy-phenolic	Solvents—excellent Alkalies—fair Dilute acids—good	Excellent	Pigmented—fair; clear—poor	Good	Medium	No	Cured by coreactant	High bakes 300 to 400°F	Spray, dip	High-bake solvent and chemical resistant coating.
Fluocarbon TFE	Solvents—excellent Alkalies—good Dilute acids—excellent	Excellent		Good	None	No	Fusion from water or solvent dispersion	Approx. 750°F	Spray, dip	High-temperature-resistant insulation for wire.
FEP CTFE		Excellent		Good	None	No	Fusion from water or solvent dispersion	500–600°F	Spray, dip	High-temperature-resistant insulation. Extrudable.
Parylene (polysylylenes)		Excellent			None		Vapor phase deposition and polymerization requiring special license from Union Carbide			Very thin, pinhole-free coatings, possible semiconductable coating.

6.12

Material	Chemical resistance						Curing	Cure temperature	Application	Uses
Phenolics	Solvents—good to excellent, Alkalies—poor, Dilute acids—good	Excellent	Fair	Poor to good	Medium	No	Cured by heat	Bake 300–500°F	Spray, dip	High-bake chemical and solvent-resistant coatings.
Phenolic-oil varnish	Solvents—poor, Alkalies—poor, Dilute acids—good to excellent	Good	Good	Poor, unless toxic-additive	Medium	Poor	Oxidation or heat		Spray, brush, dip-impregnate	Impregnation of electronic modules, quick protective coating.
Phenoxy		Good		Good			Cured by heat			Chemical resistant coating.
Polyamide (nylon)		Fair		Good		Fairly solderable				Wire coating.
Polyester	Solvents—poor, Alkalies—poor to fair, Dilute acids—good	Fair	Very good	Good	Medium	Poor	Cured by heat or catalyst	Air dry or bake 100–250°F	Spray, brush, dip	
Chlorosulfonated polyethylene (Hypalon†)		Good		Good	Low		Solvent evaporation	Air dry or low temperature bake	Spray, brush	Moisture and fungus proofing of materials.
Polyimide	Solvents—excellent, Alkalies—, Dilute acids—	Good	Good	Good	Low	Poor	Cured by heat	High bake	Dip, impregnate, wire coater	Very high temperature resistant wire insulation.
Polystyrene		Good	Good	Good	High	Dissolve with solvents.	Solvent evaporation	Air dry or low bake	Spray, dip	Coil coating, low dielectric constant, low loss in radar uses.
Polyurethane	Solvents—good, Dilute alkalies—fair, Dilute acids—good	Good	Good	Poor to good	Medium	Excellent; melts, solder-through properties	Coreactant or moisture cure	Air dry to medium bake	Spray, brush, dip	Conformal coating of circuitry, solderable wire insulation.
Silicone	Solvents—poor, Alkalies—good (dilute) poor (concentrated), Dilute acids—good	Excellent	Excellent	Good	Very low (except in O₂ atm.)	Fair to excellent. Cut and peel.	Cured by heat or catalyst	Air dry (RTV) to high bakes	Spray, brush, dip	Heat-resistant coating for electronic circuitry. Good moisture resistance.
Vinyl chloride (poly-)	Solvents—alcohol, good, Alkalies—good	Good	Pigmented—fair to good Clear—poor	Poor to good (depends on plasticizer)	Very low	Dissolve with solvents.	Solvent evaporation	Air dry or elevated temperature for speed	Spray, dip, roller coat	Wire insulation. Metal protection (especially magnesium, aluminum).
Vinyl chloride (plastisol, organisol)		Good	Good	Poor to good (depends on plasticizer)	Low	Poor	Fusion of liquid to gel	Bake 250–330°F	Spray, dip, reverse roll	Soft-to-hard thick coatings, electroplating racks, equipment.
Vinyl fluoride		Good	Excellent	Good	Very low	Poor	Fusion from water or solvent dispersion	Bake 400–500°F	Spray, roller coat	Coatings for circuitry. Long-life exterior finish.
Vinyl formal (Formvar‡)		Good	Good	Good	Medium	Poor	Cured by heat	Bake 350–500°F	Roller coat, wire coater	Wire insulation (thin coatings) coil impregnation.

This table has been reprinted from *Machine Design*, May 25, 1967. Copyright, 1967, by The Penton Publishing Company, Cleveland, OH.

* Trademark of Hercules Powder Co., Inc., Wilmington, DE.
† Trademark of E. I. du Pont de Nemours & Co., Wilmington, DE.
‡ Trademark of Monsanto Co., St. Louis, MO.

coatings. They are formulated as lacquers, enamels, and emulsions. Lacquers and baking enamels are used as automotive and appliance finishes. In both these industries acrylics are used as top coats in multicoat finish systems. Thermosetting acrylics have replaced alkyds in applications requiring greater mar resistance such as appliance finishes. Acrylic lacquers are brittle and therefore have poor impact resistance, but their outstanding weather resistance allowed them to replace nitrocellulose lacquers in automotive finishes for many years. Acrylic and modified acrylic emulsions have been used as architectural coatings and also on industrial products. These medium-priced resins can be formulated to have excellent hardness, adhesion, abrasion, chemical, and mar resistance. When acrylic resins are used to modify other resins, their properties are imparted to the resultant resin system.

Uses. Acrylics, both lacquers and enamels, were the top coats of choice for the automotive industry from the early 1960s to the middle 1980s. Thermosetting acrylics are still used by the major appliance industry. Acrylics are used in electrodeposition and have largely replaced alkyds. The chemistry of acrylic-based resins allows them to be used in radiation curing applications alone or as monomeric modifiers for other resins. Acrylic-modified polyurethane coatings have excellent exterior durability.

Alkyds. Alkyd-resin-based coatings were introduced in the 1930s as replacements for nitrocellulose lacquers and oleoresinous-based coatings. They offer the advantage of good durability at relatively low cost. These low- to medium-priced coatings are still used for finishing a wide variety of products, either alone or modified with oils or other resins. The degree and type of modification determine their performance properties. They were used extensively by the automotive and appliance industries until the 1960s. Although alkyds are used in outdoor exposure, they are not as durable in long-term exposure and their color and gloss retention is inferior to that of acrylics.

Uses. Once the mainstay of organic coatings, alkyds are still used for finishing metal and wood products. Their durability in interior exposures is generally good, but their exterior durability is only fair. Alkyd resins are used in fillers, sealers, and caulks for wood finishing because of their formulating flexibility. Alkyds have also been used in electrodeposition as replacements for the oleoresinous vehicles. They are still used for finishing by the machine tool and other industries. Alkyds have also been widely used in architectural and trade sales coatings. Alkyd-modified acrylic latex paints are excellent architectural finishes.

Epoxies. Epoxy resins can be formulated with a wide range of properties. These medium- to high-priced resins are noted for their adhesion, make excellent primers, and are used widely in the appliance and automotive industries. Their heat resistance permits them to be used for electrical insulation. When epoxy top coats are used outdoors, they tend to chalk and discolor because of inherently poor ultraviolet-light resistance. Other resins modified with epoxies are used for outdoor exposure as top coats, and properties of many other resins can be improved by their addition. Two-component epoxy coatings are used in extreme corrosion and chemical environments. Flexibility in formulating two-component epoxy-resin-based coatings results in a wide range of physical properties.

Uses. Owing to their excellent adhesion, they are used extensively as primers for most coatings over most substrates. Epoxy coatings provide excellent chemical and corrosion resistance. They are used as electrical insulating coatings because of their high electric strength at elevated temperatures. Electrical and other properties of polymers are shown in Tables 6.7–6.10. Some of the original work with powder coating was done using epoxy resins, and they are still applied using this method. Many of the primers used for coil coating are epoxy-resin-based.

TABLE 6.7 Electric Strengths of Coatings

Material	Dielectric strength, volts/mil	Comments*	Source of information
Polymer coatings:			
Acrylics.............	450–550	Short-time method	*a*
	350–400	Step-by-step method	*a*
	400–530	....	*b*
	1,700–2,500	2-mil-thick samples	Columbia Technical Corp., Humiseal Coatings
Alkyds..............	300–350	....	*b*
Chlorinated polyether	400	Short-time method	*a*
Chlorosulfonated poly-ethylene...........	500	Short-time method	*a*
Diallyl phthalate.....	275–450	....	*b*
	450	Step-by-step method	*a*
Diallyl isophthalate...	422	Step-by-step method	*a*
Depolymerized rubber (DPR)	360–380	....	H. V. Hardman Co., DPR Subsidiary
Epoxy..............	650–730	Cured with anhydride–castor oil adduct	Autonetics, Div. of North American Rockwell
Epoxy..............	1,300	10-mil-thick dip coating	
Epoxies, modified.....	1,200–2,000	2-mil-thick sample	Columbia Technical Corp., Humiseal Coatings
Neoprene...........	150–600	Short-time method	*a*
Phenolic...........	300–450	....	*b*
Polyamide..........	780	106 mils thick	
Polyamide-imide......	2,700		
Polyesters...........	250–400	Short-time method	*a*
	170	Step-by-step method	*a*
Polyethylene........	480	....	*b*
	300	60-mil-thick sample	*c*
	500	Short-time method	*a*
Polyimide...........	3,000	Pyre-M.L., 10 mils thick	*d*
	4,500–5,000	Pyre-M.L. (RC-675)	*e*
	560	Short-time method, 80 mils thick	*a*
Polypropylene........	750–800	Short-time method	*a*
Polystyrene..........	500–700	Short-time method	*a*
	400–600	Step-by-step method	*a*
	450	60-mil-thick sample	*c*
Polysulfide...........	250–600	Short-time method	*a*
Polyurethane (single component)........	3,800	1-mil-thick sample	*f*
Polyurethane (two components)/castor oil cured...........	530–1,010	....	*g*
Polyurethane (two components, 100% solids)	275	125-mil-thick sample	Products Research & Chem. Corp. (PR-1538)
	750	25 mils thick sample	
Polyurethane (single component)	2,500	2-mil-thick sample	Columbia Technical Corp., Humiseal 1A27
Polyvinyl butyral.....	400		
Polyvinyl chloride....	300–1,000	Short-time method	
	275–900	Step-by-step method	
Polyvinyl formal......	860–1,000	....	*b*
Polyvinylidene fluoride	260	Short-time, 500-volt/sec, ⅛-in. sample	*h*
	1,280	Short-time, 500-volt/sec, 8-mil sample	*h*
	950	Step by step (1-kv steps)	*h*

TABLE 6.7   Electric Strengths of Coatings (*Continued*)

Material	Dielectric strength, volts/mil	Comments*	Source of information
Polyxylylenes:			
Parylene N........	6,000	Step by step	Union Carbide Corp.
	6,500	Short time	Union Carbide Corp.
Parylene C........	3,700	Short time	Union Carbide Corp.
	1,200	Step by step	Union Carbide Corp.
Parylene D.......	5,500	Short time	Union Carbide Corp.
	4,500	Step by step	Union Carbide Corp.
Silicone..............	500	Sylgard 182	Dow Corning Corp.
Silicone..............	550–650	RTV types	General Electric & Stauffer Chemical Co. bulletins
Silicone..............	800	Flexible dielectric gel	Dow Corning Corp.
Silicone..............	1,500	2-mil-thick sample	Columbia Technical Corp., Humiseal 1H34
TFE fluorocarbons....	400	60-mil-thick sample	c
	480	Short-time method	a
	430	Step by step	a
Teflon TFE dispersion coating	3,000–4,500	1–4-mil-thick sample	E. I. du Pont de Nemours & Co.
Teflon FEP dispersion coating	4,000	1.5-mil-thick sample	E. I. du Pont de Nemours & Co.
Other materials used in electronic assemblies:			
Alumina ceramics.....	200–300	....	b
Boron nitride........	900–1,400		
Electrical ceramics....	55–300	....	b
Forsterite...........	250	....	b
Glass, borosilicate.....	4,500	40-mil sample	c
Steatite..............	145–280	....	b

* All samples are standard 125 mils thick unless otherwise specified.
[a] *Insulation,* Directory Encyclopedia Issue, no. 7, June–July, 1968.
[b] *Mater. Eng.,* Materials Selector Issue, vol. 66, no. 5, Chapman-Reinhold Publication, mid-October, 1967–1968.
[c] Kohl, W. H.: "Handbook of Materials and Techniques for Vacuum Devices," p. 586, Reinhold Publishing Corporation, New York, 1967.
[d] Learn, J. R., and M. P. Seegers: Teflon-Pyre-M. L. Wire Insulation System, *13th Symp. on Tech. Progr. in Commun. Wire and Cables,* Atlantic City, NJ, Dec. 2–4, 1964.
[e] Milek, J. T.: Polyimide Plastics: A State of the Art Report, *Hughes Aircraft Rep.* S-8, October, 1965.
[f] *Hughson Chemical Co.* Bull. 7030A.
[g] *Spencer-Kellogg* (Division of Textron, Inc.) Bull. TS-6593.
[h] *Pennsalt Chemicals Corp. Prod. Sheet* KI-66a, Kynar Vinylidene Fluoride Resin, 1967.

Polyesters.   Polyesters are used alone or modified with other resins to formulate coatings ranging from clear furniture finishes, replacing lacquers, to industrial finishes, replacing alkyds. These moderately priced finishes permit the same formulating flexibility as alkyds but are tougher and more weather-resistant. There are basically two types of polyesters, two-component and single-package. The former are cured using peroxides which initiate free-radical polymerization, while the latter, sometimes called oil-free alkyds, are self-curing, usually at elevated temperatures. It is important to realize that, in both cases, the resin formulator can adjust properties to meet most exposure conditions. Polyesters are also applied as powder coatings.

TABLE 6.8 Volume Resistivities of Coatings

Material	Volume resistivity at 25°C, ohm-cm	Source of information
Acrylics....	10^{14}–10^{15}	a
	$> 10^{14}$	b
	7.6×10^{14}–1.0×10^{15}	Columbia Technical Corp., Humiseal
Alkyds........................	10^{14}	a
Chlorinated polyether............	10^{15}	b
Chlorosulfonated polyethylene.....	10^{14}	b
Depolymerized rubber...........	1.3×10^{13}	H. V. Hardman, DPR Subsidiary
Diallyl phthalate.................	10^{8}–2.5×10^{10}	a
Epoxy (cured with DETA)........	2×10^{16}	c,d
Epoxy polyamide	1.1–1.5×10^{14}	
Phenolics......................	6×10^{12}–10^{13}	a
Polyamides....................	10^{13}	
Polyamide-imide................	7.7×10^{16}	e
Polyethylene...................	$> 10^{16}$	
Polyimide.....................	10^{16}–10^{18}	f
Polypropylene..................	10^{10}–$> 10^{16}$	a
Polystyrene....................	$> 10^{16}$	b
Polysulfide....................	2.4×10^{11}	g
Polyurethane (single component)...	5.5×10^{12}	h
Polyurethane (single component)...	2.0×10^{12}	h
Polyurethane (single component)...	4×10^{13}	Columbia Technical Corp.
Polyurethane (two components)....	1×10^{13}	Products Research &
	$5 \times 10^{9}(300°F)$	Chemical Corp. (PR-1538)
Polyvinyl chloride..............	10^{11}–10^{15}	b
Polyvinylidene chloride..........	10^{14}–10^{16}	a
Polyvinylidene fluoride...........	2×10^{14}	h
Polyxylylenes (parylenes).........	10^{16}–10^{17}	Union Carbide Corp.
Silicone (RTV).................	6×10^{14}–3×10^{15}	Stauffer Chemical Co., Si-O-Flex SS 831, 832, & 833
Silicone, flexible dielectric gel......	1×10^{15}	Dow Corning Corp.
Silicone, flexible, clear............	2×10^{15}	Dow Corning Corp.
Silicone.......................	3.3×10^{14}	Columbia Technical Corp. Humiseal 1H34
Teflon TFE	$> 10^{18}$	b
Teflon FEP....................	$> 2 \times 10^{18}$	b

^{a} *Mater. Eng.*, Materials Selector Issue, vol. 66, no. 5, Chapman-Reinhold Publication, mid-October, 1967–1968.
^{b} *Insulation*, Directory Encyclopedia Issue, no. 7, June–July, 1968.
^{c} Lee, H., and K. Neville: "Epoxy Resins," McGraw-Hill Book Company, New York, 1966.
^{d} Tucker, Cooperman, and Franklin: Dielectric Properties of Casting Resins, *Electron. Equip.*, July, 1956.
^{e} Freeman, J. H.: A New Concept in Flat Cable Systems, *5th Ann. Symp. on Advan. Tech. for Aircraft Elec. Syst.*, Washington, D.C., October, 1964.
^{f} Milek, J. T.: Polyimide Plastics: A State of the Art Report, *Hughes Aircraft Rep.* S-8, October, 1965.
^{g} Hockenberger, L.: *Chem.-Ing. Tech.*, vol. 36, 1964.
^{h} *Hughson Chemical Co. Tech. Bull.* 7030A; *Pennsalt Chemicals Corp. Prod. Sheet* KI-66a, Kynar Vinylidene Fluoride Resin, 1967.

TABLE 6.9 Dielectric Constants of Coatings

Coating	60–100 Hz	10^6 Hz	$>10^6$ Hz	Reference source
Acrylic	....	2.7–3.2	....	a
Alkyd	....	....	3.8 (10^{10} Hz)	b
Asphalt and tars	....	....	3.5(10^{10} Hz)	b
Cellulose acetate butyrate	....	3.2–6.2	....	a
Cellulose nitrate	....	6.4	....	a
Chlorinated polyether	3.1	2.92	....	a
Chlorosulfonated polyethylene (Hypalon)	6.19 7–10(10^3 Hz)	~5	....	E. I. du Pont de Nemours & Co.
Depolymerized rubber (DPR)	4.1–4.2	3.9–4.0	....	H. V. Hardman, DPR Subsidiary
Diallyl isophthalate	3.5	3.2	3(10^8 Hz)	a,c
Diallyl phthalate	3–3.6	3.3–4.5	....	a,c
Epoxy-anhydride—castor oil adduct	3.4	3.1	2.9(10^7 Hz)	Autonetics, Division of North American Rockwell
Epoxy (one component)	3.8	3.7	....	Conap Inc.
Epoxy (two components)	3.7	....	....	Conap Inc.
Epoxy cured with methylnadic anhydride (100:84 pbw)	3.31	....	....	d
Epoxy cured with dodecenyl-succinic anhydride (100:132 pbw)	2.82	....	....	d
Epoxy cured with DETA	4.1	4.2	4.1	e
Epoxy cured with m-phenylenediamine	4.6	3.8	3.25(10^{10} Hz)	e
Epoxy dip coating (two components)	3.3	3.1	....	Conap Inc.
Epoxy (one component)	3.8	3.5	....	Conap Inc.
Epoxy-polyamide (40% Versamid* 125, 60% epoxy)	3.37	3.08	....	e
Epoxy-polyamide (50% Versamid 125, 50% epoxy)	3.20	3.01	....	e
Fluorocarbon (TFE, Teflon)	2.0–2.08	2.0–2.08	....	E. I. du Pont de Nemours & Co.
Phenolic	....	4–11	....	a
Phenolic	5–6.5	4.5–5.0	....	c
Polyamide	2.8–3.9	2.7–2.96		
Polyamide-imide	3.09	3.07		
Polyesters	3.3–8.1	3.2–5.9	....	c
Polyethylene	....	2.3	....	a
Polyethylenes	2.3	2.3	....	c
Polyimide-Pyre-M.L.† enamel	3.8	3.8	....	f
Polyimide–Du Pont RK-692 varnish	3.8	....	....	g
Polyimide–Du Pont RC-B-24951	3.0(10^3 Hz)	....	....	g
Polyimide–Du Pont RC-5060	2.8(10^3 Hz)	2......	....	g
Polypropylene	....	2.1	....	a
Polypropylene	2.22–2.28	2.22–2.28	....	c
Polystyrene	2.45–2.65	2.4–2.65	2.5(10^{10} Hz)	b
Polysulfides	6.9	....	....	h
Polyurethane (one component)	4.10	3.8	....	Conap Inc.
Polyurethane (two components—castor oil cured)	....	2.98–3.28	....	i
Polyurethane (two components)	6.8(10^3 Hz)	4.4	....	Products Research & Chemical Corp. (PR-1538)
Polyvinyl butyral	3.6	3.33	....	a
Polyvinyl chloride	3.3–6.7	2.3–3.5	....	a
Polyvinyl chloride–vinyl acetate copolymer	3–10	....	....	a
Polyvinyl formal	3.7	3.0	....	a
Polyvinylidene chloride	....	3–5	....	a
Polyvinylidene fluoride	8.1	6.6	....	j
Polyvinylidene fluoride	8.4	6.43	2.98(10^9 Hz)	k

TABLE 6.9   Dielectric Constants of Coatings (*Continued*)

Coating	60–100 Hz	10^6 Hz	>10^6 Hz	Reference source
p-Polyxylylene:				
Parylene N	2.65	2.65	....	Union Carbide Corp.
Parylene C	3.10	2.90	....	Union Carbide Corp.
Parylene D	2.84	2.80	....	Union Carbide Corp.
Shellac (natural, dewaxed)	3.6	3.3	2.75(10^9 Hz)	*l*
Silicone (RTV types)	3.3–4.2	3.1–4.0	....	General Electric and Stauffer Chemical Cos.
Silicone (Sylgard ‡ type)	2.88	2.88	....	Dow Corning Corp.
Silicone, flexible dielectric gel	3.0	....	....	Dow Corning Corp.
FEP dispersion coating	2 1(10^3 Hz)	....	....	E. I. du Pont de Nemours & Co.
TFE dispersion coating	2.0–2.2(10^3 Hz)	....	....	E. I. du Pont de Nemours & Co.
Wax (paraffinic)	2.25	2.25	2.22(10^{10} Hz)	*l*

* Trademark of General Mills, Inc., Kankakee, IL.
† Trademark of E. I. du Pont de Nemours & Co., Wilmington, DE.
‡ Trademark of Dow Corning Corporation, Midland, MI.
   a *Mater. Eng.,* Materials Selector Issue, vol. 66, no. 5, Chapman-Reinhold Publication, mid-October 1967–1968.
   b Volk, M. C., J. W. Lefforge, and R. Stetson: "Electrical Encapsulation," Reinhold Publishing Corporation, New York, 1962.
   c *Insulation,* Directory Encyclopedia Issue, no. 7, June–July, 1968.
   d Coombs, C. F. (ed.): "Printed Circuits Handbook," McGraw-Hill Book Company, New York, 1967.
   e Lee, H., and K. Neville: "Handbook of Epoxy Resins," McGraw-Hill Book Company, New York, 1967.
   f Learn, J. R., and M. P. Seegers: Teflon-Pyre-M.L. Wire Insulation System, *13th Symp. of Tech. Progr. in Commun. Wire and Cable,* Atlantic City, NJ, Dec. 2–4, 1964.
   g *Du Pont Bull.* H65-4, Experimental Polyimide Insulating Varnishes, RC-B-24951 and RC-5060, January, 1965.
   h Hockenberger, L.: *Chem.-Ing. Tech.* vol. 36, 1964.
   i *Spencer-Kellogg* (division of Textron, Inc.) Bull. TS-6593.
   j Barnhart, W. S., R. A. Ferren, and H. Iserson: 17th ANTEC of SPE, January, 1961.
   k *Pennsalt Chemicals Corp. Prod. Sheet* KI-66a, Kynar Vinylidene Fluoride Resin, 1967.
   l Von Hippel, A. R. (ed.): "Dielectric Materials and Applications," Technology Press of MIT and John Wiley & Sons, Inc., New York, 1961.

Uses.   Two-component polyesters are well known as gel coats for glass-reinforced plastic bathtubs, lavatories, boats, and automobiles. Figure 6.2 shows a polyester gel coat applied to tub and shower units. High-quality one-package polyester finishes are used on furniture, appliances, automobiles, magnet wire, and industrial products. Polyester powder coatings are used as high-quality finishes in indoor and outdoor applications for anything from tables to trucks. They are also used as coil coatings.

Polyurethanes.   Polyurethane-resin-based coatings are extremely versatile. There are higher priced than alkyds but lower than epoxies. Polyurethane resins are available as oil-modified, moisture curing, blocked, two-component, and lacquers. Table 6.11 is a selection guide for polyurethane coatings. Two-component polyurethanes can be formulated in a wide range of hardnesses. They can be abrasion-resistant, flexible, resilient, tough, chemical-resistant, and weather-resistant. Abrasion resistance of organic coatings is shown in Table 6.12. Polyurethanes can be combined with other resins to reinforce or adopt their properties. Urethane-modified acrylics have excellent outdoor weathering properties. They can also be applied as air-drying, forced-dried, and baking liquid finishes as well as powder coating.

TABLE 6.10 Dissipation Factors of Coatings

Coating	60–100 Hz	10^6 Hz	$>10^6$ Hz	Reference source
Acrylics	0.04–0.06	0.02–0.03	....	[a]
Alkyds	0.003–0.06	....	....	[a]
Chlorinated polyether	0.01	0.01	....	[a]
Chlorosulfonated polyethylene	0.03	0.07(10^3 Hz)	....	[b]
Depolymerized rubber (DPR)	0.007–0.013	0.0073–0.016	....	H. V. Hardman, DPR Subsidiary
Diallyl phthalate	0.010	0.011	0.011	[b]
Diallyl isophthalate	0.008	0.009	0.014(10^{10} Hz)	[b]
Epoxy dip coating (two components)	0.027	0.018	....	Conap Inc.
Epoxy (one component)	0.011	0.004	....	Conap Inc.
Epoxy (one component)	0.008	0.006	....	Conap Inc.
Epoxy polyamide (40 % Versamid 125, 60 % epoxy)	0.0085	0.0213	....	[c]
Epoxy polyamide (50 % Versamid 115, 50 % epoxy)	0.009	0.0170	....	[c]
Epoxy cured with anhydride–castor oil adduct	0.0084	0.0165	0.0240	Autonetics, North American Rockwell
Phenolics	0.005–0.5	0.022	....	[a]
Polyamide	0.015	0.022–0.097	....	
Polyesters	0.008–0.041	.....	....	[a]
Polyethylene (linear)	0.00015	0.00015	0.0004(10^{10} Hz)	[d]
Polymethyl methacrylate	0.06	0.02	0.009(10^{10} Hz)	[d]
Polystyrene	0.0001–0.0005	0.0001–0.0004		
Polyurethane (two component, castor oil cure)	....	0.016 –0.036	....	[e]
Polyurethane (one component)	0.038–0.039	0.068–0.074	....	Conap Inc.
Polyurethane (one component)	0.02	.....	....	Conap Inc.
Polyvinyl butyral	0.007	0.0065		
Polyvinyl chloride	0.08–0.15	0.04–0.14	....	[a]
Polyvinyl chloride, plasticized	0.10	0.15	0.01(10^{10} Hz)	[d]
Polyvinyl chloride–vinyl acetate copolymer	0.6–0.10			
Polyvinyl formal	0.007	0.02		
Polyvinylidene fluoride	0.049	0.17	....	[f]
	0.049	0.159	0.110	[g]
Polyxylylenes:				
Parylene N	0.0002	0.0006	....	Union Carbide Corp.
Parylene C	0.02	0.0128	....	Union Carbide Corp.
Parylene D	0.004	0.0020	....	Union Carbide Corp.
Silicone (Sylgard 182)	0.001	0.001	....	Dow Corning Corp.
Silicone, flexible dielectric gel	0.0005	.....	....	Dow Corning Corp.
Silicone, flexible, clear	0.001	.....	....	Dow Corning Corp.
Silicone (RTV types)	0.011–0.02	0.003–0.006	....	General Electric
Teflon FEP dispersion coating	0.0002–0.0007	.....	....	E. I. du Pont de Nemours & Co.
Teflon FEP	<0.0003	<0.0003	....	[a]
Teflon TFE	<0.0003	....	....	[a]
Teflon TFE	0.00012	0.00005	....	Union Carbide Corp.
Other materials:				
Alumina (99.5 %)	....	0.0001	....	[h]
Beryllia (99.5 %)	....	0.0003	....	[h]
Glass silica	0.0006	0.0001	0.00017(10^{10} Hz)	[d]
Glass, borosilicate	....	0.013–0.016	....	Corning Glass Works
Glass, 96 % silica	....	0.0015–0.0019	....	Corning Glass Works

[a] *Mach. Des.*, Plastics Reference Issue, vol. 38, no. 14, Penton Publishing Co., 1966.
[b] *Insulation*, Directory, Encyclopedia Issue, no. 7, June–July, 1968.
[c] Lee, H. and K. Neville: "Handbook of Epoxy Resins," McGraw-Hill Book Company, New York, 1967.
[d] Mathes, K.: Electrical Insulation Conference, 1967.
[e] *Spencer-Kellogg* (Division of Textron, Inc.) *Tech. Bull.* TS-6593.
[f] Barnhart, W. S., R. A. Ferren, and H. Iserson: 17th ANTEC of SPE, January, 1961.
[g] *Pennsalt Chemicals Corp. Prod. Sheet* KI-66a, Kynar Vinylidene Fluoride Resin, 1967.
[h] *Mach. Des.*, Design Guide, Sept. 28, 1967.

FIGURE 6.2 Polyester gel coats are used to give a decorative and protective surface to tub shower units which are made out of glass-fiber-reinforced plastics. (*Courtesy of Owens-Corning Fiberglas Corp.*)

Uses. Polyurethanes have become very important finishes in the transportation industry, which includes aircraft, automobiles, railroads, trucks, and ships. Owing to their chemical resistance and ease of decontamination from chemical, biological, and radiological warfare agents, they are widely used for painting military land vehicles, ships, and aircraft. They are used on automobiles as coatings for plastic parts and as clear top coats in the newer basecoat-clearcoat finish systems. Low-temperature baking polyurethanes are used as mar-resistant finishes for products that must be packaged while still warm. Polyurethanes are finding their way into many more coating applications.

Polyvinyl Chloride. Polyvinyl chloride (PVC) coatings, commonly called vinyls, are noted for their toughness, chemical resistance, and durability. They are available as solutions, dispersions, and lattices. Properties of vinyl coatings are listed in Table 6.13. They are applied as lacquers, plastisols, organisols, and lattices. PVC coating powders have essentially the same properties as liquids. PVC organisol, plastisol, and powder coatings have limited adhesion and require primers.

Uses. Vinyls have been used in various applications including beverage and other can linings, automobile interiors, and office machine exteriors. They are also used as thick-film liquids and as powder coatings for electrical insulation. Owing to their excellent chemical resistance, they are used as tank linings and as rack coatings in electroplating shops. Typical applications for vinyl coatings are shown in Fig. 6.3. Vinyl-modified acrylic latex trade sale paints are used as trim enamels for exterior applications and as semigloss wall enamels for interior applications.

TABLE 6.11 Guide to Selecting Polyurethane Coatings

Property	One-component			Two-component	Lacquer
	Urethane oil	Moisture	Blocked		
Abrasion resistance	Fair–good	Excellent	Good–exc.	Excellent	Fair
Hardness	Medium	Med.–hard	Med.–hard	Soft–very hard	Soft–med.
Flexibility	Fair–good	Good–exc.	Good	Good–exc.	Excellent
Impact resistance	Good	Excellent	Good–exc.	Excellent	Excellent
Solvent resistance	Fair	Poor–fair	Good	Excellent	Poor
Chemical resistance	Fair	Fair	Good	Excellent	Fair–good
Corrosion resistance	Good	Fair	Fair	Excellent	Good–exc.
Adhesion	Good	Fair–good	Good	Excellent	Fair–good
Toughness	Poor	Excellent	Poor	Excellent	Good–exc.
Elongation	Fair	Poor	Fair–good	Good–exc.	Excellent
Tensile	....	Good	....	....	Excellent
Weatherability	....	....	....	....	....
Aliphatic	Good	Poor–fair	Poor–fair	Good–exc.	Good
Conventional	Poor–fair	Poor–fair	....	Poor–fair	Poor
Pigmented glass	High	High	High	High	Medium
Cure rate	Slow	Slow	Fast	Fast	None
Cure temp	Room temp	Room temp	300–390°F	212°F	150–225°F
Work life	Infinite	1 year	6 months	1 s–24 h	Infinite

TABLE 6.12 Abrasion Resistance of Coatings

Coating	Taber ware index, mg/1000 rev.
Polyurethane Type 1	55–67
Polyurethane Type 2 (clear)	8–24
Polyurethane Type 2 (pigmented)	31–35
Polyurethane Type 5	60
Urethane oil varnish	155
Alkyd	147
Vinyl	85–106
Epoxy-amine-cured varnish	38
Epoxy-polyamide enamel	95
Epoxy-ester enamel	196
Epoxy-polyamide coating (1:1)	50
Phenolic spar varnish	172
Clear nitrocellulose lacquer	96
Chlorinated rubber	200–220
Silicone, white enamel	113
Catalyzed epoxy, air-cured (PT-401)	208
Catalyzed epoxy, Teflon-filled (PT-401)	122
Catalyzed epoxy, bake-Teflon-filled (PT-201)	136
Parylene N	9.7
Parylene C	44
Parylene D	305
Polyamide	290–310
Polyethylene	360
Alkyd TT-E-508 enamel (cured for 45 min at 250°F)	51
Alkyd TT-E-508 (cured for 24 hr at room temperature)	70

6.5.2 Other Coating Resins

In addition to the aforementioned materials, there are a number of other important resins used in formulating coatings. These materials, used alone or as modifiers for other resins, provide coating vehicles having diverse properties.

Aminos. Resins of this type, such as urea formaldehyde and melamine, are used in modifying other resins to increase their durability. Notable among the modified resins are the superalkyds used in automotive and appliance finishes.

Uses. Melamine and urea formaldehyde resins are used as modifiers for alkyds and other resins to increase hardness and accelerate cure.

Cellulosics. Nitrocellulose lacquers are the most important of the cellulosics. They were introduced in the 1920s and used as fast-drying finishes for a number of manufactured products. Applied at low solids using expensive solvents, they will not meet air-quality standards. By modifying nitrocellulose with other resins such as alkyds and ureas, the VOC content can be lowered and properties can be increased. Other important cellulosic resins are cellulose acetate butyrate and ethyl cellulose.

Uses. Although no longer used extensively by the automotive industry, nitrocellulose lacquers are still used by the furniture industry because of their fast drying and hand-rubbing properties. Cellulose acetate butyrate has been used for coating metal in numerous applications. In 1959 one of the first conveyorized powder coating lines in the United States coated transformer lids and hand-hole covers with a cellulose acetate butyrate powder coating.

TABLE 6.13 Properties of Vinyl Coatings

Coating type	Outstanding characteristics	Mechanical properties [a]					Color and gloss [a]			Weathering [a] properties	
		Hardness	Abrasion resistance	Adhesion	Flexibility	Toughness	Film color	Color retention	Gloss	Weather resistance	Gloss retention
Solution[b]	Excellent color, flexibility, chemical resistance; tasteless, odorless	F	E	F to G	E	E	E[f]	E	G	E[f]	E
Plastisol[c]	Toughness; resilience; abrasion resistance; can be applied without solvents	F	E	E to cloth[e]	E	E	E[f]	E	F	E[f]	F to G
Organosol[d]	High solids content; excellent color, flexibility; tasteless, odorless	F	E	E to cloth[e]	E	E	E[f]	E	P to G	E[f]	G

[a] E = excellent; G = good; F = fair; P = poor.
[b] Vinyl chloride acetate copolymers; resins vary widely in compatibility with other materials.
[c] Vinyl chloride acetate copolymer and vinyl chloride resins.
[d] Vinyl chloride acetate copolymers; require grinding for good dispersions.
[e] Requires primer for use on metal.
[f] Pigmented.

FIGURE 6.3   Vinyl plastisols and organisols are used extensively for dip coating of wire products. The coatings can be varied from very hard to very soft. (*Courtesy of M&T Chemicals.*)

Chlorinated Rubber.   Chlorinated rubber coatings are used as swimming pool paints and traffic paints.

Fluorocarbons.   These high-priced coatings require high processing temperatures and therefore are limited in their usage. They are noted for their lubricity or nonstick properties due to low coefficients of friction, and for weatherability. Table 6.14 gives the coefficients of friction of typical coatings.

Uses.   Fluorocarbons are used as chemical-resistant coatings for processing equipment. They are also used as nonstick coatings for cookware, as friction-reducing coatings for tools, and as dry lubricated surfaces in many other consumer and industrial products, as shown in Fig. 6.4. Table 6.15 compares the properties of four fluorocarbons.

Oleoresinous Coatings.   Oleoresinous coatings, based on drying oils such as soybean and linseed, are slow curing. For many years prior to the introduction of synthetic resins, they were used as the vehicles in most coatings. They still find application alone or as modifiers to other resins.

Uses.   Oleoresinous vehicles are used in low-cost primers and enamels for structural, marine, architectural, and, to a limited extent, industrial product finishing.

Phenolics.   Introduced in the early 1900s, phenolics were the first commercial synthetic resins. They are available as 100 percent phenolic baking resins, oil-modified, and phenolic dispersions. Phenolic resins, used as modifiers, will improve heat and chemical resistance of other resins. Baked phenolic-resin-based coatings are well known for their corrosion, chemical, moisture, and heat resistance.

Uses.   Phenolics coating are used on heavy-duty air-handling equipment, chemical equipment, and as insulating varnishes. Phenolic resins are also used as binders for electrical and decorative laminated plastics.

TABLE 6.14 Coefficients of Friction of Typical Coatings

Coating	Coefficient of friction, μ	Information source
Polyvinyl chloride...	0.4–0.5	a
Polystyrene..	0.4–0.5	a
Polymethyl methacrylate................................	0.4–0.5	a
Nylon...	0.3	a
Polyethylene...	0.6–0.8	a
Polytetrafluoroethylene (Teflon).......................	0.05–0.1	a
Catalyzed epoxy air-dry coating with Teflon filler.......	0.15	b
Parylene N..	0.25	c
Parylene C..	0.29	c
Parylene D..	0.31–0.33	c
Polyimide (Pyre-M.L.).................................	0.17	d
Graphite..	0.18	d
Graphite–molybdenum sulfide:		
Dry-film lubricant....................................	0.02–0.06	e
Steel on steel.......................................	0.45–0.60	e
Brass on steel.......................................	0.44	e
Babbitt on mild steel................................	0.33	e
Glass on glass.......................................	0.4	e
Steel on steel with SAE no. 20 oil...................	0.044	e
Polymethyl methacrylate to self......................	0.8 (static)	e
Polymethyl methacrylate to steel.....................	0.4–0.5 (static)	e

[a] Bowder, F. P., *Endeavor*, vol. 16, no. 61, p. 5, 1957.
[b] Product Techniques Incorporated: Bulletin on PT-401 TE, Oct. 17, 1961.
[c] Union Carbide data.
[d] *DuPont Tech. Bull.* 19, Pyre-M.L. Wire Enamel, August, 1967.
[e] Electrofilm, Inc. data.

Polyamides.   One of the more notable polyamide resins is nylon, which is tough, wear-resistant, and has a relatively low coefficient of friction. It can be applied as a powder coating by fluidized bed, electrostatic spray, or flame spray. Table 6.16 compares the properties of three types of nylon polymers used in coatings. Nylon coatings generally require a primer. Polyamide resins are also used as curing agents for two-component epoxy resin coatings. Film properties can be varied widely by polyamide selection.

   Uses.   Applied as a powder coating, nylon provides a high degree of toughness and mechanical durability to office furniture. Other polyamide resins are used as curing agents in two-component epoxy-resin-based primers and top coats, adhesives, and sealants.

Polyolefins.   These coatings, which can be applied by flame spraying, hot melt, or powder coating methods, have limited usage.

   Uses.   Polyethylene is used for impregnating or coating packaging materials such as paper and aluminum foil. Certain polyethylene-coated composite packaging materials are virtually moisture proof. Table 6.17 compares the moisture-vapor transmission rates of various coatings and films. Polyethylene powder coatings are used on chemical-processing and food-handling equipment.

Polyimides.   Polyimide coatings have excellent long-term thermal stability, wear, mar and moisture resistance, as well as electrical properties. They are high-priced.

FIGURE 6.4 Nonstick feature of fluorocarbon finishes makes them useful for products such as saws, fan and blower blades, door-lock parts, sliding- and folding-door hardware, skis, and snow shovels. (*Courtesy of E. I. du Pont de Nemours & Company.*)

Uses. Polyimide coatings are used in electrical applications as insulating varnishes and magnet wire enamels in high-temperature, high-reliability applications. NEMA standards and manufacturers' trade names for various wire enamels are shown in Table 6.18. They are also used as alternatives to fluorocarbon coatings on cookware, as shown in Fig. 6.5.

Silicones. Silicone resins are high-priced and are used alone or as modifiers to upgrade other resins. They are noted for their high-temperature resistance, moisture resistance, and weatherability. They can be hard or elastomeric, baking or room-temperature curing.

Uses. Silicones are used in high-temperature coatings for exhaust stacks, ovens, and space heaters. Figure 6.6 shows silicone coatings on fireplace equipment. They are also used as conformal coatings for printed wiring boards, moisture repellants for masonry, weather-resistant finishes for outdoors, and thermal control coatings for space vehicles. The thermal conductivities of coatings are listed in Table 6.19.

6.6 APPLICATION METHODS

The selection of an application method is as important as the selection of the coating itself. Basically, the application methods for protective and decorative industrial liquid coatings are dipping, flow coating, and spraying, although some coatings are applied by brushing, rolling, and silk screening. In these times of environmental awareness and regulation, it is mandatory that coatings be applied in the most effi-

TABLE 6.15 Properties of Four Fluorocarbons

Property	Polyvinyl fluoride (PVF) $(CH_2-CHF)_n$	Polyvinyl-idene fluoride (PVF-2) $(CH_2-CF_2)_n$	Polytrifluoro-chloro-ethylene (PTFCl) $(CClF-CF_2)_n$	Polytetra-fluoro-ethylene (PTFE) $(CF_2-CF_2)_n$
Physical properties:				
Density.................	1.4	1.76	2.104	2.17–2.21
Fusing temp, °F...........	300	460	500	750
Max continuous service and temp, °F................	225	300	400	550
Coefficient of friction.......	0.16	0.16	0.15	0.1
Flammability.............	Burns	Non-flammable	Non-flammable	Non-flammable
Mechanical properties:				
Tensile strength, lb/in.^{2}....	7,000	7,000	5,000	2,500–3,500
Elongation, %............	115–250	300	250	200–400
Izod impact, ft-lb/in.......	....	3.8	5	3
Durometer hardness.......	....	80	74–78	50–65
Yield strength at 77°F, lb/in.^{2}.................	6,000	5,500	4,500	1,300
Heat-distortion temp at 66 lb/in.^{2}, °F.............	NA	300	265	250
Coefficient of linear expansion................	2.8×10^5	8.5×10^5	15×10^5	8×10^5
Modulus (tension) $\times 10^5$ lb/in.^{2}..................	2.5–3.7	1.2	1.9	0.6
Electrical properties:				
Dielectric strength, V/mil...	....	260 (0.125)	500 (0.063)	600 (0.060)
Short time, V/mil, in.....	3,400 (0.002)			
Dielectric constant, 10^3 Hz..	8.5	7.72	2.6	2.1
Arc resistance (77°F) ASTM D 495...........	NA	60	300	300
Volume resistivity Ω-cm at 50% RH 77°F............	10^{12}	10^{14}	10^{16}	10^{18}
Dissipation factor, 100 Hz...	1.6	0.05	0.022	0.0003

TABLE 6.16 Properties of Nylon Coatings

	Nylon 11	Nylon 6/6	Nylon 6
Elongation (73°F), %..........................	120	90	50–200
Tensile strength (73°F), lb/in.^{2}.................	8,500	10,500	10,500
Modulus of elasticity (73°F), lb/in.^{2}.............	178,000	400,000	350,000
Rockwell hardness............................	R 100.5	R 118	R 112–118
Specific gravity..............................	1.04	1.14	1.14
Moisture absorption, %, ASTM D 570...........	0.4	1.5	1.6–2.3
Thermal conductivity Btu/(ft^2)(h)(°F/in.).........	1.5	1.7	1.2–1.3
Dielectric strength (short time), V/mil..........	430	385	440
Dielectric constant (10 Hz).....................	3.5	4	4.8
Effect of:			
Weak acids..............................	None	None	None
Strong acids.............................	Attack	Attack	Attack
Strong alkalies...........................	None	None	None
Alcohols................................	None	None	None
Esters..................................	None	None	None
Hydrocarbons...........................	None	None	None

TABLE 6.17 Moisture-Vapor Transmission Rates per 24-h Period of Coatings and Films in $g/(mil)(in^2)$

Coating or film	MVTR	Information source
Epoxy-anhydride..................	2.38	Autonetics data (25°C)
Epoxy–aromatic amine.............	1.79	Autonetics data (25°C)
Neoprene.......................	15.5	Baer[29] (39°C)
Polyurethane (Magna X-500)........	2.4	Autonetics data (25°C)
Polyurethane (isocyanate-polyester)	8.72	Autonetics data (25°C)
Olefane,* polypropylene............	0.70	Avisun data
Cellophane (type PVD uncoated film)	134	Du Pont
Cellulose acetate (film).............	219	Du Pont
Polycarbonate...................	10	FMC data
Mylar†........................	1.9	Baer[29] (39°C)
	1.8	Du Pont data
Polystyrene.....................	8.6	Baer[29] (39°C)
	9.0	Dow data
Polyethylene film.................	0.97	Dow data (1-mil film)
Saran resin (F120)................	0.097 to 0.45	Baer[29] (39°C)
Polyvinylidene chloride............	0.15	Baer[29] (2-mil sample, 40°C)
Polytetrafluoroethylene (PTFE)......	0.32	Baer[29] (2-mil sample 40°C)
PTFE, dispersion cast.............	0.2	Du Pont data
Fluorinated ethylene propylene (FEP)	0.46	Baer[29] (40°C)
Polyvinyl fluoride.................	2.97	Baer[29] (40°C)
Teslar.........................	2.7	Du Pont data
Parylene N.....................	14	Union Carbide data (2-mil sample)
Parylene C.....................	1	Union Carbide data (2-mil sample)
Silicone (RTV 521)................	120.78	Autonetics data
Methyl phenyl silicone.............	38.31	Autonetics data
Polyurethane (AB0130–002).........	4.33	Autonetics data
Phenoxy.......................	3.5	Lee, Stoffey, and Neville[40]
Alkyd-silicone (DC-1377)...........	6.47	Autonetics data
Alkyd-silicone (DC-1400)...........	4.45	Autonetics data
Alkyd-silicone...................	6.16–7.9	Autonetics data
Polyvinyl fluoride (PT-207).........	0.7	Product Techniques Incorp.

* Trademark of Avisun Corporation, Philadelphia, PA.
† Trademark of E. I. du Pont de Nemours & Co., Wilmington, DE.

cient manner.[4] Not only will this help meet the air-quality standards, but it will also reduce material costs. The advantages and disadvantages of various coating application methods are given in Table 6.20.

Liquid spray coating equipment can be classified by its atomizing method: air, hydraulic, or centrifugal. These can be subclassified into air atomizing, airless, airless electrostatic, air-assisted airless electrostatic, rotating electrostatic disks and bells, and high-volume, low-pressure types. While liquid dip coating equipment is usually simple, electrocoating equipment is highly sophisticated using electrophoresis as the driving force. Other liquid coating methods include flow coating, which can be manual or automated, roller coating, curtain coating, and centrifugal coating. Powder coating equipment includes fluidized beds, electrostatic fluidized beds, and electrostatic spray outfits.

It is important to note that environmental and worker safety regulations can be met, hazardous and nonhazardous wastes can be reduced, and money can be saved by using compliance coatings (those that meet the VOC emission standards) in

TABLE 6.18   NEMA Standards and Manufacturers' Trade Names for Magnet Wire Insulation

Manufacturer	Plain enamel	Polyvinyl formal	Polyvinyl formal modified	Polyvinyl formal with nylon overcoat	Polyvinyl formal with butyral overcoat	Poly-amide	Acrylic	Epoxy
Thermal class.............	105°C	105°C	105°C	105°C	105°C	105°C	105°C	130°C
NEMA Standard^s.........	MW 1	MW 15	MW 27	MW 17	MW 19	MW 6	MW 4	MW 9
Anaconda Wire & Cable Co.	Plain enamel	Formvar	Hermetic Formvar	Nyform	Cement coated Formvar	....	....	Epoxy epoxy-cement coated
Asco Wire & Cable Co.......	Enamel	Formvar	....	Nyform	Formbond	Nylon	Acrylic	Epoxy
Belden Manufacturing Co...	Beld-enamel	Formvar	....	Nyclad	....	....	....	Epoxy
Bridgeport Insulated Wire Co.	....	Formvar	....	....	Quickbond	Quick-Sol	....	....
Chicago Magnet Wire Corp.	Plain enamel	Formvar	....	Nyform	Bondable Formvar	Nylon	Acrylic	Epoxy
Essex Wire Corp............	Plain enamel	Formvar	Formetex	Nyform	Bondex	....	Ensolex/ESX	Epoxy
General Cable Corp.........	Plain enamel	Formvar	Formetic	Formlon	Formeze	....	Solderable acrylic	Epoxy
General Electric Co........	....	Formex	....	....	....	Nylon	....	....
Haveg-Super Temp Div..... Hitemp Wires Co. Division Simplex Wire & Cable Co.	....	....	....	....	....	....	....	....
Hudson Wire Co............	Plain enamel	Formvar	....	Nyform	Formvar AVC	Ezsol	....	....
New Haven Wire & Cable, Inc.	Plain enamel	....	....	....	....	....	....	....
Phelps Dodge Magnet Wire Corp.	Enamel	Formvar	Hermeteze	Nyform	Bondeze	....	....	....
Rea Magnet Wire Co., Inc...	Plain enamel	Formvar	Hermetic Formvar special	Nyform	Koilset	Nylon	....	Epoxy
Viking Wire Co., Inc........	Enamel	Formvar	....	Nyform	F-Bondall	Nylon	....	....

Courtesy of Rea Magnet Wire Co., Inc.

equipment having the highest transfer efficiency (the percentage of the coating used which actually coats the product and is not otherwise wasted). The theoretical transfer efficiencies (TE) of coating application equipment are indicated in the text and in Table 6.21, where they are listed in descending order.[4]

In the selection of a coating method and equipment, the product's size, configuration, intended market, and appearance must be considered. To aid in the selection of the most efficient application method, each will be discussed in greater detail.

6.6.1   Dip Coating

Dip coating (95 to 100 percent TE) is a simple coating method where products are dipped in a tank of coating material, withdrawn, and allowed to drain in the solvent-

TABLE 6.18 NEMA Standards and Manufacturers' Trade Names for Magnet Wire Insulation (*Continued*)

Teflon	Polyurethane	Polyurethane with friction surface	Polyurethane with nylon overcoat	Polyurethane with butyral overcoat	Polyurethane with nylon and butyral overcoat	Polyester	Polyester with overcoat	Polyimide	Polyester polyimide	Ceramic, ceramic-Teflon, ceramic-silicon
200°C MW 10	105°C MW 2	105°C	130°C MW 28	105°C MW 3 (PROP)	130°C MW 29 (PROP)	155°C MW 5	155°C MW 5	220°C MW 16	180°C	180°C+ MW 7
....	Analac	....	Nylac	Cement-coated analac	Cement coated nylac	....	Anatherm D, Anatherm 200	Al 220 M.L.	Anatherm N Anamid M (amide-imide)	
....	Poly	....	Nypol	Asco bond-P	Asco bond	Ascotherm	Isotherm 200	M.L.	Ascomid	
....	Beldure	....	Beldsol	....	....	Isonel	Polyther-maleze	M.L.		
....	Polyurethane	Uniwind	Polynylon	Polybond	....	Isonel 200				
....	Soderbrite	....	Nysod	Bondable polyurethane	....	Polyester 155				
....	Soderex	....	Soderon	Soderbond	Soderbond N	Thermalex F	Polythermalex/PTX 200	Allex		
....	Enamel "G"	....	Genlon	....	....	Gentherm	Polythermaleze 200			
Teflon Temprite	....	....	....	....	....	Alkanex Isonel				
....	Hudsol	Gripon	Nypoly	Hudsol AVC	Nypoly AVC	Isonel 200	Isonel 200-A	M.L.	Isomid	
....	Impsol	....	Impsolon	....	....	Imp-200				
....	Sodereze	Gripeze	Nyleze	....	S-Y Bondeze	....	Polythermaleze 200 II	M.L.		
....	Solvar	....	Nylon solvar	Solvar koilset	....	Isonel 200	Polythermaleze 200	Pyre M.L.	Isomid	Ceroc
....	Polyurethane	....	Polynylon	P-Bondall	....	Isonel 200	Iso-poly	M.L.	Isomid Isomid-P	

rich area above the coating's surface and then allowed to dry. The film thickness is controlled by viscosity, flow, percent solids by volume, and rate of withdrawal. This simple process can also be automated with the addition of a drain-off area, which allows excess coating material to flow back to the dip tank.

Dip coating is a simple, quick method that does not require sophisticated equipment. The disadvantages of dip coating are film thickness differential from top to bottom, resulting in the so-called wedge effect; fatty edges on lower parts of products; and runs and sags. Although this method coats all surface areas, solvent refluxing can cause low film build. Light products can float off the hanger and hooks and fall into the dip tank. Solvent-containing coatings in dip tanks and drain tunnels must be protected by fire extinguishers and safety dump tanks. The fire hazard can be eliminated by using waterborne coatings.

FIGURE 6.5 Polymide coating is used as a protective finish on the inside of aluminum, stainless steel, and other cookware. (*Courtesy of Mirro Aluminum Co.*)

6.6.2 Electrocoating

Electrocoating (95 to 100 percent TE) is a sophisticated dipping method commercialized in the 1960s to solve severe corrosion problems in the automotive industry. In principle it is similar to electroplating, except that organic coatings, rather than metals, are deposited on products from an electrolytic bath. Electrocoating can be either anodic (deposition of coatings on the anode from an alkaline bath) or cathodic (deposition of coatings on the cathode from an acidic bath). The bath is aqueous and contains very little volatile organic solvent. The phenomenon called throwing power causes inaccessible areas to be coated with uniform film thicknesses. Electrocoating is gaining a significant share of the primer and one-coat enamel coatings market.

Advantages of the electrocoating method include environmental acceptability owing to decreased solvent emissions and increased corrosion protection to inaccessible areas. It is less labor-intensive than other methods, and it produces uniform film thickness from top to bottom and inside and outside on products with a complex shape. Disadvantages are high capital equipment costs, higher material costs, and more thorough pretreatment. Higher operator skills are also required.

6.6.3 Spray Coating

Spray coating (30 to 90 percent TE), which was introduced to the automobile industry in the 1920s, revolutionized industrial painting. The results of this development were increased production and improved appearance. Electrostatics, which were added in the 1940s, improved transfer efficiency and reduced material consumption. Eight types of spray-painting equipment are discussed in this section. The transfer

FIGURE 6.6 Silicone coatings are used as heat-stable finishes for severe high-temperature applications such as fireplace equipment, exhaust stacks, thermal control coatings for spacecraft, and wall and space heaters. (*Courtesy of Copper Development Association.*)

efficiencies listed are theoretical. The actual transfer efficiency depends on many variables, including the size and configuration of the product and the airflow in the spray booth.

Rotating Electrostatic Disks and Bell Spray Coating. Rotating spray coaters (80 to 90 percent TE) rely on the centrifugal force to atomize droplets of liquid as they leave the highly machined, knife-edged rim of an electrically charged rotating applicator. The new higher-rotational-speed applicators will atomize high-viscosity, high-solids coatings (65 percent volume solids and higher). Disk-shaped applicators are almost always used in the automatic mode, with vertical reciprocators, inside a loop in the conveyor line. Bell-shaped applicators are used in automated systems in the same configurations as spray guns and can also be used manually.

An advantage of rotating disk and bell spray coating is its ability to atomize high-viscosity coating materials. A disadvantage is maintenance of the equipment.

High-Volume, Low-Pressure Spray Coating. High-volume, low-pressure (HVLP) spray coaters (40 to 60 percent TE) are a development of the early 1960s and have been upgraded. Turbines rather than pumps are now used to supply high volumes of

TABLE 6.19　Thermal Conductivities of Coatings

Material	k value,* cal/(sec)(cm²) (°C/cm) × 10⁴	Source of information
Unfilled plastics:		
Acrylic..........................	4–5	*a*
Alkyd...........................	8.3	*a*
Depolymerized rubber..............	3.2	H. V. Hardman, DPR Subsidiary
Epoxy...........................	3–6	*b*
Epoxy (electrostatic spray coating)....	6.6	Hysol Corp., DK-4
Epoxy (electrostatic spray coating)....	2.9	Minnesota Mining & Mfg., No. 5133
Epoxy (Epon† 828, 71.4% DEA, 10.7%)	5.2	
Epoxy (cured with diethylenetriamine)	4.8	*c*
Fluorocarbon (Teflon TFE)..........	7.0	Du Pont
Fluorocarbon (Teflon FEP)..........	5.8	Du Pont
Nylon...........................	10	*d*
Polyester........................	4–5	*a*
Polyethylenes....................	8	*a*
Polyimide (Pyre-M.L. enamel)........	3.5	*e*
Polyimide (Pyre-M.L. varnish)........	7.2	*f*
Polystyrene......................	1.73–2.76	*g*
Polystyrene......................	2.5–3.3	*a*
Polyurethane.....................	4–5	*n*
Polyvinyl chloride.................	3–4	*a*
Polyvinyl formal..................	3.7	*a*
Polyvinylidene chloride.............	2.0	*a*
Polyvinylidene fluoride.............	3.6	*h*
Polyxylylene (Parylene N)..........	3	Union Carbide
Silicones (RTV types)..............	5–7.5	Dow Corning Corp.
Silicones (Sylgard types)...........	3.5–7.5	Dow Corning Corp.
Silicones (Sylgard varnishes and coatings)........................	3.5–3.6	Dow Corning Corp.
Silicone (gel coating)...............	3.7	Dow Corning Corp.
Silicone (gel coating)...............	7 (150°C)	Dow Corning Corp.
Filled plastics:		
Epon 828/diethylenetriamine = A.....	4	*b*
A + 50% silica...................	10	*b*
A + 50% alumina	11	*b*
A + 50% beryllium oxide..........	12.5	*b*
A + 70% silica...................	12	*b*
A + 70% alumina	13	*b*
A + 70% beryllium oxide..........	17.8	*b*
Epoxy, flexibilized = B.............	5.4	*i*
B + 66% by weight tabular alumina	18.0	*i*
B + 64% by volume tabular alumina	50.0	*i*
Epoxy, filled.....................	20.2	Emerson & Cuming, 2651 ft
Epoxy (highly filled)...............	15–20	Wakefield Engineering Co.
Polyurethane (highly filled)..........	8–11	International Electronic Research Co.
Other materials used in electronic assemblies:		
Alumina ceramic..................	256–442 (20–212°F)	*a*
Aluminum.......................	2767–5575	*a*
Aluminum oxide (alumina), 96%......	840	*j*
Beryllium oxide, 99%..............	5500	*j*
Copper..........................	8095–9334	*a*
Glass (Borosill, 7052)..............	28	*k*
Glass (pot-soda-lead, 0120)..........	18	*k*

TABLE 6.19   Thermal Conductivities of Coatings (*Continued*)

Material	k value,* cal/(sec)(cm²) (°C/cm) × 10^h	Source of information
Glass (silica, 99.8 % SiO₂).............	40	*l*
Gold..............................	7104 (20–212°F)	*a*
Kovar.............................	395	*m*
Mica..............................	8.3–16.5	*a*
Nichrome‡........................	325	*m*
Silica............................	40	*k*
Silicon nitride....................	359	*m*
Silver............................	9995 (20–212°F)	*a*
Zircon............................	120–149	*a*

* All values are at room temperature unless otherwise specified.
† Trademark of Shell Chemical Co., New York, NY.
‡ Trademark of Driver-Harris Co., Harrison, NJ.
a *Mater. Eng.*, Materials Selector Issue, vol. 66, no. 5, Chapman-Reinhold Publication, mid-October, 1967.
b Wolf, D. C.: *Proc. Nat. Electron. and Packag. Symp.*, New York, June, 1964.
c Lee, H., and K. Neville: "Handbook of Epoxy Resins," McGraw-Hill Book Company, New York, 1966.
d Davis, R.: *Reinf. Plast.*, October, 1962.
e *DuPont Tech. Bull.* 19, Pyre-M.L. Wire Enamel, August, 1967.
f *DuPont Tech. Bull.* 1, Pyre-M.L. Varnish RK-692, April, 1966.
g Teach, W. C., and G. C. Kiessling: Polystyrene, Reinhold Publishing Corporation, New York, 1960.
h Barnhart, W. S., R. A. Ferren, and H. Iserson: 17th ANTEC of SPE, January, 1961.
i Gershman, A. J., and J. R. Andreotti: *Insulation,* September, 1967.
j *American Lava Corp.* Chart 651.
k Shand, E. B.: "Glass Engineering Handbook," McGraw-Hill Book Company, 1958.
l Kingery, W. D.: Oxides for High Temperature Applications, *Proc. Int. Symp.,* Asilomar, CA, October, 1959, McGraw-Hill Book Company, New York, 1960.
m Kohl, W. H.: "Handbook of Materials and Techniques for Vacuum Devices," Reinhold Publishing Company, New York, 1967.
n "Modern Plastics Encyclopedia," McGraw-Hill, Inc., New York, 1968.

low-pressure, heated air to the spray guns. Newer versions use ordinary compressed air. The air is heated to reduce the tendency to condense atmospheric moisture and to stabilize solvent evaporation. Low atomizing pressure results in lower droplet velocity, reduced bounce-back, and reduced overspray.

The main advantage of HVLP spray coating is the reduction of overspray and bounce-back and the elimination of the vapor cloud usually associated with spray painting.

Airless Electrostatic Spray Coating.   The airless electrostatic spray coating method (70 to 80 percent TE) uses airless spray guns with the addition of a DC power source that electrostatically charges the coating droplets.

Its advantage over airless spray is the increase in transfer efficiency due to the electrostatic attraction of charged droplets to the product.

Air-Assisted Airless Electrostatic Spray Coating.   The air-assisted airless electrostatic spray coating method (70 to 80 percent TE) is a hybrid of technologies. The addition of atomizing air to the airless spray gun allows the use of high-viscosity, high-solids coatings. Although the theoretical transfer efficiency is in a high range, it is lower than that of airless electrostatic spray coating because of the higher droplet velocity.

TABLE 6.20 Application Methods for Coatings

Method	Advantages	Limitations	Typical applications
Spray.........	Fast, adaptable to varied shapes and sizes. Equipment cost is low.	Difficult to completely coat complex parts and to obtain uniform thickness and reproducible coverage.	Motor frames and housings, electronic enclosures, circuit boards, electronic modules.
Dip.........	Provides thorough coverage, even on complex parts such as tubes and high-density electronic modules.	Viscosity and pot life of dip must be monitored. Speed of withdrawal must be regulated for consistent coating thickness.	Small- and medium-sized parts, castings, moisture and fungus proofing of modules, temporary protection of finished machined parts.
Brush.........	Brushing action provids good "wetting" of surface, resulting in good adhesion. Cost of equipment is lowest.	Poor thickness control; not for precise applications. High labor cost.	Coating of individual components, spot repairs, or maintenance.
Roller.........	High-speed continuous process; provides excellent control on thickness.	Large runs of flat sheets or coil stock required to justify equipment cost and setup time. Equipment cost is high.	Metal decorating of sheet to be used to fabricate cans, boxes.
Impregnation	Results in complete coverage of intricate and closely spaced parts. Seals fine leaks or pores.	Requires vacuum or pressure cycling or both. Special equipment usually required.	Coils, transformers, field and armature windings, metal castings, and sealing of porous structures.
Fluidized bed	Thick coatings can be applied in one dip. Uniform coating thickness on exposed surfaces. Dry materials are used, saving cost of solvents.	Requires preheating of part to above fusion temperature of coating. This temperature may be too high for some parts.	Motor stators; heavy-duty electrical insulation on castings, metal substrates for circuit boards, heat sinks.
Screen-on......	Deposits coating in selected areas through a mask. Provides good pattern deposition and controlled thickness.	Requires flat or smoothly curved surface. Preparation of screens is time-consuming.	Circuit boards, artwork, labels, masking against etching solution, spot insulation between circuitry layers or under heat sinks or components.
Electrocoating	Provides good control of thickness and uniformity. Parts wet from cleaning need not be dried before coating.	Limited number of coating types can be used; compounds must be specially formulated; ionic polymers. Often porous, sometimes nonadherent.	Primers for frames and bodies, complex castings such as open work, motor end bells.
Vacuum deposition......	Ultrathin, pinhole-free films possible. Selective deposition can be made through masks.	Thermal instability of most plastics; decomposition occurs on products. Vacuum control needed.	Experimental at present. Potential use is in microelectronics, capacitor dielectrics.
Electrostatic spray	Highly efficient coverage and use of paint on complex parts. Successfully automated.	High equipment cost. Requires specially formulated coatings.	Heat dissipators, electronic enclosures, open-work grills and complex parts.

TABLE 6.21 Theoretical Transfer Efficiencies TE of Coating Application Methods

Coating method	TE, %
Autodeposition	95–100
Centrifugal coating	95–100
Curtain coating	95–100
Electrocoating	95–100
Fluidized-bed powder	95–100
Electrostatic fluidized-bed powder	95–100
Electrostatic-spray powder	95–100
Flow coating	95–100
Roller coating	95–100
Dip coating	95–100
Rotating electrostatic disks and bells	80–90
Airless electrostatic spray	70–80
Air-assisted airless electrostatic spray	70–80
Air electrostatic spray	60–70
Airless spray	50–60
High-volume, low-pressure spray	40–60
Air-assisted airless spray	40–60
Multicomponent spray	30–70
Air-atomized spray	30–40

The advantage of using the air-assisted airless electrostatic spray method is its ability to handle high-viscosity materials. An additional advantage is better spray pattern control.

Air Electrostatic Spray Coating. The air electrostatic spray coating method (60 to 70 percent TE) uses conventional equipment with the addition of electrostatic charging capability. The atomizing air permits the use of most high-solids coatings.

Air electrostatic spray equipment has the advantage of being able to handle high-solids materials. This is overshadowed by its having the lowest transfer efficiency of the electrostatic spray coating methods.

Airless Spray Coating. When it was introduced, airless spray coating (50 to 60 percent TE) was an important paint-saving development. The coating material is forced by hydraulic pressure through a small orifice in the spray gun nozzle. As the liquid leaves the orifice, it expands and atomizes. The droplets have low velocities because they are not propelled by air pressure as in conventional spray guns. To reduce the coating's viscosity without adding solvents, in-line heaters were added.

Advantages of airless spray coating are less solvent used, less overspray, less bounce-back, and compensation for seasonal ambient air temperature and humidity changes. A disadvantage is its slower coating rate.

Multicomponent Spray Coating. Multicomponent spray coating equipment (30 to 70 percent TE) is used for applying fast curing coating system components simultaneously. Since they can be either hydraulic or air-atomizing, their transfer efficiencies vary from low to medium. They have two or more sets of supply and metering pumps to transport components to a common spray head.

Their main advantage, the ability to apply fast-curing multicomponent coating, can be overshadowed by disadvantages in equipment cleanup, maintenance, and low transfer efficiency.

Air-Atomized Spray Coating. Air-atomized spray coating equipment (30 to 40 percent TE) has been used to apply protective and decorative coatings to products since the 1920s. A stream of compressed air mixes with a stream of liquid coating material, causing it to atomize or break up into small droplets. The liquid and air streams are adjustable, as is the spray pattern, to meet the finishing requirements of most products. This equipment is still being used.

The advantage of the air-atomized spray gun is that a skilled operator can adjust fluid flow, air pressure, and coating viscosity to apply a high-quality finish on most products. The disadvantages are its low transfer efficiency and ability to spray only low-viscosity coatings, which emit great quantities of VOCs to the atmosphere.

6.6.4 Powder Coating

Powder coating (95 to 100 percent TE), developed in the 1950s, is a method for applying finely divided, dry, solid resinous coatings by dipping products in a fluidized bed or by spraying them electrostatically. The fluidized bed is essentially a modified dip tank. When charged powder particles are applied during the electrostatic spraying method, they adhere to grounded parts until fused and cured. In all cases the powder coating must be heated to its melt temperature, where a phase change occurs, causing it to adhere to the product and fuse to form a continuous coating. Elaborate reclaiming systems to collect and reuse oversprayed material in electrostatic-spray powder systems boost transfer efficiency. Since the enactment of the air-quality standards this method has grown markedly.

Fluidized-Bed Powder Coating. Fluidized-bed powder coating (95 to 100 percent TE) is simply a dipping process using dry, finely divided plastic materials. A fluidized bed is a tank having a porous bottom plate, as illustrated in Fig. 6.7. The plenum below the porous plate supplies low-pressure air uniformly across the plate. The rising air surrounds and suspends the finely divided plastic powder particles, causing the powder-air mixture to resemble a boiling liquid. Products that are preheated above the melt temperature of the material are dipped in the fluidized bed, where the powder melts and fuses into a continuous coating. Thermosetting powders often require additional heat to cure the film on the product. The high transfer efficiency results from no dragout and consequently no dripping. This method is used to apply heavy coats, 3 to 10 mil, in one pass, uniformly, to complex-shaped products. The film thickness is dependent on the powder chemistry, preheat temperature, and dwell time. It is possible to build film thicknesses of 100 mil using higher preheat temperatures and multiple dips. An illustration of film buildup is presented in Fig. 6.8.

Advantages of fluidized-bed powder coating are uniform and reproducible film thicknesses on all complex-shaped product surfaces. Another advantage is a heavy coating in one dip. A disadvantage of this method is the 3-mil minimum film thickness required to form a continuous coating.

Electrostatic Fluidized-Bed Powder Coating. An electrostatic fluidized bed coater (95 to 100 percent TE) is essentially a fluidized bed having a high-voltage DC

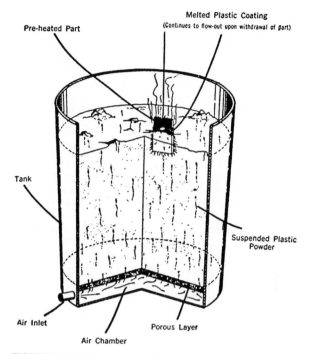

FIGURE 6.7 Illustration of fluidized-bed process principle.

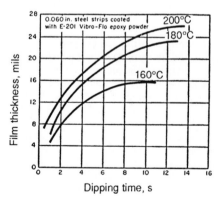

FIGURE 6.8 Effect of preheat temperature and dipping time on film build in coating a steel bar with epoxy resin.

grid installed above the porous plate to charge the finely divided particles. Once charged, the particles are repelled by the grid and repel each other, forming a cloud of powder above the grid. These electrostatically charged particles are attracted to and coat products that are at ground potential. Film thicknesses of 1½ to 5 mil are possible on cold parts, and 20 to 25 mil are possible on heated parts.

The advantage of the electrostatic fluidized bed is that small products such as electrical components can be coated uniformly and quickly. The disadvantage is that the product size is limited, and inside corners have low film thicknesses owing to the well known faraday cage effect.

Electrostatic-Spray Powder Coating. Electrostatic-spray powder coating (95 to 100 percent TE) is a method for applying finely divided, electrostatically charged plastic particles to products that are at ground potential. A powder-air mixture from a small fluidized bed in the powder reservoir is supplied by a hose to the spray gun, which has a charged electrode in the nozzle fed by a high-voltage DC power pack. In some cases the powder is electrostatically charged by friction. The spray guns can be manual or automatic, fixed or reciprocating, and mounted on one or both sides of a conveyorized spray booth. Electrostatic-spray powder coating operations use collectors to reclaim overspray. Film thicknesses of 1½ to 5 mil can be applied on cold products. If the products are heated slightly, 20- to 25-mil-thick coatings can be applied on these products. As with other coating methods, electrostatic-spray powder coating has limitations. Despite these limitations, powder coatings are replacing liquid coatings in a growing number of cases.

The advantage of this method is that coatings, using many of the resin types, can be applied in low (1½- to 3-mil) film thicknesses with no VOC emission at extremely high transfer efficiency. Disadvantages include the difficulty in obtaining less than 1-mil-thick continuous coatings and, owing to the complex powder reclaiming systems, color changes are more difficult than with liquid spray systems.

6.6.5 Other Coating Methods

Autodeposition Coating. Autodeposition (95 to 100 percent TE) is a dipping method where coatings are applied on the product from an aqueous solution. Unlike electrocoating, there is no electric current applied. Instead, the driving force is chemical.

Advantages of autophoretic coating are no VOC emissions, no metal pretreatment other than cleaning, and uniform coating thickness. This technique requires 30 percent less floor space than electrocoating, and capital equipment costs are 25 percent lower than for electrocoating. Disadvantages of autophoretic coatings are that black is the only color available, and corrosion resistance is lower than for electrocoated products.

Centrifugal Coating. A centrifugal coater (95 to 100 percent TE) is a self-contained unit. It consists of an inner basket, a dip coating tank, and exterior housing. Products are placed in the inner basket, which is dipped into the coating tank. The basket is withdrawn and spun at a speed high enough to remove excess coating material by centrifugal force. This causes the coating to be flung onto the inside of the exterior housing, from which it drains back into the dip coating tank.

The advantage of centrifugal coating is that large numbers of small parts can be coated at the same time. The disadvantage is that the appearance of the finish is a problem owing to the parts touching each other.

Flow Coating.   In a flow coater (95 to 100 percent TE), the coating material is pumped through hoses and nozzles onto the surfaces of the product, from which the excess drains into a reservoir to be recycled. Flow coaters can be either automatic or manual. Film thickness is controlled by the viscosity and solvent balance of the coating material. A continuous coater is an advanced flow coater using airless spray nozzles mounted on a rotating arm in an enclosure.

Advantages of flow coating are high transfer efficiency and low volume of paint in the system. Products will not float off hangers, and extremely large products can be painted. As with dip coating, the disadvantages of flow coating are coating thickness control and solvent refluxing.

Curtain Coating.   Curtain coating (95 to 100 percent TE), which is similar to flow coating, is used to coat flat products on conveyorized lines. The coating falls from a slotted pipe or flows over a weir in a steady stream or curtain while the product is conveyed through it. Excess material is collected and recycled through the system. Film thickness is controlled by coating composition, flow rates, and line speed.

The advantage of curtain coating is uniform coating thickness on flat products with high transfer efficiency. The disadvantage is inability to uniformly coat three-dimensional objects.

Roller Coating.   Roller coating (95 to 100 percent TE), which is used mainly by the coil coating industry for prefinishing metal coils to be later formed into products, has seen steady growth. It is also used for finishing flat sheets of material. There are two types of roller coaters, direct and reverse, depending on the direction of the applicator roller relative to the direction of the substrate movement. Roller coating can apply multiple coats to the front and back of coil stock with great uniformity.

The advantages of roller coating are consistent film thickness and elimination of painting operations at a fabricating plant. The disadvantages are limited metal thickness, limited bend radius, and corrosion of unpainted cut edges.

6.7 CURING

No dissertation on organic protective and decorative coatings is complete without mentioning film formation and cure. It is not the intent of this chapter to fully discuss the mechanisms, which are more important to researchers and formulators than to end users; but rather to show that differences exist and to aid the reader in making selections. Most of the organic coating resins are liquid, which cure or dry to form solid films. They are classified as thermoplastic or thermosetting. Thermoplastic resins dry by solvent evaporation and will soften when heated and harden when cooled. Thermosetting resins will not soften when heated after they are cured. Most thermoplastic and some thermosetting resin films are affected by solvents. Table 6.22 shows solvents that affect certain resins. Another classification of coatings is by their various film-forming mechanisms, such as solvent evaporation, coalescing, phase change, and conversion. Coatings are also classified as room-temperature curing, sometimes called air drying; or heat curing, generally called baking or force drying, which uses elevated temperatures to accelerate air drying. Thermoplastic and thermosetting coatings can be both air drying and baking.

6.7.1 Air Drying

Air drying coatings will form films and cure at room or ambient temperatures (20°C) by the mechanisms described in this section.

Solvent Evaporation.   Thermoplastic coating resins that form films by solvent evaporation are shellac and lacquers, such as nitrocellulose, acrylic, styrene-butadiene, and cellulose acetate butyrate.

Conversion.   In these coatings, films are formed as solvents evaporate, and they cure by oxidation, catalyzation, or cross-linking. Thermosetting coatings cross-link to form films at room temperature by oxidation or catalyzation. Oxidative curing of drying oils and oil-modified resins can be accelerated by using catalysts. Monomeric materials can form films and cure by cross-linking with polymers in the presence of catalysts, as in the case of styrene monomers and polyester resins. Epoxy resins will cross-link with polyamide resins to form films and cure. In the moisture-curing polyurethane resin coating systems, airborne moisture starts a reaction in the vehicle, resulting in film formation and cure.

Coalescing.   Emulsion or latex coatings, such as styrene-butadiene, acrylic ester, and vinyl acetate acrylic, form films by coalescing and dry by solvent evaporation.

6.7.2 Baking

Baking coating will form films at room temperature, but require elevated temperatures (150 to 200°C) to cure.

TABLE 6.22   Solvents that Affect Plastics

Resin	Heat-distortion point, °F	Solvents that affect surface
Acetal	338	None
Methyl methacrylate	160–195	Ketones, esters, aromatics
Modified acrylic	170–190	Ketones, esters, aromatics
Cellulose acetate	110–209	Ketones, some esters
Cellulose propionate	110–250	Ketones, esters, aromatics, alcohols
Cellulose acetate butyrate	115–227	Alcohols, ketones, esters, aromatics
Nylon	260–360	None
Polyethylene:		
High density	140–180	
Med. density	120–150	None
Low density	105–121	
Polypropylene	210–230	None
Polycarbonate	210–290	Ketones, esters, aromatics
Polystyrene (G.P. high heat)	150–195	Some aliphatics, ketones, esters, aromatics
Polystyrene (impact, heat-resistant)	148–200	Ketones, esters, aromatics, some aliphatics
ABS	165–225	Ketones, esters, aromatics, alcohol

Conversion. The cure of many oxidative thermosetting coatings is accelerated by heating. In other resins systems, such as thermosetting acrylics and alkyd melamines, the reactions do not occur below temperature thresholds of 135°C or higher. Baking coatings (those that require heat to cure) are generally tougher than air drying coatings. In some cases the cured films are so hard and brittle that they must be modified with other resins.

Phase Change. Thermoplastic coatings that form films by phase changes—generally from solid to liquid then back to solid—are polyolefins, waxes, and polyamides. Plastisols and organisols undergo phase changes during film formation. Fluidized-bed powder coatings, both thermoplastic and thermosetting, also undergo phase changes during film formation and cure.

6.7.3 Force Drying

In many cases the cure rate of thermoplastic and thermosetting coatings can be accelerated by exposure to elevated temperatures that are below those considered to be baking temperatures.

6.7.4 Reflowing

Certain thermoplastic coating films will soften and flow to become smooth and glossy at elevated temperatures. This technique is used on acrylic lacquers by the automotive industry to eliminate buffing.

6.7.5 Radiation Curing

Films are formed and cured by bombardment with ultraviolet and electron-beam radiation with little increase in surface temperature. Infrared radiation, on the other hand, increases the surface temperature of films and is therefore a baking process. Figure 6.9 shows typical radiation-curing equipment.

6.7.6 Vapor Curing

Vapor curing is essentially a catalyzation or cross-linking conversion method for two-component coatings. The product is coated with one component of the coating in a conventional manner. It is then placed in an enclosure filled with the other component—the curing agent in vapor form. It is in this enclosure that the reaction occurs.

6.8 SUMMARY

This chapter aids readers in selecting surface preparation methods, coating materials, and application methods. It acquaints them with curing methods and helps them to comply with environmental regulations.

FIGURE 6.9 Radiation curing is fast, allowing production-line speeds of 2000 ft/min. This technique takes place at room temperature, and heat-sensitive wooden and plastic products and electronic components and assemblies can be given the equivalent of a baked finish at speeds never before possible. (*Courtesy of Radiation Dynamics.*)

Coating selection is not easy, owing to the formulating versatility of modern coating materials. This versatility also contributes to one of their faults, which is the possible decline in one performance property when another is enhanced. Because of this, the choice of a coating must be based on specific performance properties and not on generalizations. This choice is further complicated by the need to comply with governmental regulations.

To apply coatings in the most effective manner, the product's size, shape, ultimate appearance, and end use must be addressed. This chapter emphasizes the importance of transfer efficiency in choosing a coating method.

To meet these requirements, product finishers have all the tools at their disposal. They can choose coatings that apply easily, coat uniformly, cure rapidly and efficiently, and comply with governmental regulations at lower costs. By applying coatings using methods that have high transfer efficiencies, finishers will not only comply with air- and water-quality standards, but they will also provide a safe workplace and decrease the generation of hazardous wastes.[3]

REFERENCES

1. C. P. Izzo, "Today's Paint Finishes: Better than Ever," *Prod. Finish. Mag.,* Oct. 1986.
2. *Paint/Coatings Dictionary,* Federation of Societies for Coatings Technology, 1978.
3. C. Izzo, "Overview of Industrial Coating Materials," in *Products Finishing Directory,* Gardner Publ., Cincinnati, OH, 1996.
4. C. Izzo, "How Are Coatings Applied," in *Products Finishing Directory,* Gardner Publ., Cincinnati, OH, 1994.

CHAPTER 7
JOINING OF PLASTICS, ELASTOMERS, AND COMPOSITES

Edward M. Petrie
Manager, Materials and Process Center
ABB Transmission Technology Institute
Raleigh, North Carolina

7.1 INTRODUCTION

This chapter provides practical information and guidance on how to join parts made from plastics, elastomers, or composites to themselves and to other substrates. A number of processes are available for joining these materials:

1. Adhesive bonding
2. Solvent cementing
3. Thermal welding
4. Mechanical bonding

All four methods are covered in this chapter. The focus, however, is on adhesive bonding, solvent cementing, and thermal welding because of the multidisciplined nature of these technologies and their usefulness for many applications involving polymeric substrates. This chapter also provides information on the formulation and properties of polymeric adhesives.

Each design engineer must determine the joining method that best suits the purpose. The methodology for this will often depend on a given assembly, the nature of the substrate materials, expected service conditions, production requirements, plant facilities, and the like.

7.1.1 Advantages and Disadvantages of Joining Methods

For the purposes of general discussion, solvent cementing and thermal welding will be considered as a type of adhesive bonding where the substrate serves also as the adhesive. Adhesive bonding (including solvent cementing and thermal welding) present several distinct advantages over conventional mechanical methods of fastening. These are summarized in Table 7.1. However, in certain applications, mechanical joining methods may be a more appropriate choice.

Physical Considerations.   An adhesive bond can provide certain physical advantages. The need to drill holes or add inserts that can weaken the structure is avoided. The loading stresses are distributed over a large area rather than being concentrated at the mechanical fastener.

The stress-distribution characteristics and inherent toughness of polymeric adhesives provide bonds with superior fatigue resistance, as shown in Fig. 7.1. Generally, in well designed joints, the adherends will fail in fatigue before the adhesive.

The most serious limitation to the use of modern polymeric adhesives is their time-dependent strength in degrading service environments such as moisture, high temperatures, or chemicals. There are polymeric adhesives that perform well at temperatures between -60 and $350°F$. But only a few adhesives can withstand operating temperatures outside that range. Adhesives can also be degraded by chemical environments and outdoor weathering. The rate of strength degradation may be accelerated by continuous stress or elevated temperatures.

Design Considerations.   Unlike rivets or bolts, adhesives produce smooth contours that are aerodynamically and aesthetically beneficial. Adhesives also offer a better strength/weight ratio than mechanical fasteners.

Adhesives can join any combination of solid materials, regardless of their shape or thickness. Materials such as plastics, elastomers, and composites can generally be joined more economically and efficiently by adhesive bonding than by other methods.

However, the adhesive joint must be carefully designed for optimum performance. Design factors must include the type of stress, environmental influences, and production methods that will be used. The strength of the adhesive joint is dependent on the type and direction of stress. Generally, adhesives perform better when stressed in shear or tension than when exposed to cleavage or peel forces.

Production Considerations.   Adhesive bonding is frequently faster and less expensive than conventional fastening methods. As the size of the area to be joined

TABLE 7.1   Advantages and Disadvantages of Adhesive Bonding

Advantages	Disadvantages
1. Provides large stress-bearing area.	1. Surfaces must be carefully cleaned.
2. Provides excellent fatigue strength.	2. Long cure times may be needed.
3. Damps vibration and absorbs shock.	3. Limitation on upper continuous operating
4. Minimizes or prevents galvanic corrosion between	temperature (generally 350°F).
dissimilar metals.	4. Heat and pressure may be required.
5. Joins all shapes and thicknesses.	5. Jigs and fixtures may be needed.
6. Provides smooth contours.	6. Rigid process control usually necessary.
7. Seals joints.	7. Inspection of finished joint difficult.
8. Joins any combination of similar or dissimilar mate-	8. Useful life depends on environment.
rials.	9. Environmental, health, and safety
9. Often less expensive and faster than mechanical	considerations are necessary.
fastening.	10. Special training sometimes required.
10. Heat, if required, is too low to affect metal parts.	
11. Provides attractive strength-to-weight ratio.	

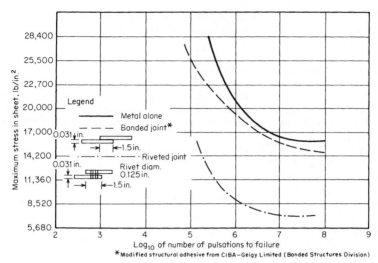

FIGURE 7.1 Fatigue strengths of aluminum-alloy specimens under pulsating tensile load.[1]

increases, the time and labor saved by using adhesives instead of mechanical fasteners become progressively greater.

All adhesives require clean surfaces to attain optimum results. Depending on the type and condition of the substrate and the bond strength desired, surface preparations ranging from a simple solvent wipe to chemical etching are necessary.

If the adhesive has multiple components, the parts must be carefully weighed and mixed. The setting operation often requires heat and pressure. Lengthy set time could make assembly jigs and fixtures necessary. Finally, the inspection of finished joints for quality control is very difficult. This necessitates strict control over the entire bonding process to ensure uniform bond quality.

The following items contribute to a hidden cost of using adhesives, and they also could lead to serious production difficulties:

- The storage life (shelf life) of the adhesive may be unrealistically short; some adhesives require refrigerated storage.
- The adhesive may begin to solidify or gel too early in the bonding process.
- Waste, safety, and environmental concerns can be essential cost factors.
- Cleanup is a cost factor, especially where misapplied adhesive may ruin the appearance of a product.
- Once bonded, samples cannot easily be disassembled; if misalignment occurs and the adhesive cures, usually the part must be scrapped.

Other Considerations.   Adhesives allow us to engineer into the product multiple functions in addition to the adhesive's prime function of holding the parts together. For example, adhesives can be made to function as electrical and thermal insulators. The degree of insulation can be varied with different adhesive formulations and fillers. Adhesives can even be made electrically and thermally conductive with silver and boron nitride fillers, respectively. Adhesives can also perform sealing functions, offering a barrier to the passage of fluids and gases. Adhesives may also act as vibration dampers to reduce the noise and oscillation encountered in assemblies.

7.1.2 Adhesive Bonding Fundamentals

Theories of Adhesion. Various theories attempt to describe the phenomena of adhesion. No single theory explains adhesion in a general way. However, knowledge of adhesion theories can assist in understanding the basic requirements for a good bond.

Mechanical Theory. The surface of a solid material is never truly smooth but consists of a maze of microscopic peaks and valleys. According to the mechanical theory of adhesion, the adhesive must penetrate the cavities on the surface and displace the trapped air at the interface.

Such mechanical anchoring appears to be a prime factor in bonding many porous substrates. Adhesives also frequently bond better to abraded surfaces than to natural surfaces. This beneficial effect may be due to:

* Mechanical interlocking
* Formation of a clean surface
* Formation of a more reactive surface
* Formation of a larger surface area

Adsorption Theory. The adsorption theory states that adhesion results from molecular contact between two materials and the surface forces that develop. The process of establishing intimate contact between an adhesive and the adherend is known as wetting. Figure 7.2 illustrates good and poor wetting of a liquid spreading over a surface.

For an adhesive to wet a solid surface, the adhesive should have a lower surface tension than the solid's critical surface tension. Tables 7.2 and 7.3 list surface tensions of common adherends and liquids.

Most organic adhesives easily wet metallic solids. But many solid organic substrates have surface tensions less than those of common adhesives. From Tables 7.2 and 7.3, it can be forecast that epoxy adhesives will wet clean aluminum or copper surfaces. However, epoxy resin will not wet a substrate having a critical surface tension significantly less than 47 dynes/cm. Epoxies will not, for example, wet either a metal surface contaminated with silicone oil or a clean polyethylene substrate.

After intimate contact is achieved between adhesive and adherend by wetting, it is believed that adhesion results primarily through forces of molecular attraction. Four general types of chemical bonds are recognized: The first three—electrostatic, covalent, and metallic—are referred to as primary bonds, and the fourth, van der Walls forces, are referred to as secondary bonds. The adhesion between adhesive and adherend is believed to be primarily due to van der Walls forces of attraction.

Electrostatic and Diffusion Theories. The electrostatic theory states that electrostatic forces in the form of an electrical double layer are formed at the adhesive-adherend interface. These forces account for resistance to separation.

The fundamental concept of the diffusion theory is that adhesion occurs through the interdiffusion of molecules in the adhesive and adherend. The diffusion theory is primarily applicable when both the adhesive and adherend are polymeric. For example, bonds formed by solvent or heat welding of thermoplastics result from the diffusion of molecules.

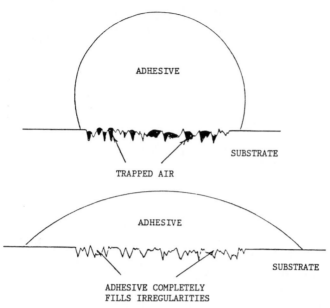

FIGURE 7.2   Illustration of poor (top) and good (bottom) wetting by adhesive spreading over a surface.

TABLE 7.2   Critical Surface Tensions of Common Plastics and Metals

Materials	Critical surface tension, dyn/cm
Acetal	47
Acrylonitrile-butadiene-styrene	35
Cellulose	45
Epoxy	47
Fluoroethylene propylene	16
Polyamide	46
Polycarbonate	46
Polyethylene	31
Polyethylene terephthalate	43
Polyimide	40
Polymethylmethacrylate	39
Polyphenylene sulfide	38
Polystyrene	33
Polysulfone	41
Polytetrafluoroethylene	18
Polyvinyl chloride	39
Silicone	24
Aluminum	≈ 500
Copper	≈ 1000

TABLE 7.3   Surface Tensions of Common Adhesives and Liquids

Material	Surface tension, dyn/cm
Epoxy resin	47
Fluorinated epoxy resin*	33
Glycerol	63
Petroleum lubricating oil	29
Silicone oils	21
Water	73

* Experimental resin; developed to wet low-energy surfaces. (Note low surface tension relative to most plastics.)

Weak-Boundary-Layer Theory.   According to the weak-boundary-layer theory, when bond failure seems to be at the interface, usually a cohesive break of a weak boundary layer is the real event.[2] Weak boundary layers can originate from the adhesive, the adherend, the environment, or a combination of any of the three.

When bond failure occurs, it is the weak boundary layer that fails, although failure seems to occur at the adhesive-adherend interface. Figure 7.3 shows examples of certain possible weak boundary layers due to moisture effects. Weak boundary layers can usually be removed or strengthened by various surface treatments.

Weak boundary layers formed from the shop environment are very common. When the adhesive does not wet the substrate as shown in Fig. 7.2, a weak boundary layer of air is trapped at the interface, causing lowered joint strength. Moisture from the air or dispersed mold release agents may also form weak boundary layers.

Requirements of a Good Bond.   The basic requirements for a good adhesive bond are cleanliness, wetting, solidification, and proper selection of adhesive and joint design.

Cleanliness.   To achieve an effective adhesive bond, one must start with a clean surface. Foreign materials such as dirt, oil, moisture, and weak oxide layers must be removed from the surface, or else the adhesive will bond to these weak boundary layers rather than the actual substrate. There are various surface preparations that remove or strengthen the weak boundary layers. Surface preparation methods for specific substrates will be discussed in a later section.

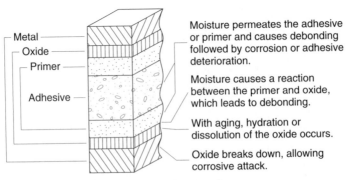

Metal
Oxide
Primer
Adhesive

Moisture permeates the adhesive or primer and causes debonding followed by corrosion or adhesive deterioration.

Moisture causes a reaction between the primer and oxide, which leads to debonding.

With aging, hydration or dissolution of the oxide occurs.

Oxide breaks down, allowing corrosive attack.

FIGURE 7.3   Examples of weak boundary layers.[3]

Wetting. While it is in the liquid state, the adhesive must wet the substrate. The result of good wetting is greater contact area between adherend and adhesive over which the forces of adhesion may act.

Solidification. The liquid adhesive, once applied, must be capable of conversion into a solid. The process of solidifying can be completed in different ways (e.g., chemical reaction by any combination of heat, pressure, and curing agents; cooling from a molten liquid to a solid; and drying due to solvent evaporation).

Adhesive Choice. Factors most likely to influence adhesive selection are listed in Table 7.4. The main areas of concern when selecting adhesives are the material to be bonded, service requirements, production requirements, and overall cost.

Joint Design. The adhesive joint should be designed to optimize the forces acting on and within the joint. Although adequate adhesive-bonded assemblies have been made from joints designed for mechanical fastening, maximum benefit can be obtained only in assemblies specifically designed for adhesive bonding.

Mechanism of Bond Degradation. Adhesive joints may fail adhesively or cohesively. Adhesive failure is interfacial bond failure between the adhesive and adherend. Cohesive failure occurs when the adhesive fractures, allowing a layer of adhesive to remain on both substrates. When the adherend fails before the adhesive, there is what is known as a cohesive failure of the adherend. The various modes of possible bond failures are shown in Fig. 7.4.

The exact cause of adhesive failure is very hard to determine, because so many factors in adhesive bonding are interrelated. However, when an adhesive bond is

TABLE 7.4   Factors Influencing Adhesive Selection

Stress	Tension
	Shear
	Impact
	Peel
	Cleavage
	Fatigue
Chemical factors	External (service-related)
	Internal (effect of adherend on adhesives)
Exposure	Weathering
	Light
	Oxidation
	Moisture
	Salt spray
Temperature	High
	Low
	Cycling
Biological factors	Bacteria or mold
	Rodents or vermin
Working properties	Application
	Bonding time and temperature range
	Tackiness
	Curing rate
	Storage stability
	Coverage

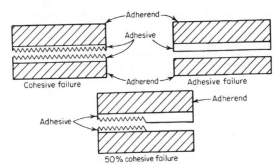

FIGURE 7.4   Cohesive and adhesive bond failures.

made, there are certain common elements at work that contribute to the weakening of all bonds. The influences of these elements are qualitatively summarized in Fig. 7.5.

If the adhesive does not wet the surface of the substrate, the joint will be inferior. Internal stresses occur in the adhesive joint because of a natural tendency of the adhesive to shrink during solidification and because of differences in physical properties between adhesive and substrate.

The coefficient of thermal expansion of adhesive and adherend should be as close as possible to limit stresses that may develop during thermal cycling or after cooling from an elevated-temperature cure. Adhesives can be formulated with various fillers to modify their thermal-expansion characteristics and limit internal stresses. A relatively elastic adhesive, capable of accommodating internal stress, may also be useful when thermal expansion differences are of concern.

The type of stress, its orientation to the joint, and the rate in which the stress is applied are important. Sustained loads can cause premature failure in service, even

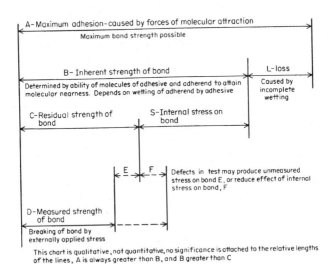

FIGURE 7.5   Relations between factors involved in adhesion.[4]

though similar unloaded joints may exhibit adequate strength when tested after aging. Most adhesives have poor strength when stresses are acting to peel or cleave the adhesive from the substrate. Many adhesives are sensitive to the rate in which the joint is stressed. Rigid, brittle adhesives sometimes have excellent tensile or shear strength but stand up very poorly under an impact test.

Operating environments are capable of degrading an adhesive joint in various ways. The adhesive may have to withstand temperature variation, weathering, oxidation, moisture, and other exposure conditions. If more than one of these factors are present, a synergistic effect could cause a rapid decline in adhesive strength.

Adhesive Classification.   Adhesives may be classified by many methods: function, chemical composition, mode of application and setting, and end use. The most common classification is by chemical composition.

The chemical composition classification broadly describes adhesives as thermosetting, thermoplastic, elastomeric, or combinations of these. There are then many chemical types within each classification. They are described in Table 7.5.

Adhesive Bonding Process.   A typical flowchart for the adhesive bonding process is shown in Fig. 7.6. The elements of the bonding process are as important as the adhesive itself for a successful end product.

Many of the adhesive problems that develop are not due to a poor choice of material or joint design but are directly related to faulty production techniques. The adhesive user must obtain the proper processing instructions from the manufacturer and follow them closely and consistently to ensure acceptable results.

7.1.3 Mechanical Joining Fundamentals

Different requirements exist for mechanically joining plastics, elastomers, and composites, and these must be considered during the design process. Fastener considerations include factors such as corrosion compatibility, fastener materials and strength, head configurations, clamping force, interference fit, and thermal expansion coefficients.

Mechanical assembly operations may be classified into two broad categories. The first category involves those associated with molded parts where the fastener can be molded into the part. The second involves those that arise in the creation of the product from basic shapes such as sheet, rod, tube, and so on. Here, separate fasteners are generally used.

Molded-In Joints.   It is often possible and desirable to incorporate fastening mechanisms in the design of the molded part itself (like a hinge or latch). Only stronger plastics are suitable for this method since the joint must survive the strain of assembly, service load, and possible repeated use. This form of fastening is suitable for lightly loaded, nonrigid assemblies where precession is not critical.

With brittle plastics, the assembly operation may cause the plastic to crack if conditions are not carefully controlled. In addition, plastics, especially thermoplastics, are subject to cold flow under stress. Under the continued stress from a forced fit, for example, the plastic may relax, causing the stress forming the joint to fall to a value insufficient to maintain the integrity of the joint.

As with adhesive bonding, it is important to remember that the thermal coefficient of expansion for most plastics is about ten times that of metals. Thus, changes in ambient temperature could cause stresses due to the dissimilar materials.

TABLE 7.5 Adhesives Classified by Chemical Composition[5]

Classification	Thermoplastic	Thermosetting	Elastomeric	Alloys
Types within group...	Cellulose acetate, cellulose acetate butyrate, cellulose nitrate, polyvinyl acetate, vinyl vinylidene, polyvinyl acetals, polyvinyl alcohol, polyamide, acrylic, phenoxy	Cyanoacrylate, polyester, urea formaldehyde, melamine formaldehyde, resorcinol and phenol-resorcinol formaldehyde, epoxy, polyimide, polybenzimidazole, acrylic, acrylate acid diester	Natural rubber, reclaimed rubber, butyl, polyisobutylene, nitrile, styrene-butadiene, polyurethane, polysulfide, silicone, neoprene	Epoxy-phenolic, epoxy-polysulfide, epoxy-nylon, nitrile-phenolic, neoprene-phenolic, vinyl-phenolic
Most used form......	Liquid, some dry film	Liquid, but all forms common	Liquid, some film	Liquid, paste, film
Common further classifications.........	By vehicle (most are solvent dispersions or water emulsions)	By cure requirements (heat and/or pressure most common but some are catalyst types)	By cure requirements (all are common); also by vehicle (most are solvent dispersions or water emulsions)	By cure requirements (usually heat and pressure except some epoxy types); by vehicle (most are solvent dispersions or 100% solids); and by type of adherends or end-service conditions
Bond characteristics...	Good to 150-200°F; poor creep strength; fair peel strength	Good to 200-500°F; good creep strength; fair peel strength	Good to 150-400°F; never melt completely; low strength; high flexibility	Balanced combination of properties of other chemical groups depending on formulation; generally higher strength over wider temp range
Major type of use......	Unstressed joints; designs with caps, overlaps, stiffeners	Stressed joints at slightly elevated temp	Unstressed joints on lightweight materials; joints in flexure	Where highest and strictest end-service conditions must be met; sometimes regardless of cost, as military uses
Materials most commonly bonded.........	Formulation range covers all materials, but emphasis on nonmetallics—esp. wood, leather, cork, paper, etc.	For structural uses of most materials	Few used "straight" for rubber, fabric, foil, paper, leather, plastics films; also as tapes. Most modified with synthetic resins	Metals, ceramics, glass, thermosetting plastics; nature of adherends often not as vital as design or end-service conditions (i.e., high strength, temp)

FIGURE 7.6   Basic steps in bonding process.

The design of joint systems in molded parts will often depend on the designer's creativity. However, careful attention must be given to the suitability for use with the plastic involved, moldability, and other features.

Mechanical Fastening.   A large variety of mechanical fasteners can be used for joining plastic parts to themselves and to each other. These include machine screws, self-tapping screws, rivets, spring clips, nuts, and other miscellaneous hardware. Designing plastic parts for use with any one fastener will depend on the fastener, the particular plastic with which it is to be used, and the functional requirements of the application.

Threaded fasteners work best on thick sections. Thread-forming screws are preferred for softer materials, while thread-cutting screws work best on harder plastics. Push-on locknuts and clips may be better for thinner sections.

If a fastener has to be removed a number of times, metal inserts are recommended. They may be molded in place, forced, glued, or expanded into molded or drilled holes, or inserted ultrasonically.

7.2   DESIGN OF JOINTS

7.2.1   Design of Adhesive Joints

Types of Stress.   Four basic types of loading stress are common to adhesive joints regardless of the substrate: tensile, shear, cleavage, and peel. Any combination of these stresses, illustrated in Fig. 7.7, may be encountered in an application.

Tensile stress develops when forces acting perpendicular to the plane of the joint are distributed uniformly over the entire bonded area. Adhesive joints show good resistance to tensile loading because all of the adhesive contributes to the strength of the joint. In practical applications, however, loads are rarely axial, and unwanted cleavage or peel stresses tend to develop.

Shear stress results when forces acting in the plane of the adhesive try to separate the adherends. Joints that are dependent upon the adhesive's shear strength are relatively easy to design and offer favorable properties. Adhesive joints are strong when stressed in shear because all of the bonded area contributes to the strength.

Cleavage and peel stresses are undesirable. Cleavage occurs when forces at one end of a rigid joint act to split the adherends apart. Peel stress is similar to cleavage but applies to a joint where one or both of the adherends are flexible. Joints loaded

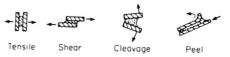

Tensile   Shear   Cleavage   Peel

FIGURE 7.7   Four basic types of adhesive stress.

in peel or cleavage offer lower strength than joints loaded in shear because the stress is concentrated at only a very small area of the total bond. The remainder of the bonded area makes no contribution to the strength of the joint. Peel and cleavage stresses should be avoided where possible.

Joint Efficiency.   The adhesive joint designer should take into consideration the following rules:

1. Keep the stress on the bond line to a minimum.
2. Design the joint so that the operating loads will stress the adhesive in shear.
3. Peel and cleavage stresses should be minimized.
4. Distribute the stress as uniformly as possible over the entire bonded area.
5. Adhesive strength is directly proportional to bond width. Increasing width will always increase bond strength; increasing the depth does not always increase strength.
6. Generally, rigid adhesives are better in shear, and flexible adhesives are better in peel.

Brittle adhesives are particularly weak in peel because the stress is localized at only a very thin line at the edge of the bond, as shown in Fig. 7.8. Tough, flexible adhesives distribute the peeling stress over a wider bond area and show greater resistance to peel.

For a given adhesive and adherend, the strength of a joint stressed in shear depends primarily on the width and depth of the overlap and the thickness of the adherend. Adhesive shear strength is directly proportional to the width of the joint. Strength can sometimes be increased by increasing the overlap depth, but the relationship is not linear. Since the ends of the bonded joint carry a higher proportion of the load than the interior area, the most efficient way of increasing joint strength is by increasing the width of the bonded area.

In a shear joint made from thin, relatively flexible adherends, there is a tendency for the bonded area to distort because of eccentricity of the applied load. This distortion, illustrated in Fig. 7.9, causes cleavage stress on the ends of the joint, and the joint strength may be considerably impaired. Thicker adherends are more rigid, and

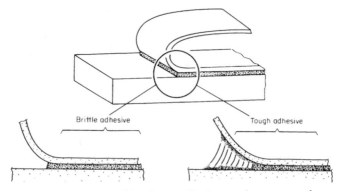

FIGURE 7.8   Tough, flexible adhesives distribute peel stress over a larger area.[6]

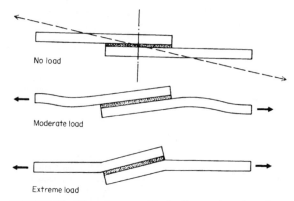

No load

Moderate load

Extreme load

FIGURE 7.9   Distortion caused by loading can introduce cleavage stresses and must be considered in joint design.[5]

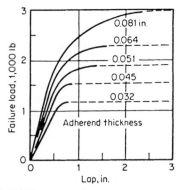

FIGURE 7.10   Interrelation of failure loads, depth of lap, and adherend thickness for lap joints with a specific adhesive and adherend.[7]

the distortion is not as much a problem as with thin-gage adherends. Figure 7.10 shows the general interrelationship between failure load, depth of overlap, and adherend thickness for a specific metallic adhesive joint.

Joint Design.   A favorable stress can be applied by using proper joint design. However, some joint designs may be impractical, expensive to make, or hard to align. The design engineer will often have to weigh these factors against optimum joint performance.

Flat Adherends.   The simplest joint to make is the plain butt joint. However, butt joints cannot withstand bending forces because the adhesive would experience cleavage stress. The butt joint can be improved by redesign in a number of ways, as shown in Fig. 7.11.

Lap joints are commonly used because they are simple to make and applicable to thin adherends, and they stress the adhesive in its strongest direction. Tensile loading of a lap joint causes the adhesive to be stressed in shear. However, the simple lap joint is offset, and the shear forces are not in line, as was illustrated in Fig. 7.9. Modifications of lap-joint design include:

1. Redesigning the joint to bring the load on the adherends in line
2. Making the adherends more rigid (thicker) near the bond area (see Fig. 7.10)
3. Making the edges of the bonded area more flexible for better conformance, thus minimizing peel

Modifications of lap joints are shown in Fig. 7.12.

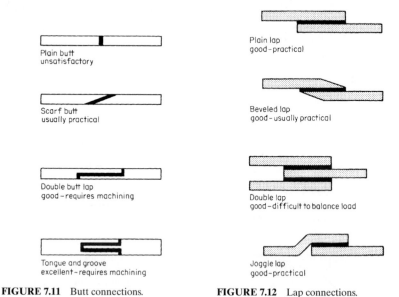

Plain butt
unsatisfactory

Plain lap
good - practical

Scarf butt
usually practical

Beveled lap
good - usually practical

Double butt lap
good – requires machining

Double lap
good – difficult to balance load

Tongue and groove
excellent – requires machining

Joggle lap
good - practical

FIGURE 7.11   Butt connections.

FIGURE 7.12   Lap connections.

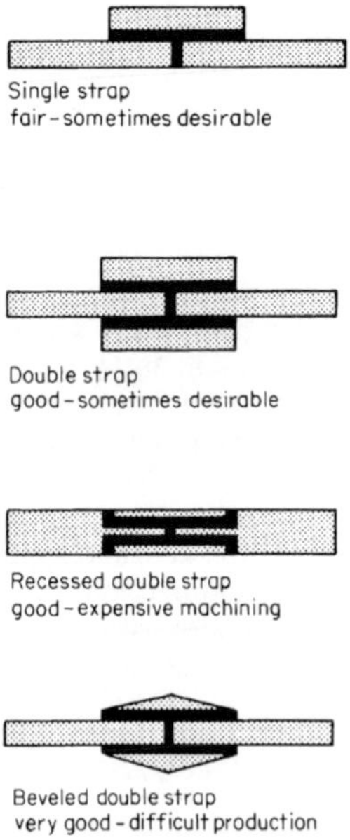

Single strap
fair – sometimes desirable

Double strap
good – sometimes desirable

Recessed double strap
good – expensive machining

Beveled double strap
very good - difficult production

FIGURE 7.13   Strap connections.

Strap joints keep the operating loads aligned and are generally used where overlap joints are impractical because of adherend thickness. Strap-joint designs are shown in Fig. 7.13. Like the lap joint, the single strap is subjected to cleavage stress under bending forces.

When thin members are bonded to thicker sheets, operating loads generally tend to peel the thin member from its base, as shown in Fig. 7.14 (top). The subsequent illustrations show what can be done to decrease peeling tendencies in simple joints.

Cylindrical Adherends.   Several recommended designs for rod and tube joints are illustrated in Fig. 7.15. These designs should be used instead of the simpler butt joint. Their resistance to bending forces and subsequent cleavage is much better, and the bonded area is larger. Unfortunately, most of these joint designs require a machining operation.

Angle and Corner Joints.   A butt joint is the simplest method of bonding two surfaces that meet at an angle. Although the butt joint has good resistance to pure tension and compression, its bending strength is very poor. Dado, L-, and T-angle joints, shown in Fig. 7.16, offer

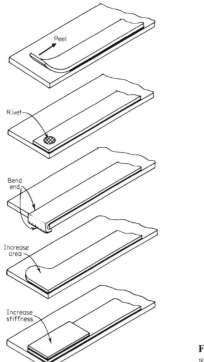

FIGURE 7.14 Minimizing peel in adhesive joints.[8]

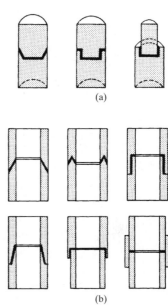

FIGURE 7.15 Recommended designs for rod and tube joints.[9] (*a*) Three joint designs for adhesive bonding of round bars. (*b*) Six joint configurations useful in adhesive-bonding cylinders or tubes.

greatly improved properties. The T design is the preferable angle joint because of its large bonding area and good strength in all directions.

Corner joints made of relatively flexible adherends such as sheet metal should be designed with reinforcements for support. Various corner-joint designs are shown in Fig. 7.17.

Flexible Substrates. Thin or flexible substrates such as polymeric film or elastomers may be joined using a simple or modified lap joint. The double strap joint is best, but it is also the most time-consuming to fabricate. The strap material should be made out of the same material as the parts to be joined, or at least have approximately equivalent strength, flexibility, and thickness. The adhesive should have the same degree of flexibility as the adherends.

If the sections to be bonded are relatively thick, a scarf joint is acceptable. The length of the scarf should be at least four times the thickness; sometimes larger scarfs may be needed.

When bonding elastomers, forces on the substrate during setting of the adhesive should be carefully controlled, since excess pressure will cause residual stresses at the bond interface.

Composites. Reinforced plastics are often anisotropic materials. Their strength properties are directional. Joints made from anisotropic substrates should be designed to stress both the adhesive and adherend in the direction of greatest

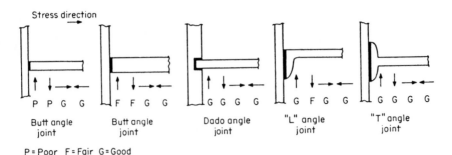

P = Poor   F = Fair   G = Good

FIGURE 7.16   Types of angle joints and methods of reducing cleavage.[8]

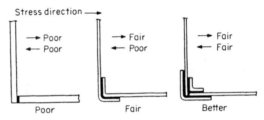

FIGURE 7.17   Reinforcement of bonded corners.

strength. Laminates, for example, should be stressed parallel to the laminations. Stresses normal to the laminate may cause the substrate to delaminate.

Single and joggle lap joints are more likely to cause delamination than scarf or beveled lap joints. The strap-joint variations are useful when bending loads are expected.

7.2.2   Design of Joints for Mechanical Assembly

There are basically two methods of mechanical assembly. The first uses fasteners such as screws or bolts; the second uses interference fit such as press fit or snap fit and are primarily used in thermoplastic applications. The optimum design for a plastic product is a one-part molding because it eliminates the need for a secondary assembly operation. However, mechanical limitations often make it necessary to join one part to another using a fastening device. A number of mechanical fastening methods are available to accomplish this task.

Mechanical fasteners and the design of parts to accommodate them are covered in detail in later sections of this chapter. The following describes the use of press-fit and snap-fit molded-in joints to achieve assembly.

Press Fit.   This technique provides joints with high strength at low cost. It is a simple and fast means for part assembly. The advisability of its use will depend on the relative properties of the two materials being assembled. It is generally used where two dissimilar materials are being assembled, the harder material usually being forced into the softer. For example, a metal shaft can be press-fitted into plastic hubs. With brittle plastics, such as the thermosets, press-fit assembly may cause the plastic to crack if conditions are not carefully controlled.

Where press fits are used, the designer generally seeks the maximum pullout force that is obtained using the greatest allowable interference between parts consistent with the strength of the plastics used. Figures 7.18 and 7.19 show calculated interference limits at room temperature for press-fitted shafts and hubs of Delrin® acetal resin and Zytel® nylon resin. These represent the maximum allowable interference based on yield point and elastic modulus data. Safety factors of 1.5 to 2 are used in most applications.

For a press-fit joint, the effects of thermal cycling and stress relaxation on the strength of the joint must be carefully evaluated. Testing of the factory-assembled parts under expected temperature cycles is obviously indicated.

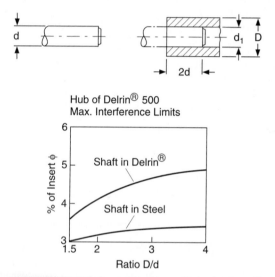

FIGURE 7.18 Maximum interference limits for Delrin® acetal.[10]

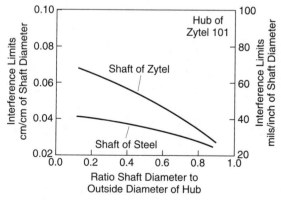

FIGURE 7.19 Theoretical interference limits for Zytel® nylon.[10] Based on yield point and elastic modulus at room temperature and average moisture conditions.

Differences in coefficient of thermal expansion can result in reduced interference due either to one material shrinking or expanding away from the other, or the creation of thermal stresses as the temperature changes.

Since plastic materials will creep or stress relieve under continued loading, loosening of the press fit, at least to some extent, can be expected. To counteract this, the designer should knurl or groove the parts. The plastic will then tend to flow into the grooves and retain the holding power of the joint.

Snap Fit. In all types of snap-fit joints, a protruding part of one component, such as a hook, stud, or bead, is briefly deflected during the joining operation and is caught in a depression (undercut) in the mating component. This method of assembly is uniquely suited to thermoplastic materials due to their flexibility, high elongation, and ability to be molded into complex shapes.

The two most common types of snap fits are those with flexible cantilevered lugs (Fig. 7.20) and those with a full cylindrical undercut and mating lip (Fig. 7.21).

Cylindrical snap fits are generally stronger but require greater assembly force than cantilevered lugs. Cylindrical snap fits require deformation for removal from the mold. Materials with good recovery characteristics are required.

In order to obtain satisfactory results, the undercut design must fulfill certain requirements:

- The wall thickness should be kept uniform.
- The snap fit must be placed in an area where the undercut section can expand freely.
- The ideal geometric shape is circular.
- Ejection of an undercut core from the mold is assisted by the fact that the resin is still at relatively high temperature.
- Weld lines should be avoided in the area of the undercut.

In the cantilevered snap-fit design, the retaining force is essentially a function of the bending stiffness of the resin. Cantilevered lugs should be designed in a way so as not to exceed allowable stresses during assembly. Also, since undercuts in the mold are used, the snap-fit material should be capable of being stripped from the mold without distortion.

Cantilevered snap fits should be dimensioned to develop constant stress distribution over their length. This can be achieved by providing a slightly tapered section

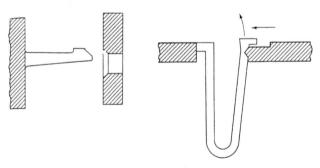

FIGURE 7.20   Snap-fitting cantilevered arms.[11]

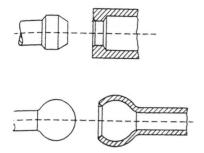

FIGURE 7.21 Undercuts for snap joints.[11]

or by adding a rib. Special care must be taken to avoid sharp corners and other possible stress concentrations.

Snap fits cannot create joints that carry a load in excess of the force necessary to make or break the snap fit. Thus, they are generally used to attach lids or covers that are meant to be disassembled in the use of the part or that will be lightly loaded. The design should be such that after the assembly, the joint should return to a stress-free condition.

Many more designs and configurations can be used with the snap-fit configuration. The individual plastic resin suppliers are suggested for design rules and guidance on specific applications.

7.3 TEST METHODS

7.3.1 Test Methods for Adhesive Bonding

A number of standard tests for adhesive bonds have been specified by the American Society for Testing and Materials (ASTM). Selected ASTM standards are presented in Table 7.6. The properties usually reported by adhesive suppliers are ASTM tensile-shear and peel strength.

Lap-Shear Tests. The lap-shear or tensile-shear test measures the strength of the adhesive in shear. It is the most common adhesive test because the specimens are inexpensive, easy to fabricate, and simple to test. This method is described in ASTM D 1002, and the standard test specimen is shown in Fig. 7.22(a). The specimen is loaded in tension, causing the adhesive to be stressed in shear until failure occurs. Since the test calls for a sample population of five, specimens can be made and cut from large test panels, illustrated in Fig. 7.22(b).

Tensile Tests. The tensile strength of an adhesive joint is seldom reported in the adhesive suppliers' literature because pure tensile stress is not often encountered in actual production. Tensile test specimens also require considerable machining to assure parallel surfaces.

ASTM tension tests are described in D 897 and D 2095 and employ bar- or rod-shaped butt joints. The maximum load at which failure occurs is recorded in pounds per square inch of bonded area. Test environment, joint geometry, and type of failure should also be recorded.

A simple cross-lap specimen to determine tensile strength is described in ASTM D 1344. This specimen has the advantage of being easy to make, but grip alignment and adherend deflection during loading can cause irreproducibility. A sample population of at least ten is recommended for this test method.

Peel Test. Because adhesives are notoriously weak in peel, tests to measure peel resistance are very important. Peel tests involve stripping away a flexible adherend from another adherend that may be flexible or rigid. The specimen is usually peeled at an angle of 90 or 180°.

TABLE 7.6   ASTM Adhesive Standards*

Aging

Resistance of Adhesives to Cyclic Aging Conditions, Test for (D 1183)
Bonding Permanency of Water- or Solvent-Soluble Liquid Adhesives for Labeling Glass Bottles, Test for (D 1581)
Bonding Permanency of Water- or Solvent-Soluble Liquid Adhesives for Automatic Machine Sealing Top Flaps of Fiber Specimens, Test for (D 1713)
Permanence of Adhesive-Bonded Joints in Plywood under Mold Conditions, Test for (D 1877)
Accelerated Aging of Adhesive Joints by the Oxygen-Pressure Method, Practice for (D 3632)

Amylaceous Matter

Amylaceous Matter in Adhesives, Test for (D 1488)

Biodeterioration

Susceptibility of Dry Adhesive Film to Attack by Roaches, Test for (D 1382)
Susceptibility of Dry Adhesive Film to Attack by Laboratory Rats, Test for (D 1383)
Permanence of Adhesive-Bonded Joints in Plywood under Mold Conditions, Test for (D 1877)
Effect of Bacterial Contamination of Adhesive Preparations and Adhesive Films, Test for (D 4299)
Effect of Mold Contamination on Permanence of Adhesive Preparation and Adhesive Films, Test for (D 4300)

Blocking Point

Blocking Point of Potentially Adhesive Layers, Test for (D 1146)

Bonding Permanency

(See Aging)

Chemical Reagents

Resistance of Adhesive Bonds to Chemical Reagents, Test for (D 896)

Cleavage

Cleavage Strength of Metal-to-Metal Adhesive Bonds, Test for (D 1062)

Cleavage/Peel Strength

Strength Properties of Adhesives in Cleavage Peel by Tension Loading (Engineering Plastics-to-Engineering Plastics), Test for (D 3807)
(See also Peel Strength)

Corrosivity

Determining Corrosivity if Adhesive Materials, Practice for (D 3310)

Creep

Conducting Creep Tests of Metal-to-Metal Adhesives, Practice for (D 1780)
Creep Properties of Adhesives in Shear by Compression Loading (Metal-to-Metal), Test for (D 2293)
Creep Properties of Adhesives in Shear by Tension Loading, Test for (D 2294)

Cryogenic Temperatures

Strength Properties of Adhesives in Shear by Tension Loading in the Temperature Range from -267.8 to $-55°C$ (-450 to $-67°F$), Test for (D 2557)

Density

Density of Adhesives in Fluid Form, Test for (D 1875)

TABLE 7.6   ASTM Adhesive Standards (*Continued*)

Durability (Including Weathering)
Effect of Moisture and Temperature on Adhesive Bonds, Test for (D 1151) Atmospheric Exposure of Adhesive-Bonded Joints and Structures, Practice for (D 1828) Determining Durability of Adhesive Joints Stressed in Peel, Practice for (D 2918) Determining Durability of Adhesive Joints Stressed in Shear by Tension Loading, Practice for (D 2919) (See also Wedge Test)
Electrical Properties
Adhesives Relative to Their Use as Electrical Insulation, Testing (D 1304)
Electrolytic Corrosion
Determining Electrolytic Corrosion of Copper by Adhesives, Practice for (D 3482)
Fatigue
Fatigue Properties of Adhesives in Shear by Tension Loading (Metal/Metal), Test for (D 3166)
Filler Content
Filler Content of Phenol, Resorcinol, and Melamine Adhesives, Test for (D 1579)
Flexibility
(See Flexural Strength)
Flexural Strength
Flexural Strength of Adhesive Bonded Laminated Assemblies, Test for (D 1184) Flexibility Determination of Hot Melt Adhesives by Mandrel Bend Test Method, Practice for (D 3111)
Flow Properties
Flow Properties of Adhesives, Test for (D 2183)
Fracture Strength in Cleavage
Fracture Strength in Cleavage of Adhesives in Bonded Joints, Practice for (D 3433)
Gap-Filling Adhesive Bonds
Strength of Gap Filling Adhesive Bonds in Shear by Compression Loading, Practice for (D 3931)
High-Temperature Effects
Strength Properties of Adhesives in Shear by Tension Loading at Elevated Temperatures (Metal-to-Metal), Test for (D 2295)
Hydrogen-Ion Concentration
Hydrogen Ion Concentration, Test for (D 1583)
Impact Strength
Impact Strength of Adhesive Bonds, Test for (D 950)
Light Exposure
(See Radiation Exposure)
Low and Cryogenic Temperatures
Strength Properties of Adhesives in Shear by Tension Loading in the Temperature Range from -267.8 to $-55°C$ (-450 to $-67°F$), Test for (D 2557)

TABLE 7.6 ASTM Adhesive Standards (*Continued*)

Nonvolatile Content

Nonvolatile Content of Aqueous Adhesives, Test for (D 1489)

Nonvolatile Content of Urea-Formaldehyde Resin Solutions, Test for (D 1490)
Nonvolatile Content of Phenol, Resorcinol, and Melamine Adhesives, Test for (D 1582)

Odor

Determination of the Odor of Adhesives, Test for (D 4339)

Peel Strength (Stripping Strength)

Peel or Stripping Strength of Adhesive Bonds, Test for (D 903)
Climbing Drum Peel Test for Adhesives, Method for (D 1781)
Peel Resistance of Adhesives (T-Peel Test), Test for (D 1876)
Evaluating Peel Strength of Shoe Sole Attaching Adhesives, Test for (D 2558)
Determining Durability of Adhesive Joints Stressed in Peel, Practice for (D 2918)
Floating Roller Peel Resistance, Test for (D 3167)

Penetration

Penetration of Adhesives, Test for (D 1916)

pH

(See Hydrogen-Ion Concentration)

Radiation Exposure (Including Light)

Exposure of Adhesive Specimens to Artificial (Carbon-Arc Type) and Natural Light, Practice for (D 904)
Exposure of Adhesive Specimens to High-Energy Radiation, Practice for (D 1879)

Rubber Cement Tests

Rubber Cements, Testing of (D 816)

Salt Spray (Fog) Testing

Salt Spray (Fog) Testing, Method of (B 117)
Modified Salt Spray (Fog) Testing, Practice For (G 85)

Shear Strength (Tensile Shear Strength)

Shear Strength and Shear Modulus of Structural Adhesives, Test for (E 229)
Strength Properties of Adhesive Bonds in Shear by Compression Loading, Test for (D 905)
Strength Properties of Adhesives in Plywood Type Construction in Shear by Tension Loading, Test for (D 906)
Strength Properties of Adhesives in Shear by Tension Loading (Metal-to-Metal), Test for (D 1002)
Determining Strength Development of Adhesive Bonds, Practice for (D 1144)
Strength Properties of Metal-to-Metal Adhesives by Compression Loading (Disk Shear), Test for (D 2181)
Strength Properties of Adhesives in Shear by Tension Loading at Elevated Temperatures (Metal-to-Metal), Test for (D 2295)
Strength Properties of Adhesives in Two-Ply Wood Construction in Shear by Tension Loading, Test for (D 2339)
Strength Properties of Adhesives in Shear by Tension Loading in the Temperature Range from −267.8 to −55°C (−450 to −67°F), Test for (D 2557)
Determining Durability of Adhesive Joints Stressed in Shear by Tension Loading, Practice for (D 2919)
Determining the Strength of Adhesively Bonded Rigid Plastic Lap-Shear Joints in Shear by Tension Loading, Practice for (D 3163)
Determining the Strength of Adhesively Bonded Plastic Lap-Shear Sandwich Joints in Shear by Tension Loading, Practice for (D 3164)
Strength Properties of Adhesives in Shear by Tension Loading of Laminated Assemblies, Test for (D 3165)

TABLE 7.6 ASTM Adhesive Standards (*Continued*)

Shear Strength (Tensile Shear Strength) (*continued*)

Fatigue Properties of Adhesives in Shear by Tension Loading (Metal/Metal), Test for (D 3166)
Strength Properties of Double Lap Shear Adhesive Joints by Tension Loading, Test for (D 3528)
Strength of Gap-Filling Adhesive Bonds in Shear by Compression Loading, Practice for (D 3931)
Measuring Strength and Shear Modulus of Nonrigid Adhesives by the Thick Adherend Tensile Lap Specimen, Practice for (D 3983)
Measuring Shear Properties of Structural Adhesives by the Modified-Rail Test, Practice for (D 4027)

Specimen Preparation

Preparation of Bar and Rod Specimens for Adhesion Tests, Practice for (D 2094)

Spot-Adhesion Test

Qualitative Determination of Adhesion of Adhesives to Substrates by Spot Adhesion Test Method, Practice for (D 3808)

Spread (Coverage)

Applied Weight per Unit Area of Dried Adhesive Solids, Test for (D 898)
Applied Weight per Unit Area of Liquid Adhesive, Test for (D 899)

Storage Life

Storage Life of Adhesives by Consistency and Bond Strength, Test for (D 1337)

Strength Development

Determining Strength Development of Adhesive Bonds, Practice for (D 1144)

Stress-Cracking Resistance

Evaluating the Stress Cracking of Plastics by Adhesives Using the Bent Beam Method, Practice for (D 3929)

Stripping Strength

(See Peel Strength)

Surface Preparation

Preparation of Surfaces of Plastics Prior to Adhesive Bonding, Practice for (D 2093)
Preparation of Metal Surfaces for Adhesive Bonding, Practice for (D 2651)
Analysis of Sulfochromate Etch Solution Used in Surface Preparation of Aluminum, Methods of (D 2674)
Preparation of Aluminum Surfaces for Structural Adhesive Bonding (Phosphoric Acid Anodizing), Practice for (D 3933)

Tack

Pressure Sensitive Tack of Adhesives Using an Inverted Probe Machine, Test for (D 2979)
Tack of Pressure-Sensitive Adhesives by Rolling Ball, Test for (D 3121)

Tensile Strength

Tensile Properties of Adhesive Bonds, Test for (D 897)
Determining Strength Development of Adhesive Bonds, Practice for (D 1144)
Cross-Lap Specimens for Tensile Properties of Adhesives, Testing of (D 1344)
Tensile Strength of Adhesives by Means of Bar and Rod Specimens, Method for (D 2095)

Torque Strength

Determining the Torque Strength of Ultraviolet (UV) Light-Cured Glass/Metal Adhesive Joints, Practice for (D 3658)

Viscosity

Viscosity of Adhesives, Test for (D 1084)
Apparent Viscosity of Adhesives Having Shear-Rate-Dependent Flow Properties, Test for (D 2556)
Viscosity of Hot Melt Adhesives and Coating Materials, Test for (D 3236)

TABLE 7.6 ASTM Adhesive Standards (*Continued*)

Volume Resistivity
Volume Resistivity of Conductive Adhesives, Test for (D 2739)

Water Absorptiveness (of Paper Labels)
Water Absorptiveness of Paper Labels, Test for (D 1584)

Weathering
(See Durability)

Wedge Test
Adhesive Bonded Surface Durability of Aluminum (Wedge Test) (D 3762)

Working Life
Working Life of Liquid or Paste Adhesive by Consistency and Bond Strength, Test for (D 1338)

* The latest revisions of ASTM standards can be obtained from the American Society for Testing and Materials, 1916 Race Street, Philadelphia, PA 19103.

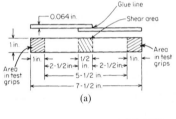

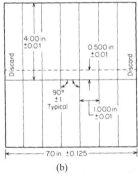

FIGURE 7.22 Standard lap-shear test specimen design. (*a*) Form and dimensions of lap-shear test specimen. (*b*) Standard test panel of five lap-shear specimens. (*From ASTM D 1002.*)

The most common types of peel test are the T-peel, Bell, and climbing-drum methods. Representative test specimens are shown in Fig. 7.23. The values resulting from each test method can be substantially different; hence it is important to specify the test method employed.

Peel values are recorded in pounds per inch of width of the bonded specimen. They tend to fluctuate more than other adhesive test results because of the extremely small area at which the stress is localized during loading.

The T-peel test is described in ASTM D 1876 and is the most common of all peel tests. The T-peel specimen is shown in Fig. 7.24. Generally, this test method is used when both adherends are flexible.

A 90° peel test, such as the Bell peel (ASTM D 3167), is used when one adherend is flexible and the other is rigid. The flexible member is peeled at a constant 90° angle through a spool arrangement. Thus, the values obtained are generally more reproducible.

The climbing-drum peel specimen is described in ASTM D 1781. This test method is intended for determining peel strength of thin metal facings on honeycomb cores, although it can generally be used for joints where at least one member is flexible.

A variation of the T-peel test is a 180° stripping test illustrated in Fig. 7.25 and described in ASTM D 903. It is commonly used when one adherend is flexible enough to permit a 180° turn near the point of loading. This test offers more reproducible results than the T-peel test because the angle of peel is maintained constant.

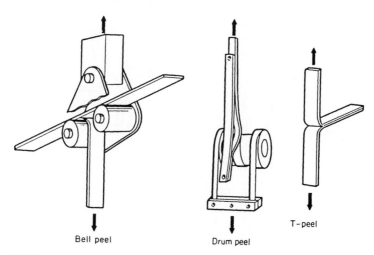

FIGURE 7.23 Common types of adhesive peel tests.[12]

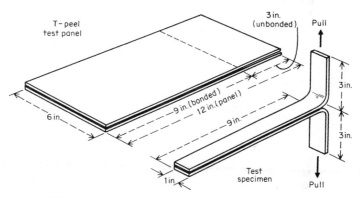

FIGURE 7.24 Test panel and specimen for T peel. (*From ASTM D 1876.*)

Cleavage Test. Cleavage tests are conducted by prying apart one end of a rigid bonded joint and measuring the load necessary to cause rupture. The test method is described in ASTM D 1062. A standard test specimen is illustrated in Fig. 7.26. Cleavage values are reported in pounds per inch of adhesive width. Because cleavage test specimens involve considerable machining, peel tests are usually preferred.

Fatigue Test. Fatigue testing places a given load repeatedly on a bonded joint. Standard lap-shear specimens are tested on a fatiguing machine capable of inducing cyclic loading (usually in tension) on the joint. The fatigue strength of an adhesive is reported as the number of cycles of a known load necessary to cause failure.

Fatigue strength is dependent on adhesive, curing conditions, joint geometry, mode of stressing, magnitude of stress, and duration and frequency of load cycling.

Impact Test. The resistance of an adhesive to impact can be determined by ASTM D 950. The specimen is mounted in a grip shown in Fig. 7.27 and placed in a standard

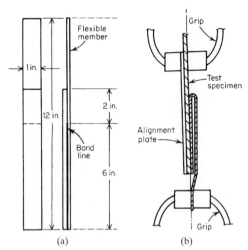

FIGURE 7.25   180° peel test specimens. (*a*) Specimen
design. (*b*) Specimen under test. (*From ASTM D 903.*)

impact machine. One adherend is struck with a pendulum hammer traveling at 11
ft/s, and the energy of impact is reported in pounds per square inch of bonded area.

Creep Test.   The dimensional change occurring in a stressed adhesive over a long
period of time is called creep. Creep data are seldom reported in the adhesive suppliers' literature because the tests are time-consuming and expensive. This is very
unfortunate, since sustained loading is a common occurrence in adhesive applica-

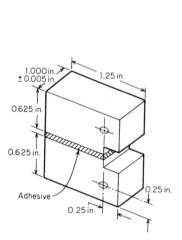

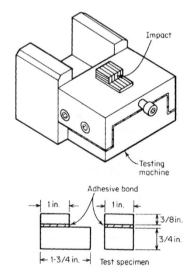

FIGURE 7.26   Cleavage test specimen. (*From*
ASTM D 1062.)

FIGURE 7.27   Impact test specimen and holding fixture. (*From ASTM D 950.*)

tions. All adhesives tend to creep, some much more than others. With weak adhesives, creep may be so extensive that bond failure occurs prematurely. Certain adhesives have also been found to degrade more rapidly when aged in a stressed rather than an unstressed condition.

Creep tests are made by loading a specimen with a predetermined stress and measuring the total deformation as a function of time or measuring the time necessary for complete failure of the specimen. Depending on the adhesive, loads, and testing conditions, the time required for a measurable deformation may be extremely long. ASTM D 2294 defines a test for creep properties of adhesives utilizing a spring-loaded apparatus to maintain constant stress.

Environmental Tests. Strength values determined by short-term tests do not give an adequate indication of an adhesive's permanence during continuous environmental exposure. Laboratory-controlled aging tests seldom last longer than a few thousand hours. To predict the permanence of an adhesive over a 20-year product life requires accelerated test procedures and extrapolation of data. Such extrapolations are extremely risky because the causes of adhesive-bond deterioration are many and not well understood.

7.3.2 Test Methods for Mechanical Fastening

Mechanical fastening tests, unlike adhesive joint tests, are often considered as subcomponent testing. The inherent characteristics of the substrate material are determined via coupon testing; whereas, assembled joints are considered structural parts or subcomponents and are designed and tested to verify manufacturing approaches and structural integrity. In many respects, simple mechanical fastener testing under uniaxial tension and compression loading resembles the tests described above for adhesive joints.

Interference Fit Joints. There is a lack of standard methods for testing integral plastic joints such as the press fit and snap fit designs. This is because the finished part must be tested in an environment that closely resembles the expected service environment. Thus, the actual assembled part and environmental conditioning tests are commonly used.

Mechanical Fasteners. In the case of mechanical fasteners, more established test methods have been developed, especially in the composites industry. Tests of tensile pull-through of thin sheets, static joint lap shear strength, and lap shear fatigue have become the three most important means of evaluating new fastener designs.

Tensile pull-through in thin sheets is performed in a fixture. With the fastener installed in two laminates, a recording of load versus deflection is made while the sheets are separated, and the fastener blind head is pulled through the laminate. The larger the blind head, the higher the load required to pull the fastener through the sheet.

Static joint lap shear tests are usually conducted on a specimen such as that depicted in Fig. 7.28. Key factors influencing lap-shear joint strengths are hole quality, fit between hole and fastener shank, blind head diameters and stiffness, and fastener preload. For highest strengths, fastener rotation in the joint should be restricted, thereby minimizing joint elongation as load is applied.

Lap-shear fatigue tests are also performed on specimens shown in Fig. 7.28 at several load levels. It has been demonstrated that fastener preload is the most influential factor affecting results because the higher the preload, the greater the number of

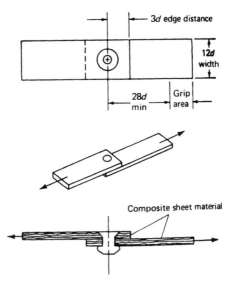

FIGURE 7.28 Lap-shear composite joint for mechanical fasteners.[13]

cycles to failure. However, if the preload is too high, crushing of the substrate can occur. High fastener preload also provides resistance to movement between the two substrates, thereby minimizing heat buildup and fretting.

One must consider several important factors when evaluating test results of fastened polymeric materials. Chief among these are: installation effects on jointed material, sheet take-up (clamp-up), fastener materials and strength, head configurations, interference fit, and sensitivity to hole quality.

7.4 SURFACE PREPARATION FOR ADHESIVE BONDING

7.4.1 Importance of Surface Preparation

Surface preparation of adherends prior to bonding is one of the most important factors in the adhesive-bonding process. Initial bond strength and joint permanence are greatly dependent on the quality of surface that is in contact with the adhesive. Prebond treatments are intended to remove weak boundary layers and provide easily wettable surfaces. As a general rule, all adherends must be treated in some manner prior to bonding.

Surface preparations can range from simple solvent wiping to a combination of mechanical abrading, chemical cleaning, and acid etching. In many low- to medium-strength applications, extensive surface preparation may be unnecessary. But, where maximum bond strength, permanence, and reliability are required, carefully controlled surface-treating processes are necessary. The following factors should be considered in the selection of a surface-preparation process:

1. The ultimate initial bond strength required
2. The degree of permanence necessary and the service environment
3. The degree and type of contamination on the adherend
4. The type of adherend and adhesive

Table 7.7 shows the effect of various metallic-surface preparations on adhesive-joint strength.

Surface preparations enhance the quality of a bonded joint by performing one or more of the following functions: (1) remove contaminants, (2) control adsorbed water, (3) control oxide formation, (4) poison surface atoms which catalyze adhesive breakdown, (5) protect the adhesive from the adherend and vice versa, (6) match the adherend crystal structure to the adhesive molecular structure, and (7) control surface roughness. Thus, surface preparations can affect the permanence of the joint as well as the initial strength.

Plastic, elastomeric, and polymeric-based composite adherends are even more dependent than metals on surface preparation. Many of these surfaces are contaminated with mold-release agents or processing additives. Such contaminants must be removed before bonding. Because of their low surface energy, polytetrafluoroethylene, polyethylene, and certain other polymeric materials are completely unsuitable for adhesive bonding in their natural state. The surfaces of these materials must be chemically or physically altered prior to bonding to improve wetting.

TABLE 7.7   Effect of Substrate Pretreatment on Strength of Adhesive-Bonded Joints

Adherend	Treatment	Adhesive	Shear strength, lb/in.^{2}	Ref.
Aluminum	As received	Epoxy	444	14
	Vapor degreased		837	
	Grit blast		1,751	
	Acid etch		2,756	
Aluminum	As received	Vinyl-phenolic	2,442	15
	Degreased		2,741	
	Acid etch		5,173	
Stainless steel	As received	Vinyl-phenolic	5,215	15
	Degreased		6,306	
	Acid etch		7,056	
Cold-rolled steel	As received	Epoxy	2,900	16
	Vapor degreased		2,910	
	Grit blast		4,260	
	Acid etch		4,470	
Copper	Vapor degreased	Epoxy	1,790	17
	Acid etch		2,330	
Titanium	As received	Vinyl-phenolic	1,356	15
	Degreased		3,180	
	Acid etch		6,743	
Titanium	Acid etch	Epoxy	3,183	18
	Liquid pickle		3,317	
	Liquid hone		3,900	
	Hydrofluorosilicic acid etch		4,005	

7.4.2 General Surface Preparation Methods

Listed here are several methods of preparing both metal and polymer substrates for adhesive bonding. The chosen method will ultimately be the process that yields the necessary strength and permanence with the least cost.

Solvent Wiping.   Where loosely held dirt, grease, and oil are the only contaminants, simple solvent wiping will provide surfaces for weak- to medium-strength bonds. Solvent wiping is widely used, but it is the least effective substrate treatment. Volatile solvents such as acetone and trichloroethylene are acceptable. Trichloroethylene is often favored because of its nonflammability. A clean cloth should be saturated with the solvent and wiped across the area to be bonded until no signs of residue are evident on the cloth or substrate. Special precautions are necessary to prevent the solvent from becoming contaminated. For example, the wiping cloth should never touch the solvent container, and new wiping cloths must be used often. After cleaning, the parts should be air-dried in a clean, dry environment before being bonded.

Vapor Degreasing.   Vapor degreasing is a reproducible form of solvent cleaning that is attractive when many parts must be prepared. It consists of suspending the adherends in a container of solvent vapor such as trichloroethylene or perchloroethylene. When the hot vapors come into contact with the relatively cool substrate, solvent condensation occurs on the surface of the part that dissolves the organic contaminants. Vapor degreasing is preferred to solvent wiping because the surfaces are continuously being washed in distilled, uncontaminated solvent. The vapor degreaser must be kept clean and a fresh supply of solvent used when the contaminants in the solvent container lower the boiling point significantly.

Modern vapor degreasing equipment is available with ultrasonic transducers built into the solvent rinse tank. The parts are initially cleaned by vapor and then subjected to ultrasonic scrubbing. The cleaning solutions and processing parameters must be optimized by test.

Abrasive Cleaning.   Mechanical methods for surface preparation include sandblasting, wire brushing, and abrasion with sandpaper, emery cloth, or metal wool. These methods are most effective for removing heavy, loose particles such as dirt, scale, tarnish, and oxide layers.

The parts should always be degreased prior to abrasive treatment to prevent contaminants from being rubbed into the surface. Solid particles left on the surfaces after abrading can be removed by blasts of clean, dry, oil-free air or by solvent wiping.

Chemical Cleaning.   Strong detergent solutions are used to emulsify surface contaminants on both metallic and nonmetallic substrates. These solutions are usually heated. Parts for cleaning are generally immersed in a well agitated solution maintained at 150 to 210°F for approximately 10 min. The surfaces are then rinsed immediately with deionized water and dried. Chemical cleaning is often used in combination with other surface treatments. Chemical cleaning by itself will not remove heavy or strongly attached contaminants such as rust or scale.

Alkaline detergents recommended for prebond cleaning are combinations of alkaline salts such as sodium metasilicate and tetrasodium pyrophosphate with surfactants included. Many commercial detergents are available.

Other Cleaning Methods.   Vapor honing and ultrasonic cleaning are efficient treating methods for small, delicate parts. In cases where the substrate is so delicate that usual abrasive treatments may be too rough, contaminants can be removed by vapor honing. This method is similar to grit blasting except that very fine abrasive particles are suspended in a high-velocity water or steam spray. Thorough rinsing after vapor honing is usually not required.

Ultrasonic cleaning employs a bath of cleaning liquid or solvent that is ultrasonically activated by a high-frequency transducer. The part to be cleaned is immersed in the liquid, which carries the sonic waves to the surface of the part. High-frequency vibrations then dislodge the contaminants. Commercial ultrasonic cleaning units are available from a number of manufacturers.

Alteration of Surfaces.   Certain treatments change the physical and chemical properties of the surface to produce greater wettability and/or a stronger surface. Specific processes are required for each substrate material. The part or area to be bonded is usually immersed in an active chemical solution for a matter of minutes. The parts are then immediately rinsed with deionized water and dried.

Chemical solutions must be changed regularly to prevent contamination and assure repeatable concentration. Tank temperature and agitation must also be controlled. Personnel need to be trained in the safe handling and use of chemical solutions and must wear the proper clothing.

Combined Methods.   More than one cleaning method is usually required for optimum adhesive properties. A three-step process that is recommended for most substrates consists of (1) degreasing, (2) mechanical abrasion, and (3) chemical treatment. The sequence of these processes is important because one wants to remove the soluble contaminants from the surface before abrasion, or else the contaminant could be driven further into the substrate by the abrading process.

7.4.3   Surface Treatment of Polymers

Treating of plastic surfaces usually consists of one or a combination of the following processes: solvent cleaning, abrasive treatment and etching, flame, hot air, electric discharge, or plasma treatments. The purpose of the treatment is either to remove or strengthen the weak boundary layer or to increase the critical surface tension.

Abrasive treatments consist of scouring, machining, hand sanding, and dry and wet abrasive blasting. The choice is generally determined by available production facilities and cost.

Chemical etching treatments vary with the type of plastic surface. The plastic resin supplier is the best guide to the appropriate etching chemicals and process. Etching processes can involve the use of corrosive and hazardous materials. The most common processes are sulfuric-dichromate etch (polyethylene and polypropylene) and sodium-naphthalene etch (fluorocarbons).

Flame, hot air, electrical discharge, and plasma treatments physically and chemically change the nature of polymeric surfaces. Treatment of certain polymeric surfaces with radio-frequency-excited inert gases greatly improves the bond strength of adhesive joints prepared from these materials. With this technique, called plasma treatment, a low-pressure inert gas is activated by an electrodeless radio-frequency discharge or microwave excitation to produce metastable species that react with the polymeric surface. The type of plasma gas can be selected to initiate a wide assortment of chemical reactions. In the case of polyethylene, plasma treatment produces a strong, wettable, cross-linked skin. Commercial instruments are available that can treat polymeric materials in this manner. Table 7.8 shows that plasma treatment results in improved plastic-joint strength with common epoxy adhesives.

7.4.4 Primers

Primers are applied and cured onto the adherend prior to adhesive bonding. They serve four primary functions individually or in combination:

1. Protection of surfaces after treatment (primers can be used to extend the time between preparing the adherend surface and bonding)
2. Developing tack for holding or positioning parts to be bonded
3. Inhibiting corrosion during service
4. Serving as an intermediate layer to enhance the physical properties of the joint and improve bond strength

Some primers have been found to provide corrosion resistance for the joint during service. The primer protects the adhesive-adherend interface and lengthens the service life of the bonded joint. Representative data are shown in Fig. 7.29.

Primers may also be used to modify the characteristics of the joint. For example, elastomeric primers are used with rigid adhesives to provide greater peel or cleavage resistance.

Metal primers are often used in bonding of elastomers to metal. They generally are solutions of chlorinated rubber and phenolic resins. The resin component provides cohesive strength and enhances adhesion to the metal. The rubber component provides toughness to the system and assists in reducing residual bondline stresses.

Certain primers can also chemically react with the adhesive and adherend to provide greater joint strengths. This type of primer is referred to as an adhesion promoter. The use of reactive silane adhesion promoters to improve the adhesion of resin to glass fibers in polymeric laminates is well known in the plastics industry.

TABLE 7.8 Typical Adhesive-Strength Improvement with Plasma Treatment; Aluminum-Plastic Shear Specimen Bonded with Epon 828–Versamid 140 (70–30) Epoxy Adhesive[19]

	Strength of bond, lb/in.^{2}	
Material	Control	After plasma treatment
High-density polyethylene	315 + 38	Greater than 3,125 + 68
Low-density polyethylene	372 + 52	Greater than 1,466 ± 106
Nylon 6	846 ± 166	Greater than 3,956 ± 195
Polystyrene	566 ± 17	Greater than 4,015 ± 85
Mylar* A	530 ± 51	1,660 ± 40
Mylar D	618 ± 25	1,185 ± 83
Polyvinylfluoride (Tedlar)*	278 ± 2	Greater than 1,280 ± 73
Lexan†	410 ± 10	928 ± 66
Polypropylene	370 ±	3,080 ± 180
Teflon* TFE	75	750

* Trademark of E. I. du Pont de Nemours & Company, Inc.
† Trademark of General Electric Co.

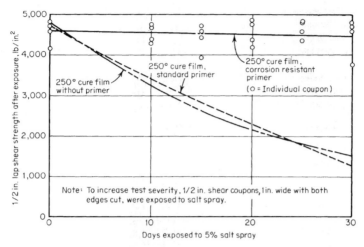

FIGURE 7.29   Effect of primer on lap-shear strength of aluminum joints exposed to 5 percent salt spray.[20]

7.4.5   Specific Surface Treatments and Characteristics

Metallic Adherends.   Table 7.9 lists common surface-treating procedures for metallic adherends. The general methods previously described are all applicable to metallic surfaces, but the processes listed in Table 7.9 have been specifically found to provide reproducible structural bonds and fit easily into the bonding operation. The metals most commonly used in bonded structures and their respective surface treatments are described more fully in the following sections.

Aluminum and Aluminum Alloys.   The effects of various aluminum surface treatments have been studied extensively. The most widely used process for high-strength, environment-resistant adhesive joints is the sodium dichromate–sulfuric acid etch, developed by Forest Product Laboratories and known as the FPL etch process. Abrasion or solvent degreasing treatments result in lower bond strengths, but these simpler processes are more easily placed into production. Table 7.10 qualitatively lists the bond strengths that can be realized with various aluminum treatments.

Copper and Copper Alloys.   Surface preparation of copper alloys is necessary to remove weak oxide layers attached to the copper surface. This oxide layer is especially troublesome because it forms very rapidly. Copper specimens must be bonded or primed as quickly as possible after surface preparation. Copper also has a tendency to form brittle surface compounds when used with certain adhesives that are corrosive to copper.

One of the better surface treatments for copper does not remove the oxide layer but creates a deeper and stronger oxide formation. Chromate conversion coatings are also used for high-strength copper joints.

Steel and Stainless Steel.   Steels are generally easy to bond, provided that all rust, scale, and organic contaminants are removed. This may be accomplished easily by a combination of mechanical abrasion and solvent cleaning.

Prepared steel surfaces are easily oxidized. Once processed, they should be kept free of moisture and primed or bonded within 4 h. Stainless surfaces are not as sensitive to oxidation as carbon steels, and a slightly longer time between surface preparation and bonding is acceptable.

TABLE 7.9　Surface Preparation for Metals

Adherend	Degreasing solvent	Method of treatment	Remarks
Aluminum and aluminum alloys	Trichloro- ethylene	1. Sandblast or 100-grit emery cloth followed by solvent degreasing	Medium- to high-strength bonds, suitable for noncritical applications
		2. Immerse for 10 min at 77 ± 6°C in a commercial alkaline cleaner or	Optimum bond strength, specified in ASTM D 2651 and MIL-A-9067. Solvent degrease may replace alkaline cleaning

	Parts by wt.
Sodium metasilicate..	30.0
Sodium hydroxide....	1.5
Sodium pyrophosphate	1.5
Nacconol NR (Allied Chemical Co.)......	0.5
Water (distilled)......	128.0

Wash in water below 65°C and etch for 10 min at 68±3°C in

	Parts by wt.
Sodium dichromate...	1
Sulfuric acid (96%, sp. gr. 1.84).........	10
Water (distilled)......	30

Rinse in distilled water after washing in tap water, and dry in air

Table continued on p. 7.35

Titanium Alloys.　Because of the usual use of titanium at high temperatures, most surface preparations are directed at improving the thermal resistance of titanium joints. Like magnesium, titanium can also react with the adhesive during cure and create a weak boundary layer.

Plastic Adherends.　Many plastics and plastic composites can be treated by simple mechanical abrasion or alkaline cleaning to remove surface contaminants. In some cases it is necessary that the polymeric surface be physically or chemically modified to achieve acceptable bonding. This applies particularly to crystalline thermoplastics such as the polyolefins, linear polyesters, and fluorocarbons. Methods used to improve the bonding characteristics of these surfaces include:

1. Oxidation via chemical treatment or flame treatment
2. Electrical discharge to leave a more reactive surface
3. Ionized inert gas, which strengthens the surface by cross linking and leaves it more reactive
4. Metal-ion treatment

Table 7.11 lists common recommended surface treatments for plastic adherends. These treatments are necessary when plastics are to be joined with adhesives. Solvent and heat welding are other methods of fastening plastics that do not require chemical alteration of the surface. Welding procedures will be discussed in another section of this chapter. As with metallic substrates, the effects of plastic surface treatments decrease with time. It is necessary to prime or bond soon after the surfaces are treated. Listed below are some common plastic materials that require special physical or chemical treatments to achieve adequate surfaces for adhesive bonding.

TABLE 7.9   Surface Preparation for Metals (*Continued*)

Adherend	Degreasing solvent	Method of treatment	Remarks
		3. Vapor degrease or solvent wipe. Immerse for 5 min at 71–82°C in Sulfuric acid (96%, sp. gr. 1.84)......... 1 gal, Chromic acid........ 45 oz, Distilled water....... 9 gal. Rinse in distilled water and dry in air	Alternative to sulfuric-dichromate etch
		4. Degrease with 50/50 solution of methyl ethyl ketone and chlorothene. Abrade lightly with mildly abrasive cleaner. Rinse in deionized water; wipe; or air dry. Etch 20 min at RT in *Parts by wt.* Sodium dichromate... 2, Sulfuric acid (96%, sp. gr. 1.84)......... 7. Rinse thoroughly in deionized water; dry at 70°C for 30 min	Room-temp etch
		5. Form a paste using sulfuric acid–dichromate solution and finely divided silica or fuller's earth. Apply; do not permit paste to dry. Time depends on degree of contamination (usually greater than 10 min at RT). Wash very thoroughly with deionized water, and air-dry	Paste form of acid etch useful when part cannot be immersed. ASTM D 2651
Brass and bronze (see also Copper and copper alloys)	Trichloroethylene	1. Etch for 5 min at 20°C in *Parts by wt.* Zinc oxide........... 20, Sulfuric acid (96%, sp. gr. 1.84)......... 460	Temperatures must not exceed 65°C when washing and drying

Table continued on p. 7.36

Fluorocarbons. Fluorocarbons such as polytetrafluoroethylene (TFE), polyfluoroethylene propylene (FEP), polychlorotrifluoroethylene (CFE), and polymonochlorotrifluoroethylene (Kel-F) are notoriously difficult to bond because of their low surface tension. However, epoxy and polyurethane adhesives offer moderate strength if the fluorocarbon is treated prior to bonding.

The fluorocarbon surface may be made more wettable by exposing it for a brief moment to a hot flame to oxidize the surface. The most satisfactory surface treatment is achieved by immersing the plastic in a bath consisting of sodium-naphthalene dispersion in tetrahydrofuran. This process is believed to remove fluorine atoms, leaving a carbonized surface that can be wet easily. Fluorocarbon films pretreated for adhesive bonding are available from most suppliers. A formulation and description of the sodium-naphthalene process may be found in Table 7.11. Commercial chemical products for etching fluorocarbons are also listed.

TABLE 7.9 Surface Preparation for Metals (*Continued*)

Adherend	Degreasing solvent	Method of treatment	Remarks
		Nitric acid (69%, sp. gr. 1.41)............ 360 Rinse in water, below 65°C, and re-etch in the acid solution for 5 min at 49°C. Rinse in distilled water after washing, and dry in air	
Chromium	Trichloroethylene	1. Abrasion. Grit or vapor blast, or 100-grit emery cloth, followed by solvent degreasing	Suitable for general-purpose bonding
		2. Etch for 1–5 min at 90–95°C in	
			Parts by wt.
		Hydrochloric acid (37%).............. 17 Water............... 20 Rinse in distilled water after cold/hot-water washing, and dry in hot air	For maximum bond strength
Copper and copper alloys	Trichloroethylene	1. Abrasion. Sanding, wire brushing, or 100-grit emery cloth, followed by vapor or solvent degreasing	Suitable for general-purpose bonding. Use 320-grit emery cloth for foil
		2. Etch for 10 min at 66°C in	For maximum bond strength. Suitable for brass and bronze. ASTM D 2651
		Parts by wt. Ferric sulfate........ 1.0 Sulfuric acid (96%)... 0.75 Water.............. 8.0 Wash in water at 20°C, and etch in cold solution of	
		Parts by wt. Sodium dichromate... 5 Sulfuric acid (96%)... 10 Water.............. 85 Etch until a bright clean surface has been obtained. Rinse in water, dip in ammonium hydroxide (sp. gr. 0.88), and wash in tap water. Rinse in distilled water, and dry in warm air	
		3. Etch for 1–2 min at 20°C in	Room-temp etch. ASTM D 2651
		Parts by wt. Ferric chloride (42% w/w solution)....... 0.75 Nitric acid (sp. gr. 1.41).............. 1.5 Water.............. 10.0 Rinse in distilled water after cold-water wash and dry in air stream at 20°C	

TABLE 7.9 Surface Preparation for Metals (*Continued*)

Adherend	Degreasing solvent	Method of treatment	Remarks
		4. Etch for 30 s at 20°C in *Parts by wt.* Ammonium persulfate 1 Water............. 4 Rinse in distilled water after cold-water wash, and dry in air stream at 20°C	Alternative etching solution to above where fast processing is required
		5. Solvent degrease. Immerse 30 s at 20°C in *Parts by vol.* Nitric acid (69%).... 10 Deionized water...... 90 Rinse in running water and transfer immediately to next solution; immerse for 1–2 min at 98°C in Ebonol C (Ethone, Inc., New Haven, Conn.) 24 oz and equivalent water to make 1 gal Rinse in deionized water, and air-dry	For copper alloys containing over 95% copper. Stable surface for hot bonding. ASTM D 2651
Gold	Trichloro-ethylene	Solvent or vapor degrease after light abrasion with a fine emery cloth	
Iron		See Steel (mild)	
Lead and solders	Trichloro-ethylene	Abrasion. Grit or vapor blast, or 100-grit emery cloth, followed by solvent degreasing	
Magnesium and magnesium alloys	Trichloro-ethylene	1. Abrasion with 100-grit emery cloth followed by solvent degreasing	Apply the adhesive immediately after abrasion
		2. Vapor degrease. Immerse for 10 min at 60–70°C in *Parts by wt.* Deionized water...... 95 Sodium metasilicate.. 2.5 Trisodium pyrophosphate............. 1.1 Sodium hydroxide.... 1.1 Nacconal NR (Allied Chemical Co.)...... 0.3 Rinse in water and dry below 60°C	Medium to high bond strength. ASTM D 2651
		3. Vapor degrease. Immerse for 10 min at 71–88°C *Parts by wt.* Water............. 4 Chromic acid........ 1 Rinse in water, and dry below 60°C	High bond strength. ASTM D 2651

TABLE 7.9   Surface Preparation for Metals (*Continued*)

Adherend	Degreasing solvent	Method of treatment	Remarks
		4. Vapor degrease. Immerse for 5–10 min at 63–80°C in *Parts by wt.* Water............... 12 Sodium hydroxide.... 1 Rinse in water. Immerse for 5–15 min at RT in *Parts by wt.* Water............... 123 Chromic acid........ 24 Calcium nitrate...... 1.8 Rinse in water, and dry below 60°C	Alternative to procedure 2
		5. Light anodic treatment and various corrosion-preventive treatments have been developed by magnesium producers (Dow 17 and Dow 7, Dow Chemical Co.)	Dow 17 preferred under extreme environmental conditions
Nickel	Trichloroethylene	1. Abrasion with 100-grit emery cloth followed by solvent degreasing	For general-purpose bonding
		2. Etch for 5 s at 20°C in nitric acid (69%, sp.gr.1.41). Wash in cold and hot water followed by a distilled-water rinse, and air-dry at 40°C	For general-purpose bonding
Silver	Trichloroethylene	Abrasion with 320-grit emery cloth followed by solvent degreasing	
Steel (stainless)	Trichloroethylene	1. Abrasion with 100-grit emery cloth, grit or vapor blast, followed by solvent degreasing	
		2. Solvent degrease and abrade with grit paper. Degrease again. Immerse for 10 min at 71–82°C in *Parts by wt.* Sodium metasilicate.. 3 Tetrasodium pyrophosphate.......... 1.5 Sodium hydroxide.... 1.5 Nacconol NR (Allied Chemical Co.)...... 0.5 Distilled water....... 138.0 Rinse in deionized water; dry in air, at 93°C. Immerse for 10 min at 85–91°C in *Parts by wt.* Oxalic acid.......... 1 Sulfuric acid (sp. gr. 1.84)............... 1 Distilled water....... 8 Rinse in deionized water; dry at 93°C for 10–15 min	Heat-resistant bond. Alkaline clean alone sufficient for general bonding. Commercial alkaline cleaners (Prebond 700, American Cyanamid) available

TABLE 7.9   Surface Preparation for Metals (*Continued*)

Adherend	Degreasing solvent	Method of treatment	Remarks

		3. Etch for 15 min at 63°C in
		
		               Parts
		              by wt.
		Sodium dichromate
		(saturated solution).. 0.30
		Sulfuric acid......... 10.0
		Remove carbon residue with nylon brush while rinsing. Rinse in distilled water and dry in warm air, at 93°C
			ASTM D 2651

		4. Etch for 2 min at 93°C in
		
		               Parts
		              by wt.
		Hydrochloric acid
		(37%)............. 20
		Orthophosphoric acid
		(85%)............. 3
		Hydrofluoric acid
		(35%)............. 1
		Rinse in warm water with a final rinse in distilled water. Dry in air below 93°C
			For maximum resistance to heat and environment. ASTM D 2651

		5. Vapor degrease for 10 min and pickle for 10 min at 20°C in
		
		               Parts
		              by vol.
		Nitric acid (69%, sp. gr. 1.41)............ 10
		Hydrofluoric acid
		(48%)............. 2
		Water............... 88
		Dry in air under 70°C
			Room-temperature etch. Treatment may be followed by passivation for 20 min in 5–10% w/v chromic acid (CrO_3) solution

Steel (mild), iron, and ferrous metals other than stainless	Trichloroethylene	1. Abrasion. Grit or vapor blast followed by solvent degreasing with water-free solvents
		
		2. Etch for 5–10 min at 20°C in
		               Parts
		              by wt.
		Hydrochloric acid
		(37%)............. 1
		Water............... 1
		Rinse in distilled water after cold-water wash, and dry in warm air for 10 min at 93°C
		
		3. Etch for 10 min at 60°C in
		               Parts
		              by wt.
		Orthophosphoric acid
		(85%)............. 1
		Ethyl alcohol
		(denatured)......... 2
		Brush off carbon residue with nylon brush while washing in running water. Rinse with distilled water, and heat for 1 h at 120°C
			Xylene or toluene is preferred to acetone and ketone, which may be moist enough to cause rusting
			
			Bonding should follow immediately after etching treatment since ferrous metals are prone to rusting. Abrasion is more suitable for procedure where bonding is delayed
			
			For maximum strength

Tin	Trichloroethylene	Solvent or vapor degrease after light abrasion with a fine emery cloth (320-grit)	

TABLE 7.9   Surface Preparation for Metals (*Continued*)

Adherend	Degreasing solvent	Method of treatment	Remarks
Titanium and titanium alloys	Trichloroethylene	1. Abrasion. Grit or vapor blast, or 100-grit emery cloth, followed by solvent degrease; or scour with a nonchlorinated cleaner, rinse, and dry 2. Etch for 5–10 min at 20°C in	For general-purpose bonding

		Parts by wt.	
	Sodium fluoride	2	
	Chromium trioxide	1	
	Sulfuric acid (96%, sp. gr. 1.84)	10	
	Water	50	

Rinse in water and distilled water. Dry in air at 93°C

3. Etch for 2 min at 20°C in

	Parts by vol.
Hydrofluoric acid (60%)	63
Hydrochloric acid (37%)	841
Orthosphosphoric acid (85%)	89

Rinse in water and distilled water. Dry in air at 93°C

Suitable for alloys to be bonded with polybenzimidazole adhesives. Bond within 10 min of treatment. ASTM D 2651

4. Etch for 10–15 min at 38–52°C in

	Parts by vol.
Nitric acid (69%)	6
Hydrofluoric acid (60%)	1
Water	20

Rinse with water and distilled water. Dry in oven at 71–82°C for 15 min

Alternative etch for alloys to be bonded with polyimide adhesives is nitric: hydrofluoric: water in ratio 5:1:27 by wt. Etch 30 s at 20°C

5. Commercial etching liquids and pastes (Plasa-Jell 107C available from Semco, South Hoover, Los Angeles, Calif. 18881)

Table continued on p. 7.41

Polyethylene Terephthalate.   A medium-strength bond can be obtained with polyethylene terephthalate plastics and films by abrasion and solvent cleaning. However, a stronger bond can be achieved by immersing the surface in a warm solution of sodium hydroxide or in an alkaline cleaning solution for 2 to 10 min.

Polyolefins, Polyformaldehyde, Polyether.   These materials can be effectively bonded only if the surface is first located. Polyethylene and polypropylene can be prepared for bonding by holding the flame of an oxyacetylene torch over the plastic until it becomes glossy or else by heating the surface momentarily with a blast of hot air. It is important not to overheat the plastic, thereby causing deformation. The treated plastic must be bonded as quickly as possible after surface preparation.

Polyolefins such as polyethylene, polypropylene, and polymethylpentene, as well as polyformaldehyde and polyether, may be more effectively treated with a sodium

TABLE 7.9   Surface Preparation for Metals (*Continued*)

Adherend	Degreasing solvent	Method of treatment	Remarks
Zinc and zinc alloys	Trichloro- ethylene	1. Abrasion. Grit or vapor blast, or 100-grit emery cloth, followed by solvent degreas- ing	For general-purpose bonding
		2. Etch for 2–4 min at 20°C in	Glacial acetic acid is an alterna- tive to hydrochloric acid

	Parts by vol.
Hydrochloric acid (37%)	10–20
Water	90–80

Rinse with warm water and dis- tilled water. Dry in air at 66– 71°C for 30 min

3. Etch for 3–6 min at 38°C in   Suitable for freshly galvanized metal

	Parts by wt.
Sulfuric acid (96%, sp. gr. 1.84)	2
Sodium dichromate (crystalline)	1
Water	8

Source: Based on Refs. 12, 21–24.

TABLE 7.10   Surface Treatment for Adhesive Bonding Aluminum[25]

Surface treatment	Type of bond
Solvent wipe (MEK, MIBK, trichloroethylene)	Low to medium strength
Abrasion of surface, plus solvent wipe (sandblasting, coarse sandpaper, etc.)	Medium to high strength
Hot-vapor degrease (trichloroethylene)	Medium strength
Abrasion of surface, plus vapor degrease	Medium to high strength
Alodine treatment	Low strength
Anodize	Medium strength
Caustic etch*	High strength
Chromic acid etch (sodium dichromate–sulfuric acid)†	Maximum strength

  * A good caustic etch is Oakite 164 (Oakite Products, Inc., 19 Rector Street, New York, NY).
  † Recommended pretreatment for aluminum to achieve maximum bond strength and weatherability:
1. Degrease in hot trichloroethylene vapor (160°F).
2. Dip in the following chromic acid solution for 10 min at 160°F:

Sodium dichromate ($Na_2Cr_2O_7 \cdot 2H_2O$)	1 part/wt.
Conc. sulfuric acid (sp. gr. 1.86)	10 parts/wt.
Distilled water	30 parts/wt.

3. Rinse thoroughly in cold, running, distilled, or deionized water.
4. Air-dry for 30 min., followed by 10 min at 150°F.

dichromate–sulfuric acid solution. This treatment oxidizes the surface, allowing bet- ter wetting. Activated-gas plasma treatment, described in the general section on sur- face treatments, is also an effective treatment for these plastics. Table 7.12 shows the tensile-shear strength of bonded polyethylene pretreated by these various methods.

Elastomeric Adherends.   Vulcanized rubber joints are often contaminated with mold release and plasticizers or extenders that can migrate to the surface. As shown

TABLE 7.11 Surface Preparations for Plastics

Adherend	Degreasing solvent	Method of treatment	Remarks
Acetal (co-polymer)	Acetone	1. Abrasion. Grit or vapor blast, or medium-grit emery cloth followed by solvent degreasing	For general-purpose bonding
		2. Etch in the following acid solution: *Parts by wt.* Potassium dichromate 75 Distilled water....... 120 Concentrated sulfuric acid (96%, sp. gr. 1.84).............. 1,500 for 10 s at 25°C. Rinse in distilled water, and dry in air at RT	For maximum bond strength. ASTM D 2093
Acetal (homo-polymer)	Acetone	1. Abrasion. Sand with 280A-grit emery cloth followed by solvent degreasing	For general-purpose bonding
		2. "Satinizing" technique. Immerse the part in *Parts by wt.* Perchloroethylene.... 96.85 1,4-Dioxane.......... 3.00 *p*-Toluenesulfonic acid.............. 0.05 Cab-o-Sil (Cabot Corp.)............. 0.10 for 5–30 s at 80–120°C. Transfer the part immediately to an oven at 120°C for 1 min. Wash in hot water. Dry in air at 120°C	For maximum bond strength. Recommended by du Pont
Acrylonitrile butadiene styrene	Acetone	1. Abrasion. Grit or vapor blast, or 220-grit emery cloth, followed by solvent degreasing	
		2. Etch in chromic acid solution for 20 min at 60°C	Recipe 2 for methyl pentane
Cellulosics: Cellulose, cellulose acetate, cellulose acetate butyrate, cellulose nitrate, cellulose propionate, ethyl cellulose	Methanol, isopropanol	1. Abrasion. Grit or vapor blast, or 220-grit emery cloth, followed by solvent degreasing 2. After procedure 1, dry the plastic at 100°C for 1 h, and apply adhesive before the plastic cools to room temperature	For general bonding purposes
Diallyl phthalate, diallyl isophthalate	Acetone, methyl ethyl ketone	Abrasion. Grit or vapor blast, or 100-grit emery cloth, followed by solvent degreasing	Steel wool may be used for abrasion

TABLE 7.11 Surface Preparations for Plastics (*Continued*)

Adherend	Degreasing solvent	Method of treatment	Remarks
Epoxy resins	Acetone, methyl ethyl ketone	Abrasion. Grit or vapor blast, or 100-grit emery cloth, followed by solvent degreasing	Sand or steel shot are suitable abrasives
Ethylene vinyl acetate	Methanol	Prime with epoxy adhesive and fuse into the surface by heating for 30 min at 100°C	
Furane	Acetone, methyl ethyl ketone	Abrasion. Grit or vapor blast, or 100-grit emery cloth, followed by solvent degreasing	
Ionomer	Acetone, methyl ethyl ketone	Abrasion. Grit or vapor blast, or 100-grit emery cloth, followed by solvent degreasing	Alumina (180-grit) is a suitable abrasive
Melamine resins	Acetone, methyl ethyl ketone	Abrasion. Grit or vapor blast, or 100-grit emery cloth, followed by solvent degreasing	
Methyl pentene	Acetone	1. Abrasion. Grit or vapor blast, or 100-grit emery cloth, followed by solvent degreasing 2. Immerse for 1 h at 60°C in <table><tr><td></td><td>*Parts by wt.*</td></tr><tr><td>Sulfuric acid (96%, sp. gr. 1.84)..........</td><td>26</td></tr><tr><td>Potassium chromate..</td><td>3</td></tr><tr><td>Water...............</td><td>11</td></tr></table>Rinse in water and distilled water. Dry in warm air 3. Immerse for 5–10 min at 90°C in potassium permanganate (saturated solution), acidified with sulfuric acid (96%, sp. gr. 1.84). Rinse in water and distilled water. Dry in warm air 4. Prime surface with lacquer based on urea-formaldehyde resin diluted with carbon tetrachloride	For general-purpose bonding Coatings (dried) offer excellent bonding surfaces without further pretreatment
Phenolic resins, phenolic melamine resins	Acetone, methyl ethyl ketone, detergent	1. Abrasion. Grit or vapor blast, or abrade with 100-grit emery cloth, followed by solvent degreasing 2. Removal of surface layer of one ply of fabric previously placed on surface before curing. Expose fresh bonding surface by tearing off the ply prior to bonding	Steel wool may be used for abrasion. Sand or steel shot are suitable abrasives. Glass-fabric decorative laminates may be degreased with detergent solution

TABLE 7.11   Surface Preparations for Plastics (*Continued*)

Adherend	Degreasing solvent	Method of treatment	Remarks
Polyamide (nylon)	Acetone, methyl ethyl ketone, detergent	1. Abrasion. Grit or vapor blast, or abrade with 100-grit emery cloth, followed by solvent degreasing	Sand or steel shot are suitable abrasives
		2. Prime with a spreading dough based on the type of rubber to be bonded in admixture with isocyanate	Suitable for bonding polyamide textiles to natural and synthetic rubbers
		3. Prime with resorcinol-formaldehyde adhesive	Good adhesion to primer coat with epoxy adhesives in metal-plastic joints
Polycarbonate, allyl diglycol carbonate	Methanol, isopropanol, detergent	Abrasion. Grit or vapor blast, or 100-grit emery cloth, followed by solvent degreasing	Sand or steel shot are suitable abrasives
Fluorocarbons: Polychlorotrifluoroethylene, polytetrafluoroethylene, polyvinyl fluoride, polymonochlorotrifluoroethylene	Trichloroethylene	1. Wipe with solvent and treat with the following for 15 min at RT: Naphthalene (128 g) dissolved in tetrahydrofuran (1 liter) to which is added sodium (23 g) during a stirring period of 2 h. Rinse in deionized water, and dry in warm air	Sodium-treated surfaces must not be abraded before use. Hazardous etching solutions requiring skillful handling. Proprietary etching solutions are commercially available (see 2). PTFE available in etched tape. ASTM D 2093
		2. Wipe with solvent and treat as recommended in one of the following commercial etchants: Bond aid.....W. S. Shamban and Co. 11617 W. Jefferson Blvd. Culver City, Calif. Fluorobond...Joclin Mfg. Co. 15 Lufbery Ave. Wallingford, Conn. Fluoroetch....Action Associates 1180 Raymond Blvd. Newark, N.J. Tetraetch....W. L. Gore Associates 487 Paper Mill Rd. Newark, Del.	
		3. Prime with epoxy adhesive, and fuse into the surface by heating for 10 min at 370°C followed by 5 min at 400°C	
		4. Expose to one of the following gases activated by corona discharge: Air (dry) for 5 min Air (wet) for 5 min Nitrous oxide for 10 min Nitrogen for 5 min	Bond within 15 min of pretreatment
		5. Expose to electric discharge from a tesla coil (50,000 V ac) for 4 min	Bond within 15 min of pretreatment

TABLE 7.11   Surface Preparations for Plastics (*Continued*)

Adherend	Degreasing solvent	Method of treatment	Remarks

Polyesters, polyethylene terephthalate (Mylar)	Detergent, acetone, methyl ethyl ketone	1. Abrasion. Grit or vapor blast, or 100-grit emery cloth, followed by solvent degreasing
		
		2. Immerse for 10 min at 70–95°C in
		
		    Parts by wt.
		    Sodium hydroxide. . . . 2
		    Water. 8
		    Rinse in hot water and dry in hot air
			For general-purpose bonding
			
			For maximum bond strength. Suitable for linear polyester films (Mylar)

Chlorinated polyether	Acetone, methyl ethyl ketone	Etch for 5–10 min at 66–71°C in
		    Parts by wt.
		    Sodium dichromate. 5
		    Water. 8
		    Sulfuric acid (96%, sp. gr. 1.84). 100
		    Rinse in water and distilled water. Dry in air
			Suitable for film materials such as Penton. ASTM D 2093

Polyethylene, polyethylene (chlorinated), polyethylene terephthalate (see polyesters), polypropylene, polyformaldehyde	Acetone, methyl ethyl ketone	1. Solvent degreasing
		
		2. Expose surface to gas-burner flame (or oxyacetylene oxidizing flame) until the substrate is glossy
		
		3. Etch in the following:
		    Parts by wt.
		    Sodium dichromate. . . 5
		    Water. 8
		    Sulfuric acid (96%, sp. gr. 1.84). 100
		    Polyethylene 60 min at 25°C or
		    and polypropylene 1 min at 71°C
		    Polyformaldehyde. . . . . .10 s at 25°C
		
		4. Expose to following gases activated by corona discharge:
		    Air (dry). . . . . .For 15 min
		    Air (wet). . . . . .For 5 min
		    Nitrous Oxide. . . . . . .For 10 min
		    Nitrogen. . . . . .For 15 min
		
		5. Expose to electric discharge from a tesla coil (50,000 V ac) for 1 min
			Low-bond-strength applications
			
			For maximum bond strength. ASTM D 2093
			
			Bond within 15 min of pretreatment. Suitable for polyolefins.
			
			Bond within 15 min of pretreatment. Suitable for polyolefins.

Polymethyl methacrylate, methacrylate butadiene styrene	Acetone, methyl ethyl ketone, detergent, methanol, trichloroethylene, isopropanol	Abrasion. Grit or vapor blast, or 100-grit emery cloth, followed by solvent degreasing	For maximum strength relieve stresses by heating plastic for 5 h at 100°C

TABLE 7.11   Surface Preparations for Plastics (*Continued*)

Adherend	Degreasing solvent	Method of treatment	Remarks
Poly-phenylene	Trichloro-ethylene	Abrasion. Grit or vapor blast, or 100-grit emery cloth, followed by solvent degreasing	
Poly-phenylene oxide	Methanol	Solvent degrease	Plastic is soluble in xylene and may be primed with adhesive in xylene solvent
Polysty-rene	Methanol, isopro-panol, deter-gent	Abrasion, Grit or vapor blast, or 100-grit emery cloth, followed by solvent degreasing	Suitable for rigid plastic
Poly-sulfone	Methanol	Vapor degrease	
Polyure-thane	Acetone, methyl ethyl ketone	Abrade with 100-grit emery cloth and solvent degrease	
Polyvinyl chloride, polyvinyl-idene chloride polyvinyl fluoride	Trichloro-ethylene, methyl ethyl ketone	1. Abrasion. Grit or vapor blast, or 100-grit emery cloth followed by solvent degreas-ing 2. Solvent wipe with ketone	Suitable for rigid plastic. For maximum strength, prime with nitrile-phenolic adhesive Suitable for plasticized material
Styrene acrylo-nitrile	Trichloro-ethylene	Solvent degrease	
Urea for-malde-hyde	Acetone, methyl ethyl ketone	Abrasion. Grit or vapor blast, or 100-grit emery cloth, followed by solvent degreasing	

Source: Based on Refs. 12, 21–24, 26.

in Table 7.13, solvent washing and abrading are common treatments for most elastomers, but chemical treatment may be required for maximum properties.

Synthetic and natural rubbers may require cyclizing with concentrated sulfuric acid until a brittle surface develops. When flexed, this surface generates hairline fractures that act as points of mechanical interlock for the adhesive. Cyclization is often used for bonding rubber to metal with epoxy adhesives.

Chlorination of vulcanized elastomers has also been recognized as being of value in improving adhesion, particularly with room-temperature-curing adhesives. The disadvantage of this procedure is the release of chlorine into the atmosphere. An industrially more acceptable process was developed that involves the use of trichloroisocyanuric acid (TCICA) in an organic solvent. This process was specifically developed to assist in the bonding of shoe soles to leather uppers.

Composites.   All of the surface-preparation procedures described above for plastic substrates are also applicable for composite substrates. The nature of the treatment

TABLE 7.12 Effects of Surface Treatments on Bonding to Polyethylene with Various Types of Adhesives[27]

Specimen No.	Control	Flame treated	Sanded	Acid treated, oven-dried at 90°C	Acid treated, oven-dried at 71°C	Acid treated, wiped, air-dried at 22°C	Acid treated, acetone-dried	Plasma treatment			
								Helium (30 s)	Helium (30 min)	Oxygen (30 s)	Oxygen (30 min)
Epoxy											
1	40	464*	186	454	428*	500*	516*	468*	...	423*	490*
2	48	480*	166	480*	440*	524*	490*	470*	...	463*	424*
3	24	452*	182	472*	532*	500*	502*	470*	...	484*	439*
4	58	486*	220	462*	460*	524*	500*	450*	...	495*	424*
5	58	520*	216	506*	424*	448*	476*		...		484*
Avg lb/in.²	46	480	195	475	457	499	497	464	...	466	445
Polyester											
1	74	502*	214	290	300	294	452*	196	284	264	480*
2	102	472*	146	290	288	416*	462*	230	396*	246	514*
3	70	430*	170	230	322	412	392	214	380	320	300
4	70	364*	188	268	318	426*	464*	160	400*	240	370
5	108	400*	178	308	256	236	200	148	372	244	484*
Avg lb/in.²	85	434	175	277	297	357	394	190	346	263	430
Nitrile-Rubber-Phenolic											
1	42	196	54	102	100	124	106	...	166	...	210
2	38	120	54	100	92	128	136	...	110	...	170
3	46	88	52	88	64	120	102	...	276	...	220
4	46	120	64	96	158	88	54	...	112	...	110
5	48	166	56	96	124	88	164	...	224	...	170
Avg lb/in.²	44	138	56	96	108	110	112	...	178	...	176

* Adherend failed rather than bond. All values in this table are based upon lap-shear strength calculated as lb/in².

TABLE 7.13 Surface Preparations for Elastomers

Adherend	Degreasing solvent	Method of treatment	Remarks
Natural rubber	Methanol, isopropanol	1. Abrasion followed by brushing. Grit or vapor blast, or 280-grit emery cloth, followed by solvent wipe	For general-purpose bonding
		2. Treat the surface for 2–10 min with sulfuric acid (sp. gr. 1.84) at RT. Rinse thoroughly with cold water/hot water. Dry after rinsing in distilled water. (Residual acid may be neutralized by soaking for 10 min in 10% ammonium hydroxide after hot-water washing)	Adequate pretreatment is indicated by the appearance of hairline surface cracks on flexing the rubber. Suitable for many synthetic rubbers when given 10–15 min etch at room temperature. Unsuitable for use on butyl, polysulfide, silicone, chlorinated polyethylene, and polyurethane rubbers
		3. Treat surface for 2–10 min with paste made from sulfuric acid and barium sulfate. Apply paste with stainless-steel spatula, and follow procedure 2, above	

		4. Treat surface for 2–10 min in
		
		                    Parts by vol.
		 Sodium hypochlorite.. 6
		 Hydrochloric acid
		    (37%)............. 1
		 Water.............. 200
		 Rinse with cold water and dry
			Suitable for those rubbers amenable to treatments 2 and 3

Butadiene styrene	Toluene	1. Abrasion followed by brushing. Grit or vapor blast, or 280-grit emery cloth, followed by solvent wipe	Excess toluene results in swollen rubber. A 20-min drying time will restore the part to its original dimensions
		2. Prime with butadiene styrene adhesive in an aliphatic solvent.	
		3. Etch surface for 1–5 min at RT, following method 2 for natural rubber.	
Butadiene nitrile	Methanol	1. Abrasion followed by brushing. Grit or vapor blast, or 280-grit emery cloth, followed by solvent wipe	
		2. Etch surface for 10–45 s at RT, following method 2 for natural rubber	
Butyl	Toluene	1. Solvent wipe	For general-purpose bonding
		2. Prime with butyl-rubber adhesive in an aliphatic solvent	For maximum strength
Chlorosulfonated polyethylene	Acetone or methyl ethyl ketone	Abrasion followed by brushing. Grit or vapor blast, or 280-grit emery cloth, followed by solvent wipe	General-purpose bonding
Ethylene propylene	Acetone or methyl ethyl ketone	Abrasion followed by brushing. Grit or vapor blast, or 280-grit emery cloth, followed by solvent wipe	General-purpose bonding

TABLE 7.13   Surface Preparations for Elastomers (*Continued*)

Adherend	Degreasing solvent	Method of treatment	Remarks
Fluoro-silicone	Methanol	Application of fluorosilicone primer (A 4040) to metal where intention is to bond unvulcanized rubber	Primer available from Dow Corning
Polyacrylic	Methanol	Abrasion followed by brushing. Grit or vapor blast, or 100-grit emery cloth followed by solvent wipe	General-purpose bonding
Polybuta-diene	Methanol	Solvent wipe	General-purpose bonding
Polychloro-prene	Toluene, methanol, isopro-panol	1. Abrasion followed by brushing. Grit or vapor blast, or 100-grit emery cloth, followed by solvent wipe 2. Etch surface for 5–30 min at RT, following method 2 for natural rubber	Adhesion improved by abrasion with 280-grit emery cloth followed by acetone wipe
Polysulfide	Methanol	Immerse overnight in strong chlorine water, wash and dry	
Polyure-thane	Methanol	1. Abrasion followed by brushing. Grit or vapor blast, or 280-grit emery cloth followed by solvent wipe 2. Incorporation of a chlorosilane into the adhesive-elastomer system. 1% w/w is usually sufficient	Chlorosilane is available commercially. Addition to adhesive eliminates need for priming and improves adhesion to glass, metals. Silane may be used as a surface primer
Silicone	Acetone or methanol	1. Application of primer, Chemlok 607, in solvent (dries 10–15 min) 2. Expose to oxygen gas activated by corona discharge for 10 min	Primer available from Hughson Chemical Company

Source: Based on Refs. 12, 21–24, 26.

will depend on the resin matrix, the joint permanence characteristics required, and the production facilities that are available.

Many surface-roughening approaches have been tried, and all have some merit. One method that has gained wide acceptance is the use of a peel ply. In this technique, a closely woven nylon or polyester cloth is used as the outer layer of the composite during lay-up. This ply is then torn or peeled away just before bonding. The tearing or peeling process fractures the resin matrix coating and exposes clean, virgin, roughened surface for the bonding process.

In the cases where the peel ply is not used, some sort of light abrasion is required to break the glazed finish on the matrix resin surface. The glaze on the matrix surface should be roughened without damaging the reinforcing fibers or forming subsurface cracks in the matrix.

Other Adherends.   Table 7.14 provides surface treatments for a variety of materials not covered in the preceding tables. Bonding to painted or plated parts requires spe-

TABLE 7.14   Surface Preparations for Materials Other Than Metals, Plastics, and Elastomers

Adherend	Degreasing solvent	Method of treatment	Remarks
Asbestos (rigid)	Acetone	1. Abrasion. Abrade with 100-grit emery cloth, remove dust, and solvent degrease 2. Prime with diluted adhesive or low-viscosity rosin ester	Allow the board to stand for sufficient time to allow solvent to evaporate off
Brick and fired non-glazed building materals	Methyl ethyl ketone	Abrade surface with a wire brush; remove all dust and contaminants	
Carbon graphite	Acetone	Abrasion. Abrade with 220-grit emery cloth and solvent degrease after dust removal	For general-purpose bonding

cial consideration. The resulting adhesive bond is only as strong as the adhesion of the paint or plating to the base material.

7.5   TYPES OF ADHESIVES

7.5.1   Adhesive Composition

Modern-day adhesives are often fairly complex formulations of components that perform specialty functions. The adhesive base or binder is the primary component of an adhesive. The binder is generally the resinous component from which the name of the adhesive is derived. For example, an epoxy adhesive may have many components, but the primary material is epoxy resin.

A hardener is a substance added to an adhesive formulation to initiate the curing reaction and take part in it. Two-component adhesive systems have one component that is the base and a second component that is the hardener. Upon mixing, a chemical reaction ensues that causes the adhesive to solidify. A catalyst is sometimes incorporated into an adhesive formulation to speed the reaction between base and hardener.

Solvents are sometimes needed to lower viscosity or to disperse the adhesive to a spreadable consistency. Often a mixture of solvents is required to achieve the desired properties.

A reactive ingredient added to an adhesive to reduce the concentration of binder is called a diluent. Diluents are principally used to lower viscosity and modify processing conditions of some adhesives. Diluents react with the binder during cure, become part of the product, and do not evaporate as does a solvent.

Fillers are generally inorganic particulates added to the adhesive to improve working properties, strength, permanence, or other qualities. Fillers are also used to reduce material cost. By selective use of fillers, the properties of an adhesive can be changed tremendously. Thermal expansion, electrical and thermal conduction, shrinkage, viscosity, and thermal resistance are only a few properties that can be modified by use of selective fillers.

TABLE 7.14 Surface Preparations for Materials Other Than Metals, Plastics, and Elastomers (*Continued*)

Adherend	Degreasing solvent	Method of treatment	Remarks
Glass and quartz (nonoptical)	Acetone, detergent	1. Abrasion. Grit blast with carborundum and water slurry, and solvent degrease. Dry for 30 min at 100°C. Apply the adhesive before the glass cools to RT	For general-purpose bonding. Drying process improves bond strength
		2. Immerse for 10–15 min at 20°C in *Parts by wt.* Sodium dichromate... 7 Water.............. 7 Sulfuric acid (96%, sp. gr. 1.84)......... 400 Rinse in water and distilled water. Dry thoroughly	For maximum strength
Glass (optical)	Acetone, detergent	Clean in an ultrasonically agitated detergent bath. Rinse; dry below 38°C	
Ceramics and porcelain	Acetone	1. Abrasion. Grit blast with carborundum and water slurry, and solvent degrease	Suitable for unglazed ceramics such as alumina, silica
		2. Solvent degrease or wash in warm aqueous detergent. Rinse and dry	For glazed ceramics such as porcelain
		3. Immerse for 15 min at 20°C in *Parts by wt.* Sodium dichromate... 7 Water.............. 7 Sulfuric acid (96%, sp. gr. 1.84)......... 400 Rinse in water and distilled water. Oven-dry at 66°C	For maximum strength bonding of small ceramic (glazed) artefacts
Concrete, granite, stone	Perchloroethylene, detergent	1. Abrasion. Abrade with a wire brush, degrease with detergent, and rinse with hot water before drying	For general-purpose bonding
		2. Etch with 15% hydrochloric acid until effervescence ceases. Wash with water until surface is litmus-neutral. Rinse with 1% ammonia and water. Dry thoroughly before bonding	Applied by stiff-bristle brush. Acid should be prepared in a polyethylene pail 10–12% hydrochloric or sulfuric acids are alternative etchants. 10% w/w sodium bicarbonate may be used instead of ammonia for acid neutralization
Wood, plywood		Abrasion. Dry wood is smoothed with a suitable emery paper. Sand plywood along the direction of the grain	For general-purpose bonding
Painted surface	Detergent	1. Clean with detergent solution, abrade with a medium emery cloth, final wash with detergent	Bond generally as strong as the paint
		2. Remove paint by solvent or abrasion, and pretreat exposed base	For maximum adhesion

Source: Based on Refs. 12, 21–24, 26.

A carrier or reinforcement is usually a thin fabric used to support a semicured (B-staged) adhesive to provide a product that can be used as a tape or film. The carrier can also serve as a spacer between the adherends and reinforcement for the adhesive.

Adhesives can be broadly classified as being thermoplastic, thermosetting, elastomeric, or an alloy blend. These four adhesive classifications can be further subdivided by specific chemical composition as described in Tables 7.15 through 7.18. The types of resins that go into the thermosetting and alloy adhesive classes are noted for high strength, creep resistance, and resistance to environments such as heat, moisture, solvents, and oils. Their physical properties are well suited for structural adhesive applications.

Elastomeric and thermoplastic adhesive classes are not used in applications requiring continuous load because of their tendency to creep under stress. They are also degraded by many common service environments. These adhesives find greatest use in low-strength applications such as pressure-sensitive tape, sealants, and hot-melt products.

7.5.2 Structural Adhesives

Epoxy. Epoxy adhesives offer a high degree of adhesion to all substrates except some untreated plastics and elastomers. Cured epoxies have thermosetting molecular structures. They exhibit excellent tensile-shear strength but poor peel strength unless modified with a more resilient polymer. Epoxy adhesives offer excellent resistance to oil, moisture, and many solvents. Low cure shrinkage and high resistance to creep under prolonged stress are characteristic of epoxy resins.

Epoxy adhesives are commercially available as liquids, pastes, and semicured (B-staged) film and solids. Epoxy adhesives are generally supplied as a 100 percent solids (nonsolvent) formulation, but some sprayable epoxy adhesives are available in solvent systems. Epoxy resins have no evolution of volatiles during cure and are useful in gap-filling applications.

Depending on the type of curing agent, epoxy adhesives can cure at room or elevated temperatures. Higher strengths and better heat resistance are usually obtained with the heat-curing types. Room-temperature-curing epoxies can harden in as little as 1 min at room temperature, but most systems require from 18 to 72 h. The curing time is greatly temperature-dependent, as shown in Fig. 7.30.

Epoxy resins are the most versatile of structural adhesives because they can be cured and coreacted with many different resins to provide widely varying properties. Table 7.19 describes the influence of curing agents on the bond strength of epoxy to various adherends. The type of epoxy resin used in most adhesives is derived from the reaction of bisphenol A and epichlorohydrin. This resin can be cured with amines or polyamides for room-temperature-setting systems; anhydrides for elevated-temperature cure; or latent curing agents such as boron trifluoride complexes for use in one-component, heat-curing adhesives. Polyamide curing agents are used in most general-purpose epoxy adhesives. They provide a room-temperature cure and bond well to many substrates including plastics, glass, and elastomers. The polyamide-cured epoxy also offers a relatively flexible adhesive with fair peel and thermal-cycling properties.

Toughened Epoxy. One of the best ways to toughen rigid epoxy adhesives is by the incorporation of a discontinuous rubbery phase such as a nitrile elastomer. This is accomplished by the addition of about 5–10% by weight of a carboxy functional

TABLE 7.15   Thermosetting Adhesives

Adhesive	Description	Curing method	Special characteristics	Usual adherends	Price range
Cyanoacrylate	One-part liquid	Rapidly at RT in absence of air	Fast setting; good bond strength; low viscosity; high cost; poor heat and shock resistance; will not bond to acidic surfaces	Metals, plastics, glass	Very high
Polyester	Two-part liquid or paste	RT or higher	Resistant to chemicals, moisture, heat, weathering. Good electrical properties; wide range of strengths; some resins do not fully cure in presence of air; isocyanate-cured system bonds well to many plastic films	Metals, foils, plastics, plastic laminates, glass	Low-med
Urea formaldehyde	Usually supplied as two-part resin and hardening agent. Extenders and fillers used	Under pressure	Not as durable as others but suitable for fair range of service conditions. Generally low cost and ease of application and cure. Pot life limited to 1 to 24 h	Plywood	Low
Melamine formaldehyde	Powder to be mixed with hardening agent	Heat and pressure	Equivalent in durability and water resistance (including boiling water) to phenolics and resorcinols. Often combined with ureas to lower cost. Higher service temp than ureas	Plywood, other wood products	Medium
Resorcinol and phenol-resorcinol formaldehyde	Usually alcohol-water solutions to which formaldehyde must be added	RT or higher with moderate pressure	Suitable for exterior use; unaffected by boiling water, mold, fungus, grease, oil, most solvents. Bond strength equals or betters strength of wood; do not bond directly to metal	Wood, plastics, paper, textiles, fiberboard, plywood	Medium
Epoxy	Two-part liquid or paste; one-part liquid, paste, or solid; solutions	RT or higher	Most versatile adhesive available; excellent tensile-shear strength; poor peel strength; excellent resistance to moisture and solvents; low cure shrinkage; variety of curing agents/hardeners results in many variations	Metals, plastic, glass, rubber, wood, ceramics	Medium
Polyimide	Supported film, solvent solution	High temp	Excellent thermal and oxidation resistance; suitable for continuous use at 550°F and short-term use to 900°F; expensive	Metals, metal foil, honeycomb core	Very high
Polybenzimidazole	Supported film	Long, high-temp cure	Good strength at high temperatures; suitable for continuous use at 450°F and short-term use at 1000°F; volatiles released during cure; deteriorate at high temperatures on exposure to air; expensive	Metals, metal foil, honeycomb core	Very high
Acrylic	Two-part liquid or paste	RT	Excellent bond to many plastics, good weather resistance, fast cure, catalyst can be used as a substrate primer; poor peel and impact strength	Metals, many plastics, wood	Medium
Acrylate acid diester	One-part liquid or paste	RT or higher in absence of air	Chemically blocked, anaerobic type; excellent wetting ability; useful temperature range −65 to 300°F; withstands rapid thermal cycling; high-tensile-strength grade requires cure at 250°F, cures in minutes at 280°F	Metals, plastics, glass, wood	Very high

TABLE 7.16   Thermoplastic Adhesives

Adhesive	Description	Curing method	Special characteristics	Usual adherends	Price range
Cellulose acetate, cellulose acetate butyrate	Solvent solutions	Solvent evaporation	Water-clear, more heat resistant but less water resistant than cellulose nitrate; cellulose acetate butyrate has better heat and water resistance than cellulose acetate and is compatible with a wider range of plasticizers	Plastics, leather, paper, wood, glass, fabrics	Low
Cellulose nitrate	Solvent solutions	Evaporation of solvent	Tough, develops strength rapidly, water-resistant; bonds to many surfaces; discolors in sunlight; dried adhesive is flammable	Glass, metal, cloth, plastics	Low
Polyvinyl acetate	Solvent solutions and water emulsions, plasticized or unplasticized, often containing fillers and pigments. Also dried film which is light-stable, water-white, transparent	On evaporation of solvent or water; film by heat and pressure	Bond strength of several thousand lb/in.² but not under continuous loading. The most versatile in terms of formulations and uses. Tasteless, odorless; good resistance to oil, grease, acid; fair water resistance	Emulsions particularly useful with porous materials like wood and paper. Solutions used with plastic films, mica, glass, metal, ceramics.	Low
Vinyl vinylidene	Solutions in solvents like methyl ethyl ketone	Evaporation of solvent	Tough, strong, transparent and colorless. Resistant to hydrocarbon solvents, greases, oils	Particularly useful with textiles; also porous materials, plastics	Medium
Polyvinyl acetals	Solvent solutions, film, and solids	Evaporation of solvent; film and solid by heat and pressure	Flexible bond; modified with phenolics for structural use; good resistance to chemicals and oils; includes polyvinyl formal and polyvinyl butaryl types	Metals, mica, glass, rubber, wood, paper	Medium
Polyvinyl alcohol	Water solutions, often extended with starch or clay	Evaporation of water	Odorless, tasteless and fungus-resistant (if desired). Excellent resistance to grease and oils; water soluble	Porous materials such as fiberboard, paper, cloth	Low
Polyamide	Solid hot-melt, film, solvent solutions	Heat and pressure	Good film flexibility; resistant to oil and water; used for heat-sealing compounds	Metals, paper, plastic films	Medium
Acrylic	Solvent solutions, emulsions and mixtures requiring added catalysts	Evaporation of solvent; RT or elevated temp (two-part)	Good low-temperature bonds; poor heat resistance; excellent resistance to ultraviolet; clear; colorless	Glass, metals, paper, textiles, metallic foils, plastics	Medium
Phenoxy	Solvent solutions, film, solid hot-melt	Heat and pressure	Retain high strength from 40 to 180°F; resist creep up to 180°F; suitable for structural use	Metals, wood, paper, plastic film	Medium

nitrile rubber into the epoxy formulation. This increases the peel strength and impact properties of several epoxy adhesives substantially without reducing their temperature or chemical resistance.

The crack-stopping effects of these dispersed rubber particles is illustrated in Fig. 7.31, which shows T-peel test results. Since the rubber does not soften the continuous

TABLE 7.17 Elastomeric Adhesives

Adhesive	Description	Curing method	Special characteristics	Usual adherends	Price range
Natural rubber	Solvent solutions, latexes and vulcanizing type	Solvent evaporation, vulcanizing type by heat or RT (two-part)	Excellent tack, good strength. Shear strength 30–180 lb/in.[2]; peel strength 0.56 lb/in. width. Surface can be tack-free to touch and yet bond to similarly coated surface	Natural rubber, masonite, wood, felt, fabric, paper, metal	Medium
Reclaimed rubber	Solvent solutions, some water dispersions. Most are black. some gray and red	Evaporation of solvent	Low cost, widely used. Peel strength higher than natural rubber; failure occurs under relatively low constant loads	Rubber, sponge rubber, fabric, leather, wood, metal, painted metal, building materials	Low
Butyl	Solvent system, latex	Solvent evaporation, chemical cross linking with curing agents and heat	Low permeability to gases, good resistance to water and chemicals, poor resistance to oils, low strength	Rubber, metals	Medium
Polyisobutylene	Solvent solution	Evaporation of solvent	Sticky, low-strength bonds; copolymers can be cured to improve adhesion, environmental resistance, and elasticity; good aging; poor thermal resistance; attacked by solvents	Plastic film, rubber, metal foil, paper	Low
Nitrile	Latexes and solvent solutions compounded with resins, metallic oxides, fillers, etc.	Evaporation of solvent and/or heat and pressure	Most versatile rubber adhesive. Superior resistance to oil and hydrocarbon solvents. Inferior in tack range; but most dry tack-free, an advantage in pre-coated assemblies. Shear strength of 150–2,000 lb/in.[2], higher than neoprene, if cured	Rubber (particularly nitrile) metal, vinyl plastics	Medium
Styrene butadiene	Solvent solutions and latexes. Because tack is low, rubber resin is compounded with tackifiers and plasticizing oils	Evaporation of solvent	Usually better aging properties than natural or reclaimed. Low dead load strength; bond strength similar to reclaimed. Useful temp range from −40 to 160°F	Fabrics, foils, plastics film laminates, rubber and sponge rubber, wood	Low
Polyurethane	Two-part liquid or paste	RT or higher	Excellent tensile-shear strength from −400 to 200°F; poor resistance to moisture before and after cure; good adhesion to plastics	Plastics, metals, rubber	Medium
Polysulfide	Two-part liquid or paste	RT or higher	Resistant to wide range of solvents, oils, and greases; good gas impermeability; resistant to weather, sunlight, ozone; retains flexibility over wide temperature range; not suitable for permanent load-bearing applications	Metals, wood, plastics	High
Silicone	Solvent solution: heat or RT curing and pressure-sensitive; and RT vulcanizing solventless pastes	Solvent evaporation, RT or elevated temp	Of primary interest is pressure-sensitive type used for tape. High strengths for other forms are reported from −100 to 500°F; limited service to 700°F. Excellent dielectric properties	Metals; glass; paper; plastics and rubber, including silicone and butyl rubber and fluorocarbons	High–very high
Neoprene	Latexes and solvent solutions, often compounded with resins, metallic oxides, fillers, etc.	Evaporation of solvent	Superior to other rubber adhesives in most respects—quickness; strength; max temp (to 200°F, sometimes 350°F); aging; resistant to light, weathering, mild acids, and oils	Metals, leather, fabric, plastics, rubber (particularly neoprene), wood, building materials	Medium

TABLE 7.18 Alloy Adhesives

Adhesive	Description	Curing method	Special characteristics	Usual adherends	Price range
Epoxy–phenolic	Two-part paste, supported film	Heat and pressure	Good properties at moderate cures; volatiles released during cure; retains 50% of bond strength at 500°F; limited shelf life; low peel strength and shock resistance	Metals, honeycomb core, plastic laminates, ceramics	Medium
Epoxy–polysulfide	Two-part liquid or paste	RT or higher	Useful temperature range −70 to 200°F, greater resistance to impact, higher elongation, and less brittleness than epoxies	Metals, plastic, wood, concrete	Medium
Epoxy–nylon	Solvent solutions, supported and unsupported film	Heat and pressure	Excellent tensile-shear strength at cryogenic temperature; useful temperature range −423 to 180°F; limited shelf life	Metals, honeycomb core, plastics	Medium
Nitrile–phenolic	Solvent solutions, unsupported and supported film	Heat and pressure	Excellent shear strength; good peel strength; superior to vinyl and neoprene–phenolics; good adhesion	Metals, plastics, glass, rubber	Medium
Neoprene–phenolic	Solvent solutions, supported and unsupported film	Heat and pressure	Good bonds to a variety of substrates; useful temp range −70 to 200°F, excellent fatigue and impact strength	Metals, glass, plastics	Medium
Vinyl–phenolic	Solvent solutions and emulsions, tape, liquid and coreacting powder	Heat and pressure	Good shear and peel strength; good heat resistance; good resistance to weathering, humidity, oil, water, and solvents, vinyl formal and vinyl butyral forms available, vinyl formal–phenolic is strongest	Metals, paper, honeycomb core	Low–medium

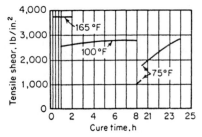

FIGURE 7.30 Characteristics of a particular epoxy adhesive under different curing time and temperature relationships.[28]

(epoxy) phase, the adhesive retains many of the inherently good properties of the rigid epoxy such as high hardness, hot strength, and high tensile-shear strength. However, it also has the excellent peel properties that were only available before this formulation with an entirely elastomeric adhesive formulation.

Epoxy Alloy. A variety of polymers can be blended and coreacted with epoxy resins to provide certain desired properties. The most common of these are phenolic, nylon, and polysulfide resins.

Epoxy-Phenolic. Adhesives based on epoxy-phenolic blends are good for continuous high-temperature service in the 350°F range or intermittent service as high as 500°F. They retain their properties over a very wide temperature range, as shown in Fig. 7.32. Shear strengths of up to 3,000 lb/in^2 at room temperature and 1,000 to 2,000 lb/in^2 at 500°F are available. Resistance to oil, solvents, and moisture is very good. Because of their rigid nature, epoxy-phenolic adhesives have low peel strength and limited thermal-shock resistance.

TABLE 7.19 Influence of Epoxy Curing Agent on Bond Strength Obtained with Various Base Materials[29]

Curing agent	Amount*	Cure cycle, h at °F	Tensile-shear strength, lb/in.^{2}					
			Polyester glass-mat laminate	Polyester glass-cloth laminate	Cold-rolled steel	Aluminum	Brass	Copper
Triethylamine	6	24 at 75, 4 at 150	1,850	2,100	2,456	1,810	1,765	655
Trimethylamine	6	24 at 75, 4 at 150	1,054	1,453	1,385	1,543	1,524	1,745
Triethylenetetramine	12	24 at 75, 4 at 150	1,150	1,632	1,423	1,675	1,625	1,325
Pyrrolidine	5	24 at 75, 4 at 150	1,250	1,694	1,295	1,733	1,632	1,420
Polyamid amine equivalent 210–230	35–65	24 at 75, 4 at 150	1,200	1,450	2,340	3,120	2,005	1,876
Metaphenylenidiamine	12.5	4 at 350	780	640	2,150	2,258	2,150	1,650
Diethylenetriamine	11	24 at 75, 4 at 150	1,010	1,126	1,350	1,420	1,135	1,236
Boron trifluoride monoethylamine	3	3 at 375	.....	.....	1,732	1,876	1,525	1,635
Dicyandiamide		4 at 350	530	432	2,680	2,785	2,635	2,550
Methyl nadic anhydride	85	6 at 350	600	756	2,280	2,165	1,955	1,835

Epoxy resin used was derived from bisphenol A and epichlorohydrin and had an epoxide equivalent of 180 to 195; the adhesives contained no filler.

* Per 100 parts by weight of resin.

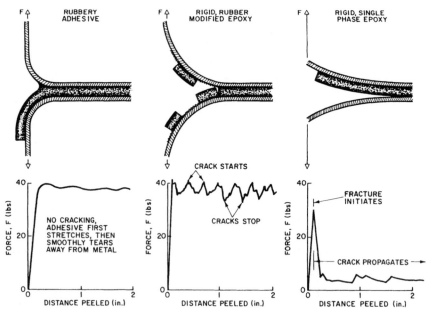

FIGURE 7.31   Failure modes and force-distance plots for T-peel specimens bonded with various adhesives.[30]

These adhesives are available as pastes, solvent solutions, and B-staged film supported on glass fabric. Cure generally requires 350°F for 1 h under moderate pressure. Epoxy-phenolic adhesives were developed primarily for bonding metal joints in high-temperature applications.

Epoxy-Nylon. Epoxy-nylon adhesives offer both excellent shear and peel strength. They maintain their physical properties at cryogenic temperatures but are limited to a maximum service temperature of 180°F.

Epoxy-nylon adhesives are available as unsupported B-staged film or in solvent solutions. A moderate pressure of 25 lb/in² and temperature of 350°F are generally required for 1 h to cure the adhesive. Because of their excellent filleting properties and high peel strength, epoxy-nylon adhesives are used to bond aluminum skins to honeycomb core in aircraft structures.

Epoxy-Polysulfide. Polysulfide resins combine with epoxy resins to provide adhesives with excellent flexibility and chemical resistance. These adhesives bond well to many different substrates. Shear strength and elevated temperature properties are poor, but resistance to peel forces and low temperatures is very good. The epoxy-polysulfide alloy is supplied as a

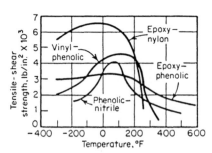

FIGURE 7.32   Effect of temperature on tensile-shear strength of adhesive alloys (substrate material is aluminum).[29]

two-part, flowable paste that cures to a rubbery solid at room temperature. A common application for epoxy-polysulfide adhesives is as a sealant.

Modified Phenolics.   Phenolic or phenol formaldehyde is also used as an adhesive for bonding wood. However, because of its brittle nature, this resin is unsuitable alone for more extensive adhesive applications. By modifying phenolic resin with various synthetic rubbers and thermoplastic materials, flexibility is greatly improved. The modified adhesive is well suited for structural bonding of many materials.

Nitrile-Phenolic.   Certain blends of phenolic resins with nitrile rubber produce adhesives useful to 300°F. On metals, nitrile phenolics offer shear strength in excess of 4,000 lb/in^2 and excellent peel properties. Good bond strengths can also be achieved on rubber, plastics, and glass. These adhesives have high impact strength and resistance to creep and fatigue. Their resistance to solvent, oil, and water is also good.

Nitrile-phenolic adhesives are available as solvent solutions and as supported and unsupported film. They require heat curing at 300 to 500°F under pressure of up to 200 lb/in^2. The nitrile-phenolic systems with the highest curing temperature have the greatest resistance to elevated temperatures during service. Because of good peel strength and elevated-temperature properties, nitrile-phenolic adhesives are commonly used for bonding linings to automobile brake shoes.

Vinyl-Phenolic.   Vinyl-phenolic adhesives are based on a combination of phenolic resin with polyvinyl formal or polyvinyl butyral resins. They have excellent shear and peel strength. Room-temperature shear strength as high as 5,000 lb/in^2 is available. Maximum operating temperature, however, is only 200°F because the thermoplastic constituent softens at elevated temperatures. Chemical resistance and impact strength are excellent.

Vinyl-phenolic adhesives are supplied in solvent solutions and as supported and unsupported film. The adhesive cures rapidly at elevated temperatures under pressure. They are used to bond metals, elastomers, and plastics to themselves or each other. A major application of vinyl-phenolic adhesive is the bonding of copper sheet to plastic laminate in printed-circuit-board manufacture.

Neoprene-Phenolic.   Neoprene-phenolic alloys are used to bond a variety of substrates. Normal service temperature is –70 to 200°F. Because of high resistance to creep and most service environments, neoprene-phenolic joints can withstand prolonged stress. Fatigue and impact strengths are also excellent. Shear strength, however, is lower than that of other modified phenolic adhesives.

Temperatures over 300°F and pressure greater than 50 lb/in^2 are needed for cure. Neoprene-phenolic adhesives are available as solvent solutions and film. During cure these adhesives are quite sensitive to atmospheric moisture, surface contamination, and other processing variables.

Polyaromatics.   Polyimide and polybenzimidazole resins belong to the aromatic heterocycle polymer family, which is noted for its outstanding thermal resistance. These two highly cross-linked adhesives are the most thermally stable systems commercially available. The polybenzimidazole (PBI) adhesive has shear strength on steel of 3,000 lb/in^2 at room temperature and 2,500 lb/in^2 at 700°F. The polyimide adhesive offers a shear strength of approximately 3,000 lb/in^2 at room temperature, but it does not have the excellent strength at 700 to 1,000°F that is characteristic of PBI. Polyimide adhesives offer better elevated-temperature aging properties than PBI. The maximum continuous operating temperature for a polyimide adhesive is 600 to 650°F, whereas PBI adhesives oxidize rapidly at temperatures over 500°F.

Both adhesives are available as supported film, and polyimide resins are also available in solvent solution. During cure, temperatures of 550 to 650°F and high pressure are required. Volatiles are released during cure that contribute to a porous, brittle bond line with relatively low peel strength.

Polyester. Polyesters are a large class of synthetic resins having widely varying properties. They may be divided into two distinct groups: saturated and unsaturated.

Unsaturated polyesters are fast-curing, two-part systems that harden by the addition of catalysts, usually peroxides. Styrene monomer is generally used as a reactive diluent for polyester resins. Cure can occur at room or elevated temperature depending on the type of catalyst. Accelerators such as cobalt naphthalene are sometimes incorporated into the resin to speed cure. Unsaturated polyester adhesives exhibit greater shrinkage during cure and poorer chemical resistance than epoxy adhesives. Certain types of polyesters are inhibited from curing by the presence of air, but they cure fully when enclosed between two substrates. Depending on the type of resin, polyester adhesives can be quite flexible or very rigid. Uses include patching kits for the repair of automobile bodies and concrete flooring. Polyester adhesives also have strong bond strength to glass-reinforced polyester laminates.

Saturated polyester resins exhibit high peel strength and are used to laminate plastic films such as polyethylene terephthalate (Mylar®). They also offer excellent clarity and color stability. These polyester types, in both solution and solid form, can be chemically cross-linked with curing agents such as the isocyanates for improved thermal and chemical stability.

Polyurethane. Polyurethane-based adhesives form tough bonds with high peel strength. Generally supplied as a two-part liquid, polyurethane adhesives can be cured at room or elevated temperatures. They have exceptionally high strength at cryogenic temperatures, but only a few formulations offer operating temperatures greater than 250°F. Like epoxies, urethane adhesives can be applied by a variety of methods and form strong bonds to most surfaces. Some polyurethane adhesives degrade substantially when exposed to high-humidity environments.

Polyurethane adhesives bond well to many substrates, including hard-to-bond plastics. Since they are very flexible, polyurethane adhesives are often used to bond films, foils, and elastomers. Moisture-curing one-part urethanes are also available. These adhesives utilize the humidity in the air to activate their curing mechanism.

Anaerobic Adhesives. Acrylate acid diester and cyanoacrylate resins are called anaerobic adhesives because they cure when air is excluded from the resin. Anaerobic resins are noted for being simple-to-use, one-part adhesives, having fast cure at room temperature and high cost. However, the cost is moderate when considering a bonded-area basis because only a small volume of adhesive is required. Most anaerobic adhesives do not cure when gaps between adherend surfaces are greater than 10 mils, although some monomers have been developed to provide for thicker bond lines.

The acrylate acid diester adhesives are available in various viscosities. They cure in minutes at room temperature when a special primer is used or in 3 to 10 min at 250°F without the primer. Without the primer, the adhesive requires 3 to 4 h at room temperature to cure.

The cyanoacrylate adhesives are more rigid and less resistant to moisture than acrylate acid diester adhesives. They are available only as low-viscosity liquids that cure in seconds at room temperature without the need of a primer. The cyanoacrylate adhesives bond well to a variety of substrates, as shown in Table 7.20 but have relatively poor thermal resistance. Modifications of the original cyanoacrylate resins have been introduced to provide faster cures, higher strengths with some plastics, and greater thermal resistance.

Thermosetting Acrylics. Thermosetting acrylic adhesives are newly developed two-part systems that provide high shear strength to many metals and plastics, as shown in Table 7.21. These acrylics retain their strength to 200°F. They are relatively rigid adhesives with poor peel strength. These adhesives are particularly noted for their weather and moisture resistance as well as fast cure at room temperature.

One manufacturer has developed an acrylic adhesive system where the hardener is applied to the substrate as a primer solution. The substrate can then be dried and stored for up to six months. When the parts are to be bonded, only the acrylic resin need be applied between the already primed substrates. Cure can occur in minutes at room temperature depending on the type of acrylic resin used. Thus, this system offers the user a fast-reacting, one-part adhesive (with primer) with long shelf life.

7.5.3 Nonstructural Adhesives

Nonstructural adhesives are characterized by low shear strength (usually less than 1000 psi) and poor creep resistance at slightly elevated temperatures. The most common nonstructural adhesives are based on elastomers and thermoplastics. Although these systems have low strength, they are usually easy to use and fast setting. Most

TABLE 7.20   Performance of Cyanoacrylate Adhesives on Various Substrates[31]

Substrate	Age of bond	Shear strength, lb/in.^{2} of adhesive bonds
Steel–steel	10 min	1,920
	48 h	3,300
Aluminum–aluminum	10 min	1,480
	48 h	2,270
Butyl rubber–butyl rubber	10 min	150*
SBR rubber–SBR rubber	10 min	130
Neoprene rubber–neoprene rubber	10 min	100*
SBR rubber–phenolic	10 min	110*
Phenolic–phenolic	10 min	930*
	48 h	940*
Phenolic–aluminum	10 min	650
	48 h	920*
Aluminum–nylon	10 min	500
	48 h	950
Nylon–nylon	10 min	330
	48 h	600
Neoprene rubber–polyester glass	10 min	110*
Polyester glass–polyester glass	10 min	680
Acrylic–acrylic	10 min	810*
	48 h	790*
ABS–ABS	10 min	640*
	48 h	710*
Polystyrene–polystyrene	10 min	330*
Polycarbonate–polycarbonate	10 min	790
	48 h	950*

* Substrate failure.

TABLE 7.21   Tensile-Shear Strength of Various Substrates Bonded with Thermosetting Acrylic Adhesives[32]

	Avg lap shear, lb/in.^{2} at 77°F		
Substrate*	Adhesive A	Adhesive B	Adhesive C
Alclad aluminum, etched.	4,430	4,235	5,420
Bare aluminum, etched.	4,305	3,985	5,015
Bare aluminum, blasted.	3,375	3,695	4,375
Brass, blasted.	4,015	3,150	4,075
302 stainless steel, blasted.	4,645	4,700	5,170
302 stainless steel, etched.	2,840	4,275	2,650
Cold-rolled steel, blasted.	2,050	3,385	2,135
Copper, blasted	2,915	2,740	3,255
Polyvinyl chloride, solvent wiped.	1,375†	1,250†	1,250†
Polymethyl methacrylate, solvent wiped.	1,550†	1,160†	865†
Polycarbonate, solvent wiped.	2,570†	960	2,570†
ABS, solvent wiped.	1,610†	1,635†	1,280†
Alclad aluminum–PVC.	1,180†		
Plywood, ⅝-in. exterior glued (lb/in.).	802†	978†	
AFG-01 gap fill (¹⁄₁₆-in.) (lb/in.).	. . . . .	1,083†	

* Metals solvent cleaned and degreased before etching or blasting.
† Substrate failure.

nonstructural adhesives are used in assembly-line fastening operations or as sealants and pressure-sensitive tapes.

Elastomer-Based Adhesives.   Natural- or synthetic-rubber-based adhesives usually have excellent peel strength but low shear strength. Their resiliency provides good fatigue and impact properties. Temperature resistance is generally limited to 150 to 200°F, and creep under load occurs at room temperature.

The basic types of rubber-based adhesives used for nonstructural applications are shown in Table 7.22. These systems are generally supplied as solvent solutions, latex cements, and pressure-sensitive tapes. The first two forms require driving the solvent or water vehicle from the adhesive before bonding. This is accomplished by either simple ambient air evaporation or forced heating. Some of the stronger and more environment-resistant rubber-based adhesives require an elevated-temperature cure. Generally, only slight pressure is required to achieve a substantial bond.

Pressure-sensitive adhesives are permanently tacky and flow under pressure to provide intimate contact with the adherend surface. Pressure-sensitive tapes are made by placing these adhesives on a backing material such as rubber, vinyl, canvas, or cotton cloth. After pressure is applied, the adhesive tightly grips the part being mounted as well as the surface to which it is affixed. The ease of application and the many different properties that can be obtained from elastomeric adhesives account for their wide use.

In addition to pressure-sensitive adhesives, elastomers go into mastic compounds that find wide use in the construction industry. Neoprene and reclaimed-rubber mastics are used to bond gypsum board and plywood flooring to wood-framing members. Often the adhesive bond is much stronger than the substrate. These mastic systems cure by evaporation of solvent through the porous substrates.

Silicone.   Silicone pressure-sensitive adhesives have low shear strength but excellent peel strength and heat resistance. Silicone adhesives can be supplied as solvent

TABLE 7.22 Properties of Elastomeric Adhesives Used in Nonstructural Applications[29]

Adhesive	Application	Advantages	Limitations
Reclaimed rubber	Bonding paper, rubber, plastic and ceramic tile, plastic films, fibrous sound insulation and weather-stripping; also used for the adhesive on surgical and electrical tape	Low cost, applied very easily with roller coating, spraying, dipping, or brushing, gains strength very rapidly after joining, excellent moisture and water resistance	Becomes quite brittle with age, poor resistance to organic solvents
Natural rubber	Same as reclaimed rubber; also used for bonding leather and rubber sides to shoes	Excellent resilience, moisture and water resistance	Becomes quite brittle with age; poor resistance to organic solvents; does not bond well to metals
Neoprene rubber	Bonding weather stripping and fibrous soundproofing materials to metal; used extensively in industry; bonding synthetic fibers, i.e., Dacron	Good strength to 150°F, fair resistance to creep	Poor storage life, high cost; small amounts of hydrochloric acid evolved during aging that may cause corrosion in closed systems; poor resistance to sunlight
Nitrile rubber	Bonding plastic films to metals, and fibrous materials such as wood and fabrics to aluminum, brass, and steel; also, bonding nylon to nylon and other materials	Most stable synthetic-rubber adhesive, excellent oil resistance, easily modified by addition of thermosetting resins	Does not bond well to natural rubber or butyl rubber
Polyiso-butylene	Bonding rubber to itself and plastic materials; also, bonding polyethylene terephthalate film to itself, aluminum foil and other plastic films	Good aging characteristics	Attacked by hydrocarbons; poor thermal resistance
Butyl	Bonding rubber to itself and metals; forms good bonding with most plastic films such as polyethylene terephthalate and polyvinylidene chloride	Excellent aging characteristics; chemically cross-linked materials have good thermal properties	Metals should be treated with an appropriate primer before bonding; attacked by hydrocarbons

solutions for pressure-sensitive application. The adhesive reaches maximum physical properties after being cured at elevated temperature with an organic peroxide catalyst. A lesser degree of adhesion can also be developed at room temperature. Silicone adhesives retain their qualities over a wide temperature range and after extended exposure to elevated temperature.

Room-temperature vulcanizing (RTV) silicone-rubber adhesives and sealants form flexible bonds with high peel strength to many substrates. These resins are one-component pastes that cure by reacting with moisture in the air. Because of this unique curing mechanism, nonporous substrates should not overlap by more than 1 in.

RTV silicone materials cure at room temperature in about 24 h. Fully cured adhesives can be used for extended periods up to 450°F and for shorter periods up to 500°F.

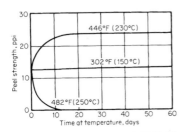

FIGURE 7.33   Peel strength of RTV silicone rubber bonded to aluminum as a function of heat aging.[33]

Figure 7.33 illustrates the peel strength of an RTV adhesive on aluminum as a function of heat aging. With most RTV silicone formulations, acetic acid is released during cure. Consequently, corrosion of metals such as copper and brass in the bonding area may be a problem. However, special formulations are available that liberate methanol instead of acetic acid during cure. Silicone rubber bonds to clean metal, glass, wood, silicone resin, vulcanized silicone rubber, ceramic, and many plastic surfaces.

Thermoplastic Adhesives.   Table 7.16 describes the most common types of thermoplastic adhesives. These adhesives are useful in the −20 to 150°F temperature range. Their physical properties vary with chemical type. Some resins like the polyamides are quite tacky and flexible, while others are very rigid.

Thermoplastic adhesives are generally available as solvent solutions, water-based emulsions, and hot melts. The first two systems are useful in bonding porous materials such as wood, plastic foam, and paper. Water-based systems are especially useful for bonding foams that could be affected adversely by solvents. When hardened, thermoplastic adhesives are very nonresistant to the solvent in which they are originally supplied.

Hot-melt systems are usually flexible and tough. They are used extensively for sealing applications involving paper, plastic films, and metal foil. Table 7.23 offers a general comparison of hot-melt adhesives. Hot melts can be supplied as (1) tapes or ribbons, (2) films, (3) granules, (4) pellets, (5) blocks, or (6) cards that are melted and pressed between the substrate. The rate at which the adhesive cools and sets is dependent on the type of substrate and whether it is preheated.

7.6   SELECTING AN ADHESIVE

There is no general-purpose adhesive. The best adhesive for a particular application will depend on the materials to be bonded, the service and assembly requirements, and the economics of the bonding operation. By using these factors as criteria for selection, the many commercially available adhesives can be narrowed down to a few possible candidates. All the desired properties are often not available in a single adhesive system. In these cases, a compromise adhesive can usually be chosen by deciding which properties are of major and minor importance.

The materials to be bonded are a prime factor in determining which adhesive to use. Some adherends such as stainless steel or wood can be successfully bonded with a great many adhesive types; other adherends such as nylon can be bonded with only a few.

7.6.1   Adhesives for Metal

The chemical types of structural adhesives for metal bonding were described in the preceding section. Since organic adhesives readily wet most metallic surfaces, the

TABLE 7.23 General Comparison of Common Hot-Melt Adhesives[34]

Property	Ethylene vinyl acetate	Ethylene ethyl acetate	Ethylene acrylic acid	Ionomer	Phenoxy	Poly-amide	Polyester	Poly-ethylene	Polyvinyl acetate	Polyvinyl butyral
Softening point, °C	40	60	70	75	100	100	......	......	65–195	130
Melting point, °C	95	90 (80–100)	......	......	......	......	267	137	......	
Crystallinity	L	L	M	L	L	L	H	H or L	L	L
Melt index	6 (2–20)	3 (2–20)	3 (0.5–400)	2	2.5	2	5	5 (0.5–20)	......	L
Tensile strength, lb/in.²	2,750	2,000	3,300	4,000	9,500	2,000	4,500	2,000	5,000	6,500
% elongation	800	700	600	450	75	300	500	150	10	100
Lap shear on metal, lb/in.²	......	......	1,700	......	3,500	1,050	......	......		
Film peel, lb/in.	1	4.5	12	12	......	......	5	2		
Cost	M	M	M	M	H	M	H	L	M–L	M
Usage	H	M	L	L	L	H	H	L	M	H

H = high. M = medium. L = low.

7.65

adhesive selection does not depend as much on the type of metal substrate as on other bonding requirements.

Selecting a specific adhesive from a table of general properties is difficult because formulations within one class of adhesive may vary widely in physical properties. General physical data for common structural metal adhesives are presented in Table 7.24. This table may prove useful in making preliminary selections or eliminating obviously unsuitable adhesives.

Nonstructural adhesives for metals include elastomeric and thermoplastic resins. These are generally used as pressure-sensitive or hot-melt adhesives. They are noted for fast production, low cost, and low to medium strength. Typical adhesives for nonstructural bonding applications were previously described. Most pressure-sensitive and hot-melt cements can be used on any clean metal surface and on many plastics and elastomers.

7.6.3 Adhesives for Plastics

The physical and chemical properties of both the solidified adhesive and the plastic substrate affect the quality of the bonded joint. Major elements of concern are the thermal expansion coefficient and glass transition temperature of the substrate relative to the adhesive. Special consideration is also required of polymeric surfaces that can change during normal aging or exposure to operating environments.

Significant differences in thermal expansion coefficient between substrates and the adhesive can cause serious stress at the plastic joint's interface. These stresses are compounded by thermal cycling and low-temperature service requirements. Selection of a resilient adhesive or adjustments in the adhesive's thermal expansion coefficient via fillers or additives can reduce such stress.

Structural adhesives must have a glass-transition temperature higher than the operating temperature to avoid a cohesively weak bond and possible creep problems. Modern engineering plastics, such as polyimide or polyphenylene sulfide, have very high glass-transition temperatures. Most common adhesives have a relatively low glass-transition temperature so that the weakest thermal link in the joint may often be the adhesive.

Use of an adhesive too far below the glass-transition temperature could result in low peel or cleavage strength. Brittleness of the adhesive at very low temperatures could also manifest itself in poor impact strength.

Plastic substrates could be chemically active, even when isolated from the operating environment. Many polymeric surfaces slowly undergo chemical and physical change. The plastic surface, at the time of bonding, may be well suited to the adhesive process. However, after aging, undesirable surface conditions may present themselves at the interface, displace the adhesive, and result in bond failure. These weak boundary layers may come from the environment or within the plastic substrate itself.

Moisture, solvent, plasticizers, and various gases and ions can compete with the cured adhesive for bonding sites. The process by which a weak boundary layer preferentially displaces the adhesive at the interface is called desorption. Moisture is the most common desorbing substance, being present both in the environment and within many polymeric substrates.

Solutions to the desorption problem consist of eliminating the source of the weak boundary layer or selecting an adhesive that is compatible with the desorbing material. Excessive moisture can be eliminated from a plastic part by postcuring or drying the part before bonding. Additives that can migrate to the surface can possibly be

TABLE 7.24 Properties of Structural Adhesives Used to Bond Metals

Adhesive	Service temp, °F Max	Service temp, °F Min	Shear strength, lb/in.^{2}	Peel strength	Impact strength	Creep resistance	Solvent resistance	Moisture resistance	Type of bond
Epoxy-amine	150	−50	3,000–5,000	Poor	Poor	Good	Good	Good	Rigid
Epoxy-polyamide	150	−60	2,000–4,000	Medium	Good	Good	Good	Medium	Tough and moderately flexible
Epoxy-anhydride	300	−60	3,000–5,000	Poor	Medium	Good	Good	Good	Rigid
Epoxy-phenolic	350	−423	3,200	Poor	Poor	Good	Good	Good	Rigid
Epoxy-nylon	180	−423	6,500	Very good	Good	Medium	Good	Poor	Tough
Epoxy-polysulfide	150	−100	3,000	Good	Medium	Medium	Good	Good	Flexible
Nitrile-phenolic	300	−100	3,000	Good	Good	Good	Good	Good	Tough and moderately flexible
Vinyl-phenolic	225	−60	2,000–5,000	Very good	Good	Medium	Medium	Good	Tough and moderately flexible
Neoprene-phenolic	200	−70	3,000	Good	Good	Good	Good	Good	Tough and moderately flexible
Polyimide	600	−423	3,000	Poor	Poor	Good	Good	Medium	Rigid
Polybenzimidazole	500	−423	2,000–3,000	Poor	Poor	Good	Good	Good	Rigid
Polyurethane	150	−423	5,000	Good	Good	Good	Medium	Poor	Flexible
Acrylate acid diester	200	−60	2,000–4,000	Poor	Medium	Good	Poor	Poor	Rigid
Cyanoacrylate	150	−60	2,000	Poor	Poor	Good	Poor	Poor	Rigid
Phenoxy	180	−70	2,500	Medium	Good	Good	Poor	Good	Tough and moderately flexible
Thermosetting acrylic	250	−60	3,000–4,000	Poor	Poor	Good	Good	Good	Rigid

eliminated by reformulating the plastic resin. Also, certain adhesives are more compatible with oils and plasticizers than others. For example, the migration of plasticizer from flexible polyvinyl chloride can be counteracted by using a nitrile-based adhesive. Nitrile-adhesive resins are capable of absorbing the plasticizer without degrading.

Thermoplastics.   Many thermoplastics can be joined by solvent or heat welding as well as with adhesives. These alternative joining processes are discussed in detail in the next section. The plastic manufacturer is generally the leading source of information on the proper methods of joining a particular plastic.

Thermosetting Plastics.   Thermosetting plastics cannot be heat- or solvent-welded. They are easily bonded with many adhesives. Abrasion is generally recommended as a surface treatment.

Plastic Foams.   Some solvent cements and solvent-containing pressure-sensitive adhesives will collapse thermoplastic foams. Water-based adhesives, based on styrene butadiene rubber (SBR) or polyvinyl acetate, and 100 percent solids adhesives are often used. Butyl, nitrile, and polyurethane adhesives are often used for flexible polyurethane foam. Epoxy adhesives offer excellent properties on rigid polyurethane foam.

7.6.3   Adhesives for Elastomers

Vulcanized Elastomers.   Bonding of vulcanized elastomers to themselves and other materials is generally completed by using a pressure-sensitive adhesive derived from an elastomer similar to the one being bonded. Flexible thermosetting adhesives such as epoxy-polyamide or polyurethane also offer excellent bond strength to most elastomers. Surface treatment consists of washing with a solvent, abrading, or acid cyclizing as described in Table 7.13.

Elastomers vary greatly in formulation from one manufacturer to another. Fillers, plasticizers, antioxidants, and the like may affect the adhesive bond. Adhesives should be thoroughly tested on a specific elastomer and then reevaluated if the elastomer manufacturer or formulation is changed.

Unvulcanized Elastomers.   Unvulcanized elastomers may be bonded to metals and other rigid adherends by priming the adherend with a suitable air- or heat-drying adhesive before the elastomer is molded against the adherend. The most common elastomers to be bonded in this way include nitrile, neoprene, urethane, natural rubber, SBR, and butyl rubber. Less common unvulcanized elastomers such as the silicones, fluorocarbons, chlorosulfonated polyethylene, and polyacrylate are more difficult to bond. However, recently developed adhesive primers improve the bond of these elastomers to metal. Surface treatment of the adherend before priming should be according to good standards.

7.6.4   Adhesives for Composites

Mechanical fasteners and/or adhesives are commonly used to join composites. Mechanical fasteners generally require drilling or machining of the composite, which could weaken the part. Adhesives that give satisfactory results on the resin matrix alone may also be used to bond composites. Three adhesives are often used to bond composites: epoxies, acrylics, and urethanes.

Surface preparation of reinforced thermosetting plastics consists of abrasion and solvent cleaning. A degree of abrasion is desired so that the glaze on the resin surface is removed, but the reinforcing fibers are not exposed.

Reinforced thermoplastic parts are generally abraded and cleaned prior to adhesive bonding. However, special surface treatment such as used on the thermoplastic resin matrix may be necessary for optimum strength. Care must be taken so that the treatment chemicals do not wick into the substrate and cause degradation.

Certain reinforced thermoplastics may also be solvent-cemented or heat-welded. However, the percentage of filler in the substrate must be limited or else the bond will be starved of resin.

7.7   METHODS OF WELDING POLYMERIC SUBSTRATES

Certain thermoplastic substrates may be joined by methods other than adhesive bonding. By careful application of solvent or heat to a thermoplastic substrate, one may liquify the surface resin and use it as the adhesive.

Table 7.25 indicates the various joining methods that are suitable for common types of plastics. Descriptions of these various bonding techniques are presented in Table 7.26, and they are discussed below.

7.7.1   Direct Heat Welding

Welding by direct application of heat provides an advantageous method of joining many thermoplastics that do not degrade rapidly at their melt temperatures. The three principal methods of direct welding are heated-tool welding, hot-gas welding, and resistance-wire welding.

Heated-Tool Welding.   In this method, the surfaces to be fused are heated by holding them against a hot surface; then the parts are brought into contact and allowed to harden under slight pressure. Electric-strip heaters, soldering irons, hot plates, and resistance blades are common methods of providing heat locally.

A simple hot plate has been used extensively with many plastics. The parts are held on the hot plate until sufficient fusible material has been developed. Table 7.27 lists typical hot-plate temperatures for a variety of plastics. A similar technique involves butting flat plastic sheets on a table next to a resistance blade that runs the length of the sheet. Once the plastic begins to soften, the blade is raised, and the sheets are pressed together and fused.

Figure 7.34 illustrates a heated tool commonly used to bond plastic sheet and film. Care must be taken especially with thin film not to apply too much pressure or heat and melt through the plastic. Table 7.28 provides heat-sealing temperature ranges for common plastic films.

Hot-Gas Welding.   An electric or gas-heated welding gun with an orifice temperature of 425 to 700°F can be used to bond many thermoplastic materials. The pieces to be joined are beveled and positioned with a small gap between them. A welding rod made of the same plastic that is being bonded is laid in the joint with a steady pressure. The heat from the gun is directed to the tip of the rod, where it fills the gap, as shown in Fig. 7.35.

TABLE 7.25 Assembly Methods for Plastics[35]

Plastic	Common assembly methods							
	Adhesives	Dielectric welding	Induction bonding	Mechanical fastening	Solvent welding	Spin welding	Thermal welding	Ultrasonic welding
Thermoplastics								
ABS	X		X	X	X	X	X	X
Acetals	X		X	X	X	X	X	X
Acrylics	X		X	X	X	X		X
Cellulosics	X				X	X		
Chlorinated polyether	X	X		X			X	
Ethylene copolymers	X							
Fluoroplastics		X						X
Ionomer							X	X
Methylpentene								X
Nylons	X		X	X	X	X		X
Phenylene oxide–based materials	X			X	X	X	X	X
Polyesters	X			X	X			
Polyamide-imide	X			X				X
Polyaryl ether	X			X				X
Polyaryl sulfone	X			X				X
Polybutylene								
Polycarbonate	X		X	X	X	X	X	X
Polycarbonate/ABS	X			X	X	X	X	X
Polyethylenes	X	X	X	X		X	X	X
Polyimide	X			X				
Polyphenylene sulfide	X			X				
Polypropylenes	X	X	X	X		X	X	X
Polystyrenes	X		X	X	X	X	X	X
Polysulfone	X			X	X			X
Propylene copolymers	X	X	X	X	X	X		
PVC/acrylic alloy	X	X	X	X			X	X
PVC/ABS alloys	X			X			X	
Styrene acrylonitrile	X	X	X	X	X	X	X	X
Vinyls	X	X	X	X	X		X	X
Thermosets								
Alkyds	X							
Allyl diglycol carbonate	X							
Diallyl phthalate	X							
Epoxies	X			X				
Melamines	X			X				
Phenolics	X			X				
Polybutadienes	X							
Polyesters	X			X				
Silicones	X			X				
Ureas	X							
Urethanes	X			X				

TABLE 7.26 Bonding or Joining Plastics: What Techniques Are Available and What Do They Offer[36]

Technique	Description	Advantages	Limitations	Processing considerations
Solvent cementing and dopes	Solvent softens the surface of an amorphous thermoplastic; mating takes place when the solvent has completely evaporated. Bodied cement with small percentage of parent material can give more workable cement, fill in voids in bond area. Cannot be used for polyolefins and acetal homopolymers	Strength, up to 100% of parent materials; easily and economically obtained with minimum equipment requirements	Long evaporation times required; solvent may be hazardous; may cause crazing in some resins	Equipment ranges from hypodermic needle or just a wiping media to tanks for dip and soak. Clamping devices are necessary, and air dryer is usually required. Solvent-recovery apparatus may be necessary or required. Processing speeds are relatively slow because of drying times. Equipment costs are low to medium

Thermal Bonding

Technique	Description	Advantages	Limitations	Processing considerations
Ultrasonics	High-frequency sound vibrations transmitted by a metal horn generate friction at the bond area of a thermoplastic part, melting plastics just enough to permit a bond. Materials most readily weldable are acetal, ABS, acrylic, nylon, PC, polyimide, PS, SAN, phenoxy	Strong bonds for most thermoplastics; fast, often less than 1 s. Strong bonds obtainable in most thermal techniques if complete fusion is obtained	Size and shape limited. Limited applications to PVCs, polyolefins	Converter to change 20 kHz electrical into 20 kHz mechanical energy is required along with stand and horn to transmit energy to part. Rotary tables and high-speed feeder can be incorporated
Hot-plate and hot-tool welding	Mating surfaces are heated against a hot surface, allowed to soften sufficiently to produce a good bond, then clamped together while bond sets. Applicable to rigid thermoplastics	Can be very fast, e.g., 4–10 s in some cases; strong	Stresses may occur in bond area	Use simple soldering guns and hot irons, relatively simple hot plates attached to heating elements up to semiautomatic hot-plate equipment. Clamps needed in all cases
Hot-gas welding	Welding rod of the same material being joined (largest application is vinyl) is softened by hot air or nitrogen as it is fed through a gun that is softening part surface simultaneously. Rod fills in joint area and cools to effect a bond	Strong bonds, especially for large structural shapes	Relatively slow; not an "appearance" weld	Requires a hand gun, special welding tips, an air source and welding rod. Regular hand-gun speeds run 6 in./min; high-speed hand-held tool boosts this to 48–60 in./min

TABLE 7.26 Bonding or Joining Plastics: What Techniques Are Available and What Do They Offer[36] (*Continued*)

Technique	Description	Advantages	Limitations	Processing considerations
		Thermal Bonding (continued)		
Spin welding	Parts to be bonded are spun at high speed, developing friction at the bond area; when spinning stops, parts cool in fixture under pressure to set bond. Applicable to most rigid thermoplastics	Very fast (as low as 1–2 s); strong bonds	Bond area must be circular	Basic apparatus is a spinning device, but sophisticated feeding and handling devices are generally incorporated to take advantage of high-speed operation
Dielectrics	High-frequency voltage applied to film or sheet causes material to melt at bonding surfaces. Material cools rapidly to effect a bond. Most widely used with vinyls	Fast seal with minimum heat applied	Only for film and sheet	Requires rf generator, dies, and press. Operation can range from hand-fed to semiautomatic with speeds depending on thickness and type of product being handled. 3–25 kW units are most common
Induction	A metal insert or screen is placed between the parts to be welded, and energized with an electromagnetic field. As the insert heats up, the parts around it melt, and when cooled form a bond. For most thermoplastics	Provides rapid heating of solid sections to reduce chance of degradation	Since metal is embedded in plastic, stress may be caused at bond	High-frequency generator, heating coil, and inserts (generally 0.02–0.04 in. thick). Hooked up to automated devices, speeds are high. Work coils, water cooling for electronics, automatic timers, multiple-position stations may also be required
		Adhesives*		
Liquids solvent, water base, anaerobics	Solvent-and-water-based liquid adhesives, available in a wide number of bases—e.g., polyester, vinyl—in one- or two-part form fill bonding needs ranging from high-speed lamination to one-of-a-kind joining of dissimilar plastics parts. Solvents provide more bite, but cost much more than similar base water-type adhesive. Anaerobics are a group of adhesives that cure in the absence of air,	Easy to apply; adhesives available to fit most applications	Shelf and pot life often limited. Solvents may cause pollution problems; water-base not as strong; anaerobics toxic	Application techniques range from simply brushing on to spraying and roller coating-lamination for very high production. Adhesive application techniques, often similar to decorating equipment, from hundreds to thousands of dollars with sophisticated laminating equipment costing in the tens of thousands of dollars. Anaerobics are generally applied a drop at a time from a special bottle or dispenser

Pastes, mastics	Highly viscous single- or two-component materials which cure to a very hard or flexible joint depending on adhesive type	Does not run when applied	Shelf and pot life often limited	Often applied via a trowel, knife or gun-type dispenser; one-component systems can be applied directly from a tube. Various types of roller coaters are also used. Metering-type dispensing equipment in the $2,500 range has been used to some extent
Hot melts	100% solids adhesives that become flowable when heat is applied. Often used to bond continuous flat surfaces	Fast application; clean operation.	Virtually no structural hot melts for plastics	Hot melts are applied at high speeds via heating the adhesive, then extruding (actually squirting) it onto a substrate, roller coating, using a special dispenser or roll to apply dots or simply dipping
Film	Available in several forms including hot melts, these are sheets of solid adhesive. Mostly used to bond film or sheet to a substrate	Clean, efficient	High cost	Film adhesive is reactivated by a heat source; production costs are in the medium-high range depending on heat source used
Pressure-sensitive	Tacky adhesives used in a variety of commercial applications (e.g., cellophane too). Often used with polyolefins	Flexible	Bonds not very strong	Generally applied by spray with bonding effected by light pressure
Mechanical fasteners (staples, screws, molded-in inserts, snap fits and variety of proprietary fasteners)	Typical mechanical fasteners are listed on the left. Devices are made of metal or plastic. Type selected will depend on how strong the end product must be, appearance factors. Often used to join dissimilar plastics or plastics to nonplastics	Adaptable to many materials; low to medium costs; can be used for parts that must be disassembled	Some have limited pull-out strength; molded-in inserts may result in stresses	Nails and staples are applied by simply hammering or stapling. Other fasteners may be inserted by drill press, ultrasonics, air or electric gun, hand tool. Special molding—i.e., molded-in-hole—may be required

* Typical adhesives in each class are: Liquids: 1. Solvent—polyester, vinyl, phenolics acrylics, rubbers, epoxies, polyamide; 2. Water—acrylics, rubber-casein; 3. Anaerobics—cyanoacrylate; mastics—rubbers, epoxies; hot melts—polyamides, PE, PS, PVA; film—epoxies, polyamide, phenolics; pressure, sensitive—rubbers.

TABLE 7.27 Hot-Plate Temperatures
to Weld Plastics[37]

Plastic	Temp, °F
ABS	450
Acetal	500
Phenoxy	550
Polyethylene LD	360
HD	390
Polycarbonate	650
PPO	650
Noryl*	525
Polypropylene	400
Polystyrene	420
SAN	450
Nylon 6, 6	475
PVC	450

* Trademark of General Electric Co.

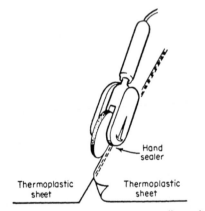

FIGURE 7.34 Hand-operated heat-sealing tool.

For polyolefins, the heated gas must be inert, since air will oxidize the surface of the plastic. After welding, the joint should not be stressed for several hours. This is particularly true for polyolefins, nylons, and polyformaldehyde. Hot-gas welding is not recommended for filled materials or substrates less than $\frac{1}{16}$ in thick. Applications are usually large structural assemblies. The weld is not cosmetically attractive, but tensile strengths 85 percent of the parent material can easily be obtained.

Resistance-Wire Welding. This method employs an electrical resistance heating wire laid between mating substrates to generate heat of fusion. After the bond has been made, the exterior wire is cut off. A similar type of process can be used to cure thermosetting adhesives when the heat generated by the resistance wire is used to advance the cure.

7.7.2 Indirect Thermal Welding

Indirect heating occurs when some form of energy is applied to the joint that acts on the plastic to cause heating at the interface or in the plastic as a whole. The applied energy is generally in the form of friction, high-frequency electric fields (dielectric), electromagnetic fields (induction), or ultrasonics.

Friction or Spin Welding. Spin welding uses the heat of friction to cause fusion at the interface. One substrate is rotated very rapidly while in touch with the other substrate so that the surfaces melt without damaging the entire part. Sufficient pressure is applied during the process to force out excess air bubbles. The rotation is then

TABLE 7.28   Heat-Sealing Temperatures for Plastic Films[38]

Film	Temp, °F
Coated cellophane	200–350
Cellulose acetate	400–500
Coated polyester	490
Poly(chlorotrifluoroethylene)	415–450
Polyethylene	250–375
Polystyrene (oriented)	220–300
Poly(vinyl alcohol)	300–400
Poly(vinyl chloride) and copolymers (nonrigid)	200–400
Poly(vinyl chloride) and copolymers (rigid)	260–400
Poly(vinyl chloride)–nitrile rubber blend	220–350
Poly(vinylidene chloride)	285
Rubber hydrochloride	225–350
Fluorinated ethylene–propylene copolymer	600–750

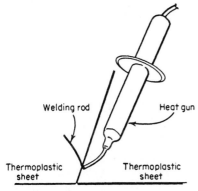

Welding rod   Heat gun

Thermoplastic sheet   Thermoplastic sheet

FIGURE 7.35   Hot-gas welding apparatus.

stopped, and pressure is maintained until the weld sets. Rotation speed and pressure are dependent on the thermoplastics being joined.

A wide variety of joints can be made by spin welding. Since the outer edges of the rotating substrate move considerably faster than the center, joints are generally designed to concentrate pressure at the center. Spin welding is a popular method of joining large-volume products, packaging, and toys.

Dielectric Welding.   Dielectric sealing can be used on most thermoplastics except those that are relatively transparent to high-frequency electric fields. It is used mostly to seal vinyl sheeting such as automobile upholstery, swimming-pool liners, and rainwear. An alternating electric field is imposed on the joint, which causes rapid reorientation of polar molecules, and heat is generated by molecular friction. The field is removed, and pressure is then applied and held until the weld cools.

Variables in the bonding operation are the frequency generated, dielectric loss of the plastic, the power applied, pressure, and time. Dielectric heating can also be used to generate the heat necessary for curing polar, thermosetting adhesives, or it can be used to quickly evaporate water from water-based adhesives.

Induction Heating.   An electromagnetic induction field can be used to heat a metal grid or insert placed between mating thermoplastic substrates. When the joint is positioned between induction coils, the hot insert causes the plastic to melt and fuse together. Slight pressure is maintained as the induction field is turned off and the joint hardens.

Electromagnetic adhesives can be made from metal-filled thermoplastics. These adhesives can be shaped into gaskets or film that will melt in an induction field. The advantage of this method is that stresses caused by large metal inserts are avoided. Table 7.29 shows compatible plastic combinations for electromagnetic adhesives.

TABLE 7.29 Compatible Plastic Combinations for Bonding with Electromagnetic Adhesives[39]

	ABS	Acetal	Acrylic	Nylon	PC	PE	PP	PS	PVC	SAN
ABS	X	..	X	..	..	..	..	X	X	
Acetal	..	X								
Acrylic	X	..	X	..	..	..	..	X	X	
Nylon	..	..	..	X						
Polycarbonate	..	..	..	..	X					
Polyethylene	..	..	..	..	..	X				
Polypropylene	..	..	..	..	..	..	X			
Polystyrene	X	..	X	..	..	..	..	X	X	
Polyvinyl chloride	X	..	X	..	..	..	..	X	X	
SAN	..	..	..	..	..	..	..	..	..	X

X = compatible combinations.

Ultrasonic Welding. During ultrasonic welding, a high-frequency electrodynamic field resonates a metal horn that is in contact with one plastic substrate. The horn vibrates the substrate sufficiently fast relative to a stationary second substrate so that great heat is generated at the interface. With pressure and subsequent cooling, a strong bond can be obtained with many thermoplastics. Rigid plastics with a high modulus of elasticity are best. Excellent results generally are obtainable with polystyrene, SAN, ABS, polycarbonate, and acrylic plastics.

Typical ultrasonic joint designs are shown in Fig. 7.36. Ultrasonics can also be used to stake plastics to other substrates and for inserting metal parts. Ultrasonic bonding is considered faster than methods of direct-heat welding.

Vibration Welding. Vibration welding is similar to ultrasonic welding except that it uses lower frequencies of vibration. In this way, very large parts can be bonded. Vibration welding has been used on large thermoplastic parts such as canisters, pipe sections, and other parts that are too large to be excited with an ultrasonic generator.

Both vibration welding and ultrasonic welding are very fast operations and applicable to high-volume assembly. The manufacturers of ultrasonic and vibration welding equipment are the same.

7.7.3 Solvent Cementing

Solvent cementing is a simple and economical method of joining noncrystalline thermoplastics. Solvent-cemented joints are less sensitive to thermal cycling than joints bonded with adhesives, and they are as resistant to degrading environments as the parent plastic. Bond strength 85 to 100 percent of the parent plastic can be obtained. The major disadvantages of solvent cementing are the possibility of stress cracking of the part and the possible hazards of using low vapor point solvents. When two dissimilar thermoplastics are to be joined, adhesive bonding is generally desirable because of solvent and polymer compatibility problems.

Solvent cements should be chosen with approximately the same solubility parameter as the plastic to be bonded. Table 7.30 lists typical solvents used to bond major plastics. It is common to use a mixture of fast-drying solvent with a less volatile solvent to prevent crazing. The solvent cement can be bodied up to 25 percent by weight with the parent plastic to fill gaps and reduce shrinkage and internal stress during drying.

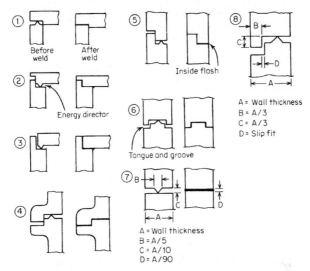

FIGURE 7.36   Typical joint designs used in ultrasonic welding.[40]

The parts to be bonded should be unstressed and annealed if necessary. The solvent cement is generally applied to the substrate with a syringe or brush. In some cases, the surface may be immersed in the solvent. After the area to be bonded softens, the parts are mated and held under pressure until dry. Pressure should be low and uniform so that the joint will not be stressed. After the joint hardens, the pres-

TABLE 7.30   Typical Solvents for Solvent Cementing[36]

Plastic	Solvent
ABS	Methyl ethyl ketone, methyl isobutyl ketone, tetrahydrofuran, methylene chloride
Acetate	Methylene chloride, acetone, chloroform, methyl ethyl ketone, ethyl acetate
Acrylic	Methylene chloride, ethylene dichloride
Cellulosics	Methyl ethyl ketone, acetone
Nylon	Aqueous phenol, solutions of resorcinal in alcohol, solutions of calcium chloride in alcohol
PPO	Trichloroethylene, ethylene dichloride, chloroform, methylene chloride
PVC	Cyclohexane, tetrahydrofuran, dichlorobenzene
Polycarbonate	Methylene chloride, ethylene dichloride
Polystyrene	Methylene chloride, ethylene ketone, ethylene dichloride, trichloroethylene, toluene, xylene
Polysulfone	Methylene chloride

These are solvents recommended by the various resin suppliers. A key to the selection of solvents is how fast they evaporate: a fast-evaporating product may not last long enough for some assemblies; too slow evaporation could hold up production.

sure is released, and an elevated-temperature bake may be necessary depending on the plastic and desired joint strength. The bonded part should not be packaged or stressed until the solvent has adequate time to escape from the joint.

7.8 METHODS OF MECHANICAL JOINING

Reliable mechanically fastened joints require:

- A firm, strong connection
- Materials that are stable in the environment
- Stable geometry
- Appropriate stresses in the parts including a correct clamping force

In addition to joint strength, mechanically fastened joints should prevent slip, separation, vibration, misalignment, and wear of parts. Well designed joints provide the above without being excessively large or heavy or burdening assemblers with bulky tools. Designing plastic parts for mechanical fastening will depend on the particular plastic and the functional requirements of the application.

7.8.1 Mechanical Fasteners

A large variety of mechanical fasteners can be used for joining plastic parts to themselves and to each other. These include machine screws, self-tapping screws, rivets, spring clips, and nuts. In general, when repeated disassembly of the product is anticipated, mechanical fasteners are used.

Machine Screws and Bolts. Parts molded of thermoplastic resin are sometimes assembled with machine screws or bolts, nuts, and washers (Figs. 7.37 and 7.38) especially if it is a very strong plastic. Machine screws are generally used with threaded inserts, nuts, and clips. They rarely are used in pretapped holes.

Molded-in inserts provide very high strength assemblies and relatively low unit cost. However, molded-in inserts could increase cycle time while the inserts are manually placed in the mold. There are four types of postmolded inserts: press-in, expansion, self-tapping, and thread-forming, and inserts that are installed by some method of heating (e.g., ultrasonic).

Particular attention should be paid to the head of the fastener. Conical heads, called flat heads, produce undesirable tensile stresses and should not be used. Bolt

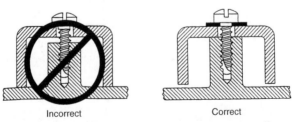

Incorrect   Correct

FIGURE 7.37 Mechanical fastening with self-tapping screws.[11]

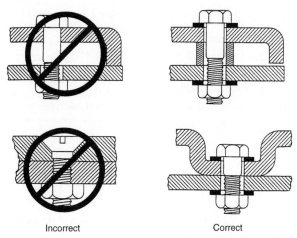

Incorrect   Correct

FIGURE 7.38   Mechanical fastening with bolts, nuts, and washers.[11]

or screw heads with a flat underside, such as pan heads, round heads, and so forth (Fig. 7.39) are preferred because the stress produced is more compressive. Flat washers are also suggested and should be used under both the nut and the fastener head.

Sufficient diametrical clearance for the body of the fastener should always be provided in the parts to be joined. This clearance can nominally be 0.25 mm (0.010 in).

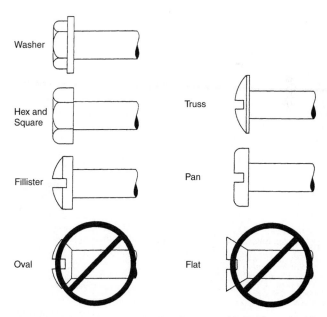

FIGURE 7.39   Common head styles of screws and bolts. Flat underside of head is preferred.[11]

When the application involves infrequent disassembly, molded-in threads can be used. Coarse threads can be molded into most materials. Threads of 32 or finer pitch should be avoided, along with tapered threads, because of excessive stress. If the mating part is metal, overtorquing will result in part failure.

Self-Threading Screws.   Self-threading screws can either be thread cutting or thread forming. To select the correct screw, the designer must know which plastic will be used and its modulus of elasticity. The advantages of using these types of screws are:

- They are generally off-the-shelf items
- Low cost
- High production rates
- Minimum tooling investment

The principal disadvantage of these screws is limited reusability; after repeated disassembly and assembly, these screws will cut or form new threads in the hole, eventually destroying the integrity of the assembly.

Thread-forming screws are used in the softer, more ductile plastics (modulus below 1380 MPa or 200,000 psi). There are a number of fasteners especially designed for use with plastics (Fig. 7.40). Thread-forming screws displace materials. This type of screw induces high stress levels in the part and is not recommended for parts made of some materials.

Thread-cutting screws are used in harder, less ductile plastics. Thread-cutting screws remove material as they are installed, thereby avoiding high stress. However, these screws should not be installed and removed repeatedly.

Blunt-tip fasteners are suitable for most commercial plastics. Harder plastics require a fastener with a cutting tip. Hardest plastics require both a piercing and drilling tip, as in these fasteners.

BLUNT

CUTTING

PIERCING

Twin lead fastener seats in two revolutions.

TWIN LEAD

For rapid installation on lightly loaded joints, some fasteners have a thread configuration that allows the screws to be pushed into place. Typical is this design. Suitable for ductile plastics, this fastener relies on plastics relaxation around the shank to form threads. The thread is helical so that it can be unscrewed, but reuse is limited.

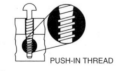
PUSH-IN THREAD

Reverse saw-tooth edges bite into the walls of the plastic.

MILFORD

Triangular configuration is another technique for capturing large amounts of plastic. After insertion, the plastic cold-flows or relaxes back into the area between the lobes. The Trilobe design also creates a vent along the length of the fastener during insertion, eliminating the "ram" effect. In some ductile plastics, pressure builds up in the hole under the fastener as it is inserted, shattering or cracking the material.

TRILOBE

Dual-height thread design boosts holding power by increasing the amount of plastic captured between threads.

HI-LO

Some specials have thread angles smaller than the 60° common on most standard screws. Included angles of 30 or 45° make sharper threads that can be forced into ductile plastics more readily, creating deeper mating threads and reducing stress. With smaller thread angles, boss size can sometimes be reduced.

SHARP
THREAD

Barbs provide holding power.

BARBED

Pushtite fastener is pushed into place and can be screwed out.

PUSHTITE

FIGURE 7.40   Thread-forming fasteners for plastics.[3]

Assembly strengths using thread-forming or self-tapping screws can be increased by reducing hole diameter in the more ductile plastics, increasing screw thread engagement, or going to a larger-diameter screw when space permits. The most common problem encountered with these types of screws is boss-cracking. This can be minimized or eliminated by increasing the size of the boss, increasing the diameter of the hole, decreasing the size of the screw, changing the thread configuration of the screw, or changing the part to a more ductile plastic.

Rivets.   Rivets provide permanent assembly at very low cost. Clamp load is limited to low levels to prevent distortion of the part. To distribute the load, rivets with large heads should be used with washers under the flared end of the rivet. The heads should be three times the shank diameter. Standard rivet heads are shown in Fig. 7.41.

Riveted composite joints should be designed to avoid loading the rivet in tension. Generally, a hole ¹⁄₆₄″ larger than the rivet shank is satisfactory for composite joints. A number of patented rivet designs are commercially available for joining aircraft or aerospace structural composites.

Spring Steel Fasteners.   Push-on spring steel fasteners (Fig. 7.42) can be used for holding light loads. Spring steel fasteners are simply pushed on over a molded stud. The stud should have a minimum 0.38 mm (0.015 in) radius at its base. Too large a radius could create a thick section, resulting in sinks or voids in the plastic molding.

7.8.2   Special Consideration for Composites

The structural efficiency of a composite structure is established, with very few exceptions, by its joints, not by its basic structure. The selection of a joining method for composites is as broad as is possible with metals, namely riveting, bolting, pinning, and bonding. Only classic welding and brazing of metals cannot be applied to composites that utilize thermosetting resin matrices. Thermoplastic and metallic matrix

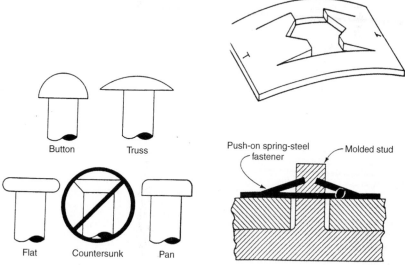

FIGURE 7.41   Standard rivet heads.[11]

FIGURE 7.42   Push-on spring steel fasteners.[11]

composites can be joined by welding or brazing. Composites are mechanically fastened in a manner similar to metals. Adherends are drilled and countersunk and joined with the fastener.

Rivets, pins, two-piece bolts, and blind fasteners made of titanium, stainless steel, and aluminum are all used for composites. Several factors should be considered:

1. Differential expansion of the fastener in the composite
2. The effect of drilling on the structural integrity of the composite as well as delamination caused by the fastener under load
3. Water intrusion between the fastener and composite
4. Electrical continuity of the composite and arching between fasteners
5. Possible galvanic corrosion at the composite joint
6. Weight of the fastening system
7. Environmental resistance of the fastening system

Table 7.31 is a performance comparison for some of the mechanical fasteners most commonly used to join composite materials.

Aluminum and stainless fasteners expand and contract when exposed to temperature extremes, as in aircraft applications. In carbon-fiber composites, contraction and expansion of such fasteners can cause changes in clamping load. Pressure within the joint is often critical.

Drilling and machining can damage composites. Several techniques exist for producing quality holes in composites. Materials reinforced with carbon, aramid, and boron fiber each require different drilling methods and tools. When composites are

TABLE 7.31 Fasteners for Advanced Composites[3]

			Suggested application			
Fastener type	Fastener material	Surface coating	Epoxy/graphite composite	Kevlar	Fiberglass	Honeycomb
Blind rivets[a]	5056 Al	None	NR	E ⎫[h]	E ⎫[h]	⎫[c]
	Monel	None	G ⎫[h]	E	E	
	A-286	Passivated	G ⎭	E	E ⎭	⎭
Blind bolts[b]	A-286	Passivated	E[h]	E ⎫[h]	E	⎫
	Alloy steel	Cadmium	NR	E ⎭	E	⎭
Pull-type lockbolts	Titanium	None	E[d]	E[c]	E[c]	G or NR[f]
Stump-type lockbolts	Titanium	None	E[d]	E[c]	E[c]	G or NR[f]
Asp fasteners	Alloy steel	Cad/Nickel	G[g]	E	E	E
Pull-type lockbolts	7075 Al	Anodize	NR	E	E	NR

NR = Not Recommended. E = Excellent. G = Good.
[a] Blind rivets with controlled shank expansion.
[b] Blind bolts are not shank expanding.
[c] Fasteners can be used with flanged titanium collars or standard aluminum collars.
[d] Use flanged titanium collar.
[e] Performance in honeycomb should be substantiated by installation testing.
[f] Depending on fastener design. Check with manufacturer.
[g] Nickel plated Asp only.
[h] Metallic structure on backside.

cut, fibers are exposed. These fibers can absorb water, which weakens the material. Sealants can be used to prevent moisture absorption in the clearance hole. Sleeved fasteners can provide fits that reduce water absorption as well as provide tightness (Fig. 7.43).

Fastener holes should be straight and round within limits specified. Normal hole tolerance is 0.075 mm (0.003 in). Interference fits may cause delamination of the composite. Holes should be drilled perpendicular to the sheet within one degree. Special sleeved fasteners can limit the changes of damage in the clearance hole and still provide an interference fit. Fasteners can also be bonded in place with adhesives to reduce fretting.

Additionally, galvanic corrosion may occur in carbon fiber composites if aluminum fasteners are used due to the chemical reaction of the aluminum with the carbon. Coating the fastener guards against corrosion but adds cost and time to assembly. Aluminum fasteners are often replaced by more expensive titanium and stainless steel fasteners in carbon fiber composite joints.

When joining composites with mechanical fasteners, special consideration must be given to creep. There are two kinds of creep: creep of the fastener hole and long-term material compression. The greater the material modulus, the lower the creep. There are mechanical ways to reinforce the hole or distribute the load so that the creep problem is minimized. For fasteners that rely on inserts, the ability of the composite under consideration to retain the fastener must be considered.

Like mechanically fastened metal structures, composites exhibit failure modes in tension, shear, and bearing, but, because of the complex failure mechanisms of composites, two further modes are possible, namely cleavage and pull-out. Figure 7.44 shows the location of each of these modes. Environmental degradation of a bolted joint after exposure to a hot, wet environment is most likely to occur in the shear and bearing strength properties. The evidence shows for fiber reinforced epoxies that temperature has a more significant effect than moisture, but in the presence of both at 127°C a strength loss of 40 percent is possible.[41]

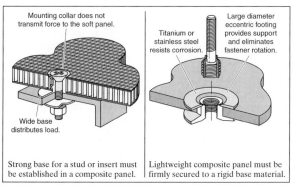

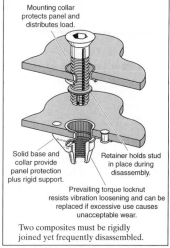

Mounting collar does not transmit force to the soft panel.	Titanium or stainless steel resists corrosion.	Large diameter eccentric footing provides support and eliminates fastener rotation.	Mounting collar protects panel and distributes load.
	Wide base distributes load.		Solid base and collar provide panel protection plus rigid support.	Retainer holds stud in place during disassembly.
			Prevailing torque locknut resists vibration loosening and can be replaced if excessive use causes unacceptable wear.
Strong base for a stud or insert must be established in a composite panel.	Lightweight composite panel must be firmly secured to a rigid base material.	Two composites must be rigidly joined yet frequently disassembled.

FIGURE 7.43   Specialty composite fasteners.[3]

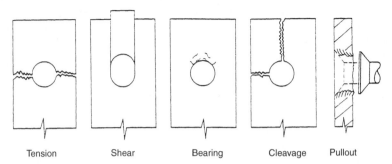

Tension   Shear   Bearing   Cleavage   Pullout

FIGURE 7.44   Modes of failure for mechanical joints in fiberglass reinforced plastics.[41]

Evidence exists to suggest that the failure behavior of thermoplastics is much the same as for thermoset composites and that similar high-joint efficiencies can be obtained with suitable consideration to the joint design, fastener type, and environmental factors.

7.9   RECOMMENDED PROCESSES FOR COMMON POLYMERIC SUBSTRATES

When adhesive bonding, thermal welding, or solvent cementing, special considerations must be taken because of the nature of the substrate and possible interactions with the adhesive or the environment. The following sections identify some of these considerations.

Acetal Homopolymer and Acetal Copolymer.   Epoxies, isocyanate-cured polyester, and cyanoacrylates are used to bond acetal copolymer (Celcon®). Generally, the surface is treated with a sulfuric-chromic acid treatment. Epoxies have shown 150–500 psi shear strength on sanded surfaces and 500–1000 psi on chemically treated surfaces.

Plasma treatment has also shown to be effective on acetal substrates. Thermal welding and solvent cementing are commonly used for bonding this material to itself.

Acetal homopolymer (Delrin®) surfaces should be chemically treated prior to bonding. This is accomplished with a sulfuric-chromic acid treatment followed by a solvent wipe. Epoxies, nitrile, and nitrile-phenolics can be used as adhesives. Acetal homopolymer surfaces are not usually solvent-cemented, but they can be heat-welded.

Acrylonitrile-Butadiene-Styrene (ABS).   The best adhesives for ABS are epoxies, urethanes, thermosetting acrylics, nitrile-phenolics, and cyanoacrylates. These adhesives have shown joint strength greater than that of the ABS substrates being bonded. ABS substrates do not require special surface treatments other than simple cleaning and removal of possible contaminants. ABS can also be bonded to itself and to certain other thermoplastics by either solvent cementing or any of the heat-welding methods.

Cellulosics (Cellulose Acetate, Cellulose Acetate Butyrate, Cellulose Nitrate, Ethyl Cellulose, and the Like).   Adhesives commonly used are epoxies, urethanes, isocyanate-cured polyesters, nitrile-phenolic, and cyanoacrylate. Only cleaning is required prior to applying the adhesive. However, cellulosic plastics may contain plasticizers. The extent of plasticizer migration and the compatibility with the adhesive must be evaluated. Cellulosic materials can also be solvent-cemented.

Fluorocarbons (PTFE, CTFE, FEP, and the Like). The combination of properties that makes the fluorocarbons highly desirable engineering plastics also makes them hard to bond and almost impossible to heat or solvent weld. For adhesive bonding, a sodium-naphthalene etch is necessary as a minimum surface preparation. Commercial sodium-naphthalene etching solutions are available (Table 7.11). The sodium-naphthalene-treated fluorocarbon surface is degraded by UV light and should be protected from direct exposure. Plasma treatment can also be used to increase the wettability of fluorocarbon surfaces. Once treated, conventional epoxy or urethane adhesives can be used.

Polyamide (Nylon). Various commercial adhesives have been used to provide bond strength on the order of 250–1000 psi. Adhesive bonding is usually considered inferior to heat welding or solvent cementing. However, priming of nylon adherends with compositions based on resorcinol formaldehyde, isocyanate modified rubber, and cationoic surfactant have been reported to provide improved joint strength. Elastomeric (nitrile, urethane), hot melt (polyamide, polyester), and reactive (epoxy, urethane, acrylic, and cyanoacrylate) adhesives have been used for bonding nylon.

Polyamide (Nylon®) parts can contain a high percentage of absorbed water. This water can create a weak boundary layer under certain conditions. Generally parts are dried to less than 0.5% water before bonding.

Polycarbonate. Polycarbonate plastics are generally joined by solvent-cementing or thermal welding methods. When adhesives are used, epoxies, urethanes, and cyanoacrylates are chosen. Adhesive bond strengths with polycarbonate are generally 1000–2000 psi. Cyanoacrylates, however, are claimed to provide over 3000 psi when bonding polycarbonate to itself.

Polycarbonates can stress crack in the presence of certain solvents. When cementing polycarbonate parts to metal parts, a room-temperature curing adhesive is suggested to avoid stress in the interface caused by differences in thermal expansion.

Polyethylene, Polypropylene, and Polymethyl Pentene. Epoxy and nitrile-phenolic adhesives have been used to bond these plastics after surface preparation. The surface can be etched with a sodium sulfuric-dichromate acid solution at elevated temperature. Flame treatment and corona discharge have also been used. However, plasma treatment has proven to be the optimum surface process for these materials. Shear strengths in excess of 3000 psi have been reported on polyethylene treated for 10 min in an oxygen plasma and bonded with an epoxy adhesive. Polyolefin materials can also be thermally welded, but they cannot be solvent-cemented.

Polyethylene Terephthalate and Polybutylene Terephthalate. Polyethylene terephthalate (PET) and polybutylene terephthalate (PBT) parts are generally joined by adhesives. Surface treatments recommended specifically for PBT include abrasion and solvent-cleaning with toluene. Gas plasma surface treatments and chemical etch have been used where maximum strength is necessary.

Solvent cleaning of PET surfaces is recommended. The linear film of polyethylene terephthalate (Mylar®) surface can be pretreated by alkaline etching or plasma for maximum adhesion, but often a special treatment is unnecessary.

Commonly used adhesives for both PBT and PET substrates are isocyanate-cured polyesters, epoxies, and urethanes. Polyethylene terephthalate cannot be solvent-cemented or heat-welded.

Polyetherimide (PEI), Polyamide-imide, Polyetheretherketone (PEEK), Polyaryl Sulfone, and Polyethersulfone (PES). These high-temperature thermoplastic materials can be bonded with epoxy or urethane adhesives. No special surface treatment is required other than abrasion and solvent cleaning. Polyetherimide (Ultem®), polyamide-imide (Torlon®), and polyethersulfone can be solvent-cemented, and ultrasonic welding is possible.

Polyimides. Polyimide parts can be bonded with epoxy adhesives. Only abrasion and solvent cleaning are necessary to treat the substrate prior to bonding. The plastic part will usually have higher thermal rating than the adhesive. Thermosetting polyimides cannot be heat-welded or solvent-cemented.

Polymethylmethacrylate (Acrylic). Epoxies, urethanes, cyanoacrylates, and thermosetting acrylics will result in bond strengths greater than the strength of the acrylic part. The surface needs only to be clean of contamination. Molded parts may stress crack when in contact with an adhesive containing solvent or monomer. If this is a problem, an anneal (slightly below the heat distortion temperature) is recommended prior to bonding. Acrylics are also commonly solvent-cemented or heat-welded.

Polyphenylene Sulfide (PPS). Adhesives recommended for polyphenylene sulfide (Ryton®) include epoxies and urethanes. Joint strengths in excess of 1000 psi have been reported for abraded and solvent-cleaned surfaces. Somewhat better adhesion has been reported for machined surfaces over as-molded surfaces. The high heat and chemical resistance of polyphenylene sulfide plastics make them inappropriate for solvent-cementing or heat-welding.

Polystyrene. Polystyrene parts are conventionally solvent-cemented or heat-welded. However, urethanes, epoxies, unsaturated polyesters, and cyanoacrylates will provide good adhesion to abraded and solvent-cleaned surfaces. Hot-melt adhesives are used in the furniture industry. Polystyrene foams will collapse when in contact with certain solvents. For polystyrene foams, a 100 percent solids adhesive or a water-based contact adhesive is recommended.

Polysulfone. Urethane and epoxy adhesives are recommended for bonding polysulfone substrates. No special surface treatment is necessary. Polysulfones can also be easily joined by solvent-cementing or thermal welding methods.

Polyvinyl Chloride (PVC). Rigid polyvinyl chloride can be easily bonded with epoxies, urethanes, cyanoacrylates, and thermosetting acrylics. Flexible polyvinyl chloride parts present a problem because of plasticizer migration over time. Nitrile adhesives are recommended for bonding flexible polyvinyl chloride because of compatibility with the plasticizers used. Adhesives that are found to be compatible with one particular polyvinyl chloride plasticizer may not work with another formulation. Solvent-cementing and thermal welding methods are also commonly used to bond both rigid and flexible polyvinyl chloride parts.

Thermoplastic Polyesters. These materials may be bonded with epoxy, thermosetting acrylic, urethane, and nitrile-phenolic adhesives. Special surface treatment is not

necessary for adequate bonds. However, plasma treatment has been reported to provide enhanced adhesion. Solvent-cementing and certain thermal welding methods
can also be used with thermoplastic polyester.

Thermosetting Plastics (Epoxies; Diallyl Phthalate; Polyesters; Melamine, Phenol,
and Urea Formaldehyde; Polyurethanes; and the Like). Most thermosetting plastics are not particularly difficult to bond. Abrasion and solvent-cleaning are generally recommended as the surface treatment. Epoxies, thermosetting acrylics, and
urethanes are the best adhesives for the purpose. Thermosetting plastic substrates
cannot be solvent-cemented or heat-welded.

7.10 EFFECT OF ENVIRONMENT

For an adhesive bond to be useful, it not only must withstand the mechanical forces
acting on it, but it must also resist the service environment. Adhesive strength is
influenced by many common environments, including temperature, moisture, chemical fluids, and outdoor weathering. Table 7.32 summarizes the relative resistance of
various adhesive types to common environments.

7.10.1 High Temperature

All polymeric materials are degraded to some extent by exposure to elevated temperatures. Not only are physical properties lowered at high temperatures, but they
also degrade due to thermal aging. Newly developed polymeric adhesives have been
found to withstand 500 to 600°F continuously. To use these materials, the designer
must pay a premium in adhesive cost and also be capable of providing long, high-
temperature cures. High-temperature adhesives are primarily reserved for metal substrates or composite substrates that are made with a high-temperature resin matrix.

For an adhesive to withstand elevated-temperature exposure, it must have a high
melting or softening point and resistance to oxidation. Materials with a low melting
point, such as many of the thermoplastic adhesives, may prove excellent adhesives at
room temperature. However, once the service temperature approaches the glass-
transition temperature of these adhesives, plastic flow results in deformation of the
bond and degradation in cohesive strength. Thermosetting materials, exhibiting no
melting point, consist of highly cross-linked networks of macromolecules. Many of
these materials are suitable for high-temperature applications. When considering
thermoset adhesives, the critical factor is the rate of strength reduction due to thermal oxidation and pyrolysis.

Thermal oxidation initiates progressive chain scission of molecules resulting in
losses of weight, strength, elongation, and toughness within the adhesive. Figure 7.45
illustrates the effect of oxidation by comparing adhesive joints aged in both high-
temperature air and inert-gas environments. The rate of strength degradation in air
depends on the temperature, the adhesive, the rate of airflow, and even the type of
adherend. Certain metal-adhesive interfaces are capable of accelerating the rate of
oxidation. For example, many structural adhesives exhibit better thermal stability
when bonded to aluminum than when bonded to stainless steel or titanium (Fig. 7.45).

High-temperature adhesives are usually characterized by a rigid polymeric structure, high softening temperature, and stable chemical groups. The same factors also
make these adhesives very difficult to process. Only epoxy-phenolic-, polyimide-,

TABLE 7.32 Relative Resistance of Synthetic Adhesives to Common Service Environments[42]

Adhesive type	Shear	Peel	Heat	Cold	Water	Hot water	Acid	Alkali	Oil, grease	Fuels	Alcohols	Ketones	Esters	Aromatics	Chlorinated solvents
Thermosetting Adhesives															
1. Cyanoacrylate	2	6	5		6	6	6	6	3	3	5	5	5	4	4
2. Polyester + isocyanate	2	2	3	2	1	3	3	2	2	2	3	2	2	6	2
3. Polyester + monomer	2	6	5	3	3	3	3	6	2	2	2	6	6	6	6
4. Urea formaldehyde	2	6	3	3	2	6	2	2	2	2	2	2	2	2	2
5. Melamine formaldehyde	2	6	3	2	2	5	2	2	2	2	2	2	2	2	2
6. Urea–melamine formaldehyde	2	6	2	2	2	2	1	1	2	2	2	2	2	2	2
7. Resorcinol formaldehyde	2	6	2	2	2	2	2	2	2	2	2	2	2	2	2
8. Phenol-resorcinol formaldehyde	2	5	2	5	3	3	2	2	2	3	1	2	2	1	2
9. Epoxy (+ polyamine)	2	5	1	4	2	6	3	6	2	2	1	6	6	2	
10. Epoxy (+ polyanhydride)	2	2	6	2	1	4	2	2	2	2	1	2	2	1	2
11. Epoxy (+ polyamide)	2	4	1	1	2	4	2	2	2	2	2	2	2	2	2
12. Polyimide	2	4	1	1	1	3	2	2	2	2	2	2	2	2	2
13. Polybenzimidazole	2	4	5	3	1	3	2	2	3	2	2	2	2	2	2
14. Acrylic	2	6	1	3	1	4	2	6	2	2	2	2	2	2	4
15. Acrylate acid diester	2	5	3	3	4	4	6	6	3	3	5	5	5	4	4
Thermoplastic Adhesives															
16. Cellulose acetate	2	6	2	3	1	6	1	2		2	4	6	6	6	6
17. Cellulose acetate butyrate	2	3	3	3	2		3	2			6	6	6	6	6
18. Cellulose nitrate	2	6	6	3	3	3	3	6	2	2	6	6	6	6	6
19. Polyvinyl acetate	2	6	3		2	3	3	3	2	2	2	6	6	6	6
20. Vinyl vinylidene	2	3	3	2	6		3		2	2	2	2	2		
21. Polyvinyl acetal	2	6	5		2	6	6	3	2	2	3	3	6	3	2
22. Polyvinyl alcohol	2	2	3		6	6	5	5	2	1	3	1	1	1	1
23. Polyamide	2	3	5	3	5	3	6	2	2	2	6	2	2	2	6
24. Acrylic	2	2	4	3	3	4			3	1	6	4	4	6	4
25. Phenoxy		3	4		3		3	2	3	2	5			6	
Elastomer Adhesives															
26. Natural rubber	2	3	3		3		3	3	6	6	2	4	4	6	6
27. Reclaimed rubber	2	3	3	3	2	6	3	3	6	6	2	4	4	6	6
28. Butyl	2	6	6	3	2	6	1	2	6	6	2	2	2	6	6
29. Polyisobutylene	6	6	3	3	2	6	1	2	6	2	2	2	2	6	6
30. Nitrile	2	3	3	2	1	5	2	6	2	1	2	6	6	3	5
31. Styrene butadiene	3	6	3		2	3	5	2	6	5	2	5	5	6	5
32. Polyurethane	3	3	6	3	1	6	3	3	6	2	3	6	6	6	3
33. Polysulfide	3	2	1	2	2	3	2	2	6	3	3	3	3	3	3
34. Silicone (RTV)	2	5	1		2	2	3	2	2	2	3	4	4	3	3
35. Silicone resin	2	2	1	2	2				2	2	3	6	6	3	3
36. Neoprene	2	3	3	3	2	4	3	6	3	2	3	6	6	6	6
Alloy Adhesives															
37. Epoxy-phenolic	1	6	1	3	2	2	2	2	3	3	2	6	6	2	6
38. Epoxy-polysulfide	2	2	6	2	1	6	2	2	2	2	2	6	6	6	6
39. Epoxy-nylon	1	1	6	2	2	2				2	3	6	6	6	6
40. Phenolic-nitrile	2	2	2	2	2		2	2		2	2	6	6	6	6
41. Phenolic-neoprene	3	3	3	2	1	6	3	2	3	2	3	6	6	6	5
42. Phenolic-polyvinyl butyral	2	3	3	2	2	3	4	2	2	2	4	6	6	6	6
43. Phenolic-polyvinyl formal	2	3	6	6	2	4	4	2	2	2	4	6	6	6	6

Key: 1. Excellent   2. Good   3. Fair   4. Poor   5. Very poor   6. Extremely poor

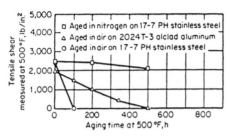

FIGURE 7.45   Effect of 500°F aging in air and nitrogen on epoxy-phenolic adhesive (HT-424).[43]

and polybenzimidazole-based adhesives can withstand long-term service temperatures greater than 350°F.

Epoxy.   Epoxy adhesives are generally limited to continuous applications below 300°F. Figure 7.46 illustrates the aging characteristics of a typical epoxy adhesive at elevated temperatures. Certain epoxy adhesives are able to withstand short terms at 500°F and long-term service at 300 to 350°F. These systems were formulated especially for thermal environments by incorporation of stable epoxy coreactants, high-temperature curing agents, and antioxidants into the adhesive.

One successful epoxy coreactant system is an epoxy-phenolic alloy. The excellent thermal stability of the phenolic resins is coupled with the adhesion properties of epoxies to provide an adhesive capable of 700°F short-term operation and continuous use at 350°F. The heat-resistance and thermal-aging properties of an epoxy-phenolic adhesive are compared with those of other high-temperature adhesives in Fig. 7.47.

Anhydride curing agents give unmodified epoxy adhesives greater thermal stability than most other epoxy-curing agents. Phthalic anhydride, pyromellitic dianhydride, and chlorendic anhydride allow greater cross-linking and result in short-term heat resistance to 450°F. Long-term thermal endurance, however, is limited to 300°F. Typical epoxy formulations cured with pyromellitic dianhydride offer 1,200–2,600 lb/in² shear strength at 300°F and 1,000 lb/in² at 450°F.

Modified Phenolics.   Of the common modified phenolic adhesives, the nitrile-phenolic blend has the best resistance to shear at elevated temperatures. Nitrile phenolic adhesives have high shear strength up to 250 to 350°F, and the strength retention on aging at these temperatures is very good. The nitrile phenolic adhesives are also extremely tough and provide high peel strength.

Silicone.   Silicone adhesives have very good thermal stability but low strength. Their chief application is in nonstructural applications such as high-temperature pressure-sensitive tape.

Attempts have been made to incorporate silicones with other resins such as epoxies and phenolics, but long cure times and low strength have limited their use.

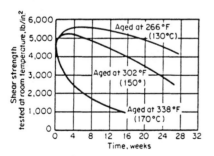

FIGURE 7.46   Effect of temperature aging on typical epoxy adhesive in air. Strength measured at room temperature.[44]

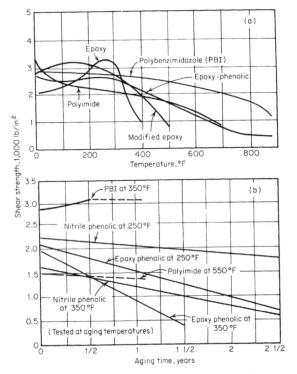

FIGURE 7.47 Comparison of (*a*) heat resistance and (*b*) thermal aging of high-temperature structural adhesives.[45]

Polyaromatics. The most common polyaromatic resins, polyimide and polybenzimidazole, offer greater thermal resistance than any other commercially available adhesive. The rigidity of their molecular chains decreases the possibility of chain scission caused by high temperatures. The aromaticity of these structures provides high bond-dissociation energy and acts as an "energy sink" to the thermal environment.

The strength retention of polyimide adhesives for short exposures to 1000°F is slightly better than that of an epoxy-phenolic alloy. However, the thermal endurance of polyimides at temperatures greater than 500°F is unmatched by other commercially available adhesives.

Polyimide adhesives are usually supplied as a glass-fabric-reinforced film having a limited shelf life. A cure of 90 min at 500 to 600°F and 15 to 200 lb/in² pressure is usually necessary for optimum properties. High-boiling volatiles can be released during cure, which causes a somewhat porous adhesive layer. Because of the inherent rigidity of this material, peel strength is low.

7.10.2 Low Temperature

The factors that determine the strength of an adhesive at very low temperatures are (1) the difference in coefficient of thermal expansion between adhesive and adherend, (2) the elastic modulus, and (3) the thermal conductivity of the adhesive. The difference in thermal expansion is very important, especially since the elastic modulus of the adhesive generally decreases with falling temperature. It is necessary that the adhesive retain some resiliency if the thermal-expansion coefficients of adhesive and adherend cannot be closely matched. The adhesive's coefficient of thermal conductivity is important in minimizing transient stresses during cooling. This is why thinner bonds have better cryogenic properties than thicker ones.

Low-temperature properties of common structural adhesives used for cryogenic applications are illustrated in Fig. 7.48. Epoxy-polyamide adhesives can be made serviceable at very low temperatures by the addition of appropriate fillers to control thermal expansion. But the epoxy-based systems are not as attractive as some others because of brittleness and corresponding low peel and impact strength at cryogenic temperatures.

Epoxy-phenolic adhesives are exceptional in that they have good adhesive properties at both elevated and low temperatures. Vinyl-phenolic adhesives maintain fair shear and peel strength at −423°F, but strength decreases with decreasing temperature. Nitrile-phenolic adhesives do not have high strength at low service temperatures because of rigidity.

Polyurethane and epoxy-nylon systems offer outstanding cryogenic properties. Polyurethane adhesives are easily processible and bond well to many substrates. Peel strength ranges from 22 lb/in at 75°F to 26 lb/in at −423°F, and the increase in shear strength at −423°F is even more dramatic. Epoxy-nylon adhesives also retain flexibility and yield 5,000 lb/in² shear strength in the cryogenic temperature range.

Heat-resistant polyaromatic adhesives also have shown promising low-temperature properties. The shear strength of a polybenzimidazole adhesive on stainless steel substrates is 5,690 lb/in² at a test temperature of −423°F, and polyimide adhesives have exhibited shear strength of 4,100 lb/in² at −320°F. These unique properties show the applicability of polyaromatic adhesives on structures seeing both very high and low temperatures.

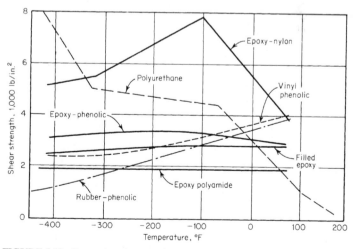

FIGURE 7.48   Properties of cryogenic structural adhesive systems.[45]

7.10.3 Humidity and Water Immersion

Moisture can affect adhesive strength in two significant ways. Some polymeric materials, notably ester-based polyurethanes, will revert, i.e., lose hardness, strength, and in the worst cases turn fluid during exposure to warm, humid air. Water can also permeate the adhesive and preferentially displace the adhesive at the bond interface. The latter mechanism is the most common cause of adhesive-strength reduction in moist environments.

The rate of reversion or hydrolytic instability depends on the chemical structure of the base adhesive, the type and amount of catalyst used, and the flexibility of the adhesive. Certain chemical linkages such as ester, urethane, amide, and urea can be hydrolyzed. The rate of attack is fastest for ester-based linkages. Ester linkages are present in certain types of polyurethanes and anhydride-cured epoxies. Generally amine-cured epoxies offer better hydrolytic stability than anhydride-cured types. Figure 7.49 illustrates the hydrolytic stability of various polymeric materials determined by a hardness measurement. Reversion is usually much faster in flexible materials because water permeates more easily.

Structural adhesives not susceptible to the reversion phenomenon are also likely to lose adhesive strength when exposed to moisture. The degradation curves shown in Fig. 7.50 are typical for an adhesive exposed to moist, high-temperature environments. The mode of failure in the initial stages of aging is usually truly cohesive. After aging, the failure becomes one of adhesion. It is expected that water vapor permeates the adhesive through its exposed edges and concentrates in weak boundary layers at the interface. This effect is greatly dependent on the type of adhesive and substrate.

Stress accelerates the effect of environments on the adhesive joint. Little data are available on this phenomenon because of the time and expense associated with stress-aging tests. However, it is known that moisture, as an environmental burden, markedly decreases the ability of an adhesive to bear prolonged stress. Figure 7.51 illustrates the effect of stress aging on specimens exposed to relative humidity cycling

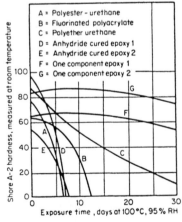

FIGURE 7.49 Hydrolytic stability of potting compounds. Materials showing rapid hardness loss will soften similarly after 2 to 4 years at ambient temperatures in high-humidity tropical climate.[46]

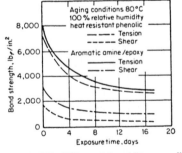

FIGURE 7.50 Effect of humidity on adhesion of two structural adhesives to stainless steel.[47]

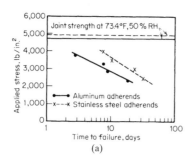

(a)

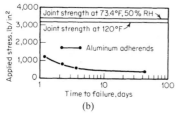

(b)

FIGURE 7.51   Time to failure versus stress for two adhesives in warm high-humidity environment. (*a*) One-part heat-curing modified epoxy. (*b*) Flexibilized amine-cured epoxy.

from 90 to 100 percent and simultaneous temperature cycling from 80 to 120°F. The loss of load-bearing ability of a certain flexibilized epoxy adhesive (Fig. 7.51) is exceptional. The stress on this particular adhesive had to be reduced to 13 percent of its original strength in order for the joint to last a little more than 44 days in the test environment.

7.10.4   Outdoor Weathering

The most detrimental factors influencing adhesives aged outdoors are heat and humidity. Thermal cycling, ultraviolet radiation, and cold are relatively minor factors. The reasons why warm, moist climates degrade adhesive joints were presented in the last section.

When exposed to weather, structural adhesives rapidly lose strength during the first six months to 1 year. After 2 to 3 years, the rate of decline usually levels off, depending on the climate zone, adherend, adhesive, and stress level. The following generalizations are of importance in designing an adhesive joint for outdoor service:

1. The most severe locations are those with high humidity and warm temperatures.
2. Stressed panels deteriorate more rapidly than unstressed panels.
3. Stainless steel panels are more resistant than aluminum panels because of corrosion.
4. Heat-cured adhesive systems are generally more resistant than room-temperature-cured systems.
5. With the better adhesives, unstressed bonds are relatively resistant to severe outdoor weathering, although all joints will eventually exhibit some strength loss.

7.10.5   Chemicals and Solvents

There is no best adhesive for universal chemical environments. As an example, maximum resistance to bases almost axiomatically means poor resistance to acids. It is relatively easy to find an adhesive that is resistant to one particular chemical environment. It becomes more difficult to find an adhesive that will not degrade in two widely differing chemical environments. Generally, adhesives that are most resistant to high temperatures have the best resistance to chemicals and solvents.

The temperature of the immersion medium is a significant factor in the aging properties of the adhesive. As the temperature increases, more fluid is generally absorbed by the adhesive, and the degradation rate increases.

From the rather limited information reported in the literature, it may be summarized that:

1. Chemical-resistance tests are not uniform in concentrations, temperature, time, and properties measured.
2. Generally, chlorinated solvents and ketones are severe environments.
3. High-boiling solvents, such as dimethylformamide and dimethyl sulfoxide, are severe environments.
4. Acetic acid is a severe environment.
5. Amine curing agents for epoxies are poor in oxidizing acids.
6. Anhydride curing agents are poor in caustics.

7.11 PROCESSING AND QUALITY CONTROL

Processing and quality control are usually the final considerations in the design of an adhesive-bonding system. These decisions are very important, however, because they alone may (1) restrict the degrees of freedom in designing the end product, (2) determine the types and number of adhesives that can be considered, (3) affect the quality and reproducibility of the joint, and (4) affect the total assembly cost.

7.11.1 Measuring and Mixing

When a multiple-part adhesive is used, the concentration ratios have a significant effect on the quality of the joint. Strength differences caused by varying curing-agent concentration are most noticeable when the joints are tested at elevated temperatures or after exposure to water or solvents. Exact proportions of resin and hardener must be weighed out on an accurate balance or in a measuring container for best adhesive quality and reproducibility.

The weighed-out components must be mixed thoroughly. Mixing should be continued until no color streaks or density stratifications are noticeable. Caution should be taken to prevent air from being mixed into the adhesive through overagitation. This can cause foaming of the adhesive during heat cure, resulting in porous bonds. If air does become mixed into the adhesive, vacuum degassing may be necessary before application.

Only as much adhesive as the job requires should be mixed before the adhesive begins to cure. The working life of an adhesive is defined as the period of time during which an adhesive remains suitable for use after mixing with a catalyst. Working life is decreased as the ambient temperature increases and as the batch size becomes larger. One-part and some heat-curing, two-part adhesives have very long working lives at room temperature, and application and assembly speed or batch size are not critical.

For a large-scale bonding operation, hand mixing is costly, messy, and slow, and repeatability is entirely dependent on the operator. Equipment is available that can meter, mix, and dispense multicomponent adhesives on a continuous or shot basis.

7.11.2 Application of Adhesives

The selection of an application method depends primarily on the form of the adhesive: liquid, paste, powder, or film. Table 7.33 describes the advantages and limitations realized in using each of the four basic forms. Other factors influencing the application method are the size and shape of parts to be bonded, the total area where the adhesive is to be applied, and production volume and rate.

TABLE 7.33 Characteristics of Various Adhesive-Application Methods[49]

Application method	Viscosity	Operator skill	Production rate	Equipment cost	Coating uniformity	Material loss
Liquid						
Manual, brush or roller........	Low to medium	Little	Low	Low	Poor	Low
Roll coating, reverse, gravure........	Low	Moderate	High	High	Good	Low
Spray, manual, automatic, airless, or external mix......	Low to high	Moderate to high	Moderate to high	Moderate to high	Good	Low to high
Curtain coating........	Low	Moderate	High	High	Good to excellent	Low
Bulk						
Paste and mastic......	High	Little	Low to moderate	Low	Fair	Low to high
Powder						
Dry or liquid primed......	......	Moderate	Low	High	Poor to fair	Low
Dry Film........	......	Moderate to high	Low to high	Low to high	Excellent	Lowest

7.11.3 Bonding Equipment

After the adhesive is applied, the assembly must be mated as quickly as possible to prevent contamination of the adhesive surface. The substrates are held together under pressure and heated if necessary until cure is achieved. The equipment required to perform these functions must provide adequate heat and pressure, maintain the prescribed pressure during the entire cure cycle, and distribute pressure uniformly over the bond area. Of course, many adhesives require only simple contact pressure at room temperature, and extensive bonding equipment is not necessary.

Pressure Equipment.   Pressure devices should be designed to maintain constant pressure on the bond during the entire cure cycle. They must compensate for thickness reduction from adhesive flow or thermal expansion of assembly parts. Thus, screw-actuated devices like C-clamps and bolted fixtures are not acceptable when a constant pressure is important. Spring pressure can often be used to supplement clamps and compensate for thickness variations. Dead-weight loading may be applied in many instances; however, this method is sometimes impractical, especially when heat cure is necessary.

Pneumatic and hydraulic presses are excellent tools for applying constant pressure. Steam or electrically heated platen presses with hydraulic rams are often used for adhesive bonding. Some units have multiple platens, thereby permitting the bonding of several assemblies at one time.

Large bonded areas such as on aircraft parts are usually cured in an autoclave. The parts are mated first and covered with a rubber blanket to provide uniform pressure distribution. The assembly is then placed in an autoclave, which can be pressurized and heated. This method requires heavy capital-equipment investment.

Vacuum-bagging techniques can be an inexpensive method of applying pressure to large parts. A film or plastic bag is used to enclose the assembly, and the edges of the film are sealed airtight. A vacuum is drawn on the bag, enabling atmospheric pressure to force the adherends together. Vacuum bags are especially effective on large areas because size is not limited by equipment.

Heating Equipment.   Many structural adhesives require heat as well as pressure. Most often the strongest bonds are achieved by an elevated-temperature cure. With many adhesives, trade-offs between cure times and temperature are permissible. But generally, the manufacturer will recommend a certain curing schedule for optimum properties.

If, for example, a cure of 60 min at 300°F is recommended, this does not mean that the assembly should be placed in a 300°F oven for 60 min. It is the bond line that should be at 300°F for 60 min. Total oven time would be 60 min plus whatever time is required to bring the assembly up to 300°F. Large parts act as a heat sink and may require substantial time for an adhesive in the bond line to reach the necessary temperature. Bond-line temperatures are best measured by thermocouples placed very close to the adhesive. In some cases, it may be desirable to place the thermocouple in the adhesive joint for the first few assemblies being cured.

Oven-heating is the most common heating technique for bonded parts, even though it involves long curing cycles because of the heat-sink action of large assemblies. Ovens may be heated with gas, oil, electricity, or infrared units. Good air circulation within the oven is mandatory to prevent nonuniform heating.

Heated-platen presses are good for bonding flat or moderately contoured panels when faster cure cycles are desired. Platens are heated with steam, hot oil, or electricity and are easily adapted with cooling-water connections to further speed the bonding cycle.

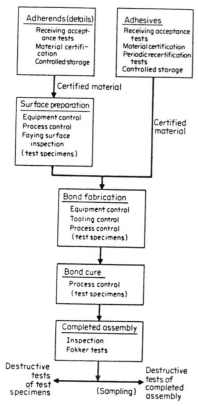

Adhesive-Thickness Control. It is highly desirable to have a uniformly thin (2- to 10-mil) adhesive bond line. Starved adhesive joints, however, will yield exceptionally poor properties. Three basic methods are used to control adhesive thickness. The first method is to use mechanical shims or stops that can be removed after the curing operation. Sometimes it is possible to design stops into the joint.

The second method is to employ a film adhesive that becomes highly viscous during the cure cycle, preventing excessive adhesive flow-out. With supported films, the adhesive carrier itself can act as the shims. Generally, the cured bond-line thickness will be determined by the original thickness of the adhesive film.

The third method of controlling adhesive thickness is to use trial and error to determine the correct pressure-adhesive viscosity factors that will yield the desired bond thickness.

7.11.4 Quality Control

A flow chart of a quality-control system for a major aircraft company is illustrated in Fig. 7.52. This system is designed to ensure reproducible bonds and, if a substandard bond is detected, to

FIGURE 7.52 Flowchart of quality control system for adhesive bonding.[50]

make suitable corrections. Quality control should cover all phases of the bonding cycle, from inspection of incoming material to the inspection of the completed assembly. In fact, good quality control will start even before receipt of materials.

Prebonding Conditions. The human element enters the adhesive-bonding process more than in other fabrication techniques. An extremely high percentage of defects can be traced to poor workmanship. This generally prevails in the surface-preparation steps but may also arise in any of the other bonding steps. This problem can be largely overcome by proper motivation and education. All employees from design engineer to laborer to quality-control inspector should be somewhat familiar with adhesive-bonding technology and aware of the circumstances that can lead to poor joints.

The plant's bonding area should be as clean as possible prior to receipt of materials. The basic approach to keeping the assembly area clean is to segregate it from the other manufacturing operations, either in a corner of the plant or in isolated rooms. The air should be dry and filtered to prevent moisture or other contaminants from gathering at a possible interface. The cleaning and bonding operations should be separated from each other. If mold release is used to prevent adhesive flash from sticking to bonding equipment, it is advisable that great care be taken to assure that the release does not contaminate the adherends. Spray mold releases, especially silicone release agents, have a tendency to migrate to undesirable areas.

Quality Control of Adhesive and Surface Treatment.   Acceptance tests on adhesives should be directed toward assurance that incoming materials are identical from lot to lot. The tests should be those that can quickly and accurately detect deficiencies in the adhesive's physical or chemical properties. ASTM lists various test methods that are commonly used for adhesive acceptance. Actual test specimens should also be made to verify strength of the adhesive. These specimens should be stressed in directions that are representative of the forces that the bond will see in service, i.e., shear, peel, tension, or cleavage. If possible, the specimens should be prepared and cured in the same manner as actual production assemblies. If time permits, specimens should also be tested in simulated service environments like high temperature and humidity.

Surface preparations must be carefully controlled for reliable production of adhesive-bonded parts. If a chemical surface treatment is required, the process must be monitored for proper sequence, bath temperature, solution concentration, and contaminants. If sand- or grit-blasting is employed, the abrasive must be changed regularly. An adequate supply of clean wiping cloths for solvent cleaning is also mandatory. Checks should be made to determine if cloths or solvent containers have become contaminated.

The specific surface preparation can be checked for effectiveness by the water-break free test. After the final treating step, the substrate surface is checked for a continuous film of water that should form when deionized water drops are placed on the surface.

After the adequacy of the surface treatment has been determined, precautions must be taken to ensure that the substrates are kept clean and dry until bonding. The adhesive or primer should be applied to the treated surface as quickly as possible.

Quality Control of the Bonding Process.   The adhesive metering and mixing operation should be monitored by periodically sampling the mixed adhesive and testing it for adhesive properties. A visual inspection can also be made for air entrapment and degree of mixing. The quality-control engineer should be sure that the oldest adhesive is used first and that the specified shelf life has not been exceeded.

During the actual assembly operation, the cleanliness of the shop and tools should be verified. The shop atmosphere should be controlled as closely as possible. Temperature in the range of 65 to 90°F and relative humidity from 20 to 65 percent are best for almost all bonding operations.

The amount of the applied adhesive and the final bond-line thickness must also be monitored because they can have a significant effect on joint strength. Curing conditions should be monitored for heat-up rate, maximum and minimum temperature during cure, time at the required temperature, and cool-down rate.

Bond Inspection.   After the adhesive is cured, the joint can be inspected to detect gross flaws or defects. This inspection procedure can be either destructive or nondestructive in nature. Destructive testing generally involves placing samples in simulated or accelerated service and determining if they have similar properties to a specimen that is known to have a good bond and adequate service performance. The causes and remedies for faults revealed by such mechanical tests are described in Table 7.34. Nondestructive testing (NDT) is far more economical, and every assembly can be tested if desired. Great amounts of energy are now being devoted to improve NDT techniques.

TABLE 7.34   Faults Revealed by Mechanical Tests

Fault	Cause	Remedy
Thick, uneven glue line	Clamping pressure too low	Increase pressure. Check that clamps are seating properly
	No follow-up pressure	Modify clamps or check for freedom of moving parts
	Curing temperature too low	Use higher curing temperature. Check that temperature is above the minimum specified throughout the curing cycle
	Adhesive exceeded its shelf life, resulting in increased viscosity	Use fresh adhesive
Adhesive residue has spongy appearance or contains bubbles	Excess air stirred into adhesive	Vacuum-degas adhesive before application
	Solvents not completely dried out before bonding	Increase drying time or temperature. Make sure drying area is properly ventilated
	Adhesive material contains volatile constituent	Seek advice from manufacturers
	A low-boiling constituent boiled away	Curing temperature is too high
Voids in bond (i.e., areas that are not bonded), clean bare metal exposed, adhesive failure at interface	Joint surfaces not properly treated	Check treating procedure; use clean solvents and wiping rags. Wiping rags must not be made from synthetic fiber. Make sure cleaned parts are not touched before bonding. Cover stored parts to prevent dust from settling on them
	Resin may be contaminated	Replace resin. Check solids content. Clean resin tank
	Uneven clamping pressure	Check clamps for distortion
	Substrates distorted	Check for distortion; correct or discard distorted components. If distorted components must be used, try adhesive with better gap-filling ability
Adhesive can be softened by heating or wiping with solvent	Adhesive not properly cured	Use higher curing temperature or extend curing time. Temperature and time must be above the minimum specified throughout the curing cycle. Check mixing ratios and thoroughness of mixing. Large parts act as a heat sink, necessitating larger cure times

Nondestructive Testing Procedures

Visual Inspection.   A trained eye can detect a surprising number of faulty joints by close inspection of the adhesive around the bonded area. Table 7.35 lists the characteristics of faulty joints that can be detected visually. The most difficult defect to be found by any way are those related to improper curing and surface treatment. Therefore, great care and control must be given to surface-preparation procedures and shop cleanliness.

Sonic Inspection.   Sonic and ultrasonic methods are at present the most popular NDT technique for use on adhesive joints. Simple tapping of a bonded joint with a

TABLE 7.35 Visual Inspection for Faulty Bonds

Fault	Cause	Remedy
No appearance of adhesive around edges of joint or adhesive bond line too thick	Clamping pressure roo low Starved joint Curing temperature too low	Increase pressure. Check that clamps are seating properly Apply more adhesive Use higher curing temperature. Check that temperature is above the minimum specified
Adhesive bond line too thin	Clamping pressure too high Curing temperature too high Starved joint	Lessen pressure Use lower curing temperature Apply more adhesive
Adhesive flash breaks easily away from substrate	Improper surface treatment	Check treating procedure; use clean solvents and wiping rags. Make sure cleaned parts are not touched before bonding
Adhesive flash is excessively porous	Excess air stirred into adhesive Solvent not completely dried out before bonding Adhesive material contains volatile constituent	Vacuum-degas adhesive before application Increase drying time or temperature Seek advice from manufacturers
Adhesive flash can be softened by heating or wiping with solvent	Adhesive not properly cured	Use higher curing temperature or extend curing time. Temperature and time must be above minimum specified. Check mixing

coin or light hammer can indicate an unbonded area. Sharp, clear tones indicate that adhesive is present and adhering to the substrate in some degree; dull, hollow tones indicate a void or unattached area. Ultrasonic testing basically measures the response of the bonded joint to loading by low-power ultrasonic energy.

Other NDT Methods. Radiography (X-ray) inspection can be used to detect voids or discontinuities in the adhesive bond. This method is more expensive and requires more skilled experience than ultrasonic methods.

Thermal-transmission methods are relatively new techniques for adhesive inspection. Liquid crystals applied to the joint can make voids visible if the substrate is heated. This test is simple and inexpensive, although materials with poor heat-transfer properties are difficult to test, and the joint must be accessible from both sides. An infrared inspection technique has also been developed for detection of internal voids and nonbonds. This technique is somewhat expensive, but it can accurately determine the size and depth of the flaw.

The science of holography has also been used for NDT of adhesive bonds. Holography is a method of producing photographic images of flaws and voids using coherent light such as that produced by a laser. The major advantage of holography is that it photographs successive "slices" through the scene volume. A true three-dimensional image of a defect or void can then be reconstructed.

7.11.5 Environmental and Safety Concerns

Four primary safety factors must be considered in all adhesive bonding operations: toxicity, flammability, hazardous incompatibility, and equipment.

All adhesives, solvents, chemical treatments, and the like must be handled in a manner preventing toxic exposure to the workforce. Methods and facilities must be provided to assure that the maximum acceptable concentrations of hazardous materials are never exceeded. These values are prominently displayed on the material's Material Safety Data Sheet (MSDA), which must be maintained and available for the workforce.

Where flammable solvents and adhesives are used, they must be stored, handled, and used in a manner preventing any possibility of ignition. Proper safety containers, storage areas, and well-ventilated workplaces are required.

Certain adhesive materials are hazardous when mixed together. Epoxy and polyester catalysts, especially, must be well understood prior to departing from the manufacturers' recommended procedure for mixing. Certain unstabilized solvents, such as trichloroethylene and perchloroethylene, are subject to chemical reaction on contact with oxygen or moisture. Only stabilized grades of solvents should be used.

Certain adhesive systems, such as heat-curing epoxy and room temperature curing polyesters, can develop very large exothermic reactions on mixing. The temperature generated during this exotherm is dependent on the mass of the material being mixed. Exotherm temperatures can get so high that the adhesive will catch fire and burn. Adhesive products should always be applied in thin bond lines to minimize the exotherm until the chemistry of the product is well understood.

Safe equipment and proper operation is, of course, crucial to a workplace. Sufficient training and safety precautions must be installed in the factory before the bonding process is established.

REFERENCES

1. Powis, C. N., "Some Applications of Structural Adhesives," in D. J. Alner (ed.), *Aspects of Adhesion*, vol. 4, University of London Press, London, 1968.

2. Bikerman, J. J., "Causes of Poor Adhesion," *Ind. Eng. Chem.*, Sept. 1967.

3. *Mach. Des.*, Nov. 17, 1988.

4. Reinhart, F. W., "Survey of Adhesion and Types of Bonds Involved," in J. E. Rutzler and R. L. Savage (eds.), *Adhesion and Adhesives Fundamentals and Practices*, Soc. of Chemical Industry, London, 1954.

5. Merriam, J. C., "Adhesive Bonding," *Mater. Des. Eng.*, Sept. 1959.

6. Rider, D. K., "Which Adhesives for Bonded Metal Assembly," *Prod. Eng.*, May 25, 1964.

7. Perry, H. A., "Room Temperature Setting Adhesives for Metals and Plastics," in J. E. Rutzler and R. L. Savage (eds.), *Adhesion and Adhesives Fundamentals and Practices*, Soc. of Chemical Industry, London, 1954.

8. Koehn, G. W., "Design Manual on Adhesives," *Mach. Des.*, Apr. 1954.

9. "Adhesive Bonding Alcoa Aluminum," Aluminum Co. of America, 1967.

10. "General Design Principles," *Design Handbook for Dupont Engineering Polymers*, Dupont Polymers, Wilmington, DE.

11. "Engineering Plastics," *Engineered Materials Handbook*, vol. 2, ASM International, Metals Park, OH, 1988.

12. DeLollis, N. J., *Adhesives for Metals Theory and Technology*, Industrial Press, New York, 1970.

13. "Composites," *Engineered Materials Handbook*, vol. 1, ASM International, Metals Park, OH, 1987.

14. Chessin, N. and V. Curran, "Preparation of Aluminum Surface for Bonding," in M. J. Bodnar (ed.), *Structural Adhesive Bonding,* Interscience, New York, 1966.

15. Muchnick, S. N., "Adhesive Bonding of Metals," *Mech. Eng.,* Jan. 1956.

16. Vazirani, H. N., "Surface Preparation of Steel and Its Alloys for Adhesive Bonding and Organic Coatings," *J. Adhesion,* July 1969.

17. Vazirani, H. N., "Surface Preparation of Copper and Its Alloys for Adhesive Bonding and Organic Coatings," *J. Adhesion,* July 1969.

18. Walter, R. E., D. L. Voss, and M. S. Hochberg, "Structural Bonding of Titanium for Advanced Aircraft," in *Proc. Nat. SAMPE Tech. Conf.,* vol. 2, *Aerospace Adhesives and Elastomers,* 1970.

19. Bersin, R. L., "How to Obtain Strong Adhesive Bonds via Plasma Treatment," *Adhesives Age,* Mar. 1972.

20. Krieger, R. B., "Advances in Corrosion Resistance of Bonded Structures," in *Proc. Nat. SAMPE Tech. Conf.,* vol. 2, *Aerospace Adhesives and Elastomers,* 1970.

21. Cagle, C. V., *Adhesive Bonding Techniques and Applications,* McGraw-Hill, New York, 1968.

22. Guttmann, W. H., *Concise Guide to Structural Adhesives,* Reinhold, New York, 1961.

23. "Preparing the Surface for Adhesive Bonding," Bull. G1-600, Hysol Div., Dexter Corp.

24. Landrock, A. H., *Adhesives Technology Handbook,* Noyes Publ., Park Ridge, NJ, 1985.

25. *Adhesive Bonding Aluminum,* Reynolds Metals Co., 1966.

26. Schields, J., *Adhesives Handbook,* CRC Press, Boca Raton, FL, 1970.

27. Devine, A. T. and M. J. Bodnar, "Effects of Various Surface Treatments on Adhesive Bonding of Polyethylene," *Adhesives Age,* May 1969.

28. Austin, J. E. and L. C. Jackson, "Management: Teach Your Engineers to Design Better with Adhesives," *SAE J.,* Oct. 1961.

29. Burgman, H. A., "Selecting Structural Adhesive Materials," *Electrotechnol.,* June 1965.

30. Bolger, J., "Structural Adhesives: State of the Art," in G. L. Schneberger (ed.), *Adhesives in Manufacturing,* Marcel Dekker, New York, 1983.

31. "Three New Cyanoacrylate Adhesives," Leaflet R-206A, Eastman Chemical Co.

32. "TAME, A New Concept in Structural Adhesives," Bull. GPC-72-AD-3, B. F. Goodrich General Products Co.

33. "Information about Silastic RTV Silicone Rubber," Bull. 61-015a, Dow Corning.

34. Bruno, E. J. (ed.), *Adhesives in Modern Manufacturing,* Soc. of Manufacturing Engineers, 1970, p. 29.

35. "Engineer's Guide to Plastics," *Mater. Eng.,* May 1972.

36. Trauernicht, J. O., "Bonding and Joining, Weigh the Alternatives, Part 1, Solvent Cements, Thermal Welding," *Plast. Technol.,* Aug. 1970.

37. Gentle, D. F., "Bonding Systems for Plastics," in D. J. Almer (ed.), *Aspects of Adhesion,* vol. 5, University of London Press, London, 1969.

38. Mark, H. F., N. G. Gaylord, and N. M. Bihales (eds.), *Encyclopedia of Polymer Science and Technology,* vol. 1, Wiley, New York, 1964, p. 536.

39. "Electromagnetic Bonding—It's Fast, Clear, and Simple," *Plast. World,* July 1970.

40. "How to Fasten and Join Plastics," *Mater. Eng.,* Mar. 1971.

41. "Joining of Composites," in A. Kelley (ed.), *Concise Encyclopedia of Composite Materials,* The MIT Press, Cambridge, 1989.

42. Weggemans, D. M., "Adhesive Charts," in *Adhesion and Adhesives,* vol. 2, Elsevier, Amsterdam, 1967.

43. Krieger, R. B. and R. E. Politi, "High Temperature Structural Adhesives," in D. J. Almer (ed.), *Aspects of Adhesion,* vol. 3, University of London Press, London, 1967.

44. Burgman, H. A., "The Trend in Structural Adhesives," *Mach. Des.*, Nov. 21, 1963.

45. Kausen, R. C., "Adhesives for High and Low Temperature," *Mater. Eng.*, Aug.–Sept. 1964.

46. Bolger, J. C., "New One-Part Epoxies Are Flexible and Reversion Resistant," *Insulation*, Oct. 1969.

47. Falconer, D. J. et al., "The Effect of High Humidity Environments on the Strength of Adhesive Joints," *Chem. Ind.*, July 4, 1964.

48. Sharpe, L. H., "Aspects of the Permanence of Adhesive Joints," in M. J. Bodnar (ed.), *Structural Adhesive Bonding*, Interscience, New York, 1966.

49. Carroll, K. W., "How to Apply Adhesive," *Prod. Eng.*, Nov. 22, 1965.

50. Smith, D. F. and C. V. Cagle, "A Quality Control System for Adhesive Bonding Utilizing Ultrasonic Testing," in M. J. Bodnar (ed.), *Structural Adhesive Bonding*, Interscience, New York, 1966.

CHAPTER 8
PLASTICS IN PACKAGING

Ruben J. Hernandez, PhD
Associate Professor, School of Packaging,
Michigan State University
East Lansing, Michigan

Plastics are used extensively in packaging due to their outstanding physical, mechanical, and chemical properties. Plastics are readily available; versatile; easy to process, either alone or in combination with other materials; and relatively low in cost. The use of plastics has grown faster than that of any other group of materials in the packaging industry. The vigorous growth has been driven by a material substitution trend, the flexibility in design, and the wide range in product protection that plastics afford. In the last decade there has been a strong expansion of the use of plastics especially in the food and pharmaceutical industries. Market specialty niches like snack foods, baby foods, aseptic food packages, microwavable foods, and modified atmosphere packaging are examples of new applications for plastics. Tables 8.1 and 8.2 list the U.S. demand of resins for flexible and rigid packaging. Overall, the use of plastics is expected to grow by more than 4.0 percent between 1993 and 1998. Similar trends are expected in Europe and Japan.

Plastics are used in packaging in diverse forms: single films, multilayer flexible structures, sheets, coatings, adhesives, foams, laminations, and rigid or semirigid containers. There is a great variety of products delivered and distributed around the nation and world using plastics. Solids, liquids, chemicals, alkalis, acids, electronics, hardware, foods, beverage, and health care are just a few of the many products. In the United States alone, nearly 1 million tons per year of plastics are used in the packaging industry, with half of this amount going into films and coatings.

The most widely used plastics in flexible packaging are LDPE, the very fast growing LLDPE, PP, and HDPE. HDPE is widely used for rigid containers also, fulfilling almost 50 percent of the total demand. However, the fastest growing resin for rigid containers is PET and its copolymers, with an expected annual average growth rate of 9 percent. Other resins such as EVOH, VDC copolymers, and nylons have specialized applications as high-barrier materials.

New technologies are continuously improving existing resin properties as well as generating entirely new resins like metallocenes and polyethylene naphthalate. These technologies widen the processing capability of single and composite structures in the production of flexible and rigid containers, while at the same time increasing the number of applications. The increasing sophistication in the use of

TABLE 8.1 Plastics Demand in Flexible Packaging (Thousand Metric Tons)*

Resin	1993	1998	Growth (%)[†]
LDPE	1648	1690	0.5
LLDPE	708	1135	9.9
PP	334	386	2.9
HDPE	393	543	6.7
PVC	114	102	−2.1
PS	114	143	4.7
TP polyesters	114	182	9.8
EVA	36	41	2.4
Nylon	39	45	3.3
PVA	23	30	5.4
Other	109	129	3.5
Total	3632	4226	4.0

* 1 metric ton = 2203 lb.
[†] Annual average growth rate.
Source: G. O. Shroeder, *Modern Plastics Encyclopedia,* McGraw-Hill, NY, 1995, p. A-35.

TABLE 8.2 Plastics Demand in Rigid Packaging (Thousand Metric Tons)*

Resin	1993	1998	Growth (%)[†]
HDPE	2100	2506	3.6
PS	776	885	2.7
PP polyester	713	1103	9.0
PP	493	608	4.3
PVC	95	79	−3.6
LLDPE	127	191	8.4
LDPE	36	32	−2.6
Polyurethanes	55	66	3.9
Other	195	236	3.9
Total	4595	5707	4.4

* 1 metric ton = 2203 lb.
[†] Annual average growth rate.
Source: G. O. Shroeder, *Modern Plastics Encyclopedia,* McGraw-Hill, NY, 1995, p. A-36.

plastics requires a better quantitative treatment of the barrier characteristics of these materials as well as the interactions that occur between the plastic package and the product. The interest in package/product interactions has recently increased with the use of recycled plastics, most noticeably recycled PET for food applications. PET's 28 percent recycling rate and new processes for cleaning the recycled resin and making bottles available have prompted the Food and Drug Administration to clear the use of conventionally recycled postconsumer PET in direct-contact food packaging.

Plastics, however, have some limitations as packaging materials. These limitations are associated with: (1) relatively low temperature of use; (2) mass transfer characteristics that allow molecular exchange between a plastic container, its product, and the external environment; (3) tendency to environmental stress cracking; and (4) viscoelastic behavior that may produce physical distortion under external forces and heat. The correct evaluation of the mass transfer phenomena in a package is crucial for the optimization of package protection and product quality. These aspects determine the design and selection of plastic for food and pharmaceutical packages. This chapter describes the plastics commonly used in the packaging industry; examines their principle properties, characteristics, and evaluations; and studies the fundamental aspects of mass transfer in plastic package systems.

8.1 PACKAGING PLASTICS

8.1.1 Polyethylene

Polyethylene (PE) is a group of polymers resulting from the polymerization of ethylene by chain reaction. Polyethylene can be either a homopolymer or a copolymer and either linear or branched. Homopolymer PE is based mostly on ethylene monomer. In PE copolymer, on the other hand, ethylene can be copolymerized with short alkenes or with compounds having polar functional groups, such as vinyl acetate (VA), acrylic acid (AA), ethyl acrylate (EA), methyl acrylate (MA), or vinyl alcohol (VOH); see Fig. 8.1. When the molar percent of the comonomer is less than 10%, the polymer can be classified as either copolymer or homopolymer.

Branched polyethylenes are nonlinear, thermoplastic, partially crystalline homopolymers or copolymers of ethylene. They are fabricated at high pressure and temperature by a free-radical polymerization process that produces ramification in the main backbone chain. The polymerization of ethylene under these conditions

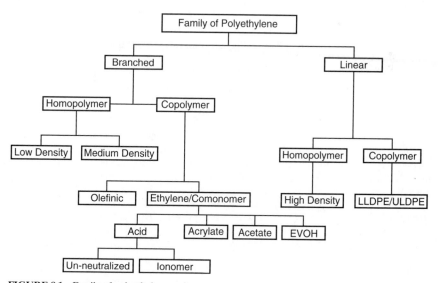

FIGURE 8.1   Family of polyethylene resins.

produces a polymer made of a combination of large molecules with different back-bone lengths, variable side-chain lengths, and many degrees of side-chain branching. This molecular architecture prevents the polymer from reaching a high percent of crystallinity. Nevertheless, branched polyethylene may show crystallinity between 30 and 50 percent and the density ranges from 0.910 to 0.950 g/cm³. Low crystallinity gives a distinctive clear and transparent appearance to LDPE. The molecular weight distribution of LDPE is controlled by the addition of a comonomer such as propylene or hexene during the polymerization process. Several types of branched polyethylene are commercially available depending on the reaction conditions and the type and amount of comonomer.

Linear polyethylene is made of long chains without major branching that tend to crystallize because the polyethylene molecule stereoregularity is not hindered by branches. Linear PE has, indeed, a high percent of crystallinity, ranging from 70 to 90 percent. High-density PE has a density between 0.950 and 0.970 g/cm³. Therefore, linear PE is normally a highly crystalline polymer with high density, while branched PE is a polymer of relatively low crystallinity with density ranging from 0.89 to 0.94 g/cm³.

Branched Homopolymer Polyethylene

Low-Density Polyethylene. LDPE is a branched thermoplastic polyethylene with density values ranging from 0.915 to 0.942 g/cm³ and melt flow index (MFI) between 0.2 and 20 g/10 min. Chain-branching yields desirable characteristics such as: clarity, flexibility, sealability, and ease of processing. The actual values of these properties depend on the balance of the molecular weight, molecular weight distribution, and branching.

Due to its rheological characteristics, LDPE can be processed as blown film, cast film, extrusion coating, extrusion molding, or blow molding. Film is the single largest production form of LDPE; in the United States, 55 percent of the total LDPE produced is made into films with thicknesses under 12 mils (300 microns). LDPE has a low melting-temperature range (98–115°C) and, therefore, is an easily sealable material, a property of great value in flexible packaging. When compared with other plastics, LDPE provides an excellent barrier to water. On the other hand, it shows one of the highest permeability values for oxygen, carbon dioxide, and organic vapors.

Containers, bags for food and clothing, industrial liners, vapor barriers, agricultural films, household products, and shrink- and stretch-wrap films are commonly fabricated from LDPE. Other packaging applications for LDPE are: bakery items, snacks, produce, durable consumer goods, textiles, and industrial items.

Medium-Density Polyethylene. MDPE, with density between 0.930–0.945 g/cm³ and melt-flow index ranging from 0.02 to 20 g/10 min, is somewhat stronger, stiffer, and less permeable than LDPE. MDPE is processed analogously to LDPE, though usually at slightly higher temperatures. The major competitor material of both MDPE and LDPE is LLDPE. LLDPE provides superior strength at any given density. Nevertheless, LDPE is preferred for high-clarity films and coating substrates.

Suppliers of branched PE are: American Polymers, Bamberger Polymers, Chevron, Dow Plastics, DuPont, Eastman, Exxon, Mobil, Monmouth, Novacor Chemicals, Quantum Chemical, RSG Polymers, Rexene, Union Carbide, Washington Penn Plastics, and Westlake Plastics.

Branched Copolymers of Polyethylene. Ethylene monomer can be copolymerized with either alkenes of four to eight carbons or monomers containing polar functional groups like vinyl acetate, acrylic acid, and vinyl alcohol. Branched ethylene/alkene copolymers are equivalent to LDPE, since in commercial practice a certain amount of propylene and hexene are always added to help control the average molecular weight. Inserting polar comonomers in branched ethylene copolymers increases their flexibility, widens the range of heat-sealing temperatures, improves the barrier properties, and lowers their crystallinities below those of homopolymers.

Ethylene Vinyl Acetate (EVA). The properties of EVA, a random copolymer of ethylene and vinyl acetate, depend on the content of vinyl acetate. EVA resins show better flexibility, toughness, and heat sealability than LDPE. The content of vinyl acetate (VA) in the copolymer ranges from 5 percent to 50 percent, although for optimal applications in food-packaging applications VA content should range from 5 percent to 20 percent. As the percent of VA increases, it happens that:

1. The crystallinity of EVA decreases, and, in contrast to PE, the density increases.

2. EVA becomes clearer, more flexible at low temperatures, and more resistant to impact.

3. At a 50 percent of VA and above, EVA is totally amorphous and transparent.

The presence of VA enhances the intermolecular bonds between chains and, therefore, increases the adhesion strength and tackiness of EVA with respect to LDPE. As the molecular weight increases, the viscosity, toughness, heat-seal strength, hot tack, and flexibility increase. Because of its excellent adhesion and ease of processing, EVA is used in extrusion coating, heat sealing layers with PET, cellophane, and biaxially oriented PP films (20 percent VA) for cheese wrap and medical packages. However, EVA has limited thermal stability and low melting temperature, and, therefore, must be processed at relatively low temperatures. Nevertheless, this aspect can be advantageous if toughness is required at low temperature. For this reason, EVA is a good choice for ice bags and stretch wrap for refrigerated meat and poultry.[1]

Suppliers of EVA are: AT Plastics, Chevron, DuPont, Exxon, Federal Plastics, Mobil, Quantum Chemical, Polymers, Rexene, A. Schulman, Union Carbide, and Westlake Plastics.

Ethylene Acrylic Acid (EAA). Copolymerizing ethylene with acrylic acid (AA) produces copolymers containing carboxyl groups along the main and side chains of the molecule. EAA copolymers are flexible thermoplastics with chemical resistance and barrier properties similar to those of LDPE. EAA is superior to LDPE in strength, toughness, hot tack, and adhesion. Clarity and strength adhesion increase with AA content, while the heat-seal temperature decreases by a few degrees. EAA uses include blister packaging and as extrusion-coating tie layer between aluminum foil and other polymers.

Films of EAA are used in skin packaging, adhesive lamination, and flexible packaging of meat, cheese, snack foods, and medical products. Extrusion-coating applications of EAA include coated paperboard, aseptic cartons, composite cans, toothpaste tubes, and food packages. FDA regulations permit the use of ethylene acid copolymers containing up to 25 percent acrylic acid and 20 percent methylacrylic acid in direct food applications.[2]

Dow Chemical is the main supplier of EAA.

Ionomers. Ionomers are unique plastics because they combine both covalent and ionic bonds in the polymer chain. They are produced by the neutralization of EAA (or similar copolymers) with cations such as Na^+, Zn^{++}, or Li^+. Ionomers show better transparency, toughness, and higher melt strength than the unneutralized copolymer. In general, sodium ion types are better in optical, hot tack, and oil resistance. Zinc ionomers are more inert to water and have better adhesion properties in coextrusion and extrusion coating of foil.

Random cross-linking between the chains, produced by the ionic bonds, yields solid-state properties usually associated with very high molecular-weight materials. But contrary to true cross-linking, the ionic bond in ionomers, strong at low temperatures, weakens as the temperature increases. Therefore, on heating, ionomers behave like normal thermoplastic materials with normal processing temperatures ranging from 175° to 290°C. Ionomers show great pinhole resistance and have low barrier properties.

In packaging, ionomers are used in coating applications, forming heat-seal layers with nylon, PET, LDPE, PVDC, and composite structures. They are commonly used in coextrusion lamination and extrusion coating. In addition, ionomers adhere very well to aluminum foil. Ionomers are used in packaging where formability, toughness, adhesion, and visual appearance are important. Ionomers are highly resistant to oils and aggressive products and provide reliable seals over a broad range of temperatures. Uses of ionomers in food packaging include: frozen food (fish and poultry), cheese, snack foods, fruit juice, wine, water, oil, margarine, nuts, and pharmaceutical products. Heavy-gauge ionomer films serve as skin packaging for hardware and electronic products because of their great adhesion to paperboard. Ionomers can resist forceful impacts at temperature as low as –90°C (lower than LDPE).[3]

Suppliers of Ionomers are DuPont and Exxon. DuPont supplies ionomers under the trade name Surlyn.

Linear Polyethylene. Linear polyethylene comprises a group of the following polymers: ultra-low-density PE (ULDPE), linear low-density PE (LLDPE), high-density PE (HDPE), and high-molecular-weight HDPE (HMW-HDPE). Physical properties of commercially available linear PE are markedly dependent on the average molecular weight, molecular weight distribution, and resin density; see Table 8.3. Common applications of linear polyethylene are illustrated in Fig. 8.2. Linear PE is produced in gas-phase low-pressure processes in the presence of transition catalysts such as Zigler-Natta (titanium and aluminum) or Phillips (chromium oxide). For instance, the Union Carbide Unipol process produces linear polyethylene based on these transition catalysts. More recent technologies employ single-site catalysts

TABLE 8.3   Effects of Density, MW, and MWD on Linear
PE Properties

Property	Density	Average molecular weight	Molecular weight distribution
Chemical resistance	I	I	0
Permeability	D	d	0
ESCR	D	I	0
Tensile	I	I	0
Stiffness	I	i	d
Toughness	D	I	D
Melt strength	—	I	I

As variable increases, property undergoes a/n: I = increase; D = decrease; i = slight increase; d = slight decrease; 0 = no significant effect.

Source: J. W. Taylor, *Modern Plastics Encyclopedia Handbook,* McGraw-Hill, NY, 1994, p. 29.

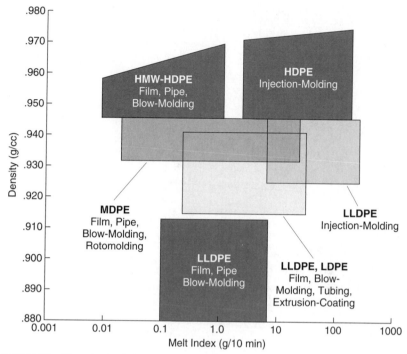

FIGURE 8.2 Uses of polyethylene as a function of density and melt index. (Reprinted with permission from P. W. Manders, *Modern Plastics Encyclopedia '95*, McGraw-Hill Inc., 1994.)

(metallocenes, or constrained-geometry catalysts) to produce polymers with more controlled molecular configurations and very narrow molecular weight distributions. These new materials, in addition to improving mechanical properties, have low levels of extractables (an important characteristic in packaging for flavor-sensitive products). Polyolefins plastomers are produced by Dow Chemical using single-site catalysts under the trade name Affinity.

LLDPE. LLDPE is characterized by narrow molecular weight distributions, a linear structure with very short branches, and low density values (0.916–0.940 g/cm³). Due to its molecular linearity, LLDPE is more crystalline and, therefore, stiffer than LDPE. This results in a melting point that is 10–15°C higher than that of LDPE. LLDPE has better tensile strength, puncture resistance, tear properties, and elongation than LDPE. Because of its greater crystallinity, LLDPE is more hazy and glossy than LDPE. Packaging uses of LLDPE are stretch/cling film, grocery sacks, shopping bags, and heavy-duty shipping sacks.

HDPE. High-density PE is a linear thermoplastic made of long chains with little branching, 65–90 percent crystallinity, and density ranging from 0.940 to 0.965 g/cc. The high percent of crystallinity gives HDPE higher moisture and oxygen barriers (see Figs. 8.3 and 8.4), better chemical resistance, and more opacity than LDPE. Blow-molding applications of HDPE include industrial chemical drums and containers for milk, detergent, bleach, juice, and water. Thin-walled dairy containers and closures are made by injection-molding. Cosmetic containers, pharmaceutical bottles, shampoo, and deodorant containers are made by injection blow-molding. Both blown

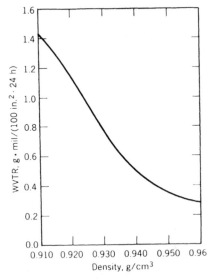

FIGURE 8.3 Effect of density on water-vapor transmission rate in HDPE. (Reprinted with permission from M. A. Smith, *The Wiley Encyclopedia of Packaging Technology,* John Wiley and Sons, Inc., New York, 1986.)

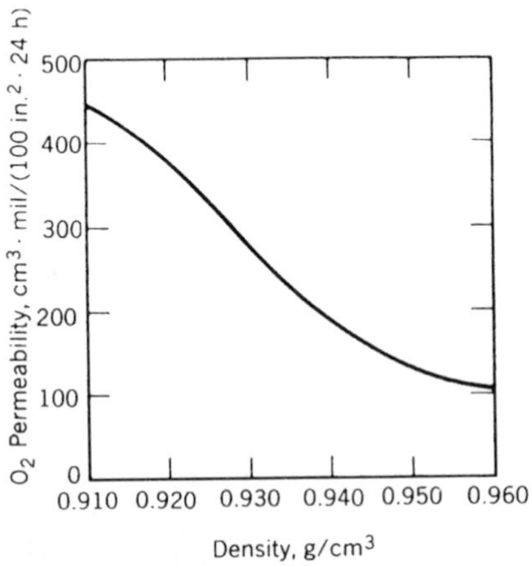

FIGURE 8.4 Effect of density on oxygen permeability in HDPE. (Reprinted with permission from M. A. Smith, *The Wiley Encyclopedia of Packaging Technology,* John Wiley and Sons, Inc., New York, 1986.)

and cast films are frequently utilized in flexible packaging applications. HDPE replaces glassine for cereal, cracker, and snack-food packaging. It is also used for produce bags and wrap for delicatessen products. HDPE strongly resists a wide range of chemical compounds: water-based products, medium-molecular-weight aliphatic hydrocarbons, alcohols, ketones, and dilute acids and bases. However, HDPE should not be used with aromatic hydrocarbons such as benzene, which penetrate the polymer. Compared to other polymers, HDPE is an excellent moisture barrier but shows high permeability to gases like oxygen and CO_2 as well as organic vapors.

Suppliers of linear PE, in addition to the ones listed as producers of branched PE, are: Federal Plastics, Hoechst Celanese, M.A. Polymers, Novacor, Paxon, A. Shulman, and Solvay Polymers.

8.1.2 Polypropylene

Polypropylene (PP) is a group of thermoplastic polymers based on the polymerization of the propylene monomer. They are commercially available as PP homopolymers and PP random copolymers. The latter is produced by the addition of a small amount of ethylene (2–5 percent) during the polymerization process. Thermoplastic PP polymers are characterized by their low density (0.89–0.92 g/cm³) and high resistance to chemical and mechanical fatigue. PP resins are frequently employed in films and rigid containers.

PP Homopolymer. Depending on the catalyst and polymerization conditions, the molecular structure of the resulting polymer consists of the three different types of stereo configurations for vinyl polymers: isostatic, syndiotactic, and atactic. Isotactic PP is synthesized using Ziegler-Natta catalysts under controlled conditions of temperature and pressure. Industrial processes are designed to minimize the production of the atactic PP, a lower-value, noncrystalline tacking by-product used mainly in adhesives.

Isostatic PP (iso-PP). This is the most common form of PP. The placement of all the methyl groups on the same side of polymer backbone produces a stereoregular molecule and, consequently, a highly crystalline material. Because of its crystalline nature, isostatic PP is fairly opaque and has good solvent resistance. Clarity in iso-PP can be increased by the addition of chemical agents to prevent the growth of large crystals that give the opaqueness to PP. Compared to both LDPE and HDPE, PP has lower densities, higher melting-point temperatures, and higher stiffness values (higher modulus). These properties determine unique applications for the PP homopolymer. For example, a higher value of stiffness and ease of stretching make PP homopolymers very suitable for wrapping and stretching applications, while the higher heat resistance allows a container made of this material to be autoclavable. On the other hand, iso-PP is more sensitive than PE to heat and light oxidative degradation. An oxidative degradation process during melt processing of PP produces chain scissions that reduce the molecular weight, which in turn increases flow rate. For this reason, antioxidant compounds are added to the resin during processing to control oxidation. Antistatic agents are commonly incorporated for packaging applications to dissipate static charge.

Iso-PP has desirable rheological properties, a wide range of flow rate, and good processing behavior. With molecular-weight averages commonly in the range of 200,000–600,000 daltons, the melt-flow index ranges from 0.5 to 50 g/10 min. Broad MWD materials are easy to process by injection molding. The density of PP, ranging from 0.89 to 0.91 g/cm^3, is one of the lowest among plastics. This low density produces films with high values of yield. Yield y (or area factor) is calculated as $y = 1/(\rho \cdot l)$, where ρ and l are the resin density and film thickness, respectively. PP shows excellent moisture-barrier properties like other polyolefins. Films are produced by both blown and casting methods. Biorientation improves the film's optical appearance and strength; see Table 8.4. Oriented PP films are used to wrap compact-disc boxes, toys, games, hardware items, frozen foods, and cigarette packaging.

PP films are commercially available with different coatings (acrylic and PVDC) to improve heat sealing, barrier, and optical characteristics, as well as to decrease the coefficient of friction. Also, metallized PP films are available to provide extremely low values of permeability to gases and vapor.[4] PP films are well suited for bag-in-box

TABLE 8.4 Effect of Orientation on PP Properties

	Nonoriented PP	Oriented PP
WVTR g mil/m^2 day at 90% RH and 100°F	15	6
Stiffness	Very low	High, similar to cellophane
Propagated tear strength	High	Very low CD, very high MD
Heat sealabilty	Yes, 350–450°F	No, film distorts
Density	0.902	No change
Optics	Very good	Excellent
Surface adhesivity to inks, etc.	Low	Low
Oxygen permeability, cc mil/m^2 day atm	3700	2500

applications in cereals, crackers, soup mixes, and stand-up pouches. PP is also an excellent material to fabricate injection-molded closures for HDPE, PET, and glass bottles.

PP Random Copolymer. Random copolymer PP typically contains 1.5 to 7 percent ethylene, by weight, as a comonomer. The random placement of ethylene in the molecular chain prevents the stereoregularity of the chains and high values of crystallinity as seen in iso-PP. Therefore, polymers with low crystallinity, resulting in better clarity and flexibility, and lower melting points are obtained. Random copolymers with densities ranging 0.89–0.90 g/cm^3 are slightly lighter than homopolymers. They are tougher and have lower temperature impact than homopolymer PP. Random copolymers show good chemical resistance to acids, alkalies, alcohols, and low-boiling hydrocarbons (nonaromatic hydrocarbons). Oriented films can be used as shrink wrap for toys and audio products. Moisture-barrier properties are also good. For example, at 100°F and 90 percent RH, permeance is 0.6 g · mil/100 in^2 · 24 h. PP random copolymers are processed as films and by blow and injection molding. Packaging applications include: medical, food and bakery products, produce, and clothing. The 7 percent ethylene copolymer is used as a heat-seal layer in food packaging.

Suppliers of PP are: American Polymers, Amoco Chemical, Aristech Chem, Bamberger Polymers, ComAlloy, Epsilon, Chevron, Federal Plastics, Ferro Corporation, Fina, M.A. Polymers, Monmouth, Montell Polyolefins, Phillips, Quantum Chemical, RSG Polymers, Rexene, A. Schulman, Shell, Shuman, Solvay Polymers, Thermofil, and Washington Penn Plastics.

8.1.3 Polyvinyl Chloride (PVC)

PVC is a homopolymer of vinyl chloride. Eighty percent of commercial PVC in packaging is produced by chain-reaction polymerization using a suspension method. Other methods are emulsion and solution polymerization. Chain-reaction polymerization requires initiators to produce free radicals, and then the reaction proceeds until the chain is terminated. The predominant configuration of the monomer in the polymer chain follows a head-to-tail alignment to yield a syndiotactic polymer.

PVC resins start decomposing at temperatures as low as 100°C (212°F) with the generation of HCl corrosive vapors. PVC without plasticizers (rigid PVC) has a glass-transition temperature (T_g) of 82°C. This high T_g value makes PVC difficult to process into useful products. The addition of plasticizers decreases the glass-transition temperature of PVC, thus decreasing processing temperature. Liquid plasticizers permit the production of a flexible film of PVC with moderate values of oxygen permeability. Plasticizers also permit the production of blow-molded bottles, blown films, and flexible materials to be produced. Therefore, the manufacture of a wide variety of products from PVC is possible, then, because it is miscible with a range of plasticizers. DOA, or di(2-ethyl hexyl)adipate, is one of the most common plasticizers used in PVC. Stabilizers such as Ca/Zn salts, to avoid the decomposition of PVC and the corresponding product of HCl, are also employed during compounding. When used for food applications, these stabilizers must have FDA clearance.

PVC films show good clarity, fair barrier properties, high puncture resistance, and great sealability. PVC films are also tough and resilient. For these reasons, PVC films are extensively used in food packaging, particularly for red and fresh meats. The oxygen permeability of PVC film is well suited to maintaining the oxygen requirements of the meat. This is necessary to keep the red color of the meat and its appearance of freshness. In the U.S., chilled poultry and tray-packed poultry parts are packaged with PVC stretch films. PVC is also used to wrap fresh fruits and vegetables. Other food applications include bottles for milk, dairy products, and edible oil, as well as blister packages for fish and produce. Nonfood packaging uses of PVC include vacuum-formed blisters and bottles for toiletries, cosmetics, and detergents. Medical packaging applications include tubing and bags for parenteral products as well as for blood and intravenous solutions.

In the polymerization process of PVC, less than 100 percent of vinyl chloride monomer (VCM) is converted to polymer. This means that relatively high values of VCM may remain unreacted and trapped in the resin. For this reason, the resin is subsequently submitted to a process to remove the VCM by repeated applications of vacuum. Currently, the industry produces PVC with extremely low levels of VCM in the resin, and the amount of VCM that might migrate to food is well below the sensitivity of common analytical methods (lower than 1 ppm). Due to its chlorine content, PVC has been the center of an environmental controversy in the 1990s. Several European countries have banned the use of all PVC packaging because of the fear that, during incineration of solid wastes, HCl gas and chlorinated organic compounds (in which dioxins can be found) are generated and released into the environment. Although there is available technology that eliminates the emission of such unwanted compounds (Japan incinerates a large portion of its solid waste), in many European countries, the use of PVC in food packaging is still very common.

Suppliers of flexible unfilled PVC are: Alpha Gary, Borden, Colorite, Novatec, Rimtec, A. Shulman, Shintec, Synergistecs, Teknor Apex, Tenneco, Union Carbide, Vi-Chem, and Vista Chemical.

8.1.4 Vinylidene Chloride Copolymers

Vinylidene chloride homopolymer and copolymers are known as Saran, a registered trademark of Dow Chemical, the company that developed them during the 1930s. They are commonly, though wrongly, referred as PVDC. The polymers produced are all based on vinylidene chloride (VDC) and comonomers such as vinyl chloride (VC), acrylates (methyl acrylate), and vinyl nitriles. VDC homopolymer has a melting point of 388°C to 401°C, but it decomposes at 205°C. These conditions make VDC homopolymer difficult to process. By copolymerization, the melting point of the resin is decreased to a range of 140–175°C, making melt processing feasible. Saran polymer contains 2–10 percent plasticizer (for example, dibutyl sebacate or diisobutyl adipate), and heat stabilizers. Molecular weight ranges from 65,000–150,000 daltons. The most notable attributes of Saran copolymers are their extremely low permeability to gases and liquids, and chemical resistance comparable to EVOH resins. Oxygen-permeability values range from 0.5 to 15.0 cc · mil/m^2 · day · atm.

Saran is available in the following forms: F-resins (with acrylonitrile as copolymer) used as solvent-soluble polymer for barrier coating; aqueous emulsion latexes for barrier coatings; and extrusion resins, which are melt-processable in rigid multilayer coextruded containers, extrusion of films, and sheets. F-resins include F-239 and F-278 types. These resins are used for coating plastic films such as cellophane and polyester. Resin F-310 is used for paper coating. The heat-sealing temperature of F-resins are in the 100°–130°C range (Dow Chemical Co., Form 190-305-1084). Latexes are applied to coat paper, paperboard, and plastic films such as PP and PE. Also PET, PVC, PS, and PE rigid containers can be coated with latexes. Barrier properties of latexes are similar to those of F-resins (Dow Form 190-309-1084). Extrusion resins are used for flexible packaging in monolayer and multilayer (coextruded or laminated) structures for meat and other food applications. For nonrefrigerated foods in rigid containers, extrusion resins can be coextruded with PP or PS resins. Extrusion resins provide poorer barriers than do F-resins or latexes (Dow Form 190-320-1084). Saran HB films (with vinyl chloride as comonomer) are better barriers than F-resins (Dow Form 500-1083-586).

Several processing methods are possible with Saran resins: extrusion, coextrusion, laminating resin, and latex coating. Also, injection molding, blown extrusion film, and cast film are common industry processes for these resins. The main applications of Saran resins are in food packaging as barrier materials to moisture, gases, flavors, and odors. Monolayer films are widely used in household wrap. Multilayer films, generally coextrusions with polyolefins, are used to package meat, cheese, and other moisture- or gas-sensitive foods. The structures usually contain 10–20 percent of VDC copolymer and are commonly used as shrinkable films to provide a tight barrier around the food product. VDC copolymers are used as barrier layers in semirigid thermoformed containers. PVDC coatings are used to improve barrier properties of paper and paperboard, cellophane, plastic films, and even semirigid containers such as PET bottles. Other industrial applications of monolayer films include laminations for unit dose packaging and pack liners for moisture-, oxygen-, and solvent-sensitive products in pharmaceutical and cosmetic packaging.[5]

The supplier of saran resins is Dow Plastics.

8.1.5 Polystyrene (PS)

New copolymerization methods, additives, rubber modification, and blending have made polystyrene (PS) a versatile packaging material. Polystyrenes are hydrophobic, nonhygroscopic, and can be processed by extrusion and thermoforming. Three types of PS are available: general purpose, impact PS, and foams.

General-Purpose Polystyrene (GPPS).   Although referred to as "crystal" PS, these are totally amorphous materials showing no melting temperature. They are therefore highly transparent, with excellent optical properties. Crystal PS linear polymer has a T_g value ranging from 74 to 105°C, which makes it brittle and stiff at room temperature. There are three grades of GPPS: high heat, medium flow, and high flow (or easy flow). High-heat resins have high molecular weight, contain few or no additives, and are brittle. They are used as extruded foams and thermoformed materials for electronic packaging, injection-molded jewel boxes, high-quality cosmetic containers, and boxes for compact discs.

High-flow resins have low molecular weight and usually contain 3–4 percent mineral oil as an additive. This makes crystal PS more flexible (less brittle) with a lower distortion temperature. Typical applications include disposable medical ware, dinner ware, and coextruded sheets for thermoformed packaging.

Medium-flow resins have intermediate molecular weight, with 1–2 percent mineral oil as an additive. These resins are used in blow-molded bottles and coextruded materials for food and medical packaging.

High-Impact Polystyrene (HIPS).   HIPS contains particulates of rubber that are added to enhance impact resistance. This produces an opaque material easy to process that can be thermoformed. Typical food-packaging applications are: tubs for refrigerated dairy products, serving-size cups, lids, plates, and bowls. Limiting factors for HIPS are: low heat resistance, high oxygen permeability, low UV light stability, and low resistance to oil and chemicals. According to Toebe et al.,[6] containers made of HIPS have intense flavor-scalping action on foods.

Expandable PS (EPS). Foam form is a form of crystal PS supplied as a partially expanded bead. EPS foam has good shock-absorbing and heat-insulation characteristics. Application in food packaging includes egg cartons and meat trays.[7]

Suppliers of PS are: A&E Plastics, American Polymers, Amoco Chemical, Bamberger Polymers, BASF, Chevron, Dow Plastics, Federal Plastics, Fina, Hudsman, Mobil, Chemical, RSG Polymers, and A. Schulman.

8.1.6 Ethylene Vinyl Alcohol

Introduced in 1970 in Japan, ethylene vinyl alcohol (EVOH) is produced by a controlled hydrolysis of ethylene vinyl acetate copolymer. The hydrolytic process transforms the vinyl acetate group in vinyl alcohol, VOH ($CH_2 = CHOH$). The presence of OH in the backbone chain, equivalent of substituting a certain number of H atoms in a polyethylene chain, has several noticeable effects on EVOH resins. First, the OH group, which is highly polar, increases the intermolecular forces between polymer chains and, at the same time, makes the resin hydrophilic. Second, the OH group is small enough to yield a polymer with high crystallinity, even if it is randomly distributed in the chain. This yields an outstanding barrier to permeants that is sensible to humidity. If the percent of vinyl alcohol is equivalent to 100 percent, polyvinyl alcohol, PVOH, is obtained.

Contrary to PE, PVOH has exceptional gas- and odor-barrier properties (the lowest permeability of any polymer available). PVOH is difficult to process and is water-soluble. When the percent of VOH in EVOH ranges from 52 to 70 percent, the ethylene-vinyl alcohol copolymers obtained combine the processability and water resistance of polyethylene and the gas- and odor-barrier characteristics of PVOH. EVOH copolymers are highly crystalline, and their processing and barrier properties vary with respect to the equivalent percent of ethylene. When the ethylene percent is near 30, the gas and organic vapor barrier is exceptionally high, but processing conditions become more difficult. When ethylene content increases, water-barrier characteristics and processability improve.

The most important characteristic of EVOH is its outstanding oxygen- and odor-barrier properties. A layer of EVOH in packaging structures provides high retention of flavors and protection to oxidation of the food product packaged. EVOH also provides a very high resistance to oils and organic vapors. This resistance decreases somewhat as the polarity of the penetrating compound increases. For example, the resistance to linear and aromatic hydrocarbons is outstanding, but for ethanol and methanol it is low (it may absorb up to about 12 percent of ethanol). As indicated, the hydroxyl group OH makes the polymer hydrophilic, attracting water molecules. EVOH films in equilibrium with a humid environment show higher oxygen permeability values (see Fig. 8.5). This poses an interesting challenge in the design of high-barrier packages, since external nonhydrophilic layers are necessary to protect the EVOH oxygen barrier characteristics.

EVOH can be coextruded in numerous combinations with PE or PP, laminated or coated to several substrates including PET, PE, nylons, and the like. EVOH can also be extruded into films and processed in blow molding, injection molding, and coextrusion blow-molding. The FDA has cleared the use of EVOH resins for direct food contact for VOH content up to 80 percent. Applications in packaging include flexible and rigid containers. Typical applications are: ketchup and barbecue sauce bottles, jelly preserves, vegetable juice, mayonnaise containers, and meat packages. Nonfood applications include packaging of solvents and chemicals.[8]

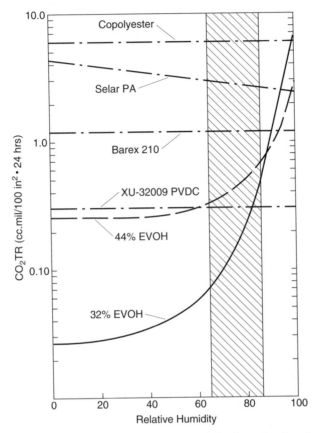

FIGURE 8.5 Oxygen permeability of selected resins as function of relative humidity. (Reprinted with permission from EVAL Company of America.)

Producers of EVOH resins are: Eval Company of America (EVALCA), Nippon Goshei Japan, and Kurary Co Japan.

8.1.7 Nylons

Nylons are linear thermoplastic condensation polyamides that contain the amide group (—CONH—) as a recurring part of the chain. In general they are clear, thermoformable, strong, and tough over a broad range of temperatures. Nylons show excellent chemical resistance and are good barriers to gas, oil, and aromas. They are, however, moisture-sensitive materials like EVOH. When left in normal environmental conditions, nylons can easily absorb 6–8 percent of their weight in water. The amount of water sorbed by a nylon sample as a function of relative humidity is described by a sorption isotherm curve.[9] For most packaging applications, nylons are used in film form as a single component or in multilayer structures.

Nylons are polymers with strong H-bonding between the $C\!=\!O$ and HN groups of different chains. These high intermolecular forces combined with high crystallinity yield tough, high-melting thermoplastic materials—Nylon 6,6 has a melting point of 516°F (269°C). Nylons have good puncture resistance, impact strength, and temperature stability. However, the flexibility of the aliphatic portion in the chain permits film orientation that enhances strength. Amorphous nylons can be produced by the copolymerization of two different acids. For example, isophthalic and terephthalic acids are used to produce the amorphous nylon Selar PA, which is commercialized by DuPont. Oxygen permeability increases with moisture content.

Nylons are melt-processable using conventional extrusion equipment. Films can be produced by either cast-film processing or blown-film processing. During film production, diverse degrees of crystallinity are obtained depending on the temperature-quenching rate. When the cooling rate is increased, a less crystalline nylon is obtained since the polymer is not given sufficient time to form crystals. The increase in amorphousness produces a more transparent and more easily thermoformable film. Nylons are used in coextrusion with other plastic materials, providing strength and toughness to the structure. Polyolefins are commonly used in nylon coextrusions to provide heat salability, moisture, and low cost. Nylons are used in extrusion-coat paperboard to obtain heavy-duty paperboard.

The blow-molding process is used with nylon resins to produce industrial containers, moped fuel tanks, and oil reservoirs. Thermoformed nylons are employed for disposable medical devices, meat and cheese packaging, and thermoform/fill/seal packaging. For most applications, nylons are combined with other materials, such as LDPE, ionomer, and EVA, to add a moisture barrier and heat sealability. Multilayer films containing a nylon layer are used principally in vacuum packing bacon, cheese, bologna, hot dogs, and other processed meats. Polyvinylidene chloride copolymer coatings on nylons are available to improve oxygen-, moisture vapor-, and grease-barrier properties. Biaxial orientation of nylon (BON) films provides increased crack resistance and better mechanical and barrier properties.

Suppliers of molding and extrusion nylons 6 and 6,6 include: Allied Signal, BASF, Bamberger Polymers, Bayer Corp., ComAlloy, Custom Resins, DSM Engineering Plastics, DuPont, Federal Plastics, Hoechst-Celanese, Monsanto, Nylon Corporation of America, Polymers International, A. Shulman, Texapol Thermofil, and Wellman.

8.1.8 Polyethylene Terephthalate (PET)

PET, a linear semicrystalline thermoplastic homopolymer, has become a very important packaging material as for both food-packaging films and carbonated-beverage bottles. Copolymerization with other monomers produces polyester resins with different degrees of crystallinity, including an amorphous material used for containers and trays. The great acceptance of PET as a carbonated-beverage packaging material is due to its toughness, clarity, capability of being oriented, and reasonable cost, as well as the development of high-speed bottle-processing technology. Compared to glass, PET containers are lightweight and shatter-resistant. They provide an acceptable barrier, and they are recyclable. PET is produced by the condensation of terephthalic acid and glycol. Its T_g is around 78°C, and its melting temperature is 270°C. Crystallinity of PET ranges between 20 and 40 percent, and its density is between 1.30 and 1.40 g/cm^3.

The presence of moisture during the extrusion of PET reverts the condensation reaction and produces some degree of depolymerization. Prior to processing, moisture content should be less than 0.005 percent to minimize hydrolytic breakdown and loss of properties. Values of acetaldehyde in the resins are extremely low.

Films can be cast using chill roll, while injection-stretch blow-molding is used to produce bottles for carbonated beverages. PET is used for packaging food, distilled spirits, carbonated soft drinks, noncarbonated beverages, and toiletries. Typical food products include: mustard, pickled foods, peanut butter, spices, edible oil, syrups, and cocktail mixers. PET is extensively used for extrusion coating and extrusion into film and sheet. Its crystalline form (CPET) is the basic material for oven-ware containers. Biaxially oriented PET is used in meat and cheese packaging.

Thermoplastic Copolyesters. The name *copolyester* is applied to those polyesters whose synthesis is carried out by using more than one glycol and/or diacid. The copolyester chain is less regular than the homopolymer (some of those are amorphous), and the degree of crystallinity is lower. PCTA is a polymer of cyclohexanedimethanol and terephthalic/isophthalic acids. It is an amorphous polyester designed primarily for film forming and sheeting in food, pharmaceutical packaging, and general blister-packaging applications.

Suppliers of unfilled PET include: DuPont, Eastman Chemical, Hoechst Celanese, MA Polymers, A. Schulman, Shell, Texapol, and Wellman.

8.1.9 Polyethylene Naphthalene (PEN)

Polyethylene naphthalate, a relatively new polyester, has a great potential as a resin for bottles. PEN resin is a clear material, although less clear than PET, with enhanced barrier values to oxygen (a permeability coefficient that is five times lower than PET), carbon dioxide, and other chemicals. It has a higher glass-transition temperature and is stronger and stiffer than PET. This makes it more suitable for hot filling and an excellent material for carbonated beverages. Bottles made of PEN provide the product with additional ultraviolet protection. PEN bottles can be returnable, refillable, and recyclable. PEN resins can be processed by blow molding, injection molding, and extrusion thermoforming.

8.1.10 Polycarbonate

Polycarbonate (PC) is a glassy, amorphous thermoplastic material. It offers an excellent balance of toughness and clarity. Its heat-deflection temperature is about 130°C, and its glass-transition temperature is 149°C. Polycarbonate has good potential for packaging applications. Commonly, polycarbonate is produced by the reaction of bisphenol-A and carbonyl chloride.

Toughness is the most relevant property of PC. Being tough and clear makes PC a material well suited for reusable bottles, particularly 19-liter (5-gallon) water bottles and one-gallon milk bottles. Systems with washing stations have been developed for reusing PC bottles. Polycarbonate films are odorless, tasteless, and do not become stained through normal contact with natural or synthetic coloring agents.

PC has good resistance to fruit juices, aliphatic hydrocarbons, and aqueous solutions of ethanol. It is, however, attacked by some solvents such as acetone and dimethyl ethyl ketone. Since PC is FDA-approved, food-contact applications include microwave, oven-ware, and food storage containers. Other applications include hot-filling, modified atmospheric packaging, rigid packaging to substitute for PVC, high gloss for paper, and as a barrier for fruit-juice cartons. PC finds applications in packaging for medical devices—it can be sterilized by commercial sterilization techniques such as ethylene oxide, autoclave sterilization, and gamma sterilization.

PC can be processed by injection molding, extrusion, coextrusion, and blow molding. Coextrusions with EVAL or polyamides are carried out with the help of adhesives. PC can be laminated or coextruded to PP, PE, PET, PVC, and PVDC. PC are hydrophilic polymers and at ambient conditions can reach moisture levels of 0.35 percent.

Suppliers of unfilled molding PC are: Albis, American Polymers, Ashley Polymers, Bamberger Polymers, Bayer Corporation, Dow Plastics, Federal Plastics, GE plastics, RMC Polymers Resources, Progressive Polymers, and Shuman.

A list of selected properties of plastics commonly used in packaging is presented in Table 8.5.

8.2 PROPERTIES OF PACKAGING PLASTICS

A description of the most relevant properties of plastics packaging and their test methods is presented in this section. For easy reference they are grouped under these headings: (1) morphology and density; (2) thermophysical; (3) mechanical; (4) barrier; (5) surface and adhesion; (6) optical; and (7) electrical.

8.2.1 Morphology and Density

Morphology.   Polymer morphology refers to the presence, shape, arrangement, and physical state of amorphous and crystalline regions that are generally found coexisting in a polymer sample. Polymers used in packaging are either homogenous amorphous solids or, more frequently, heterogenous phase semicrystalline solids. A semicrystalline polymer can be thought of as small crystalline regions embedded in an amorphous phase. LDPE, HDPE, PET, Nylon 6, and EVOH are examples of semicrystalline polymers, while general-purpose polystyrene, butadiene-styrene copolymers, some PET copolymers, and Nylon 6I/6T are amorphous polymers.

The actual morphology of a polymer depends primarily on three factors: chemical composition, degree of polymerization, and chain conformation. Other variables affecting the final physical state of a polymer sample include the thermomechanical history and processing methods.

Polymer molecules tend to seek out an arrangement in the lowest energy state possible (lowest Gibbs free energy). The lowest energy level that a compound can achieve is a crystal form. Although crystallization tends to occur naturally, crystallization takes place in small regions of a polymer (in the order of 1×10^{-9} m), if at all. The ability of a polymer to crystallize is largely determined by the regular placement of atoms in the chain. Polymers made of symmetrical unsaturated monomers such as polyethylene and polyvinylidene chloride crystallize easily. Asymmetric polymers such as polypropylene (PP) crystallize only within regular configurations (isotactic or syndiotactic), while atactic PP is amorphous. But for the most part, asymmetric polymers with an atactic configuration can crystallize if the subtituents are small and polar as in the case of EVOH and PAN. On the other hand, polymers with asymmetric, repeating units and regular configurations do not crystallize if the substituent is too bulky as is the case of PS. Normally, step-reaction polymers that are synthesized from difunctional monomers containing alcohol, acid, or amines can crystal-

TABLE 8.5   Selected Properties of Packaging Plastics

	LDPE	HDPE	LLDPE	>12% VA EVA	Ionomer
Density, g/cc	0.91–0.925	0.945–0.967	0.918–0.923	0.94	0.94–0.96
Yield, in²/lb·mil × 10⁻³	30	29.0	30.0	29.5	28.6–29.5
Tensile strength, kpsi	1.2–2.5	3.0–7.5	3.5–8.0	3–5	3.5–5.5
Elongation at break, %	225–600	10–500	400–800	300–500	300–600
Impact strength, kg·cm	7–11	1–3	8–13	11–15	6–11
Elmendorf tear strength, g/mil	100–400	15–300	80–800	50–100	15–150
WVTR, g·mil/100 in²·day @ 100°F and 90% R.H.	1.2	0.3–0.65	1.2	3.9	1.3–2.1
Oxygen transmission rate, cm³·mil/100 in²·day·atm @ 77°F and 0% R.H.	250–840	30–250	250–840	515–645	226–484
CO₂ permeability, cm³·mil/100 in²·day·atm @ 77°F and 0% R.H.	500–5000	250–645	500–5000	2260–2900	626–1150
Resistance to grease and oil	varies	good	good	varies	good
Dimensional change at high R.H., %	0	0	0	0	0
Haze, %	4–10	25–50	6–20	2–10	1–15
Light transmission, %	65	N/A	N/A	55–75	85
Heat-seal temperature range, °F	250–350	275–310	250–350	150–300	225–300
Service temperature range, °F	–70–180	–60–250	–60–180	–60–140	–150–150
Tensile modulus, 1% secant, kpsi	20–40	125	25	8–20	10–50

* Some PVC data depend on plasticizer content.
† Some EVOH data depend on ethylene content.
Source: Osborn, K. R. and W. A. Jenkins, *Plastic Films, Technology and Packaging Applications,* Technomic Publishing Co., Lancaster, PA, 1992, p. 233.

lize. These polymers produce highly ordered chains, since the difunctionality of the monomers forces the chain to grow in only one isomeric configuration. This is the case, for instance, with Nylon 6 and Nylon 6,6. When the difunctional monomers contain aromatic and cyclohexane rings, only polymers with substitutions in the 1 and 4 positions, like PET and PC, are crystallizable. A 1,3 linkage in the ring increases the randomness of the chain and hence minimizes crystallinity. Copolymers of step-reaction polymers combining 1,4 and 1,3 substitutions will produce amorphous polymers as in the case of Nylon 6I/6T and PCTA. When the step-reaction polymerization process includes monomers with three and four functional groups, the resulting polymer will form a tridimensional network. This is the case in adhesives such as epoxies and polyurethanes, which are both amorphous and thermoset materials.[10]

Amorphous polymeric materials and inorganic glasses do not show melting points. But amorphous materials do have a glass-transition temperature T_g defined as the freezing (on cooling) or thawing (on heating) of the micro-Brownian motion of chain segments 2–50 carbon atoms in length.[11] At low temperature, an amorphous polymer is glassy, hard, and brittle, but as the temperature increases, the polymer becomes rubbery, soft, and elastic. There is a smooth transition in the polymer's properties from a solid to a flow melt. At the glass-transition temperature, properties like specific volume, enthalpy, shear modulus, and permeability show significant changes. See Fig. 8.6.

Oriented PET	OPP	PVC*	PS	PVDC	Nylon	EVOH†	BON
1.4	0.905	1.21–1.37	1.05	1.6–1.7	1.14	1.12–1.19	1.14
20–22	30.6	20–22.5	26	16.2–16.8	23.5–24.5	23–24	23.5–24.5
25	25–30	2–16	5.0–8.0	8–20	7–18	1.2–1.7	25–30
70–100	60–100	5–500	2–3	40–100	250–500	220–280	70
25–30	5–15	12–20	N/A	10–15	4–6	N/A	N/A
13–80	4–6	N/A	N/A	10–20	20–50	N/A	N/A
1.3	0.3–0.4	2.8	5.0	.05–0.3	24–26	High	12
5	110	5–1500	100–200	0.08–1.7	2.6	0.01	2
N/A	240–285	50–13,500	N/A	0.04–10	4.7	N/A	N/A
good	good	good	good	good	good	good	good
0	0	0	0	0	1.3	0	0
4	3	1–2	0–1	2	2	N/A	1–2
88	80	90	90	80–88	N/A	N/A	N/A
275	200–300 (coated)	280–340	194–212	250–300	350–500	N/A	N/A
–100–400	–60–250	–20–200	–80–175	0–275	–75–400	N/A	–100–400
700	350	350–600	330–475	50–150	N/A	300–385	250–300

Woodward[12] describes seven common crystalline morphologies in polymers: faceted single lamellas, nonfaceted lamellas, dendritic structures, sheaflike lamellar ribbons, spherulite arrays, fibrous structures, and epitaxial lamellar overgrowths on microfibrils. Spherulites are complex, ordered aggregations of submicroscopic crystals. PE crystal spherulites, for example, are about 10 nm thick. The spherulites are separated from one another by small amorphous regions called micelles. The spherulites are larger than the wavelength of visible light, producing light-scattering, which makes the polymer opaque. Plastic materials with high crystallinity are opaque, while plastics with low degrees of crystallinity are transparent or clear. Amorphous materials are totally transparent.

Process conditions associated with the cooling rate of polymer melts, blown stretching, and film orientation enhance the polymer's natural anisotropy. Unlike low-molecular-weight compounds, polymers show anisotropism in both amorphous and crystalline phases because of the strong covalent bonds in the backbone chain and weak intermolecular forces. This anisotropic behavior increases upon molecular orientation. Unbalanced films, therefore, show different values of tensile and tear strength along machine direction than they do along cross-machine direction. In PVC and PET stretch-blown molded containers, molecular orientation plays an important role in the mechanical, barrier, and optical properties of the container. For a particular molecular weight and stretch ratio, chain orientation increases with strain rate and lower temperatures. Imbalances in orientation can be detected by

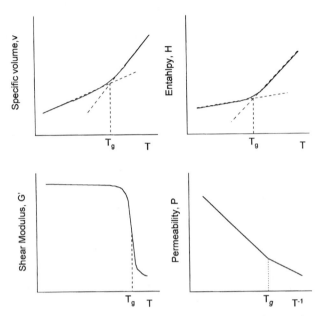

FIGURE 8.6   Variation in specific volume v, enthalpy H, shear modulus G', and permeability P, near the glass-transition temperature.

optical measurements such as birefringence.[13] Several important polymer properties depend on the morphology of the polymer. As crystallinity increases, density, permeability, opacity, strength, and heat-sealing temperatures increase, while blocking, clarity, tear strength, impact strength, toughness, ductility, ultimate elongation, and heat-sealing temperature ranges decrease.

Degree of Crystallinity.   Degree of crystallinity is the ratio of crystalline region to amorphous region in a polymer sample. It can be expressed as a volume or mass ratio. Degree of crystallinity is primarily determined by X-ray scattering. In practice, however, this is a tedious and expensive method. The crystallinity of a polymer sample can also be determined by the density gradient method, ASTM D 1505. In this method two solutions, A and B, are prepared with density in the range of interest. Solution A, with the lowest density, and solution B with the highest density, are combined in a glass tube to form a vertical column of liquid in which the density varies linearly from the bottom to the top (see Fig. 8.7). The column is calibrated with glass beads of known density. Plastic samples are dropped in the column, and the sample will settle at the corresponding density value. The density of the plastic is calculated from the position of the sample and calibrated beads using the calibration scale. Based on the two-phase model,[14] the degree of crystallinity of a polymer sample is related to its density by:

$$a_v = \frac{\rho - \rho_a}{\rho_c - \rho_a}$$

$$a_m = a_v \frac{\rho_c}{\rho}$$

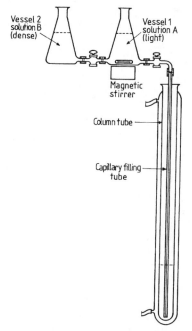

FIGURE 8.7 Density gradient column. (Reprinted with permission from R. P. Brown, *Handbook of Plastics Test Methods,* G. Godwin, Ltd., 1981.)

where a_v and a_m are the volume and mass crystallinity, respectively; and ρ, ρ_c, and ρ_a are the density of the sample, crystallinity, and amorphousness, respectively. Percent of crystallinity is a_v or a_m multiplied by 100.

Polymer Orientation. Molecules in polymer film or sheet are oriented into more orderly morphology in response to external stress above T_g. Orientation tends to increase crystallinity (although not always in the strict sense of the word; for instance, oriented PP is clearer than nonoriented PP, but the molecules are more compact) and decrease permeability. In an oriented film, a large fraction of the molecules tend to line up in the direction of stretch. The result of uniaxial molecular orientation is a substantial increase in strength and toughness in the direction of the stretch (but decrease in strength in the perpendicular direction). Polymers can be unoriented, uniaxially oriented, or biaxially oriented. The films that are made to fabricate bags are generally uniaxially oriented to improve tensile strength. Films used for pouches are biaxially oriented because it is expected the tensile forces may act in both directions, and, if shrinking occurs, the pouch should shrink in both directions. Biaxial orientation can be balanced (stretching the same amount in both directions) or unbalanced.

8.2.2 Thermophysical Properties

Relaxation Temperatures. Consider a completely amorphous polymer, or a polymer that does not crystallize, even when cooled from the melt at a very slow rate. One effect of cooling the melt is a decrease in the degree of thermal agitation of the molecular segments. As the cooling process continues, the rate of segmented movement becomes more and more sluggish until the segmental movement finally stops. At this point, the glassy state of the amorphous polymer is reached. The temperature at which this glassy state takes place is called the glass-transition temperature T_g. This process is associated with the limitation of segmental chain mobility as defined below. The glassy state consists, then, of "frozen" entangled chain molecules with a complete absence of stereoregularity and motion. This absence of stereoregularity is typical of liquids.

As seen in Fig. 8.6, the change in the specific volume as a function of temperature shows a smooth curve from the melt state to the glassy state. However, there is a change in the slope near T_g. In the glass region, the specific volume is much larger than the value corresponding to the crystal phase. This is as if additional or free volume has been trapped between the entangled and frozen polymeric chain. The reduction in specific volume as the temperature decreases below T_g is associated with molecular movement of the total molecules moving with respect to each other, and it is determined by intermolecular force. As the free volume increases, density decreases and properties like permeability increases.

The motions associated with T_g results from the semicooperative actions involving torsional oscillation and/or rotations around the backbone bonds in a given chain as well as in neighboring chains. Torsional motion of side groups around the axis connecting them to the main chain may also be involved.[11] T_g involves, then, both intrachain and interchain segmental motions. In addition to T_g, other relaxation temperatures like T_β, T_{ll}, and T_{lp} are characteristics of amorphous polymers.[11] Unlike T_g, T_β is associated mostly with intrachain subgroup motions (2–10 consecutive chain atoms). T_{ll} is a weak transition-relaxation temperature about 1.2 times greater than T_g that is associated with the thermal disruption of intermolecular segment-to-segment contacts know as segmental melting. T_{ll} marks the onset of the true liquid state of an amorphous polymer. At a temperature 30 to 50°K above T_{ll}, the molecule chains contain enough energy to break the rotational barrier that marks the true liquid nature of an amorphous polymer. This relaxation has been identified as T_{lp}. According to Boyer,[11] physical aging of polymers occurs at all temperatures between T_β and T_g (the higher the temperature, the faster the aging process). The effect of aging can be erased by heating the polymer above T_g. At a given temperature, toughness is associated with the presence of one or more relaxation temperatures below that particular temperature.

An amorphous polymer, whether pure or coexisting with crystalline regions, can be a glassy material (brittle) or a rubberlike substance (soft). The glass-transition temperature separates a quasi-liquid state (above T_g) behavior from a glassy behavior (below T_g). If an amorphous polymer is at a temperature below T_g, it will be brittle and show aspects of glassy materials. As the temperature of the sample increases above T_g, the polymer will exhibit leathery behavior, and there will be a loss in the elastic modulus and a decrease in its barrier properties. T_g marks the lower limit temperature for melt flow, resiliency, and polymer molecular orientation. Wetting, tackiness, and adhesion appear to be fully developed at T_{ll}.

For many polymers, it has been determined experimentally the ratio of the glass-transition temperature to the melting temperature: $T_g/T_m \approx 0.6$ (both temperatures in degrees Kelvin).[11] For example, for polypropylene, $T_g = -19°C$ and $T_m = 176°C$, so $T_g/T_m = 0.57$. T_g can be calculated by group-contribution methods.[15]

Melting Temperature. The melting temperature T_m is a true transition temperature. This means that, at T_m, both the liquid and solid phase have the same free energy. While most semicrystalline polymers show a melting range temperature, pure amorphous polymers do not have T_m at all. Similarly to T_g, T_m can be estimated from contribution groups. Plastics show T_m as low as 275°K for polyisobutylene, and as high as 728°K in polyethylene terephthalamide.[15, 16] ASTM methods D 2117 and D 3418 describe methods for measuring T_m.

Heat Capacity. Heat capacity, or specific heat, is the amount of energy needed to change a unit of mass of a material one degree in temperature. The heat capacity of plastics, which are obtained at constant pressure, are temperature-dependent, especially near the glass-transition temperature. The heat-capacity values of polymers at 25°C vary from 0.9 to 1.6 J/gK for amorphous polymers and from 0.96 to 2.3 J/gK for crystalline polymers. In a semicrystalline polymer, the heat capacity of the amorphous phase is larger than the heat capacity in the crystalline phase. This implies that the heat capacity values depend on the percent of the polymer's crystallinity. Reliable data regarding the heat capacity of amorphous and crystalline phases are available for only a limited number of polymers.[15,17] Heat-capacity values for polymers can be found in the review by Wunderlich.[18] Usual techniques for measuring specific heat are differential thermal analysis (DTA) and differential scanning calorimetry (DSC).

Heat of Fusion.   The heat of fusion ΔH_m is the energy involved during the formation and melting of crystalline regions. For semicrystalline polymers, the energy of fusion is proportional to the percent of crystallinity. Amorphous polymers or amorphous polymer regions do not have heat of fusion, since amorphous structures show a smooth transition from the viscous liquid amorphous state to the true liquid state. Experimental values of crystalline heat of fusion for common packaging plastics vary from 8.2 kJ/mol for polyethylene to 43 kJ/mol for Nylon 6,6.[15] ASTM D 3417 describes a method for measuring the heat of fusion and crystallization of a polymer by differential scanning calorimetry (DSC).

Thermal Conductivity.   Thermal conductivity is the parameter in Fourier's law that relates the flow of heat to the temperature gradient. In specific terms, thermal conductivity k is a measure of a material's ability to conduct heat. The thermal conductivity of a polymer is the amount of heat conducted through a unit of thickness per unit of area, time, and degree of temperature. Thermal conductivity values control the heat-transfer process in applications such as container-forming, heat-sealing, cooling and heating of packaging, and sterilization processes. Plastics have values of k much lower than metals and are therefore poor heat conductors. Thermal conductivity for plastics ranges from 3×10^{-4} cal/s cm°C for PP to 12×10^{-4} cal/s cm°C for HDPE. For aluminum, k is 0.3 cal/s cm°C, and for steel it is 0.08 cal/s cm°C.[19] For plastic foams, values of k are much lower than those of unfoamed plastic. This is due to the presence of air trapped in the cellular structure of the plastic foam. Plastics with low thermal conductivity values do not conduct heat well. Enhanced by low thermal conductivity of air, foams are excellent insulating materials. In addition foams are very attractive cushioning materials. Plastic fillers may increase the thermal conductivity of plastics. Methods for measuring k are given in ASTM D 4351, C 518, and C 177.

Thermal Expansion Coefficient.   The coefficient of linear (or volume) thermal expansion is the change of length (volume) per unit of length (volume) per degree of temperature change at constant pressure: $\beta = (1/L)(dL/dT)p$ and $\alpha = (1/V)(dV/dT)p$. Units of α and β are $°K^{-1}$ or $°F^{-1}$. Compared to other materials, polymers have high values of thermal expansion coefficients. While metals and glass have values in the range 0.9 to 2.2 $°K^{-1}$, polymers[19] range from 5.0 to 12.4 $°K^{-1}$. Thermal expansion coefficients can be measured by thermomechanical analysis (TMA). ASTM D 696 describes a method using a quartz dilatometer, while ASTM E 831 describes the determination of the linear thermal expansion of solid materials. Volume contraction of a container from the molding operation temperature down to room temperature is called shrinkage, and its measurement is described in the ASTM D 955, D 702, and D 1299.

8.2.3 Mechanical Properties

Bursting Strength.   Bursting strength is the hydrostatic pressure given in psi (or pascal) required to rupture a flat material (film or sheet) when the pressure is applied at a controlled increasing rate through a circular rubber diaphragm that is 30.48 mm (1.2 in) in diameter. *Points bursting strength* is the pressure expressed in psi. ASTM method D 774 describes the measurement of the bursting strength of plastic films.

Dimensional Stability. Dimensional stability refers to how well a structure maintains its dimensions under changing temperature and humidity conditions. ASTM D 1204 describes a standard method for linear dimensional changes of flexible thermoplastic films and sheets at elevated temperatures. Dimensional stability is a property relevant in any flexible-material converting process. During printing, for example, even a small change in dimensions may lead to serious problems in holding a print pattern. In a flexible structure, dimensional stability may produce different changes in the machine and transverse directions.

Folding Endurance. This is a measure of the material's resistance to flexure or creasing. Folding endurance is greatly influenced by the polymer's glass-transition temperature and the presence of plasticizers. The ASTM D 2176 describes the procedure to determine the number of folds necessary to break the sample film.

Impact Strength. Impact strength is the material's resistance to breakage under a high-velocity impact. Widely used impact tests are: for rigid materials, Izod (D 256A) and Charpy (D 256); and for flexible structures, dart-drop impact (ASTM D 4272) and pendulum impact resistance (ASTM D 3420). A free-falling-dart method for polyethylene films is described in ASTM D 1709. Unlike low-speed uniaxial tensile tests, the pendulum impact test measures the resistance of film to impact puncture, simulating high-speed end-use applications. Dart drop measures the energy lost by a moderate-velocity blunt impact passing through the film. Both pendulum and impact tests measure the toughness of a flexible structure.

Pinhole Flex Test. Pinhole flex resistance is the property of a plastic film to resist the formation of pinholes during repeated folding. A related test is the folding endurance test. Films having a low value of pinhole flex resistance will easily generate pinholes at the folding line under repeated flexing. The test is described by the standard ASTM F 456.

Tensile Properties. The mechanical behavior of a polymer can be evaluated by its stress-strain tensile characteristics (see Fig. 8.8). The stress is measured in force per area and expressed in unit of pressure. The strain is the dimensionless fractional length increase.

Modulus of elasticity E is the elastic ratio between the stress applied and the strain produced, giving the material's resistance to elastic deformation. The tensile modulus gives also a measure of the material's stiffness: the larger the modulus, the more brittle the material. For example, E of LDPE is 250 mPa, while for crystal PS, it is 2500 mPa. Comparatively, values of tensile modulus in polymers (1.9×10^3 MPa for nylon) are much lower than for glass (55×10^3 MPa) or mild steel (210×10^3 MPa).[16,20]

Elastic elongation is the maximum strain under elastic behavior. Ultimate strength or tensile strength is the maximum tensile stress the material can sustain. Ultimate elongation is the strain at which the sample ruptures.

Toughness is the energy a film can absorb before rupturing, and it is measured by the area under the stress-strain curve. Brittleness is the lack of toughness.[21] Amor-

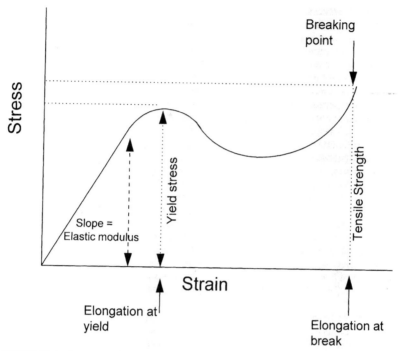

FIGURE 8.8 Stress-strain curve.

phous and semicrystalline polymers become brittle when cooled below their glass-transition temperature. Tests for tensile properties are described in the ASTM D 882, for flexural strength in ASTM D 790, and for flexural modulus in ASTM D 790M.

Tear Strength. The measurement of tear strength evaluates the energy absorbed by a film and sheeting sample during the tear initiation and propagation. Two standard methods are available: ASTM methods D 1004 and D 1922 (see Fig. 8.9). The former is designed to measure the force to initiate tearing. It covers the determination of the tear resistance of a flexible plastic sample at a very low rate of loading, 51 mm/min. ASTM D 1922 refers to propagation of a tear after being initiated by a small hole in the sample. The value of tear strength in a film depends on the orientation stretching ratio and whether the measurement is performed along or across the machine direction.

Abrasion Resistance. Abrasion is a difficult property to define as well as to measure. It is normally accepted that abrasion depends on the polymer's hardness and resilience, frictional forces, load, and actual area of contact. ASTM method D 1044 evaluates the resistance of transparent plastics to one kind of surface abrasion by measuring its effects on the transmission of light. Another test method to evaluate abrasion, ASTM 1242, measures the volume lost by two different types of abrasion machines: loose abrasion and bonded abrasion.

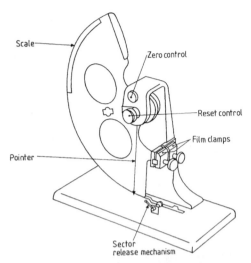

FIGURE 8.9   Tear-testing apparatus of the Elmendorf type. (Reprinted with permission from R. P. Brown, *Handbook of Plastics Methods,* G. Godwin Ltd., 1981.)

8.2.4   Barrier Properties

Barrier properties of materials indicate their resistance to diffusion and sorption of molecules. A good barrier polymer has low values of both diffusion (D) and solubility (S) coefficients. Since the permeability coefficient P is a derived function of D and S, a high-barrier polymer has a low value of P. The diffusion coefficient is a measure of how fast a penetrant will move within the polymer, while the solubility coefficient gives the amount of the penetrant taken (or sorbed) by the polymer from a contacting phase per unit of sorbate concentration. Both diffusion and solubility can be applied to the reverse process of sorption, that is, the migration of compounds from the polymer to a surrounding media. Several factors influence the effective value of diffusion and solubility coefficients in polymers: (1) chemical compositions of the polymer and permeant; (2) polymer morphology (diffusion and sorption occur mainly through the amorphous phase and not through crystals); (3) temperature (as temperature increases, diffusion increases while solubility decreases); (4) glass-transition temperature; and (5) the presence of plasticizers and fillers. A detailed discussion of barrier properties is presented in Sec. 8.3 of this chapter.

Diffusion Coefficient.   In Fick's law, the diffusion coefficient D is a parameter that relates the flux of a penetrant in a medium to its concentration gradient. A diffusion coefficient value is always given for a particular molecule-polymer pair. For solid polymers, the diffusion coefficient values of a large number of low-molecular-mass substances may range from 1×10^{-8} to 1×10^{-14} cm^2/s. The diffusion theory states that diffusion is an activated phenomenon that follows Arrhenius' law. In addition to temperature, penetrant concentration and plastizers also affect the value of the diffusion coefficient. Methods for the determination of D are discussed in Sec. 8.3.

Permeability Coefficient. The permeability coefficient P combines the effect of the diffusion and solubility coefficients. The barrier characteristic of a polymer is commonly associated with its permeability coefficient value. The well known relationship $P = DS$ holds when D is concentration-independent and S follows Henry's law. Standard methods for measuring the permeability of organic compounds are not yet available. ASTM E96 describes a method for measuring the water vapor transmission rate. ASTM D1434 covers a method for the determination of oxygen permeability.

Solubility Coefficient. The solubility coefficient is the amount of sorbate per unit of mass per unit of pressure. The simplest solubility coefficient is defined by Henry's law of solubility, which is valid at low concentration values. The solubility of CO_2 in PET at high pressure is described by combining Henry's and Langmuir's laws.

8.2.5 Surface and Adhesion

Adhesive Bond Strength. Adhesive bond strength between an adhesive and a solid substrate is a complex phenomenon. It is controlled (at least in part) by the values of surface tension, solubility parameters, and adhesive viscosity. To obtain good wettability and adhesion between a polymeric substrate and an adhesive, the surface tension of the adhesive must be lower than that of the substrate. Usually, the difference between the two values must be at least 10 dynes/cm. The similarity in solubility parameters between the two phases indicates the similarity of the intermolecular forces between the two phases. For good compatibility, the values of the solubility parameters must be very close. Low viscosity in the adhesive is necessary for good spreadability of adhesive and wettability of the substrate.[22]

Cohesive Bond Strength. Cohesive bond strength is the force within the adhesive itself when bonding two substrates. The cohesive bond strength depends mainly on the intermolecular forces of the adhesive, molecular mass, and temperature.

Blocking. Blocking is the tendency of a polymer film to stick to itself upon physical contact. This effect is controlled by the adhesion characteristic of the polymer. Blocking is enhanced by surface smoothness and by pressure on the films present in stacked sheets or compacted rolls. Blocking can be measured by the perpendicular force needed to separate two sheets, and it can be minimized by incorporating additives such as talc in the polymer film. ASTM D 1893, D 3354, and Packaging Institute Procedure T 3629 present methods to evaluate blocking.

Friction. The coefficient of friction (COF) is a measure of the friction forces between two surfaces. It characterizes a film's frictional behavior. The COF of a surface is determined by the surface adhesivity (surface tension and crystallinity), additives (slip, pigment, and antiblock agents), and surface finish. Cases in which the material's COF values require careful consideration include: film passing over free-running rolls, bag forming, the wrapping of film around a product, and the stacking of bags and other containers. In addition to the intrinsic variables affecting a material's COF, environmental factors such as machine speed, temperature, electrostatic buildup, and humidity also have considerable influence on its final value. The static COF is associated with the force needed to start moving an object. It is usually higher than the kinetic COF, which is the force needed to sustain movement. Determination of static COF is described in TAPPI standard T 503 and ASTM D 1894. Thompson has studied the effect of additives on the COF values of polypropylene.[23]

Heat Sealing.   An important property for wrapping, bag making, or sealing a flexible structure is the heat sealability of the material. At a given thickness, heat-sealing characteristics of flexible web material are determined by the material's composition (which controls strength), average molecular mass (controlling sealing temperature and strength), molecular mass distribution (setting temperature range and molecular entanglement), and the thermal conductivity (controlling dwell time).[24] Tests normally conducted to evaluate the heat sealability of a polymeric material are the cold peel strength (ASTM F 88) and the hot tack strength.[25] Hot tack is the melt strength of a heat seal without mechanical support. The hot adhesivity is associated with the molecular entanglement of the polymer chains, viscosity, and intermolecular forces of the material.

Surface Tension.   In both solids and liquids, the forces associated with inside molecules are balanced because each molecule is surrounded by like molecules. On the other hand, molecules at the surface are not completely surrounded by the same type of molecules, generating, therefore, unbalanced forces. At the surface these molecules show additional free surface energy. The intensity of the free energy is proportional to the intermolecular forces of the material. The free surface energy of liquids and solids is called surface tension. It can be expressed in mJ/m^2 or dyne/cm. Values of surface tension in polymers range from 20 dyne/cm for Teflon to 46 dyne/cm for Nylon 6,6. The measurement of surface tension by contact angle measurement is covered by ASTM D2578. Several independent methods are available to estimate the surface tension of liquids and solid polymers including the parachor.

When two condensed phases are in close contact, the free energy at the interface is called interfacial energy. Interfacial energy and surface energy in polymeric materials control adhesion, wetting, printing, surface treatment, and fogging.

Wettability.   Adhesion and printing operations on a plastic surface depend on the value of the substrate wettability or surface tension. A measure of the wettability of a surface is given by a material's surface tension as described in ASTM D 2578.

8.2.6   Optical Appearance

Among the most important optical properties of polymers are: absorption, reflection, scattering, and refraction. Absorption of light takes place at the molecular level, when the electromagnetic energy is absorbed by group of atoms. If visible light is absorbed, a color will appear; however, most polymers show no specific absorption with visible light and are, therefore, colorless. Reflection is the light that is remitted on the surface. It depends on the refractive indices of air and the polymer. Scattering of light is caused by optical inhomogeneities reflecting the light in all directions. Refraction is the change in direction of light due to the difference between the polymer and air refraction indices. Optical appearance properties are of two types: optical morphological properties, which correlate with transparency and opacity; and optical surface properties, which produce the specular reflectance and attenuated reflectance.[15, 26] Transparency, opacity, and gloss of a polymer are mainly determined by the polymer morphology.

Gloss.   Gloss is the percentage of incident light that is reflected at an angle equal to the angle of the incident rays (usually 45°). It is a measure of the ability of a surface to reflect the incident light. If the specular reflectance is near zero, the surface is said to be matte. A surface with high reflectance has a high gloss, which produces a sharp image of any light source and gives a pleasing sparkle. Surface roughness, irregularities, and scratches all decrease gloss. Test method ASTM D 2457 describes the determination of gloss.

Haze.   Haze is the percentage of transmitted light that, in passing through the sample, deviates by more than 2.5° from an incident parallel beam. The appearance of haze is caused by light being scattered by surface imperfections and nonhomogeneity. The measurement of haze is described in ASTM D 1003.

Transparency and Opacity.   Transmittance is the percent of incident light that passes through the sample. It is determined by the intensity of the absorption and scattering effects. The absorption in polymers is insignificant, so, if the scattering is zero, the sample will be transparent. An opaque material has low transmittance and, therefore, large scattering power. The scattering power of a polymer results from morphological inhomogeneities and/or the presence of crystals. An amorphous homogeneous polymer such as "crystal" polystyrene will have little or no scattering power and, therefore, will be transparent. A highly crystalline polymer such as HDPE will be mostly opaque. Transmittance can be determined according to standard ASTM D 1003. A transparent material has a transmittance value above 90 percent.

8.2.7   Electrical Properties

Electrical conductivity, dielectric constant, dissipation factor, and triboelectric behavior are electrical properties of polymers subject to low electric field strength. Materials can be classified as a function of their conductivity (κ) in $(\Omega/cm)^{-1}$ as follows: conductors, $0-10^{-5}$; dissipatives, $10^{-5}-10^{-12}$; and insulators, 10^{-12} or lower. Plastics are normally nonconductive materials. The relative dielectric constant of insulating materials (ϵ) is the ratio of the capacities of the parallel plate condenser with and without the material between the plates. A correlation between the dielectric constant and the solubility parameter (δ) is given by the relationship $\delta \approx 7.0\epsilon$.[15] There is also a relation between resistivity R (inverse of conductivity) and the dielectric constant at 298°K: $\log R = 23 - 2\epsilon$.[15] Values of ϵ for polymers are presented in Refs. 15 and 17. When two polymers are rubbed against each other, one becomes positively charged and the other negatively charged. Whether a polymer becomes positive or negative depends on the electron donor-acceptor characteristic of the polymer. A triboelectric series is a listing of polymers according to their charge intensity. The polymers are ordered from more negatively charged polymers (electron acceptors) to neutral polymers to more positively charged polymers (electron donors). The charge of polymer films also takes place by friction during industrial operations such as form-fill-seal. A brief triboelectric series is presented in Table 8.6, where polymers are listed as a function of the polymer's dielectric constant. Hydrophilic polymers that absorb water become more conductive because their dielectric constant increases. Standard methods for measuring the triboelectric charge of films and foams are, at the time of this writing, under consideration by the ASTM Committee D10.[27]

TABLE 8.6   Brief Triboelectric Series

Polymer	Dielectric constant
Negative:	
Polypropylene	2.2
Polyethylene	2.3
Polystyrene	2.6
Neutral:	
Polyvinylchloride	2.8
PVDC	2.9
Polyacrylonitrile	3.1
Positive:	
Cellulose	3.7
Nylon 6,6	4.0

Source: Van Kevelen, D. W., *Properties of Polymers,* 3d ed., Elsevier, New York, 1990.

8.3   MASS TRANSFER IN POLYMERIC PACKAGING SYSTEMS

8.3.1   Packaging Interactions

An ideal packaging material, in addition to containing and enclosing a product, provides an inert separating phase between the product and the environment. With an inert material, there should be no molecular exchange of oxygen, carbon dioxide, water, ions, product ingredients, and packaging-material components between the product and the package. In other words, a truly inert packaging material will show no interaction with the packaged product. Such a packaging material may not be available, at least at ambient temperature, or may not be of reasonable cost. Although inorganic glasses approach this ideal barrier concept, helium and hydrogen diffuse through glass. In addition to that, inorganic glass components like sodium can leach out into an aqueous solution. On the other hand, metals are good barrier materials, but metallic ions tend to dissolve and/or react with both the product and the environment through corrosion.

Due to their chemical nature, polymeric materials have the tendency to dissolve molecules of gases, vapors, and other low-molecular-weight substances in a higher degree than inorganic glasses. Dissolved penetrant molecules diffuse through the polymer via an activated process of random walks promoted by Brownian motion of the polymer chains. The diffusion process is driven by thermodynamic forces tending to equilibrate the penetrant's chemical potential, or, more commonly, concentration. A diffusant molecule may transfer between adjacent phases because its tendency to equilibrate its chemical potential. When a polymeric phase is in direct contact with a solid, liquid, or gas phase, a penetrant can enter of leave the polymer. Plastic materials, then, are not inert packaging materials. When they are in direct contact with a packaged product, it is certain that an exchange of substances between product and package will occur. We refer to this phenomenon as an interaction between the polymer, product, and environment. The degree of interaction depends on the type of plastic material. For a particular diffusant molecule, both the diffusivity and solubility of a particular penetrant in the polymer will vary depending on the composition of the polymer and contacting phase. See Fig. 8.10. As indicated in Sec. 8.2.4, both the diffusion coefficient and solubility coefficient are the two fundamental parameters controlling the molecular interactions in packaging systems.

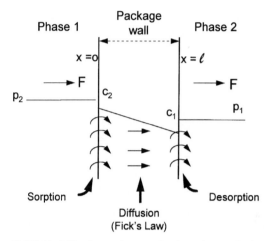

FIGURE 8.10 Interaction mechanisms in a plastic package.

8.3.2 Type of Interactions

The mass transfer of low-molecular-weight compounds has been of interest to scientists and engineers since 1829, when Thomas Graham performed the first diffusion experiments with natural membranes. In 1855, Adolf Fick derived the fundamental equations of diffusion. Mass-transfer processes (together with physical and mechanical properties) provide the rationale for packaging design and modeling a product's shelf life and quality. Molecular interactions in packaging systems start from the moment the package contacts the product during production, and extend throughout the entire package's shelf life. For easy reference, these interactions can be classed as permeation, migration, and sorption.

Permeation. Permeation is the diffusional molecular exchange of gases, vapors, or liquids (called permeants) across a plastic packaging material with the exclusion of perforations and cracks. A permeation process may significantly impact a product's shelf life, since the product may gain or lose components and/or develop unwanted chemical reactions. The loss of water or CO_2 by liquid products, the uptake of moisture by dry products, or the oxidation of oxygen-sensitive products affects the product's composition and therefore affects shelf life. Other consequences of permeation also include the transfer of airborne contaminants or volatile components from secondary packages (such as corrugated boxes) into the product. These permeants may produce off-flavors in certain flavor-sensitive foods.

While the permeation process may have a negative impact in product quality in many cases, there are instances in which a permeation process is not only beneficial but also necessary. Such is the case with modified atmospheric packaging where selected rates of O_2 and CO_2 need to permeate through the package.[28]

Migration.   Migration is the transfer of substances originally present in the plastic material into a packaged food. These particular substances are called migrants. Examples of migrants include residual monomers, solvents, remaining catalysts, and polymer additives. Migration may affect the product's sensorial quality as well as its toxicological characteristics, since it incorporates undesirable components from the package into the food. The presence of potential migrant substances in the packaging material is the subject of legal control by the Food and Drug Administration. Vinyl chloride in PVC, styrene in PS, and acrylonitrile in PAN are well known examples of residual monomer migrants. These cases were an object of intense investigations in the 1970s. Resins are commercially produced with residual monomer contents below those allowed by the FDA.

However, even if virgin resins contain a very low level of monomers and additives, migrating substances that may potentially impart off-flavors to products can be developed during converting operations of packaging materials. Excessive temperature-time processing conditions of resins may oxidize and degrade the resin. This may occur during extrusion coating, blown film production, injection molding, blow molding, or even heat-seal operations. Adhesives, inks, pigments, printing solvents, and printing pretreatment ingredients are potential sources of packaging contamination during lamination operations. Residual solvents like toluene, hexane, or pentanol from packages and coupons can produce unwanted odor and taste in food products. The presence of recycled and regrinded resins are also potential sources of migrants in the finished package if not sufficiently cleaned. Other sources of migrant substances from plastic containers may arise when contaminated equipment and lines are used to process the resin. Similarly, impure air used to blow films can result in a contaminated package.

The sensorial threshold limit may vary widely depending on the substance and product. It is not unusual that amounts as little as a few parts per billion may be sufficient to reject a product. The extend of a migration process depends on the initial concentration of the migrant substance in the plastic container and the migrant partition coefficient between the plastic and the food.

Sorption.   Sorption, also called scalping, is the uptake of food components, such as flavor, aroma, or colorant compounds (called sorbates) by the package material. Many sorption studies of food components have been reported.[29–32, 36] Similarity in chemical structure between the sorbate and polymers enhances sorption.[15] As the molecular weight distribution (MWD) of the polymer increases, sorption also increases. Metallocenes, or in-site polyolefins, which are characterized by narrow MWD, are intended to better reduce the sorption of volatile compounds compared to LDPE. Sorption studies of benzaldehyde, citral, and ethyl butyrate showed that sorption levels by ionomers are similar to the sorption of LDPE.[33] As in the case of migration, the extent to which the sorption process occurs depends on the initial concentration of the sorbate in the food and the migrant partition coefficient between the plastic and the food. The dynamic of the diffusion process for permeation, migration, and sorption can be calculated from the corresponding solutions to Fick's second law.

8.3.3 Diffusion and Sorption Equations

The fundamental driving force that prompts a molecule to diffuse within the polymer or between a polymer and a surrounding phase is, according to the solution theory, the tendency to equilibrate the species' activity. In packaging, polymers are generally contacted by gas (air) and/or liquid phases. In the case of multilayer structures where several layers of polymers are in direct contact, a layer is contacted by at least one adjacent-layer solid phase. Therefore, any mobile molecular species that is not in thermodynamic equilibrium within the phase will tend to equilibrate its activity value as indicated by Fig. 8.10. In a packaging system, mass-transfer sorption and migration processes involves the diffusion of substances between two adjacent phases: polymer phase and a surrounding liquid or gas. The diffusant substance must diffuse within each phase and move across the interphase. In a permeation process, the permeant needs to move across the two interphases of the package's wall. The maximum concentration of a substance retained by the polymer, from a contacting gas or liquid phase, is the solubility. Solubility is controlled by the equilibrium thermodynamics of the system. From solution theory the activity of a species a can be represented by the activity coefficient ξ and concentration c:

$$a = \xi c \tag{8.1}$$

In most packaging situations, ξ is approximately 1, and concentration replaces activity.

Fick's laws of diffusion quantitatively describe permeation, migration, and sorption processes in packaging systems.[34] In isotropic phases, the diffusion theory states that the rate of transfer of a diffusing substance per unit of area F is given by Fick's first law:

$$F = -D \frac{\partial c}{\partial x} \tag{8.2}$$

where c is the penetrant concentration in the polymer, x is the direction of the diffusion, and D is the molecular diffusion coefficient. Quantity in F and c is expressed in the same quantity unit: mass or volume of gas at standard temperature and pressure. The fundamental equation for unsteady-state, one-dimensional diffusion in an isotropic phase is Fick's second law:

$$\frac{\partial c}{\partial t} = D \frac{\partial^2 c}{\partial x^2} \tag{8.3}$$

where t is time. In systems where the diffusant concentration is relatively low, the diffusion coefficient in Eqs. 8.2 and 8.3 is assumed to be independent of both penetrant concentration and polymer relaxations. Diffusion processes in packaging generally involve low values of diffusant concentration. Also, the diffusion is perpendicular to the flat surface of the package with a negligible amount diffusing through the edges. The concentration of penetrants in polymers c (especially polymers above their glass-transition temperature and penetrants at low pressure) is, in many cases, well described by the linear isotherm, Henry's law of solubility:

$$c = kp \tag{8.4}$$

where p is the partial pressure of penetrant and k is Henry's law constant, which is the solubility coefficient and commonly represented by S. The solubility coefficient S is constant with the pressure p for gases such as O_2 and CO_2 up to one atmosphere.

For glassy polymers and high-pressure penetrants like in the cases of CO_2 in PET, a nonlinear Langmuir–Henry's law model is followed:

$$c = kp + \frac{C'_H bp}{1 + bp} \tag{8.5}$$

where C_H' is the Langmuir capacity constant and b is the Langmuir affinity constant.[35]

In closed systems and when there is a partition process of the penetrant between two phases, the partition coefficient K at equilibrium is defined as:

$$K = \frac{c_f^*}{c_p^*} \tag{8.6}$$

where c_f^* and c_p^* are the sorbate or migrant equilibrium concentration in the packaged product and polymer, respectively.

For organic vapors, Henry's law is valid, in most cases only at very low pressure and penetrant concentrations in the order of milligrams of penetrant per liter of air (ppm). However, the actual applicability range of Henry's law depends on the particular organic penetrant/polymer under consideration. The Flory-Huggins equation applies to high solvent activity in polymers.[36]

8.3.4 Diffusion Across a Single Sheet

Steady-State Diffusion: Permeability.   Consider the case of a plane sheet of thickness l that is contacted from both sides with a penetrant at different concentration values as indicated in Fig. 8.10. At the surface $x = 0$, the penetrant concentration $c = c_2$; and at $x = l$, $c = c_1$. Applying these condition to Eq. 8.2, the penetrant flow rate F across any section of the sheet is given by:

$$F = -D\frac{dc}{dx} = D\frac{c_2 - c_1}{l} \tag{8.7}$$

In permeability experiments, however, the partial pressure in the gas phase surrounding the sheet is easier to measure than the penetrant concentration c in the polymer. Substituting p for c and P for D, Eq. 8.7 can be written now as

$$F = P\frac{p_1 - p_2}{l} \tag{8.8}$$

where P is the permeability coefficient. Notice that Eq. 8.8 is actually the definition of permeability. A material having a P value for a particular permeant is considered to be a good barrier, only if a small quantity of permeant is transferred. Conversely, a high permeability value indicates a low barrier material. The diffusant flow rate F is given by

$$F = \frac{q}{At} \tag{8.9}$$

where q is the quantity of permeant flowing across the sheet in time t, and A is the sheet area exposed to the permeant. The quantity q can be expressed in mass, volume, or moles.

By combining Eqs. 8.5 and 8.6 we obtain

$$P = \frac{ql}{At\Delta p} \tag{8.10}$$

where Δp is $p_2 - p_1$. Eq. 8.10 is a simple but very useful design equation for packages at steady state.

Water vapor transmission rate $WVTR$ relates to P by

$$P = \left(\frac{q}{At}\right)\frac{l}{\Delta p} = WVTR \frac{l}{\Delta p} \qquad (8.11)$$

EXAMPLE: *The WVTR of a film 25 microns thick measured according to ASTM (100°F and 90 percent RH) is 0.1 g/day m², calculate P.*

SOLUTION: *The saturation vapor pressure of water at 100°F is 49.7 mmHg, since $p_2 = 0$, $\Delta p = p_1 - p_2 = 49.7 \times 0.9 = 44.73$ mmHg. $P = WVTR (l/\Delta p) = 0.1 \times 25/44.73 = 0.056$ g·μm/m²·d·mmHg; $P = 4860 \times 10^{-18}$ kg·m/m²·s·Pa = 4860 attosecond (as).*

The permeability coefficient P is a derived parameter based on two more fundamental parameters: the diffusion and solubility coefficients. When the polymer diffusion coefficient D is independent of the permeant concentration, and the penetrant dissolves in the polymer according to Henry's law of solubility, Eqs. 8.7 and 8.8 yield the well known relationship:

$$P = DS \qquad (8.12)$$

The permeability of a polymer, therefore, depends on the diffusion and solubility coefficients, which can be related by a simple expression like Eq. 8.12 or alternatively by a more complex one. For instance, the solubility of CO_2 in PET follows the Langmuir–Henry's law model, Eq. 8.5, and P is given as[37]

$$P = kD_D + \frac{C_H'bD_H}{1 + bp} \qquad (8.13)$$

where D_D and D_H are the diffusion coefficients for the Henry's and Langmuir's laws' populations, respectively.

Figure 8.10 illustrates the mechanism of the permeation process, which involves three steps: (1) the permeant is dissolved in the polymer interphase; (2) the permeant diffuses within the polymer film from the side of high concentration toward the low-concentration side; and (3) the permeant diffuses out from the opposite polymer interphase. These steps are always present in any system regardless of whether D and S follow Fick's and Henry's laws, respectively.

A good barrier material has a low value of combined diffusion-coefficient and solubility-coefficient values. Preferably, both D and S should be low. For instance, polyethylene is an excellent barrier to water because water has very low solubility- and diffusion-coefficient values.

The plastic industry offers a large number of polymeric structures to cover a wide range of barrier characteristics to satisfy the diverse needs within the packaging industry. There are high-barrier materials available to protect a product from oxygen, water, or organic vapors, as well as structures with high permeability values for oxygen or carbon dioxide that are needed, for example, in modified atmosphere packaging of produce.

Variables Affecting Permeability

Structure-Permeability Relationships. The chemical structure of the polymer's constitutional unit ultimately determines the polymer barrier behavior. Chemical composition, polarity, stiffness of the polymer chain, bulkiness of side- and backbone-chain groups, and degree of crystallinity, significantly impact the sorption and diffusion of penetrants. Salame[38] has developed a comprehensive semiempirical correlation of polymer structure and gas permeability based on the permachor parameter, provided the permeant does not swell the polymer. The permachor concept is based on the molecular forces holding the polymer together, cohesive energy density, and free-volume fraction of the polymer.

The permeability P is related to permachor π by

$$P = Ae^{-s\pi} \tag{8.14}$$

where A and s are constants based on the permeant gas. Table 8.7 presents values of A, s, and π for selected gases and polymers. From Eq. 8.14 and Table 8.6, the permeability ratio for CO_2 and O_2 for any polymer at 25°C is calculated as

$$\frac{P_{CO_2}}{P_{O_2}} = \frac{[Ae^{-s}]_{CO_2}}{[Ae^{-s}]_{O_2}} \approx 6.2 \tag{8.15}$$

which is very close to experimental values. Figure 8.11 shows a plot of gas permeability and permachor values. Unfortunately, the permachor concept has not been yet extended to organic permeants.

Effect of Temperature. Considering that the diffusion, sorption, and permeability are activated processes, D, S, and P are related to the temperature by an Arrhenius-type equation:

$$\Gamma = \Gamma_o e^{-(E_\Gamma/RT)} \tag{8.16}$$

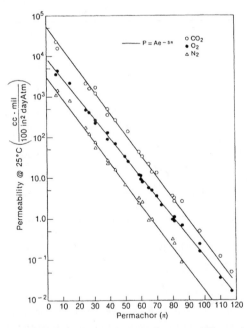

FIGURE 8.11 Correlation of gas permeability and permachor. (Reprinted with permission from M. Salame, *Polym. Eng. Sci.* 26:22, 1543, 1986.)

where Γ represents either D, S, or P; Γ_o is the respective pre-exponential term; E_Γ is the respective activation energy (sorption enthalpy for S), R is the gas constant, and T is temperature in degrees Kelvin. Equation 8.16 is valid over a relatively small range of temperatures, which should not include the polymer's glass-transition temperature. In particular for P, polymers show values of E_p larger at temperatures above T_g than below T. Equation 8.16 can be used to estimate the permeability coefficient at a desired temperature from a known value.

EXAMPLE: *Calculate the permeability coefficient of oxygen through PET at 50°C (P_2) considering that the permeability coefficient at 25°C (P_1) is 220 cc·mil/m²·d·atm.*

SOLUTION: P_2 *is related to* P_1 *by the following equation,*

$$P_2 = P_1 \, e^{(E_p/R)[(1/T_1) - (1/T_2)]} \tag{8.17}$$

which allows calculation of the permeability value at a temperature T_2 *if the value of* P *is given at* T_1 *and* E_p *is known.* E_p = 43.8 *KJ/mol* = 10,500 *cal/g mol[17]; exp[(10,500/1.987)] (298.2⁻¹ − 323.2⁻¹) = 3.94;* $P_2 = P_1 \times 3.94 = 220 \times 3.94 = 866$ *cc·mil/m²·d·atm.*

Humidity.   Hydrophilic polymers such as polyamides and EVOH absorb water from humid air. For example, water-sorption isotherms for nylon have been determined by Hernandez and Gavara.[39] The presence of water in a hydrophilic polymer affects the permeability of oxygen, carbon dioxide, organic vapor, flavor, and aroma compounds. An increase in moisture content increases oxygen permeability in EVOH and Nylon 6 (Fig. 8.5), while, for amorphous nylon, permeability tends to decrease (Fig. 8.12).

Morphology.   In the development of diffusion, sorption, and permeation equations (Eqs. 8.2 to 8.13), it is assumed that the polymer phase is homogeneous and isotropic; that is, an amorphous polymer. The presence of a crystalline microphase

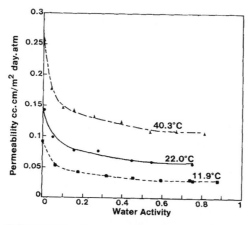

FIGURE 8.12   Oxygen permeability in Nylon 6I/6T. (Reprinted with permission from R.J. Hernandez, *J. Food Eng.* 22, 495, 1994.)

complicates this assumption considerably and makes the diffusion process in semicrystalline polymers more complex. Semicrystalline polymers consist of a microcrystalline phase dispersed in an amorphous phase. The dispersed crystalline phase decreases the sorption of penetrants whenever the crystal conformations produce regions of higher density than the amorphous polymer. A closer atomic packing tends to exclude relatively large molecules such as CO_2 or O_2. For this reason it is generally accepted that gases and vapors are normally sorbed, and are therefore able to diffuse, only in the polymer's rubbery or amorphous phase. The dispersed microcrystals are impermeable to penetrant diffusion and create a more tortuous path for the diffusing molecule. Additionally, the microcrystalline phase also acts as a tridimensional crosslinking agent increasing the nonisotropism of the polymer. The combined decrease in sorption and diffusion contributes, then, to a lower permeability.

The solubility S of a semicrystalline polymer S, having a crystalline volume fraction of α and an amorphous phase solubility S_a is given by

$$S = (1 - \alpha)S_a \tag{8.18}$$

One of the earliest evidences of Eq. 8.18 was reported by Van Amerongen.[40] The work of Michaels[41] and coworkers on PET and PE supported the model presented in Eq. 8.18. Puleo et al.[42] reported the gas sorption and permeation in a semicrystalline polymer for which the crystal phase has a lower density than the amorphous phase. At 100 percent crystallinity, the sorption of CO_2 and CH_4 was about 25–30 percent of the solubility of the amorphous phase. The diffusion coefficient in the amorphous phase D_a has been shown to be related to the diffusion coefficient in the semicrystalline phase D by

$$D_a = D\beta\tau \tag{8.19}$$

where β is a chain immobilization factor and τ is a geometric impedance factor. Both β and $\tau > 1$. Michaels et al. applied Eq. 8.19 to semicrystalline PET.[43] However Eq. 8.19 does not work well for annealed polymers.[44]

Other Factors Affecting Permeation. Chain orientation, permeant concentration, plasticizers, and fillers affect permeability. Chain orientation in general decreases the permeability to gases (see Table 8.4). Permeant concentration of gases below one atmosphere of pressure in general does not affect the permeability coefficient. However, strong effects have been observed in the permeability of organic compounds. For instance, the permeability of organic vapors such as aromas, flavors, and solvents in general are strongly dependent on concentration.[45] The addition of plasticizers, usually, but not always, increases the permeability. Film thickness itself does not affect permeability, diffusion, or solubility of a penetrant, provided the polymer morphology is not affected. However, polymer film produced in different thicknesses may have different morphology due to different cooling conditions during processing. The molecular weight of a polymer has been found to have little effect on permeability, except in the very low range of molecular weight. Inorganic mineral fillers such as talc, $CaCO_3$, or TiO_2, used as much as 40 percent, affect the permeability of a film. When coupling agents, such as titanates, are used to improve the interfacial bond between polymer and filler, the permeability to gases and vapors decreases. The absence of bonding agents may increase permeability (see Table 8.7).

TABLE 8.7   Selected Values of A and s for
Permachor Equation at 25°C

Gas	A cc-m/m²d atm	s
O_2	3.48	0.112
N_2	1.18	0.121
CO_2	21.69	0.122

These values are good for both glassy and rubber
materials only at 25°C.
   Source: Salame, M., "Prediction of Gas Barrier Properties of High Polymers," *Polymer Eng.
Sci.,* 26: 22, 1543, 1986.

8.3.5   Diffusion Across Composite Structure

Many packaging materials and containers are often produced by the combination
of more than one plastic material. Common configurations include multilayer
structures, two-phase polymer blends, and plastics with fillers. While multilayer
structures offer the lowest permeability for a given composition, the performance
of two phase-blends depends on which polymer provides the continuous phase. The
effectiveness of a filler to enhance or diminish the barrier value of a polymer
depends on the adhesivity and wetting interphase characteristics of the polymer-
filler pair.

Multilayer Configuration.   Multilayer structures of two or more polymers are
commonly produced by coextrusion, adhesion lamination, coating, and extrusion-
coating. In a lamination, the tie layer contribution to the structure barrier characteristics is often neglected, although in some cases it may have some impact.

   Plane Configuration.   Consider a plane multilayer structure, the layers of which
are perpendicular to the permeation flow. This can be referred to as a structure in
series.

   Figure 8.13 illustrates the case of a three-layer structure during a steady-state
permeation process. Applying Fick's first law to each layer and considering that the
flow rate F is the same through each layer, the following can be concluded:

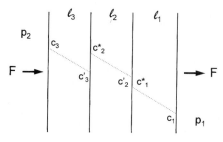

FIGURE 8.13   Permeability through a plane
multilayer structure.

$$F = D_1 \frac{c_1^* - c_1}{l_1} \tag{8.20}$$

$$F = D_2 \frac{c_2^* - c_2'}{l_2} \tag{8.21}$$

$$F = D_3 \frac{c_3 - c_3'}{l_3} \tag{8.22}$$

where l_n = thickness of layer n
D_n = diffusion coefficient of the penetrant in each layer
c^* and c' = the penetrant concentrations at the layer's interphases

The partition coefficients between the layers are calculated as

$$K_{12} = \frac{c_1^*}{c_2'} \tag{8.23}$$

$$K_{23} = \frac{c_2^*}{c_3'} \tag{8.24}$$

When the concentrations at the interphases are eliminated from Eqs. 8.20 to 8.24, the following expression is obtained:

$$F_{ss} = \frac{c_3 K_{23} K_{12} - c_1}{\dfrac{l_3}{D_3} K_{23} K_{12} + \dfrac{l_2}{D_2} K_{12} + \dfrac{l_1}{D_1}} \tag{8.25}$$

which gives the steady-state flux value F_{ss} as a function of external concentrations c_1 and c_3, partition coefficients, layer thickness, and diffusion coefficients. Equation 8.25 can be extended to n layers[44] as

$$F_{ss} = \frac{c_n K_{(n-1)n} \cdots K_{12} - c_1}{\dfrac{l_n}{D_n} K_{(n-1)n} \cdots K_{12} + \cdots + \dfrac{l_3}{D_3} K_{23} K_{12} + \dfrac{l_2}{D_2} K_{12} + \dfrac{l_1}{D_1}} \tag{8.26}$$

where $K_{(n-1)n}$ = partition coefficient of penetrant between two contacting layers. Consider two practical cases in a three-layer structure.

Case 1. Layers 1 and 3 are the same polymer. Therefore, $D_1 = D_3$ and $K_{12} = 1/K_{23} \equiv K$. Equation 8.25 reduces to

$$F_{ss} = \frac{c_3 - c_1}{\dfrac{l_1 + l_3}{D_1} + \dfrac{l_2 K}{D_2}} \tag{8.27}$$

Similarly in Eqs. 8.20 to 8.22, an apparent diffusion coefficient D_{eff} of the whole structure can be defined as

$$F_{ss} \equiv D_{eff} \frac{c_3 - c_1}{l_T} \tag{8.28}$$

where l_T is the total thickness of the structure.

From Eqs. 8.27 and 8.28, the D_{eff} is related to the individual D and K by

$$D_{eff} = \frac{l_T}{\dfrac{l_1 + l_3}{D} + \dfrac{l_2 K}{D_2}} \tag{8.29}$$

Further, assuming that Henry's law of solubility applies at both external layers inter-phases in contact with the permeant gas phase, the effective permeability P_{eff} of the multilayer is

$$P_{eff} = SD_{eff} = \frac{Sl_T}{\dfrac{l_1 + l_3}{D} + \dfrac{l_2 K}{D_2}} \tag{8.30}$$

When the partition coefficient K favors the external layer, i.e. the permeant preferentially dissolves much better in the external layers rather than in the middle layer, $K >> 1$. Additionally, if $K/D_2 >> l/D$, then P_{eff} is controlled by the permeant solubility in the external layer S, the diffusion coefficient of the middle layer D_2, and K:

$$P_{eff} \approx \frac{3SD_2}{K} \tag{8.31}$$

For the middle layer to be a high barrier, the permeant diffusion coefficient D_2 must be very small, and its solubility must also be low.

Case 2. (*a*) First consider that both external layers have equal thickness: $l_1 = l_3 \equiv l$, and $l_2 < l$. If $K = 1$ (similar preferential sorption of permeant in both polymers) and D_2 is large, Eq. 8.30 simplifies to $P_{eff} \approx SD$. This indicates, as expected, that the middle layer does not contribute at all to the barrier because the middle layer is similar to the external layers.

(*b*) If $l_2 = \epsilon l$ ($\epsilon < 1$) and $\epsilon K >> 1$, assuming that $D \approx D_2$, then Eq. 8.30 becomes $P_{eff} \approx 2SD_2/\epsilon K$, and the middle layer is now a high barrier material. Additionally, if $D_2 < D$, then $P_{eff} \approx (2 + \epsilon)SD_2/\epsilon k$, which is the best combination for a middle-layer barrier.

When the values of permeability for each layer are known instead of the diffusion and partition coefficients, P_{eff} is given by

$$P_{eff} = \frac{\sum l_i}{\sum \dfrac{l_i}{P_i}} \tag{8.32}$$

where P_i is the permeability coefficient of each layer.

Cylindrical and Spherical Multilayer Structures. For a cylindrical structure with concentric parallel layers of radius $R_0 < R_1, \ldots < R_n$, P_{eff} is[34]

$$\ln \frac{R_n/R_o}{P_{eff}} = \ln \frac{R_1/R_o}{P_1} + \ln \frac{R_2/R_1}{P_2} + \ldots + \ln \frac{R_n/R_{n-1}}{P_n} \tag{8.33}$$

while for a multilayer spherical shell,

$$\frac{(1/R_o - 1/R_n)}{P_{eff}} = \frac{(1/R_o - R_1)}{P_1} + \frac{(1/R_1 - R_2)}{P_2} + \ldots + \frac{(1/R_{n-1} - R_n)}{P_n} \tag{8.34}$$

where $l_i = R_i - R_{i-1}$, $\sum l = R_n - R_o$ and $P_i = D_i k_i$. Barrer[46] discusses the case of a laminate in which the diffusion coefficient of each layer is permeant-concentration-dependent.

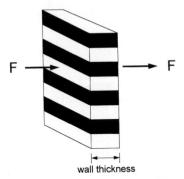

F → → F

|←——→|
wall thickness

FIGURE 8.14   Permeability in parallel structures.

Parallel Structures. Although multi-layer structures parallel to flow are not common in packaging, the case of two materials, or phases, is illustrated in Fig. 8.14. The effective permeability is given by

$$P_{eff} = \phi_1 P_1 + \phi_2 P_2 \qquad (8.35)$$

where ϕ is the volume fraction of each polymer. This is the least efficient way to improve barrier properties by combining two materials.

Immiscible Blends. Immiscible polymer blends having good interphase bonding provide a continuous two-phase heterogenous medium to diffusion and sorption. A detailed study of these cases is beyond the scope of this chapter. Let us consider the simple case of a two-polymer immiscible blend in which one phase remains continuous across a composition range and the other is a discontinuous phase dispersed as spherical particles (aspect ratio of 1). The effective permeability is described by the following equation[47]:

$$P_{eff} = \frac{P_c[P_d + 2P_c - 2\phi_d(P_c - P_d)]}{[P_d + 2P_c + \phi_d(P_c - P_d)]} \qquad (8.36)$$

where P_c, P_d ϕ_c, and ϕ_d are the permeability and volume fractions of the continuous and discontinuous phase, respectively. Equation 8.36 applies to the continuous phase polymer. In an immiscible two-polymer blend, there is a value of ϕ between 0 and 1 in which a phase inversion occurs. At this point, the continuous phase becomes a discontinuous phase and vice versa. Figure 8.15 shows the values of P_{eff} calculated by Eq. 8.37 for two polymers having permeability 0.1 and 150 cm³·mil/100 in²·day·atm, respectively.

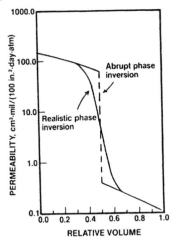

FIGURE 8.15   Calculated values of permeability in a immiscible blend. (Reprinted with permission from P. T. DeLassus, *TAPPI J. 3*, 126, 1988.)

Permeability of Polymers with Fillers. Permeability of a polymer with inorganic fillers can be estimated by Eq. 8.37,[48] provided good adhesion and wettability exist between the polymer and the filler.

$$P_{eff} = P_p \phi_p \left(1 + \frac{L}{2W} \phi_p\right) \qquad (8.37)$$

where P_p and ϕ_p are the permeability and volume fractions of the unfilled polymer, ϕ_f is the volume fraction of the filler, and W/L is the aspect ratio. Aspect ratio is the average dimension of the dispersed filler particles parallel to the plane of the film L divided by the average dimension perpendicular to the film W. When there is not good adhesion between the filler and polymer, permeability may increase in a less predictable form by diffusion through interphase microvoids. Table 8.8 shows values of oxygen permeability of LDPE filled with calcium carbonate.[49]

TABLE 8.8 Oxygen Permeability of LDPE with
Calcium Carbonate

Filler, % by volume	Surface-treated?	Oxygen permeability cc-m/m^{2}d atm
0	No	0.189
15	Yes	0.098
25	Yes	0.059
15	No	0.394
25	No	0.787

Source: Steingeser, S., G. Rubb, and M. Salame, *Encyclopedia of Chemical Technology,* vol. 3, 3d ed., 1980, p. 480.

Miscible Blends. Miscible blends include polymer blends, random copolymers, and polymer/plasticizer blends.[50] The effective permeability in a miscible blend can be described by the following expression[51]:

$$P_{eff} = \phi_1 \ln P_1 + \phi_2 \ln P_2 \tag{8.38}$$

The permeability of O_2, CO_2, and H_2O in a vinylidene chloride copolymer increases exponentially with increasing plasticizer concentration.[47]

8.3.6 Permeability Measurement

Standard methods for measuring the permeability of oxygen and carbon dioxide gas and water vapor through flat polymer structures are described in ASTM D3985-81 and E96-80, respectively. The measurement of permeability of organic vapors has been the subject of numerous investigations.[52,56] Commercial equipment for measuring the permeability of organic vapors through plastic films are now available from MAS Technologies Inc. (Zumbrota, MN) and Modern Controls Inc. (Minneapolis, MN).

Methods for determining permeability are of two types: continuous flow and lag-time or quasi-isostatic.

Continuous-flow Method. Figure 8.16[57] shows a schematic of a continuous-flow-method setup for measuring oxygen permeability in the presence of water vapor. The solution to Fick's second law from a continuous-flow permeation experiment with D independent of concentration is given by[58]

$$\frac{F_t}{F_{ss}} = \left(\frac{4}{\sqrt{\pi}}\right)\left(\sqrt{\frac{l^2}{4Dt}}\right)\sum_{1,3..}^{\infty} \exp\left(\frac{-n^2l^2}{4Dt}\right) \tag{8.39}$$

where F_t is the flow rate of the permeant through the film sample and F_{ss} is the steady-state flow. A typical curve obtained using the continuous-flow method is presented in Fig. 8.17. The experimental permeation flow up to a value of the flow ratio of 0.95 can be virtually described by the first term of Eq. 8.39:

$$\psi = \frac{F_t}{F_\infty} = \left(\frac{4}{\sqrt{\pi}}\right)X^{1/2} \exp(-X) \tag{8.40}$$

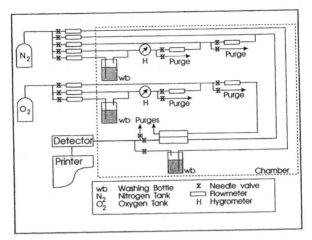

FIGURE 8.16 Continuous flow permeation method. (Reprinted with permission from R. Gavara and R. J. Hernandez. *J. Polym. Sci., Part B, 32, 2375, 1994.)*

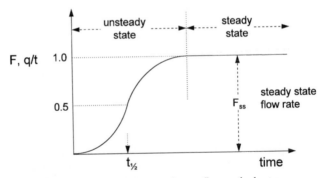

FIGURE 8.17 Flow profile in a continuous-flow method.

where $X = l/4Dt$. The Newton-Rhapson method can be used to evaluate X versus t in Eq. 8.40, and the diffusion coefficient D is obtained from the slope of the linear fitting of X^{-1} versus t for $0.05 < \psi < 0.95$.[54] The permeability coefficient P is calculated from the value of F_{ss},

$$P = \frac{F_{ss}l}{\Delta p} \tag{8.41}$$

where Δp is the partial pressure across the film during the permeation experiment. Simple constant diffusion coefficient D can be approximately calculated by

$$D = \frac{l^2}{7.20\, t_{1/2}} \tag{8.42}$$

where $t_{1/2}$ is the time required at $F_t/F_{ss} = 0.5$.

Gavara and Hernandez[59] have presented a simple method to analyze the consistency of experimental Fickian permeability data.

Lag-Time Method. Figure 8.18 shows an apparatus for measuring the permeation and sorption of oxygen in polymers films.[60] The solution of Fick's second law when the initial concentration of permeant in the film is zero is

$$Q(l) = \frac{Dc_2}{l}\left(t - \frac{l^2}{6D}\right) - \frac{2lc_2}{\pi^2}\sum_1^\infty \frac{(-1)^n}{n^2}\exp\left(\frac{-Dn^2\pi^2 t}{l^2}\right) \qquad (8.43)$$

where $Q(l)$ is the quantity of permeant that has passed at $x = l$ per unit of the membrane area during time t.[34] A plot of Eq. 8.43 is presented in Fig. 8.19, where the partial pressure is plotted versus time t. At the steady state, at very large t values, Q is given by

$$Q(l) = \frac{Dc_2}{l}\left(t - \frac{l^2}{6D}\right) \qquad (8.44)$$

The intercept of the steady-state line with the t axis, which occurs at $Q(l) = 0$, defines the lag time θ:

$$\theta = \frac{l^2}{6D} \qquad (8.45)$$

A quasi-isostatic (lag-time) method has been used by Wahid[61] for measuring organic vapor permeability through metallocene polymer films.

For a penetrant/polymer system that follows the dual-mode sorption model of Eqs. 8.5 and 8.13, the lag-time is given by the following expression:

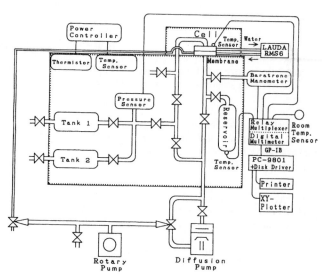

FIGURE 8.18 Lag-time permeation apparatus. (Reprinted with permission from K. Toi, H. Suzuki, I. Ikenasto, T. Ito, and T. Kasi, *J. Polym. Sci.*, Part B, 33, 777, 1995.)

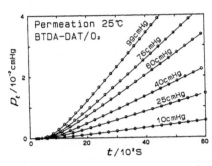

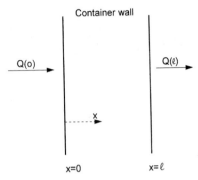

FIGURE 8.19 Plot of lag-time permeation experiment. (Reprinted with permission from K. Toi, H. Suzuki, I. Ikenasto, T. Ito, and T. Kasi, *J. Polym. Sci.*, Part B, 33, 777, 1995.)

FIGURE 8.20 Permeation through a container wall of thickness l. At $x = 0$, total quantity entered is $Q(0)$; at $x = l$, total quantity permeated is $Q(l)$.

$$\theta = \frac{l^2}{6D_D}\,[1 + f\,(C'_H, k_D, D_H, D_D, b, p)] \qquad (8.46)$$

where f is a complex function.[62]

8.3.7 Permeant Transport and Packaging Shelf Life

The barrier to the diffusion of small molecules provided by a container is of primary concern for protecting food, cosmetic, and pharmaceutical products.

Simplified Shelf-Life Calculation. Consider a permeation process through a container with a homogeneous wall thickness l, as illustrated in Fig. 8.20. In many applications, the steady-state permeation Eq. 8.10 is an acceptable approximation to correlate a product's shelf life t with package area A, wall thickness l, and environmental conditions of pressure and temperature. This approximation is valid for thin films, and the time to reach steady-state flow rate t_{ss} is very small compared with the package shelf life.

EXAMPLE: *Estimate the film thickness l for a plastic pouch containing an oxygen-sensitive product, provided that not more than 5 cc of oxygen can be allowed to permeate the package during six months. Assume the pouch area A = 400 cm², and the material permeability P = 165 cc μm/m² day atm.*

SOLUTION: *Since this is an oxygen-sensitive product, the headspace partial pressure of oxygen is near zero, $p_1 \approx 0$ atm, and Δp across the package can be considered constant during the shelf life. From Eq. 8.10, $l = PtA p_2/q = 165 \times 180 \times 0.04 \times 0.21/5 = 50$ microns. If the diffusion coefficient of the material is 5×10^{-10} cm²/s, the time to reach the steady-state flow is 3 times the lag time. By Eq. 8.45, $t_{ss} = 6.9$ hours, which is small compared to 180 days. Fig. 8.21 shows plots of $t_{ss} = 3\Theta$, calculated for different thickness and diffusion-coefficient values.*

When Δp changes during the package's shelf life as in the case of moisture-sensitive products, Eq. 8.10 can be written as

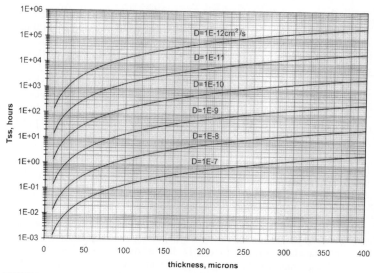

FIGURE 8.21 Values of t_{ss} calculated for different thicknesses and diffusion coefficient values; $t_{ss} = 3\Theta$ (Eq. 8.45).

$$t = \frac{l}{PA} \int \frac{dq}{p_e - p(M)} \tag{8.47}$$

where p_e is the water partial pressure outside the package, and $p(M)$ is the package's internal partial pressure in equilibrium with the product.[63] The finite-differences method has been applied by Kim et al.[64] to calculate the uptake of water by solving simultaneously the diffusion equation in both the product and package.

Gaining and Losing Permeant. The unsteady-state equation describing the mass transfer in Fig. 8.20 can be written for both sides of the film, $x = 0$ and $x = l$. The resulting equations have practical interest. A packaged product that loses a component by permeation through the package will start losing it at the internal side of the package, $x = 0$. If the product uptakes permeant from the environment, the permeant will be incorporated into the product only after the permeant reaches $x = l$ and has totally penetrated the packaging wall. The distinction of these two cases is especially relevant when the permeant diffusion coefficient is very small and/or the packaging wall is relatively thick. To evaluate the quantity of permeant at $x = 0$ and $x = l$, Fick's second law needs to be solved subject to the following conditions: the initial permeant concentration in the packaging wall is zero, and for $t > 0$, one side, $x = 0$, will be at c_2; and the other, $x = l$, will be at $c_1 = 0$. The quantity permeated at $x = l$, $Q(l)$ is already given by Eq. 8.43, while the quantity entering the plane at $x = 0$, $Q(o)$ is[43]

$$Q(o) = \frac{c_2}{3} + c_2 D t l^2 - \frac{2c_2}{\pi^2} \sum_{1}^{\infty} \frac{1}{n^2} \exp\left(\frac{-n^2\pi^2 Dt}{l^2}\right) \tag{18.48}$$

Figure 8.22 is a plot of Eqs. 8.43 and 8.48. After a long time, the steady-state permeation is reached, and both lines, $Q(o)$ and $Q(l)$, become parallel. At steady state, the difference between $Q(o)$ and $Q(l)$ is the quantity of permeant sorbed by the polymer;

that is, $lc_2/2$. The intersection of the asymptotic steady state of $Q(l)$ with Dt/l^2-axis is $D\Theta/l^2 = 1/6$, where Θ is the lag-time of Eq. 8.45. The asymptotic lines on both curves indicate the pseudosteady-state solution given by Eq. 8.10 and have the same slope, $F = Q/t$, at both sides of the container wall. As indicated in Fig. 8.22, the pseudosteady-state solution agrees within less than 1 percent of Eq. 8.43 at $t > 2.64\Theta$, or at a value of $Dt/l^2 > 0.44$. However, it is generally recommended to consider steady-state flow at $Dt/l^2 > 0.50$, equivalent to 3 times the lag time Θ or 3.6 times $t_{1/2}$ in Eq. 8.42. $Dt/l^2 = 0.5$ can be used to estimate the time needed for a permeant to reach the steady-state permeation rate from the moment a product containing the permeant is put in a package and sealed.

EXAMPLE: *If $D = 1 \times 10^{-10}$ cm^2/s and the packaging thickness $l = 25$ microns (1 mil), $t = l^2/2D = (25 \times 10^{-4})^2/2 \times 10^{-10} = 31,125$ sec $= 8.68$ hours. It will take 8.68 hours to reach the steady-state permeation rate t_{ss}. Other values of t_{ss} as a function of D and thickness l are presented in Fig. 8.21.*

A plot of $Q(o)$ of Eq. 8.48 as a function of thickness is presented in Fig. 8.23. From this figure it can be seen that the value of $Q(o)$ at a given time t depends on both the thickness and the diffusion coefficient D. For a given D the loss of a permeant decreases with thickness to a value l_{min} beyond which there is no real increase of barrier power. This value is[65]

$$l_{min} = (3Dt)^{1/2} \tag{8.49}$$

Figure 8.23 also shows the pseudosteady-state Eq. 8.10, wrongly indicating that $Q(o)$ continues decreasing below the minimum thickness value l_{min}.

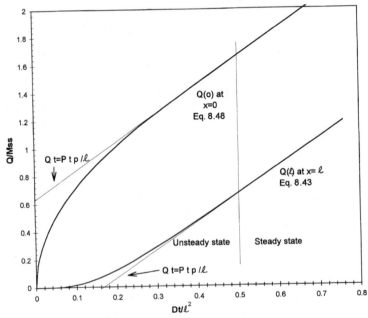

FIGURE 8.22 Quantity permeated at $x = 0$ and $x = l$. (Adapted from P. Masi and D. Paul, *J. Memb. Sci.* 12, 137, 1982.)

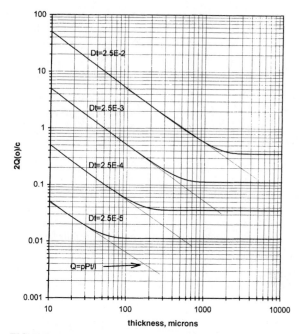

FIGURE 8.23 Plot of quantity permeated at $Q(o)$ as a function of diffusion coefficient and container thickness. (Adapted from P. Masi and D. Paul, *J. Memb. Sci.* 12, 137, 1982.)

Loss of Pressurized CO₂ Through PET. The transport of pressurized gases in glassy polymers such as PET and PAN exhibit dual sorption behavior. Masi and Paul[66] carried out simulations based on the dual-mobility sorption model of Eq. 8.13 as well as the simple sorption model of Henry's law, Eq. 8.10. Comparison of these two models is shown in Fig. 8.24, which shows that the loss at $Q(o)$ is greatly overestimated by the simple Henry's law model. However, at $Q(l)$ the difference is not very great. Also, Henry's law model predicts a larger amount of sorbed CO_2 than the dual-mobility model. The total amount of permeant sorbed by the polymer at a given time is $Q(o) - Q(l)$.

8.3.8 Migration and Sorption in a Semi-Infinite Polymer Plane

Consider the packages of Fig. 8.25, in which a product (or contacting phase) is in direct contact with a polymeric material. From the moment the product is put in contact with the package, substances from the product will be sorbed by the polymer and, conversely, components from the plastic will migrate into the product. Selected sorption and migration cases are reviewed now.

Since the processes are controlled by diffusion, the following assumptions are made:

1. The diffusion process is described by Fick's second laws, Eq. 8.3.

2. The migration and sorption is limited to a single component that is initially homogeneously distributed in the polymer or in the contacting product, respec-

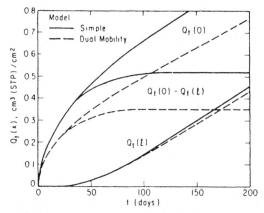

FIGURE 8.24   Comparison of quantity permeated at $x = 0$ and $x = l$ of simple model Eq. (8.10) and dual mobilized model Eq. (8.13). (Reprinted with permission from P. Masi and D. Paul, *J. Mem. Sci.* 12, 137, 1982.)

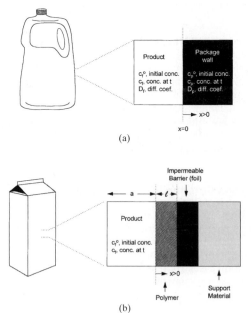

FIGURE 8.25   (*a*) Single-wall plastic package. (*b*) Package with a impermeable foil layer.

tively. This also means that the diffusion of one adjuvant is not affected by the presence of other adjuvants.

3. For a given plastic material, the diffusion coefficient of the adjuvant is only a function of temperature.

4. There is no swelling of the polymer by any sorbed component, or, if swelling does occur, the above assumptions still hold.

5. A flat sheet can be considered of infinite thickness if less than 40 percent of any migrant originally contained in the sheet is extracted by a contacting media.

For easy reference, D_p is the diffusion coefficient of the adjuvant in the polymer, and D_f is the diffusion of the adjuvant in the contacting phase. In solving Fick's second law, the initial conditions are

$$C_p = C_p^\circ \text{ at } t = 0, x > 0 \qquad C_f^\circ = 0 \text{ at } t = 0, x < 0 \qquad (8.50)$$

and the boundary condition is

$$F = D_p\left(\frac{\partial c_p}{\partial x}\right) \text{at } x = 0 \qquad (8.51)$$

which gives the rate of adjuvant transferred at the polymer/product interphase.

Transient Migration

Less than 40 Percent of the Original Migrant is Transferred to a Contacting Well Mixed Liquid. Consider a single polymer layer with an initial migrant concentration C_p°, at $t = 0$ for $x > 0$, as in Fig. 8.25(A). The migrant substance in the polymer can be residual polymerization reactants, processing additives, converting operation residuals, or contaminants in a single layer of postconsumer recycled resin. The contacting product with the polymer is considered to be very large with respect to the amount of migrant to be transferred and initially contains no migrant, that is, $C_f^\circ = 0$ for $x < 0$ at any time. This case can be simulated by a diffusional process in a semi-infinite/infinite domain[43] with an infinite mass-transfer coefficient, like a well agitated liquid. The solution to Fick's law gives the amount of migrant transferred from the polymer to the liquid[43]

$$M = 2C_p^\circ\left(\frac{D_p t}{\pi}\right)^{1/2} \qquad (8.52)$$

where M is the total quantity of migrant lost from the polymer per unit area of containing phase at time t and D_p is the diffusion coefficient of migrant in the polymer. From Eq. 8.52, it follows that a plot of M/C_p° versus $(Dt)^{1/2}$ is a straight line. This is confirmed in Fig. 8.26, which presents experimental data of styrene monomer migrated from a polystyrene into contacting oils at 40°C.[67,68]

Equation 8.52 can be written to give the quantity of migrant transferred per unit of mass of product $W_f = M/\zeta$ as a function of dimensionless group $D_p t/l^2$.

$$\frac{\zeta W_f}{c_p^\circ l} = \frac{2}{\pi^{1/2}}\sqrt{\frac{D_p t}{l^2}} \qquad (8.53)$$

where ζ is the ratio of the mass of contacting phase/package area. For food products, the Food and Drug Administration uses a $\zeta = 1.55$ g/cm^2. A plot of Eq. 8.53 is shown in Fig. 8.27. Equations 8.52 and 8.53 show that the diffusion coefficient of migrant in the polymer is a key parameter.

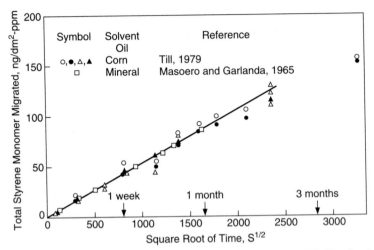

FIGURE 8.26 Migration of styrene monomer into solvent oil at 40°C. (Reprinted with permission from R. C. Reid et al., *Ind. Eng. Chem. Prod. Res. Dev.*, 19 (4), 580, 1980.)

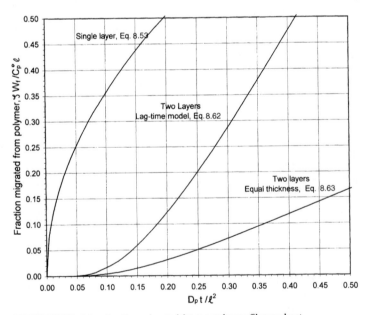

FIGURE 8.27 Mass fraction migrated from a polymer film or sheet.

EXAMPLE: *Calculate the concentration of a migrant in a foodstuff that is in direct contact with a monolayer of recycled polymeric package when $D_p t / l^2 = 0.1$, $C_p^\circ = 23$ $\mu g/cm^3$, and thickness $l = 50\ \mu m$.*

SOLUTION: $\zeta W_f / C_p^\circ l = 2(D_p t/l^2)^{1/2}/\pi^{1/2} = 0.3568;$ $W_f = 0.3568 \times C_p^\circ \times l/\zeta = 0.3568 \times 23 \times 0.005/1.55 = 26.5\ \mu g/kg = 26.5\ ppb$. *This corresponds to 36 percent of the original amount of migrant in the polymer, for $\zeta = 1.55\ g/cm^2$. Therefore, the assumption of a semi-infinite polymer is acceptable. Further, if $D_p = 1 \times 10^{-12}\ cm^2/s$, calculate how long will it take to transfer the calculated 36 percent of migrant: $t = 0.1 \times l^2/D_p = 0.1 \times 0.005^2/1 \times 10^{-12} = 2.5 \times 10^6 s = 28.9\ days$.*

Equation 8.53 represents the worst possible scenario to estimate migration from the plastic. Since the contacting phase is well mixed, the difference in concentration between the polymer and liquid is the highest possible. Also, Eq. 8.53 does not include a partition coefficient value that would slow the migration process. Therefore, this case gives the shorter possible time for migration, or, for a given time, the larger amount of migrant transferred from the plastic. Calculated values from this case can be used to estimate an upper bound of migrant leaving the polymer in the transient state. The effect of temperature in Eq. 8.52 was studied by Markelov and coworkers in the migration of acrylonitrile in commercial copolymers.[69]

Semi-Infinite Polymer Contacted by a Well Mixed Finite Amount of Liquid Phase of Volume V_f. As is illustrated in Fig. 8.25(b), the amount of contacting liquid is given by $V_f = Aa$. Since the liquid is considered to be in agitation, there is no concentration gradient of the migrant into the liquid and no effect of the mass-transfer coefficient. To solve this case, the following additional boundary condition applies:

$$a \frac{dc_f}{dt} = D_p \frac{\partial c_p}{\partial x} \text{ at } x = 0 \tag{8.54}$$

which indicates that the flux of migrant is transferred entirely into the contacting phase with no gradient in concentration in the latter. The quantity migrated by unit of area M in time t given in dimensionless form is[67]

$$Y = \frac{M}{aKC_p^\circ} = [1 - \exp(Z^2) \operatorname{erfc} Z] \tag{8.55}$$

where $Z = (D_p t)^{1/2}/aK$ and Y is the ratio of M to the maximum quantity of migrant transferred at equilibrium, aKC_p°. $Y = 1$ at equilibrium, otherwise $0 < Y < 1$. K is the partition coefficient of the migrant between the contacting phase and polymer as defined in Eq. 8.6. K is considered to be independent of migrant concentration, which is a very realistic assumption for very low adjuvant concentration values commonly found in packaging migration studies. A plot of Eq. 8.55 is presented in Fig. 8.28, and a table of erfc (error function) is given in Table 8.9. If $Z < 0.05$ (large liquid volume, small migration time t, or very low D), Eq. 8.55 reduces to Eq. 8.52. The assumption of an infinitely thick polymer is fully acceptable if the quantity of migrant diffused into the liquid is less than 40 percent. Therefore, for $Y < 0.40$, the polymer thickness l is not a significant variable.[67]

EXAMPLE: *Consider a similar system as in the previous example, but now the contacting phase has a limited volume Aa with $a = 2.5\ cm$. Estimate the time to transfer 25 percent of the total migrant that would potentially migrate at equilibrium (infinite time).*

SOLUTION: *From Eq. 8.55 or Fig. 8.28, at $Y = 0.25$, $Z = 0.23$, if $K = 1 \times 10^{-2}$ and $D_p = 1 \times 10^{-10}\ cm^2/s$, $t = (Z \cdot a \cdot K)^2/D_p = 3.3 \times 10^5\ s = 3.8\ days$.*

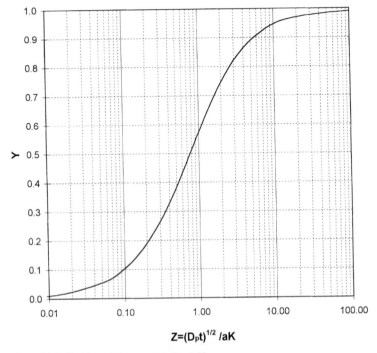

$$Z=(D_p t)^{1/2} / aK$$

FIGURE 8.28   Plot of Y versus Z in Eq. 8.55.

Semi-Infinite Polymer Contacted by a Semi-Infinite Immobile Phase.   In this case, mixing in the product is insignificant or does not exist at all. The mass transfer takes place only by molecular diffusion. This can be the case with a viscous liquid such as honey or a solid frozen meal in a plastic container. In addition to the initial conditions of Eq. 8.50 and 8.51, the following boundary condition describes the system:

$$D_p \frac{\partial c_p}{\partial x} = D_f \frac{\partial c_f}{\partial x} \text{ at } x = 0 \tag{8.56}$$

which indicates that the migrant diffusion coefficient D_f in the immobile product is important and that all the diffusant leaving the polymer enters the contacting phase. When the contacting phase is very large, $c_f = 0$ at any time, the solution is[67]

$$M = \frac{2c_p^o}{1+\beta} \left(\frac{D_f t}{\pi} \right)^{1/2} \tag{8.57}$$

where $\beta = K(D_f/D_p)^{1/2}$. If $\beta > 100$, Eq. 8.57 reduces to Eq. 8.52 with an error of less than 1 percent. Large β values might result from large K (the migrant is very soluble in the contacting phase), high values of D_f, or very small D_p. However, if K values are very small (no preference of the liquid phase by the migrant), β may be also very small. Equation 8.57 reduces, then, to

$$M = 2C_p^o K \left(\frac{D_f t}{\pi} \right)^{1/2} \tag{8.58}$$

TABLE 8.9 Error Function

χ	erfc χ	χ	erfc χ	χ	erfc χ
0	1.0	0.7	0.322199	1.8	0.010909
0.05	0.943628	0.75	0.288844	1.9	0.007210
0.1	0.887537	0.8	0.257899	2.0	0.004678
0.15	0.832004	0.85	0.229322	2.1	0.002979
0.2	0.777297	0.9	0.203092	2.2	0.001863
0.25	0.723674	0.95	0.179109	2.3	0.001143
0.3	0.671373	1.0	0.157299	2.4	0.000689
0.35	0.620618	1.1	0.119795	2.5	0.000407
0.4	0.571608	1.2	0.089686	2.6	0.000236
0.45	0.524518	1.3	0.065992	2.7	0.000134
0.5	0.479500	1.4	0.047715	2.8	0.000075
0.55	0.436677	1.5	0.033895	2.9	0.000041
0.6	0.396144	1.6	0.023652	3.0	0.000022
0.65	0.357971	1.7	0.016210		

which indicates that the diffusion coefficient in the polymer no longer affects the mass transfer.

Semi-Infinite Polymer Contacted by a Limited Volume of Immobile Phase. The additional boundary condition for this case is $D_f(\partial c_f/\partial x) = 0$ at $x = a$, and the corresponding solution is

$$Y = \frac{2\beta z}{\pi^{1/2}(1 + \beta)} [1 + F(\beta, z)] \qquad (8.59)$$

where $F(\beta,z)$ is a function of β and z[67]

$$F(\beta, z) = \frac{2}{1 + \beta} \sum_{n=1}^{\infty} \left(\frac{\beta - 1}{1 + \beta}\right)^{n-1} \exp\left(\frac{-n^2}{\beta^2 z^2}\right) \left[1 - E\left(\frac{n}{\beta z}\right)\right] \qquad (8.60)$$

where $E(n/\beta z) = \pi^{1/2}(n/\beta z) \exp(n/\beta z)^2 \operatorname{erfc}(n/\beta z)$.

Equation 8.59 is plotted in Fig. 8.29. For values of $\beta > 5$, the mixing does not have any effect in the mass transfer, and Eq. 8.59 reduces to Eq. 8.55. The linear portion of Fig. 8.29 is given by the following expression:

$$Y = \frac{2\beta Z}{\pi^{1/2}(1 + \beta)} \qquad (8.61)$$

EXAMPLE: *If* $Z = 1$, $\beta = 0.22$, *and* $Y = 0.203$, *then 20.3 percent of the migrant has passed to the contacting phase.*

Contaminant Transferred from a Coextruded Recycled Polymer and a Virgin Layer. The migration of contaminants from a single-layered recycled polymer is described by Eq. 8.52 for low migrant-concentration levels. Consider now the migration from a recycled layer coextruded with a virgin layer of the same polymer as illustrated in Fig. 8.30, where r and l are the thicknesses of the recycled and virgin layers, respectively. The thickness r can be expressed as a function of l: $r = \gamma l$, where $\gamma > 0$. The migrating substances must diffuse out from the recycled layer through the virgin layer and into

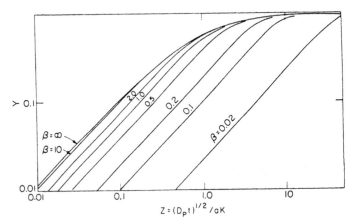

FIGURE 8.29 Migration from a semi-infinite polymer contacted by a limited volume of immobile phase, Eq. 8.59. (Reprinted with permission from R. C. Reid et al., *Ind. Eng. Chem. Prod. Res. Dev.*, 19 (4), 580, 1980.)

the food-contacting phase. It is further assumed that there is no direct diffusion from the recycled layer into the opposite side. Two solutions are possible:

(1) Lag-time approach. Here the virgin polymer layer is considered a membrane having a constant contaminant concentration C_p^o at the recycled polymer interphase and zero at the contacting phase. The corresponding solution is given by Eq. 8.43, the lag-time equation for permeation,[70] that can be rewritten in dimensionless form as

$$\frac{\zeta W_f}{c_p^o l} = \frac{Dt}{l^2} - \frac{1}{6} - \frac{2}{\pi^2} \sum_1^\infty \frac{(-1)^n}{n^2} \exp\left(\frac{n^2\pi^2 D_p t}{l^2}\right) \qquad (8.62)$$

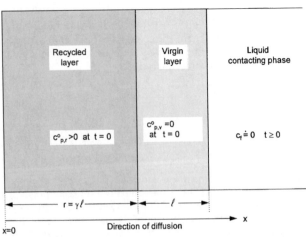

FIGURE 8.30 Migration from a layer of recycled resin and coextruded with virgin layer of the same resin.

As expected, this equation grossly overestimates the amount of migrant transferred into the foodstuff for $D_p t / l^2 > 0.052$, because it assumes that, at the recycled/virgin polymer interphase, C_p^o remains always constant. For comparison, Eq. 8.62 is plotted together with Eq. 8.53 in Fig. 8.27.

(2) Continued diffusion approach. This approach considers that the contaminant continuously diffuses through the recycled and virgin layers with the corresponding decrease in concentration at the recycled/virgin interphase. Figure 8.31 illustrates the concentration profile within the coextruded structure for $r = l$. The migrant diffuses out only at the virgin resin/product interphase, and the contacting phase is well mixed. This implies a mass-transfer coefficient of infinite value. The corresponding solution is[71]

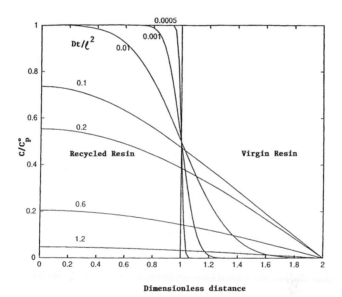

FIGURE 8.31 Concentration profile of a migrant within a coextruded recycled/virgin structure in contact with a well agitated liquid. (Reprinted with permission from S. Laoubi et al., *Pack. Techn. and Sci.* 8, 249, 1995.)

$$\frac{\zeta W_f}{c_r^o l} = 1 - \frac{8}{\pi^2} \frac{\gamma+1}{\gamma} \sum_{0}^{\infty} \frac{(-1)^n}{(n+1)^2} \sin\left[\frac{(2n+1)\pi\gamma}{2(\gamma+1)}\right] \exp\left[\frac{(2n+1)^2\pi^2}{4(\gamma+1)^2} \frac{D_p t}{l^2}\right] \quad (8.63)$$

where C_r^o is the initial concentration of the contaminant in the recycled layer of thickness $r = \gamma l$. Eq. 8.63 is plotted in Fig. 8.27 for $\gamma = 1$; that is, both the virgin and recycled layer have the same thickness l. Equation 8.63 gives the transferred fraction of migrant for a given value of $D_p t / l^2$.

EXAMPLE: *Calculate the maximum migrant concentration in a mixed foodstuff in direct contact with a coextruded structure made of a recycled and virgin resin of equal thickness* $r = l = 25$ *μm, during a 30-day period. The diffusion coefficient* $D_p = 1 \times 10^{-12}$ *cm²/s, and* $c_r^o = 23$ *μg/cm³.*

SOLUTION: *At* $D_p t/l^2 = 1 \times 10^{-12} \times 30 \times 24 \times 3600/(0.0025)^2 = 0.415$. *From Fig. 8.27, for the line corresponding to Eq. 8.63,* $\zeta W_f/c_f^0 l = 0.125$; *therefore,* $W_f = 0.125 \times 23 \times 0.0025/1.55 = 0.0046$ μg/kg = 4.6 ppb. *This corresponds to 12.5 percent of the original amount of migrant in the polymer.*

Equation 8.63 still overestimates the amount of migrant in the contacting phase, since it is assumed that all migrant diffuses only in the direction of the contacting phase, which is perfectly mixed.

Transient Sorption

Sorption at One Side of the Film. Foodstuff constituents in polymeric packages such as the one in Fig. 8.25(b) lose aroma through sorption. The penetration of the sorbate takes place on only one side of the polymer and on the inside of the package. The polymer layer is of thickness *l*, and its initial concentration $C = C_p^0$. Assuming that the sorbant concentration in the contacting phase remains essentially constant during the sorption process (that is, the amount sorbed is very small compared with the total amount originally in the food), the solution giving the total amount of sorbed compound after a very large time (to infinity) is[34]

$$\frac{M}{M_\infty} = 1 - \frac{8}{\pi^2} \sum_{n=0}^{\infty} \frac{1}{(2n+1)^2} \exp\left[(2n+1)^2 \pi^2 \frac{D_p t}{4l^2}\right] \tag{8.64}$$

The corresponding solution for short times is

$$\frac{M}{M_\infty} = 2\left(\frac{Dt}{l^2}\right)^{1/2} \left[\pi^{-1/2} + 2\sum_{n=0}^{\infty} (-1)^n \operatorname{ierfc} \frac{nl}{D_p^{-1/2}}\right] \tag{8.65}$$

Sorption from Both Sides. The polymer layer of thickness *l* is contacted by both sides when it is immersed in a fluid phase of constant sorbant concentration. The amount sorbed is

$$\frac{M}{M_\infty} = 1 - \frac{8}{\pi^2} \sum_{m=0}^{\infty} \frac{1}{(2m+1)^2} \exp\left[(2m+1)^2 \pi^2 \frac{D_p t}{l^2}\right] \tag{8.66}$$

Equation 8.66 is commonly used to determine the diffusion coefficient of a sorbant in a polymer film. The film is exposed on both sides to a flow of sorbant carried in a gas stream, and the increase in mass is continuously monitored.[39,72] The corresponding solution for short times of Eq. 8.66 is

$$\frac{M}{M_\infty} = 2\left(\frac{Dt}{l^2}\right)^{1/2} \left[\pi^{-1/2} + 2\sum_{n=0}^{\infty} (-1)^n \operatorname{ierfc} \frac{nl}{Dp^{-1/2}}\right] \tag{8.67}$$

Reduced Sorption Equations. Because the first term in the series is dominant, for small times, Eq. 8.66 is reduced to

$$\frac{M}{M_\infty} = \frac{4}{\pi^{1/2}} \left(\frac{Dt}{l^2}\right)^{1/2} \tag{8.68}$$

with a relative error of 1 percent for $0.5 < M/M_\infty \le 0.63$, and 0.1 percent for $M/M_\infty \le 0.50$. Similarly, for long times, when $M/M_\infty \ge 0.55$, Eq. 68 is reduced to

$$\frac{M}{M_\infty} = 1 - \frac{8}{\pi^2} \exp\left(\frac{-\pi^2 Dt}{l^2}\right) \tag{8.69}$$

with a relative error of 0.1 percent.

For $M/M_\infty = 0.5$, Eq. 8.68 can be further simplified to the well known expression

$$D = 0.049 \frac{l^2}{t_{0.5}} \qquad (8.70)$$

Equations 8.64 to 8.70 also describe a desorption process.

Sorption from a Stirred Solution of Limited Volume. When a plane sheet or film of thickness $2l$ is immersed in a limited volume of a solution of thickness $2a$ containing the sorbate, the sorbate concentration falls with time due to the sorption process by polymer. The space occupied by the film is $-l \leq x \leq l$, and the liquid $-(l + a) \leq x \leq -l, l \leq x \leq l + a$. Initial conditions are $c = 0$ for $-l < x < l$ at $t = 0$.

The boundary condition is given by

$$aK \frac{\partial c}{\partial t} = \pm D \frac{\partial c}{\partial x} \qquad x = \pm l, t > 0 \qquad (8.71)$$

where K is the partition coefficient ($K = c_f^*/c_p^*$) and c_f^* and c_p^* are the equilibrium concentrations of the adjuvant in the contacting phase and polymer, respectively. The solution of Fick's second law subject to these conditions is[34]

$$\frac{M}{M_\infty} = \sum_{n=1}^{\infty} \frac{2\alpha(\alpha+1)}{(1+\alpha+\alpha^2 q_n^2)} \exp\left(\frac{-Dq_n^2 t}{l^2}\right) \qquad (8.72)$$

where M is the quantity of diffusant entering the sheet or film, $q_n s$ are the nonzero positive roots of $\tan q_n = -\alpha q_n$, and $\alpha = aK/l$. The fraction uptake U is

$$U = \frac{M_\infty}{2ac_f^\circ} = \frac{V_p c_f^*}{V_f c_f^\circ} = \frac{1}{1+\alpha} \qquad (8.73)$$

where V_f is the volume of the contacting phase and V_p is the volume of plastic in the container.

8.3.9 Migration and Sorption at Equilibrium

All diffusion processes involving the transfer of a diffusant from or to a polymeric material asymptotically reaches a steady condition at long times. In any closed system that has reached this condition, there is not effective mass transfer between the polymer and the surrounding phases, the adjuvant concentrations are at equilibrium, and the transferred material is the maximum possible amount. In reference to Fig. 8.25(b), consider a sorption or migration process that has reached equilibrium. V_f is the volume of contacting phase, V_p is the volume of plastic in the container, and c_f^* and c_p^* are the equilibrium concentrations of the adjuvant in the contacting phase and polymer, respectively. For the case of sorption of a substance from the food phase by the polymer, the mass balance at equilibrium is

$$^sC_p^* V_p + c_f^* V_f = c_f^\circ V_f \qquad (8.74)$$

and for a migration process,

$$^mC_p^* V_p + c_f^* V_f = c_p^\circ V_p \qquad (8.75)$$

The equilibrium concentration $^sC_f^*$ for a sorption process and $^mC_f^*$ for a migration process are expressed as a function only of adjuvant initial concentration (in the polymer of food), the partition coefficient K, and the ratio of polymer and food volume,

$$^sC_f^* = \frac{C_f^{\circ}}{K^{-1}\dfrac{V_p}{V_f} + 1} \qquad (8.76)$$

$$^mC_f^* = \frac{C_p^{\circ}}{K^{-1} + \dfrac{V_f}{V_p}} \qquad (8.77)$$

where the s and m superscripts at the right of C refer to sorption and migration processes, respectively. Figures 8.32 and 8.33 plot $^mC_f^*$ and $^sC_f^*$, respectively, for common values of K and V_f/V_p.

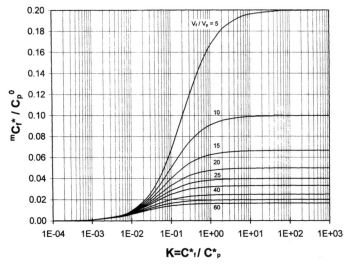

FIGURE 8.32 Equilibrium concentration in a migration process.

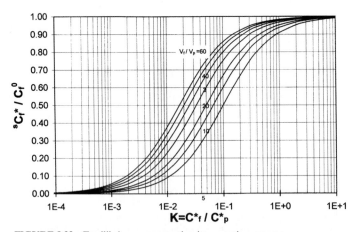

FIGURE 8.33 Equilibrium concentration in a sorption process.

REFERENCES

1. Landvatter, G. R., *Modern Plastics Encyclopedia Handbook*, McGraw-Hill Co., Inc., NY, 1995, p. 39.
2. Mergenhagen, L. K., *Modern Plastics Encyclopedia Handbook*, 1994, p. 37.
3. Statz, R. J., *Modern Plastics Encyclopedia Handbook*, 1994, p. 41.
4. *Mobil OPP Films Product Characteristics*, 1994.
5. Clark, D. L., *Modern Plastics Encyclopedia Handbook*, 1994, p. 75.
6. Toebe, J., H. Hoojjat, R. J. Hernandez, J. Giacin, and B. Harte, "Interaction of Flavor Components from an Onion/Garlic Flavored Sour Cream with HIPS," *Packaging Technology and Science*, 3, 133, 1990.
7. Schwarz, R. A., *Modern Plastics Encyclopedia Handbook*, 1994, p. 55.
8. Foster, R. H., *Modern Plastics Encyclopedia Handbook*, 1994, p. 40.
9. Hernandez, R. J., "Effect of Water Vapor on the Transport Properties of Oxygen through Polyamides Packaging Materials," *J. Food Eng.*, 22, 495, 1994.
10. Eisenberg, A., "The Glassy State," in *The Physical Properties of Polymer*, 2d ed., American Chemical Society, Washington DC, 1993.
11. Boyer, R. F., "Transitions and Relaxations in Amorphous and Semicrystalline Organic Polymers and Copolymers," reprint from *Encyclopedia of Polymer Science and Technology Supplement*, vol. 2, John Wiley & Sons, Inc., NY, 1977, pp. 745–839.
12. Wooward, A. E., *Understanding Polymer Morphology*, Hanser Publishers, New York, 1995.
13. Birley, A. W., B. Haworth, and J. Batchelor, *Physics of Plastics Processing Properties and Materials Engineering*, Hanser Publishers, New York, 1992.
14. Peterlin, A., "Polymer Morphology," in *Encyclopedia of Polymer Science and Engineering*, vol. 10, 1987.
15. Van Krevelen, D. W., *Properties of Polymers*, 3d ed., Elsevier, New York, 1990.
16. *Guide to Plastics Properties and Specification Charts*, Modern Plastics, McGraw-Hill Inc., New York, 1987.
17. Brandrup, J. and E. Immergut, eds., *Polymer Handbook*, 3d ed., Wiley-Interscience, New York, 1989.
18. Wunderlich, B., S. Cheng, and K. Loufakis, *Encyclopedia of Polymer Science and Engineering*, vol. 16, New York, 1989, pp. 787–807.
19. Rosato, D., *Rosato's Plastic Encyclopedia and Dictionary*, Hanser Publisher, New York, 1993.
20. Brady, G. S. and H. R. Clauser, *Materials Handbook*, 13th ed., McGraw-Hill Inc., New York, 1991.
21. Nielsen, L. A. and R. F. Landel, *Mechanical Properties of Polymers and Composites*, 2d ed., Marcel Dekker, New York, 1993.
22. Packham, D. E., *Handbook of Adhesives*, Wiley and Sons, New York, 1993.
23. Thompson, K. I., *TAPPI Journal*, 9, 1988, p. 157.
24. Spink, J. W., R. J. Hernandez, and J. R. Giacin, "Correlation of Heat Sealing and Hot Tack Parameters with Molecular Properties of EVA Heat Sealant Copolymers," TAPPI Proceedings Polymer Laminations and Coatings Conference, 1991, p. 597.
25. Theller, H., "Heat sealability of Flexible Web Materials in Hot-Bar Sealing Applications," *J. Plastic Film and Sheeting*, 5:1, 66, 1989.
26. Bor, M. and E. Wolf, *Principle of Optics*, 6th ed., Pergamon Press, Oxford, 1980.
27. Hall, C. W., S. Singh, and Burgess, "Use of a Proposed Test Procedure for Quantifying the Triboelectric Charging Propensity of Packaging Films," *J. Testing and Evaluation*, 21, 57, 1992.

28. Cameron, A. C., R. M. Beaudry, N. Banks, and M. Yelanich, "Modified-Atmosphere Packaging of Blue Berry Fruit: Modeling Respiration and Package Oxygen Partial Pressure as Function of Temperature." *J. Amer. Soc. Hort. Sci.,* 119:3, 534, 1994.

29. Ikegami, T. Nagashima, M. Shimoda, Y. Tanaka, and Y. Osajima, "Sorption of Volatile Compounds in Aqueous Solutions by EVA Films," *J. Food Sci.,* 56, (2), 500, 1991.

30. Koszinoswski, J., "Diffusion and Solubility of n-Alkanes in Polyolefines," *J. Applied Polym. Sci.,* 31, 1805, 1986.

31. Tseng, D. T., R. Matthews, J. Gregory, C. Wei, and R. Littell, "Sorption of Ethyl Butyrate and Octanal Constituents of Oranges Essences by Polymeric Absorbents," *J. Food Sci.,* 59, (4), 801, 1993.

32. Nielsen, T. J., "Limonene and Mircene Sorption into Refillable Polyethylene Terephthalate, and Washing Effects on Removal of Sorbed Compounds," *J. Food Sci.,* 59, (1), 227, 1994.

33. Kwapong, O. Y. and J. H. Hotchkiss, "Comparative Sorption of Aroma Compounds by Polyethylene and Ionomer Food-Contact Plastics," *J. Food Sci.,* 52, (3), 761, 1987.

34. Crank, J., *The Mathematics of Diffusion,* 2d ed., Clarendon Press, Oxford, 1975.

35. Michaels A. S., W. R. Wieth, and J. A. Barrie, "Diffusion of gases in Poly(ethylene Terephthalate)." *J. Appl. Phys.,* 34:13, 1963.

36. Berens, A. R., "The Solubility of Vinyl Chloride in PVC," *Angew. Makromol. Chem.,* 47, 85, 1985.

37. Frisch, H. C., "Sorption and Transport in Glassy Polymers—A Review," *Polym. Eng. Sci.,* 20, (1), 2, 1980.

38. Salame, M., "Prediction of Gas Barrier Properties of High Polymers," *Polym. Eng. Sci.,* 26: 22, 1986, pp. 1543–6.

39. Hernandez, R. J. and R. Gavara, "Sorption and Transport of Water in Nylon-6 Films," *J. Polym. Sci.: Part B: Polym. Phys,* 32, 2367, 1994.

40. Van Amerongen, G. J., "Solubility, Diffusion, and Permeation in Gutta-Percha," *J. Polym. Sci.,* 2, 318, 1947.

41. Michaels, A. S., W. R. Vieth, and J. A. Barrier, "Solution of Gases in Polyethylene Terephthalate," *J. Appl. Phys.,* 34, (1), 1963, pp. 1–12.

42. Puleo, A. C., D. R. Paul, and P. K. Wong, "Gas Sorption and Transport in semicrystalline poly(4-methyl-1-pentene)," *Polymer,* 30, 1989, pp. 1357–66.

43. Michaels, A. S., W. R. Vieth, and J. A. Barrier, "Diffusion of Gases in Polyethylene Terephthalate," *J. Appl. Phys.,* 34, (1), 1963, pp. 13–30.

44. Veith, W. R., *Diffusion In and Through Polymers,* Hanser Publishers, New York, New York, 1991.

45. Meares, P., "Transient Permeation of Organic Vapors Through Polymers Membranes," *J. Appl. Polym. Sci.,* 9, 917, 1965.

46. Barrer, R. M., "Diffusion and Permeation in Heterogeneous Media," in *Diffusion in Polymers,* Crank and Park (eds.), Academic Press, New York, 1968, pp. 165–217.

47. DeLassus, P. T., "Barrier Expectations for Polymer Combinations," *TAPPI Journal,* March, 216, 1988.

48. Nielsen, L. E., "Models for the Permeability of Filled Polymer Systems," *J. Macromol. Sci. (Chem.),* A1(5):929, 1967.

49. Steingiser, S., G. J. Rubb, and M. Salame, *Encyclopedia of Chemical Technology,* vol. 3, 3d ed., 1980, p. 480.

50. Bernabeo, A. E., W. S. Creasy, and L. M. Robeson, *J. Polym. Sci.: Polym. Chem. Ed.,* 13: 1979, 1975.

51. Paul, D. R., *J. Mem. Sci.* 18: 75, 1984.

52. Pye, D. G., M. M. Moher, and M. Panar, "Measurement of Gas Permeability of Polymers, II. Apparatus for Determination of the Permeability of Mixed Gases and Vapors," *J. Appl. Polym. Sci.,* 20, 1921, 1976.

53. Niebergall, W., A. Humeid, and W. Blochl, "The Aroma Permeability of Packaging Films and Its Determination by Means of a Newly Developed Measuring Apparatus," *Lebesn. Wiss U. Technol.,* 11, (1), 1, 1978.

54. Hernandez, R. J., J. R. Giacin, and A. L. Baner, "The Evaluation of the Aroma Barrier Properties of Polymer Films," *J. Plastic Film and Sheeting,* 2, 187, 1986.

55. Zobel, M. G., "The Odour Permeability of Polypropylene Packaging Films," *Polymer Testing,* 5, (2), 153, 1985.

56. Hilton, B. W. and S. Y. Nee, "Permeability of Organic Vapors Through Packaging Films," *Ind. Eng. Chem. Prod. Res. Dev.,* 17, (1), 80, 1978.

57. Gavara, R. and R. J. Hernandez, "The Effect of Water on the Transport of Oxygen Through Nylon-6 Films," *J. Polym. Sci. Part B,* 32, 2375, 1994.

58. Pasternak, R. A., J. F. Scimscheimer, and J. Heller, "A Dynamic Approach to Diffusion and Permeation Experiments," *J. Polym. Sci. Part A-2,* 8:467, 1978.

59. Gavara, R. and R. J. Hernandez, "Consistency Test for Continuous Flow Permeability Experimental Data," *J. Plastic and Film Sheeting,* 9, 126, 1993.

60. Toi, K., H. Suzuki, I. Ikemoto, T. Ito, and T. Kasai, "Permeation and Sorption for Oxygen and Nitrogen into Polyimide Membrane," *J. Polym. Sci: Part B: Polym. Phys.,* 33, 777, 1995.

61. Wahid, M. A., "Permeation of 2-Nonanone Vapor Through LLDPE Affinity Films as Applied to Modified Atmosphere Packaging," M.S. Thesis, Michigan State University, 1996.

62. Paul, D. R. and W. J. Koros, "Effect of Partially Immobilized Sorption on Permeability and Diffusion Lag-Time," *J. Polym. Sci.: Polym. Phys. Ed.,* 14, 675, 1976.

63. Mizrahi, S. and M. Karel, "Moisture Transfer in a Packaged Product in Isothermal Storage," *J. Food Processing and Preservation,* 1, 225, 1977.

64. Kim, J., R. J. Hernandez, and G. Burgess, "Application of the Finite Difference Method to Estimate the Moisture Content of a Pharmaceutical Tablet in a Blister Package," *J. Packag. Technol. and Sci.,* Submitted.

65. Lee, L. M., *Modern Plastics,* 53, (12), 66, 1976.

66. Masi, P. and D. R. Paul, "Modeling Gas Transport in Packaging Applications," *J. Mem. Sci.* 12, 137, 1982.

67. Reid, R. C., K. R. Didman, A. D. Schwope, and D. E. Till, "Loss of Adjuvants from Polymer Films to Food or Food Simulants," *Ind. Chem. Eng. Prod. Res. Dev.,* 19, (4), 580, (1980).

68. Goydan, R., A. D. Schwope, R. C. Rear, and G. Cramer, "High Temperature Migration of Antioxidants from Polyolefins," *Food Addit. Contam.,* 7, 323, 1990.

69. Markelov, M., M. M. Alger, T. D. Lickly, and E. M. Rosen, "Migration Studies of Acrylonitrile from Commercial Copolymers," *Ind. Eng. Chem. Res.,* 31, 2140, 1992.

70. Begley, T. H. and H. C. Hollifield, "Recycled Polymers in Food Packaging: Migration Considerations," *Food Technology,* 11, 109, 1993.

71. Laoubi, S., A. Feigenbaum, and J. M. Vergnaud, "Effect of Functional Barrier Thickness for a Food Package with Recycled Polymer," *Packaging Technol. and Sci.,* 8, 249, 1995.

72. Berens, A. R., "Diffusion of Gases and Vapors in Rigid PVC," *J. Vinyl Technol.,* 1:1, 8, 1979.

CHAPTER 9
ELASTOMERS AND ENGINEERING THERMOPLASTICS FOR AUTOMOTIVE APPLICATIONS

Ronald Toth
Chrysler Corporation
Auburn Hills, Michigan

9.1 INTRODUCTION[1,2]

The word *rubber* refers to a class of polymeric materials comprising vulcanized cis-1,4-polyisoprene (natural rubber) as well as several cross-linked synthetic polymers that exhibit elastic mechanical properties. The properties that most easily distinguish a rubber from an engineering thermoplastic are elastic memory, high value of ultimate elongation, and very low value of elastic modulus. Elastic memory is the ability of a material to quickly recover to essentially its original shape after removal of the deforming force. Ultimate elongation is the amount of stretch imparted in a material at its yield point—200 percent or more for natural rubber and synthetic elastomers. Elastic modulus is a ratio of stress to strain as the material is deformed under a dynamic load.

In this chapter, we will use the word *elastomer* in a generic sense to include all synthetic rubbers. The word *rubber* will refer to specific materials, such as natural rubber or nitrile rubber, where rubber is historically a part of the name of the material. The word *plastic* will specifically denote engineering thermoplastics plus two selected thermoset plastic-material types with significant automotive applications.

There is an additional category of thermoplastic materials that has elastic memory similar to rubber and the processing characteristics of traditional engineering thermoplastics. These materials are known as thermoplastic elastomers (TPE). TPE use is growing in many industries, including automotive. A large portion of this growth is at the expense of traditional thermoset elastomers in carefully selected applications.

The major focus of this chapter will be on thermoset elastomers. The selection principles developed are applicable to both elastomers and engineering thermoplastics.

9.2 ELASTOMERS AND PLASTICS

Engineers generally think of elastomers and plastics as significantly different types of materials. There are many practical reasons for doing so. For example, the mechanical and functional requirements of parts made from these materials are quite different; and plastics and elastomers could not be readily used interchangeably. Also, the processing technologies for plastics and elastomers are significantly different. Elastomers are thermoset materials, requiring heat for the vulcanization process, which develops a network of cross links between the molecular chains. Plastics require heat only for flow in processing. The technical vocabularies and ways of thinking in the elastomer and plastic industries are quite different. Further, plastic-materials technology resides primarily in the polymer, while elastomer-materials technology, within any given polymer class, is profoundly dependent on the details of specific compound formulation. Finally, plastics can have the image of high-tech, space-age materials. In contrast, elastomers, which require elaborate recipes and have a manufacturing tradition reaching back to the early days of the industrial revolution, can seem more akin to the products of black art.

Despite these important differences, elastomers, plastics, and the resins used in composite materials are fundamentally very similar.

9.2.1 Similarities of Elastomers and Plastics

Except for silicones, all plastics and elastomers are carbon-based polymers, made up from the linking of one or more monomers into long molecular chains. In fact, many of the same monomers are found in both thermoplastic and elastomeric polymers. Examples are styrene, acrylonitrile, ethylene, propylene, and acrylic acid and its esters. Prior to vulcanization, an elastomer is in a thermoplastic state.

The harsh automotive environment can present a number of challenges from heat, oxidation, ultraviolet (UV) light, and chemical attack caused by aggressive fluids. Because of the chemical similarities between elastomers and plastics, these materials are susceptible to many of the same types of chemical attack. Therefore, many of the same material-selection principles come into play for both plastics and elastomers.

9.2.2 Differences Between Elastomers and Plastics[1,3,4]

An engineering thermoplastic is a rigid solid at room temperature. An uncured elastomer can be a soft, pliable gum, or a leathery, flexible solid. The fundamental property that accounts for the differences among elastomers and the differences between elastomers and engineering thermoplastics is molecular mobility.

This distinction refers to a modulus of the material—a ratio between an applied force and the amount of resulting deformation. Deformation is greatest when the molecular resistance to motion is the least. The factors that influence the resistance to motion include various intermolecular attractions, crystallinity, the presence of side chains, and physical entanglements of the molecular strands. The cumulative effect of these factors determines the glass-transition temperature (T_g) of the polymer. Below this temperature, a plastic or elastomeric polymer is a supercooled liquid and behaves in many ways like a rigid solid. Above this temperature, a cross-linked elastomer will display rubber-like properties. The behavior of a thermoplastic material above its glass-transition temperature will depend on its level of crystallinity. A noncrystalline (amorphous) polymer will display a large decrease in modulus at the glass-transition temperature. The modulus is then relatively insensitive to temperature between the glass-transition and melting temperatures. A partially crystalline plastic will display a relatively small modulus change at the glass-transition temperature, followed by a steadily decreasing modulus as the temperature increases.

Elastomers in the uncured state have glass-transition temperatures well below room temperature. Engineering thermoplastics have glass-transition temperatures that fall in a wide temperature range, extending substantially above and below room temperature.

The presence of cross links is the major physical difference between a cured elastomer and an engineering thermoplastic. The cross links anchor the elastomer molecular chains together at a cross-link density such that many molecular units exist between cross-link sites. Thus, the resistance to deformation is increased only slightly. However, the presence of the cross links increases the resilience of the elastomer, causing it to spring back to its original shape when the deforming stress is removed. Resilience is the percentage of energy returned per cycle of rapid deformation upon removal of the stress. A related concept is hysteresis, the percentage of energy lost, determined as one minus the resilience. A practical result of hysteresis in a mechanical system is damping, which reduces the magnitude of a system's vibration. In principle, most amorphous thermoplastic materials could be converted into elastomers, in some appropriate temperature range, by the controlled addition of a limited number of cross links.

In addition to resilience, there is another practical result from the cross linking of elastomer molecules. Creep resistance and compression set resistance are related phenomena describing the dimensional stability of fabricated articles over time under a compressive load. The presence of cross links helps in maintaining a desired geometry. Engineering thermoplastics, which contain no cross links, are subject to creep, or cold flow, under load. They also have poor compression set resistance. Set is a permanent deformation that occurs under a load. In sulfur-cured elastomers, set also results from the breaking of existing, less stable cross links, followed by the formation of new, more stable cross links that lock the molecules into the distorted geometry. Cross-linked elastomers can vary significantly in their set resistance. Compounding, choice of vulcanization system, and degree of cross linking can profoundly affect set resistance.

9.3   WHY USE ELASTOMERS?

Elastomers are used in a wide variety of automotive applications because they offer a varied combination of physical characteristics—resiliency, flexibility, extensibility, and durability—that is unmatched by other classes of material. Other useful properties of elastomers include resistance to abrasion; resistance to aqueous fluids, including nonoxidizing acids, and other environmental factors; ease of fabrication by molding or extrusion processes; and energy absorption. Further, the properties of elastomers can be significantly modified by compounding. Thus, elastomeric materials can be, and frequently are, tailor-made for specific applications.

9.4   WHY USE PLASTICS?

Plastics are also used in a wide variety of automotive applications. They compete with other materials based on weight savings, design flexibility, parts consolidation, and an ease of fabrication that translates into cost savings in the finished part. Plastics were originally used as alternatives to other materials in applications that benefit from their advantageous performance characteristics and processing properties. Now plastics are frequently chosen in preference to other materials when new parts are designed.

9.5 THE IMPORTANCE OF MATERIAL SELECTION

There is great variety in the functions that rubber and plastic materials perform in the automobile. The exterior, interior, and under-hood environments all differ in their effects on materials. The challenge in materials selection is to pick a cost-effective material that meets all of the functional requirements of the part in its application environment.

There are at least 18 different elastomers used in automotive components, and almost as many thermoplastics. They range from commodity materials such as natural rubber and polypropylene sold in truckload quantities for under $1 per pound, to sophisticated specialty polymers such as fluorosilicones and low-temperature-resistant fluorocarbon elastomers selling for over $50 per pound.

9.5.1 Functionality

Elastomers perform a variety of functions, including static-load bearing, dynamic energy transfer, vibration damping, and fluid containment. While the great variety of available polymer types offers the opportunity to specifically tailor an elastomer to its application, it also presents an even greater number of opportunities for inappropriate material selection. Elastomers vary not only in price but also with regard to all of the advantageous physical characteristics already mentioned. They also vary widely with regard to their chemical make-up, polarity, thermal stability, low-temperature properties, dynamic properties, adhesion, hardness, and set resistance. The choice of an elastomer for a specific application must be based on the chemical type of the polymer and the specific set of physical characteristics required by the application.

The ability to function in the intended use is the most important factor in elastomer selection. The selected material must have the right combination of mechanical and chemical properties to perform its designed function.

9.5.2 Reliability and Durability

Reliability is the assurance that a component or a material will perform its intended function in a particular operating environment. The environmental factors include both high and low temperature extremes, exposure to fluids of various polarities, physical stresses from impact and vibration, oxygen, ozone (from the atmosphere or electrical components), UV light (from sunlight), and electrical fields. For example, the functions of a radiator hose and a fuel filler neck hose are virtually identical— fluid transport. However, the fluids differ so greatly in their polarity and useful temperatures that different elastomers must be selected for the two types of hoses.

Durability results from continuously reliable performance over an extended time period. Durability and reliability must be considered because many elastomer articles are vital to the operation of the vehicle. Often they are inaccessible and must therefore continue to function reliably for the life of the vehicle. Other articles must perform consistently with recommended maintenance schedules.

When an elastomer or plastic part fails, the failure is often the result of chemical deterioration of the organic polymer, usually caused by oxidation. The polymer can react with ozone, with free radicals caused by the action of ultraviolet light, or at elevated temperatures with oxygen from the atmosphere or air trapped during mixing. In any polymer, the chemical attack can occur in the molecular chain in one of two ways. If the polymer chain is severed, the molecular weight is reduced, and mechanical properties deteriorate. If two or more polymer chains are joined, the result is an uncontrolled amount of cross linking, which can cause embrittlement, shrinkage, and cracking. In plastics, molecular-weight reduction, accompanied by a loss in impact strength, is the most common effect of oxidation.

Elastomers can also fail in another way: from attack at the cross-link sites. This reaction, known as reversion or devulcanization, causes a loss in elastomeric properties.

9.5.3 Regulatory and Safety Considerations

Certain automotive parts perform functions that are vital not only to the performance of the vehicle but also to the safety of the occupant. Other parts are important in meeting regulations to control emissions and other potential sources of pollution. These parts must be made of materials that will not only survive but continue to function for the life of the vehicle with absolute reliability. An obvious example is brake-line hose. Many other elastomers in steering, suspension, and fuel-delivery applications also contribute to safe vehicle operation.

The consideration of regulatory effects and safety implications is significant because of the serious outcomes that would result from failure. The engineer must be aware of any regulatory or safety implications that may affect a part.

One aspect of material selection is intelligent and informed risk taking. These risks must be viewed much more conservatively when regulations and safety are involved. Tables 9.1 and 9.2 point out some affected elastomeric components.

9.5.4 Cost

Frequently, choices will occur among several candidate elastomers or plastics, all of which are capable of meeting functional, reliability, and durability requirements. The material selection should then be made on the basis of cost. Since compound formulation and material processing significantly affect the cost of finished elastomeric articles, they must be considered along with the relative cost of the various polymers. The automotive industry is extremely competitive and highly cost-conscious. Therefore, a significant factor in proper material selection is a consideration of cost-effectiveness. It can be a delicate balancing act when material cost is weighed against functionality, reliability, and durability.

9.6 ENGINEERING THE AUTOMOBILE

The engineer responsible for designing or approving elastomeric or plastic components does not need to have an intimate knowledge of these materials. However, a

TABLE 9.1 Rubber Components with Regulatory Considerations

System	Component	Regulation
Emission sensor	Hoses Tubing Seals	Tailpipe emissions
Fuel delivery	Seals Fuel filler hose	Fuel vapor emissions
Fuel delivery	Fuel supply hose	Vehicle safety requirements
Brakes	Hoses hydraulic vacuum antilock	Vehicle safety requirements

basic understanding of the various polymer types and their respective strengths and weaknesses will be helpful at every stage of the product life cycle, from concept to performance evaluation and failure analysis. Even though the engineer is responsible for the part, a materials expert will probably be involved in making the material selection. The engineer needs to be able to describe the functional requirements of the part and the details of the application environment.

The manufacturing supply chain from polymer source to engineer can have several links. It is quite possible, for example, that a molded elastomer article buried deep in a component will be designed by someone in the supply chain. This person may not have a clear understanding of the conditions that the article will be exposed to. A clear communication of part requirements all along the supply chain is vital to proper part design and material selection.

TABLE 9.2 Rubber Components with Safety Implications

System	Components
Brakes	Hoses hydraulic vacuum antilock Seals Cups
Power steering	Hoses pressure return Seals
Suspension	Motor mounts Tie rod end seals Bushings
Fuel delivery	Fuel supply hose Pump seals Injector O-rings

9.6.1 The Automotive Environment

The automobile is a torture chamber for organic materials. Even a simple exterior decorative part is exposed to constant abuse from widely fluctuating temperatures, acid rain, ozone and other atmospheric pollutants, and ultraviolet light.

Under the hood, the situation is even worse. Engines, exhaust manifolds, catalytic converters, and brakes can all add heat to the under-hood compartment. Electrical components can cause an increase in ozone concentration under the hood. Current styling concepts have led to smaller engine compartments with reduced capacity for efficient air flow. Under-hood components must therefore operate at elevated temperatures, sometimes in excess of 400°F.

Over the past several years, automotive fluids have become much more aggressive. Type SH engine oils can contain higher levels or more aggressive types of amines as dispersing agents for foreign matter in the oil. These amines can react with fluorocarbon elastomers, causing embrittlement. Reformulated gasoline contains oxygenated hydrocarbons, which alter the polarity of the fuel blend.

9.6.2 Specific Elastomeric Components by Function

Each functional elastomeric component has a set of performance and quality requirements and an operating environment. This section will briefly describe several types of automotive components, along with their typical requirements and environments.

Energy Management. Although it may not be the primary function, most automotive elastomeric components manage energy in some way. Among these are specific components that have energy management as their primary function. Elastomers manage energy by returning it to the energy source (resilience), dissipating it as heat (hysteresis, damping), or transferring it to a movable component. The usual model for describing the stress-strain relationships in elastomers is a dashpot in combination with a spring. This model represents the viscoelastic nature of elastomers; they behave in some respects like solids and in other respects like liquids. Which effect predominates in a particular situation is determined by the temperature of the system, the molecular structure of the polymer, and the shape and size of the part.

Damping and Isolation.[2,3] The practical effect of resilience in a vibrating system is the decoupling of the vibration source from other components that are attached via the resilient material. Decoupling occurs when the vibrational energy is returned to the source rather than being transmitted through the elastomer. This phenomenon, known as isolation, occurs when the forced vibration of the system is well above (three times or more) the natural frequency of the isolator. The natural frequency is a function of the shape of the isolator and the modulus of the material of construction.

Isolation is the most important function of engine mounts. The vibration that results from running an automobile engine is an approximately steady-state oscillation. Engine mounts isolate this oscillation and prevent its transfer to the passenger compartment. Since natural rubber is the most resilient elastomer, it is the preferred material for engine-mount applications. Although it is somewhat less resilient, polychloroprene is also used in cases where the operating temperatures exceed the upper temperature limit of natural rubber. In contrast, a material with low resilience would directly transmit a large portion of the vibrational energy.

Damping is significant for events that have a low vibration frequency, occur infrequently, or have low energy. The most significant example is road vibration. Body mounts that connect the vehicle body to the chassis absorb and dissipate this energy. Butyl rubber is ideal for this application because its high hysteresis gives it the ability to absorb and dissipate large amounts of energy.

Many molded rubber parts serve as bumpers and fillers in various locations throughout the automobile. Deck-lid over-slam bumpers are typical examples. Since these are often less critical applications, there is considerable latitude in material choice. An ethylene-propylene-based rubber (EPDM) is frequently used in these applications because of its heat and ozone resistance. Other low-cost elastomers require protective additives that can cause staining or discoloration on adjacent parts.

Energy Transfer. The most obvious example of a part used for energy transfer is the belt. The belt takes energy from a rotating pulley and transfers it to other rotating pulleys that drive various components such as air conditioners. Belts come in a variety of designs—flat, V-shaped, or multigrooved, depending on the specific application. In each case, the belt is composed of an elastomeric matrix combined with fabric, a longitudinal array of cords, or both. Since the belt is rapidly and continuously flexing as it is routed around the pulleys, hysteresis is an important consideration. First, any energy dissipated will reduce the efficiency of the system. Even more importantly, since the action of a moving belt is continuous and intense, hysteresis heat buildup can affect the mechanical properties of the belt material or lead to thermal degradation. Polychloroprene has been used for many years in automotive belts. Rising under-hood temperatures have driven the use of materials with greater thermal stability, such as hydrogenated nitrile (HNBR).

Material Transport. In the automobile, material transport typically involves the movement of a fluid from a reservoir to an active component. The reservoir may simply be a storage device, as in a fuel tank or washer-fluid bottle, or it may provide cooling, as in the radiator or transmission oil cooler.

Ducts. An automotive duct generally has a diameter of three to six inches, a length of up to three feet, and side walls with ribs or a convoluted structure. The convolutes maintain strength when wall thicknesses are small and increase the flexibility of the item. Ducts typically transport air that is at or near atmospheric pressure. Although the application generally is not demanding, the construction of a duct can be complex, since both ends must attach to other components. For processing reasons, designs of this type can be achieved more easily using thermoplastic elastomer (TPE) materials than with elastomers. Because of the variety of end-use environments, both elastomers and TPE materials are used.

Tubing and Connectors. Tubing is a flexible, hollow extrusion with a round cross section. It is used for transporting liquids or vapors under conditions of moderate pressure or to maintain vacuum. For an elastomeric tube construction, the burst strength is equivalent to the tensile strength of the material. Other important quality considerations for tubing include concentricity of the inner and outer diameters and consistency of wall thickness, both around the circumference and along the length of the tube. Material selection is dictated by ambient operating conditions and resistance to the fluid being transported. An example is the washer-fluid tube, which transports the fluid from the reservoir to the spray nozzle. EPDM is a good choice because it is inexpensive and highly resistant to water-alcohol mixtures. It is also the typical choice for vacuum tubing.

Dual extrusions are used when the inside and outside environments are significantly different. For example, a nitrile inner tube might be specified for its resistance to hydrocarbon fluids. Since nitrile has only moderate ozone resistance, it can be combined with a tube cover made from a compatible ozone-resistant elastomer such as chlorinated polyethylene. Dual extrusions can also be used to control part cost.

Fluorocarbon elastomer has excellent fluid resistance and can be used for tubing that handles flexible fuel mixtures. A considerable cost savings results when the fluoroelastomer is dual-extruded with a cover made from a less costly elastomer such as ethylene acrylic copolymer.

Connectors are molded parts that attach tubes or hoses together or to other components. Quality considerations include dimensional accuracy and stability. The same material selection factors used for tubing also apply to connectors.

Hoses. Hoses are similar to tubes but more complex. They are constructed from two extruded elastomeric layers with a fiber reinforcement to impart burst strength and resistance to kinking and collapse under vacuum. The inner layer is essentially a tube. After it is extruded, the fiber reinforcement is applied in a knit, braid, or spiral pattern around the tube exterior. The second elastomer layer, or cover, is then extruded around this assembly, and the completed hose is vulcanized. The vulcanization is typically carried out in a steam autoclave. The tube and cover need not be made from the same elastomer. The different elastomers must, however, have compatible cure systems and good adhesion to each other as well as to the reinforcing fiber. Different elastomers are selected when the inner and outer service requirements are different. For example, the tube material may be exposed to a particular fluid, while the cover material is exposed to hot air and ozone. Elastomer pairs are selected to optimize the hose performance under these conditions. Fiber selection is determined by compatibility with and adhesion to the elastomers, tensile strength, and thermal stability. The selected fiber material generally has an upper service temperature well above that of the elastomers. Typical reinforcements include cotton, rayon, polyester, polyamide, polyvinyl alcohol, and polyaramid fibers. Certain hoses, such as those used in brake systems and power steering, must be designed to withstand high internal pressures with absolute reliability.

Government regulations, especially in California, are mandating greatly reduced hydrocarbon emissions from all parts of the vehicle. New improved hose designs include a barrier layer of a fluorine-containing elastomer or plastic material with high permeation resistance. This layer is used in addition to, or instead of, the tube elastomer. Because of the expense of the barrier materials and the extra processing, barrier hoses carry a high price tag.

Fluid and Vapor Sealing. [5] Sealing applications are among the most important elastomer uses. They include a wide variety of parts and designs. Some of these applications, such as O-rings and lip seals, contain only the elastomer. In others, the elastomer is a binder for other materials. Certain gaskets are made from rubber-coated metal sheet. A controlled amount of swell in the contained fluid is often beneficial and can compensate for dimensional changes caused by compression set and creep in the gasket materials.

For static seals, the operating temperature and resistance to the fluids being contained are the primary considerations in material selection. Other factors include compatibility with or adhesion to other materials in the component. For dynamic seals, abrasion resistance is also an important consideration.

Gaskets. Gaskets fall under the general category of static seals. Materials used in gaskets include paper, solid metals, cork, and elastomers, either alone or in various combinations. A typical gasket is made from a single or multilayer flat sheet material. Sealing is achieved by deformation under load, which accommodates irregularities in the two joining surfaces. Therefore, the gasket material must have a combination of strength and compressibility that is compatible with the range of pressures used for sealing at all operating temperatures, along with resistance to the fluids that are contained in the system. Static seals must also accommodate some relative motion between the joining surfaces resulting from vibration, differential thermal expansion, and externally applied forces. Nitrile rubber (NBR), styrene-butadiene copolymer (SBR), and silicone have many automotive static-sealing applications.

Dynamic Seals. Dynamic seals provide containment for fluids or greases against a moving part. These parts typically have a round cross section. Examples are rotating drive shafts and the reciprocating pistons found in automatic transmissions. Precise dimensional control of O-rings, D-rings, and lip seals is essential to maintaining seal integrity against a moving part. Therefore, elastomer swell must be controlled at a low level, and the elastomer must be both highly abrasion-resistant and nonabrasive to the moving part. Depending on the fluids and operating temperatures, automotive dynamic seals are made from NBR, acrylic, ethylene-acrylic copolymer, and fluoroelastomers.

Body Sealing. Body-sealing components are typically made of an extruded elastomer, often in combination with other materials. Door and deck lid seals fill the build variations between these parts and the mating body openings. Their function is to keep water and dust out of the passenger compartment and the trunk. Door seals also prevent air leakage and the associated wind noise. An important design consideration for door seals is the balance between seal integrity and door-closing effort. Other sealing components include glass run channels and belt-line seals. Body seals vary in complexity from simple extrusions to coextrusions of dense (not foamed) and foamed elastomer, possibly combining relatively harder and softer versions of the same dense elastomer. The extrusion may incorporate metal or plastic reinforcing carriers, flocking, or various kinds of surface coatings. EPDM is the most common body-sealing elastomer due to its excellent resistance to oxidation and UV light. In the past, body seals have been made from SBR and polychloroprene.

9.6.3 Specific Plastic Components by Function

Typically, elastomers are used in applications where other material types are not appropriate. In contrast, plastics are almost always in competition with other candidate materials. After all factors are weighed, the decision can still be based as much on intangible factors as it is on technical or cost-driven reasons.

Structural and Mechanical Applications. When plastic is chosen for a structural application, it is often as a replacement for metal. In general, a one-for-one replacement of metal with plastic will yield a weight savings. Also, plastics can easily be formed into shapes that are difficult to achieve with metals. By designing the part in an attractive shape that favors plastic forming, the engineer can achieve a savings in cost and weight along with a cosmetic improvement. In addition, a molded plastic part can be made with a wide variety of colors and surface textures to coordinate with and enhance an aesthetic theme. This presents an additional cost-savings opportunity over metals, since metal parts often require painting or coating to achieve a desired appearance and to protect the metal from corrosion.

Plastics should also be considered when a new part is conceived. The design engineer has the ability to choose among a wide variety of materials, including plastics, metals, wood, and pressed-fiber products. Plastics can successfully compete in these applications because they are cost-effective and lightweight, and they can be designed to incorporate multiple functions.

Appropriate plastics for consideration in these applications are acetals, polyamides, and specialty thermoplastics. Selection among these materials will be based on mechanical requirements, and the temperature and chemical end-use environ-

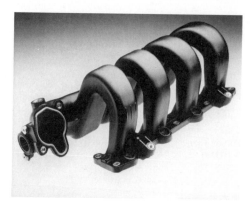

FIGURE 9.1   Exhaust manifold.

ment for the part in question. Figure 9.1 illustrates plastic use in an underhood application: the exhaust manifold for the Dodge and Plymouth Neon. The selected material is glass-reinforced polyamide (nylon).

Exterior Applications.   Plastics in exterior applications are also substitutes for or alternatives to metals. The advantages of plastics are in design flexibility, parts consolidation, and ease of assembly. Plastics are also superior to metals in dent and corrosion resistance. The most prominent mechanically functional exterior plastic application is the door handle. Other types of exterior parts include body panels and exterior ornamentation. In both of these broad categories, specific parts may be either molded in black or painted to match the color, surface finish, and gloss level of adjacent painted metal body panels. Painting involves a significant cost penalty over molding in color. It is used on high-line vehicles to enhance a styling theme or to achieve product differentiation.

Several different plastics are used in exterior applications, including acrylonitrile-butadiene-styrene polymer (ABS), polycarbonate and polycarbonate blends, acrylic-styrene-acrylonitrile terpolymer (ASA), thermoplastic olefins (TPO), and polyamides (nylon). Material selection is made on an application- and vehicle-specific basis, driven by the trade-off between styling requirements and total system cost. Table 9.3 indicates several plastic exterior applications, along with appropriate material choices.

Transparencies.   Automotive applications for plastic transparencies are in head-lamp and tail-lamp lenses and overlays for the instrument-panel cluster, message center, and radio face. Plastic transparencies have been able to replace glass because they offer distinct advantages. The specific gravity of most plastics is about one half that of glass, so there is a significant weight savings. Also, plastics have a higher level of transparency, so more light is available for illumination. They also offer the opportunity for molded-in details, as seen in lamp lenses. Most automotive transparencies are made of acrylic or polycarbonate.

Interior Decorative Parts.   The main function of automotive interior trim parts is to mask the details of auto assembly. Plastics also offer the opportunity to enhance styling concepts with molded shapes and surface textures that would be difficult, expensive, or impossible to duplicate in other materials. Materials use in interior applications include ABS and polypropylene.

TABLE 9.3 Selected Exterior Applications and Plastic Materials

Exterior application	Material	Potential advantages	Disadvantages
Fascia	Thermoplastic olefin (TPO)	Mold in color (dark) available Low cost (unpainted) Paintable Ease of assembly	Molded in color only available with grained surface Painting drives up cost
Fascia	Polyurethane (RIM)	Dimensional control Body colors available Weight savings Can use recycled material Low molding pressures	Must be painted Painting can cost up to $40
Hood and deck lid	Sheet molding compound (SMC)	Design flexibility Low cost tooling	Material cost higher than metal
Fender	Thermoplastic polyester (PET)	Design flexibility Surface appearance	May require off-line paint operation Complex assembly
Mirror housing Cowl screen Belt molding	ABS	Body colors available	Requires painting
Mirror housing Cowl screen Belt molding	ASA	Mold in color (black) Low cost	Black only Scratch and mar concerns
Mirror housing Cowl screen Belt molding	Polycarbonate blends	Mold in color (black) Low cost Paintable	Painting drives up cost
Belt molding	PVC	High gloss mold in color	Black only
Glazing encapsulation	PVC	High gloss mold in color Low cost	Requires additional adhesive High molding pressure
Window encapsulation	Polyurethane (RIM)	Low molding pressure Design flexibility	High cost Can require painting

Containers. Plastics offer several practical advantages over other materials for use as radiator-overflow reservoirs, washer-fluid containers, and gas tanks. Plastics combine flexibility with resilience and are therefore less fragile than glass and more resistant to puncture than metals. Permeability is a major concern, and this must be addressed by appropriate material selection.

9.7 AUTOMOTIVE ELASTOMERS

Elastomers are named for the polymers that they contain, since the polymer is the major determinant of the elastomer's mechanical and chemical properties. These properties can be modified, within limits, by a variety of additives. The capability for modification gives the compounder the ability to fine-tune the elastomer to the specific requirements of the application.

One method of classifying and specifying elastomers is by arranging them according to their useful temperature range and resistance to hydrocarbon oil. SAE J200[6] and ASTM D2000,[7] both of which are described in Sec. 9.10.1, use this method. The accompanying chart shows the placement of automotive elastomers in the J200 classification system. In general, polymers with higher temperature resistance and greater oil resistance also have higher cost. Table 9.4 lists common elastomers and their standard polymer designations.[8]

9.7.1 Polymers[9,10]

The stability and performance characteristics of an elastomer depend on the chemical nature of the polymer chain. The most significant chemical features of the polymer are the elements in the polymer chain, residual double bonds (unsaturation) in the polymer chain, and the presence and type of side-chain groups.

Polymers with no double bonds in the backbone chain, such as EPDM, exhibit the greatest stability toward heat and oxidation. Double bonds in the polymer chain are weak links for attack by heat and oxidizing chemicals. Side chains can shield the double bond to some extent, with a corresponding increase in polymer stability. Moreover, side chains of various chemical types can affect the polarity of the polymer and

TABLE 9.4 Common Automotive Elastomers

Polymer	Elements in the polymer chain	ASTM D 1418 Polymer Symbol	SAE J200 Type and Class
Butyl	C	IIR	CA, DA
Bromobutyl	C	BIIR	CA, DA
Chlorobutyl	C	CIIR	CA, DA
Chlorinated polyethylene	C	CM	BC
Chlorosulfonated polyethylene	C	CSM	CE
Epichlorohydrin	C, O	CO, ECO, GECO	CH
Ethylene-acrylic copolymer	C	AEM	EE
Ethylene-propylene-diene terpolymer	C	EPDM	AA, BA, CA, DA
Ethylene-vinyl acetate	C	EVM	EC to EJ
Fluoroelastomer	C	FKM	HK
Fluorosilicone	Si, O	FVMQ	FK
Natural rubber	C	NR	AA
Nitrile	C	NBR	BF, BG, BK, CH
Polyacrylate	C, O	ACM	DF, DH
Polychloroprene	C	CR	BC, BE, DE
Styrene-butadiene copolymer	C	SBR	AA, BA
Silicone	Si, O	MQ, VMQ, PVMQ	FC, FE

therefore also influence resistance to oil swell. For certain polymers, such as poly-chloroprene, side chains play a further role in the cross-linking chemical reaction.

Many commercial elastomers are blends of two polymers. The purpose of blending is to overcome a weakness in one of the polymers. To be blended successfully, the polymers must have similar polarity. To be cross linked together, they must have compatible vulcanization systems. The risk in forming polymer blends is that the resulting polymer will retain all of the worst properties of the original polymers. Further treatment of polymer blends is beyond the scope of this chapter.

The Big Three.   Three common elastomers—natural rubber, nitrile rubber, and EPDM—have wide applicability in the automotive industry. They also illustrate the variability in polymer types described in the preceding section. Because of their differences in chemical makeup, their use applications generally do not overlap. The engineer should consider these three materials first for most automotive applications. They are described in Table 9.5.

TABLE 9.5   The Big Three of Automotive Elastomers

Elastomer	Strengths	Weaknesses
Natural rubber	Tensile strength Flex fatigue life Vibration damping Low temperature flexibility Low cost	Ozone attack Heat attack Oil swell

Nitrile rubber	Oil resistance
	Moderate cost
		Ozone attack
		Heat attack
		Brittle at low temperature
		  (depending on acrylonitrile
		  content)

EPDM	Ozone resistance Heat resistance Low cost	Oil swell Low flex fatigue life Difficult to adhere Can be compounded for poor properties

Natural Rubber (NR).[2]   Natural rubber (NR) is derived from the latex found under the bark of the rubber tree, *Hevea brasiliensis*. Chemically, the polymer is 100 percent cis-polyisoprene. This chemical structure gives the rubber molecule a geometric regularity that allows for a very stable close packing known as crystallinity. Unvulcanized NR can crystallize spontaneously during storage, especially at reduced temperatures. It will also crystallize rapidly when stretched due to the application of a tensile load. In either case, the crystallization is reversible. When it occurs from cold storage, the crystallinity may be reversed, or melted, by warming the rubber. Strain-induced crystallinity reverses when the stress is removed. The relatively high tensile strength and tear resistance of natural rubber both result from the crystallization that occurs when the rubber is stretched. NR offers the best combination of resilience, tensile strength, fatigue life, and abrasion resistance of any elastomer. It is the material of choice for applications such as engine mounts, where the perfor-

TABLE 9.6   Automotive Applications
for Natural Rubber

Application	Reasons for use
Engine mounts	Isolation
	Resilience
	Flex fatigue life
	Set resistance
Suspension system	Resilience
Tie rod end seals	Flex fatigue life
Bushings	Set resistance
Isolators	
Bumpers and grommets	Set resistance
	Tear strength
Miscellaneous molded parts	Low cost

mance is determined by these dynamic properties. Most other automotive applications require greater heat and fluid resistance than can be obtained with NR. Table 9.6 lists typical applications for natural rubber.

Natural rubber is generally vulcanized with sulfur by a variety of techniques. The least expensive methods yield many polysulfidic linkages in which the cross links contain a chain of linked sulfur atoms. The more expensive, efficient vulcanization systems result in a greater proportion of monosulfidic linkages. The vulcanized elastomer then has improved heat resistance and set resistance, and the formulation is better able to accommodate certain stabilizers such as antioxidants.

Natural rubber contains carbon-to-carbon double bonds (unsaturation) in the polymer backbone. It is therefore subject to attack from heat, oxidizing agents, ozone, and UV light. In automotive applications, natural rubber is always compounded with carbon black. This imparts UV resistance and causes a significant increase in tensile strength and other mechanical properties. Oxidation resistance is not affected by carbon black.

Ethylene-Propylene-Diene Terpolymer (EPDM).   EPDM technology dates from the early 1960s. It is a synthetic terpolymer of ethylene, propylene, and one of three dienes: dicyclopentadiene, norbornadiene, or hexadiene. One of the diene's double bonds is a reaction site for polymerization. It becomes saturated as it is incorporated into the polymer backbone. The other double bond is then located on a pendant side-chain, not in the polymer back bone. EPDM can then be cross-linked with sulfur via this residual double bond. Different commercial grades of EPDM are made by varying the ethylene-to-propylene ratio and the type and amount of diene monomer. These factors are of interest to the compounder and can affect the processibility of the compound as well as the performance properties of the vulcanized elastomer.

Since EPDM has no unsaturation in the polymer chain, it is inherently resistant to attack from oxidizing agents, ozone, and UV light. It has the greatest thermal stability and therefore the highest operating temperature of all elastomers except the expensive specialty polymers. The upper temperature limit and set resistance can be further increased by using a peroxide cure mechanism. This will significantly affect part cost because the curing package is more expensive, and also because the use of peroxide limits the amount of extending oils that can be incorporated into the formulation.

EPDM is a highly cost-effective material. First, it has the lowest specific gravity of any elastomer, so it yields more parts per pound of material. Second, although the cost per pound for EPDM is higher than for natural rubber or SBR, EPDM is considerably cheaper than most other elastomers. Finally, EPDM can accept a much greater loading of extending oils and mineral fillers than any other elastomer. This offers the opportunity to manufacture compounds in which the polymer content is only a small fraction of the total formulation. Thus, EPDM is uniquely susceptible to the production of off-specification material due to the excessive incorporation of inexpensive fillers. Excess filler content will reduce the mechanical properties of the compound and lead to inferior set resistance.

The weakness of EPDM is its vulnerability to hydrocarbon fluids and greases. The polymer will rapidly absorb a huge quantity of these materials, resulting in a volume increase of up to several hundred percent and a dramatic loss of mechanical properties. Although EPDM has reasonably good resilience, it is not a good high-temperature alternative to NR in dynamic applications. EPDM has relatively low tensile strength and tear resistance, leading to poor flex fatigue life. Table 9.7 lists typical applications for EPDM.

Nitrile Rubber (NBR) and Derivatives. Shortages of natural rubber during the two World Wars drove the development of synthetic elastomers during the decades from 1920 to 1950. Nitrile rubber was developed in the 1930s, mainly in Germany. NBR is a copolymer of acrylonitrile and butadiene. The acrylonitrile monomer introduces polarity into the polymer chain, making NBR resistant to hydrocarbon

TABLE 9.7 Automotive Applications for EPDM

Application	Reasons for use

Body sealing
  Door seals
  Glass run seals
  Deck lid seals
  Body plugs and grommets
  Body sealing (foam)
	Set resistance
	Weathering stability
	Nonstaining
	Nonblooming

Cooling system
  Heater hoses
  Radiator hoses
  Coolant overflow tubing
  Coolant bypass hose
  Radiator gaskets
	Set resistance
	Heat resistance
	Water resistance

Vacuum tubing	Ozone resistance Heat resistance Low cost
Tail pipe hangers	Heat resistance

Brake system applications
  Hydraulic brake hose
  Seals
	Brake-fluid resistance
	Ozone resistance
	Heat resistance

Windshield washer tubing	Washer fluid resistance Ozone resistance Heat resistance Set resistance Nonstaining Low cost
Air supply ducts	Low cost

fluids and greases. The polymer chain also has residual unsaturation from the butadiene. Therefore, NBR is only moderately stable toward oxidation, ozone, and heat. The acrylonitrile functionality does offer some protection, so NBR is superior to NR, especially in thermal stability. The stability can be further increased by the addition of antioxidants and antiozonants.

NBR is generally used where resistance to hydrocarbon fluids and greases at moderate temperatures is of primary importance. Typical applications include oil seals, oil cooler hoses, and fuel-line hoses. At high temperatures, NBR is subject to heat-induced cross linking, leading to embrittlement and loss of elastomeric properties. Due to the increasing severity of automotive-application environments, NBR is sometimes replaced by specialty elastomers. Table 9.8 lists typical applications for nitrile rubber.

TABLE 9.8 Automotive Applications for Nitrile Rubber

Application	Reasons for use
Engine gaskets	Oil resistance
	Temperature resistance
	Set resistance
Fuel system	Fuel resistance
Filler neck hose inner	Set resistance
Vent hose inner	
Oil and grease seals	Oil resistance
	Temperature resistance
	Set resistance
Fluid- and vapor-resistant tubing	Oil resistance
	Temperature resistance
Manifold pressure sensor	Set resistance
Vapor cannister	

NBR is available in several grades, with acrylonitrile content varying between 15 and 50 percent. As the nitrile content increases, resistance to hydrocarbons also increases, as does tensile strength and heat resistance. However, resilience, low-temperature flexibility, and set resistance all decrease. These results are consistent with what one would expect from considerations of polarity.

Resistance to heat, oxidation, and ozone can also be increased by the addition of protective additives. Polyvinyl chloride (PVC) plastic is often incorporated into NBR by the polymer supplier. The PVC improves processibility and causes a dramatic improvement in ozone resistance while also increasing tensile strength. However, PVC is capable of releasing small amounts of chloride ion, probably as hydrochloric acid. In vapor-tubing applications that are in sequence with delicate electronic sensors, PVC/NBR blends must be avoided, since the chloride ion will foul the sensor.

Hydrogenated nitrile rubber (HNBR) is made from nitrile rubber polymer that is further modified in a catalytic hydrogenating process. This reaction selectively hydrogenates the unsaturation in the polymer backbone while leaving the acrylonitrile groups unaffected. HNBR retains the useful properties of NBR with the additional advantages of ozone and weathering resistance, plus an upper-temperature limit in the range of EPDM. An HNBR polymer has greater resilience than an NBR polymer at the same nitrile content.

Because the residual unsaturation following hydrogenation is low, HNBR has good resistance to sulfur and sulfur-containing compounds. It also has good high-temperature set resistance when peroxide-cured. HNBR is suitable for fuel and oil seals, hoses, and belts. It is being evaluated for use in engine mounts. Because HNBR is a very expensive polymer, it is only used in circumstances where the end-use requirements are severe and long part-life is essential.

Carboxylated nitrile rubber (XNBR) incorporates up to 10 percent of a third comonomer with organic acid functionality. XNBR has dramatically improved abrasion resistance and strength when compared to NBR. When XNBR is blended with up to 25 percent of polybutadiene (BR), a significant further increase in abrasion resistance occurs, along with improvements in resilience and low-temperature flexibility. However, tensile strength and oil resistance are both reduced by the addition of BR. XNBR can be difficult to process, and it requires special formulations to avoid sticking to mixer surfaces and premature vulcanization (scorch).

Other Polymers. Other polymers are used when there is a specific need to go outside the use ranges of NR, NBR, and EPDM. These needs are generally caused by combinations of environmental factors, such as heat plus oil resistance, for example. Some of these polymers like polychloroprene have a rich automotive history but limited current use. Others such as fluoroelastomers will probably be used more extensively as operating conditions become more severe.

Polymers Based on Carbon or Carbon and Oxygen

Butyl Rubber (IIR). Butyl rubber is a copolymer of isobutylene and isoprene. IIR has a very low level of residual unsaturation, making it relatively more stable toward heat and oxidation than natural rubber. It also has good resistance to ozone and weathering. It is comparable to NR in oil-swell susceptibility. The polymer has excellent flexibility but very low resilience due to its molecular structure. Therefore, it absorbs a great deal of any mechanical energy that is put into it. This energy absorption is responsible for butyl rubber's vibration-damping properties, which are maintained over a broad temperature range. Butyl rubber is therefore an excellent choice where high vibration-damping is required, as in body mounts and suspension bumpers. Butyl is also the best choice when low permeation to air and moisture is required. It serves as an excellent barrier to these two fluids. Because of its low level of unsaturation, IIR is relatively difficult to cure, requiring long cycles at high temperature.

There are two commercially available halogenated butyl rubber derivatives: bromobutyl (BIIR) and chlorobutyl (CIIR). The halogen atoms are incorporated into the polymer on the isoprene units. These derivatives have chemical properties that allow the polymers to be covulcanized with other elastomers more readily than IIR. The end-use properties and suitable applications are similar to those for IIR. A typical application is as the cover material for air-conditioning hoses.

Ethylene-Acrylic Copolymer (AEM). Copolymers of ethylene can be elastomeric if the other comonomer is capable of preventing the crystallinity that occurs naturally in polyethylene. Acrylate esters are effective comonomers in this respect. The newest varieties of AEM contain only the ethylene and acrylic monomers and must be peroxide-cured. The original AEM polymer also contains a small amount of a third comonomer that introduces reactive sites for cross linking. This variety of AEM does not respond well to peroxide cure systems. Organic amines are used as vulcanizing agents through the carboxyl group of the third comonomer. Amine-cured AEM requires an additional postcure heat cycle to fully cross link the polymer, since undercure causes a significant reduction in mechanical properties. In either case, the polymer backbone contains no residual unsaturation, so it has excellent resistance to heat, ozone, and oxidation. It also has fair to good hydrocarbon oil resistance and low-temperature properties. AEM has special compounding requirements and can be difficult to process. Automotive uses for AEM include boots and dynamic fluid seals.

Ethylene-Vinyl Acetate Copolymer (EVM). Like AEM, EVM owes its elastomeric properties to the absence of crystallinity in the copolymer. EVM can only be cured by peroxides. The resulting elastomer has excellent thermal stability, even surpassing EPDM. It has outstanding set resistance at elevated temperatures but poor set resistance below room temperature. Similarly, the best resilience is obtained at elevated temperatures. Resistance to ozone, oxidation, and weathering are excellent. Resistance to hydrocarbon fluids is strongly dependent on the vinyl acetate content. Higher levels of vinyl acetate increase the polarity of the polymer and therefore increase the hydrocarbon resistance. When properly compounded, EVM has levels of volume resistivity and dielectric constant suitable for low-voltage insulation and cable sheathing. An important automotive use is in wire insulation. EVM has excellent adhesion to metals and can be difficult to mold.

Polyacrylate (ACM). The polymer backbone of a polyacrylate elastomer contains alkyl or alkoxy esters of acrylic acid plus a small amount of a reactive co-monomer that introduces sites for cross linking. Because of its fully saturated backbone, the polymer has excellent resistance to heat, oxidation, and ozone. It is one of the few organic elastomers with higher heat resistance than EPDM. ACM is resistant to aliphatic hydrocarbons and petroleum-derived lubricants and greases. It is also resistant to the sulfur additives found in some varieties of gear lubricant and the amine additives of engine oils. It is not highly resistant to automotive fuels due to their aromatic content. ACM is not stable towards continuous exposure to hot water or steam and can also be affected by other polar solvents.

The polymers are available from a number of sources worldwide. The various polymers have differing cure systems, although most are based on amine chemistry. Because of these different cure systems, ACM polymers from different sources cannot be used interchangeably in compound formulations. Relatively long cure times or a postcure operation in a hot-air oven are required to develop optimum properties.

Typical automotive uses include seals for transmissions, oil pans, valve stems, and crankshafts.

Styrene-Butadiene Copolymer (SBR). SBR is one of the oldest synthetic elastomers, and it is still among the most important. It was developed as a replacement for natural rubber, and its general properties are similar to those of NR. It offers a slight advantage in thermal stability, and it is sometimes blended with NR for that reason. It is, however, inferior to NR in tear strength, low-temperature flexibility, and hysteresis. Therefore, NR-SBR blends must be properly formulated to achieve the correct balance of properties. SBR is superior to NR in several of its processing properties. It also has excellent abrasion resistance and resistance to hydraulic brake fluid. Molded SBR parts have a strong tendency to bloom due to the presence of added stabilizers. The bloom creates an unsightly appearance and can cause migration staining when in contact with paint. Therefore, SBR is not generally used in readily visible locations or in contact with painted surfaces. SBR's primary automotive use is in passenger car tires. This application accounts for the great majority of worldwide SBR production. It is also used in suspension components and auto-body interior grommets and as cups and diaphragms in hydraulic brake systems.

Carbon-Based Polymers Containing Chlorine.[11] The incorporation of halogen atoms onto a polymer molecule has three major effects: oil resistance is increased, flammability is decreased, and the halogen atoms offer alternative sites for polymerization with nonsulfur-containing cure systems. In each case the chlorine is attached to a carbon atom in the polymer backbone.

Chlorinated Polyethylene (CM). Unmodified polyethylene cannot be converted into a suitable elastomer because of its molecular crystallinity. Chlorination of the polyethylene polymer chain disrupts the crystallinity and allows the polymer to become elastomeric upon vulcanization. CM can be vulcanized by peroxides or by certain nitrogen-containing organic compounds that cross link via the chlorine atoms.

CM has good heat resistance and excellent resistance to ozone. It has reasonably good resistance to fuels and oils. Applications include hoses, protective boots, and flexible dust shields.

Chlorosulfonated Polyethylene (CSM). The chlorosulfonation of polyethylene yields a polymer with varying amounts of chlorination on the polymer chain, similar to that found in CM. The resulting polymer also contains a relatively small number of sulfonyl chloride groups. Although the chemistry and manufacturing technology for CM and CSM are quite different, the end-use properties of CM and CSM are similar, with CSM having greater oil resistance. Because of its high dielectric strength, CSM is used in wire insulation and spark-plug boots. It also offers a greater latitude in cross linking methods via the sulfonyl chloride groups.

CSM and CM have base polymer prices in the same range. Historically, CSM has been more cost-effective because its cure systems are less expensive. Currently the cost of CSM is increasing, and the polymer has been on allocation due to supply shortages. These factors will probably favor CM in the near term as engineers search for alternative materials to CSM.

Epichlorohydrin Rubber (CO, ECO, GECO). The epichlorohydrin polymer group includes the homopolymer (CO), the copolymer with ethylene oxide (ECO), and terpolymers (GECO). These elastomers have excellent thermal stability and resistance to ozone, weathering, and hydrocarbon fuels. Oil-swell resistance is in the range found with nitrile and polyacrylate elastomers. The various polymers exhibit widely differing performance levels with regard to resilience and low-temperature flexibility. The ECO version has dynamic properties similar to those found with natural rubber. Applications include seals and tubes in air-conditioning and fuel systems. Epichlorohydrin polymers are susceptible to reversion (devulcanization) in the presence of oxidized fuels (sour gasoline). Therefore, they should not be used in the fuel systems of fuel-injected vehicles.

Polychloroprene (Neoprene, CR).[12] Polychloroprene is a very good general-purpose elastomer with many commercial applications. While CR is not usually the absolute best polymer choice with respect to any given physical or mechanical property, it is often a very good choice. Further, it offers a good balance of physical properties, often making it a good, cost-effective compromise selection. CR is more expensive than NR or EPDM but considerably less expensive than specialty synthetic polymers.

Automotive uses for CR include timing belts, accessory drive belts, and engine mounts. CR is among the best synthetic polymers in rebound resilience and offers an operating temperature at least 35°F higher than NR. It is the next logical choice for dynamic components when application temperatures are outside the range of NR. As automotive application environments drive material selection to the best material for each given application, CR use may continue to diminish.

Carbon-Based Polymers Containing Fluorine (Fluoroelastomers, FKM). FKM elastomers have outstanding chemical resistance and thermal stability due to the presence of fluorene in the polymer. Most varieties of FKM are copolymers or terpolymers of vinylidene fluoride with other fluorine-containing monomers, typically hexafluoropropylene (HFP). Several grades are available, with differing properties determined by the fluorene content. In general, as the amount of fluorene increases, chemical resistance also increases at the expense of low-temperature performance. FKM is particularly useful because it has excellent resistance to hydrocarbon fluids at temperatures far above the useful range of NBR as well as very good compression-set resistance. Specialty grades are available, with superior low-temperature performance.

All grades of FKM are extremely expensive, so they are only used when the service requirements of the application disqualify other elastomers. Typical applications include shaft seals, valve stem seals, fuel hoses, and injector O-rings. Except for fully submerged hoses such as those used in fuel tanks, the FKM layer of a hose is usually a thin veneer to control cost by minimizing material use.

FKM does have one chemical weakness. Copolymers of vinylidene fluoride and HFP are subject to attack from amines such as those used as engine-oil additives. There are three commercially available amine-stable polymers that incorporate tetrafluoroethylene along with other comonomers. These polymers have molecular arrangements that eliminate the reactive site for amine attack. They are under evaluation for automotive applications.

Silicone-Based Elastomers. Silicone polymers are formed by the condensation polymerization of organosilanes. The resulting polymers have an alternating sequence of silicone and oxygen atoms in the backbone chain. There is no unsaturation, so silicone polymers have unsurpassed ozone resistance. They also offer a unique combination of thermal and weathering stability, low-temperature flexibility, and compression set resistance. Silicones are highly energy-absorbing over a wide temperature and frequency range.

The simplest silicone polymer is dimethylpolysiloxane, commonly called methyl silicone (MQ). The replacement of a small number of methyl groups with other organic groups can have a significant effect on performance characteristics. Methyl-vinyl silicone (VMQ) offers an alternative curing route with improved set resistance through the unsaturated vinyl side-chains. Methyl-phenyl silicone (PMQ) has improved low-temperature flexibility. Methyl-phenyl-vinyl silicone (PVMQ) incorporates the characteristics of both the vinyl and phenyl varieties.

Silicone is used in applications where thermal and chemical stability are important. Oil-pan gaskets fit in this category. Silicone elastomers have excellent electrical properties and are used in ignition-wire jackets and insulation and in spark-plug boots.

Fluorosilicone elastomers (FVMQ) are partially fluorinated. Like other silicones, they have a polymer chain based on silicone and oxygen. They have excellent low-temperature properties and outstanding resistance to hydrocarbon fluids, and they are highly impermeable to engine oils. They are used in sealing applications with severe service requirements.

9.7.2   The Importance of Compounding[9,10]

Material decisions usually involve trade-offs, and nowhere is this more evident than in the compounding of elastomers. Through compounding, the rubber chemist is able to optimize selected properties or fine-tune the formulation to meet a desired balance of properties. For example, SAE J200 contains several tables of elastomer properties. The materials described in any particular table vary widely with regard to hardness, tensile strength, extensibility, and other material characteristics. A material that meets any specified set of requirements is designed through the formulation of the compound.

Every ingredient in the formulation, from the polymer to the stabilizing additives, is selected for a particular reason. Also, it is important to consider the interactions that are likely to occur among the various ingredients. The selected compound must meet all functional part requirements as well as the fabricator's processing needs while also supplying reasonable economics for both the supplier and the customer. In this section, we will consider the various ingredients in an elastomer formulation and their functions in the finished compound. Table 9.9 indicates some important compounding considerations.

TABLE 9.9 Compounding Trade-Offs

An improvement in	Can be at the expense of
Tensile strength hardness	Extensibility Dynamic properties
Dynamic properties	Thermal stability Compression set resistance
Compression set resistance	Flex fatigue strength Resilience
Oil resistance	Low temperature flexibility
Abrasion resistance	Resilience
Damping	Resilience Compression set resistance
Balanced properties	Processability
Optimized properties	Higher cost

Polymers. The first choice is in the selection of the specific polymer. Even when a particular polymer type, such as EPDM or polychloroprene, is specified, there are a number of commercial grades that can be selected. These grades can differ in molecular weight, comonomer type and ratio, the presence of additional cure-site monomers, polarity, crystallinity, included additives, and straight-chain versus branched structure. These variables will influence finished properties as well as such processing properties as viscosity, mixing time, moldability, extrudability, and response to vulcanizing agents.

Vulcanization Systems. Vulcanization is the process that transforms the polymer from a thermoplastic to an elastomer. The load-bearing capability of the material, determined by properties like set resistance, hardness, and modulus, are developed during vulcanization. Other changes that occur during vulcanization include increases in tensile strength and resilience and decreases in extensibility and solvent swell.

The major types of vulcanizing (cross linking, curing) systems are based on sulfur and peroxides. Other specific curatives exist for particular elastomers. A vulcanizing system can contain the cross-linking agent along with activators, accelerators, and retardants. This combination of chemicals is selected to achieve desired end-use properties and a manageable cure rate. If the cure is too slow, then the cycle times will be too long for efficient processing. If it is too fast, the polymer can set up prematurely, resulting in mold-packing problems and high scrap rates. In the rubber industry, premature vulcanization is known as scorch.

Sulfur vulcanization, the original process discovered by Charles Goodyear in 1839, can be used with any diene-based polymer. The residual double bonds in the polymer backbone become the active cure sites. However, direct curing with elemental sulfur is slow and difficult to control. For these reasons, a number of coagents are available that accelerate the cure and influence the vulcanization chemistry.

Elemental sulfur will bridge the polymer molecules via a combination of mono-, di-, and polysulfidic linkages. Monosulfidic cross links are the most thermally stable and result in the greatest compression set resistance available with sulfur systems. Di- and polysulfidic linkages break and reform with relative ease. As this happens, a greater proportion of cross links is converted to the monosulfidic variety. If this conversion takes place while the finished article is under stress, a permanent deformation will occur. This deformation contributes to poor set resistance.

A variety of organic nitrogen and sulfur-containing organic compounds offer more effective and efficient curing chemistry. They act either as accelerators to control the reaction rate or as sulfur donors that form the cross links. A cure package that contains both sulfur and selected coagents will result in a compound that cures with an acceptable molding cycle and also develops the desired end-use properties. Many accelerated sulfur-curing systems require zinc in order to be effective. Stearic acid is also an aid to effective curing. These two materials are typically included in sulfur vulcanization systems. Sulfur curing is commonly used with natural rubber, SBR, NBR, and EPDM.

Peroxide curing systems do not require double bonds in the polymer backbone, so they offer a cure method that is appropriate for polymers that cannot be sulfur-cured. The peroxide is heat-activated and decomposes to form free radicals. The free radicals are highly reactive towards organic materials and can cause polymer scission as well as cross linking. Free radicals can also react with other compound ingredients, such as plasticizers, stabilizers, and processing aids. Therefore, the types and amounts of these ingredients must be carefully controlled.

Diene-based polymers can also be peroxide-cured. The resulting elastomers have more stable cross links and can also have a higher cross-link density than is attainable when the same polymers are cross-linked with sulfur. This results in improved thermal stability and compression set resistance but reduced tensile strength, resilience, and extensibility.

Peroxide-cured elastomers are generally more expensive than sulfur-cured elastomers. The peroxide cure systems are more costly than sulfur systems, and the elastomer compound must be polymer-rich. Plasticizers and process aids cost less than polymers, but their use must be reduced or eliminated in peroxide-cured systems. For these reasons, peroxide-cured elastomers can be more difficult to process and fabricate. Peroxide curing is used with EPDM, CM, CSM, AEM, FKM, silicones, and fluorosilicones.

Other vulcanizing systems take advantage of specific chemically reactive groups on the polymer. They can be naturally occurring, such as the allylic chlorines that are present in about 1.5 percent of polymerized chloroprene molecules or specifically added cure-site monomers that alter the reactivity of the polymer. An example is the carboxyl functional group in ethylene–acrylic acid copolymer, which permits the use of amine curing systems.

Metal oxides are used with polychloroprene and chlorosulfonated polyethylene, polymers that contain a reactive chlorine atom. In polychloroprene, this leads to a linkage via an unsaturated four-carbon chain, accompanied by the conversion of some of the metal oxide to a chloride. In chlorosulfonated polyethylene, the reaction leads to a monosulfidic linkage, accompanied by the formation of water and metallic chlorides and sulfates.

Fluoroelastomers can be cured by either organic diamines or bisphenol systems containing metal-oxide acid acceptors. These are necessary because the bisphenol-cure reaction gives off hydrofluoric acid as a by-product.

Processing Aids, Plasticizers, and Extenders.   Several types of waxes, oils, esters, and low-molecular-weight polymers can be added to the elastomer formulation. They must be chosen for compatibility with the polymer and their effect on performance properties and processing characteristics. Processing aids serve several functions: as viscosity adjustors for mixing; as release agents for mixing and molding; and as flow modifiers for molding and extruding. Plasticizers aid in mixing and also enhance flexibility and resilience, especially at low temperatures. Extenders are used to lower the cost of the formulation and to increase the capacity of the polymer to accept fillers.

Fillers.   Large amounts of carbon black, minerals, talc, and various clays and silicates are incorporated in elastomer formulations. Any of these fillers will increase the hardness, stiffness, and abrasion resistance of the vulcanized compound. They can also affect mixing efficiency as well as flow and molding characteristics. Fillers with relatively large particle size (above 1 micron) extend and dilute the polymer but will not greatly influence the other physical properties. However, extremely finely divided materials (.01 to .1 microns) will act as reinforcement, significantly increasing tensile strength and tear resistance. Carbon black is particularly effective in reinforcing natural rubber and most synthetic organic elastomers. There are dozens of commercial grades of carbon black with differing particle size, surface characteristics, and degree of particle aggregation. Carbon black selection will profoundly affect the mechanical properties, resilience, and fatigue life of the elastomer. Fillers will affect the effectiveness of the vulcanization systems, so the accelerators must be adjusted accordingly.

Stabilizers.   Most elastomers need protection from oxidation, both during mixing and vulcanization and during service life. Oxidation is a free-radical process similar to peroxide curing. Thus, it can lead to excessive cross linking and embrittlement or chain scission and softening. In either case, the ability of an elastomer to perform its designed function will be impaired. Oxidation is retarded by the addition of antioxidant stabilizers. These are chemicals that react with and consume the hydrocarbon and hydroperoxy radicals that are formed during the oxidation reactions. Antioxidants are destroyed during these reactions. Therefore, although they extend the service life of the elastomer to several years or decades, the life is still finite.

Diene-based elastomers are subject to attack from ozone that is present in the atmosphere or generated by electrical components. The attack occurs at unsaturation points of the polymer on an exposed surface of the part. Unless the part is stretched, the damage will not penetrate below the surface. When the part is stressed, ozone attack is rapid and dramatic. Ozone resistance can be achieved by adding a wax that migrates to the surface and acts as a protective layer, by adding chemicals that scavenge the ozone, or both.

9.8 AUTOMOTIVE ENGINEERING THERMOPLASTICS[4,13,14]

Engineering thermoplastics are synthetic materials that can be used in load-bearing and structural applications or as mechanical components such as machine parts. The useful properties of thermoplastics derive mainly from their polymer molecules. Although fillers, additives, and stabilizers are important in improving performance, they generally do not play the essential role that nonpolymeric ingredients play in elastomer formulations. Plastics are usually processed by injection molding, blow molding, or extrusion. The polymer is purchased in a moldable form, usually as pellets. Additives and colorants are frequently blended into the polymer as received. The polymer portion of the plastic formulation is also known as the resin.

The molecular structures of thermoplastic polymers can be either amorphous or semicrystalline. Crystallinity can increase the rigidity, temperature resistance, and general chemical resistance of the polymer. However, crystalline materials are relatively brittle, leading to lower impact strength. Amorphous polymers are more dimensionally stable and less prone to warpage. However, they are susceptible to stress cracking by solvents. Reinforcement with glass fibers or annealing of the processed part can cause an improvement in stress-cracking resistance.

This section lists some of the more common engineering thermoplastics used in automotive applications. It is not intended to be all-inclusive. Several additional thermoplastics that are not structurally capable are also included because they fill important automotive needs.

9.8.1 General-Purpose Engineering Thermoplastics

There are many opportunities for the materials in this class to compete with each other. The material-selection process should include a consideration of structural strength and stability, weatherability, and surface characteristics such as friction and abrasion resistance as well as material cost.

Acetals. The simplest acetal is polyoxymethylene (POM). It has a highly crystalline molecular structure, resulting in hardness and rigidity. Compared to other crystalline polymers, acetals have good dimensional stability in molded parts. The fatigue life and creep resistance are superior to that found in polyamides. Like polyamides, acetals also have low surface friction. The alternating carbon-oxygen backbone has good resistance to ozone and fair to good resistance to UV light. The UV resistance can be improved by painting or by filling the polymer with carbon black or other UV-protecting materials. Acetals have many automotive applications, including fuel-system components and moving parts such as gears, structural components in the suspension system, and bushings. A set of opposed gears made from acetal and polyamide will run quietly with low friction.

Acrylics. Acrylics are polymers of acrylic or methacrylic acids and their esters. Although there are many possibilities for polymers and copolymers within this group, the most common acrylic in automotive applications is polymethyl methacrylate. It offers a level of transparency that meets or exceeds that of glass. It also accepts coloring agents very easily and is available in a wide variety of transparent colors. The acrylic polymer structure is fully saturated and offers excellent resistance to UV light and outdoor weathering. Acrylic is suitable for interior and exterior lens applications, and has been used in tail-lamp lenses for several decades.

Acrylic-Styrene-Acrylonitrile Terpolymer (ASA).   ASA is an amorphous polymer that is similar to ABS in most of its processing variables and performance characteristics. The acrylic content lends weatherability and UV stability to parts made from ASA. ASA is also resistant to hydrocarbon oils and greases and salt solutions. Polar organic fluids, including esters, ketones, and chlorinated solvents will attack ASA.

ASA is compatible with other polymers such as PC and PVC and can be blended into or coextruded with them. Like other hard plastics, ASA can cause squeaks and rattles when in contact with other plastic or metal surfaces. Because of its UV stability, ASA does not require painting. It can be molded in color for exterior applications. Automotive applications include interior and exterior trim parts, mirror housings, and body side moldings. Figure 9.2 illustrates an ASA mirror housing used on Dodge trucks.

FIGURE 9.2   Mirror housing.

Acrylonitrile-Butadiene-Styrene Polymer (ABS).   ABS is a polymer blend of styrene-acrylonitrile copolymer (SAN) with polybutadiene rubber that has been graft-polymerized with SAN. Various commercial grades of ABS are made by altering the ratios of these materials in the copolymer and the blend, along with the molecular weight. ABS has good chemical resistance and thermal resistance. However, the double bonds in the butadiene backbone reduce the UV light and weathering stability. Therefore, ABS must be painted or coated in most automotive applications. It is used for interior trim parts, such as door panels and instrument panel surfaces, and exterior parts such as mirror housings, cowl screens, and belt moldings. ABS is not usually used in load-bearing or impact-sensitive applications.

Polyamide (nylon, PA).   Polyamide, commonly known as nylon, is formed from the copolymerization of an organic diamine with a dicarboxylic acid, or from the homopolymerization of an organic monomer with both acid and amine functionality. The resulting polymers are highly crystalline, offering stiffness, strength, and heat resistance. Polyamide is a workhorse material in the automotive and other industries. Approximately one-half of all molded PA parts are used in automotive or other transport applications. The various PA polymers are named according to the number of atoms in the carbon chains of the amine and acid, respectively. For example, polyamide 6/12 is a copolymer of a six-carbon diamine and a twelve-carbon dicarboxylic acid.

PA is a relatively expensive thermoplastic with excellent resistance to chemicals, including oils, greases, and hydrocarbon fluids. It can be used in load-bearing components and exterior parts if UV stabilizers are added. Molding temperatures are high, and the resin must be dried before processing, because it is moisture absorbing. It is otherwise very easy to process and will mold into geometries with very thin cross sections. Water absorption will also occur in finished parts, causing dimensional changes and altering physical properties. The loss of moisture in a dry climate can lead to embrittlement. These factors should be considered when a part is designed. PA is selected where the use environment or the mechanical demands on the part are too severe for other thermoplastics. Typical applications include underhood components, exterior parts, and mechanical parts such as gears, cams, and bearings. Figure 9.3 illustrates the use of PA in the fuel-filler door and housing for Dodge and Plymouth Neon.

FIGURE 9.3   Fuel-filler door and housing.

Polycarbonate (PC).   Most commercial grades of polycarbonate are polyesters of carbonic acid with bisphenol A. PC offers excellent impact resistance along with a good balance of properties over a wide temperature range. It also has excellent creep resistance. However, it has only fair solvent resistance, and it scratches easily and will yellow when weathered unless protected by UV-absorbing additives. Automotive uses include headlamp lenses, radio knobs and faces, and other interior components such as instrument-panel top covers. The polymer is recyclable, and interior parts have been made from 100 percent recycled material. Because of its vulnerability to stress-induced cracking, PC is not used in load-bearing applications.

PC is also blended with ABS or polybutylene terephthalate (PBT) to form plastic alloys. The alloys with ABS remain UV-sensitive and must be painted when used in exterior applications. These materials remain ductile over a wide temperature range and can be used for energy management in components such as knee blockers. The blend with PBT has improved UV stability and can be used without paint when molded in black color.

Polypropylene (PP). Polypropylene is a highly cost-effective thermoplastic material with low specific gravity. It has good stiffness and tensile strength along with excellent fatigue resistance, making PP an ideal material for living hinge designs. Polypropylene has poor weathering stability and is particularly susceptible to UV radiation. Resistance to UV deterioration can be improved by the use of protective additives. Filling with glass fibers significantly increases tensile strength, impact strength, flexural modulus, and deflection temperature under load, with a corresponding reduction in elongation. Glass-filled PP can be used in load-bearing applications. Mineral-filled grades are used in miscellaneous molded articles to reduce cost and prevent warpage. The proper use of these reinforcing materials raises PP from a commodity thermoplastic to a material that competes with higher-performance engineering materials.

Polypropylene has a relatively narrow processing window. Special care must be taken to prevent residual stresses from the molding operation in order to avoid warping of finished parts. Automotive applications include interior trim parts, air-conditioning ducts, and underhood components. Interior parts can be molded in color. PP is resistant to contacts and spills of all underhood fluids, including battery acid.

Polyvinyl Chloride (PVC). PVC is an extremely versatile material with many commercial uses. Unplasticized PVC is extremely rigid. Typically, it is blended with another more flexible polymer, such as ABS, to overcome brittleness. Plasticized PVC is flexible and nonbrittle. It has excellent weathering stability, retaining mechanical properties and color, and is therefore suitable for exterior automotive applications. Body side moldings and encapsulated glazings are two examples. PVC is more cost-effective than RIM urethane for glazing encapsulation. However, the high molding temperatures and pressures limit the types and shapes of glazings that can be accommodated. PVC is also used as jacketing for electrical wire.

Thermoplastic Olefins (TPO). TPO is an injection-moldable blend of polypropylene with an ethylene-propylene rubber (EP or EPDM). With the rubber modification of the PP, TPO can also be considered as a variety of thermoplastic elastomer (TPE). It offers potential assembly advantages, since the parts can be heat-staked and ultrasonically welded. TPO is used in grills and fascias. In either case, the part may be molded in black or other dark color, or it may be painted to match the body color. Due to the softness of the material, parts molded in color should also have a grained or textured surface to hide mars. Painted surfaces are capable of high gloss. The painting process is very expensive and can add as much as $40 to the production cost of the part.

TPO is used in the vehicle interior in many of the same applications as polycarbonate—instrument-panel top covers and knee blockers. It is also used in airbag doors.

Thermoplastic Polyesters. Polyesters are a family of materials formed from the reaction of difunctional alcohols (glycols) with dicarboxylic acids. Two polyesters used in the automotive industry are polyethylene terephthalate (PET) and polybutylene terephthalate (PBT). PET and PBT both combine good UV stability with mechanical properties in the same range as PA. The polyesters have better dimension stability due to low moisture absorption, while PA has superior impact strength.

Dimensional stability and mechanical properties are both improved by the addition of glass fibers and mineral fillers. PET is appropriate for housings, racks, mirror parts, and latch mechanisms. It can be painted to match metal body panels and has been used in fenders. The more expensive PBT can be used in similar applications as well as in grills, wheel covers, and door handles. It has very good dielectric strength, making it an excellent choice for ignition components and electrical connectors.

9.8.2 Specialty Thermoplastics

High-end engineering thermoplastic materials are characterized by superior long-term heat resistance. They also have overall physical and mechanical properties that are typically higher than those found in most general-purpose thermoplastics when tested at elevated temperatures, although not necessarily when tested at room temperature. They are correspondingly more expensive and are therefore used only when the end-use environment requires a higher level of performance or stability.

Polyether-imide (PEI).   Polyether-imides are noncrystalline polymers made up of alternating aromatic ether and imide units. This molecular structure causes rigidity, strength, and impact resistance in fabricated parts over a wide temperature range. Even without reinforcement, PEI is among the strongest thermoplastics. The addition of glass or carbon fiber reinforcement further increases both strength and rigidity. PEI also has outstanding resistance to creep, even at high stress levels and elevated temperatures. PEI has better chemical resistance than most noncrystalline polymers and is therefore suitable for use in the engine compartment. Typical applications include connectors and MAP sensors.

Polyphenylene Sulfide (PPS).   Polyphenylene sulfide is a straight-chain polymer with a backbone composed of alternating sulfur atoms and benzene rings (phenyl groups). The sulfur-to-carbon attachments occur at the para positions; that is, at the 1 and 4 positions, diametrically opposed on the six-carbon ring. The polymer has excellent resistance to solvents and chemicals, inherent flame resistance, exceptional thermal stability, and good electrical insulating properties. PPS is among the easiest mold-filling polymers and is available in a variety of melt viscosities. However, it can cause molding problems due to flash and out-gassing. Filled grades are used in automotive applications for improved stiffness, strength, and processibility. Ultraviolet-light stability is good and can be increased to excellent by the addition of carbon black.
  Automotive applications for PPS include molded components in electrical, cooling, fuel-handling, and emission-control systems.

Polyphthalamide (PPA).   PPA is a polyamide-type copolymer in which the dicarboxylic acid is phthalic acid. It offers general chemical resistance similar to PA, along with greater strength, stiffness, thermal stability, friction-wear resistance, and dimensional stability. It is less moisture-absorbent than other PA polymers. Like other polyamides, PPA is affected by UV light. The addition of carbon black helps to maintain physical properties by protecting the polymer.
  PPA is used when service requirements are beyond the range of general-purpose PA. Examples are automatic transmission components, fuel-pump flanges, fuel rails and quick connects, cam sensors, and PCV valves.

Polysulfones.   Polyaryl (PAS) and polyether (PESV) sulfones have excellent thermal stability and mechanical strength but are still easily melt-processed. Sulfones are amorphous polymers, and when unfilled are transparent. Since the sulfones have excellent processing properties, they can be molded to precise tolerances. In addition to thermal stability, the sulfones offer dimensional stability and creep resistance, both of which are maintained for long periods at temperatures up to 180°C.

Polytetrafluoroethylene (PTFE).   As its name suggests, polytetrafluoroethylene is a straight-chain carbon-based polymer similar to polyethylene with fluorine atoms instead of hydrogen atoms attached to the carbons. PTFE has a low coefficient of friction and is, for all practical purposes, inert. It is highly resistant to water, chemicals and solvents, UV light, and ozone. PTFE is not melt-processible in the same way as other thermoplastics. Molded articles are made by a process of powder compression, followed by sintering, similar to the molding of ceramic goods. Extruded articles pass from the extrusion die through a series of heating zones, which also sinter the polymer.

Automotive applications take advantage of the low surface friction and chemical stability of PTFE. It is used in seals and rings for transmission and power steering systems, and in seals for shafts, compressors, and shock absorbers. Because of its inertness and impermeability, PTFE is used in fuel lines and as a barrier layer in elastomeric fuel hoses.

Thermoplastic Polyurethane (TPU).   TPU is a higher-priced alternative to TPO and is used in many of the same applications. Due to its lower thermal expansion, TPU offers better dimensional stability. Exterior applications include fascias and body side moldings. TPU is also used for instrument-panel skin, as an alternative to PVC. It offers a softer feel and greater durability. TPU is made without plasticizer and does not contribute to fogging on the windshield inner surface.

9.8.3   Selected Thermoset Plastics

In contrast to the other materials mentioned in this section, thermoset plastics are cross-linked thermoset polymers.

Polyurethanes (Urethanes, PUR).   Polyurethanes comprise an extremely versatile class of materials. Although they are primarily composed of only two major ingredients, a polyisocyanate and a polyol, urethanes offer a modulus range from highly flexible to extremely rigid in both dense and foamed configurations. Controllable factors that determine these properties are the specific chemistries of the two components, the degree of cross linking, foam density, and the presence of open or closed cell structure. Urethane foam is highly energy-absorbing and resilient. A molded rigid foam is used in the suspension system to manage the severe jolts and bouncing that can result from potholes. Flexible foam applications include carpet underlay, seat cushions, and sound-absorbing headliners.

Dense urethane parts for exterior applications are made by the reaction injection-molding process (RIM). The two components are pumped through a mix head and then into the mold. Both the isocyanate and the polyol have relatively low viscosities and can therefore be injected at low pressures and low temperatures. The low viscosities also allow easy molding of complex geometries. Thick and thin cross sections within a part like ribs, bosses, slots, and various kinds of inserts can all be easily accommodated. The RIM process is well understood and very forgiving of processing variability.

The two most prominent applications for RIM parts are as fascias and as glazing encapsulation. For fascias, the composition can be either filled or unfilled. Unfilled parts offer weight savings and a higher gloss-level surface. Filled parts can be made with tuned dimensional control to match adjacent panels or the overall requirements of the vehicle build. Hollow glass spheres can be used in the filler system to save weight. The system also accommodates reground, previously used RIM urethane fascias, offering an opportunity to increase recycled content in the vehicle. RIM fascias require painting for protection from UV.

RIM urethane is suitable for a greater variety of glazing encapsulations than the competitive material, PVC. The low molding temperatures will accommodate laminated glass constructions, and the low pressures allow both laminated glass and a greater degree of three-dimensional shape in the glass part. However, RIM is more expensive than PVC encapsulation. Traditional RIM encapsulations require painting or an applied coating. A more expensive UV-stable variety that eliminates this need is also used.

Sheet Molding Compound (SMC).   SMC is both highly cross-linked and highly filled. The resin (polymer) component of SMC is generally polyester. A more expensive material, vinyl ester, is also used when greater toughness is required for applications such as bumpers. SMC is composed of the ester resin, fiberglass and mineral fillers, viscosity modifiers, catalysts, and mold-release agents. It is received in sheet form, suitable for compression molding. The molded product combines high modulus with high strength and is therefore suitable for body panels such as hoods and deck lids that require this combination of physical properties. Strength and rigidity can also be increased by molding in reinforcing ribs and bosses. Other features can also be molded in, offering the opportunity for parts reduction and simplified assembly. SMC material is considerably more expensive than metal. However, tooling for the manufacturing of SMC parts involves much smaller capital expense than tooling for sheet steel. SMC is therefore ideal for selected body panels on vehicles with limited production volumes.

9.9   SELECTION PROTOCOL FOR AUTOMOTIVE ELASTOMERS AND PLASTICS

A simple method for making a material selection can be based on prior knowledge and experience of material characteristics combined with knowledge of the application requirements and environment. The known material chosen for a similar application is often a good choice or at least will offer a good starting point. The final selection can then be based on additional constraints imposed by the operating environment, such as temperature requirements or exposure to aggressive fluids. Historically, the move away from natural rubber and polychloroprene in many automotive applications has followed this route.

If the specialized knowledge and experience is lacking, or if the application is new and unique, a systematic selection protocol becomes necessary. One such protocol is to use a set of decision criteria to specifically eliminate inappropriate candidates from the available selection set. Certainly this is not the only possible selection method. Also a different group, or different ordering, of decision criteria than that suggested here may lead to more original or creative solutions to material selection problems. However, the automotive engineer must be quite conservative in his or her material selections. Considerations of durable part performance—often driven by government regulation, customer satisfaction, and competitive economics—emphasize the importance of a conservative approach. And above all, occupant safety must never be compromised.

The selection protocol suggested here considers four factors: survival in the application environment, durability of function in the application environment, material availability, and cost. The goal at each step is to effectively narrow the selection set by eliminating from consideration the largest number of candidate polymers. The factors are presented in an order designed to narrow the set most efficiently.

9.9.1 Selection Factor One—Survival in the Application Environment

The ability of an organic material to survive in the application environment can be judged by experience or by using tabulated information. The tables in this section contain information that will be helpful in making these judgments, based on resistance to various environmental factors. These factors include extremes of temperature, UV light, ozone, reactive chemicals, vibrational energy, and impact and static loading. Tables 9.10, 9.11, and 9.12 indicate the resistance of elastomers to heat, ozone, and hydrocarbon fluids, respectively.

Making the proper judgment involves a thorough knowledge of all of the environmental factors that can affect the part as well as the performance characteristics of the materials. Eliminate from consideration any materials that will not meet the application requirements for the part.

9.9.2 Selection Factor Two—Performance in the Intended Application

Given the constraints of survival, an elastomer should be selected based on its ability to perform the required function of the component. Since elastomers as a class

TABLE 9.10 Thermal Resistance of Elastomers

| | Approximate continuous upper temperature limits | |
Elastomer	Degrees Fahrenheit	Degrees Centigrade
Fluorosilicone	525	275
Silicone	525	275
Ethylene acrylic	350	175
Polyacrylate	350	175
Ethylene vinyl acetate	325	160
EPDM	300	150
Epichlorohydrin (CO)	300	150
Butyl	285	140
Chlorinated polyethylene	275	135
Chlorosulfonated polyethylene	275	135
Epichlorohydrin (ECO)	275	135
Halogenated butyl	275	135
Hydrogenated nitrile	275	135
Nitrile	235	110
Polychloroprene	225	105
SBR	175	70
Natural rubber	158	70

TABLE 9.11 Ozone Resistance of Elastomers

Elastomer	Ozone resistance	Polymer-chain features
EPDM	Excellent	No unsaturation
Epichlorohydrin	Excellent	No unsaturation
Silicone	Excellent	Si-O bonds
Fluorosilicone	Excellent	Si-O bonds
Ethylene acrylic	Excellent	No unsaturation
Ethylene vinyl acetate	Excellent	No unsaturation
Chlorinated polyethylene	Excellent	No unsaturation
Chlorosulfonated polyethylene	Excellent	No unsaturation
Polyacrylate	Excellent	No unsaturation
Polychloroprene	Good	Chlorine atoms protect polymer chain
Butyl, halobutyl	Good	Low unsaturation
Nitrile	Fair	Nitrile groups offer some protection
SBR	Poor	Highly unsaturated
Natural rubber	Poor	Highly unsaturated

have many properties in common like extensibility and flexibility, this choice is seldom definitive. There are, however, differences in degree among the various elastomers with regard to their functional capabilities. The relevant properties are resilience, tensile strength, flex fatigue strength, dielectric strength, set resistance, abrasion resistance, and fluid resistance. Table 9.13 presents a cross section of elastomeric functional applications and the appropriate polymer properties.

For plastics used in load-bearing applications, the heat deflection temperature is an important performance characteristic. It indicates the temperature at which the material begins to soften enough to deform when subjected to a specified load under standard test conditions (ASTM D 648). Reinforcement, especially by glass fibers, will increase the rigidity of the material, thus raising the heat-deflection tempera-

TABLE 9.12 Resistance of Elastomers to Hydrocarbon Fluids

Elastomer	SAE J200 Class	Hydrocarbon fluid resistance
Fluorosilicone	K	Good
Polyacrylate	J	Good
Hydrogenated nitrile	H to K	Good
Epichlorohydrin	G to H	Good
Nitrile	F to K	Good
Ethylene acrylic	F to K	Good
Silicone	F to K	Fair to good
Chlorosulfonated polyethylene	F	Fair
Chlorinated polyethylene	F	Fair
Polychloroprene	F	Fair to poor
Ethylene vinyl acetate	C to J	Poor to good
Butyl	A to B	Poor
Natural rubber	A	Poor
SBR	B	Poor
EPDM	A	Very poor

TABLE 9.13 Representative Functions of Automotive Elastomers

Function	Properties	Part	Material
Energy management • eliminate engine vibration	Isolation	Engine mount	NR
Energy management • eliminate mirror shake	Damping	Bracket pad	IIR
Energy transport • actuate pullies	Abrasion resistance Resilience Thermal stability	Timing belt	CR HNBR
Energy transport • actuate pullies	Abrasion resistance Resilience	Accessory drive belt	CR
Material transport • coolant	Water resistance Coolant resistance Set resistance	Coolant hose	EPDM
Material transport • fuel	Fuel resistance Permeation resistance Elongation Set resistance	Fuel filler Neck hose	NBR (inner) FKM (veneer)
Body sealing	Set resistance Weathering stability Nonstaining	Door seal Deck lid seal Glass run seal	EPDM
Static fluid-sealing	Set resistance Controlled swell in fluid	Engine head Gasket	NBR
Dynamic fluid-sealing	Set resistance Abrasion resistance Low swell in fluid	Transmission Lip seal	AEM
Electrical insulation	Electrical resistance Oil resistance Set resistance	Spark plug Boot	CSM

ture. The heat-deflection temperature is a thermal mechanical property, not an indication of thermal stability.

The accompanying tables present a rough ordering of the various elastomers with regard to some of these properties. Each of these properties is highly sensitive to both compounding and fabricating variables. The selected material must therefore be specified in such a way that the significant performance properties are assured.

9.9.3 Selection Factor Three—Material Availability

Ideally, the engineer would be able to take material availability for granted. In the real world, however, supply shortfalls, delays, or bottlenecks can cause a number of serious problems, especially in an industry that relies on just-in-time delivery.

Risk Factors in Material Availability. New materials can be unexpectedly withdrawn from the market for a variety of reasons. If a material has a single source, a strike or production disruption can shut off the supply. Materials from foreign sources are subject to shipping delays and pricing uncertainty due to currency fluctuations.

The best scenario for recovery from a material-supply disruption is the quick and easy substitution of another available material with similar performance characteristics. However, this will involve at least an accelerated qualification program with some associated level of risk, a probable cost increase, or both. Other scenarios are worse, including the possibility of a plant shutdown.

Managing Risk. Conservative practice dictates eliminating from consideration any material, the uninterrupted supply of which is uncertain. Disruption of the natural rubber supply during World War II was the primary driving force for the development of synthetic elastomers, both in North America and in Europe. At the time of this writing, chlorosulfonated polyethylene (CSM) is on allocation, and automotive engineers are forced to use alternate materials or blends.

9.9.4 Selection Factor Four—Cost

When the selection set has been narrowed to a small number of choices, it is appropriate to bring in cost considerations. This factor is listed last in order to eliminate any possibility of making an inappropriate choice by putting cost before performance or safety.

9.10 SPECIFICATION AND TESTING OF ELASTOMERS[3]

The laboratory tests that are generally run on elastomers are the original properties of tensile strength, elongation, hardness, tear strength, compression set, and the same set of tests run after aging in air or in a standard fluid at a specified temperature. The reasons for performing this group of tests are simple and pragmatic. The results are measurable, they are generally accepted in the industry, and compounds can be formulated to meet requirements expressed in terms of these tests.

Tensile-strength measurements are made on a tensile test machine such as an Instron. Because elastomers are viscoelastic materials, the stress-strain relationship is nonlinear except at small values of elongation. Modulus values, if they are measured at all, are taken at specific levels of elongation, typically 100 or 300 percent. The relevant physical properties actually obtained are tensile strength, as determined by the tensile load at break, and the percentage of elongation at break, also known as the ultimate elongation. To obtain meaningful and reproducible information, it is important to run these tests under controlled temperature and humidity conditions, using test specimens that are die cut, generally from molded slabs. The condition of the cutting edge on the die and the orientation of the cut sample in the molded slab will significantly affect the resulting measurements.

Hardness is usually measured in terms of the elastomer material's resistance to indentation. The measuring device is a spring-loaded small diameter probe such as a Shore Durometer. The resulting values generally fall between 40 and 80 on the Shore A scale and are used as a rough relative measure of stiffness. These measurements are quite operator-sensitive and frequently vary by as much as five points.

Tear-strength measurements are made using the tensile test machine and a test specimen that has a curved contour and a notch cut to a specific depth at the midpoint of the curved edge, or a 90° interior angle. Tear strength is an important use property in many rubber applications. It can also be important in the fabrication of complex molded shapes. A material or formulation with poor hot tear strength can be difficult to demold with acceptable scrap rates.

Compression set is determined as follows[15]:

1. Compress a disc-shaped specimen of known thickness by a given percentage of its thickness.
2. Hold the specimen in the compressed state at a specified time and temperature.
3. Remove the specimen from the compression fixture and allow it to relax at standard laboratory conditions.
4. Remeasure the specimen thickness.

Since elastomers are not perfectly elastic, there is some deformation due to creep and other factors. A low state of cure also contributes to the deformation. If the test is run at elevated temperature, some of the original cross links sever. New cross links form, resulting in a permanent deformation of the specimen.

The test results are reported in terms of the total amount of dimensional change, as a percentage of the total deformation. Thus, a compression set value of 20 percent indicates that the specimen actually recovered 80 percent of the compressed distance. A low value of compression set indicates superior performance in this test.

These tests measure basic material properties. Their use is most appropriate for defining a material and measuring process variability. However, they are at best proxies for performance criteria in the application. Increasing customer expectations and the use of product quality as a marketing tool are driving the automotive industry to develop new sets of test methods. These functional test methods will more closely relate to the actual performance requirements in the use environment.

9.10.1 SAE J200 Line Call-Out System

The Society of Automotive Engineers (SAE) has a recommended practice for identifying elastomeric materials for automotive uses. The practice is described in SAE document J200.[6] The same protocol is described by the American Society for Testing Materials (ASTM) in the document ASTM D2000.[7] The SAE J200/ASTM D2000 classification system provides a unified method of describing elastomeric materials based on a standard set of physical properties. These properties, along with the associated test methods and conditions, are presented in a shorthand technique known as a line call-out. The line call-out consists of a specifically ordered alphanumeric character string, where every character refers to a material property, test method, or test temperature.

In its simplest form, a line call-out will specify the basic requirements for the material. This information includes the type, class, and grade of the elastomer, along with its hardness and tensile strength. The type designation is a measure of heat-aging resistance. The class designation is a measure of oil-swell resistance. The grade designation is a single-digit number indicating whether the line call-out system allows additional requirements to be imposed on the specified material. A grade of 1 indicates that only the basic requirements are appropriate for the material. Any other grade indicates that a defined set of additional requirements, known as suffix requirements, can be used to complete the line call-out specification. The SAE J200 document contains tables that describe the allowable suffix requirements.

To illustrate the system we will dissect a line call-out example and describe the meaning of each element in sequence.

Example Line Call-Out.   SAEJ200M3BC614A14B14EO34F17. Elastomers that meet this line call-out include polychloroprene and chlorinated polyethylene.

"SAEJ200" indicates that the following alphanumeric string is an SAE J200 line call-out material description.

"M" indicates that the property values are expressed in metric-system units. Only metric units should be used in constructing a line call-out. However, the information tables in J200 contain both metric and English-system values to accommodate line call-outs written earlier than 1979.

"3" indicates that this is a grade-3 material. Therefore, the line call-out will contain suffix requirements, and those requirements will be selected from the appropriate table in SAE J200.

"BC" indicates the type and class of the material. Type B materials are evaluated by a heat-exposure test of 70 hours at 100°C. Class C materials exhibit a volume swell of no more than 120 percent after 70 hours immersion in a specified test oil at 100°C. The class designations associated with specific polymers are based on testing in ASTM #3 test oil. Because of OSHA considerations, this oil is no longer available. The substitute oil is designated IRM 903. Its effects on elastomers are similar to those caused by ASTM #3 oil, but the two oils are not completely equivalent. The volume swell values found using the two test oils are generally within a few percentage points.[16]

"6" indicates that the hardness measured using a Shore A durometer tester is 60 ± 5 points.

"14" indicates that the minimum tensile strength for the material is 14 MPa (2031 psi).

According to the table covering BC materials in SAE J200, the basic requirements for an M3BC614 material include a minimum elongation at break of 350 percent and a maximum compression set of 80 percent determined using test method ASTM D 395, a 22-hour test at 100°C. To meet the heat-age requirement, the following property changes are allowed: hardness ± 15 points; tensile strength ± 30 percent; ultimate elongation, 50 percent maximum. The table also indicates that the only suffix grades available at these hardness and tensile-strength values are 3 and 6. Grades 2, 4, and 5 are available, but at other values of hardness and tensile strength.

Since this is a grade-3 material, we can find additional characteristics in the table containing information on suffix requirements for BC materials. Other tables in the document indicate the meanings assigned to the various suffix characters. Table 9.14 indicates the suffix letters that are available in the J200 system.

For suffix A14, "A" indicates an additional heat-aging requirement. To meet the additional heat-age requirement, the following property changes are allowed: hardness ± 15 points; tensile strength ± 15 percent; ultimate elongation, 40 percent maximum. "1" indicates test method ASTM D 573, with a 70-hour heat-aging period. "4" indicates a test temperature of 100°C.

For suffix B14, "B" indicates an additional compression set requirement maximum compression set of 35 percent. "1" indicates test method ASTM D395, Method B. "4" indicates a test temperature of 100°C.

For suffix EO34, "EO" indicates an additional oil-exposure requirement. To meet the oil-exposure requirement, the following property changes after exposure are allowed: tensile strength, 60 percent; and ultimate elongation, 50 percent maximum. The maximum volume swell is 100 percent. "3" indicates test method ASTM 471, with a 70-hour immersion period. "4" indicates a test temperature of 100°C.

For suffix F17, "F" indicates low-temperature resistance. The sample must pass ASTM D 2137, method A. "1" indicates a 3-minute cold-exposure period. "7" indicates a test temperature of –40°C.

TABLE 9.14 Suffix Letters for SAE J200

Letter	Indicated test
A	Resistance to heat aging
B	Compression set resistance
C	Resistance to ozone or weathering
D	Compression—deflection
EA	Fluid resistance in water
EF	Fluid resistance in fuel
EO	Fluid resistance in oils
F	Low-temperature resistance
G	Resistance to tear
H	Resistance to flexing
J	Resistance to abrasion
K	Adhesion
M	Flammability
N	Resistance to impact
P	Resistance to staining
R	Resilience
Z	Additional special requirement (specific to the application)

The advantages of the line call-out system are that it supplies a set of standard-ized requirements and a unified communication tool to the rubber industry. Any competent rubber molder or extruder should be able to supply a rubber article that meets a line call-out. The formulations that comply with J200 are generic materials that should be readily available and cost-effective. The material characterizations use only ASTM test methods, so all requirements are clearly understandable, and third-party laboratories can be used as needed.

The disadvantages of this system arise from its low level of specificity. As in our example, more than one polymer may satisfy the line call-out requirements. Also, the specified material properties may not be sufficient to meet the requirements of a given application.

9.10.2 Specific Material Standards

The SAE J200 line call-out system is adequate for generic or commodity parts. It is not always adequate to describe elastomer materials used in motor-vehicle applica-tions. A specific material standard should be considered if any of the following con-ditions are met:

- The component has more stringent functional requirements than those included in J200.
- Part function is critical to vehicle performance.
- The part involves specialized material or processing technology.

The specific material standard should contain a description of the part and its function along with a detailed list of material characteristics and mechanical prop-erties, or performance criteria, as appropriate. These descriptions should include the appropriate test methods and the expected test results.

The process of developing a new part or modifying a part to meet new performance criteria or a harsher environment emphasizes the need for an open exchange of information between the automotive release engineers and their counterparts in the supplier community.

9.11 CONCLUSION—A LOOK TO THE FUTURE

Several trends will influence the use of automotive polymeric materials in the future.

1. Continuing competitive pressure will emphasize the need to make materials choices that are both reliable and cost-effective. The challenge is to achieve the reliability without overengineering the part. Therefore, it is necessary to avoid the use of expensive materials that have performance properties that exceed the requirements of the intended application.

2. Functional test methods will be developed that more closely predict the performance of materials and components in real-world applications. Current specifications are often descriptive and may be written around a particular material. This type of standard is more useful as a control tool than as a material-selection guide and verification.

3. Material recyclability will become an increasingly more important consideration. Some ways of addressing the recycling issue are marking parts for material identification, designing the parts for ease of disassembly, selecting materials based on their potential to be recycled, and incorporating recycled materials where appropriate. The challenge in material selection is to find materials that meet recycling goals without compromising other design and functional requirements.

4. TPE growth is projected at 6 percent per year, compared to 1 to 2 percent for elastomers.[17] Much of this growth is at the expense of traditional thermoset elastomers. Although the cost per pound is higher for most TPE materials, TPE processing is much more cost-effective. In many cases the overall economics favor TPE use. An additional advantage arises in the area of recyclability. Elastomers, like all thermoset materials, present a considerable challenge to recycling. TPE materials have the same recycling potential as thermoplastics.

5. New material developments will occur, especially in the area of elastomers. In recent decades there have been few fundamental elastomeric polymer developments. Now there are new technologies available, especially in catalyst systems. Also, several strategic business moves in the polymer industry have caused a realignment of research and development resources. These changes should lead to the development of new cost-effective polymers with superior performance characteristics.

REFERENCES

1. Eirich, Frederick R., *Science and Technology of Rubber,* Academic Press, New York, 1978.

2. Naunton, W. J. S., *What Every Engineer Should Know About Rubber,* London, British Rubber Development Board, 1954.

3. *Handbook of Molded and Extruded Rubber,* The Goodyear Tire and Rubber Co., Akron, 1969.

4. *Engineered Materials Handbook: Vol. 2 Engineering Plastics,* ASM International, 1988.

5. *Seals and Sealing Handbook,* The Trade and Technical Press, London, 1986.

6. SAE J200, "Classification System for Rubber Materials," Soc. Auto. Eng., Warrendale, PA, 1992.

7. ASTM D2000, "Standard Classification System for Rubber Products in Automotive Applications," Am. Soc. for Testing and Materials, Pittsburgh, PA, 1994.

8. ASTM D 1418, "Standard Practice for Rubber and Rubber Latices—Nomenclature."

9. Babbit, R. O., *The Vanderbilt Rubber Handbook,* R. T. Vanderbilt Co., 1990.

10. *Manual for the Rubber Industry,* English ed., Bayer A. G., 1995.

11. Bhowmick, A. K. and H. L. Stephens, *Handbook of Elastomers,* Marcel Dekker, New York, 1988.

12. Murray, R. M. and D. C. Thompson, *The Neoprenes,* E. I. du Pont de Nemours and Co., 1963.

13. *Mod. Plast.,* MPE Suppl., 1990.

14. Richardson, Terry L., *Industrial Plastics: Theory and Application,* Delmar Publishers, Inc., 1989.

15. ASTM D395, "Standard Test Methods for Rubber Property—Compression Set," Am. Soc. for Testing and Materials, Pittsburgh, PA, 1994.

16. ASTM ES 27, "Emergency Standard Practice for Rubber—Establishing Replacement Immersion Reference Oils for ASTM No. 2 and No. 3 Immersion Oils as Used in Test Method D 471," Am. Soc. for Testing and Materials, Pittsburgh, PA, 1994.

17. *Rubber World,* vol. 212, no. 4, 1995, pp. 12–13.

CHAPTER 10
DESIGN AND PROCESSING OF PLASTIC PARTS

John L. Hull
Hull Corporation
Hatboro, Pennsylvania

10.1 INTRODUCTION

The creation of a plastic part requires a series of conscious decisions regarding type of plastic, method of production, design of mold or tooling, and selection of machine or process. Reaching these decisions requires information as to the intended usage of the part and the conditions of environment (temperature, moisture, exposure to harsh atmospheres, physical and electrical requirements). Furthermore, in most cases, the cost to manufacture the part is a major consideration and is dependent on the choice of materials, the manufacturing process, and, of course, the quantities to be produced per shift, month, or year.

In this chapter, several production processes are discussed broadly, outlining the basics of each process, the equipment involved, the plastic materials for which the process is feasible, and some typical products made by each particular process. In addition, the chapter describes important aspects of plastic part design and considers the closely related topic of mold design and construction.

10.2 DESIGN PROCEDURE

The first step requires a preliminary engineering drawing of the intended part, with approximate overall dimensions, section thicknesses desired, probable location of holes, ribs, bosses, and the like, as well as approximate radii of curves, corners, and so on. Tolerances are not necessary at this stage but will be incorporated before final design.

Next is a list detailing the intended requirements of the final part. Physical considerations regarding stiffness; dimensions and tolerances; impact resistance; surface hardness; compressive, tensile, and torque loads; light transmittance; and so on; all need to be quantified. If the part is intended for electrical applications, determine voltage to be encountered, current-induced temperature extremes, arc resistance and surface tracking resistance, dielectric strength, electrical frequency, and the like. List possible environmental extremes such as humidity, salt or corrosive spray, ultraviolet exposure, moisture-absorption limitations, service-temperature lows and highs, extreme gravities from acceleration or shock, and possible abrasive exposure. Note also cosmetic requirements such as color or transparency, high gloss, or special tactile surface.

At this point, select several possible materials that appear (from the data sheets) to meet all the intended requirements and also appear to be appropriate for the size of part. If you know of plastic parts similar to your intended size and application requirements, determine the materials from which they are made, and add these to your list of material candidates.

The next step is to consider the several processes that might be appropriate. Later in this chapter you will find descriptions of many processes, some of which will be worth considering. But you must realize that finding a process that will be compatible to your part requirements as well as to the materials you have tentatively selected is not always easy. Compromises are often made at this stage—compromises in your design, your requirements, and your materials. (In some cases, you may decide not to use plastics as your material of construction!)

The selection of process involves size of part. For example, you can't injection mold 50-foot lengths of garden hose; but you can readily extrude them. Ultrahigh molecular weight polyethylene (UHMWPE) has fantastic resistance to abrasion; but if your part must experience temperatures of 100°C or higher, UHMWPE will lose its physical properties rapidly.

Selecting the process requires consideration of size (injection molding a boat hull is not feasible, but hand lay-up is very practical for such an application), production volume (to build a compression mold for only 100 parts would be extremely expensive, but liquid casting for the same quantity might be ideal), and economics (using a material that costs $10 per pound would be out of the question if your two-ounce part had to sell for $1). Often, choice of process may be dictated by available process equipment already in your manufacturing facility or in your favorite custom molding house. (The part may be appropriate for either transfer molding or injection molding; so if you have transfer molding equipment in your plant but not injection presses, possibly transfer molding will be the "best" process.)

If a practical solution still eludes you, it would be wise to get suggestions from plastic material suppliers, processing-machinery suppliers, or custom molders, all of whose advice will probably be given at no charge.

And if those sources fail to lead to a practical solution, try one or more of the many plastics design consultants for some guidance. The consulting fee will generally be well spent.

10.3 PROTOTYPING

Once you have reached an apparently viable decision as to material and process (having taken manufacturing economics into the formula), you should give strong consideration to starting with an appropriate prototype. Such a prototype may be machined or cast using your selected material, with final dimensions sufficiently close to your design specifications to enable good mechanical and electrical testing. If the prototype can be produced using the selected material and process (as, for example, using a single-cavity injection mold for a part that will ultimately be produced in a multicavity injection mold), you will not only be able to perform various mechanical and electrical tests on the prototype, but you may gain valuable molding parameters and cycle rates during the prototyping. You may also discover the need to make changes in wall thickness, corner radii, hole diameters or location, gate and vent location and size, and ejector pin location and size—compromises or improvements that lead to optimum production with minimum costs.

Rapid prototyping is becoming increasingly important for bringing a new product to market in shortest possible time. To date, the newest and fastest process for producing prototype molds utilizes stereo lithography (SLA) in a special computer-operated machine that guides laser beams to create a series of flat styrene layers —two-dimensional x-y configured layers (about 0.004 inch thick) at progressive increments along the z axis of the CAD part design. These layers create a three-dimensional solid plastic part to serve as a pattern. This pattern is then placed in a container filled with liquid silicone rubber. When the rubber has cured, the pattern is removed. The resulting elastomeric cavity may be used to cast rigid parts using epoxies, polyesters, acrylics, and other plastics for further part design analysis and evaluation. Such rigid prototypes can often be produced within a week from start to finish.

If the prototype needs to be molded in an injection, compression, transfer, or blow-molding process, it is possible to create a hard cavity shell by first coating the silicone rubber cavity with a release agent and then metal spray plating the cavity surface. When the plated coating is sufficiently thick, it is removed from the elastomeric cavity and backed up with liquid epoxy and appropriate reinforcing stiffeners to create a prototype mold that, with care, can be used to mold parts in the plastic intended for final part. Time from "art to part" may be merely a few weeks, as compared to the several months generally required for conventional machined steel production molds.

A variation on the above SLA process enables the generation of a powdered metal mold for the CAD designed part. Molds made by this variation have been used as production molds and have withstood millions of cycles of injection molding.

10.4 PROCESSES FOR PRODUCING PLASTIC PARTS

10.4.1 Liquid-Plastics Processing

For prototyping or for limited production runs, liquid plastics casting offers simplicity of process, relatively low investment in equipment, and fast results. The casting materials may be thermoplastics, such as acrylics, or thermosetting plastics, such as epoxy resins. The plastics may harden by simple cooling, by evaporation of solvents, or by a polymerization or cross-linking reaction. Such plastics are often poured into open molds or cavities. Because pouring is done at atmospheric pressure, molds are simple, often made of soft metals, plaster, or other plastics.

An example of liquid casting is the fabrication of design details such as scrolls and floral or leaf patterns for furniture decoration. Such parts are often made using filled polyester resins. After curing, the parts are simply glued to the wooden bureau drawer or mirror frame. The parts can readily be finished to look like wood. Molds for such parts are often made by casting an elastomeric material over a wood or plastic model of the part. When the elastomer is removed, it generally yields a cavity in which can be cast a very faithful reproduction of the original pattern. The elastomer mold may be used over and over many times.

Another widely used industrial application is casting with liquid plastics to embed objects such as electronic components or circuits in plastic cups, cases, or shells, giving the components mechanical protection, electrical insulation, and a uniform package size. When such applications require fairly high-volume production, machines for mixing and dispensing the liquid plastics may be used for convenience and higher production rates, and curing ovens, conveyors, and other auxiliary capital equipment may be incorporated. In short, the liquid-casting operation may be a low-cost manual one, or it may be highly automated, depending on the nature of the product and the quantities required.

Casting liquid acrylic into an already-molded acrylic shape in which a coin or emblem has been placed, results, when the acrylic has hardened, in an attractive paperweight. Casting liquid epoxy into a previously molded container in which has been assembled a hybrid electronic circuit with leads protruding out of the container, results, when the liquid epoxy has cured, in an attractively packaged and protected circuit. This process is called potting when the container ("pot") remains as a permanent element of the finished part.

If bubbles or voids are a concern, liquid casting and potting may be performed in a vacuum chamber.

Refer also to Chap. 4.

10.4.2   Rotational Molding (Rotomolding or Rotational Casting)

This relatively low-cost process utilizes a closed mold in which is placed an appropriate charge of liquid or granular plastic, generally thermoplastic. The mold is then mounted in a carousel, which enables it to be rotated simultaneously in at least two axes vertical to each other. As the mold spins, the plastic inside coats the inner surface of the mold cavity to a reasonably uniform thickness. The mold is brought to a controlled melt temperature, usually by hot air, which enables the plastic to flow and fuse. Subsequently, as the mold is cooled with cold air or water spray, the plastic hardens. The mold is then opened, and the hollow part is removed. Often the hollow part is cut in half to yield two usable parts, as in thin-walled covers. Applications include storage and feed tanks (as large as a 22,500-gallon tank 12 feet wide and 30 feet high!), shipping containers, automotive instrument panels, gearshift covers, door liners, playground equipment, recreational boats, and portable toilets.

Most common thermoplastics, such as polyethylene, polyvinyl chloride, nylon, polycarbonate, acetate butyrate, and polypropylene, are suitable for the rotational molding process. In recent years, some thermosetting formulations have been found successful.

Inexpensive molds are made of cast or machined aluminum, or sheet metal. For finer details, more expensive molds may be made using electroformed or vapor-formed nickel. Release agents of nonstick coatings are generally used for ease of demolding.

10.4.3   Hand Lay-Up (Composites)

When larger plastic parts are required, and often when such parts must be rigid and robust, a process referred to as hand lay-up is used. Hand lay-up is closely related to liquid-plastic casting. A reinforcing fabric or mat, frequently fiberglass, is placed into an open mold or over a form, and a fairly viscous liquid resin is poured over the fabric to wet it thoroughly and to penetrate into the weave, ideally with little or no air entrapment. When the plastic hardens, the object is removed from the mold or form, trimmed as necessary, and is then ready for use. Many boats are produced using the hand lay-up process, from small sailing dinghies and bass boats, canoes and kayaks to large commercial fishing boats and even military landing craft. Unsaturated polyesters are most often used in this process, but epoxies and polyurethanes are also used. Although glass fibers, in mat-form or woven, dominate as the reinforcing material, carbon fiber or Kevlar (DuPont's aramid filament) is used where extreme stiffness or strength may be required.

This basic process can be automated as required, with proportioning, mixing, and dispensing machines for liquid resin preparation, using matched molds (that is, two mold halves that are closed after the reinforcing material has been impregnated with the liquid to produce a smooth uniform surface on both top and bottom of the part), conveyors, ovens, and so on.

Refer also to Chap. 2.

10.4.4 Resin Transfer Molding

The resin transfer molding (RTM) process is used principally for manufacturing fiber-reinforced composites in moderate to high volumes. The process combines the techniques of hand lay-up, liquid resin casting, and transfer molding. In practice, the fiber mat or preform (a precut and preshaped insert of the reinforcing material) is placed into the open mold, and the heated mold is closed. A liquid catalyzed resin, often epoxy or polyester, is then injected into the cavity at a modest positive pressure (5 to 100 psi) until all the interstices between the fibers are completely filled. The resulting formulation may have as much as 60 percent glass by weight. Subsequently, the resin mix reacts and hardens. Often vacuum is applied to the mold cavity to remove the air prior to and during cavity fill to minimize the possibility of air entrapment and voids. Cycle times may run several hours, particularly for large aircraft and missile components. During cure, mold temperatures are often ramped up and down to achieve optimum properties of the plastic and therefore of the finished product. Fibers often used include glass, carbon, Kevlar, or combinations of these. Very high strength-to-weight ratios are achieved in such parts, exceeding those of most metals, and complex configurations are achieved more easily than by machining and forming such high strength-to-weight ratio metals as tantalum or aluminum.

To make such parts, the press and resin transfer equipment for the RTM process is generally less expensive than that for an injection press for comparable-size components, partly because RTM clamping and injection pressures are often only a few hundred pounds per square inch. Control systems have become highly sophisticated, however, raising the cost of equipment, and the long cycle times add to the manufacturing cost of each item produced. Advanced metering and mixing machines for preparing the liquid resin are capable of controlling resin-catalyst ratios over the range of 1:1 to 200:1, and in flow rates from several ounces per minute to 100 pounds per minute (Fig. 10.1).

10.4.5 Filament Winding

For structural tubes up to several inches in diameter, and for tubes and closed tanks that may hold fluids at high pressure, a continuous filament of glass or other strong polymeric material is drawn through a liquid polyester or epoxy bath and then fed to a rotating mandrel, allowing the wetted filament to closely wind onto the man-

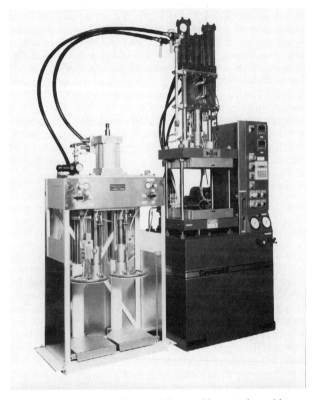

FIGURE 10.1 Liquid-silicone-molding machine complete with pro-portioning, mixing, and dispensing system for two-part resin systems. Machines similar to this are used for resin transfer molding (RTM). (Photograph provided by WABASH/MPI/Carver, Inc.)

drel, often with layering in different directions. When the predetermined number of turns has resulted in the desired thickness, the resin cures, and the mandrel is removed. The resulting tube or container has extremely high hoop strength, ideal for storing or transporting high pressure liquids or gases.

10.4.6 Thermoforming

Because thermoplastics soften with heat and harden when cooled, many items are produced using extruded thermoplastic sheet as the starting material. The sheet is heated to a temperature that softens the sheet but does not melt it. Then the sheet is shaped or formed while soft and subsequently cooled while being held in the new configuration. Often the heated sheet is forced into an open cavity by vacuum or air pressure. Most blister packages and many low-cost plastic drinking cups and containers are produced by such a thermoforming process. Acrylic (Plexiglas™) rounded canopies for small aircraft and sailplanes are frequently formed by the process, as are many advertising signs on truck doors or retail establishments. One highly successful application is the 60-inch long ABS dash panel for a low-volume street sweeper. Bike helmets and snowboards are also produced with this process.

During the process, the material often stretches, making the final wall thickness in some areas less than the original sheet thickness. With proper design and operation, the thermoforming process permits very inexpensive production of three-dimensional thin-wall parts for short or long runs. Many process variations have been developed to accomplish desired results. Such variations include drape forming, matched-mold forming, plug-assist forming, vacuum snap-back, and trapped-sheet forming. Materials suitable for the thermoforming processes include ABS, thermoplastic polyesters, polypropylene, polystyrene, acrylics, and polyvinyl chloride.

Equipment costs vary widely but are considerably lower than for comparable injection-molded parts because of the relatively low pressures and temperatures required. Highly automated machines use roll-fed stock. Residual material from trimming can be ground, blended with virgin material, and reextruded into sheet for subsequent thermoforming use.

10.4.7 Extrusion, Coextrusion, and Pultrusion

The extrusion process starts somewhat like the thermoplastic injection molding process, with a long barrel and rotating augur-type screw advancing thermoplastic crystals from a room-temperature hopper at one end, bringing the material up to melt temperature by the electrically heated barrel and also by the frictional and shearing action of the screw on the material, finally forcing the melt out the nozzle end through a precision-machined die to create the profile or shape of the continuously extruded product and cool it. Then, to cool the extrudate completely, it is generally passed through a water bath kept at a sufficiently low temperature such that the material hardens in the shape of the profile. Single-screw extrusion barrels range from ½ inch to 18 inches inner diameter, with throughput rates depending on the size and wall thickness of the profile and the material characteristics but ranging from several pounds per hour to more than five tons per hour. Twin-screw extruder throughputs can be as high as 30 tons per hour (Fig. 10.2).

In recent years, extruder screw architecture has become highly sophisticated, with special configurations for each type of extrudate. Venting in the barrel con-

FIGURE 10.2 Twin screw extruder, counterrotating, nonintermeshing, for compounding, model HTR2800-511-512-E1. Unit shown has panel open for access to barrel and heater elements. Set of extruder screws shown in front, and seven-zone temperature control panel with loss-in-weight feeder controls and main drive motor controls, stands at right of machine. (Photograph provided by Welding Engineers, Inc.)

tributes much to product quality and throughput. As with most high-speed polymer processing systems, closed-loop control systems monitor and regulate process parameters for optimum production.

Profiles may be of closed cross section, as for pipe or conduit, or they may be open sections, as for door jambs and window frames for the construction industry. Extruders can also coat wire on a continuous basis, such as vinyl-insulated wire used for appliance wiring.

Coextrusion is a process in which two different materials are extruded from two separate extruder barrels and then brought together in a complex die to achieve a laminated sheet or profile. Not many materials lend themselves to this process because temperature and chemical differences often preclude good bonding.

Successful extrusion of thermosetting materials has been achieved and put into commercial practice, but extruded thermosets represent only a tiny fraction of the amount of thermoplastics extruded annually.

Pultrusion is a process somewhat similar to extrusion in concept in that it produces a continuous profile by forcing the material through a precision die configured to the desired profile. The differences between the two processes, however, start with the fact that pultrusion generally uses thermoset plastics rather than thermoplastics, and the process *pulls* the resin and reinforcing fibers or web through the die instead of *pushing* them through. Because the profiles are heavily reinforced, the pull is on the continuous reinforcing material. Thermoplastic materials suited to the pultrusion process include PE and PEEK (Fig. 10.3).

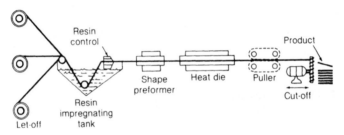

FIGURE 10.3   Schematic illustration of pultrusion process.

When thermosetting materials are used, the steps in the process involve pulling the reinforcing matrix through a tank filled with catalyzed liquid resin, often polyester, where the fibers become totally saturated. The wet matrix is then passed through a stripper to squeeze out excess liquid and sometimes to start shaping the material into a profile. The material is then pulled into a 20- to 30-inch-long heated die at a rate controlled to ensure complete curing of the resin as it passes continuously from one end of the die to the other. The material may be partially brought up to temperature prior to entering the curing die by use of high-frequency heating between the stripper and the die, thereby speeding up the curing and enabling faster pulling rates, shorter curing times, or both. Recent process variations inject the catalyzed resin directly into the die, thereby eliminating the above-described matrix saturation and stripping steps.

Perhaps the most crucial element in a pultrusion machine is the gripper mechanism, which grips the cured profile downstream of the curing die and continuously pulls at a carefully regulated rate to ensure total curing in the die, taking into account the degree of advanced cure of the catalyzed resin as it enters the die.

Applications include electric-bus duct, side rails for safety ladders, third-rail covers, walkways, structural supports in harsh chemical environments, and resilient items such as fishing rods, bicycle flagpoles, and tent poles. Common profile sections generally fall between 4 by 6 inches and 8 by 24 inches.

10.4.8 Blow Molding

Ideal for producing plastic beverage bottles and other closed shapes, the blow molding process combines elements of the extrusion process and the thermoforming process in complex, fully automated machines and mold systems to produce thin-walled holloware at fantastically high rates (Fig. 10.4).

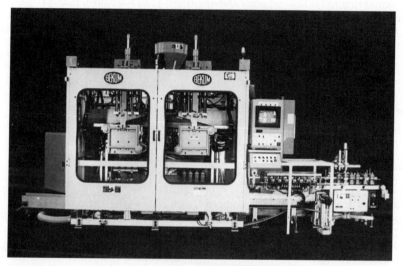

FIGURE 10.4   Bekum model H-155 double-station extrusion-blow-molding machine with oriented bottle take-out. (Photograph provided by Bekum America Corporation.)

In extrusion blow molding, a hollow tube called a parison is extruded vertically downward. Immediately after it is extruded, two halves of a mold clamp firmly over the parison, pinching the lower end closed, and forming the upper end firmly around a blowing nozzle. Compressed air or some other gas (sometimes chilled) is blown at about 100 lbs/in^2 pressure into the upper end of the parison, forcing the parison to expand rapidly in the cavity, where it cools and hardens in the shape of the cavity. The mold at the end in which air is introduced also forms the threads and neck of the hollow bottle or part. The mold is then opened, and the finished part is blown off or extracted onto a conveyor line. In many plants, blow molding is a captive operation, and the molded container is immediately transported to the labeling and then the filling station, then capped and packaged for shipment.

A variation on extrusion blow molding is injection blow molding, a two-stage process. In the first stage, parisons are injection-molded into a tubular shape complete with the threaded and formed top. These parisons may then be stored until time for the second step.

The second step involves feeding the premolded parisons into an automatic blow-molding machine where the parison is heated to the softening point and then clamped between die halves and blown as in the extrusion blow molding process. In this two-stage process, the bottom of the parison is not pinched because it is not open.

Common thermoplastic materials suited to the blow-molding process include high density polyethylene, polyvinyl chloride, polypropylene, and polyethylene terephthalate.

Although the injection blow molding process is not suited to "handleware" bottles, it is rapidly gaining favor over extrusion blow molding for bottles up to 1.5 liters and more. Advantages include the practicality of molding strong, closely toleranced necks and threads suited to childproof caps, and for molding widemouthed bottles.

In addition to bottles, a rapidly growing application for blow molding is the production of "technical" parts, such as automotive components—bumpers, ducts, and other fluid containers. For fuel tanks, a coextruded parison has polyethylene as the structural material and special barrier layers to prevent escape of polluting fumes.

Machines and molds may cost between $200,000 and $1,000,000, depending on the throughput rates and accessories.

Molds for blow molding are made from beryllium copper and aluminum because of the excellent thermal conductivity of such materials. Stainless steel and hard-chrome plated tool steels are also common.

10.4.9 Compression and Transfer Molding

Compression molding is a process very similar to making waffles. The molding compound, generally a thermosetting material such as phenolic, melamine, or urea, is placed in granular form into the lower half of a hot mold, and the heated upper half is then placed on top and squeezed down until the mold halves come essentially together, forcing the molding compound to flow into all parts of the cavity, where it finally cures, or hardens, under continued heat and pressure. The cure, or polymerization, is an irreversible chemical reaction. When the mold is opened, the part is removed and the cycle repeated. The process can be manual, semiautomatic, or fully automatic (unattended operation), depending on the equipment. Molds are generally made of hardened steel, highly polished and hard-chrome-plated, and the two mold halves, with integral electric, steam, or circulating hot oil heating provisions, are mounted against upper and lower platens in a hydraulic press capable of moving the molds open and closed with adequate pressure to make the plastic flow. Molds may be single cavity or multiple cavity, and the pressure must be adequate to provide about four tons per square inch of projected area of the molded part or parts at the mold parting surfaces. Overall cycles depend on molding material, part thickness, and mold temperature, and may be about one minute for parts of ¼ inch thickness to five or six minutes for parts of one inch thickness or larger (Fig. 10.5).

The process is generally used for high-volume production because the cost of a modern semiautomatic press of modest capacity, say 50- to 75-ton clamping force, may be as much as $50,000, and a moderately sophisticated self-contained multicavity mold may also cost $50,000. Typical applications include melamine dinnerware, toaster legs and pot handles, and electrical outlets, wall plates, and switches—parts that require the rigidity, dimensional stability, heat resistance, and electrical insulating properties typical of thermosetting compounds.

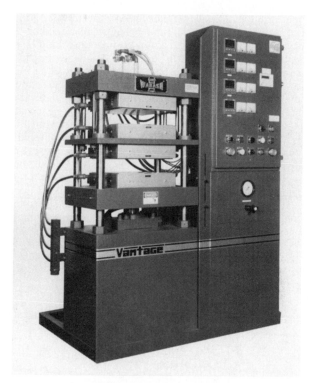

FIGURE 10.5 Vantage model V502H-18 compression molding press equipped with two heated openings for increased production. (Photo provided by WABASH/MPI/Carver, Inc.)

To simplify feeding material into the mold, the molding compound is often pre-compacted without heat into preforms, or "pills," in a specially designed automatic preformer, which compacts the granular molding compound into cylindrical or rect-angular blocks of uniform weight. To reduce the molding cycle time, the preform is often heated with high-frequency electrical energy in a self-contained unit called a preheater, arranged beside the press. The preform is manually placed between the electrodes of the preheater before each molding cycle and heated throughout in as little as 10 to 15 seconds to about 200°F, at which temperature the plastic coheres but is slightly mushy. It is then placed manually in the bottom mold cavity, and the mold-ing cycle is initiated. Through the use of preheating, the cure time may be cut in half, mold wear is reduced considerably, and the part quality is often improved.

The process of compression molding is often utilized with sheet molding com-pounds (SMC) to produce heavily reinforced composite parts. SMC is generally fiberglass mat or woven cloth, impregnated with B-staged (partially catalyzed) ther-mosetting epoxy or polyester. Following cavity fill and a suitable cure time, the press opens and the part is removed.

A related process for high-volume molding with thermosetting materials is trans-fer molding, so called because the material, instead of being compressed between the two halves of a closing mold to make it flow and fill the cavity, is placed into a

separate chamber of the mold, called a transfer pot. This transfer pot, which generally is cylindrical, is connected by small runners and smaller openings (called gates)
to the cavity or cavities. In operation, the mold is first closed and held under pressure. The preheated preform is then dropped into the pot and pressed by a plunger,
where the material liquefies from the heat of the mold and the pressure of the
plunger and flows (or is transferred) through the runners and gates into the cavity or
cavities. The plunger maintains pressure on the molding compound until the cavities
are full and the material cures. At that point, the mold is opened, the plunger is
retracted, and the part or parts, runners, and cull (the material remaining in the pot,
generally about ⅛ inch thick and having the diameter of the pot and plunger) are
removed. Because the gate is small and the cured plastic is relatively rigid, the runners and cull are readily separated from the molded parts at the part surface, leaving
a small and generally unobtrusive but visible "gate scar" (Fig. 10.6).

Transfer molding is often used when inserts are to be molded into the finished
part as, for example, contacts in an automotive distributor cap or rotor, or solenoid
coils and protruding terminals for washing machines. Whereas in compression mold-

FIGURE   10.6 Transfer/compression-molding   press,
model T-200, for high-precision applications. Press shown
features a Smart Mold™ software-based control system.
(Photo provided by Hull Corporation.)

ing such inserts might be displaced during the flow of the viscous plastic, in transfer molding the inserts are gently surrounded by a liquid flowing into the cavity at controlled rates and pressures and generally at a relatively low viscosity. Inserts are also rigidly supported by being firmly clamped at the parting line or fitted into close-toleranced holes of the cavity. When dimensions perpendicular to the parting line or parting surfaces of the mold must be held to close tolerances, transfer molding is used because the mold is fully closed prior to molding. With compression molding, parting-line flash generally prevents metal-to-metal closing of the mold halves, making dimensions perpendicular to the parting line greater by the flash thickness, perhaps by as much as 0.005 to 0.010 inch (Fig. 10.7).

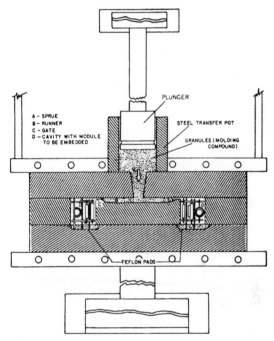

FIGURE 10.7 Schematic illustration of transfer or plunger mold.

Transfer presses and molds generally cost 5 to 10 percent more than compression presses and molds, but preheaters and preformers are the same as used in compression molding. Transfer cycle times are often slightly shorter than cycle times for compression molding, because the motion of the compound through the small runners and gates prior to its entering the cavity raises the compound temperature by frictional heat and mechanical shear, therefore accelerating the cure.

One highly significant application of transfer molding is the direct encapsulation of electronic components and semiconductor devices. Adaptation of the basic transfer molding process to successfully mold around the incredibly fragile devices and whisker wires of such items, required first the development of very soft flowing materials, generally epoxies and silicones; then modifications to conventional transfer presses to enable sensitive low-pressure control and accurate speed control (both often programmed through several steps during transfer); and finally new mold design and construction techniques to ensure close-tolerance positioning of the components in the cavities prior to material entry. It can be fairly stated that the successful development of the transfer molding encapsulation process was a large factor in the manufacture of low-cost transistors and integrated circuits.

In the automotive field, growing dependency on electronic sensors for highly sophisticated engine controls and for the increasing number of safety features leads to predictions of $2,000 average cost per car for plastics-encapsulated electronic components.

Although compression and transfer molding are used principally with thermosetting compounds, the processes are occasionally used with thermoplastic materials, often thermoplastic composites. Thermoplastic toilet seats, for example, which often have fairly thick cross sections, have been successfully compression-molded at acceptably short cure times, about 4 to 5 minutes. Molding thermoplastic requires mold cooling rather than mold heating. In compression molding such materials, therefore, the material is put in the cavities when the mold is hot enough to melt the material, enabling the melt to flow adequately to fill the cavity. Cooling water is then circulated through the mold-cooling channels to cool the plastic below its melt index. When the thermoplastic material in the cavities hardens, the mold is opened and parts removed.

As mentioned in Sec. 10.4.3, liquid transfer molding (LTM) is being increasingly used for reinforced plastics, or composites. Conventional transfer presses may be used in such applications, with the plastic being introduced into the transfer pot as a liquid catalyzed thermosetting material such as epoxy, polyester, or silicone, and caused to flow into the cavities by movement of the transfer plunger pushing against the liquid. Following cavity fill, the liquid material cures much the same way as do conventional transfer molding compounds.

10.4.10   Injection Molding—Thermosets

Similar in many respects to transfer molding is injection molding of thermosets. The process is also a closed-mold process, and the mold uses runners and gates leading to cavities in much the same way as does a transfer mold. But instead of a pot and plunger, the injection process generally uses an auger-type screw, rotating inside a long cylindrical tube called a barrel. The barrel temperature is closely controlled, usually by circulating hot water in jackets surrounding the barrel. The front of the barrel narrows down to a small opening or nozzle, which is held firmly against a mating opening in the center of one of the mold halves, called a sprue hole, which leads into the runner system at the parting surfaces of the mold. The screw and barrel are generally positioned horizontally, and the press opens left and right (as compared to the up-and-down movements traditional with compression and transfer presses), so the mold parting surface is in a vertical plane rather than a horizontal one (Fig. 10.8).

In operation, after the molded parts and runners have been removed from the open mold, the press closes the mold in preparation for the next cycle. By this time, the screw has been rotating in the barrel, conveying granular material forward from the hopper at the back end of the barrel through the screw flights. As the material is conveyed forward, it is heated by the jacketed barrel and also by the mechanical shear caused by the screw rotation in the barrel and the constant motion of the material. The material becomes a viscous, pastelike fluid by the time it reaches the

FIGURE 10.8   110-ton thermoset injection-molding machine incorporating integral deflashing system to provide one-step processing of molded parts. (Photo provided by Hull Corporation.)

nozzle end. There, it does not have enough pressure to flow through the small nozzle opening, so it exerts a pressure against the front end of the screw, forcing the screw to reciprocate back into the barrel against a controlled hydraulic pressure in a cylinder at the back end of the screw. As the charge of plasticized material accumulates at the nozzle end of the receding screw, it finally reaches the set charge weight or volume for the mold. The screw's backward motion is automatically detected by a limit switch or linear potentiometer, which then stops further backward motion and rotation. The screw is then positioned in the barrel with the correct measured charge of plasticized material between the screw tip and the nozzle end of the barrel. This plasticizing step occurs automatically in the press cycle such that it is completed by the time the mold is closed, ready for another cycle.

When the injection molding machine senses that the mold is closed and is being held closed under full pressure, the screw advances forward rapidly ("reciprocates"), during which stroke it acts as a piston, driving the plasticized charge of material through the nozzle, sprue, runners, and gates to fill the mold cavities. Fill time is generally from 1 to 3 seconds, depending on the charge mass, as compared to 10 to 30 seconds in a transfer molding operation. Frictional heat from the high-velocity flow raises the molding compound temperature rapidly, such that the material time-temperature experience assures a rapid cure in the cavity. Overall cycles of thermoset-injection processes are often half those for comparable parts produced by the transfer-molding process.

Modern thermoset-injection-molding presses are usually fully automatic and produce parts at a high rate. They are ideal for applications requiring high volume of parts at minimum cost. Machines cost about twice as much as comparable-capacity machines for transfer and compression molding. Mold costs are about the same as for transfer operations. No preforming or preheating is required, and the labor content of automatic injection molding is significantly lower than that of semiautomatic transfer and compression molding.

To achieve maximum-strength parts, a high concentration of glass or other reinforcing fibers may be mixed with the molding compound in this process. Bulk molding compounds (BMC) are often used, in which the formulation, generally polyester and glass fibers up to ½ inch in length, is puttylike in consistency. To minimize fiber breakage in BMC injection molding, the screw is often replaced with a plunger, and a special stuffing system is used to load the BMC into the barrel. Many electric switchgear components are produced with BMC injection molding.

Gas-assist injection molding is a relatively new process that enables certain types of products to be molded as hollow parts without the use of cores. The hole is produced by injecting an inert gas, generally nitrogen, into the cavity after it is partially filled with the molding compound. In some respects the process is like blow molding, but without the parison. Parts for this process are generally of relatively simple configuration: a tube or elbow closed on one or both ends. A vital requisite to the process is that the molding compound must adhere to the cavity walls as it flows into the cavity, resulting in a relatively thick contiguous "skin" on cavity walls during initial injection. At the point when cavity walls are well coated, further material injection is halted, and the inert gas is blown into the open space under high pressure. The molding compound is then forced against cavity walls, held there under high pressure until the material is hard, and then ejected (after gas pressure has been reduced to atmospheric).

Most of the common thermoset materials exhibit the mandatory (for gas-assist injection molding) laminar flow properties—phenolics, urea, melamines, thermosetting polyesters, and epoxies. Inorganic fillers and reinforcing fibers prove more successful than organic reinforcements. Parts molded with the gas-assist process are lighter than their solid equivalents by 20 percent or more. They also enjoy faster cure and, therefore, shorter cycles, and generally require less molding pressure in the cavity.

10.4.11 Injection Molding—Thermoplastics

Thermoplastic injection molding came into general practice well before thermoset injection molding, and it is the principal method for volume production of thermoplastic parts. Because thermoplastics are liquefied by heating and hardened by cooling, there are differences between thermoset and thermoplastics injection presses and molds (Fig. 10.9 and Fig. 10.10)

Molds for thermoplastics are often cooled to a temperature well above ambient. The mold temperature is controlled such that it will cool the material in the cavity at

FIGURE 10.9 Elektra 725-ton all-electric injection molding machine. (Photo provided by Cincinnati Milacron, Inc.)

FIGURE 10.10   Model VT550 550-ton toggle-type injection molding machine. (Photo provided by Cincinnati/Milacron, Inc.)

the fastest rate possible without causing it to harden before it totally fills the cavity. Molds are often made of metals more easily machined than steel, such as beryllium copper, that will afford optimum heat transfer to the molding material. Cavity surfaces are generally not chrome-plated because thermoplastics are less abrasive than thermosets. Overall cycle times, especially for thin-walled products, are often below 10 seconds.

Screw length-to-diameter ratios are often different from those for thermosets because of the critical need for very thorough mixing to achieve color and temperature uniformity. Because barrel temperatures are generally higher in thermoplastics machines to assure total melt and low-viscosity flow, barrel heating is usually electric and may be as high as 600°F or more for some of the newer engineering plastics such as nylons and liquid crystalline polymers. Also, with the significantly lower apparent viscosity of the thermoplastics at the time of injection, a check valve is often required at the nozzle end of the screw to prevent "drooling" or leakage through the nozzle prior to the injection stroke. An economic advantage not possible with thermoset injection molding is that thermoplastic scrap—sprues, runner, and rejects—may usually be reground and mixed with virgin material for reprocessing.

Applications of injection molding of thermoplastics include practically every molded thermoplastic item—picnic ware (including eating utensils); fashion buttons; refrigerator food containers; grocery-store containers for butter, yogurt, and the like; towel racks; and fan and power-tool housings are some examples. Since the thermoplastic materials are considered less dimensionally stable than thermosetting plastics because the former exhibit a property called creep, or distortion due to prolonged stress, thermoplastic selection must take into consideration the nature of any loads to be borne by the molded product. And because the heat stability (closely related to the melt index) of thermoplastics extends from about 200°F for some vinyls and styrenes to 600°F and higher for the newer engineering plastics, material selection often depends on the temperature exposure anticipated for the finished product. Where color and superior surface finish are desired, thermoplastics may offer better characteristics than do thermosets.

The costs of injection presses and molds for thermoplastics are comparable to those for thermosets, running from about $30,000 for a 25-ton press to $400,000 for a 500-ton press. Injection presses for thermoplastic molding have been built in sizes of over 5,000-ton clamp force for molds measuring 15 feet by 15 feet and more.

As with thermoset injection-molding, heavily fiber-reinforced parts are also molded from thermoplastics. Such parts still may exhibit creep during prolonged mechanical stress, but they are nevertheless considerably stronger and tougher than the nonreinforced products. Glass-reinforced nylon, for example, is used in the lost-core process for a complex air-intake manifold on a current automotive engine.

Injection-molded thermoset parts requiring hollow sections that are not suited to conventional mold configurations or to mechanical side cores may be produced using the above-mentioned lost-core process, a process borrowed from a similar metal-casting technique. In this molding process, a precisely shaped rigid metal insert is positioned in the mold cavity prior to mold closing. The injection process proceeds in the normal manner. Following ejection, the rigid insert is removed by melting the core. Obviously, a relatively low melting temperature for the core, higher than the molding temperature of the plastic but lower than the heat distortion temperature of the molded part, is mandatory for success in this process. This process is presently used for both thermosetting and thermoplastic injection-molded parts.

Reaction injection molding (RIM) is quite similar to injection molding of thermoplastics, but it is designed to process materials that combine in the press to create a chemical reaction—sometimes to produce a foamed product (such as inexpensive picnic coolers, foamed drinking cups, structural foamed containers, or sporting equipment) and sometimes to process a hybrid plastic such as polyurethane for, say, automotive bumpers. The injection process in RIM machines includes bringing two or more chemicals together at the injection nozzle as they are being injected into the mold, where the mixed chemicals rapidly blend, react, and harden.

Injection-compression molding is a process utilizing a uniquely designed mold in an injection press. The process (used with either thermoset or thermoplastic materials) requires the mold to be closed partially and then held in that position (perhaps $\frac{1}{4}$ to $\frac{1}{2}$ inch short of full closed) while the charge is delivered into the cavity. Then the mold closes fully. The special mold design prevents injected plastic from escaping at the parting line. The process minimizes the oriented flow lines and insures a part less subject to warpage or shrinkage. Reinforcing fibers are less oriented and may enable a stronger part.

Structural RIM (SRIM) uses the RIM process with a glass-reinforcing mat placed in the mold before closing and injecting the plastic. SRIM parts offer far greater strength than nonreinforced parts.

Expandable polystyrene (EPS) is an extremely low density thermoplastic material (0.2 to 10.0 lbs per ft^3) that is produced using the RIM process plus live steam injected into the mold cavity as soon as the styrene and blowing agent have been introduced into the cavity. The steam enters the cavity through a number of small holes in the cavity wall to cause rapid expansion of the material. Gases generated escape through the perforated cavity walls. The cavity is chilled prior to mold opening.

Principal EPS applications are in packaging of consumer electronic appliances needing protection against shock and vibration during handling and shipment. Other applications include disposable food containers, drinking cups, flotation systems for small boats and canoes, and energy-absorbing barriers in automobiles.

One thermoplastic used in RIM is a rubber-toughened nylon that has proven exceptionally ideal for a fender on a Caterpillar™ tractor. The rugged fender weighs only 26 pounds, is 101 inches long and 33 inches wide, and possesses high-impact resistance.

10.5   ASSEMBLY AND MACHINING GUIDELINES

Many molded plastic parts are complete and usable by themselves, such as melamine dinnerware, or closet hooks, and the like. But often plastic parts need to be joined to other plastic or metal parts to become functional.

10.5.1

Bonding plastics to plastics has become a well developed art through use of adhesives and "welding." For optimum strength in such bonding applications, the design of the joint where the two pieces are to be bonded must provide physical means for the plastic parts to take the loads rather than relying on the bonding alone. Tongue and groove, or a molded boss fitting into a molded hole; or a stepped joint, with bonding on all the mating surfaces, insures that, so long as the two pieces remain together, shear or tension or torque loads will generally be transmitted from one piece to the other without stressing the bond itself (Fig. 10.11).

In addition to the many polymeric adhesives available today for most families of plastics, actual welding of two similar plastics is possible. Many thermoplastics materials can be welded through use of ultrasonic energy. In principle, the process is similar to metal spot welding in that energy is transferred from a "horn," vibrating ultrasonically, through a finite thickness of one of the pieces to the contacting interface of the other piece to which it is temporarily clamped. The contact area is generally established at a properly designed interfacing point. The concentration of energy at the interface melts the contacting plastic surfaces of both pieces in a few seconds. When the joint cools, the pieces are effectively welded (Fig. 10.12).

Spin welding of thermoplastics uses the mechanical friction of one part spinning against another part to generate the heat needed to melt the surfaces at point of contact. As soon as localized melting has occurred, spinning stops, the bond area is cooled, and the parts become fused. Vibration welding creates similar frictional heat for melting and fusing (Fig. 10.13).

Hot-air welding uses a concentrated blast of hot air directed toward the desired joint area at a temperature above the melt index of the plastic to achieve melting and bonding.

Solvent cementing uses a solvent to temporarily create a solution of softened plastic at the joint. The parts are then clamped together until the solvent has evaporated, thereby effecting fusion. This is a very simple method of bonding acrylics and styrenes for picture frames or desktop decorative items (Fig. 10.14).

A current automobile engine utilizes valve covers molded from glass and mineral-filled nylon 6/6 thermoplastic material, weighing half as much as the formerly used aluminum cover. Ultrasonic welding is used to weld the cover to the oil seal, and vibration welding of the baffle plate to the cover enables reduction of assembly time from three minutes for the former aluminum cover to 10 seconds for the three-piece plastic cover!

Electromagnetic welding of thermoplastics is a process using induction heating with ferromagnetically filled thermoplastics. Frequencies used range from 3 to 8 megahertz to heat the joint area for melting and fusing.

10.5.2

When two pieces of plastic must be joined together in service but must also be suited to simple disassembly, threaded joints are possible. Such joints may be plastic to

FIGURE 10.11 Structural bonding application for a reinforced plastic bumper assembly for a truck. The equipment dispenses a two-component structural adhesive onto the bumper assembly before it is integrated into the truck. In production, this is typically done robotically with a manual back-up system (shown). The system being used is a See-Flo 488 meter/mix system; the dispense gun in the picture is model 2200-250-000 dual-spool Snuf-Bak™ dispense valve that utilizes a No-Flush™ disposable static mixer nozzle. (Photo provided by Sealant Equipment and Engineering, Inc.)

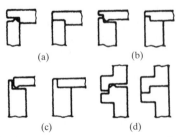

FIGURE 10.12 Basic joint variations suitable for ultrasonic welding. (*Courtesy of Branson Sonic Power Co.*)

FIGURE 10.13 Spin welding.

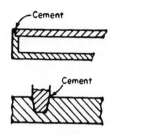

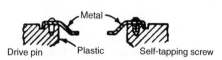

FIGURE 10.14 Assembly by solvent cement.   **FIGURE 10.15** Assembly by screws.

plastic, as in toothpaste tubes and caps, where male and female threads are molded. If metal screws are to be used, female threads can be molded in a plastic part to accept the metal screw. If such a threaded application does not have to withstand much tension in the screw, a self-tapping screw may be turned into a molded hole of appropriate diameter. Self-tapping screws designed for plastics applications are better suited than those for metal applications. Metal female-threaded inserts may be molded into plastic parts to accept metal screws in a subsequent assembly operation. Metal female-threaded inserts may also be staked into a molded hole of appropriate size, sometimes mechanically, often with heat ("heat staking") to soften the thermoplastics at the interface in order to achieve more positive bonding (Fig. 10.15 and Fig. 10.16).

It is important to locate holes for such threaded connections such that there is ample wall thickness to accommodate the concentrated stress in the immediate vicinity of the screw. Screwing a relatively thin plastic part to a more robust part could result in tearing of the thin material at the screw head if separation forces are severe. Designing a thicker section or using an ample washer or oversized screw head will yield more satisfactory results.

Screw joints are especially difficult with brittle plastics such as styrenes, acrylics, polycarbonates, and most thermosetting plastics. In such applications, slight over-tightening of the screw may result in cracking of a part or stripping of threads.

10.5.3

Machining of plastics is often necessary as, for example, in cutting and finishing formica (melamine laminate) counter tops, cutting polycarbonate sheet for glazing applications, or drilling structural shapes for attaching to supports. While such operations are commonplace when working in metals or wood, special tools and practices are necessary when machining plastics.

Brittle plastics tend to break or chip under concentrated loads such as those imposed by a sawtooth or a cutting edge of a drill bit. Softer plastics, like many ther-

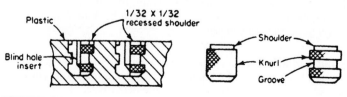

FIGURE 10.16 Molded-in inserts.

moplastics, tend to tear when local load concentrations occur. Furthermore, thermosetting plastics and thermoplastics are good thermal insulators, which means that the high energy imparted by cutting tools turns into frictional heat that, because it does not dissipate easily, quickly reaches the melting point of many plastics and the burning point of others. Essentially all plastics have heat-distortion limits where the plastics lose rigidity and strength.

When conventional woodworking and metalworking cutting and grinding tools are used on plastics, the plastics often become gummy and sticky in the cutting area, binding the cutting tool and distorting the plastic. Therefore, for machining operations on plastics, special cutting tools and special cutting techniques are necessary to achieve desired results. Cutting tools that are suited to epoxies will not be appropriate for polyethylene and the like.

When contemplating such machining operations, therefore, it is critical to contact the plastic supplier for specific recommendations as to cutting-tool configurations and speeds, and procedures often involve cooling.

Machining of Thermoplastics.   Carbide tools should be used. If a mirrorlike finish is expected, the cutters should be diamond-tipped. Diamond tools must be fed uniformly along the traverse of the cut. Sharp tools carefully ground with minimum cutting edge should be used. Removal of the molded surface layer during machining operations will affect the physical properties of the plastic material. Adding water-soluble agents to the cooling air jet improves the cooling system but requires a subsequent cleaning operation. The coolant should be selected with care, as some liquids may craze, crack, or dissolve the plastic. Glass-reinforced materials present the problem of abrasiveness, which in turn reduces cutter life.

The spiral plastic cuttings that leave the cutting edge create problems of entanglement. These problems usually can be resolved by air jets and vacuum attachments directing the chips away from the cutter. However, the type of thermoplastic to be machined will vary the techniques used.

Taps for Thermoplastics.   For tapping thermoplastics, taps with a slightly negative rake and with two or three flutes are preferred. Some plastics such as nylon may require slightly oversized taps because the resiliency of the plastic may cause it to compress during cutting, leading to swelling after the tap is removed.

Solid carbide taps and standard taps of high-speed steel with flash-chrome-plated or nitrided surfaces are necessary. During tapping, the tap should be backed out of the hole periodically to clear the threads of chips.

Reaming for Thermoplastics.   High-speed or carbide-steel machine reamers will ream accurately sized holes in thermoplastic. It is advisable to use a reamer 0.001 to 0.002 inch larger than the desired hole size to allow for the resiliency of the plastic. Tolerances as close as ±0.0005 inch can be held in through holes ¼ inch in diameter. Fluted reamers are best for obtaining a good finished surface. Reamer speeds should approximate those used for drilling. The amount of material removed per cut will vary with the hardness of the plastic. Reaming can be done dry, but water-soluble coolants will produce better finishes.

Turning and Milling.   Use tungsten carbide or diamond-tipped tools with negative back rake and front clearance. Milling cutters and end mills will remove undesired material, such as protruding gate scars, on a molded article.

Mechanical Finishing of Thermosets.   The machining of thermoset plastics must consider the abrasive interaction with tools but rarely involves the problems of melting from high-speed frictional heat. Although high-speed steel tools may be used, carbide and diamond tools will perform much better with longer tool lives. Higher cutting speeds improve machined finishes, but high-speed abrasion reduces tool life. Since the machining of thermosets produces cuttings in powder form, vacuum hoses and air jets adequately remove the abrasive chips. To prevent grabbing, tools should have an O-rake, which is similar to the rake of tools for machining brass. Adding a water-soluble coolant to the air jet will necessitate a secondary cleaning operation. The type of plastic will vary the machining technique.

Drilling Thermosets.   Drills that are not made of high-speed steel or solid carbide should have carbide or diamond tips. Also, drills should have highly polished flutes and chrome-plated or nitrided surfaces. The drill design should have the conventional land, the spiral with regular or slower helix angle (16 to 30°), the rake with positive angle (0 to +5°), the point angle conventional (90 to 118°), the end angle with conventional values (120 to 135°), and the lip clearance angle with conventional values (12 to 18°). Because of the abrasive material, drills should be slightly oversize by 0.001 to 0.002 inch.

Taps for Thermosets.   Solid carbide taps and standard taps of high-speed steel with flash-chrome-plated or nitrided surfaces are necessary. Taps should be oversize by 0.002 to 0.003 inch and have two or three flutes. Water-soluble lubricants and coolants are preferred.

Machining operations will remove the luster from molded samples. Turning and machining tools should be high-speed steel, carbide or diamond-tipped. Polishing, buffing, waxing, or oiling will return the luster to the machined part, where required.

10.6   POSTMOLDING OPERATIONS

10.6.1   Plastic Part Deflashing

Most thermoset molding operations result in some excess material, called flash, at the parting line and on molded-in inserts. It is generally necessary to remove this flash, either for cosmetic reasons or, in the case of contacts and leads extending from an encapsulated electric or electronic device, to ensure good electrical contact to the leads.

Robust parts may be tumbled randomly in a wire container to remove the flash. Slow rotation of the containers for 10 to 15 minutes, with parts gently falling against one another, will usually suffice. For more thorough flash removal, the tumbling action may be augmented by a blast of moderately abrasive material, or media, of either organic type (such as ground walnut shells or apricot pits) or polymeric type (such as small pellets of nylon or polycarbonate) directed against the tumbling parts. For more delicate parts, a deflashing system passing such components on a conveyor that holds the parts captive as they are conveyed beneath one or more directed blast nozzles, generally proves practical. Such systems have been perfected for transfer-molded electronic components, holding the lead frames captive, often temporarily masking the molded body as the devices pass through the blast area, and using as many as 24 individually positioned blast nozzles to ensure total removal of flash on a continuous basis. Such devices are magazine-fed and collected in magazines to maintain batch separation (Fig. 10.17).

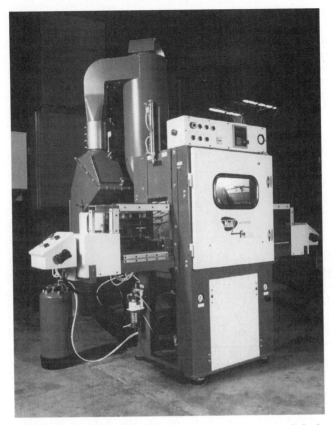

FIGURE 10.17   Side-delivery deflashing system with conveyor. It feeds, cleans, and deflashes thermoset plastic parts. Parts can be mounted on rotating spindles or indexing fixtures for maximum exposure to blast from nozzles. (Photo provided by Hull Corporation.)

Modern deflashers recycle the blast media and utilize dust collectors to minimize air pollution. The blasting chamber is effectively sealed with entry and exit ports designed to avoid dust and media escape. Chemical deflashing using solvents to remove the flash from such components is also used. In addition, water-honing deflashing has been found successful for some types of devices.

Although very simple tumbling deflashers may be built by the processor, sophisticated applications are best handled by commercial specialists who manufacture a wide variety of special and custom systems.

10.6.2   Lead Trimming and Forming

Most semiconductor devices encapsulated by transfer molding and a host of other high-volume small and fragile electronic components utilize lead frames as carriers during assembly and molding. These lead frames need to be trimmed off prior to testing, marking, and packing the devices.

Progressive trimming and forming presses and dies have been developed for this application, available both as manually fed and actuated systems and also as fully automated magazine-to-tube carrier systems.

10.6.3 Cooling Fixtures

Molded parts, both thermoplastic and thermoset, are ejected from the mold while still warm. As cooling to room temperature takes place, parts may warp or deform due in part to internal stresses or to stresses created because of uneven cooling. Such changes in shape may be minimized by placing the parts in a restraining fixture, that holds them to tolerance during final stages of cooling (Fig. 10.18).

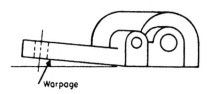

FIGURE 10.18   Product warpage.

The part design is often at fault for such deformation. Thick sections in combination with thin sections, for example, experience faster cooling of the thinner sections, resulting in localized shrinkage, which produces distortion, while the thick sections are still relatively soft. Designing parts with thin reinforcing ribs rather than thick sections often reduces or eliminates such distortion.

Another cause of internal stresses is the flow pattern as cavities are filled. In general, shrinkage is greater in a direction transverse to flow line than in the direction of flow. Part designers and mold designers need to agree on gate locations and vent locations to minimize such distortion due to flow direction. Other molding parameters, such as temperature and cavity fill pressures and rates, can be adjusted to minimize such distortion. But when all else fails, cooling fixtures may prove to be the final solution.

10.6.4 Postcure

Thermosetting plastics harden by cross linking under heat and pressure. But at the time of ejection from cavities, especially in relatively short cycles for high production, parts have not fully cross-linked and are to some extent "rubbery" at that stage of cure. They may be distorted from stresses created when ejector pins force parts from cavities or from parts piling on top of each other as they fall into a container or conveyor on being ejected. Longer cure times in the cavities may be necessary to lessen distortion from such conditions immediately following ejection.

Cross linking, or polymerization, of thermosetting plastics is generally about 90 percent completed at time of ejection, with the irreversible reaction continuing for minutes, possibly hours, and sometimes for days or months, with certain formulations. Mechanical properties of such materials may be improved by a programmed cooling following ejection, a program generally providing staged cooling in ovens instead of conventional cooling in room-temperature air. Molding compound formulators can recommend postcuring cycles where appropriate.

10.7 PROCESS-RELATED DESIGN
CONSIDERATIONS

10.7.1 Flow

As thermoplastics cool and harden into solid shapes following amorphous flow, and as thermosetting materials polymerize, changing from a viscous liquid to a shaped solid, the plastics often carry some history of their flow conditions preceding hardening. This history often adversely affects mechanical and electrical properties, dimensions, cosmetic appearance, even density of the finished parts.

Flow inside a mold cavity in injection, compression, or transfer molding should ideally be such that the cavity is completely filled while the material is still fluid. If such is the case, the cooling (for thermoplastics) or curing under sustained cavity heat (for thermosetting plastics) will proceed uniformly until the part is sufficiently rigid to withstand the rigors of ejection. But such is rarely the case.

In the case of thermoplastics, as soon as the melted material flows through the gate, it encounters a relatively cool environment and stiffens as its temperature drops to mold temperature. The material pressed against a cavity wall or flowing through restricted passage into a rib or boss, has dropped in temperature and increased in stiffness, while relatively hotter fluid continues to enter the cavity, possibly displacing some of the stiffer material on the way. The partially cooled or fully hardened material against the cool cavity walls has begun to shrink, while the later-arriving hotter material continues to enter, starting its hardening and shrinking moments later. The final amount of material passing through the gate is slowed in its flow because it is pushing the earlier-arriving material as it packs the mold during final stages of fill. The dynamic temperature change of the melt is nonuniform during cavity fill, and the final part reflects those changes, to a greater or lesser extent, depending on rate of fill and the various cavity obstructions through which or around which the flow was forced—even when cavity fill-time amounts to only a few seconds.

In extreme cases, flow lines showing earlier-hardened material adjacent to later-hardened material will be visible. Material flowing around two sides of a boss or insert in the cavity may have partially hardened before coming together on the other side, showing weld marks or lines where the material actually failed to weld to the material coming from the other side. Material flowing through a thinner section may cool so rapidly that it hardens before it reaches the far side of this section (Figs. 10.19 and 10.20).

A slow cavity fill, or a fill into a mold too chilled, will cause the plastic to stiffen before completely filling the cavity. Corners may not fill out, or detailed configurations in the cavity will not fill completely.

If the material is reinforced, such as a glass-fiber-filled nylon, relatively thin passages in the cavity may cause the low-viscosity plastic to flow into the passage but may "strain" out the glass fiber particles, resulting in a resin-rich but nonreinforced (and therefore weaker) area in the final part.

In thermosetting plastics molding, similar dynamic flow dichotomies occur because of cavity fill rate, obstructions in the cavity configuration, possible separation of resin from filler, and nonuniform cross linking during flow and cavity fill due to slightly differing time-temperature relationships, causing precure or delayed cure in various locations of the cavity during fill. Weld lines, incomplete fill, and resin-rich sections can result in reduced quality of parts and possibly lower yields of acceptable parts.

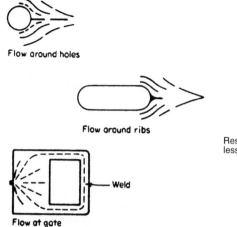

Flow around holes

Flow around ribs

Weld

Flow at gate

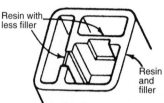

Resin with less filler

Resin and filler

FIGURE 10.19 Flow patterns around holes, bosses, ribs, and windows.

FIGURE 10.20 Flow patterns in and around thin, delicate sections.

Cavity venting—machined-in passages at one or more locations along the parting line or along ejector pins—is vital to obtaining complete cavity fill in shortest possible time. Air in the cavity must have a clear and rapid escape route. If it doesn't get out of the way in the few seconds of cavity fill because it can't find a vent or because the vent is too restricted, a void will result, or possibly a burn mark on the finished part where the highly compressed unvented air overheats according to Boyle's law—the compression of gases creating very high temperatures, enabling diesel engines to ignite fuel, and enabling air in inadequately vented mold cavities to overheat and burn plastic (Fig. 10.21 and Fig. 10.22).

Experienced part designers and mold designers need to understand the flow phenomena of various plastics during the molding process in order to minimize the occurrence of unwanted defects in the final product. But even with the "perfect"

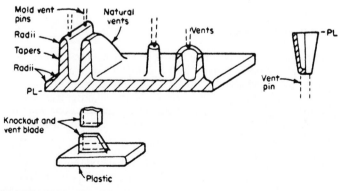

FIGURE 10.21 Ejector pins and vents.

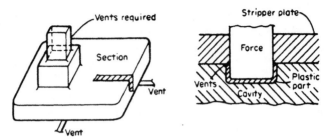

FIGURE 10.22 Parting-line vents.

design of part and the "perfect" design of mold, the processing parameters of temperatures (of melt and of mold), fill rate, fill pressures, and in-mold dwell are equally critical in achieving quality parts.

Die design for thermoplastic extruding requires extensive understanding of the melt properties of the specific material to be used as well as a full grasp of the parameters of the process. Upstream of the die, parameters of barrel temperature, screw rotational speed, and barrel venting need to be sensitively regulated for the particular plastic and die being used. Additionally, coolant flowing through the die must hold die temperature constant. Because extruding is a continuous process, these parameters must be controlled continuously to ensure a uniform product.

10.7.2

In thermoforming processes, including the blowing stage of blow molding, the plastic is not in a true fluid state but more in a relaxed elastomeric state, able to be stretched and formed and mostly stress-relieved before finally cooling to a rigid state. Parameters of temperature, pressure, and time must be closely regulated to insure that the plastic retains adequate elasticity to stretch and shape itself to the mold contours without tearing or rupturing. Thickness of material in the sheet (or in the expanded blow-molded part) must be adequate after stretching to meet the mechanical requirements of the finished part.

10.7.3 Final Part Dimensions

Physical dimensions of many processed parts must be held to fairly close tolerances to ensure proper assembly of parts into a complete structure—as, for example, molded fender panels assembled on cars, plastic screw caps for glass jars, and so on. In general, the final dimensions of the processed part will differ from the dimensions of the mold cavity or the extrusion die. Such differences are somewhat predictable but are usually unique to the specific material and to the specific process. The dimensions of a mold cavity for a polycarbonate part requiring close tolerances will often be different from dimensions of a cavity for an identical methyl methacrylate part. Similarly, the dimensions of an extrusion die for a close-toleranced vinyl profile will differ from those of a die for an identical close-toleranced polyethylene part. Both the part designer and the mold or die designer must have a full understanding of the factors affecting final dimensions of the finished product and often need to make compromises in tolerances of both part and cavity dimensions (or even in plastic-material selection) in order to achieve satisfactory results with the finished product.

The following paragraphs will address the plastic-behavior characteristics affecting dimensional tolerances.

Shrinkage of a plastic as it cools (thermoplastic) or polymerizes (thermoset) is a fact of life, often specified as parts per thousand, or, for example, as six thousandths of an inch per inch mold shrinkage. Such mold shrinkage is reasonably easy to compensate for by making the cavity proportionately larger in all dimensions.

But shrinkage in many materials is different when measured transverse to the material flow as when measured longitudinal to the flow. In reinforced or heavily filled materials, this difference is significant. Gate location and size, and multiple gates in some instances, must be considered for cavity and part design to minimize effects of such mold shrinkage.

Sink marks in a molded part often occur in relatively thick sections, usually reflecting progressive hardening of the molded or extruded or formed part from cavity wall to inside area. The outside wall hardens, while the mass of plastic in the thick section is still somewhat fluid. As this inside mass subsequently hardens (and shrinks, as most plastics will), the cured outer wall is distorted inwards, resulting in a sink mark. The best way to avoid such deformation is to avoid thick sections wherever possible. Often one or more judiciously designed thin ribs in select locations will give a part adequate strength and thickness without the need for thick sections (Fig. 10.23).

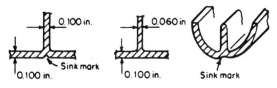

FIGURE 10.23 Sink marks.

Nonuniform hardening of the material during residence in the mold for cooling or curing generally produces internal stresses in a molded part that, after removal from the mold, may distort the part widely from the intended dimensions. Flat panels become concave, straight parts may curve, round holes may elongate, and worse. Part design and cavity design can generally accept some necessary compromises to accommodate such deformations yet still yield a part meeting its functional requirements.

Deep molded parts may require design considerations to ensure minimum stresses during the ejection phase of the molding process. Imagine a straight-walled plastic tumbler of internal and external diameters unchanging from top to bottom. As the part hardens in the cavity, it will tend to shrink around the force, or male part of the cavity. When the mold opens, the part will stay with the force. To remove it from the force will require considerable pressure, either from ejector pins or air pressure coming out the end of the force against the bottom of the tumbler or from an ejection ring moving longitudinally from the inner end of the force. The pressure exerted by either of these ejection methods will be considerable until the open end of the tumbler finally slips off the end of the force.

To minimize such ejection stresses, forces for deep molded parts are designed with an appropriate "draft" or taper, up to 5° in some cases, such that very slight

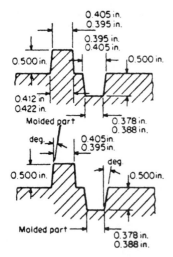

FIGURE 10.24 Dimensioning tapers.

movement of the molded part with respect to the force will suddenly free the part from its strong grip on the force, and the remainder of the ejection stroke exerts almost no stress on the part. Such draft is advisable on all plastic parts, even those with depths of only ¼ inch, to minimize ejection pressures and to prevent possible localized damage where the knock-out pins push against the plastic (Fig. 10.24).

Parting lines on molded parts require special consideration in part and mold design, especially where two molded parts must come together as, for example, on each half of a molded box with hinged opening (Fig. 10.25 and Fig. 10.26).

In compression molding or injection-compression of thermoset or thermoplastic parts, the mold is fully closed only after the material has been placed into the partially closed cavity. More often than not, some material is forced out of the cavity onto the land area before the mold is fully closed, metal to metal. In effect, then, the mold closing is halted short of full close, perhaps by as much as 0.005 inch or more. Such overflow hardens, leaving flash on the molded part. Under these cir-

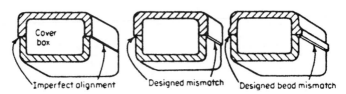

FIGURE 10.25 Mismatch parting lines.

FIGURE 10.26 Designed mismatch.

cumstances, the molded part dimension perpendicular to the mold parting surface will be at least 0.005 inch greater than intended. The tolerance of such compression-molded parts is kept very wide in deference to the inherent characteristics of the process.

When an assembly of two molded parts is ultimately required, even if the materials and the molding processes are the same, it will be virtually impossible to achieve a perfect match where the parts come together. Slight variations in shrinkage or warpage will yield an easily noticeable or "feelable" mismatch. Intentionally designing mating surfaces with an overlap or a ridge enables ingenious camouflaging of the nonuniformity of mating areas in the final assembly of the two contacting parts.

In designing molded parts, and molds to produce them, it is necessary to consciously determine how the part will be removed from the mold cavity or force and to maintain positive control of the part during mold opening, such that it is ejected as intended. This positive control is especially vital in automatic molding.

Assuming that the decision has been made that the molded part must be ejected from the moving half of the mold (as opposed to the fixed half), then it is necessary to make provisions that the part will not remain in the fixed half of the mold during mold opening but will invariably remain with the moving half.

One common way to accomplish such positive part control is to provide undercuts in the cavity or force of the moving half. These undercuts will enable plastic to flow into them and harden there before the mold is opened. On opening, the hardened plastic in the correctly designed and sized undercuts will hold the molded part in the moving mold-half during opening stroke. After mold opening, the ejector pins or mechanism in the moving half of the mold will then have to push hard enough to allow the molded part to distend sufficiently to be pushed off the undercuts and away from the moving half of the mold.

If undercuts are not practical, the fixed mold-half may be provided with spring-loaded or mechanically actuated hold-down pins, which are ejector pins that assist the molded part to leave the fixed mold-half and to follow the moving mold-half during "breakaway" and initial travel, perhaps ¼ inch or more.

The part designer and the mold designer need to consider this aspect of the molding process and agree on how to ensure proper travel of the part to guarantee controlled ejection.

10.8 MOLD CONSTRUCTION AND FABRICATION

This chapter is not intended to cover the broad field of mold and die design and construction, but the plastic-part designer needs to be aware of the several aspects of mold design that can affect the cost of mold construction. With such knowledge, the part designer may be able to achieve the desired finished product with lower mold costs, faster deliveries, lower processing costs (shorter cycles and few, if any, post-molding costs), less stringent processing parameters, and minimum mold maintenance, as well as longer mold life.

10.8.1

For the basic compression, transfer, and injection-molding processes, a wide variety of mold types may be considered. Decisions as to the optimum type will often be based on the production volume anticipated and the allowable final part cost, including mold maintenance and amortization costs as well as hourly cost rates for molding machine and labor.

10.8.2

If production quantities are as low as a few hundred parts, single-cavity molds may be feasible, even recognizing the longer time period to produce parts one at a time and the increased labor and machine time to produce the required number of parts. Single-cavity molds can be of the hand-molding type, having no mechanical ejection mechanism, no heating or cooling provisions, but requiring a set of universal heating/cooling plates bolted into the press, between which the hand mold is placed and removed each cycle. Such hand molds may be of two-plate or three-plate construction, depending on configuration of parts to be molded. If the part is relatively small, hand molds can be multicavity yet still be light enough that an operator can manually place them into and out of the press each cycle without physical strain.

Although hand molds may be made of soft metals such as aluminum or brass and the cavities may produce acceptable parts, such metals quickly develop rough surfaces and rapidly lose their practicality after a few dozen cycles. It is best to use conventional metals as used for production molds for the respective process. Molds then may be used not only for prototyping but also for modest production while waiting for full-size production tooling.

10.8.3

Production molds are generally multicavity, and they have integral heating or cooling provisions and ejection systems. And when they follow single-cavity hand molds, cavity dimensions may be fine-tuned, and vents and gates can be repositioned, based on experience with the single-cavity molds.

Family molds are multicavity molds that mold one or more sets of a group of parts that are required to make up a complete assembly of the finished product. A base, a cover, and a switch, for example, may be needed for a limit switch assembly. A family mold of, say, 36 cavities may be constructed that will yield 12 assemblies each cycle.

Production molds may also be made using a standard mold base that will accept, say, 12 identically sized cavity inserts. If the component to be produced is a small box with lid and attachable cover, in, say, 12 different sizes, each cavity insert could contain the cavities for one size box and cover. The complete mold would produce 12 boxes and covers each cycle. If sales of any one size require greater quantities than are required for one or more of the other sizes, a second cavity insert could be made to fit into the mold base, enabling twice as many of the faster-selling size each cycle.

For family molds to be successful, all parts should have approximately the same wall thickness so that molding cycles may be optimum for all sizes.

10.8.4

Some molded parts have configurations that require portions of the cavity to be removed in order for the part to be ejected. Solenoid coil bobbins, as an example, consist of a cylindrical body around which wire will be wound, with two large flat flanges at each end of the cylindrical body to keep the wire contained within the length of the body. Additionally, there is a hole through the length of the cylindrical body to accommodate the plunger of the solenoid assembly.

Such bobbins may be molded with the cylindrical body axis parallel to the parting surface of the mold. The cavity of the bottom half has the half-round shape to mold half of the body of the bobbin as well as two thin slots to mold half of each flange. The upper mold-half is almost identical to produce the other half of the bobbin. To mold the hole through the body requires a cylindrical metal mandrel, with its outside diameter equal to the inside diameter of the coil body. This removable mandrel is manually placed in the matching half-round shape on the parting line. When the mold is closed, it effectively seals around the mandrel, leaving an open cavity into which the plastic will be injected in an injection or transfer-molding machine. Following hardening of the plastic, the molded part with the mandrel is removed from the mold, the mandrel is pushed out of the bobbin with a simple manually actuated fixture, leaving the finished thin-walled bobbin intact. The mandrel is replaced in the mold for the next cycle.

For fully automatic cycles, molds for such a part may be constructed with cam-actuated or hydraulically actuated side cores that serve as the above-described mandrel. In each cycle, after the mold is closed and prior to injection of material, the side core is automatically actuated into place. Following the cycle, the side core is retracted automatically prior to mold opening and part ejection.

10.8.5

Many plastic parts are produced with molded-in inserts, such as a screwdriver with plastic handle. Molds for such items are designed to accept and hold in the correct location the steel shaft with the flat blade or Phillips head end away from the cavity, and the other end, often knurled or with flats (to insure that in use, the handle, when rotated, won't slip around the shaft) protruding into the handle cavity. After the mold is closed with insert in place, plastic is injected or transferred into the cavity, where it surrounds the shaft end and fills out the cavity to achieve its shape as a handle. Following hardening of the plastic, the mold opens and the finished part is removed.

In many insert-molding operations, fully automatic molding becomes possible when mechanisms are installed to put the insert into place before each cycle and to remove the insert with molded part following each cycle.

When inserts are to be molded into plastic parts, close coordination is obviously needed between part designer and mold designer and manufacturer.

10.9 SUMMARY

Although this chapter has touched on many aspects of part design and processing, it has not covered a myriad of special considerations that may arise in the real world. The less-experienced part designer is advised to consult with others in the field as he develops his part design, selects the material, and chooses the optimum process for his production requirements.

CHAPTER 11
RECYCLING OF PLASTIC MATERIALS

Susan E. Selke
Professor, School of Packaging
Michigan State University
East Lansing, Michigan

11.1 INTRODUCTION

In recent years there has been a significant increase in the amount of plastic recycling that is being done, both in the United States and around the world. It was estimated in 1994 that there were about 1000 plastics recycling firms in Europe, 400 in North America, and a handful in Japan. The overall plastics recycling rate that year was estimated at 4.5 percent in the United States and 7.5 percent in Europe.[1] The amount of postconsumer plastics recycled in the United States has increased dramatically in the last 15 years, from only about 20,000 tons in 1980 to about 680,000 tons in 1993 (see Fig. 11.1).[2] However, there is a fair amount of uncertainty about how much recycling actually occurs, with estimates from different sources sometimes varying widely. According to the American Plastics Council (APC), 1994 recycling rates were about 21 percent for plastic bottles and about 17 percent for plastic containers—substantial growth since 1989, when the recycling rate was just over 4 percent.[3,4] Total tonnage recycled in 1994 in the United States, by APC estimate, was 1 billion pounds.[5] For some types of bottles, the rate was much larger—nearly 49 percent for PET soft-drink bottles and nearly 26 percent for natural HDPE bottles.[4] More than 15,000 communities in the United States are now reported to collect plastics for recycling.[6]

Recovery of process scrap has a slightly longer history, becoming routine in many industries in the 1970s when increasing oil prices made it economically attractive. For uncontaminated process scrap (clean, single-resin material), recovery is now routine. Use of this regrind is not generally classified as recycling since it is considered a normal part of material usage. Considerable quantities of contaminated scrap do still go to disposal, but little accurate quantitative information is available about these materials. What can be said is that recovery of preconsumer plastics is also increasing rapidly, as the economic value embodied in these materials is recognized, and the infrastructure for converting them to a usable form is developed.

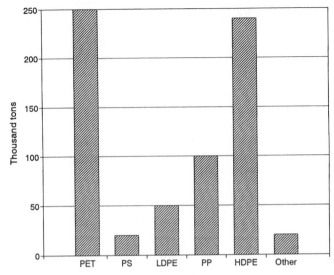

FIGURE 11.1   U.S. plastics recycled, 1993.[2]

In this chapter, our discussion will focus on recycling of postconsumer plastic materials. Routine use of regrind will not be covered. Postconsumer plastics can be defined as those materials that have fulfilled their intended use. Thus, a laundry-detergent bottle that has been filled, bought, and emptied is a postconsumer bottle, while a bottle that for any reason (defective production, production overrun, or other) never reached the consumer would be considered preconsumer scrap. Recycling of preconsumer scrap will be covered to a limited extent. The discussion will also concentrate on plastics recycling in the United States, though there will be limited discussion of similarities and differences around the world.

11.1.1 Solid-Waste Concerns

To understand the characteristics and growth of plastics recycling, it is helpful to first look at the reasons why pressure to recycle plastics developed. In the late 1980s, a number of U.S. communities began to face problems in the disposing of their municipal waste. Waste management had progressed from primary reliance on open dumps and uncontrolled incineration, with its attendant sanitary and air-pollution concerns, to almost total reliance on "sanitary landfills," where the refuse was covered with a layer of dirt at the end of each working day.

While these early sanitary landfills significantly reduced problems associated with vermin, odors, blowing trash, and air pollution, it was discovered that they too often resulted in contamination of groundwater. Growing public concern about water contamination, associated with places like Love Canal, made landfills an unpopular neighbor. Federal and state regulatory standards were tightened, which meant many existing landfills no longer met legal requirements and therefore had to close. At the same time, many of these first-generation sanitary landfills reached their designed capacity and closed. New regulatory standards made it harder to site new landfills because of geological considerations. Public unwillingness to accept new landfills was an even greater siting problem. It was not uncommon for operators to get sites approved by state agencies and then be held up in court for years by suits filed by citizen groups.

This situation led, in the late 1980s, to predictions that as many as a fourth of major U.S. cities would be out of garbage-disposal capacity within five years. Escalating disposal fees wreaked havoc with municipal budgets. Parts of Long Island, New York, saw disposal fees skyrocket from $5 per ton in 1984 to $150 per ton in 1987.[7] And then, the Mobrow set sail. This garbage scow left Islip, Long Island, in March 1987, loaded with city refuse. Six months and 6,000 miles later, after receiving worldwide media attention as it voyaged from port to port in search of a dumping ground, it ended up back in Long Island, where the garbage was eventually incinerated. All of a sudden it seemed that everyone was talking about the garbage crisis.

Since 1987, a number of things have happened (see Fig. 11.2). We have built more incinerators. We do more recycling. And we have fewer, but larger, landfills for municipal solid waste. The crisis atmosphere has largely disappeared. Disposal costs have stabilized and even decreased in some areas. We also, as a public, are more conscious of our waste. The public and our elected officials have come to see recycling as something positive that can be done, not only to reduce waste-disposal problems, but also to produce other environmental benefits.

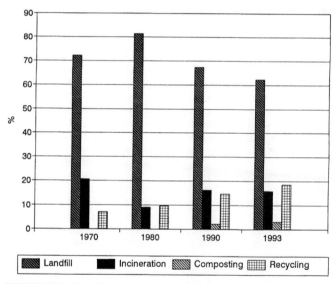

FIGURE 11.2   Handling of MSW in the United States.[2]

Elected officials have also learned that they can, at least sometimes, use legislation to promote recycling. Much of the development of plastics recycling can be fairly characterized as occurring, in large measure, because of threats of restrictive legislation. This is one of the reasons why much of the recent development of plastics recycling has focused on recovery of postconsumer materials. The concern of elected officials has been with the disposal of the materials their constituents care about and which in many cases they bear responsibility for—municipal solid waste, or MSW (see Fig. 11.3). MSW is not a well-defined term. It refers to the ordinary

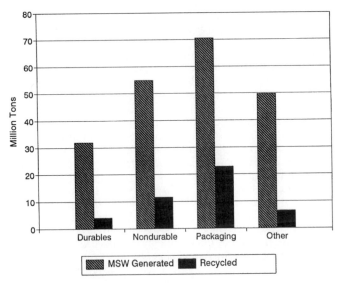

FIGURE 11.3   Products in U.S. MSW, 1993.[2]

garbage we all throw away in our homes, plus that associated with office and retail operations and institutions. It does not include industrial scrap, mining waste, construction debris, agricultural waste, and the like. As usually defined, MSW also does not include scrap automobiles, though it does include tires.

Because of the public focus on recycling as a partial solution to the problem of disposal of municipal solid waste, the type of recycling that is seen as "real" recycling is recycling of postconsumer goods. Preconsumer items are, for the most part, not part of the MSW stream, and recycling of them is thus viewed, in some sense, as not legitimate recycling.

This becomes particularly significant when legislation imposes requirements for inclusion of recycled content, or when claims of recycled content are made for marketing purposes. If an article claims "25 percent recycled content" and the public believes this means 25% postconsumer content, then the public is being misled if the recycled content is, for example, process scrap. This distinction underlies the differentiation between pre- and postconsumer recycled materials. The case is clear for differentiating between regrind, which is routinely used, and postconsumer scrap, which is both more difficult to use and represents true diversion of material from the waste stream. However, for other preconsumer materials, differential treatment is harder to justify. A bottle of milk that was pulled from distribution because of a refrigeration breakdown in a dairy is no different in characteristics from the one put in the recycling bin by a consumer. The piece of vinyl-covered wire from an automobile assembly plant is surely headed for the waste stream if it does not reach a recycler who is capable of fairly sophisticated separation and recovery operations. Many other examples could be found of preconsumer plastics that were routinely disposed of until growth in plastics-recycling capability permitted them to be diverted for processing and eventual reuse. Current federal guidelines and state regulations permit these materials to be counted as recycled content, but only if they are clearly differentiated from postconsumer materials. Thus, a label may bear a claim such as "contains at least 30 percent recycled content, at least 20 percent pre-consumer recycled content." Because of the primary concern of lawmakers with municipal, rather than industrial, waste, most recycling-related legislation is directed towards encouraging recycling of postconsumer materials.

11.1.2   Plastics in Municipal Solid Waste

The amount of plastics in municipal solid waste is increasing as use of plastics in our society increases. New resin developments permit applications of plastics where they were not suitable before. Regardless of this, the amount of plastics in municipal solid waste is less than many consumers realize. When measured by weight, plastics amounted to only 9.3 percent of municipal solid waste generated in 1993 and 11.5 percent of the MSW that went to disposal (see Fig. 11.4).[2] The difference between these two values reflects the subtraction of the MSW that is recycled or composted. Of course, when the majority of MSW is disposed in landfills, weight is not the most appropriate measure of the plastics contribution. Because plastics have a lower density than many other materials in MSW, their contribution by volume is considerably larger than their contribution by weight. In 1993, it was estimated that plastics accounted for 21.2 percent by volume of total MSW generated and 23.9 percent of MSW discarded (see Fig. 11.5).[2] The total amount of plastic generated in MSW was estimated at 19.3 million tons; plastic discarded was 18.6 million tons.[2]

Of that amount, about 6.3 million tons originated in durable goods, 3.4 million tons in nondurables, and 8.4 million tons in packaging. Six hundred eighty thousand tons of plastic from packaging materials were recycled, 150 thousand tons from durable goods, and 20 thousand tons from nondurables, according to EPA estimates for 1993.[2] Figs. 11.6 and 11.7 illustrate the sources and types of plastic in municipal solid waste.

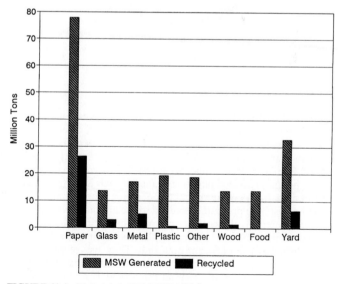

FIGURE 11.4   Materials in U.S. MSW, 1993.[2]

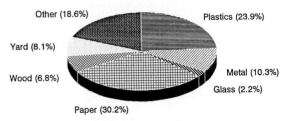

Other (18.6%)   Plastics (23.9%)
Yard (8.1%)
Wood (6.8%)   Metal (10.3%)
Glass (2.2%)
Paper (30.2%)

FIGURE 11.5   Volume (%) of materials in U.S. MSW, 1993.[2]

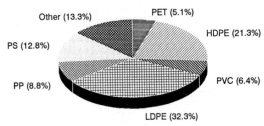

Packaging (43.3%)   Durables (32.7%)
Nondurables (24.0%)

FIGURE 11.6   Plastics in U.S. MSW by end use, 1993.[2]

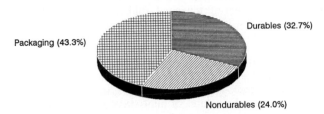

Other (13.3%)   PET (5.1%)
PS (12.8%)   HDPE (21.3%)
PP (8.8%)   PVC (6.4%)
LDPE (32.3%)

FIGURE 11.7   Plastics in U.S. MSW by resin type, 1993.[2]

11.1.3 Benefits of Recycling

An obvious benefit of recycling is that it reduces the requirement for disposal of waste materials. Items that are recycled are, by definition, diverted from the waste stream. In many, though not all, cases, another obvious benefit of recycling is cost reduction. Use of regrind, for example, became routine because of the monetary savings it provided. Similarly, certain plastics industries for years have relied on a combination of off-spec and recycled plastics because of their lower price. More recently, the desire to benefit from consumer preferences for recycled material coupled, in some cases, with legislative pressures have led to the anomalous situation of recycled plastic sometimes being worth more per pound than virgin resin.

Less obvious benefits from recycling of plastics result from the fact that use of recycled resin displaces use of virgin materials. Thus, use of recycled plastics can result in significant energy savings, since in general more energy is used to make a plastics article than to recover and remold the article. For example, Fenton[8] calculated the total energy requirement for a low-density polyethylene grocery bag to be 1400 kJ, while a bag with 50 percent recycled content required only 1164 kJ, for a savings of nearly 17 percent. A recent DOE report concluded that recycling PET products such as soft-drink and ketchup bottles requires only about ⅓ the energy needed to produce PET from virgin materials.[9]

Similarly, recycling processes generally produce fewer environmental effluents than do processes that produce virgin resins, so use of recycled plastics usually results in a decrease in air and water pollution.

Obviously, use of recycled plastics also results in less depletion of natural resources.

Historically, the major motivation for efforts by lawmakers to increase plastics recycling has been the decrease in disposal demands that results, while the major motivations for the business community have been the cost savings and marketing advantages that can be gained.

11.1.4 Contamination Issues

One of the major stumbling blocks in increasing the recycling of plastic materials is contamination. Contamination issues can be grouped into three general categories: contamination that raises safety concerns, contamination with nonplastics that reduces performance, and contamination with other plastics that reduces performance.

Contamination that Raises Safety Concerns.   For certain applications, possible or actual change in behavior of plastic materials due to impacts during prior use cycles can be particularly problematic. For example, it is probably safe to conclude that recycled plastics will never be used for implantable medical devices. It is highly unlikely that recycled plastics will be used for the packaging of sensitive drugs. Other examples, of course, could also be cited where the small but real risk of unacceptable performance or release of some damaging substance, coupled with the critical nature of the application, is likely to rule out the use of recycled plastics.

For less critical applications, such as use of recycled packaging for food products, only a few years ago the conventional wisdom was that recycled plastics should not even be considered, if for no other reason than the belief that the Food and Drug Administration (FDA) would never approve. In this area, there have been dramatic changes. One of the earliest applications of recycled plastic for packaging of food products was recycled polyethylene terephthalate (PET) in egg cartons. The physical barrier of the egg shell provided a degree of added protection. Next came use of recycled plastic in buried inner layers of packaging, such as a recycled polystyrene (PS) clamshell used for hamburgers, in which the contact between the food and the recycled plastic was mediated by a layer of virgin plastic that acted as at least a partial barrier. Next, repolymerized PET was used in direct contact with food (blended with virgin material). The nature of the repolymerization process, with its crystallization steps, provided assurance that any impurities present would be removed. Most recently, use of recycled plastics in food contact moved another step forward with FDA approval of a physical reprocessing system for PET, which, along with control over incoming material, was determined to provide a sufficient level of safety to permit direct food contact. More detailed discussion of these processes and applications for recycled plastics will follow.

The concern over use of recycled plastics in food contact falls in two general areas. First is concern about biological contaminants. In most cases, the processing steps for production of plastic food-packaging materials provide a sufficient heat history to destroy any disease-producing organisms. Thus, this is not a major concern.

The more relevant concern is directed at the possible presence of hazardous substances in the recycled feedstock. FDA regulations require food packagers to ensure that the materials they use are safe for food contact, that they do not contain substances that might migrate into the food and cause deleterious effects on human health. Recycled resins often, by their very nature, have a somewhat unknown history. What if, for example, someone put some insecticide, or some gasoline, or weed killer, or any of a myriad of hazardous substances into a soft-drink bottle and then later turned in the bottle for recycling? How can we prevent that container from contaminating new plastic packages? What we have seen in the last few years, in this as in other areas, is movement in the FDA away from absolute prohibitions and towards a more reasonable evaluation of risk.

Contamination with Nonplastics that Reduces Performance.   Inclusion of nonplastic contaminants in recycled resins can affect both processing of the material and performance of the products manufactured from these materials. Presence of remnants of paper labels, for example, can result in black specks in plastic bottles, detracting from their appearance and rendering them unsuitable for some applications. These paper fragments can also build up in the screens in the extruder, resulting in greater operating pressures (and energy use) and requiring more frequent screen changes. Presence of solid inclusions in the polymer can adversely affect the physical performance of the molded parts, resulting in premature failure. Mechanical properties can be decreased to the extent that thicker sections are required to attain the desired performance.

Contamination with Other Plastics that Reduces Performance.   Just as contamination with nonplastics can reduce the performance of the end product, contamination of one resin with another can result in diminished performance. This can result from a variety of factors.

One of the most fundamental problems is that many polymers are mutually insoluble. Thus, a blend of resins is likely to consist, on a microscopic scale, of domains of one resin imbedded in a matrix of the other resin. While this sometimes results in desirable properties, more often it does not. To further complicate matters, the actual morphology, and thus the performance, will be strongly dependent not only on the composition but also on the processing conditions.

Another problem arises from differences in melting temperatures. When PET is contaminated with PVC, for example, the PVC decomposes at the PET melt temperatures, resulting in black flecks in the clear PET. Only a small amount of PVC contamination can render useless a large quantity of recovered PET. On the opposite side, at PVC-processing temperatures, PET flakes fail to melt, resulting in solid inclusions in the PVC articles that cause them to fail. Again, a small amount of PET contamination can render recovered PVC unusable.

More subtle problems can arise, even from very similar resins. When injection-molded HDPE base cups from soft-drink bottles were contaminated by newly developed, blow-molded HDPE base cups, the recovered HDPE consisted of a blend of a high-melt flow resin with a low-melt flow resin that neither blow molders nor injection molders found usable.

Mixing resins of different colors can also be a problem. Laundry-detergent bottle producers were able to fairly easily incorporate unpigmented milk-bottle HDPE in their detergent bottles in a buried inner layer, but found it much more difficult to use recycled laundry detergent-bottles because the color tended to show through the thin pigmented layer, especially in lighter colored bottles. Motor-oil bottlers who used black bottles had no such problems. As a general rule, the lighter the color of the plastic article, the more difficult it is to incorporate recycled content, and conversely, the lighter the color of the plastic article, the easier it is to find a use for it when it is recycled.

11.1.5 Quality Issues

The issue of color contamination addressed above is closely related to more general issues of the quality of recovered plastic resins. It is well known that plastics undergo chemical changes during processing and use that ultimately lead to deterioration in properties. In fact, much of the history of plastics is related to the development of appropriate stabilizing agents to prevent this degradation. We routinely stabilize plastics against thermo-oxidative degradation that would otherwise occur during processing. We know that some resins are much more sensitive than others. Depending on the amount of stabilizer initially present, the history of the resin, and the type of resin, a recycled resin may or may not require additional stabilizer in order to be successfully utilized.[10]

Similarly, plastics that are designed to be used outdoors must, in general, be stabilized against photodegradation. Recycled materials are likely to need additional stabilizer to retain adequate performance.

When regrind began to be a common ingredient in plastics processing in the late 1970s, much effort was devoted to studying the effects of multiple processing cycles on polymer performance. For many polymers, three major types of chemical reactions occur. First is oxidation. Reaction of the polymer structure with oxygen results in the incorporation of oxygen-containing structures in the polymer, with concomitant changes in properties and increased potential for further reactions. Either with or without oxidation, chain cleavage can also occur. This results in a decrease in molecular weight, with a consequent decrease in many performance properties. Chain cleavage can be followed by cross linking, the forming of new molecular bonds that increases molecular weight and also changes properties. In some polymers, one or the other of these reactions predominates. In others, such as polyethylene, the effects of one tend to be balanced by the effects of the other. Some molecular structures are much more reactive than others. Polypropylene, for example, is significantly more susceptible to photo-oxidation than is polyethylene. Further, for some materials it is feasible to upgrade the material during reprocessing (such as solid-stating of recycled PET), while for others it is not.

In summary, the general rule is that recycled polymers will have somewhat different properties than virgin polymers. These changes are usually detrimental and range in nature from virtually unnoticeable to major. Just as not all polymers are equally sensitive, not all properties are equally sensitive. It is possible, for example, for a recycled HDPE, compared to virgin, to have virtually the same tensile strength but significantly decreased Izod impact strength.[11]

11.1.6 Collection of Materials

For a material to get recycled, it must be collected, processed into a usable form, and finally supplied to an end market. For postconsumer materials, the most difficult part of this process may be getting the material collected in the first place. Industrial scrap is "owned" by the industrial entity that produced it. If this entity cannot get it recycled, the owner will have either to dispose of it or pay some other business entity to do so. For much consumer scrap, there is little or no monetary incentive for its owner, the individual consumer, to direct it into a recycling system. Further, industrial scrap tends to be concentrated, with substantial amounts of material in relatively few locations. Postconsumer materials are typically very diffuse. It is estimated that the typical consumer generates only about 35 pounds of plastic bottles per year.[12] Thus, a fairly elaborate collection infrastructure is needed to get this material gathered together in quantities that make processing it economically viable.

A number of systems have been developed to accomplish the task of collection. Among the most well developed in the United States are bottle-deposit systems.

Bottle-Deposit Systems.   Nine states (see Fig. 11.8) have a deposit system on certain types of beverage containers. The consumer pays a deposit, usually 5 cents, when buying the container and then receives a refund of that fee when the bottle is returned to a designated collection point. In most cases, any retailer that sells beverages is obligated to accept the returns and refund the deposit. In many cases, the retailer receives a handling fee from the distributor, the originator of the deposit, to offset the costs of managing the system. In most states, the deposit was originally restricted to carbonated soft drinks and beer, though it has since been expanded in several states. In all cases, the original motivation for the deposit was primarily to reduce litter. Deposits have proved to be a powerful incentive to consumers to return the covered containers, with redemption rates routinely exceeding 90 percent. Thus, deposit legislation resulted in the collection of large numbers of polyethylene terephthalate (PET) soft-drink bottles. This, in turn, spurred the development of effective reprocessing systems for these bottles and end markets for the recovered resin. The existence of bottle-deposit legislation is in large part responsible for the successful development of PET recycling, and for the fact that, in the United States, PET has a much higher recycling rate than any other plastic. Even with the growth in curbside recycling in nondeposit states, deposit systems accounted for over 35 percent of the PET collected in 1994.[13]

Connecticut	Delaware
Iowa	Maine
Massachusetts	Michigan
New York	Oregon
Vermont	

FIGURE 11.8   States with bottle-deposit legislation.

California more recently introduced a system of refund values for beverage containers designed specifically to encourage recycling. It differs from true deposit systems in three significant ways. First, the consumer is not charged a separate deposit. The refund value affects the price of the product, but the consumer does not see this reflected explicitly. Second, containers can be returned only to designated redemption centers, so return of containers is significantly less convenient. Third, the size of the refund value (currently 2.5 cents) is less than the typical deposit (5 cents). Recycling rates in California for containers covered by this legislation are higher than the national average but lower than those in deposit states.

Maine extended its early deposit law in an explicit attempt to increase recycling. Maine now has deposits in place on most beverages with the exception of milk.

From a recycling perspective, the most significant aspect of bottle-deposit legislation is that, in most cases, the financial incentive provided does an excellent job of getting people to return their empty plastic bottles to the appropriate place. The most negative aspect is that the per-container costs of managing these systems, as they are currently designed, are higher than the costs of alternative collection systems.

Other Deposit Systems. The idea behind beverage-bottle deposits has also been applied to other products. Automobile batteries are subject to deposits in many states. While the primary motivation is to avoid the introduction of lead into landfills and incinerators, these systems have been very successful at facilitating the recycling of the polypropylene (PP) battery cases.

Buy-Back Systems. Another way to put a monetary incentive into the return for recycling of plastic materials is similar to the California bottle-refund system. Recycling centers have been set up in a number of places that pay consumers for the materials that they bring to the center, generally on a price-per-pound basis. These centers have been very successful for collection of aluminum beverage cans but have also been used to facilitate the recycling of plastic beverage bottles. They typically consist of only one, or at most a few, locations in a metropolitan area. A variation that has been successful in some areas is utilization of reverse-vending machines, in which consumers insert the bottles and receive a reedemable receipt or coupon for the refund value. For plastic bottles, however, these have been used primarily in deposit states.

Drop-Off Systems. Drop-off systems function much like buy-back systems, consisting of centralized locations for people to bring in their recyclable materials. The difference, of course, is that no monetary compensation is offered. This difference also permits these centers to operate unattended, in some cases. Drop-off systems encompass a wide range of designs, including barrels in supermarkets for people to place their plastic grocery sacks, roving multimaterial collection centers coming to a location once a month, permanent multimaterial centers in a centralized location in a community (or an out-of-the-way location), collection bins in apartment-building laundry rooms, and even sophisticated garbage and recyclables chutes in high-rise apartment buildings.

Drop-off systems, especially in their simplest designs, can offer the lowest cost collection options. Their major drawbacks are high rates of contamination and low rates of collection.

Collection Systems. A general rule of recycling is that the easier you make it for the consumer, the higher the rate of participation—and the higher the rate of diversion of material from the waste stream—you will achieve. The major method in the United States for collecting plastics for recycling from single-family dwellings, outside of deposit systems, is curbside recycling. By going to the consumer to get the materials instead of asking the consumer to go to the recycling point, significantly higher rates of collection can be achieved. These systems also differ in design but fall into three general categories. Collection of commingled recyclables refers to systems where the participant places all the recyclables together, usually in a container provided by the operator of the system. Other systems require consumers to separate the recyclables by type, and thus use multiple containers, usually provided by the consumer, for set-out of the materials. Many systems are hybrid types, with most materials collected in a commingled form, and others collected separately. A common system is a bin for commingled bottles and cans, with newspapers bundled separately. Virtually all of these systems collect multiple materials.

Systems in which consumers set out commingled recyclables at the curb can be further divided into three categories, depending on how the materials are handled in the collection vehicle.

First, in a few communities, recycled materials are placed into bags (generally blue) and collected in the same vehicles as the garbage, standard compactor trucks. The waste is then sorted to retrieve the blue bags, and sometimes other readily identifiable recyclable materials. While some of these systems seem to work reasonably well, others have experienced significant contamination problems. Even without losses due to contamination, the yield of recyclables in general is lower than in systems that provide separate collection, simply because not all the bags are recovered intact. One frequently encountered problem is contamination of newspapers with broken glass. Some systems, therefore, request that newspaper be bagged separately from the other recyclables. The largest city to try this type of collection system is Chicago. This long-delayed system began operation in December 1995.

The second category includes systems that use a separate truck, or at least a separate compartment, for commingled recyclables. The recyclables are then delivered to a sorting facility called a MRF (materials recovery facility), where they are separated by material type (and for plastics, perhaps by resin type as well, though this may occur at a separate facility). While the first-generation MRFs relied almost exclusively on hand sorting, modern MRFs are becoming increasingly mechanized. The major advantages of this system are efficiency in the time on route and in the filling of the vehicle. Major disadvantages are the need for a dedicated sorting facility and sometimes high residual levels of unwanted materials.

The third category includes systems in which the commingled collectibles are sorted at truckside into several categories. These may or may not require further processing at a MRF, depending on the materials included. The major advantage of this system is the quality control that can be practiced by the driver, coupled with ongoing education of consumers. If the householder puts an unacceptable item in the bin, he or she will find it left there, ideally with an explanatory flyer, so they can do better the next time. Another advantage is the avoidance of a requirement for a dedicated processing facility. The major disadvantages are increased time per stop and the potential for the truck filling one compartment and therefore having to leave the route and off-load, even though other compartments may not be full.

The general recommendation is that truckside sorting works well for moderate to small-sized communities, and commingled collection and MRF work best for large communities.

General rules of thumb for effective design of curbside recycling systems are: collect the recyclables in commingled form and require little if any preparation beyond cleaning; provide a readily identifiable container for use by the householder in putting out the recyclables; collect recyclables weekly on the same day as garbage collection; and put considerable effort into ongoing education and publicity efforts.

Providing a container is particularly important. The container serves several functions. First, it is a convenience for the householder, providing a useful place to deposit the recyclables. It is a visible reminder of the importance of the recycling program. This is key when a program is initiated—a mailing telling about the program can get easily lost in the junk mail, but it is hard to ignore a large plastic bin. Further, the bins serve as a potent source of peer pressure. When everyone else has their recycling bin at the curb and you do not, you are likely to feel like a bad citizen, which hopefully will be incentive for you to participate in recycling the next time. According to some reports, participation rates average 70 to 80 percent for curbside programs that provide containers, and only 30 to 40 percent in programs that do not.[14]

Another difference between recycling programs is whether they are voluntary or mandatory. The majority of curbside programs are voluntary, but several states and a number of municipalities have instituted mandatory programs. There seems to be general agreement that mandatory programs increase participation if enforcement efforts are included. If no enforcement takes place, results are not as clear. Typically, enforcement activities involve a series of warnings, ending in refusal to pick up the garbage for a period of time. While fines and even jail terms may be permitted by the ordinances, they are seldom employed.

As of 1994, it was estimated that approximately 4,000 curbside collection programs in the United States included plastic bottles.[12] A significant problem often encountered when plastics are added to a curbside program is the space the containers take up in the truck, relative to their value. Programs urging consumers to compact the plastic bottles by stepping on them before placing them in the recycling bin can provide a significant benefit. There has also been considerable investigation of on-truck compacting equipment. Results have been mixed, since extra handling is required, and the compactor takes up valuable space on board the truck. For trucks that collect commingled recyclables, there has also been investigation of compacting the entire metal, glass, and plastic fraction. In this case, the presence of plastic bottles turns out to be of value in helping minimize glass breakage.[14] Shredding or chipping the plastic on the truck has not been practiced, in large part because of the lack of reliable methods for separating chipped plastics by resin type.

For plastics, curbside-collection systems typically concentrate on PET and HDPE bottles, which represent about 95 percent of the plastic bottles generated by a typical household.[12] Costs per bottle collected are generally intermediate between drop-off and deposit systems, as are participation and diversion rates.

According to a survey conducted by the Council of Packaging in the Environment (COPE), 56 percent of U.S. consumers claim to recycle "very often." Fifty-one percent of the people who recycle do so through curbside collection programs, compared to 31 percent through drop-off centers.[15]

Australia has a unique collection system for PET beverage bottles. In Victoria and New South Wales, a private entrepreneur called a bottle merchant provides plastic bags to households for collection of the bottles, picking up the bags every two weeks. Glass and sometimes PVC bottles are also included. The bottle merchant sorts the containers and sells refillable bottles to the stores and one-way bottles to container manufacturers. Fees are set by a semigovernment association and are subsidized by soft-drink companies.[16]

Mixed Waste Processing.   Another approach to recycling plastics and other materials is not to ask consumers to do any special sorting or preparation but instead to recover recyclables from the garbage stream. The advantage of these systems is that, since they do not require any particular cooperation by consumers, they have the potential to recover the largest amount of recyclables.

The major disadvantage is the high cost of such systems and the low quality of the collected materials. The U.S. Bureau of Mines began experimenting with mixed-waste sorting facilities in the 1970s. Techniques employed drew largely on the mineral-processing industries and included size reduction and various types of size- and density-based sortation methods. For plastics, the obvious result of such processing is the production of a mixed stream of plastics, with the resultant problems. Residual contamination is also a major concern. While research on this type of recovery continues, the vast majority of plastics is recovered through source-separation-based programs, where the "free" labor of the individuals who keep the recyclable plastics separate from the garbage is crucial, both in terms of economics and quality.

11.2 POLYETHYLENE TEREPHTHALATE (PET) RECYCLING

PET is the most recycled plastic, largely due to recycling of soft-drink bottles. Four companies control nearly 75 percent of all PET recycling in the United States. The largest is Wellman, with about 37 percent. Pure Tech International is estimated to control 15 percent; Image Industries, 13 percent; and Martin Color-Fi, 8 percent.[17] Packaging is the largest source of PET in municipal solid waste (see Fig. 11.9).

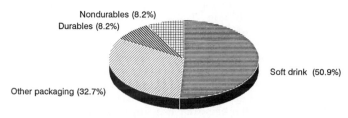

Nondurables (8.2%)
Durables (8.2%)
Soft drink (50.9%)
Other packaging (32.7%)

FIGURE 11.9   PET in U.S. MSW, 1993.[2]

11.2.1 History of Soft-Drink-Bottle Recycling

As mentioned above, soft-drink-bottle recycling got its start with the introduction of deposit legislation, which resulted in collection of significant volumes of material and recognition of the economic value embedded in them. Recycling rates for soft-drink bottles have continued to grow, reaching a recycling rate of 48.6 percent in 1994.[6] The recycling rate in deposit states is over 90 percent.[16]

At the same time, PET usage in packaging applications other than soft-drink bottles began to grow. These "custom" bottles now account for nearly 37 percent of all PET bottles.[18] In nondeposit states, it is relatively simple to add collection of custom PET bottles to curbside systems that are already collecting PET beverage bottles. In deposit states, collection of custom PET bottles is less common, since the deposit bottles, which still represent the major source of PET in household waste, are not part of the curbside system.

The overall recycling rate for PET bottles was estimated to be 34 percent in 1994. Five hundred sixty-five million pounds of PET bottles were collected for recycling.[13]

11.2.2 PET Soft-Drink-Bottle Recycling Processes

Two major approaches arose to processing PET beverage bottles into usable resin. One method focuses on separating the bottle components as the first step. Mechanical devices remove the closure (cap) and HDPE base cup from the PET body. The various components are then processed separately. While these systems work well on intact bottles, they sometimes have difficulty in handling bottles that have been baled, even when a reinflation step is added to try to return the bottles to their initial shape.

The major method for recycling PET (see Fig. 11.10), as developed at the Rutgers University Center for Plastics Recycling Research, starts with shredding the whole bottle—closure, base cup, and all. The shredding is sometimes done in two steps to facilitate achieving the particle size necessary for subsequent processing. Bottles typically arrive at the processor in baled form. Separation by color may have been achieved before baling. While some systems for automatic color sorting of previously baled bottles are available; typically the color sorting is done by hand before baling, or not at all.

After size reduction, the chipped plastic (about ⅜ inch in size)[19] is fed into an air cyclone, where light materials such as label fragments are blown off and heavy materials such as rocks and glass are also removed. Next the plastic is washed, usually with hot water and detergent, to remove product residues and separate labels and base cups that are attached to PET bottle fragments by the hot-melt glues used in their application. The hot wash is sometimes preceded by a prewash in cold water. Screening and rinsing typically follows to remove the dirty residue. The wash solutions are generally filtered and reused several times.[19]

Next, HDPE (from base cups) and PP (from caps and labels) are separated from the PET and aluminum (from caps) by flotation or hydrocyclones. The HDPE/PP and PET/aluminum streams are both dried. The HDPE/PP stream is not further separated. As long as the PP concentration is under about 5 percent, the material is marketable, but problems in performance are encountered if it rises above this level.[19] Any ethylene vinyl acetate (EVA) present from cap liners also stays in the HDPE stream at this point.

The PET and aluminum are typically passed through an electrostatic separator. In this operation, a static charge is applied to the particles, and they are then placed on an oppositely charged rotating drum. The aluminum particles, being conductive, lose their charge quickly and are flung off the drum, while the nonconducting PET particles cling to the drum and are scraped off at a later point in its revolution. Typically, several separation passes are required to get the high level of purity for PET that is needed (at least 99.9 %). The aluminum stream commonly has high levels of contamination with PET. The recovered aluminum can be sold as-is or can be further purified.

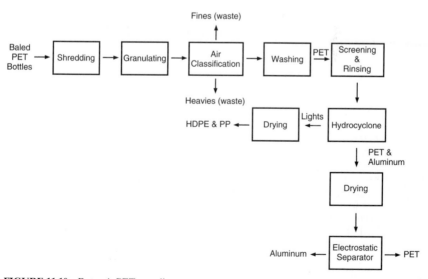

FIGURE 11.10   Rutger's PET recycling process.

Carpco Inc. manufactures a metal detector and separator that can further reduce the aluminum content in the cleaned PET flakes. The flakes are flowed on a conveyor, and the metal detector directs a jet of air to remove aluminum particles from the belt.[16]

Johnson Controls' Reco process (see Fig. 11.11) is designed for PET soft-drink bottles without metal caps. The bottles are washed (after a sorting step) in a 70–100°C hot-water bath, causing the bottle to shrink and liberate the base cups, labels, and many of the PP caps. Vibrating screens with apropriately sized holes then separate the labels and caps, and then the base cups, from the bottles. The bottles and base cups are separately granulated, washed, and rinsed. A hydrocyclone is used to separate residual light materials from the PET stream. The plastic is then dried, and the PET is passed through a metal detector.[19]

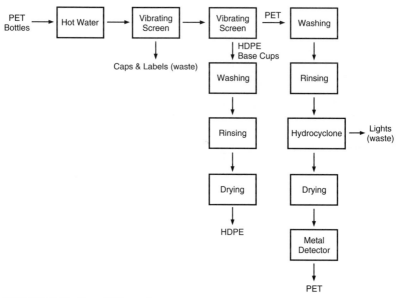

FIGURE 11.11   Reco PET recycling process.

In the Pure-Tech process, whole bottles (preferably without caps) are washed in hot water and a cleaning agent. During the 30-minute wash and rinse cycle, the labels and base cups come loose and are removed mechanically or by hand sortation. The base cups and bottles are separately granulated. The PET flakes may be subjected to a further wash and rinse step before drying.[19] This process is claimed to be unique in its ability to disperse polyvinylidene chloride (PVDC) coatings.[16]

Western Environmental Plastics uses cryogenic grinding for the PET, resulting in the adhesives being reduced to a fine powder, which can then easily be removed from the PET flake by screening.[16]

When these processes were initially developed, nearly all closures on beverage bottles were aluminum, and nearly all had HDPE base cups. Since that time, base cups have begun to disappear with the introduction of new bottle designs that can maintain stability without need of a base cup. At the same time, aluminum caps have begun to disappear, being replaced by polypropylene caps. This has serious implications for the HDPE stream from beverage-bottle recycling, which is decreasing in volume and increasing in level of contamination. There is currently no commercial system for separating HDPE and PP from each other in these materials. Since one of the major markets for recovered base-cup material has been new base cups, the amount of PP contamination in this stream has grown to the extent that it is producing problems for some users. When we reach the point where no bottles have HDPE base cups and all closures are PP, then the products of PET recycling will change from the current HDPE, PET, and aluminum streams, to PP and PET. The major product is now, and will remain, PET, in terms of both volume and value. However, other trends in soft-drink-bottle manufacture are not as encouraging. New metallicized labels are reported to clog extruder screens and discolor PET pellets. Polystyrene labels have a density close to that of PET and consequently are difficult to separate from PET. It remains to be seen whether these will be significant detriments to PET recycling.[13]

The number of recycling facilities for PET has grown dramatically. In the United States, there are, as of 1995, 24 companies operating 27 plants in 18 states, plus three firms in Canada. Eight of the U.S. companies had been in operation for only one year or less, and only six had been in operation for six or more years.[13]

Costs for producing recycled PET by such processes were about 37 cents per pound in 1994, with 7 cents per pound paid for the baled feedstock. With virgin resin selling for 55 to 65 cents per pound, this provided a clear incentive for use of recycled material.[1]

A major development in PET recycling occurred in 1994, when Johnson Controls received FDA nonobjection status for the use of physically reprocessed PET for direct food contact. Until that time, only repolymerized PET (see below) or virgin PET could be used for food and drink containers. While the process is proprietary, it reportedly involves control over the source of materials (redeemed beverage bottles from deposit states) plus a high-temperature, high-intensity washing step.

PET recycling in Europe is less developed. Switzerland collects about 75 percent of PET containers generated, the Netherlands collects about 95 percent of PET beverage containers, and Italy also has a robust PET recycling program, but the overall European recycling rate for PET bottles is only 4 percent.[13]

Retech opened a plant in the Netherlands in 1986 to recycle PET bottles collected primarily from Germany and Holland under a deposit system. In 1990 the plant was expanded, and PET bottles from other European countries were added. European PET beverage bottles are clear PET only, with a clear thermoformed PET base cup. Thus they are simpler to recycle than the traditional American design.[16]

Wellman International also operates a PET recycling facility in the Netherlands.[16]

11.2.3   Other PET Recycling

Other PET Bottles.   As mentioned above, use of PET for other packaging applications has grown substantially. PET from nonbeverage-bottle-packaging applications has been successfully incorporated into recycling systems designed for PET beverage bottles. In fact, at least in part the growth in PET packaging use can be attributed to its favorable environmental image, which in turn is due in large part to its perceived recyclability. However, while custom PET bottles are compatible with beverage-bottle recycling programs, other PET packaging is not. Bottles are estimated to make up 92 percent of rigid PET packaging, with over ⅔ of these being beverage bottles.[13]

Bottles such as those currently used for ketchup, which consist of PET along with buried inner layers of ethylene vinyl alcohol, can be successfully recycled along with pure PET bottles. The ethylene vinyl alcohol content is a small fraction of the bottle, and this substance is mostly removed during the recycling process, since the container is manufactured without adhesive tie layers.

The EPA estimated the recycling rate for custom PET bottles in the United States in 1993 as 3.6 percent (see Fig. 11.12).[2]

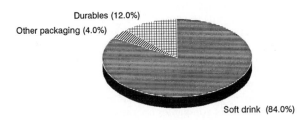

Durables (12.0%)

Other packaging (4.0%)

Soft drink (84.0%)

FIGURE 11.12   PET recycled in the United States, 1993.[2]

PET Film.   Over 50 percent of the PET film produced in the world is used in photographic film, including X-ray film.[20] Since these materials generally are coated with silver, there has long been a potent economic incentive for the recovery of these materials. In fact, silver from film has been recovered since the early 1900s, when cellulosics were used as the base film.[20] Recovered PET, therefore, is obtained as a byproduct of silver recovery. This PET is generally coated with PVDC. Gemark is reported to have a proprietary process to remove the PVDC.[16]

The 1989 rate for recycling of PET film was about 10 percent, with total recovery about 70 million pounds per year, most of this from X-ray film.[16]

Other PET Products.   In principle, many other PET products could be recycled, but the infrastructure to do so has not yet been developed. The U.S. EPA estimated recovery of PET from durable goods in 1993 as 30 thousand tons. No information was provided on the source of this material.[2]

11.2.4   Tertiary Recycling of PET

Recovered PET can be chemically broken down into small molecular species, purified, and then repolymerized. Recycling processes that use the recycled material as a chemical feedstock are generally classified as tertiary recycling. (Primary recycling is defined as applications producing the same or similar products, while secondary recycling produces products with less demanding specifications).[21] The two major processes for tertiary recycling of PET are glycolysis and methanolysis. Both processes produce PET that is essentially chemically identical to virgin resin and has been approved by the U.S. Food and Drug Administration for food contact applications. At the present time, PET produced by these processes is more costly than virgin resin. Its use is typically limited to 25 percent of the resin content for economic reasons.

In 1991, Goodyear obtained a letter of no objection from the U.S. Food and Drug Administration for the use of its Repete tertiary recycled PET in food contact applications. This process (which was later sold to Shell Chemical Co.) uses glycolysis to break down PET into monomers that are then purified by crystallization and repolymerized. In tests using model contaminants, the contaminants were removed down to a 50–100 ppm level.[19]

That same year, both Eastman Chemical Co. and Hoechst-Celanese Corp. received letters of no objection from FDA for their methanolysis-based PET depolymerization processes.[19]

Freeman Chemical Corp. uses glycolysis to process PET bottles and film into aromatic polyols used for urethane and isocyanurate manufacture.[19]

Other investigators have examined the use of glycolized PET reacted with unsaturated dibasic acids or anhydrides to form unsaturated polyesters. Recycled PET film is generally used since it is lower in cost than PET from bottles. Products include glass-fiber-reinforced applications such as bath tubs, shower stalls, and boat hulls. U.S. companies involved include Ashland Chemical, Alpha Corp., Ruco Polymer Corp., and Plexmar.[16]

One interesting use of unsaturated polyesters is in polymer concrete. One of the major advantages of these materials is their very fast cure times, facilitating repair of concrete structures without major delay in returning them to service. Polymer concrete can also be used for making precast structures. Basing polymer concrete materials on recycled PET reportedly leads to 5 to 10 percent cost savings and comparable properties compared to polymer concrete based on virgin materials. However, they are about 10 times the cost of portland cement concrete. Nonetheless, it is reported that the savings in labor cost can more than offset the added material cost.[22]

A substantial number of patents have been issued for these and other processes for depolymerizing PET from various sources.

11.2.5 Properties of Recycled PET

Recycled PET (physically processed) in general retains very favorable properties. Some reduction in intrinsic viscosity is common, but this can be reversed by solidstating. PET is very susceptible to damage from contamination with PVC. The densities of PET and PVC are overlapping, making them difficult to separate by mechanical means. Under PET processing conditions, PVC decomposes, leaving black flecks in the resin as well as causing other problems.

Residual adhesives from attachment of labels and base cups are a common contaminant concern in recycled PET. Some of the adhesive residue can become trapped in the PET granules and is not removed by washing. Since these adhesives often contain rosin acids and ethylene vinyl acetate, when the PET is extruded, the rosin acids plus acetic acid from hydrolysis of the ethylene vinyl acetate can catalyze the hydrolysis of the PET itself. A similar problem can be caused by residues of caustic soda or alkaline detergents from the wash step. In both cases, substantial loss in molecular weight can result. In addition, the adhesive residues tend to darken at PET extrusion temperatures, causing discoloration of the resin.[16]

Repolymerized PET, produced by glycolysis or methanolysis, is identical in properties to virgin PET.

11.2.6 Markets for Recycled PET

One of the earliest large-volume uses for recycled PET was as polyester fiberfill for applications such as ski jackets and sleeping bags. The range of applications has grown enormously in the last several years, now including items as diverse as carpet, automobile distributor caps, produce trays, and soft-drink bottles. Fiber applications remain the largest market (see Table 11.1). Half the polyester carpet made in the U.S. is currently made from recycled PET.[23] Recycled PET fiber has even entered the clothing market.[1]

TABLE 11.1 Uses of Recycled PET, 1990[19]

	Million lbs
Fiber	165
Extruded strapping	12
Alloys and compounds	11
Extruded sheet	2
Bottles and containers	1
Chemical conversion	10

One-hundred-percent-recycled PET containers reached the marketplace in the United States in 1988, when Proctor & Gamble began test marketing Spic and Span in recycled PET. National distribution followed in 1990 and soon spread to other nonfood containers.[16]

One of the major breakthroughs in use of recycled PET has been its approval by the U.S. FDA for use in direct food contact. The first approval was in 1989 for use in egg cartons.[16] Next came approval of repolymerized PET (as discussed above) for a wide variety of food uses, including soft-drink bottles and food jars.

Another option for food products is use of recycled PET as a buried inner layer. Continental PET Technologies received a letter of no objection from the U.S. FDA in 1993 for a coinjected multilayer PET bottle. It has a one-mil (.001-inch) layer of virgin PET between the recycled PET and the container contents. It is suitable for use with a variety of products, including soft drinks and non-hot-filled juices.[24] This approach is currently being used for soft-drink bottles in Australia, New Zealand, and Switzerland. In Australia, where the multilayer bottles were introduced in 1994, nearly 18 million lb/yr of PET were being recycled in this manner in 1995. In Switzerland, where the bottles were introduced in 1995, the technology used for the preforms for these bottles is based on a sequential injection-molding process with moveable cores. Only the United States, Australia, New Zealand, Switzerland, and Sweden permit multilayer recycled-content packaging for food products.[24–26]

Wellman, Inc., received a nonobjection letter from the U.S. FDA in 1995 for the use of conventionally recycled PET sheet in a multilayer configuration for applications such as poultry trays, salad trays, fresh fruit and vegetable trays, instant and regular coffee and tea containers, bags for cereals with no surface fat, and airline catering trays. They had received nonobjection status in 1994 for packaging for washable fruits and vegetables. These materials also have a least a 1-mil-thick virgin layer between the recyclate and the food.[27]

As discussed above, a major development was the approval in 1994 of Johnson Controls' physically reprocessed resin, Supercycle, for direct food contact. The material is now available for both container and film applications.

Other markets for recycled PET include high-performance engineering alloys and compounds. Suppliers of these materials include DuPont, General Electric,

Allied-Signal, MRC Polymers, and MA Industries, among others. Many of these are designed for the automotive industry, and others are targeted at office machines and other industries.[16]

At the present time, markets for recycled PET exceed the supply. The growth in use of PET, which is estimated to increase 143 percent over 1994 usage by 2000, and a current shortage of virgin resin production has led to increased demand for recovered material. Processing capacity for recycled resin currently exceeds supply by a significant margin.[13]

Prices for delivered truckload quantities of baled PET bottles in the United States are currently in the range of $.27 to $.35 per pound, with virgin resin selling for $.85 per pound or more. Recycled PET that is suitable for food contact is reportedly selling for 5 to 10 percent more than virgin PET, as of fall 1995.[13]

11.3 HIGH-DENSITY POLYETHYLENE (HDPE) RECYCLING

HDPE is the second most recycled plastic. In the United States, recycling of HDPE is much more fragmented than recycling of PET. It is estimated that the six largest HDPE recyclers control only about 35 percent of the market. The two largest are Graham Recycling and Union Carbide, each with about 7 percent of HDPE recycling.[17]

Sources of HDPE in municipal solid waste are shown in Fig. 11.13.

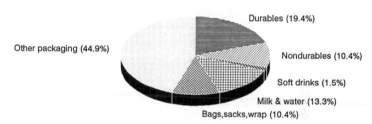

FIGURE 11.13   HDPE in U.S. MSW, 1993.[2]

11.3.1 Recycling of HDPE Milk Bottles

The first postconsumer HDPE material to be recycled in significant quantities was the HDPE milk bottle. These bottles are composed of a high-quality fractional melt-index homopolymer HDPE that is generally unpigmented, meaning that recovered resin is suitable for a wide variety of uses. A major early use was in agricultural drainage pipe. While other unpigmented HDPE bottles are available (dish-detergent bottles, for example), milk bottles had the additional advantage that nearly all plastic milk bottles were made from HDPE, while for other products, other resins were also used. This simplified the collection and sorting task, especially in the years before the SPI plastics identification code was used (see Sec. 11.11).

Recycling of natural HDPE bottles was estimated at nearly 26 percent for 1994.[28]

11.3.2 Recycling of Pigmented HDPE

After the introduction of the SPI coding system, it became possible to target a greater variety of HDPE containers for recycling. The growth in markets for recycled resin and in consumer preference for products with recycled content fueled this development. Pigmented HDPE bottles are now widely collected in curbside and drop-off recycling programs.

One type of pigmented HDPE bottle that is not accepted in many recycling programs is motor-oil bottles. Oil residuals in these containers have been associated with degradation during processing, clogging of flakes during shipment, and especially with water pollution during cleaning, requiring special (and expensive) handling of the waste water, since oil is classified as a hazardous waste. Technologies for centrifugal and cryogenic evacuation of oil residuals and techniques for separating the oil and water from the washing step are being tested and may result in increased acceptance of motor-oil bottles in the future. In fact, Phillips 66 reported that its tests show its oil bottles can be recycled and reused up to 20 times.[29]

It is reported that a plant for processing HDPE from bottles into clean flake costs 20 to 25 cents per pound to operate, including inbound and outbound freight. As of May 1995, natural baled bottles were selling for more than 30 cents per pound compared to mixed-color baled bottles for 15 to 20 cents per pound, down somewhat from the record highs set in March 1995. Natural postconsumer recyclate (PCR) was selling for 49 to 58 cents per pound, compared to 50 cents per pound for virgin homopolymer blowmolding resin.[30] This was a drastic change from 1994, when an oversupply of virgin resin depressed the value of recyclate to the point that it was lower than processing costs.[31] However, by August 1995, prices had fallen below 20 cents per pound in most of the United States.[32] Such drastic price swings have been characteristic of recycled-material markets, for other materials as well as for plastics.

Capacity for recycling HDPE increased by 121 percent between 1993 and 1995. Forty percent of HDPE-bottle recycling firms have been in business for three years or less as of 1995. Most of these firms handle both natural homopolymer and pigmented copolymer bottles. Some also process injection-molded items. Generally they process these streams separately from one another, however. About half of the plastics reclaimers also produce recycled-plastic products.[30]

In 1994, 550 million pounds of postconsumer HDPE resin were sold. HDPE bottle production that year was 2,300 million lbs.[30] In 1993, 450 million pounds of HDPE were recycled.[33] The overall recycling rate for HDPE containers in 1993 was estimated at 11 percent.[33] Pigmented HDPE bottles achieved a 10.8 percent recycling rate in 1994.[28]

11.3.3 Other HDPE Recycling

Some recovery of HDPE film occurs along with LDPE when retail bags are collected for recycling.

Suits made of Tyvek (a DuPont trademark) used in hospitals and some industries are being recycled to a small degree, mostly from hospitals. A small amount of Tyvek from envelopes and other applications is also being recycled. Markets are typically plastic lumber or pipe.[34]

Some recycling of injection-molded HDPE containers is also taking place, as mentioned above. Much of this effort in the United States is in Wisconsin, where a state law threatens bans on HDPE containers unless recovery programs are put in place.[33]

In Ontario, HDPE, LDPE, and PP tubs are being collected, along with HDPE bottles and other plastic items.[31]

The U.S. Postal Service included an evaluation of recyclability in a pallet test it conducted. The Post Office reported that damaged twin-sheet thermoformed HDPE pallets could be ground into pellets and extruded into sheets to make new pallets. Structural foam HDPE pallets can also be recycled, but not back into new pallets in significant amounts.[35]

The U.S. EPA estimated that, in 1993, 10 thousand tons of HDPE were recovered from durable goods.[2] No information on the source of these materials was presented. Ten thousand tons of HDPE bags, sacks, and wraps were also recycled.

11.3.4 Markets for Recycled HDPE

As mentioned above, a major early market for recycled HDPE was agricultural drainage pipe. Pipe continues to be a significant market, but a number of new markets have developed as well. Most plastic laundry-detergent bottles now have a three-layer structure that incorporates recycled HDPE in the middle layer. Most motor-oil bottles contain a blend of recycled and virgin HDPE, with others having a multilayer structure.[29] Recycling bins and other household products also often used recycled HDPE. Traffic cones and barriers, flower pots, and toys are other applications. Tables 11.2 and 11.3 show sources and end uses of recycled HDPE.

A potential market with great promise is use of recycled HDPE in railroad ties. These ties are now being tested on rail lines in Chicago. Since 15 million railroad ties are replaced in the United States every year, this is a very large potential market. Early results are said to be very positive, but a complete test will take 10 years.[36]

TABLE 11.2 Sources of Postconsumer HDPE Recyclate, 1993[33]

Product	%
Natural bottles	54.4
Pigmented bottles	28.9
Base cups	7.8
Other (mostly film)	8.9

TABLE 11.3 End Uses of Recycled HDPE[30]*

	Market	
	1992 (%)	1994 (%)
Bottles	17	27
Drainage pipe	17	17
Film	7	8
Pallets	6	7
Plastic lumber	7	6
Other	46	34

* Some totals do not add to 100 due to rounding.

11.3.5 Properties of Recycled HDPE

The properties of recycled HDPE depend, of course, on the properties of the virgin resin and the history of the product. In general, mechanical and flow properties are retained fairly well. Impact strength and environmental-stress crack resistance sometimes decrease.

HDPE is known to sometimes absorb components from products that it has contained. This may present problems for future applications. Butyric acid, for instance, is responsible for much of the odor of rancid milk, and has significant solubility in HDPE.[37] Burying the recycled layer between layers of virgin material was found to be effective in preventing migration of this odorous contaminant into subsequent products. It is reported that properly designed and operated washing facilities are capable of removing these residues.[19] On the other hand, tests focusing on absorption of consumer products such as gasoline, cleaner, insecticide, cooking oil, and other products found that the standard washing processes were not able to remove organic contaminants from HDPE to the degree that would be necessary to make them suitable for food contact.[38]

11.4 LOW-DENSITY POLYETHYLENE (LDPE) RECYCLING

Low-density polyethylene (LDPE) and linear low-density polyethylene (LLDPE) are used for many of the same applications and are often blended together. Recycling processes seldom attempt to differentiate between these two materials, and indeed the SPI coding system does not differentiate them. Therefore, in this discussion LDPE recycling should be interpreted to refer both to LDPE and to LLDPE.

Sources of LDPE in municipal solid waste are shown in Fig. 11.14.

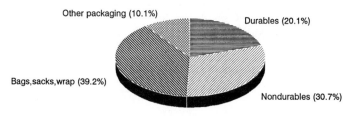

FIGURE 11.14   LDPE in U.S. MSW, 1993.[2]

11.4.1 Recycling of LDPE Stretch Film

The earliest significant recycling of LDPE was directed at stretch film used for wrapping pallets of goods. Large quantities of this material can accumulate at distribution centers like retail warehouses. While handling film is generally considered more difficult than containers, this material can be successfully recovered. The major early market was trash bags. Additional markets now include stretch film and merchandise bags.

Because of the low bulk density of the collected material, it is generally baled before it is transported. This reduces the volume by as much as 90 percent. Many recyclers of stretch film require the film to be dry, clean, and free of metal, paper, tape, or labels. Others have incorporated washing facilities and are able to handle more contaminated material.

11.4.2 Other LDPE Recycling

Retailers in many parts of the United States participate in collection programs for plastic merchandise bags, most of which are LDPE. It is estimated that over 14,000 locations in the United States have drop-off sites for plastic bag recovery.[39] Participants include grocery stores, retail stores, dry cleaners, and others. The Plastic Bag Information Clearinghouse offers a toll-free number for consumers to find out where their bags will be accepted.[39]

A recent survey conducted by the Plastic Bag Association determined that 70 percent of U.S. consumers claim to recycle their plastic bags regularly. It is estimated that 10 to 15 percent of all plastic bags are recycled, some after multiple uses.[40]

Little recycling of plastic bags or film at curbside is occurring in the United States. Even those programs that collect recyclables in plastic bags often landfill the bags themselves.[39] Plano, Texas; Glencoe, Illinois; and San Jose, California are among the few U.S. communities to collect plastic bags at curbside. Plano collects HDPE bags only. San Jose collects grocery, produce, and dry-cleaning bags, and requires them to be dry and clean. Plastic food packaging is not accepted. These communities report that a significant fraction of the bags are being discarded due to contamination. Baled bags are exported to China, where they are further processed.[39]

Contamination is one of the major problems with plastic-bag recycling. Consumers often leave paper receipts in the bags. Bags containing printing result in darker colors in the recycled material. In addition, the multiple resins used in plastic bags (HDPE, LDPE, and LLDPE) cannot generally be sorted cost effectively.[40]

Arkansas Plastics Recycling Inc. is recycling agricultural LDPE, including films and irrigation pipe, into trash bags.[41]

In Canada, the Plastic Film Manufacturers Association of Canada (PFMAC) has provided funding for collection of plastic bags and film in curbside recycling collection programs, along with a commitment to purchase the recovered film. Over 100 communities in Ontario and five in Quebec are currently participating, with many others scheduled to begin when the program is expanded. While most of the targeted film is LDPE (and linear low-density polyethylene, LLDPE), it also includes HDPE, PP, and PVC. In addition to buying the collected material, PFMAC pays $40 per metric ton collected to the participating communities and helps cover promotion costs. Materials collected include grocery, dry-cleaning and merchandise bags, overwraps, and frozen food, bread, produce, and sandwich bags. To simplify handling, households are instructed to stuff all the film into one bag, tie it closed, and place it at the curb with the recycling bin. The collection crew places the film plastics into bulk bags hanging on the side or end of the trucks. At the processing facility, the films are ground, washed, dried, and blended one-to-one with postindustrial scrap. End uses include garbage bags, merchandise and grocery bags, irrigation pipe, plastic envelopes, plastic lumber, industrial wrap, pallet covers, and agricultural film, among others.[39]

The overall recovery rate for plastic film in the United States is about 2 percent.[39] The U.S. EPA estimated recovery of LDPE bags, sacks, and wraps at 40 thousand tons in 1993. They also estimated 10 thousand tons were recovered from durable goods, but gave no information about the source of these materials.[2]

On the other hand, in the United Kingdom, PE film is reported to be the most widely recycled plastic material. Most of this material is postcommercial and agricultural film. In 1990, it amounted to about 132 million lbs, with most of it being processed by British Polythene Industries.[42]

In 1994 or 1995, the Packaging and Industrial Films Association in Nottingham, England, began a program to collect used polyethylene silage film from farmers. Six months after start-up, they had achieved a collection rate of 10 percent.[43]

The Colgate-Palmolive Co. and the Coex Coalition of SPI sponsored a study on recycling of multilayer stand-up toothpaste tubes, which contain barrier and adhesive layers buried between LDPE layers. The 13-week study demonstrated that the tubes could be successfully ground, blended with recycled HDPE milk bottles, and used in the core layer of household-cleaner bottles. However, there were no plans to continue collection beyond the study period.[44]

11.4.3 Markets for Recycled LDPE

Major markets for recycled LDPE include merchandise and grocery bags, and especially trash bags. As was previously discussed, some is also used in plastic lumber, agricultural film, envelopes, and other applications. Rubbermaid is a major purchaser of recycled LDPE stretch wrap for production of housewares.[45]

11.5 POLYSTYRENE (PS) RECYCLING

Polystyrene, especially expanded polystyrene, was widely perceived as an environmental "bad guy" in the mid 1980s. The triple problems of CFCs leading to ozone depletion, litter, and the perception that PS took up a lot of landfill space led to frequent calls for banning this material, and in some cases actually to the passage of legislation. Berkeley, California, for example, banned the use of expanded polystyrene in fast-food restaurants. In response to these pressures, the polystyrene industry developed recycling programs for this material.

Sources of PS in municipal solid waste are shown in Fig. 11.15.

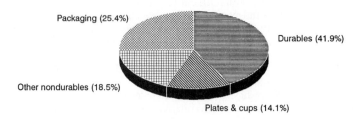

Packaging (25.4%)

Durables (41.9%)

Other nondurables (18.5%)

Plates & cups (14.1%)

FIGURE 11.15 PS in U.S. MSW, 1993.[2]

11.5.1 Recycling of Expanded PS

One of the major developments in promoting recycling of expanded PS in the United States was the founding of the National Polystyrene Recycling Company. This industry-funded organization was set up to manage the recycling of PS by building a network of facilities across the United States to process the material and working with major generators of EPS to collect the material for recycling. As part of this effort, McDonald's set up collection programs in many of their restaurants. Other major participants included school and business cafeterias. Although McDonald's discontinued most use of EPS, efforts with these other businesses have continued.

The NPRC currently operates plants in Corona, California; Chicago, Illinois; and Bridgeport, New Jersey.

Not long after the formation of the National Polystyrene Recycling Company with its emphasis on food packaging, independent foam-cushioning-material manufacturers formed their own organization, the Association of Foam Packaging Recyclers (AFPR), to promote collection and recycling of EPS cushioning materials. AFPR, founded in 1991, now has a recycling network of over 200 EPS packaging manufacturers and other collection sites. Many members offer reverse distribution systems to their customers to facilitate collection of the material. Most of the recycled material is used to make new cushioning.[45]

A typical recycling process for expanded polystyrene is as follows. Baled material is unbaled and placed on a conveyor or screening table. Hand-sorting and screening is used to remove obvious contaminants. Next, the PS is granulated and washed with hot water with detergent. An alternative process is to wash and grind simultaneously, using hot water without detergent. The washed slurry is then screened and rinsed, and the PS fluff is surface-dried in a centrifugal drier. A significant amount of water is still contained in the material at this point. A variety of methods have been employed to remove this water, including squeezing out the water physically using a screw press or a ram compactor, vacuum-vent extrusion, and convection heat drying. Twin-screw extrusion reportedly results in the highest quality products, in terms of reduction in black specks and in odor. It also permits adding modifiers before pelletizing. The last step is pelletization.[46]

Because of the very low bulk density of expanded polystyrene, properly densifying it is often the key to successful transport of the material. Densification generally involves total or partial collapse of the cell structure rather than compaction alone. Pressure alone is sometimes used, but one of the principle options is the use of heat to soften the polymer and collapse the cells. For food-contaminated EPS, densification may trap the contamination in the polymer, so this process may not be desirable until the material is cleaned. A variety of companies offer densification equipment for EPS.[46] Commercially available densifiers can readily produce a 60:1 or larger density increase.[45]

As of 1993, one estimate claimed at least 50 million pounds of EPS shape-packaging, insulation, and loose-fill were being recovered. About half was ground into fluff and used as a filler by EPS molders, with the rest repelletized and reused in loose-fill, shapes, and moldings.[47] The recycling rate for postconsumer EPS foam packaging was reported to be 10.9 percent by R. W. Beck, which estimated that 23.3 million pounds of EPS packaging were collected in 1993.[48]

Very little collection of polystyrene is done at curbside. One of the exceptions is Mississauga, Ontario, where polystyrene containers and packaging, both foam and nonfoam, are accepted.[31] Some communities in the United States, mostly on the east coast, also collect EPS containers and sheet.[14]

In addition to recycling efforts, manufacturers of polystyrene loose-fill have promoted the reusability of the material. The Plastic Loosefill Council offers a nationwide toll-free hotline that refers callers to the nearest drop-off collection site. This number is widely publicized by direct-mail companies in the form of an insert in packages. As of 1994, about 3400 collection sites were in operation.[49]

Efforts are also being directed at investigating the recycling of structural-foam PS used in business-machine parts.[50]

11.5.2 Recycling of Other PS

Little recycling of nonfoam polystyrene is currently occurring, with the exception of PS cutlery, which is collected along with EPS food service items. As previously mentioned, Mississauga and Brampton, Ontario, collect nonfoam as well as foam meat and bakery trays, cups, and fast-food containers, among other items.[31]

Eastman Kodak recycles disposable cameras, which are primarily polystyrene. The internal frame and chassis, which are also polystyrene, are retrieved intact and reused. The PS camera bodies are ground, mixed with virgin resin, and used to make new camera bodies. Other camera companies are also recycling their cameras.[51]

A study by Philips of Hamburg, Germany, was done of recycling non-flame-retardant PS from TV sets. They concluded there was no significant reduction in the properties of the material.[52]

The U.S. EPA's estimate of PS recovery for 1993 was 20 thousand tons from nondurable goods and a negligible amount from packaging and other sources. EPS and other grades were not distinguished.[2]

11.5.3 Markets for Recycled PS

Recycled EPS is reported to have properties very similar to commercial injection-molding grades of materials.[105]

Major markets for recycled PS include houseware items and cushioning material. Amoco offers a foam insulation board for residential construction consisting of at least 50 percent recycled content, along with a similar board for commercial construction. Both materials contain at least 25 percent postconsumer PS.[53]

Recycled PS foam has also been approved for food-contact packaging, first in egg cartons and later in packaging for washable fruits and vegetables. PS foam packaging with a buried inner layer of recyclate has also been used for fast-food containers.

A newly developed process for EPS recycling encapsulates it with phenolic resins and molds it into insulating foam board. The phenolic resins increase the fire-retardant and insulating properties of the EPS. The board is also denser than regular EPS board and has greater structural strength. Other mixed plastics reportedly can also be incorporated into the board.[54]

Undensified PS foam can also be used for stuffing for flower vases, fill for lawn furniture, a soil lightener for plant nurseries, and a drainage medium for ground water. Investigations of foam-filled concrete are under way.[46]

11.6 POLYPROPYLENE (PP) RECYCLING

In contrast to the materials discussed above, nonpackaging uses are the major source of recycled PP (see Figs. 11.16 to 11.18). Sources of PP in municipal solid waste are shown in Fig. 11.19.

11.6.1 Recycling of PP Automotive Battery Cases

The most prevalent recycling of polypropylene is the recovery of automotive battery cases. In a number of states in the United States, disposal of lead-acid batteries in landfills or incinerators is prohibited, and several states have imposed deposits

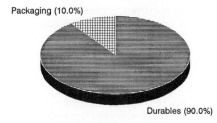

FIGURE 11.16 Recycling in PP in the United States, 1993.[2]

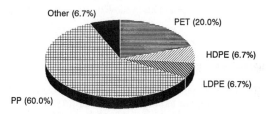

FIGURE 11.17 Recycling of plastics in durable goods in the United States, 1993.[2]

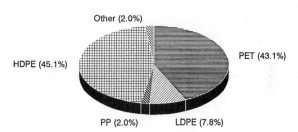

FIGURE 11.18 Recycling of plastics in packaging in the United States, 1993.[2]

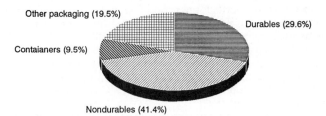

FIGURE 11.19 PP in U.S. MSW, 1993.[2]

on these batteries. When collected, the batteries are usually reprocessed, with the primary goal being recovery of the lead. Recovery of the PP cases is a secondary benefit.

In 1990, it was estimated that 100 million lbs of impact-modified virgin PP were used to manufacture lead-acid battery cases. PP makes up about 7 percent by weight of the recoverable material in a lead-acid battery. The largest U.S. battery-case reclaimers include M.A. Industries, Inc.; Exide Corp.; Witco Corp.; and Johnson Controls. The major market for the recovered materials is new battery cases. Other applications include a variety of injection-molded articles.[42]

In Germany, recovery of PP battery cases began in 1984, producing about 3,000 metric tons per year of polypropylene. The batteries are first crushed, and then the light fractions (PP and impurities) are separated from the heavy fractions (lead and other components). The polypropylene at this stage is about 97 percent pure. It is further size-reduced, dried, and sent through a cyclone separator, after which its purity is increased to 99.5 percent. This material is compounded to user specifications and then pelletized.[55]

11.6.2 Other PP Recycling

Little recovery of other polypropylene is presently occurring. While tests have shown recycled PP packaging, for example, to be a useful material, the collection and processing infrastructure has not been developed. In large part, this is related to the relatively small fraction of easily recovered PP packaging, compared to HDPE and PET. PP packaging is about evenly divided between containers, film, and closures.

Polymer Resource Group in Baltimore processes recovered PP containers, which they collect by way of a national recycling network. Containers processed include soap and sports-drink containers and multilayer ketchup bottles. Some of this resin is being used by Colgate-Palmolive for Murphy's Oil Soap bottles.[56]

In a pilot project in Germany, PP from coffee machines was recovered with better than 99 percent purity.[52]

Mayer Cohen Industries in England is reported to recycle PP film.[16]

The U.S. EPA estimated PP recycling in 1993 at 100 thousand tons, with 90 thousand tons coming from durable goods and 10 thousand tons from packaging.[2]

11.7 POLYVINYL CHLORIDE (PVC) RECYCLING

11.7.1 Recycling of PVC Bottles

PVC bottles are fairly widely collected for recycling in Europe, where they are the major container used for bottled water. Both use and recovery of PVC bottles are much less common in the United States. Few curbside programs collect PVC bottles, and those that do are generally programs that collect all types of plastic bottles. The amount of PVC packaging, including bottles, recycled in the United States in 1992 was reported to be 10 million pounds. While this was a fivefold increase over 1991, it represented a recycling rate of only slightly more than 1 percent.[57] The EPA's recycling estimate for 1993 reported only negligible PVC recycling of any materials in municipal solid waste.[2]

In Europe, waste PVC bottles are purchased by GECOM (Groupe d'Etude pour le Conditionnement Moderne) in France. An organization called GEPP, also in

France, began collecting PVC bottles from households in the late 1970s. Recovery rates reportedly reached 20–47 percent in the areas served. Collection was primarily through drop-off sites.[58]

More recent collection systems incorporate automatic sorting using X-ray detection after collection using reverse-vending machines. These machines are installed in stores and accept both PET and PVC bottles. The bottles are flattened by heated rollers, scraped off the roller with a knife, and stored and bagged in the machine. At the processing plant, the bottles pass individually under an X-ray detector to distinguish PET from PVC.[58]

A typical European processing system for PVC starts with magnetic separation and sieving to remove large wastes. The material is dried, light and heavy impurities are removed, and the material is then ground to 20 mm and screened. Next it is ground again, this time to 10 mm, washed, and separated by gravity. Glass and PE have been removed at this point, but some PET contamination remains. Aluminum is removed electrostatically. The PET and PVC are finally separated by further very fine grinding. The PET does not grind as fine as the PVC, so it is removed by screening the coarser fraction. In Australia, cryogenic grinding has been used for this purpose.[58]

11.7.2 Recycling of Other PVC

In the United States, PVC automotive trim was one of the first successful PVC recycling processes. The 1994 edition of the Vinyl Institute's directory of companies involved in PVC recycling reportedly listed more than 175 companies. Types of PVC handled range from postconsumer bottles and other packaging to consumer durables to postindustrial scrap.

IBM initiated an innovative program in which packages for circuit boards can be sent back, postage paid, to be reused two to three times and then recycled.[57]

PVC greenhouse film is recycled in Japan under a subsidized system, reaching a 40 percent recovery rate, though reportedly the system is in jeopardy.[1]

Vinyl windows, roller blinds, sheet products, and vinyl profile manufacturing waste are being recycled in Germany.[59] Scrap vinyl pipe is being recycled in the Netherlands.[60] Vinyl flooring is also being recovered in Europe.[58]

A U.S. carpet company, Collins & Aikman Floorcoverings, announced in 1995 that it had perfected a process for recycling PVC-backed carpet and carpet tile. The carpet is cut into thin strips, granulated, and then mixed with postconsumer and postindustrial LDPE in a ratio of seven parts carpet to three parts LDPE. The material is then mixed and heated, extruded, and immersed in chilled water for cooling and setting. Current products are parking stops and industrial flooring. Other products, including highway soundwall barriers, marine bulkheads, and backing for new carpet, are under development.[61]

Pilot projects for collection of vinyl-siding scrap are being carried out. One recent project involved construction by Habitat for Humanity. An ongoing pilot program in North Carolina involves mobile-home manufacturers.[62]

PVC and PE wire and cable jacketing are being recovered to some degree. As of 1989, about six companies in the United States were involved in this business. Reconstituted Plastics Ltd. in the United Kingdom sells a rising-current separator system to separate copper, PVC, and HDPE from wire and cable. The U.S. Bureau of Mines investigated electrostatic separation of such materials, and Plastic Recovery Systems of Ohio has investigated flotation techniques for these streams.[42] The presence of lead stabilizers in this material is a significant complication.[58]

11.7.3 Markets for Recycled PVC

Uses for recycled PVC include pipe, packaging, household items, toys, bedliners for pickup trucks, urinal screens, furniture, and a variety of other products.

A new use for recycled PVC is in clothing and accessories. Recycled PVC mineral-water bottles are being used in fanny packs, as well as, combined with wool, in sweaters.[63]

In the United States, in 1993 the Vinyl Institute published a directory of companies manufacturing products from recycled PVC. The Institute also offers a program of financial assistance for companies wishing to manufacture products from recycled PVC.[64]

11.8 RECYCLING OF COMMINGLED PLASTICS

A common problem associated with attempts to increase recycling of plastics is the difficulty of separating various resins from each other. Therefore, a variety of processes have been developed that do not require separation by resin type, being designed to handle mixed resin streams, or commingled plastics. Many of these processes are also designed to handle substantial amounts of nonplastic impurities.

11.8.1 Recycling Processes

Four basic types of processes have been developed specifically for processing commingled postconsumer and postindustrial plastics. These are the Klobbie intrusion process, the Reverzer process, continuous extrusion, and compression molding.[19]

The first process for recycling commingled plastics patented in the United States was the Mitsubishi Reverzer, developed in Japan. It was a short-time, high-temperature, high-shear machine that could handle mixed plastics containing up to 50 percent filler as long as a substantial fraction of the plastics mix was polyolefin. Commingled waste plastics were first softened in a hopper and then mixed in a screw. Products were formed by flow molding, extrusion, or compression molding, and could be produced in a variety of shapes. Products were designed to have large cross sections and relatively low strength demands. The equipment was never successfully marketed.[19]

The intrusion processes originated with the Klobbie process, developed in the Netherlands in the 1970s. This system consisted of an extruder coupled to several long linear molds mounted on a rotating turret. After being filled, the molds were rotated into a tank of cooling water. No extrusion nozzle or screen pack was used on the extruder. As the plastic cooled and shrank in the mold, cooling water entered the gap between the product and the mold. Products were forced out of the mold by air pressure. Since this process is considered to be a cross between injection molding and extrusion, it is often termed an intrusion process. Klobbie also patented use of foaming agents in this equipment.[19]

Continuous extrusion processes can be regarded as variations of the Klobbie process. They continuously extrude molten plastic into cooled dies, similar to processes for continuous pipe manufacture. Tri-Max of New York produces a glass-reinforced product with a structural foam core from postconsumer commingled plastics by continuous extrusion.[65]

The earliest successful technology for compression molding commingled plastics was the Recycloplast process, developed in Germany. It typically uses 70 percent postconsumer plastic scrap and 30 percent industrial scrap. Materials are heated to 180–200°C in a plastificator, with kneading providing mixing and friction heat. The resultant paste is either extruded into roll-shaped loaves that are then compression-molded in hydraulic presses or pelletized for future use. It produces products such as pallets, grates, and benches. Facilities using this process are still in operation. Current products have improved strength and appearance compared to the original Recycloplast products.[19]

Another compression-molding process was the British Regal Converter system, in which thermoplastics containing up to 50 percent nonplastic contaminants were granulated, distributed on a steel belt, melted in an oven, and then compacted into a continuous board.

More recent processes include the ET/1 produced by Advanced Recycling Technology of Brakel, Belgium. A variation of the Klobbie process, it has a short, high-speed adiabatic screw and accommodates feeds containing at least 50–60 percent polyolefin. Molds, which can be up to 12 feet long, are mounted on a revolving turret that is cooled in a water bath. A variety of profiles can be formed. Reportedly over 40 of these machines are in operation worldwide.[19] Work at Rutgers University showed that the properties of plastic lumber produced in an ET/1 could be improved by adding 10–30 percent recycled PS to mixed postconsumer bottles.[19]

Superwood International, Ltd., an Irish company, has a similar process. Blowing agents or fillers can be added to vary the end product characteristics. Feedstock is industrial scrap or postconsumer plastics. Plants using this technology have operated in England, Ireland, Canada, and the United States. The largest Superwood machine is reported to have an output 2.5 times that of the original Klobbie. After the company went into receivership in 1991, two Canadian companies purchased the North American rights to the relevant patents.[19]

Another Klobbie variant is the process developed by Hammer's Plastic Recycling in Iowa.[19] This process uses closed molds, a heated nozzle, and a screen pack, all of which increase the molding pressure compared to the Klobbie design.[65]

WormserKunststoffe Recycling GmbH in Germany processes postconsumer plastics into wood substitutes using a roller extruder that is capable of handling up to 15 percent contamination by nonpolymer materials.[19]

C.A. Greiner and Sohne in Austria have a continuous extrusion process for commingled postconsumer plastics, producing a variety of hollow profiles.[19]

Northwestern University's BIRL polymer reclamation center has developed a process called solid-state shear extrusion pulverization. It uses a twin-screw extruder to convert mixed plastics and scrap rubber into a uniform fine powder. They believe the powder will have a number of applications, especially in packaging.[66]

Other similar processes have also been developed. All have in common the production of relatively large thick-walled items.

Investigators in Italy have studied the use of mixed-plastic wastes in blends with low-density polyethylene, and high-density polyethylene as a filler. They found that the properties of the blended materials were very similar to those of the matrix up to a mixed plastic content of 25 to 50 percent. Elongation at break was an exception, decreasing rapidly even at very low concentrations of mixed plastic.[67]

11.8.2 Markets for Recycled Materials

As mentioned above, the major markets for commingled recycled plastics are large thick-walled items, where the presence of inhomogeneity and even large amounts of impurities can be tolerated.

The majority of these applications fall in the general categories of plastic lumber or concrete substitutes. The plastic lumber is generally designed for outdoor applications where treated lumber is used. The plastic has a significantly longer lifetime than the treated wood, since it does not rot. Thus, even though initial cost is higher, the lifecycle cost may be less than wood. Such applications include landscape timbers, boat docks, retaining walls, fences, park benches, and the like.

Investigators at the Center for Plastics Recycling Research at Rutgers University have engaged in extensive study of the properties of plastic lumber produced in an ET/1 as a function of the composition of the commingled plastics. The products in general have a solid skin, while the core has numerous voids, varying in size. Nonpolymer impurities also tend to migrate to the core of the material. The volume fraction of voids averages about 10 percent. Properties of the lumber also vary depending on position along the length of the profile.[65]

In one set of studies, the curbside tailings used for the plastic lumber were modified by the addition of recycled polystyrene. The polystyrene was found to significantly increase the compressive modulus, yield stress, ultimate stress, and flexural modulus of the lumber. However, when different sources of the recycled PS were used, results were quite different, showing deterioration in properties rather than improvement.[65]

Applications in which commingled plastics replace concrete include parking stops and bases for heavy machinery.

There are also applications in which commingled plastics can be used directly, without going through the type of processing equipment described above. Primary among these is the use of chipped commingled plastics as replacement for some of the aggregate in concrete or asphalt. Though these applications have been studied since at least the mid 1980s, they remain experimental.

Investigators at Rutgers University have used curbside tailings, the residual plastic when all plastic bottles are targeted and the high-value PET and natural HDPE bottles are removed, as the substrate for a postconsumer regrind resin designed to be suitable for blow molding containers. The process reportedly uses compatibilizers and other additives to improve properties. Melt filtration, drying, and devolatilization are also important components of the process.[19]

Evco Plastics in Wisconsin has also used commingled plastics from a collection program targeted at all plastic containers. This company produces injection-molded office desktop products from blended PCR with the help of a state rebate paying the difference between virgin resin and the PCR price, reportedly about $.70 per pound, and state financial support for part of the cost of the equipment.[33]

Hettinga Technologies of DeMoines, Iowa, has licensed a technology to processors in the United States and Europe called controlled-density molding (CDM). It reportedly accomodates a range of plastics, including mixed industrial and postconsumer scrap, producing panels with a thin, smooth outer skin around an inner core. A blowing agent is incorporated into the panels, and a range of densities can be achieved. Decorative finishes can also be applied in the mold.[68]

11.9 RECYCLING OF THERMOSETS

The processes for recycling thermoplastics usually involve remelting and reforming of the material. This approach obviously cannot be applied to thermosets. Recycling of thermosets is, in general, considerably less developed than recycling of thermoplastics. Two major approaches can be taken.

One alternative is to chemically break the thermosets down into small molecules that can then be repolymerized or used in other ways.

In 1995, the Fraunhofer Institute in Germany announced the development of an amine-based process for recycling thermosets, including polyurethane, into intermediate compounds that can either be mixed with an activator to restart the cross-linking process or be used as high-grade additives for paints and coatings or as raw materials for pharmaceutical, cosmetic, or medical products.[69]

The second approach, and the one most often employed, is to grind up the thermoset and use it as a filler in a new thermoset article. At present, very little recycling of thermosets is occuring, especially for postconsumer materials.

There is some use of regrind of scrap sheet-molding compound (SMC) from auto production as a filler in new automobile parts.[19] The SMC Automotive Alliance of Southfield, Michigan began recycling bad SMC parts and flashing in the fall of 1994. Materials are shipped to Toronto, where they are pulverized, cleaned, and separated into fiber and dust. The fiber fractions can be reused as fiber, and the dust is used to replace calcium carbonate as a filler in SMC. SMC parts with 10 to 25 percent regrind are reported to be stronger and as much as 10 percent lighter. No commercial use of postconsumer SMC has been reported, largely because the collection infrastructure has not yet been developed.[70]

Though little actual recycling of thermosets yet occurs, a considerable amount of research is being carried out in this area, and numerous patents have been granted.

Polyurethane Recycling. Polyurethane recycling is the most advanced of thermoset recycling processes. Most of the recovered material is industrial scrap. It is estimated that, in the United States, 125,000 to 130,000 tons of flexible polyurethane foam scrap, plus an additional 60,000 to 70,000 tons of imported scrap, are processed into carpet underlay, furniture padding, and gymnasium mats.[71]

The first step in recycling the foam is to separate it from contaminants. It is then chopped into pieces of suitable size and coated with a binder. The foam and binder mixture is placed in a mold, compressed, and then cured with heat and steam.[72]

Alternatively, the foam can be cryogenically ground to a fine powder, mixed with polyol, and handled on typical foam-processing equipment. This use of recycled scrap is reported to lower foam costs by about 3.5 percent.[72]

A Japanese company has operated a commercial plant for recycling of rigid polyurethane foam by hydrolysis and glycolysis to polyols, using a process licensed from Upjohn.[19] Rigid foam particles have been bonded into PUR particle boards and used for marine-furniture and gymnasium-flooring systems in Europe.[71]

Scrap from reaction injection molding (RIM) materials is being used for automotive mud guards, along with postconsumer bumpers, by a Japanese company. RIM materials are also being used, mixed with rubber chips, for pavements for athletic fields, tennis courts, and golf-course roads. They are claimed to have improved hardness, tensile strength, and wear resistance compared to pure rubber pavement. RIM recyclate is also used in Germany for under-floor heating systems.[71]

RIM scrap can also be recycled back into RIM parts. A two-stage size-reduction process is used, first requiring a granulator and then an impact disc mill, to reduce the RIM to a fine powder. The powder is then mixed with the polyol. It is reported that properties of the RIM elastomer produced with regrind are identical to those produced without it, including excellent surface finish of painted parts. Cost savings is estimated at about 5 percent.[72]

Compression molding of RIM scrap can be done without added binder. The material flows and knits together under the pressure and shear of molding. Properties are not as good as the virgin polymer but are still sufficient for applications like seat shells, and prices are claimed to be competitive with thermoplastics.[72]

Extrusion of RIM scrap in single or twin-screw extruders into profiles and shapes is reported to yield a material with less cross-linking that has uses in tubing, profiles, floor tiles, and body side moldings.[72]

Flexible polyurethane foams can be hydrolyzed to small molecules by high-pressure steam. Recovered polyol can be used in foam manufacture. Polyurethane can also be glycolyzed to polyols. Because these polyols have fairly low molecular weights, they are most suitable for reuse in rigid foams.[72]

Virtually all of these applications use manufacturing scrap rather than postconsumer materials.

11.10   RECYCLING OF AUTOMOTIVE PLASTICS

There is a long history of recycling automobiles, both for component parts and for scrap metal. The rest of the car has generally been disposed of as automotive shredder residue (ASR). As the plastic content of automobiles has grown and the amount of metal has decreased, the amount of ASR per car has increased. Automotive companies have begun to feel pressure to render automobiles more recyclable. This has been especially true in Germany, where the philosophy that the generator of the material is responsible for its ultimate disposal, which was first applied to packaging, is spreading to consumer goods and has already been applied to automobiles.

In the United States, Chrysler, Ford, and General Motors joined together in 1992 to form the Vehicle Recycling Partnership (VRP) to promote the development of technology to recover, reuse, and dispose of materials from scrapped motor vehicles. The United Kingdom has the Automotive Consortium on Recycling and Disposal (ACORD). Germany has the PRAVDA project, which is intended to set up an infrastructure for recycling all types of cars sold in Germany. France has a voluntary system of cooperation for automotive recycling, which may be the model for a European Union system. In fact, automobile manufacturers around the world are cooperating with each other in promoting recycling of automobiles, including their plastic components.[73]

Plastics found in automobiles include polypropylene, polyvinyl chloride, polyethylene terephthalate, polybutylene terephthalate, polycarbonate, nylon, ABS, polyurethane, polyphenylene oxide, and unsaturated polyesters, among others.

11.10.1   Dismantling of Automobiles for Recycling

One major approach to recovery of plastics from automobiles is to use disassembly to remove targeted parts before the car is shredded. The Society of Automotive Engineers has developed codes to identify the resin types in plastic parts, and the major U.S. automotive companies have committed to using these codes to mark plastic components for recycling. Since up to 40 different types of plastic are used in a single car, there are also efforts to simplify recycling by reducing the number of resins used, and by selecting compatible resins when more than one resin is used in a single part.[74] Nissan has reportedly increased its polypropylene content as a fraction of the total plastics in their automobiles from 34 percent in 1988 to 49 percent in 1990, at least partly in an effort to aid recycling.[75]

DuPont has used recycled plastic car fenders along with soft-drink bottles in a depolymerization pilot plant for PET.[75]

A demonstration project in Europe focused on Nylon 6 and Nylon 6-6 from old radiators, wheel covers, and radiator fans, demonstrating that it could be recycled and maintain acceptable properties.[75]

Polypropylene bumpers are being recycled in Japan by a process that uses a water-based organic salt solution to remove the paint. The repelletized and remolded PP is reported to have more than 95 percent of its original physical properties.[75]

Another PP bumper process shreds the bumpers into 4 mm particles; sends them through a cone press, where they are compressed and sheared to weaken the interface between the PP and the paint; and then removes the paint by contact with pins in a circular plate rotating at a high speed in a pin mill. This system is marketed on a turnkey basis by Itochu Corp. of Tokyo.[76]

Fiat operates a network of automobile-dismantling centers in Italy. These centers also accept Renault and BMW cars. They handled 50,800 cars in 1994. Among the products, polypropylene bumpers were converted into air ducts for new cars, and seat foam was converted into carpet underlay for buildings.[73]

11.10.2   Recycling of Automotive Shredder Residue (ASR)

Auto shredders typically handle 80 to 90 percent automobile hulks, with the remainder of their feedstock being obsolete appliances (white goods). After the ferrous and nonferrous metal streams are removed, the remainder is referred to as automotive shredder residue (ASR) or fluff. It is a heterogeneous mixture of plastics, glass, rubber, fiber, metals, and dirt, with the plastics component typically 20 percent or more by weight. It is estimated that 2.27–2.72 million metric tons of ASR are generated in the United States each year.[77]

Currently, virtually all of this material is landfilled. Research efforts are underway to enable it to be recycled, but for the most part it is still at early stages of development. These efforts are complicated by the variable composition of ASR, which differs not only from facility to facility but also from day to day.

Argonne National Laboratory has investigated the use of solvent extraction to recover plastics from ASR. The residue is first dried and then subjected to mechanical separation of the fines and the polyurethane foam. The foam is then cleaned using organic solvents and water and detergent solutions. The remaining plastics-rich stream contains primarily ABS, PVC, PP, and PE. Hexane is first used to remove automotive fluids from the ASR. Then solvents are used to dissolve out the target plastics. Two approaches were investigated. One used selective solvents to dissolve out the plastics in series. The second approach used a single solvent to dissolve all the plastics of interest at an elevated temperature and then employed different solvents to separate the mixed plastics. Finally, the plastics are recovered by cooling, using an antisolvent, or evaporation, depending on the characteristics of the solution. In all cases, the solvents are regenerated and reused.[78]

The University of Detroit Mercy is also investigating recycling of ASR. Instead of attempting to sort the fluff into components, they are adding binding agents and forming composite polymers. By characterizing the shredder operations, they can identify the types of uses that are suitable for the ASR from that shredder. Potential uses they have identified range from railroad ties to cushioning materials. Some materials have three times the compression strength of concrete, while others have the pliability of foam rubber. The binding agents used are reported to cost as little as 30 cents a pound.[79]

11.10.3 Uses of Recycled Plastics in Automobiles

Automotive manufacturers have generally been in favor of using recycled plastics in cars, but only if quality does not suffer and price is competitive. However, Ford has adopted a minimum 25 percent recycle-content target on a global basis and has reportedly doubled its use of postcommercial recyclate.[80] Ford expects to use more than 15,000 tons of recycled plastic in 1995.[73]

Several companies, including Bayer AG, Dow Plastics, DuPont, GE Plastics Hoechst AG, and Monsanto are now offering recycled-content automotive grades of plastics.[80]

American Commodities, Inc. of Flint, Michigan, takes scrap plastics from car bumpers, fascias, quarter panels, and splash shields and manufactures a diverse product line for automakers and their suppliers. Recycled resins they make available include polycarbonate/polybutylene terephthalate blends, acrylics, polyphenylene oxide, and Nylon 66. Costs are 10 to 40 percent less than virgin resin, and performance is near-virgin.[81]

11.11 RECYCLING OF OTHER THERMOPLASTICS

Recycling of other thermoplastics is in general less developed than recycling of the commodity thermoplastics discussed above. Much of what is recycled is industrial scrap rather than postconsumer material. Where postconsumer material is recycled, the source of the material is usually businesses rather than individual consumers. One of the major challenges is the diversity of resins and resin formulations, the complexity of the items (which often contain metals and a variety of plastics, often painted or coated), and the relatively low volumes generated by individual sources. In the United States, the American Plastics Council has been a sponsor of research projects intended to facilitate recycling of plastics from durable goods.[82,83]

11.11.1 Nylon

Some nylon from carpeting is being recycled. A DuPont plant depolymerizes nylon from industrial and commercial carpets using an ammonolysis process. It is used for production of resins for car parts with 25 percent recycled content. Operation is currently being scaled up from pilot plant scale.[84]

11.11.2 Polycarbonate

Polycarbonate from bumpers, water bottles, compact discs, computer housings, and telephones is being recycled to some extent. General Electric began buying back polycarbonate from 5-gallon water bottles several years ago.[85] In 1994, at least six new polycarbonate resins based on recycled materials were introduced. Most had 25 percent pre- or postconsumer recycled content. Material properties are reported to be similar to virgin materials, except for a 50 percent reduction in impact strength and a 33 percent reduction in tensile elongation. They are also not recommended for applications requiring strict color tolerances. Prices for the recycled material are about half that for virgin resin.[86]

The first polycarbonate CD recycling facility in Europe was being built in Dormagen, Germany, in 1995. It will reportedly separate polycarbonate from aluminum coatings, protective layers, and printing; blend it; and sell it for various uses.[87]

Polycarbonate from water-cooler bottles is reportedly being recycled into headlamp housings for Ford automobiles. Chrysler is using PC recyclate for instrument-panel covers.[73]

GE Plastics, in cooperation with Appliance Recycling Centers of America, Inc. (ARCA), announced in 1995 the initiation of a pilot buyback recycling program for both large and small quantities of polycarbonate scrap as well as other plastic resins generated by its customers in Rochester, New York. Included are instrument panels, fenders, and other polycarbonate parts from used automobiles that were manufactured with GE resins. If the pilot program is successful, GE plans to expand the program nationally.[88]

GE is also involved in recycling of Noryl, a PC/PS blend. Used computer housings of Noryl have been blended with virgin resin and molded into roofing shingles being used by McDonald's restaurants.[85]

11.11.3 ABS

Hewlett-Packard recycles computer equipment. This manufacturer is using some of the recovered ABS in manufacture of printers, the outer casings of which incorporate a minimum of 25 percent recycled content.[89]

A pilot project in Germany recycled ABS from vacuum cleaners, recovering it with better than 99 percent purity. A manufacturer of computers, automatic teller machines, and similar products has a recycling system for its machines, which recovers primarily PC/ABS blends. The recovered material is blended with virgin and used in new equipment housings.[52]

AT&T for a long time recovered housings from old telephones. The recovered material, primarily ABS, was blended with virgin by compounders and sold for a variety of purposes.[52] In the United Kingdom, British Telecom recycles telephones, recovering ABS and other materials.[85]

11.11.4 Others

Hoechst Celanese announced in 1993 the introduction of a family of resins based on reclaimed engineering thermoplastics. The source of the material is industrial scrap from customers and their suppliers, postconsumer parts returned by end users, and polymer waste from Hoechst Celanese's manufacturing operations. They planned to supply the materials either as 100 percent recycled pellets or in blends customized to the customer's specifications. Resins included are acetal, polyester, polyphenylene sulfide, Nylon 66, and liquid crystal polymer.[90]

11.12 SORTING, SEPARATION, AND COMPATIBILIZATION OF PLASTICS

As was discussed above, sorting of plastics by resin type is a common prerequisite to effective plastics recycling. One fundamental reason for this is that most plastics are mutually insoluble. Thus, blends of plastics generally consist of domains of one material embedded in a matrix of the other material. The exact morphology is highly dependent not only on the amounts of each component present but also on the processing conditions. In many cases, the result is a significant decrease in performance properties.

Additionally, the plastics found in postconsumer waste streams may have significantly different melting points. Therefore some may stay solid during processing, as in the case of PET exposed to polyolefin processing conditions, or some may decompose during processing, as is the case for PVC exposed to PET processing conditions.

Therefore, except for processes designed to handle commingled plastics, processes must be designed either to accept a limited range of resin types that are reasonably compatible, or to provide for separation by resin type. Such separation techniques can be divided into three major categories: macrosorting, microsorting, and molecular sorting.

11.12.1 Macrosorting Techniques

Macrosorting techniques are those that rely on sorting of whole (or nearly whole) containers. Most current commercial sorting processes for separation of plastics by resin type rely on hand sorting. Many accept only containers with certain SPI codes and give consumers instructions such as "#1 and #2 bottles only" (PET and HDPE). See Fig. 11.20.

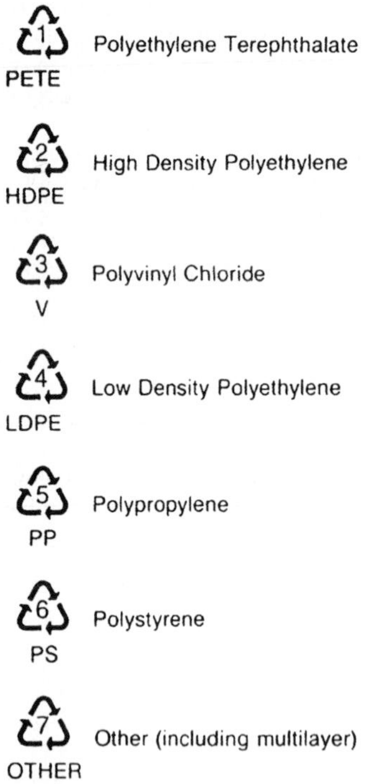

The SPI coding system itself has been a source of controversy. The impetus for designing this system was the effort of several states to mandate some form of plastics identification to make it possible for consumers or processors to easily determine what type of plastic a container was made of and whether it was acceptable for recycling in a particular recycling program. Faced with the very real potential of several states adopting mutually conflicting laws, the Society of the Plastics Industry (SPI) came up with a voluntary coding system. This system, with which many consumers are now familiar, consists of a triangle formed by three chasing arrows with a number inside and code letters underneath, denoting the type of plastic. Fairly immediately, the idea was attacked by officials in a few states who said the chasing arrows connoted to consumers that the containers bearing them were recyclable, when often this was not the case. Regardless of these early objections, many states proceeded to legally require

♲ 1 PETE	Polyethylene Terephthalate
♲ 2 HDPE	High Density Polyethylene
♲ 3 V	Polyvinyl Chloride
♲ 4 LDPE	Low Density Polyethylene
♲ 5 PP	Polypropylene
♲ 6 PS	Polystyrene
♲ 7 OTHER	Other (including multilayer)

FIGURE 11.20   SPI code.

the SPI coding systems on certain rigid containers. Though the law differs from state to state, it most often applies to bottles 16 ounces or larger, and other plastic containers 8 ounces or larger, up to some maximum size. Thirty-nine states in the United States now have this type of legislation.[91] Many packaging and some product manufacturers soon began to apply the SPI code to articles for which it was not required, such as plastic bags, lids, and blister packages. The concern about misleading of consumers continued, fueled by the proliferation of the SPI code. In addition, some representatives of the plastics recycling industry wanted modification of the code to better represent the complexities of the plastic resins. They had concerns such as the same code being used to denote blow-molding and injection-molding grades of HDPE, which are not compatible for recycling. As a result, a group was formed of industry representatives and representatives of the environmental community to try to come up with a modification of the system that would be mutually satisfactory. Despite many months of effort, they ultimately failed to reach agreement.[92]

Internationally, the International Standards Organization (ISO) developed a considerably more complicated plastics identification code. The American Society for Testing and Materials (ASTM) Recycled Plastics Subcommittee announced in 1995 an agreement to recommend the ISO code rather than the SPI code for plastics identification.[93]

The initial consumer sorting (source separation) is supplemented by employee hand sorting, either at truckside or at a reclamation facility. Hand sorting is also used to remove containers that are excessively dirty. Hand sorting is generally employed to separate out the nonplastic components of the recycle stream as well. Sorting rates of 500 to 600 lb/hr of plastic containers per sorting person are generally attainable if two types of readily identifiable containers are being removed from a relatively uncontaminated stream.[19] However, the rate drops and purity of the sorted material declines significantly as the complexity of the sorting task increases. The cost and inefficiency of these processes has led to efforts to develop automated macrosorting systems.

These systems typically rely on some type of radiative absorption serving as an identifying signal for the polymer type. Thus, the distinctive absorption of chlorine can be used to identify PVC or PVDC (polyvinylidene chloride). The simplest systems transmit at only one wavelength and are designed for a specific separation. More complex systems transmit at a variety of wavelengths and can separate a number of different resins. Some can also separate a single resin type by color.

The process generally relies on sending containers in single file past the sensor, which is coupled to an air blast to direct the targeted containers off the conveyor at the appropriate location. While such systems have been demonstrated to perform reasonably well, their high cost has limited their use.

One of the first automated sorting systems for plastic containers in the United States was tested at Rutgers University. It used X-ray fluorescence to detect the chlorine atoms in PVC and thereby distinguish PVC from PET and HDPE. Light-emitting diodes coupled with photocell receptors were used to distinguish clear PET bottles from translucent HDPE bottles by the difference in reduction of light intensity. A second-generation system design incorporated the use of a universal product code (UPC) scanner coupled to a computer. When the UPC code can be read, the computer can look up the appropriate resin type and color, as long as that information has been entered in memory.[19]

Commercially available systems include the X-ray fluorescence-based system, as described above, which is offered by Asoma Instruments, Inc., to separate PET and PVC.[19]

Tecoplast, in Italy, uses X-ray radiation to separate PET from PVC, plus an optical detector to separate clear from colored PET.[94]

Magnetic Separation Systems has a system designed to separate PET, PVC, HDPE, and PP. It uses a combination of infrared, optical, and X-ray scanning for resin identification, plus a machine vision system for color identification. It is reportedly suitable for crushed or uncrushed containers. This system is currently operating in a plastics recovery facility (PRF) in Salem, Oregon. It handles 2 to 3 bottles per second, and was financed by a $1 million grant from the American Plastics Council (APC).[95]

Automation Industrial Control has developed a system that separates PET, HDPE, LDPE, PVC, PP and PS. It uses an infrared detector for resin identification and a color camera for color identification.[19]

Bruker Analytische Messtechnik GmbH in Germany also has an infrared sorting system, said to be able to identify 30 different resin types, requiring 3–4 seconds to make the identification.[96] The P/ID 28, manufactured by Bruker Instruments, Inc., Massachusetts, will be the centerpiece of a pilot facility at a new plastics-recycling research center to open in Berkeley, California. The center is funded by the American Plastics Council and MBA Polymers, Inc. The facility will also include a three-stage air classification system, a wet grinding system, and a series of hydrocyclones, along with a paint and coating removal system using high-pressure water.[97]

National Recovery Technologies, Inc., has a system for separation of PVC bottles from HDPE and PET.[19]

TOA Electronics of Tokyo has an infrared sorting system that is reportedly capable of identifying any type of plastic, no matter how dirty the material is, with 99.9 percent accuracy. The system is adaptable to containers, film and sheet, and other materials. The main problem at this time is speed. The system currently requires about two seconds per piece to make the identification and operates at 200 kg/hr. Unlike other systems designed for installation at a plastics processing facility, the company sees this system as being used at the point of return to sort materials being turned in by consumers, much like reverse vending machines.[96]

A novel hand-held device for identifying automotive polymers is being tested by Ford in Germany. It is about the size of a flashlight and distinguishes between nylon, polyethylene, ABS, and polypropylene by analyzing their electrostatic properties. An upgraded version that can distinguish between eight polymers is in development, as is a nonportable system that can identify up to 250 plastic types using infrared light.[98]

Other efforts have been directed towards improving the ease of identification of containers by providing them with built-in signals. Continental Container Corporation has developed a marking system for containers with easily detectable invisible ink. Eastman Chemical Company and Bayer have developed techniques for incorporating an easily detectable marker chemical into resins during their manufacture.[19]

11.12.2 Microsorting Techniques

The techniques described above generally do not work effectively enough for small pieces of plastic. Thus, they cannot be applied to plastics that have been granulated. For these plastics, a microsorting technique is required.

Most microsorting techniques rely on differences in the density of different polymers. One of the oldest of such technologies was the float/sink tanks used in the first generation of PET beverage-bottle recycling processes. Here, PET and HDPE were separated by submerging the chipped plastic in water. HDPE floated and was scraped off the top of the tank, while PET sank and was scraped off the bottom. Many such facilities have now been replaced by hydrocyclones, which can do the same job of separation with a much more efficient use of space.

When the plastics to be separated are both either heavier or lighter than water, traditional float/sink options will not work. It is possible to do the separation using a fluid that has a density between the components of interest, but this often involves handling salt solutions, alcohol solutions, organic solvents, or other such streams, all of which present their own set of problems.

The only current commercial use of salt solutions for density-based separations is for the separation of ABS from metal in recycling of telephone handsets.[19] No commercial use is currently being made of organic solvents or alcohol mixtures for density-based separations of recycled plastic.

Research at Michigan State University has been directed at use of glass microbubbles to form a medium with an effective density between HDPE and PP to permit separation of these resins.

Other research is directed towards using supercritical fluids. Mixtures of carbon dioxide and sulfur hexafluoride near the critical point have been used to separate HDPE from LDPE and PP, PS from PVC and PET, and even natural HDPE from colored HDPE. By varying both pressure and composition, the density of these fluids can be adjusted to within 0.001 g/cc of the desired point in a range of 0.7 to 1.7 g/cc at room temperature.[19]

Froth flotation is employed commercially for the separation of PET and aluminum. The PET/aluminum mixture is placed in a strong caustic bath. The caustic attacks the aluminum, generating bubbles of hydrogen gas, which cause the aluminum to float, separating it from the PET.[19]

For separation of polymers, froth flotation relies on differences in surface energy that result in differences in wetting characteristics. In some cases, additives are used that affect one polymer more than another. The surface tension of the froth bath is also manipulated, with the result that, when air is dispersed into the suspension of plastic flakes, the bubbles selectively adhere to one component, causing it to float and thus permitting it to be skimmed off.[19]

A number of electrostatic-separation processes have been developed. They rely on the difference in electron affinity of the plastics. Charged plastics are dropped between a pair of oppositely charged electrodes, where one type is attracted to the positive electrode and the other to the negative one.[19]

Other approaches rely on size differences between particles of different resins. Under appropriate conditions, for example, PVC will be ground to a fine powder, while PET will stay in much larger flakes. Screening methods can then separate the two components. Alternatively, size and shape differences can be exploited in a hydrocyclone for separation.

Polymers can also be separated by using differences in their softening point. Gradual heating is employed to cause some polymers to soften and cling to a belt, while those that do not stick are removed.

There is also some use of optical sorting techniques, much as were described for macrosorting. It is reported that as of 1993 there were 11 color sorting systems operating in the United States, mostly for removing caps from natural HDPE. These systems claim to reduce the commonly encountered contamination level of 3 percent down to less than 0.1 percent on a single pass at a throughput of 5,000 pounds per hour. However, color-sorting of flake from mixed-copolymer containers requires several passes and yields a rate of only 800 pounds per hour.[33] Two such systems are those offered by Simco-Ramic of Medford, Oregon, and Massen Vision Systems GmbH of Germany.[99]

Magnetic Separation Systems has developed an X-ray-based system to separate PVC flake from other postconsumer plastics in flake form, known as the Vydar system.[36]

11.12.3 Molecular Sorting

Even microsorting techniques cannot separate polymer blends. For such materials, and possibly for at least some multilayer structures, even granulated plastics will consist of particles containing more than one resin. The only effective way to separate such materials is to break the polymer down to the molecular level by dissolution.

An appropriate choice of solvents and conditions can be used to selectively dissolve and selectively recover plastics. Typical processes use either a number of different solvents or one solvent at several different temperatures. A common drawback of such recycling systems is the potential for solvent retention in the recovered polymer and the relatively high cost associated with use, control, and recovery of organic solvents.

In the process being investigated at Rensselaer Polytechnic Institute, xylene is used as the solvent, with polystyrene being removed at 25°C, LDPE at 75°C, HDPE at 105°C, and PP at 120°C. Solutions containing 5 to 10 percent polymer are heated and then flash-evaporated, recovering about 98 percent of the solvent. More solvent is removed during extrusion of the polymer. The residual plastic from the xylene leaching is treated with another solvent to separate PVC and PET. It is reported that polymer purity is greater than 98 percent, with the contaminating polymers not presenting processing problems since they are and remain microdispersed. While the polymer is in solution, it can be further treated to remove contaminants or to add desired components.[19]

The Argonne National Laboratory has investigated the use of solvents to recover plastics from automotive shredder residue. Polyurethane foam is first removed, followed by the application of various solvents to remove ABS, PVC, and mixed polyethylenes. Hexane or acetone rinses are used to remove soluble oils.[19]

Xylene has been used to recover PVC from wire cables. Methylisobutylketone has also been used for PVC separation.[19]

11.12.4 Compatibilization

Significant efforts are being made in the area of reducing problems caused by mixtures of incompatible plastics by rendering them more compatible. Compatibilizers are typically additives that will adhere to both the components of interest and thus provide interfacial strength that improves the performance of the finished article. Most often, these are block or graft copolymers.

Rutgers recently reported the use of commingled plastics in the manufacture of containers, with compatibilizers the factor that made this development possible.[19]

11.13   THERMAL RECYCLING

11.13.1 Pyrolysis

An alternative to traditional recycling methods is to convert the polymers, using heat, into small molecules that can then be used as a chemical feedstock or alternatively can be used as a liquid or gaseous fuel. These processes are termed pyrolysis. While the chemical industry has generally regarded these as recycling processes, the environmental community has typically viewed them as akin to incineration and thus not "real" recycling.

All pyrolysis processes have in common exposing the feedstock to high temperatures in an oxygen-poor atmosphere so that significant combustion does not occur. Some processes are designed to be used for mixed waste streams, while others are designed to accommodate a plastics-only stream or even a stream containing only certain types of plastics.

Depending on the feedstock and the pyrolysis conditions, but especially on the particular polymer or polymers involved, a range of different products can be obtained. Some polymers pyrolize largely to monomers, while others give virtually no monomer at all. In general, the more nearly pure the product stream, the more likely it is that it can be used as a chemical feedstock.

Polymethyl methacrylate and styrene/methacrylate copolymers pyrolize essentially to monomers. Nylon 6 carpet scrap is reported to pyrolyze 85 percent to monomers.[19]

Pyrolysis of polyolefins such as polyethylene generally gives a very complex mixture of products, including alkanes, olefins, and dienes. These are usually suitable only for use as fuel because of the high cost of purification and the relatively low value of the species being produced.

Small facilities for the recovery of diisocyanates by pyrolysis are believed to be operating in France and Japan.[19]

BASF recently announced the start-up of a pilot plant facility for pyrolysis of mixed polymer waste. It employs a mild thermal hydrocracking of polymer feed in a polymer melt at 300°C to lower the viscosity and cause PVC to lose chlorine in the form of hydrochloric acid. Next, the melt is steam-cracked at 400 to 450°C, yielding a mixture of hydrocarbons, including fuel gases, olefins, and aromatics. Distillation is reported to yield a high-quality naphtha in a yield of 45 percent. A commercial plant was expected to start up in 1996 that would be capable of handling about 10 percent of Germany's mixed polymer waste, 300,000 metric tons per year.[100] However, in mid-1995, these plans were placed on hold, apparently as a result of unsuccessful negotiations about who would pay for the cost of commercializing the process, estimated at several hundred million dollars,[101] and it now appears that the plant will not be built.

11.13.2   Other Processes

Research on hydrogenation of plastics waste is currently being undertaken in Germany. A slurry-phase hydrogenation at temperatures up to 500°C and pressures of 20 Mpa was carried out of PVC cable waste, PE cable waste, and polyester and phenolic resins from used car bodies, mixed with old motor oil, industrial-waste oil, and chlorobenzene from waste-oil refining. The products consisted of gases, liquids, and solid residue.[100]

11.14   DESIGN ISSUES

11.14.1   Life-cycle Analysis

Increasingly, attempts to make design decisions with environmental effects in mind are relying on life-cycle assessment (LCA). This is a tool for evaluating the total environmental impact of a product or process, from initiation to ultimate disposal. ISO 14000, a global standard for manufacturing management practices, is expected to incorporate LCA as a measure of environmental accountability.[80]

The origins of life-cycle analysis can be found in the energy analyses of the 1970s and in the philosophy of waste minimization. When oil prices increased in the 1970s, and industry became concerned with saving energy to decrease costs, it was obvious that simple analyses like "don't use plastics because they're made from oil" were often inaccurate in terms of the total energy picture. Thus, when systems were investigated to see, for example, whether glass or plastic bottles were more energy-efficient, it was necessary to include the manufacture, transport, and disposal of the packages in order to reach a valid conclusion. Similarly, when concerns about toxic materials grew, and efforts began to be devoted to minimizing the use and disposal of these materials for both environmental and cost-saving reasons, it was obvious that to make a change at one point of a process that increased the generation of toxic materials at some other point, was not necessarily a positive step.

Out of these influences, then, grew the philosophy of life-cycle assessment. With this approach, the environmental impact of a product or process is calculated based on the whole life system of that product or process, from generation of raw materials through ultimate disposal. Of course, some bounds must be drawn—otherwise, you could go on forever calculating the impact of the machines that made the machines that made the. . . . A useful guideline is to draw the boundaries at the point at which the size of the influence is about the same size as the uncertainty in the data. It surely does not make sense to go beyond that point.

Life-cycle assessment is still a developing technique. Early efforts at life-cycle assessment frequently employed unstated assumptions and tended to lump diverse environmental outputs together on the basis of simple measures such as weight, in order to reach what often appeared to be self-serving conclusions. The Society of Environmental Toxicology and Chemistry (SETAC) convened a workshop in 1990 to examine the state of the art of lifecycle assessment and make recommendations for its further development.[102]

Since that time, the U.S. Environmental Protection Agency has sponsored considerable work on further development of life-cycle assessment as a tool for making environmental decisions. This work is ongoing. Groups in Europe are also actively working on further development of life-cycle assessment.

11.14.2 Design for Recycling

Design of plastic materials for recycling is accomplished in large part by paying attention to some simple rules. First, toxic constituents should be eliminated. Second, a single plastic material should be used whenever possible. Third, if multiple materials are used, they should be chosen and designed to be either easily separable or compatible without separation. Finally, recycled materials should be used whenever possible, since, without end uses, the whole recycling system falls apart. Important corollaries are that new products or product changes that interfere with existing successful recycling systems should be avoided, and that, if there is a choice of two materials, the one that is most widely recycled is generally preferable (other things being equal).

Recently, the U.S. EPA's region 5 and the states of Wisconsin and New York sponsored a report on improving the economics of plastic-container recycling. A number of design changes were recommended to facilitate the recycling process.[103]

Several design changes addressed the issue of polymer compatibility. These include a recommendation that caps, closures, and spouts on HDPE bottles be made compatible with the HDPE so that there is no need to manually remove caps during the process to render the resin suitable for high-value end uses such as film and bottle markets. PVC film labels should be used only on PVC containers. All layers of multilayer plastic bottles should be compatible enough to permit use of the postconsumer resin in high value end markets. HDPE base cups should be phased out on PET bottles.

Other recommendations dealt with the issue of color. They recommended that caps on natural HDPE bottles should not be pigmented. Pigments should not bleed from the label during processing. Printing, with the exception of date coding, should not be applied directly on unpigmented plastic containers.

Another set of recommendations dealt with processing issues. Label adhesives should be water-dispersible during processing. Metallized labels should not be used on plastic bottles with a specific gravity greater than one. Aluminum innerseals on plastic bottles should pull off completely. Aluminum caps on plastic bottles should be phased out. PVC resins should not be used in bottles for products that are also packaged in bottles made of other resins that look like PVC (presumably PET).

These recommendations also called for the development of a low-volume, low-cost automated sorting system for detecting PVC and called for further investigation of the need for MRFs to vertically integrate by installing wash-and-flake systems for plastics.[103]

11.14.3 Environmental Certification Programs

Another influence on design is the existence of programs to certify, in some fashion, the "environmental goodness" of a product or package. While recycling is certainly not the only criteria, inclusion of recycled content is sometimes the basis for such certification.

Programs include Green Cross and Green Seal in the United States, the Blue Angel program in Germany, the EcoMark in Japan, and the Environmental Choice label in Canada.

In all these programs, manufacturer participation is voluntary, with a fee generally charged to cover the cost of the program. The impetus for the manufacturer is the marketing appeal of the label.

11.14.4 Green Marketing

Marketing of products that are presented as having some positive environmentally related attribute has become big business. Much of the interest in products and packaging with recycled content is related to their consumer appeal. When this phenomenon started, there was a great deal of misleading advertising. As a result, legal action was taken in some high-profile cases in which companies were charged with misleading the consumer. Several states in the United States now have regulations on the use of environmental terminology, and the U.S. Dept. of Commerce has issued voluntary guidelines in this area. Among the important guidelines are that claims of recycled content should be specific as to minimum amount and should distinguish between preconsumer and postconsumer recycled content.

11.15   THE FUTURE OF PLASTICS RECYCLING

It seems clear that plastics recycling is continuing to grow. There are several variables that influence how much and how fast this will continue.

11.15.1   The Role of Legislation

An important variable in predicting the future of plastics recycling is the role of legislation. In the late 1980s, much legislative attention was given to proposals for packaging regulation of various types such as taxes and bans. While few of these were actually enacted, they did serve to get the attention of the targeted industries. More recently, efforts have been directed more specifically at encouraging recycling rather than trying to get rid of undesirable products or packages.

Florida and a few other states have looked at using differential taxes to encourage recycling. In the early 1990s, Florida imposed an advance disposal fee on containers that were not recycled at a 50 percent rate in the state, unless they contained a minimum recycled content. Among other effects, this influenced soft-drink companies to use PET bottles with 25 percent recycled resin content in that state. In October 1995, that legislation expired. Some have predicted that the result will be beverage companies going back to using 100 percent virgin resin, especially since they were paying a price premium for recycled resin. Since the containers were not labeled with recycled content, consumers will not be able to tell any difference.

A more common approach to encouraging recycling is mandates for recycled content or recycling rates. This approach was proven to be successful for newsprint. When a vast oversupply of newsprint resulted from many new curbside recycling programs being started, particularly on the east coast, prices for collected newsprint actually became negative. The imposition of required recycled content in newsprint encouraged the paper industry to invest in increased capacity for processing the recovered paper. Prices recovered dramatically, and paper recycling rates rose substantially.

In California, as of January 1995, most nonfood plastic containers are required either to incorporate 25 percent recycled content or be source-reduced, refillable, or recycled at a specified rate. Food and cosmetic containers are currently exempted, but that exemption expires for most at the beginning of 1997.[91] Since California represents 10 percent of the U.S. market, this legislation has had a significant effect on a national scale. Some industry sources say it is more cost-effective for companies selling in California to include recycled content across the board than to manufacture differently for that state.[12]

California also requires, as of January 1993, that all plastic trash bags sold in the state have postconsumer recycled content. The initial requirement was for bags one mil in thickness or greater to contain at least 10 percent postconsumer materials. Beginning in January 1995, the postconsumer content increased to 30 percent, and bags 0.75 mil in thickness or greater are included. The postconsumer content requirement is for an annual aggregate, not for each bag.

Oregon imposed legislation similar to that in California, requiring 25 percent recycled content in rigid plastic containers unless these containers met a 25 percent recycling rate or were reusable. In response to this legislation, the plastics industry put a considerable investment into increasing plastics recycling in Oregon. The state's official forecast for plastic recycling in Oregon for 1995, 32 percent, is currently high enough, in fact, to exempt all plastic containers from the recycled content requirement. However, some plastics-industry representatives have voiced concern that the forecast is overly optimistic. In 1995, the Oregon legislature exempted plastic packages containing solid and semisolid food products from this requirement. These are estimated to compose 25 percent of all rigid plastic containers.[104] While beverage containers are not included, the recycling rate for soft-drink containers in this deposit state exempts them from recycled-content requirements in any case. In fact, the 32 percent rate forecast includes 15 percent from bottle-bill PET bottles, 8 percent from HDPE milk-bottle collection, and 9 percent from other types of plastic containers (see Fig. 11.21).[105]

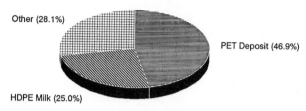

Other (28.1%)

PET Deposit (46.9%)

HDPE Milk (25.0%)

FIGURE 11.21   Plastics recycling in Oregon, 1995 estimate.[105]

Other states are approaching the task of increasing recycling by imposing landfill bans. In many states, landfill disposal of various kinds of materials such as automobile batteries and yard waste is prohibited. Massachusetts, New Hampshire, Rhode Island, and Wisconsin have extended this idea to recyclables, prohibiting their disposal.[91]

In the United States, the federal government has been a very inactive player in the recycling arena. The leadership role has fallen to the states. Thus, the current political climate in Washington with its decreased interest in regulation does not necessarily translate into decreased legislation regarding recycling. There does, however, appear to be a moderation in the demand for increased recycling, at least for the present.

Outside the United States, there have also been concerted efforts to promote recycling of plastics as well as other materials.

Canada has a national policy to reduce disposal of packaging waste by waste reduction and recycling. The National Packaging Protocol called for a 20 percent reduction by the end of 1992, which was met, and a 50 percent reduction by the end of 2000, a significantly more difficult task. Regulation was called for if the target is not met by voluntary means. Various provinces have their own targets and approaches, and negotiations about how to achieve this goal are still going on.

Germany is well known to have the most stringent requirements for packaging disposal in the world. The business entity that introduces a package into commerce is responsible for its ultimate recovery and recycling, and there are stringent targets for 80 percent collection and 80 to 90 percent recycling of collected materials. Most companies have opted to cope with consumer packaging by subscribing to the Duales System Deutschland (DSD), the members of which pay a fee in order to be allowed to participate in the Green Dot recovery system. The system is named for the green dot on the package that certifies participation in the system. Participating companies pay a fee that allows their packaging to be collected by the DSD, which will ensure that the legal requirements are met for recovery of the material. Participants thus avoid having to have their own recovery system. Nonparticipating companies are essentially frozen out of retail commerce, since retailers do not want the burden of handling the returned packages.

France and other European countries have adopted similar systems, but most permit incineration as at least a partial alternative to recycling. The European Union system will be close to the French system, but the issue of how it will affect Germany is not yet totally resolved. Clearly, making disposal the manufacturer's responsibility is at the heart of these systems.

Germany, as mentioned earlier, is now extending this philosophy beyond packaging to automobiles and other durable goods. This is at least partially responsible for the great increase in concern about recyclability on the part of European automotive manufacturers.

11.15.2   The Electronic Media

The task of linking suppliers and buyers of recycled material together in the United States may be facilitated by the opening of the Chicago Board of Trade Recycling Exchange. It provides an electronic bulletin board to link buyers and sellers of recycled materials. PET and HDPE are the only plastics listed initially.[106]

Another service available is the National Materials Exchange Network (NMEN), which provides an electronic bulletin board on the World Wide Web, allowing users to list materials and search for available materials. It is free to users.[107]

A variety of more local bulletin-board systems also exist. RecycleNet in Colorado was established in 1992. This free service is primarily for Colorado residents but can also be accessed by nonresidents.[108]

InfoCycle is operated by the California Department of Conservation. It has on-line discussions as well as files on a variety of topics.[108]

PenCycle was begun in 1990 and covers other environmental topics as well as recycling and waste management. Subscribers also have access to newsgroups on the Internet. Access is free, but there is a small charge for participation in the discussion groups.[108]

The Global Recycling Network (GRN) on the World Wide Web offers general reference material as well as a marketplace for recyclable goods. They opened in April 1995 and already have over 500 users for the materials marketing service, which requires a monthly fee. Other services are offered at no charge.[108]

Recycler's World in Canada is another World Wide Web site that opened in 1995. It has free listings of materials to buy, sell, or trade, plus an on-line news service and other services. This site reportedly averaged 150 calls per day in mid-1995.[108]

As can be seen, all of these services are very new, so their impact has yet to be fully evaluated. They offer great potential for facilitating recycling by enabling suppliers and users of recycled plastics to know of each others' existence, interest, and abilities. They may prove to be a powerful influence.

We have come a long way in making plastics recycling happen. The road ahead, however, is much longer.

REFERENCES

1. Leaversuch, R., "Recycling faces reality as bottom line looms," *Modern Plastics,* July 1994, pp. 48D–50.

2. U.S. Environmental Protection Agency, *Characterization of Municipal Solid Waste in the United States: 1994 Update,* EPA530-R-94-042, Washington, D.C., 1994.

3. Siegler, T., "Collecting plastic bottles more efficiently," *Resource Recycling,* Sept. 1994, pp. 27–42.

4. Apotheker, S., and J. Powell, "How are we doing? The 1994 report and more," *Resource Recycling*, July 1995, pp. 34–47.

5. King, R., "Plastic bottle recycling rate rises 21%," *Plastics News*, June 5, 1995, p. 1.

6. Woods, R., "APC, NAPCOR Studies Show Increases in Plastic Container Recycling Rates," *Waste Age's Recycling Times*, June 13, 1995, p. 6.

7. "Garbage: It Isn't the Other Guy's Problem Anymore," *Business Week*, 154, May 25, 1987, pp. 150–151.

8. Fenton, Robert W., "Reuse versus recycling: a look at grocery bags," *Resource Recycling*, vol. 11, no. 3, March 1992, pp. 105–110.

9. Miller, C., "DOE Report: Recycle Plastic and Metal, Refill Glass, Burn Paper," *Waste Age's Recycling Times*, March 7, 1995, p. 10.

10. Sitek, F. A., "Restabilization upgrades post-consumer recyclate," *Modern Plastics*, Oct. 1993, pp. 64–68.

11. Pattanakul, C., et al., "Properties of Recycled High Density Polyethylene Milk Bottles," *Antec '88*, Society of Plastics Engineers, 1988, pp. 1802–1804.

12. Siegler, T., "Collecting plastic bottles more efficiently," *Resource Recycling*, Sept. 1994, pp. 27–42.

13. Apotheker, S., "The bottle is the bottleneck," *Resource Recycling*, Sept. 1995, pp. 27–34.

14. Moore, W. P., "Collection and Separation," in R. Ehrig (ed.), *Plastics Recycling: Products and Processes*, Hanser Pub., Munich, 1992, pp. 17–44.

15. McCreery, P., "Consumers Recycling More Often, Study Finds," *Waste Age's Recycling Times*, March 21, 1995, p. 7.

16. Milgrom, J., "Polyethylene Terephthalate (PET)," in R. Ehrig (ed.), *Plastics Recycling: Products and Processes*, Hanser Pub., Munich, 1992, pp. 45–72.

17. McCreery, P., "Study Forecasts Wide Expansion in Use of Recovered Plastic," *Waste Age's Recycling Times*, March 7, 1995, p. 7.

18. "Resins 1995," *Modern Plastics*, Jan. 1995, pp. 37–68.

19. Bisio, A. L., and M. Xanthos (eds.), *How to Manage Plastics Waste: Technology and Market Opportunities*, Hanser Pub., Munich, 1994.

20. DeWinter, W., "Poly(ethylene terephthalate) Film Recycling," in F. LaMantia (ed.), *Recycling of Plastic Materials*, ChemTec Pub., Ontario, 1993, pp. 1–15.

21. Leidner, J., *Plastics Waste: Recovery of Economic Value*, Marcel Dekker, Inc., New York, 1981.

22. Rebeiz, K., D. Fowler, and D. Paul, "Recycling Plastics in Construction Applications," *J. of Resource Management and Tech.*, vol. 21, no. 2, 1993, pp. 76–81.

23. Sutherland, G., D. Cearley, and M. Tormey, "Market prices for recyclables: A five-year history," *Resource Recycling*, Aug. 1995, pp. 51–60.

24. Bakker, M., "Using recycled plastics in food bottles: The technical barriers," *Resource Recycling*, May 1994, pp. 59–64.

25. Myers, J., "Coca-Cola closes the loop with multi-layer PET bottle," *Modern Plastics*, April 1995, pp. 28–30.

26. Rabasca, L., "Coca-Cola Introduces Recycled PET Bottle in Switzerland," *Waste Age's Recycling Times*, Feb. 21, 1995, p. 1.

27. Ford, T., "FDA clears Wellman's recycled sheet," *Plastics News*, Aug. 28, 1995, p. 6.

28. Rabasca, L., "NAPCOR May Expand Mission To Promote PET and HDPE Recycling," *Waste Age's Recycling Times*, Aug. 22, 1995, p. 11.

29. Leaversuch, R., "Raising PCR content in oil bottles reshapes a market and a process," *Modern Plastics*, Feb. 1993, pp. 18–19.

30. Powell, J., "HDPE bottle recycling: The demand glut continues," *Resource Recycling*, May 1995, pp. 23–28.

31. Jones, K., and R. Warner, "Maximizing plastics diversion: Lessons learned," *Resource Recycling,* Jan. 1995, pp. 14–21.

32. McCreery, P., "Drastic Drop in HDPE Prices Continues," *Waste Age's Recycling Times,* Aug. 22, 1995, p. 8.

33. Apotheker, S., "High hopes for HDPE containers," *Resource Recycling,* May 1994, pp. 51–58.

34. Malloy, M., "Indiana Company Setting Up Program To Recycle DuPont Plastic Garments," *Waste Age's Recycling Times,* Sept. 5, 1995, p. 26.

35. Bregar, B., "U.S. Postal Service extols plastic pallets," *Plastics News,* Aug. 21, 1995, p. 9.

36. White, K., "Markets for Recycled Plastic Increase with New Technology, Capacity, and Innovative End Use," *Waste Age's Recycling Times,* Jan. 24, 1995, p. 6.

37. Hernandez, R., C. Lai, and S. Selke, "Butyric Acid Retention of Post-Consumer Milk Bottles," *Antec '88,* Society of Plastics Engineers, 1988, pp. 1805–1808.

38. Podborny, W., "Maintaining Packaging Quality with HDPE Post-Consumer Resin," *Packaging Technology & Engineering,* May 1995, pp. 66–68.

39. Apotheker, S., "Film at 11: A picture of curbside recovery efforts for plastic bags," *Resource Recycling,* May 1995, pp. 35–40.

40. McCreery, P., "Survey Polls Consumers on Plastic Bag Recyclability," *Waste Age's Recycling Times,* April 4, 1995, p. 10.

41. "Trash bags to be made from reclaimed agricultural film," *Modern Plastics,* Oct. 1993, p. 17.

42. Perlson, B. D., and C. C. Schababerle, "Polyolefins," in R. Ehrig (ed.), *Plastics Recycling: Products and Processes,* Hanser Pub., Munich, 1992, pp. 73–108.

43. "Program Recycles Farm Film in U.K.," *Modern Plastics,* August 1995, p. 17.

44. Ford, T., "Colgate tests toothpaste tube's recyclability," *Plastics News,* March 7, 1994, p. 32.

45. Leibovitz, H., "Cost-Effective Systems for Recycling Plastics," *NAPMInsights,* March 1994, pp. 42–44.

46. Mackey, G. A., R. C. Westphal, and R. Coughanour, "Polystyrene (PS)," in R. Ehrig (ed.), *Plastics Recycling: Products and Processes,* Hanser Pub., Munich, 1992, pp. 109–129.

47. Leaversuch, R., "EPS reuse gets lift from densification," *Modern Plastics,* Dec. 1993, pp. 75–76.

48. "Post-consumer EPS Foam Packaging Third Most Recycled Plastic Product," *Packaging Technology & Engineering,* Nov./Dec. 1994, p. 16.

49. "Consumers Prove They Know More Than Peanuts About Polystyrene Fill," *Packaging Technology & Engineering,* March, 1994, p. 14.

50. Grelle P. F., and O. Khennache, "Recycling Structural Foam Polystyrene: What Goes Around Comes Around," *Innovations in Plastics III,* Rochester Section RETECH, SPI, Sept. 15–16, 1993.

51. "Eastman Kodak recycles 50 million cameras," *Plastics News,* Aug. 14, 1995, p. 10.

52. Myers, J., "Recyclers of appliances, durables looking to Germany's proposals," *Modern Plastics,* March 1995, pp. 14–15.

53. "Amoco uses recycled PS in board," *Plastics News,* Aug. 14, 1995.

54. Ford, T., "Idea for recycling EPS gains support," *Plastics News,* June 12, 1995, p. 38.

55. Heil, K., and R. Pfaff, "Quality Assurance in Plastics Recycling by the Example of Polypropylene; Report on the experience gathered with a scrap battery recycling plant," in F. LaMantia (ed.), *Recycling of Plastic Materials,* ChemTec Pub., Ontario, 1993, pp. 171–185.

56. "Murphy scrubs virgin PP bottle in favor of PCR," *Packaging Digest,* March 1995.

57. The Vinyl Institute, *Environmental Briefs,* vol. 4, no. 2, Aug. 1993.

58. Carroll, W. F., Jr., R. G. Elcik, and D. Goodman, "Polyvinyl Chloride (PVC)," in R. Ehrig (ed.), *Plastics Recycling: Products and Processes,* Hanser Pub., Munich, 1992, pp. 131–150.

59. The Vinyl Institute, *Environmental Briefs,* vol. 5, no. 3, Nov. 1994.

60. The Vinyl Institute, *Environmental Briefs,* vol. 4, no. 3, Dec. 1993.

61. Wahlquist, C., "Company Rolls Out Carpet Recycling Process," *Waste Age's Recycling Times,* March 21, 1995, p. 6.

62. Glaz, S., "Raising the Roof on Vinyl Recycling," *Waste Age's Recycling Times,* July 25, 1995, p. 4.

63. The Vinyl Institute, *Environmental Briefs,* vol. 6, no. 1, June 1995.

64. "Vinyl Institute Promotes Production Expansion of Recycled Vinyl Products," *Waste Age's Recycling Times,* May 30, 1995, p. 10.

65. VanNess, K. E., and T. J. Nosker, "Commingled Plastics," in R. Ehrig (ed.), *Plastics Recycling: Products and Processes,* Hanser Pub., Munich, 1992, pp. 187–229.

66. White, K., "Research Center to Demonstrate New Commingled Plastic Processing System," *Waste Age's Recycling Times,* Oct. 4, 1994, p. 9.

67. LaMania, F. P., C. Perrone, and E. Bellio, "Blends of Polyethylenes and Plastics Waste. Processing and Characterization," in F. LaMantia (ed.), *Recycling of Plastic Materials,* ChemTec Pub., Ontario, 1993, pp. 83–98.

68. Mapleston, P., "Housing may be built from scrap in low-pressure process," *Modern Plastics,* Sept. 1995, p. 21.

69. "Thermoset reclaim bets on chemical process," *Modern Plastics,* April 1995, p. 13.

70. Diem, R., "Group fosters recycling of SMC auto parts," *Plastics News,* June 26, 1995, p. 10.

71. Weigand, E., et al., "Polyurethane recycling processes are reaching commercial maturity," *Modern Plastics,* Nov. 1993, pp. 71–75.

72. Farrissey, W. J., "Thermosets," in R. Ehrig (ed.), *Plastics Recycling: Products and Processes,* Hanser Pub., Munich, 1992, pp. 231–262.

73. Mapleston, P., "Auto sector's recycling goals keep plastics on hot seat," *Modern Plastics,* May 1995, pp. 48–58.

74. Bregar, B., "Obstacles numerous in auto recycling," *Plastics News,* March 25, 1992, pp. 8–9.

75. "Automotive Plastic Recycling R&D Shifts Into High Gear," *R&D Magazine,* March 1993, p. 32.

76. "Process recovers painted and coated PP bumpers," *Modern Plastics,* Feb. 1993, p. 27.

77. Bhakta, P., "Recent Technology and Trends in Automotive Recycling," *JOM,* Feb. 1994, pp. 36–39.

78. Jody, B., E. Daniels, P. Bonsignore, and N. Brockmeier, "Recovering Recyclable Materials from Shredder Residue," *JOM,* Feb. 1994, pp. 40–43.

79. Gargaro, P., "Institute recycles inseparable auto fluff," *Plastics News,* Sept. 4, 1995, p. 6.

80. Leaversuch, R., "Plastics and the Environment: A Level playing field at last," *Modern Plastics,* Aug. 1995, pp. 44–51.

81. Ford, T., "ACI renews resin to near-virgin quality," *Plastics News,* July 31, 1995, p. 7.

82. Biddle, M., and M. Fisher, "An Overview of the Recycling of Plastics from Durable Goods," in *Society of Plastics Engineers International Plastics Recycling Conference,* Schaumbure, IL, Nov. 3–4, 1994.

83. Biddle, M., and M. Fisher, "A Mechanical Recycling Process for the Recovery of Plastics from End-of-life Computer and Business Equipment," in *Society of Plastics Engineers International Plastics Recycling Conference,* Schaumbure, IL, Nov. 3–4, 1994.

84. Ford, T., "Recycled nylon resin output up," *Plastics News,* Sept. 11, 1995.

85. Burnett, R. H., and G. A. Baum, "Engineering Thermoplastics," in R. Ehrig (ed.), *Plastics Recycling: Products and Processes,* Hanser Pub., Munich, 1992, pp. 151–168.

86. Studt, T., "Polycarbonate Research Focuses on Flow, Alloys, and Recycling," *R&D Magazine,* July 1994, pp. 26–30.

87. "CD Recycling Plant is Europe's First," *Modern Plastics,* Sept. 1995, p. 13.

88. White, K., "GE Plastics Begins Buyback Program Aimed at Plastic Auto Scrap," *Waste Age's Recycling Times,* March 7, 1995, p. 8.

89. Ford, T., "Hewlett-Packard printers use recycled ABS," *Plastics News,* Aug. 7, 1995, p. 40.

90. Lindsay, K., "Engineering thermoplastics are formulated from recycled resins," *Modern Plastics,* March 1993, p. 84.

91. Thompson Publishing Group, *Environmental Packaging: U.S. Guide to Green Labeling, Packaging and Recycling,* Washington, D.C., 1995.

92. Rabasca, L., "SPI Board Stands by Chasing Arrows Code," *Waste Age's Recycling Times,* Oct. 18, 1994, pp. 3–4.

93. Rabasca, L., "ASTM Looks at Standards for Quality Assurances for Recycled Plastic," *Waste Age's Recycling Times,* Feb. 7, 1995, p. 12.

94. Sereni, E., "Techniques for Selection and Recycle of Post-Consumer Bottles," in F. LaMantia (ed.), *Recycling of Plastic Materials,* ChemTec Pub., Ontario, 1993, pp. 99–109.

95. Powell, J., "The PRFect solution to plastic bottle recycling," *Resource Recycling,* Feb. 1995, pp. 25–27.

96. Ford, T., "New sorting systems use infrared light," *Plastics News,* June 5, 1995, p. 6.

97. Wahlquist, C., "Plastics Recycling Research Center to Open in Berkeley, California," *Waste Age's Recycling Times,* March 21, 1995, p. 10.

98. Colvin, R., "Sorting mixed polymers eased by hand-held unit," *Modern Plastics,* April 1995.

99. Leaversuch, R., "Color-sortation technologies improve PCR quality and value," *Modern Plastics,* June 1993.

100. Krause, H. H., and J. M. L. Penninger, "Polymer Waste from Nuisance to Resource," in H. H. Krause and J. M. L. Penninger (eds.), *Conversion of Polymer Wastes & Energetics,* ChemTec Publishing, Ontario, 1994.

101. Ford, T., "BASF's waste-to-feedstock plan in limbo," *Plastics News,* June 5, 1955, p. 8.

102. SETAC, *A Technical Framework for Life-Cycle Assessment,* Society of Environmental Toxicology and Chemistry, Washington, D.C., 1991.

103. Rabasca, L., "EPA Issues Preliminary Report on Plastic Container Recycling," *Waste Age's Recycling Times,* April 4, 1995, p. 9.

104. McCreery, P., "Ore. Senate Exempts Food Containers from Recycling Law," *Waste Age's Recycling Times,* April 4, 1995, p. 1.

105. White, K., "Oregon Expects to Surpass 25% Plastic Container Recycling Rate in 1995," *Waste Age's Recycling Times,* Jan. 10, 1995, p. 1, 10.

106. "CBOT Recyclables Exchange To Link Nation's Traders," *Reusable News,* U.S. Environmental Protection Agency, EPA530-N-95-005, Summer/Fall 1995, p. 1.

107. Goff, J., "Recyclable Materials Listed on Internet," *Waste Age's Recycling Times,* May 30, 1995, p. 10.

108. Riggle, D., "Online Connections for Recyclers and Composters," *BioCycle,* Sept. 1995, pp. 33–35.

FURTHER READING

Selke, Susan E., *Packaging and the Environment; Alternatives, Trends and Solutions,* rev. ed., Technomic Publishing Co., Lancaster, PA, 1994.

APPENDIX A
GLOSSARY OF TERMS AND DEFINITIONS

A-stage Stage in which thermosetting reactants are mixed, but at which the polymerization reaction has not yet begun.

abhesive Film or coating, such as a mold release, applied to a surface to prevent adhesion or sticking.

ablative plastics Plastics or resins, the surface layers of which decompose when surface is heated, leaving a heat-resistant layer of charred material. Successive layers break away, exposing a new surface. These plastics are especially useful in applications such as outer skins of spacecraft, which heat up to high temperatures on reentry into the earth's atmosphere.

abrasion resistance Capability of a material to withstand mechanical forces such as scraping, rubbing, or erosion, that remove material from the surface.

ABS plastics Abbreviated phrase referring to acrylonitrile-butadiene-styrene copolymers; elastomer-modified styrene.

absorption Penetration of one substance into the mass of another, such as moisture or water absorption of plastics.

accelerated test Test in which conditions are intensified to obtain critical data in shorter time periods, such as accelerated life testing.

accelerator Chemical used to speed up a reaction or cure. Term is often used interchangeably with promoter. For example, cobalt naphthanate is used to accelerate the reaction of certain polyester resins. An accelerator is often used along with a catalyst, hardener, or curing agent.

activation Process, usually chemical, of modifying a surface so that coatings will more readily bond to that surface.

activator Chemical material used in the activation process. (*See* **activation.**)

addition reaction or polymerization Chemical reaction in which simple molecules (monomers) are added to each other to form long-chain molecules without forming by-products.

additive Substance added to materials, usually to improve their properties. Prime examples are plasticizers, flame retardants, or fillers added to plastic resins.

adhere To cause two surfaces to be held together by adhesion.

adherend   Body held to another body by an adhesive.

adhesion   State in which two surfaces are held together by interfacial forces, which may consist of valence forces or interlocking action, or both.

adhesive   Substance capable of holding materials together by surface attachment.

adhesive, anaerobic   Adhesive that sets only in the absence of air; for instance, one that is confined between plates or sheets.

adhesive, contact   Adhesive that is apparently dry to the touch and will adhere to itself instantaneously upon contact; also called contact-bond adhesive or dry-bond adhesive.

adhesive, heat-activated   Dry adhesive film that is rendered tacky or fluid by application of heat or heat and pressure to the assembly.

adhesive, pressure-sensitive   Viscoelastic material that in solvent-free form remains permanently tacky. Such material will adhere instantaneously to most solid surfaces with the application of very slight pressure.

adhesive, room-temperature-setting   Adhesive that sets in the temperature range of 20 to 30°C (68 to 86°F), in accordance with the limits for standard room temperature as specified in ASTM Methods D 618, Conditioning Plastics and Electrical Insulating Materials for Testing.

adhesive, solvent   Adhesive having a volatile organic liquid as a vehicle.

adhesive, solvent-activated   Dry adhesive film that is rendered tacky just prior to use by application of a solvent.

aging   Change in properties of a material with time under specific conditions.

air vent   Small gap in a mold to avoid gases being entrapped in the plastic part during the molding process.

airless spraying   High-pressure spraying process in which pressure is sufficiently high to atomize liquid coating particles without air.

alcohols   Characterized by the fact that they contain the hydroxyl (—OH) group, they are valuable starting points for the manufacture of synthetic resins, synthetic rubbers, and plasticizers.

aldehydes   In general, volatile liquids with sharp, penetrating odors that are slightly less soluble in water than the corresponding alcohols. The group (—CHO), which characterizes all aldehydes, contains the most active form of the carbonyl radical and makes the aldehydes important as organic synthetic agents. They are widely used in industry as chemical building blocks in organic synthesis.

aliphatic hydrocarbon   *See* **hydrocarbon**.

allowables   Statistically derived estimate of a mechanical property based on repeated tests. Value above which at least 99 percent of the population of values is expected to fall with a confidence of 95 percent.

alloy   Blend of polymers, copolymers, or elastomers under controlled conditions.

alpha particle   Heavy particle emitted during radioactive decay consisting of two protons and two neutrons bound together. It has the lowest penetration of the various emitted particles and will be stopped after traversing through only a few centimeters of air or a very thin solid film.

ambient temperature   Temperature of the surrounding cooling medium, such as gas or liquid, that comes into contact with the heated parts of the apparatus.

amine adduct   Products of the reaction of an amine with a deficiency of a substance containing epoxy groups.

amorphous plastic   Plastic that is not crystalline, has no sharp melting point, and exhibits no known order or pattern of molecule distribution.

anhydride   Organic compound from which water has been removed. Epoxy resins cured with anhydride-curing agents are generally characterized by long pot life, low exotherm during cure, good heat stability, and good electrical properties.

antioxidant   Chemical used in the formulation of plastics to prevent or slow down the oxidation of material exposed to the air.

antistatic agents   Agents that, when added to the molding material or applied onto the surface of the molded object, make it less conducting (thus hindering the fixation of dust).

aramid   Generic name for highly oriented organic material derived from a polyamide but incorporating an aromatic ring structure.

arc resistance   Time required for an arc to establish a conductive path on the surface of an organic material.

areal weight   Weight of fabric or prepreg per unit width.

aromatic amine   Synthetic amine derived from the reaction of urea, thiourea, melamine, or allied compounds with aldehydes, that contains a significant amount of aromatic subgroups.

aromatic hydrocarbon   *See* **hydrocarbon**.

aspect ratio   Ratio of length to width for a flat form, or of length to diameter for a round form such as a fiber.

assembly   Group of materials or parts, including adhesive, that has been placed together for bonding or that has been bonded together.

atactic   State when the radical groups are arranged heterogeneously around the carbon chain. (*See also* **isotactic**.)

atomic oxygen resistance   Ability of a material to withstand atomic oxygen exposure. This is related to the use of plastic and elastomers in space applications.

autoclave   Closed vessel for conducting a chemical reaction or other operation under pressure and heat.

autoclave molding   After lay-up, the entire assembly is placed in a steam autoclave at 50 to 100 lb/in^2. Additional pressure achieves higher reinforcement loadings and improved removal of air.

average molecular weight   Molecular weight of most typical chain in a given plastic. There will always be a distribution of chain sizes and, hence, molecular weights in any polymer.

B-allowable   Statistically derived estimate of a mechanical property based on numerous tests; value above which at least 90 percent of the population of values is expected to fall with a confidence of 95 percent.

B-stage   Intermediate stage in the curing of a thermosetting resin. In this stage, resins can be heated and caused to flow, thereby allowing final curing in the desired shape. The term *A-stage* is used to describe an earlier stage in the curing reaction, and the term *C-stage* is sometimes used to describe the cured resin. Most molding materials are in the B stage when supplied for compression or transfer molding.

bag molding   Method of applying pressure during bonding or molding in which a flexible cover, usually in connection with a rigid die or mold, exerts pressure on the material being molded, through the application of air pressure or drawing of a vacuum.

bagging   Applying an impermeable layer of film over an uncured part and sealing the edges so that a vacuum can be drawn.

Barcol hardness   Hardness value obtained using a Barcol hardness tester, which gages hardness of soft materials by indentation of a sharp steel point under a spring load.

basket weave   Weave where two or more warp threads cross alternately with two or more filling threads. The basket weave is less stable than the plain weave but produces a flatter and stronger fabric. It is also a more pliable fabric than the plain weave, and a certain degree of porosity is maintained without too much sleaziness, but not as much as with the plain weave.

benzene ring   Basic structure of benzene, which is a hexagonal, six-carbon-atom structure with three double bonds; also basic aromatic structure in organic chemistry. Aromatic structures usually yield more thermally stable plastics than do aliphatic structures. (*See also* **hydrocarbon.**)

binder   Resin or plastic constituent of a composite material, especially a fabric-reinforced composite.

blister   Raised area on the incompletely hardened surface of a molding caused by the pressure of gases inside it.

bismaleimide   Type of polyimide that cures by addition rather than a condensation reaction; generally of higher temperature resistance than epoxy.

blind fastener   Fastener designed for holding two rigid materials, with access limited to one side.

blow molding   Method of fabrication of thermoplastic materials in which a parison (hollow tube) is forced into the shape of the mold cavity by internal air pressure.

blowing agent   Chemical that can be added to plastics and generates inert gases upon heating. This blowing, or expansion, causes the plastic to expand, thus forming a foam; also known as foaming agent.

bond   Union of materials by adhesives; to unite materials by means of an adhesive.

bond strength   Unit load, applied in tension, compression, flexure, peel, impact, cleavage, or shear, required to break an adhesive assembly, with failure occurring in or near the plane of the bond.

boron fibers   High-modulus fibers produced by vapor deposition of elemental boron onto tungsten or carbon cores. Supplied as single strands or tapes.

boss   Projection on a plastic part designed to add strength, to facilitate alignment during assembly, to provide for fastenings, and so on.

bridging   Suspension of tensioned fiber between high points on a surface, resulting in uncompacted laminate.

bulk density   Density of a molding material in loose form (granular, nodular, and the like), expressed as a ratio of weight to volume (for instance, g/cm^3 or lb/ft^3).

bulk rope molding compound   Molding compound made with thickened polyester resin and fibers less than ½ in. Supplied as rope, it molds with excellent flow and surface appearance.

capacitance (capacity)   That property of a system of conductors and dielectrics that permits the storage of electricity when potential difference exists between the conductors. Its value is expressed as the ratio of the quantity of electricity to a potential difference. A capacitance value is always positive.

carbon, fibers   Fiber produced by the pyrolysis of organic precursor fibers such as rayon, polyacrylonitrile (PAN), or pitch, in an inert atmosphere.

cast   To embed a component or assembly in a liquid resin, using molds that separate from the part for reuse after the resin is cured. Curing or polymerization takes place without external pressure. (*See* **embed, pot**.)

catalyst   Chemical that causes or speeds up the cure of a resin, but does not become a chemical part of the final product. Catalysts are normally added in small quantities. The peroxides used with polyesters are typical catalysts.

catalytic curing   Curing by an agent that changes the rate of the chemical reaction without entering into the reaction.

caul plate   Rigid plate contained within vacuum bag to impart a surface texture or configuration to the laminate during cure.

cavity   Depression in a mold that usually forms the outer surface of the molded part; depending on the number of such depressions, molds are designated as single-cavity or multicavity.

centrifugal casting   Fabrication process in which the catalyzed resin is introduced into a rapidly rotating mold where it forms a layer on the mold surfaces and hardens.

charge   Amount of material used to load a mold for one cycle.

chlorinated hydrocarbon   Organic compound having hydrogen atoms and, more important, chlorine atoms in its chemical structure. Trichloroethylene, methyl chloroform, and methylene chloride are chlorinated hydrocarbons.

circuit board   Sheet of copper-clad laminate material on which copper has been etched to form a circuit pattern. The board may have copper on one (single-sided) or both (double-sided) surfaces. Also called printed-circuit board or printed-wiring board.

cleavage   Imposition of transverse or "opening" forces at the edge of adhesive bond.

coat   To cover with a finishing, protecting, or enclosing layer of any compound (such as varnish).

coefficient of thermal expansion (CTE)   Change in unit length or volume resulting from a unit change in temperature. Commonly used units are 10^{-6} cm/cm/°C.

cohesion   State in which the particles of a single substance are held together by primary or secondary valence forces. As used in the adhesive field, the state in which the particles of the adhesive (or the adherend) are held together.

cold flow (creep)   Continuing dimensional change that follows initial instantaneous deformation in a nonrigid material under static load.

cold-press molding   Molding process where inexpensive plastic male and female molds are used with room-temperature-curing resins to produce accurate parts. Limited runs are possible.

cold pressing   Bonding operation in which an assembly is subjected to pressure without the application of heat.

compaction   In reinforced plastics and composites, application of a temporary press bump cycle, vacuum, or tensioned layer to remove trapped air and compact the lay-up.

composite   Homogeneous material created by the synthetic assembly of two or more materials (a selected filler or reinforcing elements and compatible matrix binder) to obtain specific characteristics and properties.

compound   Some combination of elements in a stable molecular arrangement.

compression molding   Technique of thermoset molding in which the molding compound (generally preheated) is placed in the heated open mold cavity and the mold is closed under pressure (usually in a hydraulic press), causing the material to flow and completely fill the cavity, with pressure being held until the material has cured.

compressive strength   Maximum compressive stress a material is capable of sustaining. For materials that do not fail by a shattering fracture, the value is arbitrary, depending on the distortion allowed.

condensation polymerization   Chemical reaction in which two or more molecules combine with the separation of water or other simple substance.

condensation resins   Any of the alkyd, phenol-aldehyde, and urea-formaldehyde resins.

conductivity   Reciprocal of volume resistivity.

conformal coating   Insulating coating applied to printed-circuit-board wiring assemblies that covers all of the components and provides protection against moisture, dust, and dirt.

copolymer   *See* **polymer.**

corona resistance   Resistance of insulating materials, especially plastics, to failure under the high-voltage state known as partial discharge. Failure can be erosion of the plastic material, decomposition of the polymer, or thermal degradation, or a combination of these three failure mechanisms.

coupling agent   Chemical or material that promotes improved adhesion between fiber and matrix resin in a reinforced composite, such as an epoxy-glass laminate or other resin-fiber laminate.

crazing   Fine cracks that may extend in a network on or under the surface or through a layer of a plastic material.

creep   Dimensional change with time of a material under load following the initial instantaneous elastic deformation; time-dependent part of strain resulting from force. Creep at room temperature is sometimes called cold flow. See ASTM D 674, Recommended Practices for Testing Long-Time Creep and Stress-Relaxation of Plastics under Tension or Compression Loads at Various Temperatures.

cross linking   Process where chemical links set up between molecular chains of a plastic. In thermosets, cross linking makes one infusible supermolecule of all the chains, contributing to strength, rigidity, and high-temperature resistance. Thermoplastics (like polyethylene) can also be cross-linked (by irradiation or chemically through formulation) to produce three-dimensional structures that are thermoset in nature and offer improved tensile strength and stress-crack resistance.

crowfoot satin   Type of weave having a 3-by-1 interlacing; that is, a filling thread floats over the three warp threads and then under one. This type of fabric looks different on one side than on the other. Such fabrics are more pliable than either the plain or the basket weave and, consequently, are easier to form around curves. (*See* **four-harness satin**.)

crystalline melting point   Temperature at which the crystalline structure in a material is broken down.

crystallinity   State of molecular structure referring to uniformity and compactness of the molecular chains forming the polymer and resulting from the formation of solid crystals with a definite geometric pattern. In some resins, such as polyethylene, the degree of crystallinity indicates the degree of stiffness, hardness, environmental stress-crack resistance, and heat resistance.

cull   Material remaining in a transfer chamber after the mold has been filled. Unless there is a slight excess in the charge, the operator cannot be sure the cavity is filled. The charge is generally regulated to control the thickness of the cull.

cure   To change the physical properties of a material (usually from a liquid to a solid) by chemical reaction, by the action of heat and catalysts, alone or in combination, with or without pressure.

curing agent   *See* **hardener**.

curing temperature   Temperature at which a material is subjected to curing.

curing time   In the molding of thermosetting plastics, the time it takes for the material to be properly cured.

cycle   One complete operation of a molding press from closing time to closing time.

damping   In a material, the ability to absorb energy to reduce vibration.

decorative laminates   High-pressure laminates consisting of a phenolic-paper core and a melamine-paper top sheet with a decorative pattern.

deflashing   Any finishing technique used to remove the flash (excess unwanted material) from a plastic part; examples are filing, sanding, milling, tumbling, and wheelabrating. (*See* **flash**.)

deflection temperature   Formerly called heat-distortion temperature (HDT).

degas   To remove gases, usually air, from liquid resin mixture, usually achieved by placing mixture in a vacuum. Entrapped gases or voids in a cured plastic can lead to premature failure, either electrically or mechanically.

delamination   Separation of the layers of material in a laminate, either locally or in a large area. Can occur during cure or later during operational life.

denier   Numbering system for fibers or filaments equal to the weight in grams of a 9000-meter-long fiber or filament. The lower the denier, the finer the yarn.

design allowables   Tested, statistically defined material properties used for design. Usually refers to stress or strain.

dessicant   Substance that will remove moisture from materials, usually due to absorption of the moisture onto the surface of the substance; also known as a drying agent.

diallyl phythalate   Ester polymer resulting from reaction of allyl alcohol and phthalic anhydride.

dielectric constant (permittivity, specific inductive capacity)   That property of a dielectric that determines the electrostatic energy stored per unit volume for unit potential gradient.

dielectric loss   Time rate at which electric energy is transformed into heat in a dielectric when it is subjected to a changing electric field.

dielectric-loss angle (dielectric-phase difference)   Difference between 90° and the dielectric-phase angle.

dielectric-loss factor (dielectric-loss index)   Product of the dielectric constant and the tangent of the dielectric-loss angle for a material.

dielectric-phase angle   Angular difference in phase between the sinusoidal alternating potential difference applied to a dielectric and the component of the resulting alternating current having the same period as the potential difference.

dielectric-power factor   Cosine of the dielectric-phase angle (or sine of the dielectric-loss angle).

dielectric sensors   Sensors that use electrical techniques to measure the change in loss factor (dissipation) and in capacitance during cure of the resin in a laminate. This is an accurate measure of the degree of resin cure or polymerization.

dielectric strength   Voltage that an insulating material can withstand before breakdown occurs, usually expressed as a voltage gradient (such as volts per mil).

differential scanning   Calorimetry measurement of the energy absorbed (endotherm) or produced (exotherm) as a resin system is cured.

diluent   Ingredient usually added to a formulation to reduce the concentration of the resin.

diphenyl oxide resins   Thermoplastic resins based on diphenyl oxide and possessing excellent handling properties and heat resistance.

dissipation factor (loss tangent, loss angle, tan δ, approximate power factor)   Tangent of the loss angle of the insulating material.

doctor blade   Straight piece of material used to spread and control the amount of resin applied to roving, tow, tape, or fabric.

domes   In a cylindrical container, that portion that forms the integral ends of the container.

dry   To change the physical state of an adhesive on an adherend through the loss of solvent constituents by evaporation or absorption, or both.

drying agent   *See* **dessicant**.

ductility   Ability of a material to deform plastically before fracturing.

E-glass   Family of glasses with low alkali content, usually under 2.0 percent, most suitable for use in electrical-grade laminates and glasses. Electrical properties remain more stable with these glasses due to the low alkali content. Also called electrical-grade glasses.

eight-harness satin   Type of weave having a 7-by-1 interlacing; that is, a filling thread floats over seven warp threads and then under one. Like the crowfoot weave, it looks different on one side than on the other. This weave is more pliable than any of the others and is especially adaptable where it is necessary to form around compound curves, such as on radomes.

elastic limit   Greatest stress a material is capable of sustaining without any permanent strain remaining when the stress is released.

elasticity   Property of a material by virtue of which it tends to recover its original size and shape after deformation. If the strain is proportional to the applied stress, the material is said to exhibit Hookean or ideal elasticity.

elastomer   Material that at room temperature can be stretched repeatedly to at least twice its original length and, upon release of the stress, will return with force to its approximate original length. Plastics with such or similar characteristics are known as elastomeric plastics. The expression is also used when referring to a rubber (natural or synthetic) reinforced plastic, as in elastomer-modified resins.

electric strength (dielectric strength, disruptive gradient)   Maximum potential gradient that a material can withstand without rupture. The value obtained for the electric strength will depend on the thickness of the material and on the method and conditions of test.

electrode   Conductor of the metallic class through which a current enters or leaves an electrolytic cell, at which there is a change from conduction by electrons to conduction by charged particles of matter, or vice versa.

elongation   Increase in gage length of a tension specimen, usually expressed as a percentage of the original gage length. (*See also* **gage length**.)

embed   To encase completely a component or assembly in some material—a plastic, for instance. (*See* **cast, pot**.)

encapsulate   To coat a component or assembly in a conformal or thixotropic coating by dipping, brushing, or spraying.

engineering plastics   Plastics, the properties of which are suitable for engineered products. These plastics are usually suitable for application up to 125°C, well above the thermal stability of many commercial plastics. The next higher grade of plastics, called high-performance plastics, is usually suitable for product designs requiring stability of plastics above 175°C.

environmental-stress cracking   Susceptibility of a thermoplastic article to crack or craze under the influence of certain chemicals, aging, weather, or other stress. Standard ASTM test methods that include requirements for environmental stress cracking are indexed in Index of ASTM Standards.

epoxy   Thermosetting polymers containing the oxirane group; mostly made by reacting epichlorohydrin with a polyol such as bisphenol A. Resins may be either liquid or solid.

eutectic   Mixture, the melting point of which is lower than that of any other mixture of the same ingredients.

exotherm   Characteristic curve of resin during its cure that shows heat of reaction (temperature) versus time. Peak exotherm is the maximum temperature on this curve.

exothermic   Chemical reaction in which heat is given off.

extender   Inert ingredient added to a resin formulation chiefly to increase its volume.

extrusion   Compacting of a plastic material and forcing of it through an orifice.

fabric   Planar structure produced by interlacing yarns, fibers, or filaments.

failure, adhesive   Rupture of an adhesive bond such that the separation appears to be at the adhesive-adherend interface.

fiber washout   Movement of fiber during cure because of large hydrostatic forces generated in low-viscosity resin systems.

fiberglass   Individual filament made by attenuating molten glass. A continuous filament is a glass fiber of great or indefinite length; a staple fiber is a glass fiber of relatively short length (generally less than 17 in).

filament winding   Process for fabricating a composite structure in which continuous reinforcements (filament, wire, yarn, tape, or other), either previously impregnated with a matrix material or impregnated during the winding, are placed over a rotating and removable form or mandrel in a previously prescribed way to meet certain stress conditions. Generally the shape is a surface of revolution and may or may not include end closures. When the right number of layers is applied, the wound form is cured and the mandrel removed.

fill   *See* **weft**.

filler   Material, usually inert, that is added to plastics to reduce cost or modify physical properties.

film adhesive   Thin layer of dried adhesive; also class of adhesives provided in dry-film form with or without reinforcing fabric, that are cured by heat and pressure.

finish, fiber   Mixture of materials for treating glass or other fibers to reduce damage during processing or to promote adhesion to matrix resins.

fish paper   Electrical-insulation grade of vulcanized fiber in thin cross section.

flame retardants   Materials added to plastics to improve their resistance to fire.

flash   Extra plastic attached to a molding along the parting line. Under most conditions it would be objectionable and must be removed before the parts are acceptable.

flexibilizer   Material that is added to rigid plastics to make them resilient or flexible. It can be either inert or a reactive part of the chemical reaction. Also called a plasticizer in some cases.

flexural modulus   Ratio, within the elastic limit, of stress to the corresponding strain. It is calculated by drawing a tangent to the steepest initial straight-line portion of the load-deformation curve and calculating by the following equation:

$$E_B = \frac{L^3 m}{4bd^3}$$

where E_B = modulus
  b = width of beam tested
  d = depth of beam
  m = slope of tangent
  L = span, inches

flexural strength Strength of a material in bending expressed as the tensile stress of the outermost fibers of a bent test sample at the instant of failure.

fluorocarbon Organic compound having fluorine atoms in its chemical structure. This structure usually lends chemical and thermal stability to plastics.

four-harness satin Fabric, also named crowfoot satin because the weaving pattern when laid out on cloth design paper resembles the imprint of a crow's foot. In this type of weave there is a 3-by-1 interlacing; that is, a filling thread floats over the three warp threads and then under one. This type of fabric looks different on one side than on the other. Fabrics with this weave are more pliable than either the plain or basket weave and, consequently, are easier to form around curves. (*See* **crowfoot satin**.)

fracture toughness Measure of the damage tolerance of a matrix containing initial flaws or cracks. G_{1c} and G_{2c} are the critical strain energy release rates in the 1 and 2 directions.

gage length Original of that portion of the specimen over which strain is measured.

gate Orifice through which liquid resin enters mold in plastic molding processes.

gel Soft rubbery mass that is formed as a thermosetting resin goes from a fluid to an infusible solid. This is an intermediate state in a curing reaction, and a stage in which the resin is mechanically very weak.

gel coat Resin applied to the surface of a mold and gelled prior to lay-up. The gel coat becomes an integral part of the finished laminate and is usually used to improve surface appearance and so on.

gel point The point in a curing reaction at which gelatin begins. (*See* **gel**.)

gelation Point in resin cure when the viscosity has increased to a point where the resin barely moves when probed with a sharp point.

glass-transition point Temperature at which a material loses its glasslike properties and becomes a semiliquid. (*See also* **glass-transition temperature**.)

glass-transition temperature Temperature at which a plastic changes from a rigid state to a softened state. Both mechanical and electrical properties degrade significantly at this point, which is usually a narrow temperature range, rather than a sharp point, as in freezing or boiling.

glue line (bond line) Layer of adhesive that attaches two adherends.

glue-line thickness Thickness of the fully dried adhesive layer.

glycol Alcohol containing two hydroxyl (—OH) groups.

graphite fibers   High-strength, high-modulus fibers made by controlled carbonization and graphitization of organic fibers, usually rayon, acrylonitrile, or pitch.

hand lay-up   Process of placing in position (and working) successive plies of reinforcing material or resin-impregnated reinforcement on a mold by hand.

hardener   Chemical added to a thermosetting resin for the purpose of causing curing or hardening. Amines and acid anhydrides are hardeners for epoxy resins. Such hardeners are a part of the chemical reaction and a part of the chemical composition of the cured resin. The terms hardener and curing agent are used interchangeably. Note that these can differ from catalysts, promoters, and accelerators. (*See* **catalyst, promoter, accelerator.**)

hardness   *See* **indentation hardness.**

heat-deflection temperature   *See* **heat-distortion point.**

heat-distortion point   Temperature at which a standard test bar deflects 0.010 in under a stated load of either 66 or 264 lb/in^2. See ASTM D 648, Standard Method of Test for Deflection Temperature of Plastics under Load.

heat sealing   Method of joining plastic films by simultaneous application of heat and pressure to areas in contact. Heat may be supplied conductively or dielectrically.

helical pattern   Pattern generated when a filament band advances along a helical path, not necessarily at a constant angle except in the case of a cylinder, in a filament-wound object.

high-frequency preheating   Plastic to be heated forms the dielectric of a condenser to which a high-frequency (20 to 80 MHz) voltage is applied. Dielectric loss in the material is the basis. The process is used for sealing vinyl films and preheating thermoset molding compounds.

high-performance plastics   In general, plastics that are suitable for use above 175°C. (*See also* **engineering plastics.**)

homopolymer   Polymer resulting from polymerization of a single monomer. (*See also* **monomer.**)

honeycomb   Manufactured product of resin-impregnated sheet material or metal foil, formed into hexagonal cells. Skins are bonded to top and bottom surfaces to achieve strength.

hot-melt adhesive   Thermoplastic adhesive compound, usually solid at room temperature, that is heated to a fluid state for application.

hydrocarbon   Organic compound having hydrogen and carbon atoms in its chemical structure. Most organic compounds are hydrocarbons. Aliphatic hydrocarbons are straight-chained hydrocarbons, and aromatic hydrocarbons are ringed structures based on the benzene ring. Methyl alcohol, trichloroethylene, and the like are aliphatic; benzene, xylene, toluene, and the like are aromatic.

hydrolysis   Chemical decomposition of a substance involving the addition of water.

hydrophilic   Materials having a tendency to absorb water or to be wetted by water.

hydrophobic   Materials having a tendency to repel water; usually materials exhibiting a low surface energy, measured by wetting angle.

hydroxyl group   Chemical group consisting of one hydrogen atom plus one oxygen atom.

hygroscopic   Tending to absorb moisture.

immiscible   Two fluids that will not mix to form a homogeneous mixture, or that are mutually insoluble.

impact strength   Strength of a material when subjected to impact forces or loads.

impregnate   To force resin into every interstice of a part. Cloths are impregnated for laminating, and tightly wound coils are impregnated in liquid resin using air pressure or vacuum as the impregnating force.

in-situ joint   Joint between a composite and another surface that is formed during cure of the composite.

indentation hardness   Hardness evaluated from measurements of area or indentation depth caused by pressing a specified indentation into the material surface with a specified force.

inhibitor   Chemical added to resins to slow down the curing reaction. Inhibitors are normally added to prolong the storage life of thermosetting resins.

injection molding   Molding procedure whereby a heat-softened plastic material is forced from a cylinder into a cavity that gives the article the desired shape. Used with all thermoplastic and some thermosetting materials.

inorganic chemicals   Chemicals, the structure of which is based on atoms other than the carbon atom.

insulation, electrical   Protection against electrical failure in an electrical product.

insulation resistance   Ratio of applied voltage to total current between two electrodes in contact with a specified insulator.

insulator   Material that provides electrical insulation in an electrical product.

interpenetrating network (IPN)   Two or more polymers that have been formed together so that they penetrate each other in the final polymer form.

isotactic   Molecules that are polymerized in parallel arrangements of radicals on one side of the carbon chain. (*See also* **atactic**.)

Izod impact strength   Measure of the toughness of a material under impact as measured by the Izod impact test.

Izod impact test   One of the most common ASTM tests for testing the impact strength of plastic materials.

joint   Location at which two adherends are held together with a layer of adhesive.

joint, lap   Joint made by placing one adherend partly over another and bonding together the overlapped portions.

joint, scarf   Joint made by cutting away similar angular segments of two adherends and bonding the adherends with the cut areas fitted together.

joint, starved Joint that has an insufficient amount of adhesive to produce a satisfactory bond.

Kevlar Trademark for a group of DuPont aromatic polyimides that are frequently used as fibers in reinforced plastics and composites. Major characteristics are low thermal expansion, light weight, and good electrical properties, coupled with stiffness in laminated form. One special application area is in high-performance circuit boards requiring low x-y axis thermal expansion.

laminae Set of single plies or layers of a laminate (plural of *lamina*).

laminate To unite sheets of material by a bonding material, usually with pressure and heat (normally used with reference to flat sheets); product made by so bonding.

latent curing agent Curing agent that produces long-time stability at room temperature but rapid cure at elevated temperature.

lay-up As used in reinforced plastics, the reinforcing material placed in position in the mold; resin-impregnated reinforcement; process of placing the reinforcing material in position in the mold.

leno weave Locking-type weave in which two or more warp threads cross over each other and interlace with one or more filling threads. It is used primarily to prevent shifting of fibers in open fabrics.

limited-coordination specification (or standard) Specification (or standard) that has not been fully coordinated and accepted by all the interested activities. Limited-coordination specifications and standards are issued to cover the need for requirements unique to one particular department. This applies primarily to military agency documents.

liquid-crystal polymer (LCP) Polymers that spontaneously order themselves in the melt, allowing relatively easy processing at relatively high temperatures. They are characterized as rigid rods. Kevlar and Nomex are examples, as is Xydar thermoplastic.

liquid injection molding Fabrication process in which catalyzed resin is metered into closed molds.

loss angle *See* **dissipation factor**.

loss tangent *See* **dissipation factor**.

macerate To chop or shred fabric for use as a filler for a molding resin; molding compound obtained when so filled.

mandrel Form around which resin-impregnated fiber is wound to make pipes, tubes, or vessels by the filament-winding process.

mat Reinforcing material composed of randomly oriented short, chopped fibers. Manufactured in sheet or blanket form, and commonly used as alternative to woven fabric (especially glass) in fabrication of laminated plastic forms.

matched metal molding Method of molding reinforced plastics between two close-fitting metal molds mounted in a hydraulic press.

matrix Essentially homogeneous material in which the fiber system of a composite resides.

matrix manipulations Mathematical method of relating stresses and strains.

mechanical properties   Material properties associated with elastic and inelastic reactions to an applied force.

melamines   Thermosetting resins made from melamine and formaldehyde and possessing excellent hardness, clarity, and electrical properties.

melt   Molten plastic, in the melted phase of material during a molding cycle.

microcracks   Cracks formed in composites when thermal stresses locally exceed strength of matrix. These cracks generally do not penetrate or cross fibers.

micrometer (micron)   Unit of length equal to 10,000 Å, 0.0001 cm, or approximately 0.000039 in.

mock leno weave   Open-type weave that resembles a leno and is accomplished by a system of interlacings that draws a group of threads together and leaves a space between the next group. The warp threads do not actually cross each other as in a real leno and, therefore, no special attachments are required for the loom. This type of weave is generally used when a high thread count is required for strength and, at the same time, the fabric must remain porous.

modifier   Chemically inert ingredient added to a resin formulation that changes its properties.

modulus of elasticity   Ratio of unidirectional stress to the corresponding strain (slope of the line) in the linear stress-strain region below the proportional limit. For materials with no linear range, a secant line from the origin to a specified point on the stress-strain curve or a line tangent to the curve at a specified point may be used.

moisture absorption   Amount of water pickup by a material when that material is exposed to water vapor. Expressed as percent of original weight of dry material.

moisture resistance   Ability of a material to resist absorbing moisture, either from the air or when immersed in water.

moisture vapor transmission   Rate at which moisture vapor passes through a material at specified temperature and humidity levels. Expressed as grams per mil of material thickness per 24 h per 100 in^2.

mold   Medium or tool designed to form desired shapes and sizes; to process a plastics material using a mold.

mold release   Lubricant used to coat a mold cavity to prevent the molded piece from sticking to it, and thus to facilitate its removal from the mold. Also called release agent.

mold shrinkage   Difference in dimensions, expressed in inches per inch, between a molding and the mold cavity in which it was molded, both the mold and the molding being at room temperature when measured.

molecular weight   Sum of the atomic masses of the elements forming the molecule.

monofilament   Single fiber or filament.

monomer   Small molecule that is capable of reacting with similar or other molecules to form large chainlike molecules called polymers.

multilayer printed circuits   Electric circuits made on thin copper-clad laminates, stacked together with intermediate prepreg sheets and bonded together with heat and pressure. Subsequent drilling and electroplating through the layers result in a three-dimensional circuit.

necking   Localized reduction of the cross-sectional area of a tensile specimen that may occur during loading.

NEMA standards   Property values adopted as standard by the National Electrical Manufacturers Association.

notch sensitivity   Extent to which the sensitivity of a material to fracture is increased by the presence of a disrupted surface such as a notch, a sudden change in section, a crack, or a scratch. Low notch sensitivity is usually associated with ductile materials and high notch sensitivity with brittle materials.

nuclear radiation resistance   Ability of a material to withstand nuclear radiation and still perform its designated function.

nylon   Generic name for all synthetic polyamides. These are thermoplastic polymers with a wide range of properties.

olefin   Family of unsaturated hydrocarbons with the formula C_nH_n, named after the corresponding paraffins by adding "ene" or "ylene" to the stem; for instance, ethylene. Paraffins are aliphatic hydrocarbons. (*See* **hydrocarbon**.)

oligomer   Polymer containing only a few monomer units, such as a dimer or trimer.

orange peel   Undesirably uneven or rough surface on a molded part, resembling the surface on an orange.

organic   Composed of matter originating in plant or animal life, or composed of chemicals of hydrocarbon origin, either natural or synthetic. Used in referring to chemical structures based on the carbon atom.

orthotropic   Having three mutually perpendicular planes of elastic symmetry.

parting agent   *See* **mold release**.

paste   Adhesive composition having a characteristic plastic-type consistency; that is, a high order or yield value such as that of a paste prepared by heating a mixture of starch and water and subsequently cooling the hydrolyzed product.

peel   Imposition of a tensile stress in a direction perpendicular to the adhesive bond line, to a flexible adherend.

peel strength   Strength of an adhesive in peel; expressed in pounds per inch of width.

penetration   Entering of one part or material into another.

permanence   Resistance of a given property to deteriorating influences.

permeability   Ability of a material to allow liquid or gaseous molecules to pass through a film.

permittivity   *See* **dielectric constant**.

phenolic   Synthetic resin produced by the condensation of an aromatic alcohol with an aldehyde, particularly of phenol with formaldehyde.

phenylsilane   Thermosetting copolymer of silicone and phenolic resins; furnished in solution form.

pitch fibers   Fibers made from high-molecular-weight residue from the destructive distillation of coal or petroleum products.

plain weave   The most simple and commonly used weave, in which the warp and filling threads cross alternately. Plain woven fabrics are generally the least pliable, but they are also the most stable. This stability permits the fabrics to be woven with a fair degree of porosity without too much sleaziness.

plastic   Material containing an organic substance of large molecular weight that is solid in its final condition and that, at some earlier time, was shaped by flow. (*See* **resin, polymer**.)

plastic deformation   Change in dimensions of an object under load that is not recovered when the load is removed; opposed to elastic deformation.

plasticity   Property of plastics that allows the material to be deformed continuously and permanently without rupture upon the application of a force that exceeds the yield value of the material.

plasticize   To soften a material and make it plastic or moldable by means of either a plasticizer or the application of heat.

plasticizer   Material incorporated in a resin formulation to increase its flexibility, workability, or distensibility. The addition of a plasticizer may cause a reduction in melt viscosity, lower the temperature of second-order transition, or lower the elastic modulus of the solidified resin.

plastisols   Mixtures of vinyl resins and plasticizers that can be molded, cast, or converted to continuous films by the application of heat. If the mixtures contain volatile thinners, they are also known as organosols.

Poisson's ratio   Absolute value of the ratio of transverse strain to axial strain resulting from a uniformly applied axial stress below the proportional limit of the material.

polyacrylonitrile (PAN)   Synthetic fiber used as base material or precursor in manufacture of certain carbon fibers.

polyesters   Thermosetting resins produced by reacting unsaturated, generally linear, alkyd resins with a vinyl-type active monomer such as styrene, methyl styrene, or diallyl phthalate. Cure is effected through vinyl polymerization using peroxide catalysts and promoters or heat to accelerate the reaction. The resins are usually furnished in liquid form.

polyimide   High-temperature resins made by reacting aromatic dianhydrides with aromatic diamines.

polymer   Compound formed by the reaction of simple molecules having functional groups that permit their combination to proceed to high molecular weights under suitable conditions. Polymers may be formed by polymerization (addition polymer) or polycondensation (condensation polymer). When two or more monomers are involved, the product is called a copolymer. Also, any high-molecular-weight organic compound, the structure of which consists of a repeating small unit. Polymers can be plastics, elastomers, liquids, or gums and are formed by chemical addition or condensation of monomers. (*See also* **addition reaction or polymerization, condensation polymerization**.)

polymerize   To unite chemically two or more monomers or polymers of the same kind to form a molecule with higher molecular weight.

pot To embed a component or assembly in a liquid resin, using a shell, can, or case that remains as an integral part of the product after the resin is cured. (*See* **embed, cast.**)

pot life Time during which a liquid resin remains workable as a liquid after catalysts, curing agents, promoters, and the like, are added; roughly equivalent to gel time; sometimes also called working life.

power factor Cosine of the angle between the voltage applied and the current resulting.

precursor PAN or pitch fibers from which carbon graphite fiber is made.

preform Pill, tablet, or biscuit used in thermoset molding. Material is measured by volume, and the bulk factor of powder is reduced by pressure to achieve efficiency and accuracy.

preheating Heating of a compound prior to molding or casting in order to facilitate operation, reduce cycle, and improve product.

premix Molding compound prepared prior to and apart from the molding operations and containing all components required for molding: resin, reinforcement fillers, catalysts, release agents, and other compounds.

prepolymer Polymer in some stage between that of the monomers and the final polymer. The molecular weight is, therefore, also intermediate. As used in polyurethane production, reaction product of a polyol with excess of an isocyanate.

prepreg Ready-to-mold sheet that may be cloth, mat, or paper-impregnated with resin and stored for use. The resin is partially cured to a B-stage and supplied to the fabricator, who lays up the finished shape and completes the cure with heat and pressure. (*See also* **B-stage.**)

pressure-bag molding Process for molding reinforced plastics in which a tailored flexible bag is placed over the contact lay-up on the mold, sealed, and clamped in place. Fluid pressure, usually compressed air, is placed against the bag, and the part is cured.

primer Coating applied to a surface prior to the application of an adhesive to improve the performance of the bond.

printed-circuit board *See* **circuit board.**

printed-circuit laminates Laminates, either fabric- or paper-based, covered with a thin layer of copper foil and used in the photofabrication process to make circuit boards.

printed-wiring board *See* **circuit board.**

promoter Chemical, itself a weak catalyst, that greatly increases the activity of a given catalyst.

proportional limit Greatest stress a material can sustain without deviating from the linear proportionality of stress to strain (Hooke's law).

pultrusion Reversed "extrusion" of resin-impregnated roving in the manufacture of rods, tubes, and structural shapes of a permanent cross section. The roving, after passing through the resin dip tank, is drawn through a die to form the desired cross section.

qualified products list (QPL)   List of commercial products that have been pretested and found to meet the requirements of a specification, especially government specifications.

reactive diluent   As used in epoxy formulations, a compound containing one or more epoxy groups that functions mainly to reduce the viscosity of the mixture.

refractive index   Ratio of the velocity of light in a vacuum to its velocity in a substance; also ratio of the sine of the angle of incidence to the sine of the angle of refraction.

regrind   Excess or waste material in a thermoplastic molding process that can be reground and mixed with virgin raw material, within limits, for molding future parts.

reinforced molding compound   Plastic to which fibrous materials such as glass or cotton have been added to improve certain physical properties such as flexural strength.

reinforced plastic   Plastic with strength properties greatly superior to those of the base resin, resulting from the presence of reinforcements in the composition.

reinforced thermoplastics   Reinforced molding compounds in which the plastic is thermoplastic.

relative humidity   Ratio of the quantity of water vapor present in the air to the quantity which would saturate it at any given temperature.

release agent   *See* **mold release**.

release paper   Impermeable paper film or sheet that is coated with a material to prevent adhering to prepreg.

resin   High-molecular-weight organic material with no sharp melting point. For general purposes, the terms *resin, polymer,* and *plastic* can be used interchangeably. (*See* **polymer**.)

resistivity   Ability of a material to resist passage of electric current, either through its bulk or on a surface. The unit of volume resistivity is the ohm-centimeter, and the unit of surface resistivity is ohms per square.

rheology   Study of the change in form and flow of matter, embracing elasticity, viscosity, and plasticity.

rigidsol   Plastisol having a high elastic modulus, usually produced with a crosslinking plasticizer.

Rockwell hardness number   Number derived from the net increase in depth of impression as the load on a penetrator is increased from a fixed minimum load to a higher load and then returned to minimum load. Penetrators include steel balls of several specified diameters and a diamond-cone penetrator.

rotational casting (or molding)   Method used to make hollow articles from thermoplastic materials. Material is charged into a hollow mold capable of being rotated in one or two planes. The hot mold fuses the material into a gel after the rotation has caused it to cover all surfaces. The mold is then chilled and the product stripped out.

roving   Collection of bundles of continuous filaments, either as untwisted strands or as twisted yarns. Rovings may be lightly twisted, but for filament winding they are generally wound as bands or tapes with as little twist as possible.

rubber   Elastomer capable of rapid elastic recovery; usually natural rubber, Hevea. (*See* **elastomer**.)

runners   All channels in the mold through which molten or liquid plastic raw materials flow into mold.

S glass   Glass fabric made with very high tensile strength fibers for high-performance-strength requirements.

sandwich construction   Panel consisting of some lightweight core material bonded to strong, stiff skins on both faces. (*See also* **honeycomb**.)

separator ply   *See* **shear ply**.

set, mechanical   Strain remaining after complete release of the load producing the deformation.

set, polymerization   To convert an adhesive into a fixed or hardened state by chemical or physical action, such as condensation, polymerization, oxidation, vulcanization, gelation, hydration, or evaporation of volatile constituents.

shape factor   For an elastomeric slab loaded in compression, ratio of loaded area to force-free area.

shear   Action or stress resulting from applied forces that causes two contiguous parts of a body or two bodies to slide relative to each other in a direction parallel to their plane of contact.

shear ply   Low-modulus layer, rubber, or adhesive interposed between metal and composite to reduce differential shear stresses.

shear strength   Maximum shear stress a material is capable of sustaining. In testing, the shear stress is caused by a shear or torsion load and is based on the original specimen dimensions.

sheet molding compound   Compression-molding material consisting of glass fibers longer than ½ in and thickened polyester resin. Possessing excellent flow, it results in parts with good surfaces.

shelf life   Time a molding compound can be stored without losing any of its original physical or molding properties.

Shore hardness   Procedure for determining the indentation hardness of a material by means of a durometer. Shore designation is given to tests made with a specified durometer instrument.

silicones   Resinous materials derived from organosiloxane polymers, furnished in different molecular weights, including liquids and solid resins.

sink mark   Depression or dimple on the surface of an injection-molded part due to collapsing of the surface following local internal shrinkage after the gate seals. May also be an incipient short shot.

sizing agent   Chemical treatment containing starches, waxes, and the like, that is applied to fibers, making them more resistant to breakage during the weaving process. The sizing agent must be removed after weaving, as its presence would cause delamination and moisture pickup problems if it remained in the final laminate made from the woven fiber.

slipping   Lateral movement of tensioned fiber on a surface to a new unanticipated fiber angle.

slush molding   Method for casting thermoplastics in which the resin in liquid form is poured into a hot mold where a viscous skin forms. The excess slush is drained off, the mold is cooled, and the molding is stripped out.

solvent   Any substance, usually a liquid, that dissolves other substances.

specific heat   Ratio of a material's thermal capacity to that of water at 15°C.

specific modulus   Young's modulus divided by material density.

specific strength   Ultimate tensile strength divided by material density.

spiral flow test   Test method for measuring flow properties of a resin wherein the resin flows along a spiral path in a mold in the molding press. The variation in flow for different resins or different molding conditions can be compared. Flow is expressed in inches of flow in a standard spiral flow mold.

spray-up   Process in which fiber reinforcement is wetted with resin applied from a spray gun. The fiber is fed through a chopper and into a stream of resin that is sprayed onto a form or into a mold.

stabilizers   Chemicals used in plastics formulation to assist in maintaining physical and chemical properties during processing and service life. A specific type of stabilizer, known as an ultraviolet stabilizer, is designed to absorb ultraviolet rays and prevent them from attacking the plastic.

stacking sequence   Sequence of laying plies into mold. Different stacking sequences have a great effect on off-axis mechanical properties.

starved area   Part of a laminate or reinforced plastic structure in which resin has not completely wetted the fabric.

storage life   *See* **shelf life**.

strain   Deformation resulting from a stress, measured by the ratio of the change to the total value of the dimension in which the change occurred; unit change, due to force, in the size or shape of a body referred to its original size or shape. Strain is nondimensional but frequently expressed in inches per inch or centimeters per centimeter.

strength, dry   Strength of an adhesive joint determined immediately after drying under specified conditions or after a period of conditioning in the standard laboratory atmosphere.

strength, wet   Strength of an adhesive joint determined immediately after removal from a liquid in which it has been immersed under specified conditions of time, temperature, and pressure.

stress   Unit force or component of force at a point in a body acting on a plane through the point. Stress is usually expressed in pounds per square inch.

stress relaxation   Time-dependent decrease in stress for a specimen constrained in a constant strain condition.

substrate   Material upon the surface of which an adhesive-containing substance is spread for any purpose, such as bonding or coating; broader term than adherend. Material upon the surface of which a circuit is formed.

surface preparation   Physical and/or chemical preparation of an adherend to render it suitable for adhesive joining.

surface resistivity   Resistance of a material between two opposite sides of a unit square of its surface. Surface resistivity may vary widely with the conditions of measurement.

syntactic foams   Lightweight systems obtained by the incorporation of prefoamed or low-density fillers in the systems.

tack   Property of an adhesive that enables it to form a bond of measurable strength immediately after adhesive and adherend are brought into contact under low pressure.

tan δ   *See* **dissipation factor**.

tensile strength   Maximum tensile stress a material is capable of sustaining. Tensile strength is calculated from the maximum load during a tension test carried to rupture and the original cross-sectional area of the specimen.

thermal conductivity   Ability of a material to conduct heat; physical constant for the quantity of heat that passes through a unit cube of a material in a unit of time when the difference in temperatures of two faces is 1°C.

thermal stress cracking   Crazing and cracking of some thermoplastic resins resulting from overexposure to elevated temperatures.

thermoforming   Process of creating a form from a flat sheet by combinations of heat and pressure, which first soften the sheet and then form the sheet into some three-dimensional shape. This is one of the simplest, most economical plastic forming processes. There are numerous variations of this process.

thermoplastic   Plastics capable of being repeatedly softened or melted by increases in temperature and hardened by decreases in temperature. These changes are physical rather than chemical.

thermoset   Material that will undergo, or has undergone, a chemical reaction by the action of heat, catalysts, ultraviolet light, and the like, leading to a relatively infusible state that will not remelt after setting.

thermosetting   Classification of resin that cures by chemical reaction when heated and, when cured, cannot be remelted by heating.

thinner   Volatile liquid added to an adhesive to modify the consistency or other properties.

thixotropic   Material that is gel-like at rest but fluid when agitated.

time, assembly   Time interval between spreading of the adhesive on the adherend and application of pressure or heat, or both, to the assembly.

time, curing   Period of time during which an assembly is subjected to heat or pressure, or both, to cure the adhesive.

tow   Untwisted bundle of continuous fibers. Commonly used in reference to synthetic fibers, particularly carbon and graphite, but also glass and aramid. A tow designated as 12K has 12,000 filaments.

tracking   Conductive carbon path formed on surface of a plastic during electrical arcing. (*See also* **arc resistance**.)

transfer molding   Method of molding thermosetting materials in which the plastic is first softened by heat and pressure in a transfer chamber and then forced or transferred by high pressure through suitable sprues, runners, and gates into a closed mold for final curing.

transformation   Mathematical (tensor analysis) method of obtaining stress from strain values or vice versa, or to find angular properties of a laminate.

transverse isotropy   Having essentially identical mechanical properties in two directions but not the third.

transverse, properties   Properties perpendicular to the axial (x,1 or 1,1) direction. May be designated as Y or Z, 2 or 3 directions.

twill weave   Basic weave characterized by a diagonal rib or twill line. Each end floats over at least two consecutive picks, permitting a greater number of yarns per unit area than a plain weave while not losing a great deal of fabric stability.

twist   Spiral turns about its axis per unit length for a textile strand; expressed as turns per inch.

ultrasonic bonding   Bonding of plastics by vibratory mechanical pressure at ultrasonic frequencies due to melting of plastics being joined, heat being generated as frictional heat.

ultraviolet   Shorter wavelengths of invisible radiation that are more damaging than visible light to most plastics.

ultraviolet stabilizers   Additives mixed into plastic formulations for the purpose of improving resistance of plastic to ultraviolet radiation.

undercured   State of a molded article that has not been adequately polymerized or hardened in molding process, usually due to inadequate temperature-time-pressure control in molding process.

vacuum bag   Impermeable film applied to outside of lay-up to facilitate conformability to mold form and air removal during cure.

vacuum-bag molding   Process for molding reinforced plastics in which a sheet of flexible transparent material is placed over the lay-up on the mold and sealed. A vacuum is applied between sheet and lay-up. The entrapped air is mechanically worked out of the lay-up and removed by the vacuum, and the part is cured.

vacuum-injection molding   Molding process where, using a male and a female mold, reinforcements are placed in the mold, a vacuum is applied, and a room-temperature-curing liquid resin is introduced which saturates the reinforcement.

vent   Small opening placed in a mold for allowing air to exit mold as molding material enters. This eliminates air holes, voids, or bubbles in finally molded part.

Vicat softening point   One standard test for measuring temperature at which a thermoplastic will soften, involving the penetration of a flat-ended needle into the plastic under controlled conditions.

viscoelastic   Characteristic mechanical behavior of some materials that is a combination of viscous and elastic behaviors.

viscosity   Measure of the resistance of a fluid to flow (usually through a specific orifice).

void   Air bubble that has been entrapped in a plastic part during molding process. (*See also* **vent**.)

volume resistivity (specific insulation resistance) Electrical resistance between opposite faces of a 1-cm cube of insulating material, commonly expressed in ohm-centimeters. The recommended test is ASTM D 257-54T.

vulcanization Chemical reaction in which the physical properties of an elastomer are changed by causing it to react with sulfur or other cross-linking agents.

vulcanized fiber Cellulosic material that has been partially gelatinized by action of a chemical (usually zinc chloride) and then heavily compressed or rolled to required thickness, leached free from the gelatinizing agent, and dried.

warp Fibers that run lengthwise in a woven fabric. (*See also* **weft**.)

warpage Dimensional distortion in a plastic object after molding.

water absorption Ratio of the weight of water absorbed by a material to the weight of the dry material.

water-extended polyester Casting formulation in which water is suspended in the polyester resin.

weave Pattern in which a fabric is woven. There are standard patterns, usually designated by a style number.

weft Fibers that run perpendicular to warp fibers; sometimes also called fill or woof. (*See also* **warp**.)

wet lay-up Reinforced plastic structure made by applying a liquid resin to an in-place woven or mat fabric.

wetting Ability to adhere to a surface immediately upon contact.

woof *See* **weft**.

working life *See* **pot life**.

woven fabric Flat sheet formed by interwinding yarns, fibers, or filaments. Some standard fabric patterns are plain, satin, and leno.

woven roving Heavy glass-fiber fabric made by the weaving of roving.

x-y axis Directions parallel to fibers in a woven-fiber-reinforced laminate. Thermal expansion is much lower in the *x-y* axis, since this expansion is more controlled by the fabric in the laminate. (*See also* **z axis**.)

yield value (yield strength) Lowest stress at which a material undergoes plastic deformation. Below this stress, the material is elastic; above it, the material is viscous. Also, stress at which a material exhibits a specified limiting deviation from the proportionality of stress to strain.

Young's modulus Ratio of normal stress to corresponding strain for tensile or compressive stresses at less than the proportional limit of the material.

z axis Direction perpendicular to fibers in a woven-fiber-reinforced laminate; that is, through the thickness of the laminate. Thermal expansion is much higher in the *z* axis, since this expansion is more controlled by the resin in the laminate. (*See also* **x-y axis**.)

APPENDIX B
SOME COMMON ABBREVIATIONS USED IN THE PLASTICS INDUSTRY[1]

ABS	acrylonitrile butadiene styrene
AI	artificial intelligence
AMC	alkyd molding compound
AMS	alpha methyl styrene
ANSI	American National Standards Institute
ASTM	American Society for Testing and Materials
ATH	aluminum trihydrate
BMC	bulk molding compound
CA	cellulose acetate
CAB	cellulose acetate butyrate
CAD	computer-aided design
CAM	computer-aided manufacturing
CAN	cellulose acetate nitrate
CAP	cellulose acetate propionate
CIM	computer-integrated manufacturing
CNC	computer numerical control
CPE	chlorinated polyethylene
CTE	coefficient of thermal expansion
CTFE	chlorotrifluoroethylene
CVD	chemical vapor deposition
DAIP	diallyl isophthalate
DAP	diallyl phthalate
DGEBA	diglycidyl ether of bisphenol A
DODISS	Department of Defense Index of Specifications and Standards

---

[1] Several of these abbreviations are also used for elastomers.

DOP	dioctyl phthalate
EB	ethyl benzene
EC	ethyl cellulose
ECTFE	ethylenechlorotrifluoroethylene
EDM	electrical-discharge machine
EMA	ethylene methyl acrylate
EMI	electromagnetic interference
EP	epoxy
EPR	ethylene propylene rubber
EPS	expanded polystyrene
ETFE	ethylenetetrafluoroethylene
EVA	ethylene vinyl acetate
EVOH	ethylene vinyl alcohol
FDA	Food and Drug Administration
FEA	finite-element analysis
FEP	fluoroethylene propylene copolymer
FR	fiber-reinforced; flame-retardant
FRP	fiber-reinforced plastic
HDPE	high-density polyethylene
HDT	heat-deflection temperature; heat-distortion temperature
HIPS	high-impact polystyrene
IPN	interpenetrating polymer network
JIT	just in time
LCP	liquid-crystal polymer
LDPE	low-density polyethylene
LED	light-emitting diode
LIM	liquid injection molding
LLDPE	linear low-density polyethylene
LMC	low-pressure molding compound
LOI	limiting oxygen index
MA	maleic anhydride
MDPE	medium-density polyethylene
MEK	methyl ethyl ketone
MEKP	methyl ethyl ketone peroxide
MF	melamine formaldehyde
MVT	moisture vapor transmission
MW	molecular weight
MWD	molecular-weight distribution
NEAT	nothing else added to it
NR	natural rubber

OSHA	Occupational Safety and Health Act
PA	polyamide (nylon)
PAI	polyamide imide
PAN	polyacrylonitrile
PAS	polyarylsulfone
PB	polybutylene
PBI	polybenzimidazol
PBT	polybutylene terephthalate
PC	polycarbonate
PCTFE	polychlorotrifluoroethylene
PE	polyethylene
PEEK	polyetheretherketone
PEI	polyetherimide
PEKK	polyetherketoneketone
PES	polyethersulfone
PET	polyethylene terephthalate
PF	phenol formaldehyde
PFA	perfluoroalkoxy resin
PHR	parts per hundred resin
PI	polyimide
PMDA	pyromellitic dianhydride
PMMA	polymethyl methacrylate (acrylic)
PMP	polymethyl pentene
PP	polypropylene
PPE	polyphenylene ether
PPO	polyphenylene oxide
PPS	polyphenylene sulfide
PPSS	polyphenylene sulfide sulfone
PS	polystyrene
PTFE	polytetrafluoroethylene
PUR	polyurethane
PVA	polyvinyl alcohol
PVDC	polyvinylidene chloride
PVDF	polyvinylidene fluoride
PVF	polyvinyl fluoride
RFI	radio-frequency interference
RIM	reaction injection molding
RLP	reactive liquid polymer
RP	reinforced plastics
RRIM	reinforced reaction molding

RTM	resin transfer molding
RTV	room-temperature vulcanizing
SAN	styrene acrylonitrile
SBR	synthetic butyl rubber
SBS	styrene butadiene styrene
SI	silicone
SMA	styrene maleic anhydride
SMC	sheet molding compound
SPC	statistical process control
SPE	Society of Plastics Engineers
SPI	Society of the Plastics Industry
SQC	statistical quality control
SRIM	structural reaction injection molding
TCE	trichloroethylene
TDI	toluene diisocyanate
TFE	polytetrafluoroethylene
TPE	thermoplastic elastomer
UF	urea formaldehyde
UHMWPE	ultrahigh-molecular-weight polyethylene
UL	Underwriters Laboratories
UV	ultraviolet
VA	vinyl acetate
WVT	water vapor transmission

APPENDIX C
IMPORTANT PROPERTIES FOR DESIGNING WITH PLASTICS*

For most product-design engineers, designing with plastics presents difficult problems, often resulting in missed opportunities or less than optimum plastic or plastic-containing products. The primary reason is that most design engineers are not experienced with plastics, since their education and training have prepared them for designing with metals. Further, the chemical nature of plastics coupled with the extremely large number of plastics makes it often difficult for designers without specific training to properly understand and differentiate between kinds of plastics. The myriad of grades and formulations only complicates the problem.

Based on this fundamental industry problem, the requirement exists for a well-categorized listing of plastic resins and compounds and their important properties for use by design engineers. This table, adapted from *Modern Plastics Encyclopedia* '96, well meets this need. Coupled with the discussions on these materials in the text of this handbook, this comprehensive data table will provide the product design engineer with useful guidelines ranging from processing pressures and temperatures used for molding plastic materials to part shrinkage during molding and on to all of the important mechanical, thermal, and physical properties needed for part design. Also listed are the ASTM test methods used in determining these properties, for those who need further understanding of these properties and their limitations. Lastly, major suppliers are provided for each of the materials, along with a listing of addresses. This will enable source selection and help in obtaining more detailed data for each of the plastic materials listed.

Recommendation is made to the reader to subscribe to the monthly published *Modern Plastics* magazine. In addition to including an annual updated copy of the *Modern Plastics Encyclopedia*, the monthly magazine will keep designers abreast of the constant stream of new developments in this field. This comprehensive set of data and information will be invaluable for design engineers with interest in taking advantage of the product opportunities available with plastic materials.

---

* The following table is reprinted from the *Modern Plastics Encyclopedia* '96 (November 1995 issue), pp. B-146 to B-206, Resins and Compounds section. The *Encyclopedia* is produced annually by *Modern Plastics* magazine, McGraw-Hill, 1221 Avenue of the Americas, New York, NY 10020.

ABS

Materials

| | | | | Flame-retarded grades, molding and extrusion | | | | Injection molding grades |
| | | | | | | | | |

	Properties	ASTM test method	Extrusion grade	ABS	ABS/PVC	ABS/PC	ABS/Nylon	ABS/PC injection molding and extrusion	Heat-resistant
Processing	1a. Melt flow (gm./10 min.)	D1238	0.4-1.0	1.2-1.7; 6	1.9				1.1-1.8
	1. Melting temperature, °C. T_m (crystalline)								
	$\quad T_g$ (amorphous)			88-120	110-125				110-125
	2. Processing temperature range, °F. (C = compression; T = transfer; I = injection; E = extrusion)		E: 350-500	C: 350-500 I: 380-500	370-410	I: 425-520	I: 460-520	I: 460-540 E: 450-500	C: 325-500 I: 475-550
	3. Molding pressure range, 10^3 p.s.i.			8-25		10-20	8-25	10-20	8-25
	4. Compression ratio		2.5-2.7	1.1-2.0	2.0-2.5	1.1-2.5	1.1-2.0	1.1-2.5	1.1-2.0
	5. Mold (linear) shrinkage, in./in.	D955	0.004-0.007	0.004-0.008	0.003-0.006	0.004-0.007	0.003-0.010	0.005-0.008	0.004-0.009
Mechanical	6. Tensile strength at break, p.s.i.	D638[b]	2500-8000	3300-8000	5800-6500	5800-9300	4000-6000	5800-7400	4800-7500
	7. Elongation at break, %	D638[b]	20-100	1.5-80		20-70	40-300	50-125	3-45
	8. Tensile yield strength, p.s.i.	D638[b]	4300-6400	4000-7400	4300-6600	7700-9000	4300-6300	3500-8500	4300-7000
	9. Compressive strength (rupture or yield), p.s.i.	D695	5200-10,000	6500-7500		11,000-11,300			7200-10,000
	10. Flexural strength (rupture or yield), p.s.i.	D790	4000-14,000	6200-14,000	7900-10,000	12,000-14,500	8800-10,900	8700-13,000	9000-13,000
	11. Tensile modulus, 10^3 p.s.i.	D638[b]	130-420	270-400	325-380	350-455	260-320	350-380	285-360
	12. Compressive modulus, 10^3 p.s.i.	D695	150-390	130-310		230			190-440
	13. Flexural modulus, 10^3 p.s.i. 73° F.	D790	130-440	300-600	320-400	350-400	250-310	290-375	300-400
	200° F.	D790							
	250° F.	D790							
	300° F.	D790							
	14. Izod impact, ft.-lb./in. of notch ($1/8$-in. thick specimen)	D256A	1.5-12	1.4-12	3.0-18.0	4.1-14.0	15-20	6.4-12.0	2.0-6.5
	15. Hardness Rockwell	D785	R75-115	R100-120	R100-106	R115-119	R93-105	R95-120	R100-115
	Shore/Barcol	D2240/ D2583			Shore D-73				
Thermal	16. Coef. of linear thermal expansion, 10^{-6} in./in./°F.	D696	60-130	65-95	46-84	67	90-110	62-72	60-93
	17. Deflection temperature under flexural load, °F. 264 p.s.i.	D648	170-220	158[b]; 181; 190-225 annealed	169-200 annealed	180-220	130-150	210-240	220-240 annealed 181-193[g]
	66 p.s.i.	D648	170-235	210-245 annealed		195-244	180-195	220-265	230-245 annealed
	18. Thermal conductivity, 10^{-4} cal.-cm./ sec.-cm.^{2}-°C.	C177							4.5-8.0
Physical	19. Specific gravity	D792	1.02-1.08	1.16-1.21	1.13-1.25	1.17-1.23	1.06-1.07	1.07-1.15	1.05-1.08
	20. Water absorption ($1/8$-in. thick specimen), % 24 hr.	D570	0.20-0.45	0.2-0.6		0.24		0.15-0.24	0.20-0.45
	Saturation	D570							
	21. Dielectric strength ($1/8$-in. thick specimen), short time, v./mil	D149	350-500	350-500	500	450-760		430	350-500
	SUPPLIERS[a]		Albis; American Polymers; Ashley Polymers; Bamberger Polymers; Diamond Polymers; Dow Plastics; Federal Plastics; Monsanto; RSG Polymers; A. Schulman; Shuman	Albis; Ashley Polymers; ComAlloy; Diamond Polymers; DSM; Federal Plastics; GE Plastics; Monsanto; Polymer Resources; RSG Polymers; RTP; A. Schulman; Shuman	ComAlloy; Novatec; A. Schulman; Vista Chemical	Bayer Corp.; Diamond Polymers; Dow Plastics; GE Plastics; Polymer Resources; Progressive Polymers; RTP	Monsanto	Bayer Corp.; Diamond Polymers; Dow Plastics; Monsanto; Polymer Resources; Progressive Polymers; RTP	American Polymers; Ashley Polymers; Diamond Polymers; Dow Plastics; Federal Plastics; GE Plastics; Monsanto; Polymer Resources; RSG Polymers; RTP; A. Schulman; Shuman

ABS (Cont'd)

	Injection molding grades (Cont'd)								EMI shielding (conductive)		
	Medium-impact	High-impact	Platable grade	Transparent	20% glass fiber-reinforced	30% glass fiber-reinforced	20% long glass fiber-reinforced	40% long glass fiber-reinforced	6% stainless steel fiber	7% stainless steel fiber	10% stainless steel fiber
1a.	1.1-1.8	1.1-18	1.1								
1.	102-115	91-110	100-110	120	100-110	100-110	100-110	100-110	100-110	100-110	100-110
2.	C: 325-350 I: 390-525	C: 325-350 I: 380-525	C: 325-400 I: 350-500	455-500	C: 350-500 I: 350-500	I: 400-460	I: 400-460	I: 400-460	I: 400-460	I: 400-460	I: 400-460
3.	8-25	8-25	8-25		15-30						
4.	1.1-2.0	1.1-2.0	1.1-2.0								
5.	0.004-0.009	0.004-0.009	0.005-0.008	0.009-0.067	0.001-0.002	0.002-0.003	0.001-0.002	0.001	0.004-0.006	0.004	0.004-0.006
6.	5500-7500	4400-6300	5200-6400	5000	10,500-13,000	13,000-16,000	13,000	16,000	6300-9200	6000	7100
7.	5-60	5-75		20	2-3	1.5-1.8	2.0	1.5	3.8	3.8	2.5
8.	5000-7200	2600-5900		6700	7000						
9.	1800-12,500	4500-8000			13,000-14,000	15,000-17,000	14,000	17,000			
10.	7100-13,000	5400-11,000	10,500-11,500	10,000	14,000-17,500	17,000-19,000	20,000	25,000	8700-11,000	10,000-12,000	12,100
11.	300-400	150-350	320-380	290	740-880	1000-1200	900	1000	300-400		400
12.	200-450	140-300			800						
13.	310-400	179-375	340-390		650-800	1000	850	1100	280-410	430	500
14.	3.0-9.6	6.0-10.5	4.0-8.3	1.5-2.0	1.1-1.4	1.2-1.3	2.0	2.5	1.2-1.4	1.0-1.1	1.4
15.	R102-115	R85-106	R103-109	R94	M85-98, R107	M75-85	M85-95	M90-100			
16.	80-100	95-110	47-53	60-130	20-21						
17.	194-220 annealed	205-215 annealed; 192 unannealed	190-222 annealed	194	210-220	215-230	210	215	190	185-200	190
	215-225 annealed	210-225 annealed	215-222 annealed	207	220-230	230-240	225	225			
18.					4.8						
19.	1.03-1.06	1.01-1.05	1.04-1.07	1.08	1.18-1.22	1.29	1.23	1.36	1.10-1.28	1.12	1.14
20.	0.20-0.45	0.20-0.45		0.35	0.18-0.20	0.3	0.2	0.2	0.4	0.4	
21.	350-500	350-500	420-550		450-460						
	Albis; American Polymers; Ashley Polymers; Bamberger Polymers; Diamond Polymers; Dow Plastics; Federal Plastics; GE Plastics; Monsanto; Polymer Resources; RSG Polymers; RTP; A. Schulman; Shuman	Albis; American Polymers; Ashley Polymers; Bamberger Polymers; Diamond Polymers; Dow Plastics; Federal Plastics; GE Plastics; Monsanto; Polymer Resources; RSG Polymers; RTP; A. Schulman; Shuman	American Polymers; Ashley Polymers; Bamberger Polymers; Diamond Polymers; Dow Plastics; Federal Plastics; GE Plastics; Monsanto; RSG Polymers; A. Schulman	BASF; GE Plastics; Monsanto; A. Schulman	Albis; Ashley Polymers; ComAlloy; DSM; Ferro Eng. Therm.; LNP; M.A. Polymers; Polymer Resources; RTP; A. Schulman; Thermofil	Albis; Ashley Polymers; ComAlloy; DSM; Ferro Eng. Therm.; LNP; M.A. Polymers; Polymer Resources; RTP	DSM	DSM	Federal Plastics; Ferro Eng. Therm.; Hoechst Celanese; LNP; RTP	DSM; Ferro Eng. Therm.; Hoechst Celanese; LNP; RTP	Hoechst Celanese; RTP

			ABS (Cont'd)				Acetal		
			EMI shielding (conductive) (Cont'd)			Rubber-modified			
Properties	ASTM test method	20% PAN carbon fiber	20% graphite fiber	40% aluminum flake	Injection molding and extrusion grades	Homo-polymer	Copolymer	Impact-modified homo-polymer
Processing									
1a. Melt flow (gm./10 min.)	D1238					1-20	1-90	0.5-7.0
1. Melting temperature, °C. T_m (crystalline)						172-184	160-175	175
T_g (amorphous)		100-110						
2. Processing temperature range, °F. (C = compression; T = transfer; I = injection; E = extrusion)		I: 415-500	I: 420-530	I: 400-550		I: 380-470	C: 340-400 I: 360-450	I: 380-420
3. Molding pressure range, 10^3 p.s.i.		15-30				10-20	8-20	6-12
4. Compression ratio						3.0-4.5	3.0-4.5	
5. Mold (linear) shrinkage, in./in.	D955	0.0005-0.004	0.001	0.001		0.018-0.025	0.020 (Avg.)	0.012-0.019
Mechanical									
6. Tensile strength at break, p.s.i.	D638[b]	15,000-16,000	15,200-15,800	3300-4200	5400-7100	9700-10,000		6500-8400
7. Elongation at break, %	D638[b]	1.0-2.0	2.0-2.2	1.9-5	20-30	10-75	15-75	60-200
8. Tensile yield strength, p.s.i.	D638[b]				6100	9500-12,000	8300-10,400	5500-7900
9. Compressive strength (rupture or yield), p.s.i.	D695	17,000	16,000-17,000	6500		15,600-18,000 @ 10%	16,000 @ 10%	7600-11,900 @ 10%
10. Flexural strength (rupture or yield), p.s.i.	D790	23,000-25,000	23,000	6200	8900-13,000	13,600-16,000	13,000	5800-10,000
11. Tensile modulus, 10^3 p.s.i.	D638[b]	1800-2000	1660	370	310	400-520	377-464	190-350
12. Compressive modulus, 10^3 p.s.i.	D695					670	450	
13. Flexural modulus, 10^3 p.s.i. 73° F.	D790	890-1800	1560	400-600	260-380	380-490	370-450	150-350
200° F.	D790					120-135		50-100
250° F.	D790					75-90		33-60
300° F.	D790							
14. Izod impact, ft.-lb./in. of notch (⅛-in. thick specimen)	D256A	1.0	1.3	1.4-2.0	3.12-7.34	1.1-2.3	0.8-1.5	2.0-17
15. Hardness Rockwell	D785	R108		R107	90-105	M92-94, R120	M75-90	M58-79
Shore/Barcol	D2240/ D2583							
Thermal									
16. Coef. of linear thermal expansion, 10^{-6} in./in./°F.	D696	18	20	40		50-112	61-110	92-117
17. Deflection temperature under flexural load, °F. 264 p.s.i.	D648	215-225	216	190-212	181-212	253-277	185-250	148-185
66 p.s.i.	D648	225-230	240	220		324-342	311-330	293-336
18. Thermal conductivity, 10^{-4} cal.-cm./sec.-cm.^{2}.°C.	C177	9.6				5.5	5.5	
Physical									
19. Specific gravity	D792	1.13-1.14	1.17	1.54-1.61	103-119	1.42	1.40	1.32-1.39
20. Water absorption (⅛-in. thick specimen), % 24 hr.	D570	0.17	0.15	0.23		0.25-1	0.20-0.22	0.30-0.44
Saturation	D570					0.90-1	0.65-0.80	0.75-0.85
21. Dielectric strength (⅛-in. thick specimen), short time, v./mil	D149					400-500 (90 mil)	500 (90 mil)	400-480 (90 mil)
SUPPLIERS[a]		DSM; Ferro Corp.; Ferro Eng. Therm.; LNP; RTP	Albis; LNP; RTP; Thermofil	ComAlloy; Thermofil	RSG Polymers	DuPont; RTP; Shuman	Ashley Polymers; BASF; Hoechst Celanese; RTP; A. Schulman; Texapol	DuPont

Acetal (Cont'd)

	Impact-modified copolymer	Mineral-filled copolymer	Extrusion and blow molding grade (terpolymer)	Copolymer with 2% silicone, low wear	UV stabilized copolymer	20% glass-reinforced homo-polymer	25% glass-coupled copolymer	40% long glass fiber-reinforced	21% PTFE-filled homo-polymer	2-20% PTFE-filled copolymer	1.5% PTFE-filled homo-polymer
1a.	5-26		1.0	9	2.5-27.0	6.0			1.0-7.0		6
1.	160-170	160-175	160-170	160-170	170	175-181	160-180	190	175-181	160-175	175
2.	I: 360-425	I: 360-450	E: 360-400	I: 360-450	I: 360-450	I: 350-480	I: 365-480	380-430	I: 370-410	I: 350-445 / I: 325-500 / E: 360-500	I: 400-440
3.	8-15	10-20		10-20	10-20	10-20	8-20	8-12	10-20	8-20	
4.	3.0-4.5	3.0-4.0	3.0-4.0	3.0-4.0	3.0-4.0		3.0-4.5	3.0-4.0			3.0-4.5
5.	0.018-0.020	0.015-0.019	0.02	0.022	0.022	0.009-0.012	0.004 (flow) / 0.018 (trans.)	0.003-0.010	0.020-0.025	0.018-0.029	
6.		6400-11,500				8500-9000	16,000-18,500	17,400	6900-7600	8300	10,000
7.	60-300	5-55	67	60	30-75	6-12	2-3	1.3	10-22	30	13
8.	3000-8000	6400-9800	8700	7400	8800-9280	7500-8250	16,000		6900-7600	8300	10,000
9.			16,000		16,000	18,000 @ 10%	17,000 @ 10%	20,400	13,000 @ 10%	11,000-12,600	
10.	7100	12,500-13,000	12,800	12,000	13,000	10,700-16,000	18,000-28,000	27,000	11,000	11,500	13,900
11.	187-319	522-780			410-910	900-1000	1250-1400	1700	410-420	250-280	450
12.					450						
13.	120-300	430-715	350	350	375-380	600-730	1100	1430	340-380	310-360	430
						300-360			110-120		150
						250-270			80-85		95
14.	1.7-4.7	0.9-1.2	1.7	1.1-1.4	1.12-1.5	0.5-1.0	1.0-1.8	6.9	0.7-1.2	0.5-1.0	1.0
15.	M35-70; R110	M83-90	M84	M75	M80-85	M90	M79-90, R110		M78, M110	M79	M93
16.	130-150	80-90				33-81	17-44		75-113	52-68	
17.	132-200	200-279	205	230	221-230	315	320-325	320	210-244	198-225	277
	308-318	302-325	318	316	316-320	345	327-331		300-334	280-325	342
18.									4.7		
19.	1.29-1.39	1.48-1.64	1.41	1.4	1.41	1.54-1.56	1.58-1.61	1.72	0.15-1.54	1.40	1.42
20.	0.31-0.41	0.20	0.22	0.21	0.22	0.25	0.22-0.29		0.20	0.15-0.26	0.19
	1.0-1.3	0.8-0.9	0.8	0.8	0.8	1.0	0.8-1.0		0.72	0.5	0.90
21.						490 (125 mil)	480-580		400-460 (125 mil)	400-410	450 (90 mils)
	BASF; Hoechst Celanese; Texapol	BASF; Hoechst Celanese	Hoechst Celanese	Hoechst Celanese; RTP	BASF; Hoechst Celanese	ComAlloy; DSM; DuPont; Ferro Corp.; Ferro Eng. Therm.; LNP; RTP; Thermofil	BASF; ComAlloy; DSM; Ferro Eng. Therm.; Hoechst Celanese; LNP; RTP; Texapol; Thermofil	Hoechst Celanese	Adell; DSM; DuPont; Ferro Eng. Therm.; LNP; RTP	ComAlloy; DSM; Ferro Corp.; Ferro Eng. Therm.; Hoechst Celanese; LNP; RTP; Texapol; Thermofil	DuPont

						Acetal (Cont'd)		Acrylic
							EMI shielding (conductive)	Sheet
Properties	ASTM test method	Chemically lubricated homo-polymer	UV stabilized homo-polymer	UV stabilized, 20% glass-filled homo-polymer	Extrusion grade homo-polymer	30% carbon fiber	10% PAN carbon fiber, 10% PTFE-filled copolymer	Cast
Processing								
1a. Melt flow (gm./10 min.)	D1238	6	1-6	6	1-6			
1. Melting temperature, °C. T_m (crystalline)		175	175	175	175	166	163-175	
T_g (amorphous)								90-105
2. Processing temperature range, °F. (C = compression; T = transfer; I = injection; E = extrusion)		I: 400-440	I: 400-440	I: 400-440	E: 380-420	I: 350-400	I: 350-410	
3. Molding pressure range, 10^3 p.s.i.						10-20		
4. Compression ratio						3-4		
5. Mold (linear) shrinkage, in./in.	D955					0.003-0.005	0.002	1.7
Mechanical								
6. Tensile strength at break, p.s.i.	D638[b]	9500	10,000	8500	10,000	7500-11,500	12,000	66-11,000
7. Elongation at break, %	D638[b]	40	40-75	12	40-75	1.5-2	1.3	2-7
8. Tensile yield strength, p.s.i.	D638[b]	9500	10,000	8500				
9. Compressive strength (rupture or yield), p.s.i.	D695							11,000-19,000
10. Flexural strength (rupture or yield), p.s.i.	D790	13,000	14,100-14,300	10,700	14,100-14,300	12,500-16,500	14,000	12,000-17,000
11. Tensile modulus, 10^3 p.s.i.	D638[b]	450	400-450	900	400-450	1350	1300	450-3100
12. Compressive modulus, 10^3 p.s.i.	D695							390-475
13. Flexural modulus, 10^3 p.s.i. 73° F.	D790	400	420-430	730	420-430	1100-1200	1000	390-3210
200° F.	D790	130	130-135	360	130-135			
250° F.	D790	80	90	270	90			
300° F.	D790							
14. Izod impact, ft.-lb./in. of notch (1/8-in. thick specimen)	D256A	1.4	1.5-2.3	0.8	1.5-2.3	0.7-0.8	0.7	0.3-0.4
15. Hardness Rockwell	D785	M90	M94	M90	M94			M80-102
Shore/Barcol	D2240/ D2583							
Thermal								
16. Coef. of linear thermal expansion, 10^{-6} in./in./°F.	D696							50-90
17. Deflection temperature under flexural load, °F. 264 p.s.i.	D648	257	257-264	316	257-264	320	320	98-215
66 p.s.i.	D648	329	334-336	345	334-336	325	325	165-235
18. Thermal conductivity, 10^{-4} cal.-cm./ sec.-cm.^{2}.°C.	C177							4.0-6.0
Physical								
19. Specific gravity	D792	1.42	1.42	1.56	1.42	1.43-1.53	1.49	1.17-1.20
20. Water absorption (1/8-in. thick specimen), % 24 hr.	D570	0.27	0.25	0.25	0.25	0.22-0.26	0.25	0.2-0.4
Saturation	D570	1.00	0.90	1.00	0.90			
21. Dielectric strength (1/8-in. thick specimen), short time, v./mil	D149	400 (125 mils)	500 (90 mils)	490 (125 mils)	500 (90 mils)			450-550
SUPPLIERS[a]		DuPont	DuPont	DuPont	DuPont	DSM; Hoechst Celanese; LNP; RTP	DSM; Ferro Eng. Therm.; LNP; RTP; Texapol	Aristech; AtoHaas; Cyro; DuPont; ICI Acrylics

| | Acrylic (Cont'd) | | | | | | | Acrylonitrile | | | |
| | Sheet (Cont'd) | Molding and extrusion compounds | | | | | | | | | |
	Coated	Acrylic/PC alloy	PMMA	MMA-styrene copolymer	Impact-modified	Heat-resistant	Acrylic multipolymer	Molding and extrusion	Extrusion	High-impact extrusion	Injection
1a.		3-4	1.4-27	1.1-24	1-11	1.6-8.0	2-14		3	3	12
1.								135			
	90-110	140	85-105	100-105	80-103	100-165	80-105	95			
2.		I: 450-510 E: 430-480	C: 300-425 I: 325-500 E: 360-500	C: 300-400 I: 300-500	C: 300-400 I: 400-500 E: 380-480	C: 350-500 I: 400-625 E: 360-550	I: 400-500 E: 380-470	C: 320-345 I: 410 E: 350-410	380-420	380-410	380-420
3.		5-20	5-20	10-30	5-20	5-30	5-20	20	25	25	20
4.			1.6-3.0			1.2-2.0		2	2-2.5	2-2.5	2-2.5
5.		0.004-0.008	0.001-0.004(flow) 0.002-0.008(trans.)	0.002-0.006	0.002-0.008	0.002-0.008	0.004-0.008	0.002-0.005	0.002-0.005	0.002-0.005	0.002-0.005
6.	10,500	8000-9000	7000-10,500	8100-10,100	5000-9000	9300-11,500	5500-8200	9000			
7.	3	58	2-5.5	2-5	4.6-70	2-10	5-28	3-4	3-4	3-4	3-4
8.			7800-10,600		5500-8470	10,000		7500	9500	7500	9500
9.	18,000		10,500-18,000	11,000-15,000	4000-14,000	15,000-17,000	7500-11,500	12,000	12,000	11,500	12,000
10.	16,000	11,300-12,500	10,500-19,000	14,100-16,000	7000-14,000	12,000-18,000	9000-13,000	14,000	14,000	13,700	14,000
11.	450	320-350	325-470	430-520	200-500	350-650	300-430	510-580	500-550	450-500	500-550
12.	450		370-460	240-370	240-370	450					
13.	450	320-350	325-460	450-460	200-430	450-620	290-400	500-590	490	390	480
						150-440					
						350-420					
14.	0.3-0.4	26-30	0.2-0.4	0.3-0.4	0.40-2.5	0.2-0.4	1.0-2.5	2.5-6.5	5.0	9.0	2.5
15.	M105	46-49	M68-105	M80-85	M35-78	M94-100	22-56	M72-78	M60	M45	M60
16.	40	52	50-90	60-80	48-80	40-71	44-50	66	66	66	66
17.	205	101	155-212	208-211	165-209	190-310	180-194	164	156	151	151
	225		165-225		180-205	200-315		172	170	160	166
18.	5.0		4.0-6.0	4.0-5.0	4.0-5.0	2.0-4.5	5.3	6.2	6.1	6.1	6.1
19.		1.15	1.17-1.20	1.06-1.13	1.11-1.18	1.16-1.22	1.11-1.12	1.15	1.15	1.11	1.15
20.	<0.4	0.3	0.1-0.4	0.11-0.17	0.19-0.8	0.2-0.3	0.3	0.28			
21.	500		400-500	450	380-500	400-500		220-240	220-240	220-240	220-240
	DuPont	Cyro	AtoHaas; Cyro; DuPont; ICI Acrylics; Plaskolite; RTP	Novacor	AtoHaas; Cyro; DuPont; ICI Acrylics; RTP	AtoHaas; Cyro; ICI Acrylics; Plaskolite; RTP	Cyro	BP Chemicals	BP Chemicals	BP Chemicals	BP Chemicals

			Allyl			Cellulosic		
Materials			DAP molding compounds			Cellulose acetate		Cellulose acetate butyrate
Properties	ASTM test method	Allyl diglycol carbonate cast sheet	Glass-filled	Mineral-filled	Ethyl cellulose molding compound and sheet	Sheet	Molding compound	Molding compound
Processing								
1a. Melt flow (gm./10 min.)	D1238							
1. Melting temperature, °C. T_m (crystalline)			Thermoset	Thermoset	Thermoset	135	230	230	140
T_g (amorphous)								
2. Processing temperature range, °F. (C = compression; T = transfer; I = injection; E = extrusion)			C: 290-360 I: 300-350	C: 270-360	C: 250-390 I: 350-500		C: 260-420 I: 335-490	C: 265-390 I: 335-480
3. Molding pressure range, 10^3 p.s.i.			2000-6000	2500-5000	8-32		8-32	8-32
4. Compression ratio			1.9-10.0	1.2-2.3	1.8-2.4		1.8-2.6	1.8-2.4
5. Mold (linear) shrinkage, in./in.	D955		0.0005-0.005	0.002-0.007	0.005-0.009		0.003-0.010	0.003-0.009
Mechanical								
6. Tensile strength at break, p.s.i.	D638[b]	5000-6000	6000-11,000	5000-8000	2000-8000	4500-8000	1900-9000	2600-8100
7. Elongation at break, %	D638[b]		3-5	3-5	5-40	20-50	6-70	40-88
8. Tensile yield strength, p.s.i.	D638[b]						2500-7600	1600-7200
9. Compressive strength (rupture or yield), p.s.i.	D695	21,000-23,000	25,000-35,000	20,000-32,000			3000-8000	2100-7500
10. Flexural strength (rupture or yield), p.s.i.	D790	6000-13,000	9000-20,000	8500-11,000	4000-12,000	6000-10,000	2000-16,000	1800-10,100
11. Tensile modulus, 10^3 p.s.i.	D638[b]	300	1400-2200	1200-2200				50-200
12. Compressive modulus, 10^3 p.s.i.	D695	300						
13. Flexural modulus, 10^3 p.s.i. 73° F.	D790	250-330	1200-1500	1000-1400			1000-4000	90-300
200° F.	D790							
250° F.	D790							
300° F.	D790							
14. Izod impact, ft.-lb./in. of notch (1/8-in. thick specimen)	D256A	0.2-0.4	0.4-15.0	0.3-0.8	0.4	2.0-8.5	1.0-7.8	1.0-10.9
15. Hardness Rockwell	D785	M95-100	E80-87	E60-80E	R50-115	R85-120	R17-125	R11-116
Shore/Barcol	D2240/ D2583							
Thermal								
16. Coef. of linear thermal expansion, 10^{-6} in./in./°F.	D696	81-143	10-36	10-42	100-200	100-150	80-180	110-170
17. Deflection temperature under flexural load, °F. 264 p.s.i.	D648	140-190	330-550+	320-550	115-190		111-195	109-202
66 p.s.i.	D648						120-209	130-227
18. Thermal conductivity, 10^{-4} cal.-cm./ sec.-cm.^{2}-°C.	C177	4.8-5.0	5.0-15.0	7.0-25	3.8-7.0	4-8	4-8	4-8
Physical								
19. Specific gravity	D792	1.3-1.4	1.70-1.98	1.65-1.85	1.09-1.17	1.28-1.32	1.22-1.34	1.15-1.22
20. Water absorption (1/8-in. thick specimen), % 24 hr.	D570	0.2	0.12-0.35	0.2-0.5	0.8-1.8	2.0-7.0	1.7-6.5	0.9-2.2
Saturation	D570							
21. Dielectric strength (1/8-in. thick specimen), short time, v./mil	D149	380	400-450	400-450	350-500	250-600	250-600	250-475
SUPPLIERS[a]		PPG	Cosmic Plastics; Oxy-Chem; Rogers Corp.	Cosmic Plastics; Oxy-Chem; Rogers Corp.	Dow Chem.; Federal Plastics; Kleer Kast; Montell Polyolefins	Rotuba	Albis; Eastman; Kleer Kast; Rotuba	Albis; Eastman; Rotuba

	Cellulosic (Cont'd)		Chlorinated PE	Epoxy							
				Bisphenol molding compounds			Sheet molding compound (SMC)		Novolak molding compounds		Casting resins and compounds
	Cellulose acetate proplonate molding compound	Cellulose nitrate	30-42% Cl extrusion and molding grades	Glass fiber-reinforced	Mineral-filled	Low density glass sphere-filled	Glass fiber-reinforced	Carbon fiber-reinforced	Mineral- and glass-filled, encapsulation	Mineral- and glass-filled, high temperature	Unfilled
1a.											
1.	190		125	Thermoset	Thermoset	Thermoset	Thermoset	Thermoset	Thermoset	Thermoset	Thermoset
									145-155	155-195	
2.	C: 265-400 I: 335-515	C: 185-250	E: 300-400	C: 300-330 T: 280-380	C: 250-330 T: 250-380	C: 250-300 I: 250-300	C: 250-330 T: 270-330	C: 250-330 T: 270-330	C: 280-360 I: 290-350 T: 250-380	T: 340-400	
3.	8-32	2-5		1-5	0.1-3	0.1-2	0.5-2.0	0.5-2.0	0.25-3.0	0.5-2.5	
4.	1.8-3.4			3.0-7.0	2.0-3.0	3.0-7.0	2.0	2.0		1.5-2.5	
5.	0.003-0.009			0.001-0.008	0.002-0.010	0.006-0.010	0.001	.001	0.004-0.008	0.004-0.007	0.001-0.010
6.	2000-7800	7000-8000	1400-3000	5000-20,000	4000-10,800	2500-4000	20,000-35,000	40,000-50,000	5000-12,500	6000-15,500	4000-13,000
7.	29-100	40-45	300-900	4			0.5-2.0	0.5-2.0			3-6
8.											
9.	2400-7000	2100-8000		18,000-40,000	18,000-40,000	10,000-15,000	20,000-30,000	30,000-40,000	24,000-48,000	30,000-48,000	15,000-25,000
10.	2900-11,400	9000-11,000		8000-30,000	6000-18,000	5000-7000	50,000-70,000	75,000-95,000	10,000-21,800	10,000-21,800	13,000-21,000
11.	60-215	190-220		3000	350		2000-4000	10,000	2100	2300-2400	350
12.					650					660	
13.	120-350			2000-4500	1400-2000	500-750	2000-3000	5000	1400-2400	2300-2400	
							1500-2500				
							1200-1800				
14.	0.5-No break	5-7		0.3-10.0	0.3-0.5	0.15-0.25	30-40	15-20	0.3-0.5	0.4-0.45	0.2-1.0
15.	R10-122	R95-115		M100-112	M100-M112				M115		M80-110
			Shore A60-76				B: 55-65	B: 55-65	Barcol 70-75	Barcol 78	
16.	110-170	80-120		11-50	20-60		12	3	18-43	35	45-65
17.	111-228	140-160		225-500	225-500	200-250	550	550	300-500	500	115-550
	147-250										
18.	4-8	5.5		4.0-10.0	4-35	4.0-6.0	1.7-1.9	1.4-1.5	10-31	17-24	4.5
19.	1.17-1.24	1.35-1.40	1.13-1.26	1.6-2.0	1.6-2.1	0.75-1.0	0.10	0.10	1.6-2.05	1.85-1.94	1.11-1.40
20.	1.2-2.8	1.0-2.0		0.04-0.20	0.03-0.20	0.2-1.0	1.4	1.6	0.04-0.29	0.15-0.17	0.08-0.15
										0.15-0.3	
21.	300-475	300-600		250-400	250-420	380-420			325-450	440-450	300-500
	Albis; Eastman	Chem. Development; P.D. George	Dow Plastics	Hysol; ICI Fiberite; M & T; Plaskon	Hysol; ICI Fiberite; M & T; Plaskon	Hysol; ICI Fiberite	Quantum Composites	Quantum Composites	Cosmic Plastics; Hysol; ICI Fiberite; M & T; Plaskon; Rogers	Cosmic Plastics; ICI Fiberite; Plaskon; Rogers	Ciba-Geigy; Conap; Dow Plastics; Epic Resins; Grace Specialty Polymers; Hysol; ITW Devcon; Shell; UMC

			Epoxy (Cont'd)				Ethylene vinyl alcohol	Fluoroplastics	
			Casting resins and compounds (Cont'd)						Polytetra-fluoro-ethylene
	Properties	ASTM test method	Silica-filled	Aluminum-filled	Flexibilized	Cyclo-aliphatic		Polychloro-trifluoro-ethylene	Granular
Processing	1a. Melt flow (gm./10 min.)	D1238					0.8-14.0		
	1. Melting temperature, °C. T_m (crystalline)		Thermoset	Thermoset	Thermoset	Thermoset	142-191		327
	T_g (amorphous)						49-72	220	
	2. Processing temperature range, °F. (C = compression; T = transfer; I = injection; E = extrusion)						I: 365-480 E: 365-480	C: 460-580 I: 500-600 E: 360-590	
	3. Molding pressure range, 10^3 p.s.i.							1-6	2-5
	4. Compression ratio						3-4	2.6	2.5-4.5
	5. Mold (linear) shrinkage, in./in.	D955	0.0005-0.008	0.001-0.005	0.001-0.010			0.010-0.015	0.030-0.060
Mechanical	6. Tensile strength at break, p.s.i.	D638[b]	7000-13,000	7000-12,000	2000-10,000	8000-12,000	5405-13,655	4500-6000	3000-5000
	7. Elongation at break, %	D638[b]	1-3	0.5-3	20-85	2-10	180-330	80-250	200-400
	8. Tensile yield strength, p.s.i.	D638[b]					7385-10,365	5300	
	9. Compressive strength (rupture or yield), p.s.i.	D695	15,000-35,000	15,000-33,000	1000-14,000	15,000-20,000		4600-7400	1700
	10. Flexural strength (rupture or yield), p.s.i.	D790	8000-14,000	8500-24,000	1000-13,000	10,000-13,000	230-285	7400-11,000	
	11. Tensile modulus, 10^3 p.s.i.	D638[b]							58-80
	12. Compressive modulus, 10^3 p.s.i.	D695			1-350	495	300-385	150-300	60
	13. Flexural modulus, 10^3 p.s.i. 73° F.	D790						170-200	80
	200° F.	D790						180-260	
	250° F.	D790							
	300° F.	D790							
	14. Izod impact, ft.-lb./in. of notch (⅛-in. thick specimen)	D256A	0.3-0.45	0.4-1.6	2.3-5.0		1.0-1.7	2.5-5	3
	15. Hardness Rockwell	D785	M85-120	M55-85				R75-112	
	Shore/Barcol	D2240/ D2583			Shore D65-89			Shore D75-80	Shore D50-65
Thermal	16. Coef. of linear thermal expansion, 10^{-6} in./in./°F.	D696	20-40	5.5	20-100			36-70	70-120
	17. Deflection temperature under flexural load, °F. 264 p.s.i.	D648	160-550	190-600	73-250	200-450			115
	66 p.s.i.	D648						258	160-250
	18. Thermal conductivity, 10^{-4} cal.-cm./ sec.-cm.²-°C.	C177	10-20	15-25				4.7-5.3	6.0
Physical	19. Specific gravity	D792	1.6-2.0	1.4-1.8	0.96-1.35	1.16-1.21	1.12-1.20	2.08-2.2	2.14-2.20
	20. Water absorption (⅛-in. thick specimen), % 24 hr.	D570	0.04-0.1	0.1-4.0	0.27-0.5		6.7-8.6	0	<0.01
	Saturation	D570							
	21. Dielectric strength (⅛-in. thick specimen), short time, v./mil	D149	300-550		235-400			500-600	480
	SUPPLIERS[a]		Conap; Epic Resins; Grace Specialty Polymers; Hysol; ITW Devcon	Conap; Epic Resins; Grace Specialty Polymers; ITW Devcon	Conap; Dow Plastics; Epic Resins; Grace Specialty Polymers; ITW Devcon	Ciba-Geigy; Union Carbide	Eval Co. of America	Ciba-Geigy; Elf Atochem N.A.; 3M	Ausimont; DuPont; Hoechst Celanese; ICI Americas; Tetrafluor

Materials

Fluoroplastics (Cont'd)

	Polytetra-fluoro-ethylene (Cont'd)		Fluorinated ethylene propylene		Polyvinylidene fluoride			Modified PE-TFE			
	25% glass fiber-reinforced	PFA fluoro-plastic	Unfilled	20% milled glass fiber	Molding and extrusion	Wire and cable jacketing	EMI shielding (conductive); 30% PAN carbon fiber	Unfilled	25% glass fiber-reinforced	THV-200	THV-400
1a.											
1.	327	300-310	275	262	141-178	168-170		270	270	120	150
					−60 to −20	−30 to −20					
2.		C: 625-700 I: 680-750	C: 600-750 I: 625-760	I: 600-700	C: 360-550 I: 375-550 E: 375-550	E: 420-525	I: 430-500	C: 575-625 I: 570-650	C: 575-625 I: 570-650	E: 450	E: 470
3.	3-8	3-20	5-20	10-20	2-5	1.10		2-20	2-20		
4.		2.0	2.0		3	3					
5.	0.018-0.020	0.040	0.030-0.060	0.006-0.010	0.020-0.035	0.020-0.030	0.001	0.030-0.040	0.002-0.030		
6.	2000-2700	4000-4300	2700-3100	2400	3500-7250	7100	14,000	6500	12,000	3500	3400
7.	200-300	300	250-330	5	12-600	300-500	0.8	100-400	8	600	500
8.		2100			2900-8250	4460					
9.	1000-1400 @ 1% strain	3500	2200		8000-16,000	6600-7100		7100	10,000		
10.	2000			4000	9700-13,650	7000-8600	19,800	5500	10,700		
11.	200-240	70	50		200-80,000	145-190	2800	120	1200		
12.					304-420	180					
13.	190-235	95-120	80-95	250	170-120,000	145-260	2100	200	950	12	
								80	450		
								60	310		
								20	200		
14.	2.7	No break	No break	3.2	2.5-80	7	1.5	No break	9.0		
15.					R79-83, 85	R77		R50	R74		
	Shore D60-70	Shore D64	Shore D60-65		Shore D80, 82 65-70	Shore D75		Shore D75		44	53
16.	77-100	140-210		22	70-142	121-140		59	10-32		
17.				150	183-244	129-165	318	160	410		
		166	158		280-284			220	510		
18.	8-10	6.0	6.0		2.4-3.1	2.4-3.1		5.7			
19.	2.2-2.3	2.12-2.17	2.12-2.17		1.77-1.78	1.76-1.77	1.74	1.7	1.8	1.95	1.97
20.		0.03	<0.01	0.01	0.03-0.06	0.03-0.06	0.12	0.03	0.02		
21.	320	500	500-600		260-280	260-280		400	425		
	Ausimont; DuPont; Hoechst Celanese; ICI Americas; RTP; Tetrafluor	Ausimont; DuPont; Hoechst Celanese	DuPont	RTP	Ausimont; Elf Atochem N.A.; Solvay Polymers	Ausimont; Elf Atochem N.A.; Solvay Polymers	RTP	DuPont	Ausimont; DuPont; RTP	3M	3M

		Fluoroplastics (Cont'd)		Ionomer		Ketones		
Materials						Polyaryletherketone		
Properties	ASTM test method	THV-500	PE-CTFE	Molding and extrusion	Glass- and rubber-modified; molding and extrusion	Unfilled	30% glass fiber-reinforced	40% glass fiber-reinforced
Processing								
1a. Melt flow (gm./10 min.)	D1238					4-7	15-25	15-25
1. Melting temperature, °C. T_m (crystalline)		180	220-245	81-96	81-220	323-381	329-381	329
T_g (amorphous)								
2. Processing temperature range, °F. (C = compression; T = transfer; I = injection; E = extrusion)		E: 480	C: 500 I: 525-575 E: 500-550	C: 280-350 E: 300-450	C: 300-400 I: 300-550 E: 350-525	I: 715-805	I: 715-805	I: 715-805
3. Molding pressure range, 10^3 p.s.i.			5-20	2-20	2-20	10-20	10-20	10-20
4. Compression ratio				3	3	2	2	2
5. Mold (linear) shrinkage, in./in.	D955		0.020-0.025	0.003-0.010	0.002-0.008	0.008-0.012	0.001-0.009	0.001-0.009
Mechanical								
6. Tensile strength at break, p.s.i.	D638[b]	3300	6000-7000	2500-5400	3500-7900	13,500	23,700-27,550	25,000
7. Elongation at break, %	D638[b]	500	200-300	300-700	5-200	50	2.2-3.4	2
8. Tensile yield strength, p.s.i.	D638[b]		4500-4900	1200-2300	1200-4500	15,000		
9. Compressive strength (rupture or yield), p.s.i.	D695					20,000	30,000	32,500
10. Flexural strength (rupture or yield), p.s.i.	D790		7000			18,850-24,500	34,100-36,250	40,000
11. Tensile modulus, 10^3 p.s.i.	D638[b]		240	to 60		520-580	1410-1754	1900
12. Compressive modulus, 10^3 p.s.i.	D695							
13. Flexural modulus, 10^3 p.s.i. 73° F.	D790	30	240	3-55	8-700	530	1600	2100
200° F.	D790					530	1520	2010
250° F.	D790					520	1460	1910
300° F.	D790					500	1390	1790
14. Izod impact, ft.-lb./in. of notch ($^1/_8$-in. thick specimen)	D256A		No break	7-No break	2.5-1.8 No break	1.6-2.7	1.8-1.9	2
15. Hardness Rockwell	D785		R93-95	R53		M98	M102	M102
Shore/Barcol	D2240/ D2583	54	Shore D75	Shore D25-66	Shore D43-70	Shore D86	Shore D90	
Thermal								
16. Coef. of linear thermal expansion, 10^{-6} in./in./°F.	D696		80	100-170	50-100	41-44.2	18.5-20	18
17. Deflection temperature under flexural load, °F. 264 p.s.i.	D648		170	93-100		323-338	619-662	619
66 p.s.i.	D648		240	113-125	131-400	482-582	644-662	644
18. Thermal conductivity, 10^{-4} cal.-cm./ sec.-cm.^{2}.°C.	C177		3.8	5.7-6.6		7.1		10.5
Physical								
19. Specific gravity	D792	1.78	1.68-1.69	0.93-0.96	0.95-1.2	1.3	1.47-1.53	1.55
20. Water absorption ($^1/_8$-in. thick specimen), % 24 hr.	D570		0.01	0.1-0.5	0.1-0.5	0.1	0.07	0.05
Saturation	D570					0.8	0.5	
21. Dielectric strength ($^1/_8$-in. thick specimen), short time, v./mil	D149		490-520	400-450		355	370	420
SUPPLIERS[a]		3M	Ausimont	DuPont; Exxon	DuPont; A. Schulman	Amoco Polymers; BASF	BASF; RTP	RTP

| | Ketones (Cont'd) | | | | | Liquid Crystal Polymer | | | | | |
| | Polyaryletherketone (Cont'd) | | Polyetheretherketone | | | | | | | | |
	Modified, 40% glass	30% carbon fiber	Unfilled	30% glass fiber-reinforced	30% carbon fiber-reinforced	30% glass fiber-reinforced	45% glass fiber-reinforced, HDT	30% glass fiber-reinforced, high HDT	40% glass fiber-reinforced	30% mineral-filled	Glass fiber-reinforced for SMT
1a.	15-25	15-25									
1.	324	329	334	334	334						
2.	I: 715-765	I: 715-805	I: 660-750 E: 660-720	I: 660-750	I: 660-800	I: 660-690	I: 660-730	I: 685-700	I: 590-645	I: 660-690	I: 610-680
3.	10-20	10-20	10-20	10-20	10-20	4-8		4-8		4-8	
4.	2	2	3	2-3	2	2.5-3	2.5-3	2.5-3	2.5-3	2.5-3	2.5-3
5.	0.001-0.009	0.002-0.008	0.011	0.002-0.014	0.0005-0.014	0.09		-0.01			
6.	22,500	30,000	10,200-15,000	22,500-28,500	29,800-33,000	21,750	18,200	21,025	21,240	15,950	19,600
7.	1.5	1.5	30-150	2-3	1-4	2.7	2.1	2.2	1.5	4.0	1.6
8.			13,200								
9.	33,000	33,800	18,000	21,300-22,400	25,000-34,400	15,200	10,900	12,500			
10.	34,000	40,000	16,000	33,000-42,000	40,000-48,000	24,650	22,400	25,230	27,000	18,415	25,000
11.	2250	2700		1250-1600	1860-3500	3000	2170	2600	2890		2700
12.						1000	491	770		590	
13.	2100	2850	560	1260-1600	1860-2600	1700	22.4	1800	2280	1400	1950
	1960	2790	435	1400	1820						
	1750	2480		1350	1750	900		1100			
	1500	2100	290	1100	1400	800		1100			
14.	1.2	1.5	1.6	2.1-2.7	1.5-2.1	2.4	2.0	4.2	8.3	3.0	1.8
15.	M103					M61		M63			
16.	18.5	7.9	<150°C: 40-47 <150°C: 108	<150°C: 12-22 >150°C: 44	<150°C: 15-22 >150°C: 5-44	13-37	12.5-80.2	14-36		8-22	
17.	586	634	320	550-599	550-610	493	572	563	466	455	520
	643	652			615	530					
18.	8.8			4.9	4.9		2.02				
19.	1.6	1.45	1.30-1.32	1.49-1.54	1.42-1.44	1.67	1.75	1.66	1.70	1.63	1.6
20.	0.04	<0.2	0.1-0.14	0.06-0.12	0.06-0.12	0.002	<.1				<.1
			0.5	0.11-0.12	0.06	0.05					
21.	385					740	900	>710			
	Amoco Polymers; RTP	Amoco Polymers; RTP	ICI Americas; Tetrafluor	DSM; ICI Americas; LNP; RTP; Tetrafluor	DSM; ICI Americas; LNP; RTP; Tetrafluor	DuPont; RTP	RTP	DuPont; RTP	Amoco Polymers; RTP	DuPont; RTP	Amoco Polymers; RTP

Liquid Crystal Polymer (Cont'd)

	Properties	ASTM test method	Unfilled medium melting point	30% carbon fiber-reinforced	50% mineral-filled	30% glass fiber-reinforced	30% glass fiber-reinforced, high HDT	Unfilled platable grade	PTFE-filled
Processing	1a. Melt flow (gm./10 min.)	D1238							
	1. Melting temperature, °C. T_m (crystalline)		280-421	280	327	280	355		281
	T_g (amorphous)								
	2. Processing temperature range, °F. (C = compression; T = transfer; I = injection; E = extrusion)		I: 540-770	555-600	I: 605-770	I: 555-770	I: 625-730	I: 600-620	
	3. Molding pressure range, 10^3 p.s.i.		1-16	1-14	1-14	1-14			
	4. Compression ratio		2.5-4	2.5-4	2.5-4	2.5-4	2.5-4	3-4	3:1
	5. Mold (linear) shrinkage, in./in.	D955	0.001-0.006	0-0.002	0.003	0.001			0-3
Mechanical	6. Tensile strength at break, p.s.i.	D638[b]	15,900-27,000	35,000	10,400-16,500	16,900-30,000	18,000-19,800	13,500	24,500-25,000
	7. Elongation at break, %	D638[b]	1.3-4.5	1.0	1.1-2.6	1.7-2.6	1.7-2.0	2.9	3.0-5.2
	8. Tensile yield strength, p.s.i.	D638[b]							
	9. Compressive strength (rupture or yield), p.s.i.	D695	6200-19,000	34,500	6800-7500	9900-21,000	9600		
	10. Flexural strength (rupture or yield), p.s.i.	D790	19,000-35,500	46,000	14,200-23,500	21,700-23,300	24,000		18,500-29,000
	11. Tensile modulus, 10^3 p.s.i.	D638[b]	1400-2800	5400	1500-2700	700-2400	2330	1500	1100-1600
	12. Compressive modulus, 10^3 p.s.i.	D695	400-900	4800	490-2016	470-519	447		
	13. Flexural modulus, 10^3 p.s.i. 73° F.	D790	1770-2700	4800	1250-2500	1660-2100	2000-2050	1500	1000-1400
	200° F.	D790	1500-1700						
	250° F.	D790	1300-1500						
	300° F.	D790	1200-1450						
	14. Izod impact, ft.-lb./in. of notch (¹/₈-in. thick specimen)	D256A	1.7-10	1.4	0.8-1.5	2.0-3.0	2.0-2.5	0.6	2.1-3.8
	15. Hardness Rockwell	D785	M76; R60-66	M99	82	77-87			
	Shore/Barcol	D2240/D2583							
Thermal	16. Coef. of linear thermal expansion, 10^{-6} in./in./°F.	D696	5-7	-2.65	9-65	4.9-77.7			
	17. Deflection temperature under flexural load, °F. 264 p.s.i.	D648	356-671	440	429-554	485-655	518-568	410	352-435
	66 p.s.i.	D648				400			
	18. Thermal conductivity, 10^{-4} cal.-cm./sec.-cm.²-°C.	C177	2		2.57	1.73	1.52		
Physical	19. Specific gravity	D792	1.35-1.84	1.49	1.84-1.89	1.60-1.62	1.6-1.64		1.50-1.62
	20. Water absorption (¹/₈-in. thick specimen), % 24 hr.	D570	0-<0.1	<0.1	<0.1	<0.1	<0.1	0.03	
	Saturation	D570	<0.1						
	21. Dielectric strength (¹/₈-in. thick specimen), short time, v./mil	D149	800-980		900-955	640-900	900-1050	600	
	SUPPLIERS[a]		Hoechst Celanese	Hoechst Celanese; RTP	Hoechst Celanese; RTP	Hoechst Celanese; RTP	Hoechst Celanese; RTP	Hoechst Celanese	Hoechst Celanese; RTP

	Liquid Crystal Polymer (Cont'd)		Melamine formaldehyde		Phenolic					
					Molding compounds, phenol-formaldehyde					
								Impact-modified		
	15% glass fiber-reinforced	Glass/mineral-filled	Cellulose-filled	Glass fiber-reinforced	Woodflour-filled	Woodflour-and mineral-filled	High-strength glass fiber-reinforced	Cotton-filled	Cellulose-filled	Fabric and rag-filled
1a.										0.5-10
1.	280	280	Thermoset	Thermoset	Thermoset	Thermoset	Thermoset	Thermoset	Thermoset	Thermoset
2.			C: 280-370 I: 200-340 T: 300	C: 280-350	C: 290-380 I: 330-400	C: 290-380 I: 330-390 T: 290-350	C: 300-380 I: 330-390 T: 300-350	C: 290-380 I: 330-400	C: 290-380 I: 330-400	C: 290-380 I: 330-400 T: 300-350
3.			8-20	2-8	2-20	2-20	1-20	2-20	2-20	2-20
4.	3:1	3:1	2.1-3.1	5-10	1.0-1.5		2.0-10.0	1.0-1.5	1.0-1.5	1.0-1.5
5.	0-3	0-3	0.005-0.015	0.001-0.006	0.004-0.009	0.003-0.008	0.001-0.004	0.004-0.009	0.004-0.009	0.003-0.009
6.	28,000	21,000-25,000	5000-13,000	5000-10,500	5000-9000	6500-7500	7000-18,000	6000-10,000	3500-6500	6000-8000
7.	3	1.4-2.3	0.6-1	0.6	0.4-0.8		0.2	1-2	1-2	1-4
8.										
9.		18,000-21,000	33,000-45,000	20,000-35,000	25,000-31,000	25,000-30,000	16,000-70,000	23,000-31,000	22,000-31,000	20,000-28,000
10.	30,000	31,000-34,000	9000-16,000	14,000-23,000	7000-14,000	9000-12,000	12,000-60,000	9000-13,000	5500-11,000	10,000-14,000
11.	1100-2100	2400-3100	1100-1400	1600-2400	800-1700	1000-1800	1900-3300	1100-1400		900-1100
12.		2000-2800					2740-3500			
13.	1600	2200-2900	1100		1000-1200	1200-1300	1150-3300	800-1300	900-1300	700-1300
14.	5.5	1.6-3.8	0.2-0.4	0.6-18	0.2-0.6	0.29-0.35	0.5-18.0	0.3-1.9	0.4-1.1	0.8-3.5
15.		76-79	M115-125	M115	M100-115	M90-110	E54-101 Barcol 72	M105-120	M95-115	M105-115
16.		6-8	40-45	15-28	30-45	30-40	8-34	15-22	20-31	18-24
17.	430	437	350-390	375-400	300-370	360-380	350-600	300-400	300-350	325-400
18.			6.5-10	10-11.5	4-8	6-10	8-14	8-10	6-9	9-12
19.	1.5	1.68-1.89	1.47-1.52	1.5-2.0	1.37-1.46	1.44-1.56	1.69-2.0	1.38-1.42	1.38-1.42	1.37-1.45
20.			0.1-0.8	0.09-1.3	0.3-1.2	0.2-0.35	0.03-1.2	0.6-0.9	0.5-0.9	0.6-0.8
	0.2	0.2					0.12-1.5			
21.		840	270-400 175-215 @100°C	130-370	260-400	330-375	140-400	200-360	300-380	200-370
	Hoechst Celanese; RTP	Hoechst Celanese; RTP	BIP Chemicals; CYTEC; ICI Fiberite; Patent Plastics; Perstorp; Plastics Mfg.	ICI Fiberite	Lockport Thermosets; OxyChem; Plaslok; Plastics Eng.	Lockport Thermosets; OxyChem; Plaslok; Plastics Eng.	ICI Fiberite; OxyChem; Plastics Eng.; Quantum Composites; Resinoid; Rogers	ICI Fiberite; Lockport Thermosets; Miles; OxyChem; Plaslok; Plastics Eng.; Resinoid; Rogers	ICI Fiberite; Lockport Thermosets; OxyChem; Plaslok; Plastics Eng.; Resinoid; Rogers	ICI Fiberite; OxyChem; Resinoid; Rogers

			Phenolic (Cont'd)			Polyamide		
Materials			Molding compounds, PF (Cont'd)	Casting resins		Nylon alloys	Nylon, Type 6	
			Heat-resistant					
	Properties	ASTM test method	Mineral- or mineral- and glass-filled	Unfilled	Mineral-filled	Ceramic and glass fiber-reinforced	Molding and extrusion compound	15% glass fiber-reinforced
Processing	1a. Melt flow (gm./10 min.)	D1238					0.5-10	
	1. Melting temperature, °C. T_m (crystalline)		Thermoset	Thermoset	Thermoset		210-220	220
	T_g (amorphous)							
	2. Processing temperature range, °F. (C = compression; T = transfer; I = injection; E = extrusion)		C: 270-350 I: 330-380 T: 300-350				I: 440-550 E: 440-525	520-555
	3. Molding pressure range, 10^3 p.s.i.		2-20				1-20	
	4. Compression ratio		2.1-2.7				3.0-4.0	
	5. Mold (linear) shrinkage, in./in.	D955	0.002-0.006			0.002	0.003-0.015	2-3x10^{-3}
Mechanical	6. Tensile strength at break, p.s.i.	D638[b]	6000-10,000	5000-9000	4000-9000	25,000-30,000	6000-24,000	18,900[c]; 10,200[d]
	7. Elongation at break, %	D638[b]	0.1-0.5	1.5-2.0		3-4	30-100[c]; 300[d]	3.5[c]; 6[d]
	8. Tensile yield strength, p.s.i.	D638[b]					13,100[c]; 7400[d]	
	9. Compressive strength (rupture or yield), p.s.i.	D695	22,500-36,000	12,000-15,000	29,000-34,000		13,000-16,000[c]	
	10. Flexural strength (rupture or yield), p.s.i.	D790	11,000-14,000	11,000-17,000	9000-12,000	39,000-45,000	15,700[c]; 5800[d]	
	11. Tensile modulus, 10^3 p.s.i.	D638[b]	2400	400-700			380-464[c]; 100-247[d]	798[c]; 508[d]
	12. Compressive modulus, 10^3 p.s.i.	D695					250[d]	
	13. Flexural modulus, 10^3 p.s.i. 73° F.	D790	1000-2000			1800-2150	390-410[c]; 140[d]	700[d]; 420[d]
	200° F.	D790						
	250° F.	D790						
	300° F.	D790						
	14. Izod impact, ft.-lb./in. of notch ($^1/_8$-in. thick specimen)	D256A	0.26-0.6	0.24-0.4	0.35-0.5	1.6-2.0	0.6-2.2[c]; 3.0[d]	1.1
	15. Hardness Rockwell	D785	E88	M93-120	M85-120	R-120	R119[c]; M100-105[c]	M92[c]; M74[d]
	Shore/Barcol	D2240/ D2583	Barcol 70					
Thermal	16. Coef. of linear thermal expansion, 10^{-6} in./in./°F.	D696	19-38	68	75		80-83	52
	17. Deflection temperature under flexural load, °F. 264 p.s.i.	D648	275-475	165-175	150-175	410-495	155-185[c]	374
	66 p.s.i.	D648				420-515	347-375[c]	419
	18. Thermal conductivity, 10^{-4} cal.-cm./ sec.-cm.^{2}-°C.	C177	10-24	3.5			5.8	
Physical	19. Specific gravity	D792	1.42-1.84	1.24-1.32	1.68-1.70	1.59-1.81	1.12-1.14	1.23
	20. Water absorption ($^1/_8$-in. thick specimen), % 24 hr.	D570	0.02-0.3	0.1-0.36		0.35-0.50	1.3-1.9	2.6
	Saturation	D570	0.06-0.5				8.5-10.0	8.0
	21. Dielectric strength ($^1/_8$-in. thick specimen), short time, v./mil	D149	200-350	250-400	100-250		400[c]	
	SUPPLIERS[a]		ICI Fiberite; Lockport Thermosets; OxyChem; Plaslok; Plastics Eng.; Resinoid; Rogers	Ametek, Haveg; Monsanto; Schenectady Chem.; Union Carbide	Monsanto; Schenectady Chem.	Thermofil	Adell; Albis; AlliedSignal; ALM; Ashley Polymers; BASF; Bamberger Polymers; Bayer Corp.; ComAlloy; Custom Resins; DuPont; EMS; Federal Plastics; Hoechst Celanese; Nylon Corp. of Amer.; Polymer Resources; Polymers Intl.; A. Schulman; Texapol; Thermofil; Wellman	Ashley Polymers; BASF; ComAlloy; Polymers Intl.; RTP

Polyamide (Cont'd)

Nylon, Type 6 (Cont'd)

	25% glass fiber-reinforced	30-35% glass fiber-reinforced	50% glass fiber-reinforced	30% long glass fiber-reinforced	40% long glass fiber-reinforced	50% long glass fiber-reinforced	Toughened Unreinforced	Toughened 33% glass fiber-reinforced	Flame-retarded grade 30% glass fiber-reinforced
1a.									
1.	220	210-220	220	210-220	210-271	220	210-220	210-220	210-220
2.	520-555	I: 460-550	535-575	I: 460-550	I: 460-550	I: 480-540	I: 520-550	I: 520-550	I: 520-560
3.		2-20		10-20	10-20	10-20			12-25
4.		3.0-4.0		3.0-4.0	3.0-4.0	3-4			3.0-4.0
5.	2×10^{-3}	0.001-0.005	1×10^{-3}	0.003-0.009	0.002-0.010	0.002-0.008	0.006-0.02	0.001-0.003	0.001
6.	23,200[c]; 14,500[d]	24-27,600[c]; 18,900[d]	33,400[c]; 23,200[d]	25,200-26,000[c]	30,400-31,300	35,400-36,200[c]	6500-7900[c]; 5400[d]	17,800[c]	18,800-22,000[c]
7.	3.5[c]; 5[d]	2.2-3.6[c]	3.0[c]; 3.5[d]	2.3-2.5[c]	2.2-2.3	2.0-2.1[c]	65.0-150[c]	4.0[c]	1.7-3.0[c]
8.									22,000[c]
9.		19,000-24,000[c]		24,000-32,200[c]	33,800-37,400	39,700-39,900[c]			23,000[c]
10.		34-36,000[c]; 21,000[d]		38,800-40,000[c]	45,700	53,900[c]	9100[c]	25,800[c]	28,300-30,31,000[c]
11.	1160[c]; 798[d]	1250-1600[c]; 1090[d]	2320[c]; 1740[d]	1300	1800	2200-2270[c]	290[c]; 102[d]		1200[c]-1700
12.									
13.	910[c]; 650[d]	1250-1400[c]; 800-950[d]	1700[c]; 1570[d]	1200[c]	1600	1930-2000[c]	250[c]	1110[c]	1160-1400[c]
14.	2.0	2.1-3.4[c]; 3.7-5.5[d]		4.2[c]	6.2-6.4	8.4-8.6[c]	16.4[c]	3.5[c]	1.5[d]-2.2
15.	M95[c]; M83[d]	M93-96[c]; M78[d]	M104[c]; M93[d]	M93-96	M93				
16.	40	16-80	30	22[c]					
17.	410	392-420[c]	419	420[c]	405	415	135[c]; 122	400[c]	380-400[c]
	428	420-430[c]	428	425[c]			158	430[c]	420
18.		5.8-11.4							
19.	1.32	1.35-1.42	1.55	1.4	1.45	1.56	1.07; 1.08	1.33	1.62-1.7
20.	2.3	0.90-1.2	1.5	1.3				0.86	0.5-0.6
	7.1	6.4-7.0	4.8						
21.		400-450[c]		400					
	Ashley Polymers; BASF; Bayer Corp.; RTP	Adell; Albis; AlliedSignal; ALM; Ashley Polymers; BASF; Bamberger Polymers; Bayer Corp.; Chem Polymer; ComAlloy; DSM; EMS; Ferro Corp.; Ferro Eng. Therm.; Hoechst Celanese; LNP; M.A. Polymers; Nylon Corp. of Amer.; Polymers Intl.; RTP; Texapol; Thermofil; Wellman	Ashley Polymers; BASF; Bayer Corp.; ComAlloy; RTP	Adell; ALM; DSM; Ferro Eng. Therm.; Hoechst Celanese; LNP; RTP	Adell; ALM; DSM; Ferro Eng. Therm.; Hoechst Celanese; LNP; RTP	Hoechst Celanese; RTP	Adell; Albis; AlliedSignal; Ashley Polymers; BASF; Bamberger Polymers; Bayer Corp.; Chem Polymer; Custom Resins; DSM; EMS; Ferro Eng. Therm.; Nylon Corp. of Amer.; Polymers Intl.; Progressive Polymers; A. Schulman; Texapol	Adell; AlliedSignal; BASF; Bamberger Polymers; Bayer Corp.; Chem Polymer; ComAlloy; DSM; EMS; Ferro Eng. Therm.; LNP; Nylon Corp. of Amer.; Polymers Intl.; Progressive Polymers; A. Schulman; Texapol	AlliedSignal; Bayer Corp.; ComAlloy; DSM; Ferro Eng. Therm.; Hoechst Celanese; LNP; Nylon Corp. of Amer.; Polymers Intl.; RTP; Texapol

Polyamide (Cont'd)

Nylon, Type 6 (Cont'd)

		Properties	ASTM test method	40% mineral- and glass fiber-reinforced	40% mineral-reinforced	High-impact copolymers and rubber-modified compounds	Unfilled with molybdenum disulfide	Impact-modified; 30% glass fiber-reinforced
Processing	1a.	Melt flow (gm./10 min.)	D1238			1.5-5.0		
	1.	Melting temperature, °C. T_m (crystalline)		210-220	210-220	210-220	215	220
		T_g (amorphous)						
	2.	Processing temperature range, °F. (C = compression; T = transfer; I = injection; E = extrusion)		I: 450-550	I: 450-550	I: 450-580 E: 450-550	I: 460-500	I: 480-550
	3.	Molding pressure range, 10^3 p.s.i.		2-20	2-20	1-20	5-20	3-20
	4.	Compression ratio		3.0-4.0	3.0-4.0	3.0-4.0	3.0-4.0	3.0-4.0
	5.	Mold (linear) shrinkage, in./in.	D955	0.003-0.006	0.003-0.006	0.008-0.026	1.1	0.003-0.005
Mechanical	6.	Tensile strength at break, p.s.i.	D638[b]	17,400[c]; 19,000[d]	11,000-11,300	6300-11,000[c]	11,500[c]	21,000[c]; 14,500[d]
	7.	Elongation at break, %	D638[b]	3[c]; 2-6[d]	3.0	150-270[c]	60-80	5[c]-8[d]
	8.	Tensile yield strength, p.s.i.	D638[b]	19,000-20,000		9000[c]-9500	11,500-12,300	
	9.	Compressive strength (rupture or yield), p.s.i.	D695	14,000-18,000[c]		3900[c]		
	10.	Flexural strength (rupture or yield), p.s.i.	D790	23,000-30,000[c]		5000-12,000[c]	13,000	
	11.	Tensile modulus, 10^3 p.s.i.	D638[b]	1160-1400[c]; 725[d]			440	1220[c]-754[d]
	12.	Compressive modulus, 10^3 p.s.i.	D695					
	13.	Flexural modulus, 10^3 p.s.i. 73° F.	D790	900-1300[c]; 650-996[d]	650-700	110-320[c]; 130[d]	400	1160[c]-600[d]
		200° F.	D790			60-130[c]		
		250° F.	D790					
		300° F.	D790					
	14.	Izod impact, ft.-lb./in. of notch ($^1/_8$-in. thick specimen)	D256A	0.6-4.2[c]; 5.0[d]	1.8-2.0	1.8-No break[c] 1.8-No break[d]	0.9-1.0	2.2[c]-6[d]
	15.	Hardness Rockwell	D785	R118-121[c]		R81-113[c]; M50	R119-120	
		Shore/Barcol	D2240/ D2583					
Thermal	16.	Coef. of linear thermal expansion, 10^{-6} in./in./°F.	D696	11-41		72-120		20-25
	17.	Deflection temperature under flexural load, °F. 264 p.s.i.	D648	390-405[c]	270-285	113-140[c]	200	410[c]
		66 p.s.i.	D648	410-425[c]		260-367[c]	230	428[c]
	18.	Thermal conductivity, 10^{-4} cal.-cm./ sec.-cm.^{2}.°C.	C177					
Physical	19.	Specific gravity	D792	1.45-1.50	1.45	1.07-1.17	1.17-1.18	1.33
	20.	Water absorption ($^1/_8$-in. thick specimen), % 24 hr.	D570	0.6-0.9		1.3-1.7	1.1-1.4	2.0
		Saturation	D570	4.0-6.0		8.5		6.2
	21.	Dielectric strength ($^1/_8$-in. thick specimen), short time, v./mil	D149	490-550[c]		450-470[c]		
		SUPPLIERS[a]		Adell; AlliedSignal; Ashley Polymers; BASF; Bayer Corp.; ComAlloy; DSM; EMS; Ferro Eng. Therm.; LNP; M.A. Polymers; Nylon Corp. of Amer.; RTP; A. Schulman; Texapol; Thermofil; Wellman	Albis; AlliedSignal; Ashley Polymers; Bayer Corp.; RTP; A. Schulman	Adell; AlliedSignal; Ashley Polymers; BASF; Bamberger Polymers; Bayer Corp.; Custom Resins; EMS; M.A. Polymers; MKB; Nylon Corp. of Amer.; Polymers Intl.; RTP; A. Schulman; Texapol; Wellman	Ashley Polymers; DSM; Hoechst Celanese; LNP; RTP	Adell; Albis; AlliedSignal; Ashley Polymers; BASF; Bamberger Polymers; Bayer Corp.; ComAlloy; EMS; Ferro Eng. Therm.; LNP; M.A. Polymers; Nylon Corp. of Amer.; Polymers Intl.; RTP; Texapol

Polyamide (Cont'd)

	Nylon, Type 6 (Cont'd)						Nylon, Type 66		
	EMI shielding (conductive); 30% PAN carbon fiber	Cast	Cast, heat-stabilized	Cast, oil-filled	Cast, plasticized	Cast, Type 612 blend	Molding compound	13% glass fiber-reinforced, heat-stabilized	15% glass fiber-reinforced
1a.									
1.	210-220	227-238	227-238	227-238	227-238	204-216	255-265	257	260
2.	I: 520-575						I: 500-620	I: 520-570	535-575
3.							1-25	7-20	
4.							3.0-4.0	3.0-4.0	
5.	0.001-0.003						0.007-0.018	0.005-0.009	4×10^{-3}
6.	30,000-36,000[c]	12,500	12,500	9,500	10,000	10,000	13,700[c]; 11,000[d]	15,000-17,000	18,900[c]; 11,600[d]
7.	2-3[c]	20-45	20-30	45-55	25-35	25-80	15-80[c]; 150-300[d]	3.0-5	3[c]; 6[d]
8.							8000-12,000[c]; 6500-8500[d]		
9.	29,000[c]	17,000			22,000		12,500-15,000[c] (yld.)		
10.	46,000-51,000[c]	16,500	15,500	13,500	12,500	17,500	17,900-1700[c]; 6100[d]	27,500[c]; 15,000[d]	25,600[c]; 17,800[d]
11.	2800-3000[c]	500	500	350	400	300	230-550[c]; 230-500[d]		870[c]; 653[d]
12.		325	300	275	260				
13.	2500-2700[c]	430	430	370	340	330	410-470[c]; 185[d]	700-750	720[c]; 480[d]
14.	1.5-2.8[c]	0.7-0.9	0.7-0.9	1.4-1.8	0.8-0.9	0.9-1.4	0.55-1.0[c]; 0.85-2.1[d]	0.95-1.1	1.1
15.	E70[c]	R115-125	R110-115	R105-115	R110-115	R108-120	R120[c]; M83[c]; M95-M105[d]	95M/R120	M97[c]; M87[d]
		D78-83	D76-78	D74-78	D76-78	D75-81			
16.	14.0[c]	50	45	35	45	40-45	80		52
17.	415-490[c]	330-400	330-400	330-400	330-400	330-400	158-212[c]	450-470	482
	425-505[c]	400-430	400-430	400-430	400-430	400-430	425-474[c]	494	482
18.							5.8		
19.	1.28	1.15-1.17	1.15-1.17	1.14-1.15	1.14-1.16	1.10-1.13	1.13-1.15	1.21-1.23	1.23
20.	0.7-1.0	0.3-0.4	0.3-0.4	0.3-0.4	0.3-0.4	0.3-0.4	1.0-2.8	1.1	
		5-6	5-6	2-2.5	5-6	4-5	8.5	7.1	7
21.		500-600	500-600	500-600	500-600	500-600	600[c]		
	ComAlloy; DSM; Ferro Eng. Therm.; LNP; Nylon Corp. of Amer.; RTP; Thermofil	Cast Nylons	Cast Nylons	Cast Nylons	Cast Nylons	Cast Nylons	Adell; Albis; ALM; Ashley Polymers; BASF; Bamberger Polymers; Chem Polymer; ComAlloy; DSM; DuPont; EMS; Hoechst Celanese; MRC; Monsanto; Nylon Corp. of Amer.; Polymer Resources; Polymers Intl.; Rhone-Poulenc; A. Schulman; Texapol; Thermofil; Wellman	Adell; Albis; ALM; Ashley Polymers; ComAlloy; DSM; Hoechst Celanese; LNP; RTP; A. Schulman	BASF; RTP

Polyamide (Cont'd)

Nylon, Type 66 (Cont'd)

	Properties	ASTM test method	30-33% glass fiber-reinforced	50% glass fiber-reinforced	30% long glass fiber-reinforced	40% long glass fiber-reinforced	50% long glass fiber-reinforced
Processing	1a. Melt flow (gm./10 min.)	D1238					
	1. Melting temperature, °C. T_m (crystalline)		260-265	260	260-265	257-265	260
	T_g (amorphous)						
	2. Processing temperature range, °F. (C = compression; T = transfer; I = injection; E = extrusion)		I: 510-580	555-590	I: 530-570	I: 520-570	I: 550-580
	3. Molding pressure range, 10^3 p.s.i.		5-20		10-20	10-20	10-20
	4. Compression ratio		3.0-4.0			3.0-4.0	3-4
	5. Mold (linear) shrinkage, in./in.	D955	0.002-0.006	1×10^{-3}	0.003	0.002-0.10	0.002-0.007
Mechanical	6. Tensile strength at break, p.s.i.	D638[b]	27,600[c]; 20,300[d]	33,400[c]; 26,100[d]	24,000-28,000[c]	32,800-32,900	37,200-38,000[c]
	7. Elongation at break, %	D638[b]	2.0-3.4[c]; 3-7[d]	2[c]; 3[d]	2.1-2.5[c]	2.1-2.5	2.0-2.1[c]
	8. Tensile yield strength, p.s.i.	D638[b]	25,000[c]				
	9. Compressive strength (rupture or yield), p.s.i.	D695	24,000-40,000[c]		28,000-34,200[c]	37,700-42,700	42,900-44,900[c]
	10. Flexural strength (rupture or yield), p.s.i.	D790	40,000[c]; 29,000[d]	46,500[c]; 37,500[d]	40,000-40,300[c]	49,100	54,500-57,000[c]
	11. Tensile modulus, 10^3 p.s.i.	D638[b]	1380[c]; 1090[d]	2320[c]; 1890[d]		1700-1790	2270-2300[c]
	12. Compressive modulus, 10^3 p.s.i.	D695					
	13. Flexural modulus, 10^3 p.s.i. 73° F.	D790	1200-1450[c]; 800[d]; 900	1700[c]; 1460[d]	1200-1300[c]	1560-1600	1900-1920[c]
	200° F.	D790					
	250° F.	D790				810	990
	300° F.	D790					
	14. Izod impact, ft.-lb./in. of notch ($\frac{1}{8}$-in. thick specimen)	D256A	1.6-4.5[c]; 2.6-3.0[d]	2.5	4.0-5.1[c]	6.2; 6.6[c]; 6.9	8.0-9.2[c]
	15. Hardness Rockwell	D785	R101-119[c]; M101-102[c]; M96[c]	M102[c]; M98[d]	E60[c]		
	Shore/Barcol	D2240/D2583					
Thermal	16. Coef. of linear thermal expansion, 10^{-6} in./in./°F.	D696	15-54	0.33	23.4		
	17. Deflection temperature under flexural load, °F. 264 p.s.i.	D648	252-490[c]	482	485-495	490	500
	66 p.s.i.	D648	260-500[c]	482	505	490	
	18. Thermal conductivity, 10^{-4} cal.-cm./ sec.-cm.^{2}.°C.	C177	5.1-11.7				
Physical	19. Specific gravity	D792	1.15-1.40	1.55	1.36-1.4	1.45	1.56
	20. Water absorption ($\frac{1}{8}$-in. thick specimen), % 24 hr.	D570	0.7-1.1		0.9		
	Saturation	D570	5.5-6.5	4			
	21. Dielectric strength ($\frac{1}{8}$-in. thick specimen), short time, v./mil	D149	360-500		500		
	SUPPLIERS[a]		Adell; Albis; ALM; Ashley Polymers; BASF; Bamberger Polymers; Chem Polymer; ComAlloy; DSM; DuPont; EMS; Ferro Corp.; Ferro Eng. Therm.; Hoechst Celanese; LNP; MRC; Monsanto; Nylon Corp of Amer.; Polymer Resources; Polymers Intl.; RTP; Rhone-Poulenc; A. Schulman; Texapol; Thermofil; Wellman	Ashley Polymers; BASF; ComAlloy; RTP	Ashley Polymers; DSM; Ferro Eng. Therm.; Hoechst Celanese; LNP; RTP	DSM; Ferro Eng. Therm.; Hoechst Celanese; LNP; RTP	Ashley Polymers; Hoechst Celanese; RTP

Polyamide (Cont'd)

Nylon, Type 66 (Cont'd)

	60% long glass fiber-reinforced	Toughened		Modified high-impact, 25-30% mineral-filled	Flame-retarded grade		40% glass- and mineral-reinforced	40-45% mineral-filled
		Unreinforced	15-33% glass fiber-reinforced		Unreinforced	20-25% glass fiber-reinforced		
1a.								
1.	260	240-265	256-265	250-260	249-265	200-265	250-260	250-265
2.	I: 560-600	I: 520-580	I: 530-575	I: 510-570	I: 500-560	I: 500-560	I: 510-590	I: 520-580
3.	10-20	1-20					9-20	5-20
4.				10-20			3-4	3.0-4.0
5.	0.002-0.006	0.012-0.018	0.0025-0.0045[c]	0.01-0.018	0.01-0.016	0.004-0.005	0.001-0.005	0.012-0.022
6.	40,500-41,800[c]	7000[c]-11,000; 5800[d]	10,900-20,300[c]; 14,500[d]	9000-18,900[c]	8500[c]-10,600	20,300[c]-14,500[d]	15,500-31,000[c]; 13,100[d]	14,000[c]; 11,000[d]
7.	2.0-2.1[c]	4-200[c]; 150-300[d]	4.7[c]; 8[d]	5-16[c]	4-10.0[c]	2.3-3[c]	2-7[c]	5-10[c]; 16[d]
8.		7250-7500[c]; 5500[d]			10,600			13,900
9.	47,000-47,900[c]		15,000[c]-20,000		25,000[c]		18,000-37,000[c]	15,500-22,000[c]
10.	65,000[c]	8500-14,500[c]; 4000[d]	17,400-29,900[c]	20,000[c]	14,000-15,000[c]	23,000[c]	24,000-48,000[c]	22,000[c]; 9000[d]
11.	2900-3170[c]	290[c]-123[d]	1230[c]; 943[d]		420[c]	1230[c]; 870[d]		900[c]; 500[d]
12.								370[c]
13.	2580-2600[c]	225-380[c]; 125-150[d]	479-1100[c]	600[c]-653	400-420[c]	1102	985-1750[c]; 600[d]	900-1050[c]; 400[d]
	1440						600[c]	
14.	10.0-11.0[c]	12.0[c]-19.0; 3-N.B.[c]; 1.4-N.B.[d]	>3.2-5.0	1.0-3.0[c]	0.5-1.5[c]	1.1[c]	0.6-3.8[c]	0.9-1.4[c]; 3.9[d]
15.		M60[c]; R100[c]; R107[c]; R113; R114-115[c]; M50[d]	R107[c]; R115; R116; M86[c]; M70[d]	R120; M86[c]	M82[c]; R119	M98[c]; M90[d]	M95-98[c]	R106-121[c]
16.		80	43	30		50	20-54	27
17.	505	140-175[c]	446-470	300-470	170-200	482	432-485[c]	300-438
		385-442[c]	480-495	399-460	410-415	482	480-496[c]	320-480[c]
18.							11	9.6
19.	1.69	1.06-1.11	1.2-1.34	1.28-1.4	1.25-1.42	1.3-1.51	1.42-1.55	1.39-1.5
20.		0.8-2.3	0.7-1.5	0.9-1.1	0.9-1.1	0.7	0.4-0.9	0.6-0.55
		7.2	5				5.1	6.0-6.5
21.					520	430	300-525	450[c]
	Hoechst Celanese; RTP	Adell; ALM; Ashley Polymers; BASF; Bamberger Polymers; Chem Polymer; ComAlloy; DSM; DuPont; Ferro Eng. Therm.; Hoechst Celanese; LNP; MRC; Monsanto; Nylon Corp. of Amer.; Polymers Intl.; RTP; A. Schulman; Texapol	Adell; ALM; Ashley Polymers; BASF; Bamberger Polymers; Chem Polymer; ComAlloy; DSM; DuPont; EMS; Hoechst Celanese; LNP; Monsanto; Polymers Intl.; RTP; Texapol	ALM; Ashley Polymers; ComAlloy; DSM; Ferro Eng. Therm.; LNP; Nylon Corp. of Amer.; Texapol	Ashley Polymers; BASF; ComAlloy; DSM; DuPont; EMS; Montell Polyolefins; Polymers Intl.; RTP; Texapol; Wellman	BASF; ComAlloy; DSM; DuPont; Ferro Eng. Therm.; LNP; Montell Polyolefins; Polymers Intl.; RTP; Texapol	Adell; ALM; Ashley Polymers; BASF; Chem Polymer; ComAlloy; DSM; DuPont; EMS; Ferro Eng. Therm.; Hoechst Celanese; LNP; Montell Polyolefins; MRC; Monsanto; Nylon Corp. of Amer.; RTP; Texapol; Thermofil; Wellman	Adell; Albis; ALM; Ashley Polymers; BASF; ComAlloy; DSM; DuPont; Ferro Eng. Therm.; Hoechst Celanese; LNP; Montell Polyolefins; MRC; Monsanto; Nylon Corp. of Amer.; RTP; Rhone-Poulenc; Texapol; Thermofil; Wellman

Polyamide (Cont'd)

Nylon, Type 66 (Cont'd)

EMI shielding (conductive)

	Properties	ASTM test method	30% graphite or PAN carbon fiber	40% aluminum flake	5% stainless steel, long fiber	6% stainless steel, long fiber	10% stainless steel, long fiber	50% PAN carbon fiber
Processing	1a. Melt flow (gm./10 min.)	D1238						
	1. Melting temperature, °C. T_m (crystalline)		258-265	265	260-265	260-265	257-265	260-265
	T_g (amorphous)							
	2. Processing temperature range, °F. (C = compression; T = transfer; I = injection; E = extrusion)		I: 500-590	I: 525-600	I: 530-570	I: 530-570	I: 530-570	I: 530-570
	3. Molding pressure range, 10^3 p.s.i.		10-20	10-20		5-18	5-20	
	4. Compression ratio						3.0-4.0	
	5. Mold (linear) shrinkage, in./in.	D955	0.001-0.003	0.005	0.004	0.004-0.006	0.003-0.004	0.0005
Mechanical	6. Tensile strength at break, p.s.i.	D638[b]	27,600-35,000[c]	6000[c]	10,000[c]	10,000-11,300[c]	11,500	38,000[c]
	7. Elongation at break, %	D638[b]	1-4[c]	4[c]	5.0[c]	2.9; 5.0[c]	2.6	1.2[c]
	8. Tensile yield strength, p.s.i.	D638[b]						
	9. Compressive strength (rupture or yield), p.s.i.	D695	24,000-29,000[c]	7500[c]				
	10. Flexural strength (rupture or yield), p.s.i.	D790	45,000-51,000[c]	11,700[c]	16,000[c]	16,000-38,700[c]	18,100	54,000[c]
	11. Tensile modulus, 10^3 p.s.i.	D638[b]	3200-3400[c]	720[c]	450[c]	450-1500[c]	600	5000[c]
	12. Compressive modulus, 10^3 p.s.i.	D695						
	13. Flexural modulus, 10^3 p.s.i. 73° F.	D790	1500-2900[c]	690[c]	450[c]	500-1400	500	4200[c]
	200° F.	D790						
	250° F.	D790						
	300° F.	D790						
	14. Izod impact, ft.-lb./in. of notch ($^1/_8$-in. thick specimen)	D256A	1.3-2.5[c]	2.5[c]	1.3[c]	0.7; 2.2	0.7; 1.9[c]	2.0[c]
	15. Hardness Rockwell	D785	R120[c]; M106[c]	R114[c]				
	Shore/Barcol	D2240/D2583						
Thermal	16. Coef. of linear thermal expansion, 10^{-6} in./in./°F.	D696	11-16	22				
	17. Deflection temperature under flexural load, °F. 264 p.s.i.	D648	470-500	380	285	175-480	175-480	495
	66 p.s.i.	D648	500-510	400	295			505
	18. Thermal conductivity, 10^{-4} cal.-cm./sec.-cm.^{2}-°C.	C177	24.1					
Physical	19. Specific gravity	D792	1.28-1.43	1.48	1.27	1.19-1.45	1.24	1.38
	20. Water absorption ($^1/_8$-in. thick specimen), % 24 hr.	D570	0.5-0.8	1.1	0.12			0.5
	Saturation	D570						
	21. Dielectric strength ($^1/_8$-in. thick specimen), short time, v./mil	D149						
	SUPPLIERS[a]		ComAlloy; DSM; Ferro Corp.; Ferro Eng. Therm.; LNP; Nylon Corp. of Amer.; RTP; Thermofil	Thermofil	DSM; RTP	Hoechst Celanese; RTP	Hoechst Celanese; RTP	DSM; LNP; RTP

Polyamide (Cont'd)

Nylon, Type 66 (Cont'd)

	EMI shielding (conductive) (Cont'd)				Anti-friction molybdenum disulfide-filled	Lubricated			
	30% pitch carbon fiber	40% pitch carbon fiber	15% nickel-coated carbon fiber	40% nickel-coated carbon fiber		5% silicone	10% PTFE	30% PTFE	5% Molybdenum disulfide and 30% PTFE
1a.									
1.	260-265	260-265	260-265	260-265	249-265	260-265	260-265	260-265	260-265
2.	I: 530-570	I: 530-570	I: 530-570	I: 530-570	I: 500-600	I: 530-570	I: 530-570	I: 530-570	I: 530-570
3.					5-25				
4.									
5.	0.003	0.002	0.005	0.001	0.007-0.018	0.015	0.01	0.007	0.01
6.	15,500[c]-15,600	17,500[c]; 19,250	14,000[c]	20,000[c]	10,500-13,700[c]	8500[c]	9500[c]	5500[c]	7500[c]
7.	2.0[c]	1.5[c]	1.6[c]	2.5[c]	4.4-40[c]				
8.									
9.	20,500	24,000			12,000-12,500[c]				
10.	24,800-26,000[c]	28,000[c]-29,000	21,000[c]	27,000[c]	15,000-20,300[c]	15,000[c]	13,000[c]	8000[c]	1200[c]
11.	1400-2000[c]	2250-2600[c]	1100[c]		350-550[c]				
12.									
13.	1200-1500[c]	1800-2000[c]	1000[c]	2000[c]	420-495[c]	300[c]	420[c]	460[c]	400[c]
14.	0.6-0.7[c]	0.7-0.8[c]	0.7[c]	1.0[c]	0.9-4.5[c]	1.0[c]	0.8[c]	0.5[c]	0.6[c]
15.					R119[c]				
16.	16.0-19.0	9.0-14.0			54	63.0	35.0	45.0	
17.	465-490	475-490	460	470	190-260[c]	170	190	180	185
	490-500	498-500			395-430				
18.									
19.	1.30-1.31	1.36-1.38	1.20	1.46	1.15-1.18	1.16	1.20	1.34	1.37
20.	0.6	0.5	1.0	0.8	0.8-1.1	1.0	0.7	0.55	0.55
					8.0				
21.					360[c]				
	ComAlloy; LNP; Polymers Intl.; RTP	ComAlloy; LNP; RTP	DSM; RTP	DSM; RTP	Adell; ALM; Ashley Polymers; Chem Polymer; ComAlloy; DSM; LNP; Nylon Corp. of Amer.; RTP; Texapol; Thermofil	ComAlloy; DSM; Ferro Eng. Therm.; LNP; RTP; Texapol	ComAlloy; DSM; Ferro Eng. Therm.; LNP; RTP; Texapol	ComAlloy; DSM; Ferro Eng. Therm.; LNP; RTP; Texapol	ComAlloy; DSM; RTP; Texapol

Polyamide (Cont'd)

		Nylon, Type 66 (Cont'd)	Nylon, Type 69	Nylon, Type 610			
							Flame-retarded grade
Properties	ASTM test method	Copolymer	Molding and extrusion	Molding and extrusion compound	30-40% glass fiber-reinforced	30-40% long glass fiber-reinforced	30% glass fiber-reinforced
Processing							
1a. Melt flow (gm./10 min.)	D1238						
1. Melting temperature, °C. T_m (crystalline)		200-255	205	220	220	220	220
T_g (amorphous)							
2. Processing temperature range, °F. (C = compression; T = transfer; I = injection; E = extrusion)		I: 430-500	I: 450-500 E: 425-500	I: 445-485 E: 480-500	I: 510-550	I: 510-550	I: 500-560
3. Molding pressure range, 10^3 p.s.i.		1-15	1-15	1-19			
4. Compression ratio				3-4			
5. Mold (linear) shrinkage, in./in.	D955	0.006-0.015	0.010-0.015	0.005-0.015	0.015-0.04	0.013-0.03	0.002
Mechanical							
6. Tensile strength at break, p.s.i.	D638[b]	7400-12,400[c]	8500[c]	10,150[c]; 7250[d]	22,000-26,700[c]	22,400-25,400[c]	19,000[c]
7. Elongation at break, %	D638[b]	40-150[c]; 300[d]	1125[c]	70[c]; 150[d]	4.3-4.7[c]	4.0-4.1[c]	3.5[c]
8. Tensile yield strength, p.s.i.	D638[b]						
9. Compressive strength (rupture or yield), p.s.i.	D695				20,400-21,000[c]	20,000[c]	23,000[c]
10. Flexural strength (rupture or yield), p.s.i.	D790	12,000		350[c]; 217[d]	32,700-38,000[c]	34,000-37,400[c]	28,000[c]
11. Tensile modulus, 10^3 p.s.i.	D638[b]	150-410[c]	275[c]		800[c]	950-1600[c]	
12. Compressive modulus, 10^3 p.s.i.	D695						
13. Flexural modulus, 10^3 p.s.i. 73° F.	D790	150-410[c]	290[c]		1150-1500[c]	1200-1430[c]	1230[c]
200° F.	D790						
250° F.	D790						
300° F.	D790						
14. Izod impact, ft.-lb./in. of notch ($1/8$-in. thick specimen)	D256A	0.7[c]; No break[d]	1.1[c]		1.6-2.4	3.2-4.2[c]	1.5[c]
15. Hardness Rockwell	D785	R114-119; R83[d]; M75[c]	R111[c]		E43-48[c]	E42-56[c]	M89[c]
Shore/Barcol	D2240/ D2583						
Thermal							
16. Coef. of linear thermal expansion, 10^{-6} in./in./°F.	D696						
17. Deflection temperature under flexural load, °F. 264 p.s.i.	D648	135-170[c]	135-140[c]		410-415	425	390
66 p.s.i.	D648	430[c]-440	330-340[c]		430	430	
18. Thermal conductivity, 10^{-4} cal.-cm./ sec.-cm.^{2}.°C.	C177						
Physical							
19. Specific gravity	D792	1.08-1.14	1.08-1.10		1.3-1.4	1.33-1.39	1.55
20. Water absorption ($1/8$-in. thick specimen), % 24 hr.	D570	1.5-2.0	0.5	1.4	0.17-0.19	0.17-0.21	0.16
Saturation	D570	9.0-10.0		3.3			
21. Dielectric strength ($1/8$-in. thick specimen), short time, v./mil	D149	400[c]	600[c]			500	
SUPPLIERS[a]		AlliedSignal; Ashley Polymers; BASF; Chem Polymer; DuPont; EMS; Monsanto; Nylon Corp. of Amer.; Polymers Intl.; Texapol	Monsanto	BASF; Texapol	Ashley Polymers; DSM; Ferro Eng. Therm.; LNP; RTP; Texapol	DSM; LNP	DSM; LNP; RTP

Polyamide (Cont'd)

			Nylon, Type 612							Nylon, Type 46	
			Toughened		Flame-retarded grade	Lubricated					
	Molding compound	30-35% glass fiber-reinforced	35-45% long glass fiber-reinforced	Unrein-forced	33% glass fiber-reinforced	30% glass fiber-reinforced	10% PTFE	15% PTFE, 30% glass fiber-reinforced	10% PTFE, 30% PAN carbon fiber	Extrusion	Un-reinforced
1a.											
1.	195-219	213-217	195-217	195-217	195-217	195-217	195-217	195-217	195-217	295	295
2.	I: 450-550 E: 464-469	I: 450-550	I: 510-550	I: 510-550	I: 510-550	I: 500-560	I: 510-550	I: 510-550	I: 510-550	E: 560-590	I: 570-600
3.	1-15	4-20									5-15
4.										3-4	3-4
5.	0.011	0.002-0.005	0.001-0.002				0.012	0.002-0.003	0.0013		0.018-0.020
6.	6500-8800^c	22,000^c; 20,000^d	26,000-29,000^c	5500^c	18,000^c	18,000-19,000^c	7000^c	19,500-20,000^c	28,000^c	8,500	14,400
7.	150^c; 300^d	4^c; 5^d	2.9-3.2^c	40^c	5^c	2.0-3.5^c		2.5		60	25
8.	5800-8400^c; 3100^d										
9.		22,000^c	23,000^c			15,000-21,000^c		19,000			13,000
10.	11,000^c; 4300^d	32,000-35,000^c	39,000-44,000^c	6500^c	27,000^c	28,000^c		30,500-31,000^c	42,000^c	11,500	21,700
11.	218-290^c; 123-180^d	1200^c; 900^d				1000^c-1400		1200		250	435
12.											319
13.	240-334^c; 74-100^d	1100-1200^c; 900^d	1200-1500^c	195^c	1050^c	1200-1250^c	100^c	1100-1200^c	2600^c	270	460
14.	1.0-1.9^c; 1.4-No break^d	1.8-2.6^c	4.2-6.3^c	12.5^c	4.5^c	1.0-1.5^c	1.0^c	2.5-3.0^c	2.4^c	17	1.8
15.	M78^c; M34^d; R115	M93^c; E40-50^d; R116	E40^c			M89^c					R113
	D72-80^c; D63^d										D85
16.			21.6-25.2					18.0			
17.	136-180^c	390-425^c	410-415	135	385	385-390	202^c	385	390	194	320
	311-330^c	400-430^c	420-425			400					
18.	5.2	10.2									
19.	1.06-1.10	1.30-1.38	1.34-1.45	1.03	1.28	1.55-1.60	1.13	1.42-1.45	1.30	1.10	1.18
20.	0.37-1.0	0.2	0.2	0.3	0.2	0.16	0.2	0.13	0.15	1.84	2.3
	2.5-3.0	1.85									
21.	400^c	520^c				450					673
	ALM; Ashley Polymers; DuPont; EMS; Huls America; A. Schulman; Texapol	ALM; Ashley Polymers; ComAlloy; DSM; DuPont; Ferro Corp.; Ferro Eng. Therm.; LNP; RTP; Texapol	DSM; RTP	DSM; DuPont	DSM; DuPont; LNP	ComAlloy; DSM; LNP; RTP	ComAlloy; DSM; Ferro Eng. Therm.; LNP; RTP; Texapol	ComAlloy; DSM; Ferro Eng. Therm.; LNP; RTP; Texapol	ComAlloy; DSM; LNP; RTP	DSM	DSM

| | | | Polyamide (Cont'd) | | | | | | |
| | | | Nylon, Type 46 (Cont'd) | | | | | | |
Materials	Properties	ASTM test method	Super-tough	15% glass-reinforced	15% glass-reinforced, V-0	30% glass-reinforced	30% glass-reinforced, V-0	50% glass-reinforced	50% glass and mineral-filled
Processing	1a. Melt flow (gm./10 min.)	D1238							
	1. Melting temperature, °C. T_m (crystalline)		295	295	295	295	295	295	295
	T_g (amorphous)								
	2. Processing temperature range, °F. (C = compression; T = transfer; I = injection; E = extrusion)		I: 570-600	I: 570-600	I: 570-600	I: 570-600	I: 570-600	I: 570-600	I: 570-600
	3. Molding pressure range, 10^3 p.s.i.		5-15	5-15	5-15	5-15	5-15	5-15	5-15
	4. Compression ratio		3-4	3-4	3-4	3-4	3-4	3-4	3-4
	5. Mold (linear) shrinkage, in./in.	D955	0.018-0.020	0.005-0.009	0.006-0.009	0.004-0.006	0.004-0.006	0.002-0.004	0.003-0.006
Mechanical	6. Tensile strength at break, p.s.i.	D638[b]	8,500	21,500	16,500	30,000	23,000	34,000	20,000
	7. Elongation at break, %	D638[b]	60	3	8	4	3	3	2
	8. Tensile yield strength, p.s.i.	D638[b]							
	9. Compressive strength (rupture or yield), p.s.i.	D695				33,000	34,000		
	10. Flexural strength (rupture or yield), p.s.i.	D790	11,500	31,900	27,000	43,000	34,000	50,750	34,000
	11. Tensile modulus, 10^3 p.s.i.	D638[b]	250	841	1,000	1,300	1,500	2,320	2,100
	12. Compressive modulus, 10^3 p.s.i.	D695				507	688		
	13. Flexural modulus, 10^3 p.s.i. 73° F.	D790	270	798	1125	1,200	1,300	2,030	1700
	200° F.	D790							
	250° F.	D790							
	300° F.	D790							
	14. Izod impact, ft.-lb./in. of notch (1/8-in. thick specimen)	D256A	17	1.6	.5	2.0	1.3	2.2	1.1
	15. Hardness Rockwell	D785				R120	R120		
	Shore/Barcol	D2240/ D2583				D89	D88		
Thermal	16. Coef. of linear thermal expansion, 10^{-6} in./in./°F.	D696							
	17. Deflection temperature under flexural load, °F. 264 p.s.i.	D648	194	480	480	545	545	545	545
	66 p.s.i.	D648							
	18. Thermal conductivity, 10^{-4} cal.-cm./ sec.-cm.^{2}-°C.	C177							
Physical	19. Specific gravity	D792	1.10	1.3	1.47	1.41	1.68	1.62	1.6
	20. Water absorption (1/8-in. thick specimen), % 24 hr.	D570	1.84			1.5	0.9	1.15	
	Saturation	D570							
	21. Dielectric strength (1/8-in. thick specimen), short time, v./mil	D149				863	838		
	SUPPLIERS[a]		DSM	DSM	DSM	DSM	DSM	DSM	DSM

	Polyamide (Cont'd)			Polyamide-imide					
Nylon, Type 11	Nylon, Type 12	Aromatic polyamide							
Molding and extrusion compound	Molding and extrusion compound	Amorphous transparent copolymer	Aramid molded parts, unfilled	Unfilled compression and injection molding compound	30% glass fiber-reinforced	Graphite fiber-reinforced	Bearing grade	High compressive strength	Wear resistant for speeds
1a.									
1. 180-190	160-209		275						
125-155				275	275	275	275		
2. I: 390-520 E: 390-475	I: 356-525 E: 350-500	I: 480-610 E: 520-595		C: 600-650 I: 610-700	C: 630-650 I: 610-700	C: 630-650 I: 610-700	I: 580-700		
3. 1-15	1-15	5-20		2-40	15-40	15-40	15-40		
4. 2.7-3.3	2.5-4			1.0-1.5	1.0-1.5	1.0-1.5	1.0-1.5		
5. 0.012	0.003-0.015	0.004-0.007		0.006-0.0085	0.001-0.0025	0.000-0.0015	0.0025-0.0045	3.5-6.0	3.5-6.0
6. 8000^c-9,500	5100-$10,000^c$; 8000^d	7600-$14,000^c$; $13,000^d$	$17,500^c$	22,000					
7. 300^c-400	250-390^c	40-150^c; 260^d	5^c	15	7	6	7	7	9
8.	3000-6100^c	$11,000$-$14,861^c$; $11,000^d$		27,800	29,700	26,000	22,000	23,700	17,800
9. 7300-7800		$17,500^c$; $14,000^d$	$30,000^c$	32,100	38,300	36,900		24,100	18,300
10.	1400-8100^c	$10,000$-$16,400^c$; $14,000^d$	$25,800^c$	34,900	48,300	50,700	27,100-31,200	31,200	27,000
11. 185^c	36-180^c	275-410^c; 270^d		700	1560	3220	870-1130	950	870
12. 180^c		340^c	290^c		1150	1430			
13. 44-180^c	27-190^c	306-400^c; 350^d	640^c	730	1700	2880	910-1060	1000	910
20									
14. 1.8^c-N.B.	1.0-N.B.	0.8-3.5^c; 1.8-2.7^d	1.4^c	2.7	1.5	0.9	1.2-1.6	1.2	1.3
15. $R108^c$; R80	R70-109^c; 105^d	M77-93^c	$E90^c$	E86	E94	E94	E66-E72	72	66
	D58-75^d	$D83^c$; $D85^d$							
16. 100	61-100	28-70	40	30.6	16.2	9.0	25.2-27.0	14	15
17. 104-126	95-135^c	170-268	500^c	532	539	540	532-536	534	532
300^c	158-302^c	261-330^c							
18. 8	5.2-7.3	5	5.2	6.2	8.8	12.7		3.7	
19. 1.03-1.05	1.01-1.02	1.0-1.19	1.30	1.42	1.61	1.48	1.46-1.51	1.46	1.50
20. 0.4	0.25-0.30	0.4-1.36	0.6				0.17-0.33		
1.9	0.75-1.6	1.3-4.2		0.33	0.24	0.26	0.33	0.28	0.17
21. 650-750	450^c	350^c	800^c	580	840				
Elf Atochem N.A.	ALM; Ashley Polymers; Elf Atochem N.A.; EMS; Hüls America	AlliedSignal; Ashley Polymers; Bayer Corp.; DuPont; EMS; Hüls America; Nylon Corp. of Amer.; Texapol	DuPont	Amoco Polymers	Amoco Polymers	Amoco Polymers	Amoco Polymers	Amoco Polymers	Amoco Polymers

		Polyamide-imide (Cont'd)			Polyaryletherketone				
Properties	ASTM test method	Stiffness and lubricity	Cost/ performance ratio	Injection molding grade	20% glass fiber-reinforced	30% glass fiber-reinforced	20%-30% carbon fiber-reinforced	Lubricated
1a. Melt flow (gm./10 min.)	D1238								
1. Melting temperature, °C. T_m (crystalline)				381	381	381	381	381
T_g (amorphous)									
2. Processing temperature range, °F. (C = compression; T = transfer; I = injection; E = extrusion)				I: 734-788 E: 752-806	I: 752-806	I: 752-806	I: 752-806	I: 752-806
3. Molding pressure range, 10^3 p.s.i.									
4. Compression ratio									
5. Mold (linear) shrinkage, in./in.	D955	0.0-1.5		0.012-0.015	0.009	0.006	0.001-0.02	0.006
6. Tensile strength at break, p.s.i.	D638[b]				24,700	26,800	34,800-37,700	23,200
7. Elongation at break, %	D638[b]	6	7	20-30	3.3	2.5	1.3-1.6	2.0
8. Tensile yield strength, p.s.i.	D638[b]	26,000	31,800	17,100				
9. Compressive strength (rupture or yield), p.s.i.	D695	30,300	46,700					
10. Flexural strength (rupture or yield), p.s.i.	D790	40,100	52,000	18,850	34,800	36,250	48,720-52,200	
11. Tensile modulus, 10^3 p.s.i.	D638[b]		2020	580	1320	1740	2830-3770	1986
12. Compressive modulus, 10^3 p.s.i.	D695							
13. Flexural modulus, 10^3 p.s.i. 73° F.	D790	2440	2100					
200° F.	D790								
250° F.	D790								
300° F.	D790								
14. Izod impact, ft.-lb./in. of notch ($^1/_8$-in. thick specimen)	D256A	1.0	1.5	0.90-1.16	1.46	2.1	1.22-2.06	1.31
15. Hardness Rockwell	D785	107						
Shore/Barcol	D2240/ D2583			D86	D88	D90		
16. Coef. of linear thermal expansion, 10^{-6} in./in./°F.	D696	7	7	41	25	25	10-20	32
17. Deflection temperature under flexural load, °F. 264 p.s.i.	D648	534	536	388	482	482	482	545
66 p.s.i.	D648			482	590	662	590-662	662
18. Thermal conductivity, 10^{-4} cal.-cm./ sec.-cm.^{2}-°C.	C177							
19. Specific gravity	D792	1.50		1.30	1.45	1.53	1.40-1.44	1.45
20. Water absorption ($^1/_8$-in. thick specimen), % 24 hr.	D570							
Saturation	D570		0.21	0.8	0.65	0.50		
21. Dielectric strength ($^1/_8$-in. thick specimen), short time, v./mil	D149	Conductive	490					
SUPPLIERS[a]		Amoco Polymers	Amoco Polymers	BASF	BASF; RTP	BASF; RTP	BASF; RTP	BASF; RTP

	Polybutadiene	Polybutylene			Polycarbonate						
					Unfilled molding and extrusion resins		Glass fiber-reinforced				
	Casting resin	Extrusion compound	Film grades	Adhesive resin	High viscosity	Low viscosity	10% glass	30% glass	20-30% long glass fiber-reinforced	40% long glass fiber-reinforced	50% long glass fiber-reinforced
1a.					3-10	10-30	7.0				
1.	Thermoset	126	118	90							
					150	150	150	150	150	150	150
2.		C: 300-350 I: 290-380 E: 290-380	E: 380-420	300-350	I: 560	I: 520	I: 520-650	I: 550-650	I: 590-650	I: 575-620	570-620
3.		10-30			10-20	8-15	10-20	10-30	10-30	10-20	10-20
4.		2.5	2.5		1.74-5.5	1.74-5.5					3
5.		0.003 (unaged) 0.026 (aged)			0.005-0.007	0.005-0.007	0.002-0.005	0.001-0.002	0.001-0.003	0.001-0.003	0.001-0.003
6.		3800-4400	4000	3500	9100-10,500	9100-10,500	7000-10,000	19,000-20,000	18,000-23,000	23,100	25,100-26,000
7.		300-380	350	500	110-120	110-150	4-10	2-5	1.9-3.0	1.4-1.7	1.3-1.5
8.		1700-2500	1700	600	9000	9000	8500-11,600				
9.					10,000-12,500	10,000-12,500	12,000-14,000	18,000-20,000	18,000-29,800	31,600	32,700
10.	8000-14,000	2000-2300			12,500-13,500	12,000-14,000	13,700-16,000	23,000-25,000	22,000-36,900	36,400	40,800
11.	560	30-40	30	10-15	345	345	450-600	1250-1400	1200-1500	1700	2200
12.		31			350	350	520	1300			
13.		45-50		13	330-340	330-340	460-580	1100	800-1500	1500-1700	2100
					275	275	440	960			
					245	245	420	900			
14.		No break	No break	No break	12-18 @ 1/8 in. 2.3 @ 1/4 in.	12-16 @1/8 in. 2.0 @1/4 in.	2-4	1.7-3.0	3.5-4.7	5.0	6.6
15.	R40		D45	A90	M70-M75	M70-M75	M62-75; R118-122	M92, R119	M85-95		
		Shore D55-65		D25							
16.		128-150			68	68	32-38	22-23			
17.		130-140			250-270	250-270	280-288	295-300	290-300	305	310
		215-235			280-287	273-280	295	300-305	305		
18.		5.2			4.7	4.7	4.6-5.2	5.2-7.6			
19.	0.97	0.91-0.925	0.909	0.895	1.2	1.2	1.27-1.28	1.4-1.43	1.34-1.43	1.52	1.63
20.	0.03	0.01-0.02			0.15	0.15	0.12-0.15	0.08-0.14	0.09-0.11		
					0.32-0.35	0.32-0.35	0.25-0.32				
21.	630	>450			380->400	380->400	470-530	470-475			
	Colorado Chem.; OxyChem	Shell	Shell	Shell	Albis; American Polymers; Ashley Polymers; Bamberger Polymers; Bayer Corp.; Dow Plastics; Federal Plastics; GE Plastics; MRC Polymers; Polymer Resources; Progressive Polymers; RTP; Shuman	Albis; American Polymers; Ashley Polymers; Bayer Corp.; Federal Plastics; GE Plastics; MRC Polymers; Polymer Resources; Progressive Polymers; RTP; Shuman	Albis; American Polymers; Ashley Polymers; Bayer Corp.; ComAlloy; DSM; Dow Plastics; Federal Plastics; Ferro Corp.; Ferro Eng. Therm.; GE Plastics; LNP; MRC Polymers; Polymer Resources; RTP; Thermofil	Albis; American Polymers; Ashley Polymers; ComAlloy; DSM; Ferro Corp.; Ferro Eng. Therm.; GE Plastics; LNP; MRC Polymers; Polifil; Polymer Resources; RTP; Thermofil	DSM; Hoechst Celanese; RTP	Hoechst Celanese; RTP	Hoechst Celanese; RTP

Polycarbonate (Cont'd)

			Flame-retarded grade		High-heat			Conductive poly-carbonate	EMI shielding (conductive)
Properties	ASTM test method	35% random glass mat	20-30% glass fiber-reinforced	Polyester copolymer	Poly-carbonate copolymer	Impact-modified poly-carbonate/polyester blends	6% stainless steel fiber	10% stainless steel fiber
1a. Melt flow (gm./10 min.)	D1238							
1. Melting temperature, °C. T_m (crystalline)								
T_g (amorphous)			149	160-195	160-205		150	150
2. Processing temperature range, °F. (C = compression; T = transfer; I = injection; E = extrusion)		C: 560-600	I: 530-590	I: 575-710	I: 580-660	I: 475-560		I: 590-650
3. Molding pressure range, 10^3 p.s.i.		2-3.5		8-20	8-20	15-20	10-20	
4. Compression ratio				1.5-3	2-3	2-2.5	3	
5. Mold (linear) shrinkage, in./in.	D955	0.002-0.003	0.002-0.004	0.007-0.010	0.007-0.009	0.006-0.009	0.004-0.006	0.003-0.006
6. Tensile strength at break, p.s.i.	D638[b]	18,000	14,000-20,000	9500-11,300	8300-10,000	7600-8500	9800	10,110-11,000
7. Elongation at break, %	D638[b]	2.5	2.0-3.0	50-122	70-90	120-165	4.7	4.0
8. Tensile yield strength, p.s.i.	D638[b]	18,000		8500-9800	9300-10,500	7400-8300		
9. Compressive strength (rupture or yield), p.s.i.	D695	14,800	18,000-21,000	11,500		7000		
10. Flexural strength (rupture or yield), p.s.i.	D790	30,000	21,000-30,000	10,000-13,800	12,000-14,000	10,900-12,500	14,000	16,300-17,000
11. Tensile modulus, 10^3 p.s.i.	D638[b]	1100	1000-1100	320-340	320-340		410	500
12. Compressive modulus, 10^3 p.s.i.	D695							
13. Flexural modulus, 10^3 p.s.i. 73° F.	D790	1000	900-1200	294-340	320-340	280-325	400	500
200° F.	D790							
250° F.	D790							
300° F.	D790							
14. Izod impact, ft.-lb./in. of notch ($^1/_8$-in. thick specimen)	D256A	11.8	1.8-2.0	1.5-10	1.5-12	2-18	0.8-1.7	1.1-1.7
15. Hardness Rockwell	D785		M77-85	M74-92	M75-91	R114-122		
Shore/Barcol	D2240/D2583							
16. Coef. of linear thermal expansion, 10^{-6} in./in./°F.	D696			70-92	70-76	80-95		1410
17. Deflection temperature under flexural load, °F. 264 p.s.i.	D648	290	288-305	285-335	284-354	190-250	270	270-295
66 p.s.i.	D648		295	305-365	306-383	223-265		
18. Thermal conductivity, 10^{-4} cal.-cm./ sec.-cm.^{2}.-°C.	C177			4.7-5.0	4.7-4.8	4.3		
19. Specific gravity	D792	1.39	1.36-1.45	1.15-1.2	1.14-1.18	1.20-1.22	1.28	1.26-1.35
20. Water absorption ($^1/_8$-in. thick specimen), % 24 hr.	D570		0.15-0.17	0.15-0.2	0.15-0.2	0.12-0.16		0.12
Saturation	D570					0.35-0.60		
21. Dielectric strength ($^1/_8$-in. thick specimen), short time, v./mil	D149	528		509-520	>400	440-500		
SUPPLIERS[a]		Azdel	ComAlloy; DSM; Ferro Eng. Therm.; GE Plastics; LNP; Polymer Resources; RTP	Ferro Eng. Therm.; GE Plastics; Progressive Polymers	Bayer Corp.	Eastman; GE Plastics; MRC Polymers; Progressive Polymers	Hoechst Celanese; RTP	DSM; Hoechst Celanese; LNP; RTP

	Polycarbonate (Cont'd)							Polydicyclopentadiene	Polyester, thermoplastic		
	EMI shielding (conductive) (Cont'd)			Lubricated					Polybutylene terephthalate		
	20% PAN carbon fiber	30% graphite fiber	40% PAN carbon fiber	10-15% PTFE	30% PTFE	10-15% PTFE, 20% glass fiber-reinforced	2% silicone, 30% glass fiber-reinforced	RIM solid; unfilled	Unfilled	30% glass fiber-reinforced	30% long glass fiber-reinforced
1a.											
1.							Thermoset	Thermoset	220-267	220-267	235
	150	149-150		150	150	150	150	90-165			
2.	I: 590-650	I: 540-650	I: 580-620	I: 590-650	I: 590-650	I: 590-650	I: 590-650	T: <95-<100	I: 435-525	I: 440-530	I: 480-540
3.		15-20	15-20					<0.050	4-10	5-15	10-20
4.											3-4
5.	0.001	0.001-0.002	0.0005-0.001	0.007	0.009	0.002	0.002	0.008-0.012	0.009-0.022	0.002-0.008	0.001-0.003
6.	18,000-20,000	20,000-24,000	23,000-24,000	7500-10,000	6000	12,000-15,000	16,000	5300-6000	8200-8700	14,000-19,500	20,000
7.	2.0	1-5	1-2	8-10		2		5-70	50-300	2-4	2.2
8.								5000-6700	8200-8700		
9.	18,500	19,000-26,000	22,000	110,000		11,000		8500-9000	8600-14,500	18,000-23,500	26,000
10.	27,000-28,000	30,000-36,000	34,000-35,000	11,000-11,500	7600	18,000-23,000	22,000	10,000-11,000	12,000-16,700	22,000-29,000	35,000
11.	2000	250-2150	3000-3100	340		1200		240-280	280-435	1300-1450	1400
12.									375	700	
13.	1500-1800	240-1900	2800-2900	250-300	460	850-900	900	260-280	330-400	850-1200	1300
14.	1.4-2.0	1.8	1.5-2.0	2.5-3.0	1.3	1.8-3.5	3.5	5.0-9.0	0.7-1.0	0.9-2.0	5.7
15.		R118, R119	R119						M68-78	M90	
								D72-D84			
16.	180	9	11.0-14.4	58.0		21.6-23.4	12.0	46-49 in./in./°F.	60-95	15-25	
17.	290-295	280-300	295-300	270-275	260	280-290	290	217-240	122-185	385-437	405
	300	295	300			290		239	240-375	421-500	
18.		16.9	17.3						4.2-6.9	7.0	
19.	1.28	1.32-1.33	1.36-1.38	1.26-1.29	1.39	1.43-1.5	1.46	1.03-1.04	1.30-1.38	1.48-1.54	1.56
20.	0.2	0.04-0.08	0.08-0.13	0.13	0.06	0.11	0.12	0.09	0.08-0.09	0.06-0.08	
									0.4-0.5	0.35	
21.									420-550	460-560	
	ComAlloy; DSM; Ferro Eng. Therm.; LNP; RTP	ComAlloy; DSM; LNP; RTP; Thermofil	ComAlloy; DSM; Ferro Eng. Therm.; RTP; Thermofil	ComAlloy; DSM; GE Plastics; LNP; Polymer Resources; RTP	DSM; Polymer Resources; RTP	ComAlloy; DSM; Ferro Eng. Therm.; GE Plastics; LNP; Polymer Resources; RTP	ComAlloy; DSM; Ferro Eng. Therm; LNP; RTP	BFGoodrich; Hercules	Albis; Ashley Polymers; BASF; ComAlloy; Dainippon; DuPont; GE Plastics; Hoechst Celanese; Hüls America; RTP; Texapol; Thermofil	Albis; Adell; Ashley Polymers; BASF; ComAlloy; DSM; DuPont; Ferro Corp.; Eng. Therm.; GE Plastics; Hoechst Celanese; Hüls America; LNP; Polymer Resources; RTP; Texapol; Thermofil	Hoechst Celanese

Polyester, thermoplastic (Cont'd)

Polybutylene terephthalate (Cont'd)

		Properties	ASTM test method	40% long glass fiber-reinforced	50% long glass fiber-reinforced	60% long glass fiber-reinforced	25% random glass mat	35% random glass mat	40-45% glass fiber- and mineral-reinforced	35% glass fiber- and mica-reinforced
Processing		1a. Melt flow (gm./10 min.)	D1238							
		1. Melting temperature, °C. T_m (crystalline)		235	235	235			220-228	220-224
		T_g (amorphous)								
		2. Processing temperature range, °F. (C = compression; T = transfer; I = injection; E = extrusion)		I: 470-540	I: 480-540	I: 490-530	C: 520-560	C: 520-560	I: 450-520	I: 480-510
		3. Molding pressure range, 10^3 p.s.i.		10-15	10-20	10-20	1.5-3	2-3	10-15	9-15
		4. Compression ratio		3.5-4.0	3-4	3-4			3-4	
		5. Mold (linear) shrinkage, in./in.	D955	0.001-0.008	0.001-0.007	0.001-0.006	0.0035-0.0045	0.003-0.004	0.003-0.010	0.003-0.012
Mechanical		6. Tensile strength at break, p.s.i.	D638[b]	22,900	24,000	20,000	12,000	15,000	12,000-14,800	11,400-13,800
		7. Elongation at break, %	D638[b]	1.4	1.3	1.0	2.8	2.1	2-5	2-3
		8. Tensile yield strength, p.s.i.	D638[b]				12,000	15,000		
		9. Compressive strength (rupture or yield), p.s.i.	D695	24,500	24,600	24,600		14,700	15,000	
		10. Flexural strength (rupture or yield), p.s.i.	D790	35,200	35,500	41,000	28,000	32,000	18,500-23,500	18,000-22,000
		11. Tensile modulus, 10^3 p.s.i.	D638[b]	1900	1900	2100	980	1300	1350-1800	
		12. Compressive modulus, 10^3 p.s.i.	D695						1000	
		13. Flexural modulus, 10^3 p.s.i. 73° F.	D790	1600	2200	2500	900	1200	1250-1600	1200-1600
		200° F.	D790							
		250° F.	D790							
		300° F.	D790							
		14. Izod impact, ft.-lb./in. of notch ($^1/_8$-in. thick specimen)	D256A	6.6	8.5	8.0		13.0	0.7-2.0	0.7-1.8
		15. Hardness Rockwell	D785						M75-86	M50-76
		Shore/Barcol	D2240/ D2583							
Thermal		16. Coef. of linear thermal expansion, 10^{-6} in./in./°F.	D696				29	20	1.7	
		17. Deflection temperature under flexural load, °F. 264 p.s.i.	D648	415	420	450	403	425	388-395	330-390
		66 p.s.i.	D648						408-426	410-416
		18. Thermal conductivity, 10^{-4} cal.-cm./ sec.-cm.^{2}-°C.	C177	1.72						
Physical		19. Specific gravity	D792	1.65	1.75	1.87	1.45	1.59	1.58-1.74	1.59-1.74
		20. Water absorption ($^1/_8$-in. thick specimen), % 24 hr.	D570						0.04-0.07	0.04-0.11
		Saturation	D570							
		21. Dielectric strength ($^1/_8$-in. thick specimen), short time, v./mil	D149					440	540-590	450-600
		SUPPLIERS[a]		Hoechst Celanese	Hoechst Celanese	Hoechst Celanese	Azdel	Azdel	Ashley Polymers; ComAlloy; DSM; GE Plastics; Hoechst Celanese; LNP; RTP; Texapol; Thermofil	ComAlloy; GE Plastics; Hoechst Celanese; LNP; RTP

Polyester, thermoplastic (Cont'd)

| | Polybutylene terephthalate (Cont'd) | | | | | Polyester alloy | | | PCT | |
| | | | Flame-retarded grade | | | | | | | |
	Impact-modified	50% glass fiber-reinforced	7-15% glass fiber-reinforced	30% glass fiber-reinforced	EMI shielding (conductive); 30% carbon fiber	Unfilled	7.0-30% glass fiber-reinforced	Unreinforced flame retardant	15% glass fiber-reinforced	30% glass fiber-reinforced
1a.										
1.	225	225	220-260	220-260	222					
2.	I: 482-527	I: 482-527	I: 490-560	I: 490-560	I: 430-550	I: 475-540	I: 475-540	I: 470-510	I: 565-590	I: 555-595
3.					5-20	5-17	5-17	5-17		
4.						3.0-4.0	3.0-4.0	3.0-4.0		
5.	20×10^{-3}	4×10^{-3}	0.005-0.01	0.002-0.006	0.001-0.004	0.016-0.018	0.003-0.014		0.001-0.004	0.001-0.004
6.		21,800	11,500-15,000	17,400-20,000	22,000-23,000	4900-6300	7200-12,000	5000	13,700	18,000-19,500
7.		2.5	4.0	2.0-3.0	1-3	150-300	4.5-45	50	2.0	1.9-2.3
8.	6530									
9.			17,000	18,000						
10.			18,000-23,500	30,000	29,000-34,000	7100-8700	12,400-19,000	8500	23,900	24,000-29,800
11.	276	2470	800	1490; 1700	3500					
12.										
13.			580-830	1300-1500	2300-2700	210-280	395-925	260	812	1200-1450
14.	3.3	1.4	0.6-1.1	1.3-1.6	1.2-1.5	NB	2.8-4.1	NB	0.76	1.3-1.8
15.			M79-88	M88; M90	R120	R101-R109; >R115	R104-R111; >R115	R105	M88	>R115
16.	135	25	5	1.5		20	18			20
17.	122	419	300-450	400-450	420-430	105-125	170-374	130	475	500
	248	428	400-490	425-490		180-260	379-417	240	518	>500
18.					15.8	6.9	8.3			6.9
19.	1.20	1.73	1.48-1.53	1.63	1.41-1.42	1.23-1.25	1.30-1.47	1.31	1.33	1.45
20.			0.06	0.06-0.07	0.04-0.45	0.1	0.11	0.10		0.04-0.05
	0.3	0.3								
21.			460	490		460	435		462	440-460
	BASF	Ashley Polymers; BASF	Albis; Ashley Polymers; ComAlloy; DSM; DuPont; GE Plastics; Hoechst Celanese; LNP; Polymer Resources; RTP; Texapol; Wilson-Fiberfil	Albis; Ashley Polymers; BASF; ComAlloy; DSM; DuPont; Ferro Eng. Therm; Hoechst Celanese; LNP; Polymer Resources; RTP; Texapol	ComAlloy; DSM; LNP; RTP	GE Plastics; Hoechst Celanese	Albis; GE Plastics; Hoechst Celanese	Hoechst Celanese	Eastman	Eastman; GE Plastics

Polyester, thermoplastic (Cont'd)

	Properties	ASTM test method	40% glass fiber-reinforced	27-30% glass fiber- and mineral-reinforced	40% glass fiber- and mineral-reinforced	20% glass, flame retarded	30% glass, flame retarded	40% glass, flame retarded	20% glass fiber-reinforced
			PCT (Cont'd)						PCTA
Processing	1a. Melt flow (gm./10 min.)	D1238							
	1. Melting temperature, °C. T_m (crystalline)								285
	T_g (amorphous)								92
	2. Processing temperature range, °F. (C = compression; T = transfer; I = injection; E = extrusion)		I: 565-590	I: 565-590	I: 565-590	I: 565-590	I: 565-590	I: 565-590	I: 560-590
	3. Molding pressure range, 10^3 p.s.i.								8-16
	4. Compression ratio								2.5-3.5
	5. Mold (linear) shrinkage, in./in.	D955	0.0005-0.003	0.002-0.005	0.002-0.005	0.001-0.004	0.002-0.004	0.001-0.003	0.004
Mechanical	6. Tensile strength at break, p.s.i.	D638[b]	22,000	17,500	17,000	15,400	19,000	20,600	11,000
	7. Elongation at break, %	D638[b]	2.1	2.4	1.18	1.4-2.0	1.7	1.4	2.1
	8. Tensile yield strength, p.s.i.	D638[b]							
	9. Compressive strength (rupture or yield), p.s.i.	D695							
	10. Flexural strength (rupture or yield), p.s.i.	D790	33,000	26,700	27,400	24,000	29,000	32,000	17,800
	11. Tensile modulus, 10^3 p.s.i.	D638[b]							
	12. Compressive modulus, 10^3 p.s.i.	D695							
	13. Flexural modulus, 10^3 p.s.i. 73° F.	D790	1690	1180	1550	1080	1450	1910	840
	200° F.	D790							
	250° F.	D790							
	300° F.	D790							
	14. Izod impact, ft.-lb./in. of notch ($^1/_8$-in. thick specimen)	D256A	1.5	1.0	1.0	0.9	1.0	1.4	0.7
	15. Hardness Rockwell	D785	M88	M96	R119	M96	R122	M94	122
	Shore/Barcol	D2240/ D2583							
Thermal	16. Coef. of linear thermal expansion, 10^{-6} in./in./°F.	D696							
	17. Deflection temperature under flexural load, °F. 264 p.s.i.	D648	491	482	500	442	477	489	450
	66 p.s.i.	D648	518	523	527	514	527	518	>512
	18. Thermal conductivity, 10^{-4} cal.-cm./ sec.-cm.^{2}.-°C.	C177							
Physical	19. Specific gravity	D792	1.53	1.43	1.55	1.54	1.62	1.70	1.37
	20. Water absorption ($^1/_8$-in. thick specimen), % 24 hr.	D570							
	Saturation	D570							
	21. Dielectric strength ($^1/_8$-in. thick specimen), short time, v./mil	D149	420	452	462	444	430	440	
	SUPPLIERS[a]		Eastman	Eastman	Eastman	Eastman	Eastman	Eastman	Eastman

Polyester, thermoplastic (Cont'd)

	PCTA (Cont'd) Unfilled	Polyethylene terephthalate Unfilled	15% glass fiber-reinforced	30% glass fiber-reinforced	40-45% glass fiber-reinforced	35% glass, super toughened	15% glass, easy processing	30% glass, flame retarded, V-0 1/32	15% glass, flame retarded, V-0 1/32	15-20% glass, flame retarded	30% glass, flame retarded
1a.											
1.	285	212-265		245-265	252-255	245-255	245-255	245-255	245-255		
	92	68-80									
2.	I: 299-316 E: 299-302	I: 440-660 E: 520-580	I: 540-570	I: 510-590	I: 500-590	I: 525-555	I: 525-555	I: 525-555	I: 525-555	I: 540-560	I: 540-560
3.		2-7		4-20	8-12	8-18	8-18	8-18	8-18		
4.		3.1	2-3	2-3	4	3:1	3:1	3:1	3:1	2-3	2-3
5.	0.004	0.002-0.030	0.001-0.004	0.002-0.009	0.002-0.009	0.002-0.009	0.003-0.010	0.002-0.009	0.003-0.010	0.0015-0.004	0.001-0.004
6.		7000-10,500	14,600	20,000-24,000	14,000-27,500	15,000	11,500	22,000	15,500	13,700-15,700	18,600
7.	25-250	30-300	2.0	2-7	1.5-3	6.0	6.0	2.3	2.6	2.2	1.8
8.	5900-9000	8600		23,000							
9.		11,000-15,000		25,000	20,500-24,000	11,700	13,500	25,000	25,000		
10.		12,000-18,000	20,000	30,000-36,000	21,000-42,400	21,000	13,500	32,000	23,000	20,000-23,200	27,500
11.		400-600		1300-1440	1800-1950						
12.											
13.	240-285	350-450	830	1200-1590	1400-2190	1000	525	1500	850	850-1090	1540
				520	489	360	185	620	350		
				390	320	275	155	420	220		
14.	1.5-NB	0.25-0.7	1.9	1.5-2.2	0.9-2.4	4.4	2.2	1.6	1.2	1-1.2	1.4
15.	105-122	M94-101; R111	R121	M90; M100	R118; R119	M62; R107	M58, R111	M100, R120	M88, R120	M83	M84
16.	5.8×10^{-5}	65×10^{-6}		18-30	18-21	1.5	1.0	1.1	1.0		
17.	69-95	70-150	400	410-440	412-448	428	405	435	410	383-409	425
	83-142	167	464	470-480	420-480	475	454	475	471	455-462	459
18.	5×10^{-4} or 5	3.3-3.6		6.0-7.6	10.0						
19.	1.195-1.215	1.29-1.40	1.33	1.55-1.70	1.58-1.74	1.51	1.39	1.67	1.53	1.60-1.63	1.71
20.		0.1-0.2		0.05	0.04-0.05	0.25	0.24	0.05	0.07		
		0.2-0.3									
21.	422-441	420-550	475	405-650	415-600	530	450	430	490	437-460	427
	Eastman	DuPont; Eastman; Hoechst Celanese; M.A. Polymers; A. Schulman; Shell; Texapol; Wellman	Eastman	Albis; AlliedSignal; ComAlloy; DSM; DuPont; EMS; Eastman; Ferro Eng. Therm.; GE Plastics; Hoechst Celanese; M.A. Polymers; MRC; RTP; A. Schulman; Texapol; Thermofil; Wellman	Albis; AlliedSignal; ComAlloy; DSM; DuPont; EMS; Eastman; Ferro Eng. Therm.; GE Plastics; Hoechst Celanese; RTP	DuPont	DuPont	Albis; DuPont	DuPont	Eastman	Eastman

Polyester, thermoplastic (Cont'd)

Polyethylene terephthalate (Cont'd)

Materials / Properties	ASTM test method	40% glass, flame retarded	35-45% glass fiber and mica-reinforced	30% long glass fiber-reinforced	40% long glass fiber-reinforced	50% long glass fiber-reinforced	60% long glass fiber-reinforced	EMI shielding (conductive); 30% PAN carbon fiber
Processing								
1a. Melt flow (gm./10 min.)	D1238							
1. Melting temperature, °C. T_m (crystalline)			252-255	275	275	275	270-280	
T_g (amorphous)								
2. Processing temperature range, °F. (C = compression; T = transfer; I = injection; E = extrusion)		I: 540-560	I: 500-590	I: 470-530	480-540	I: 470-530	I: 500-530	I: 550-590
3. Molding pressure range, 10^3 p.s.i.			5-20	10-20	10-20	10-20	10-20	
4. Compression ratio		2-3	4	3-4		3-4	3-4	
5. Mold (linear) shrinkage, in./in.	D955	0.001-0.004	0.002-0.007	0.001-0.008	0.001-0.005	0.001-0.008	0.001-0.007	0.001-0.002
Mechanical								
6. Tensile strength at break, p.s.i.	D638[b]	19,000	14,000-26,000	20,200	23,200	23,500	23,500	25,000
7. Elongation at break, %	D638[b]	1.5	1.5-3	1.4	1.4	1.0	0.9	1.4
8. Tensile yield strength, p.s.i.	D638[b]							
9. Compressive strength (rupture or yield), p.s.i.	D695		20,500-24,000	31,000	34,200	35,100	35,100	
10. Flexural strength (rupture or yield), p.s.i.	D790	29,300	21,000-40,000	29,300	35,400	36,500	40,100	38,000
11. Tensile modulus, 10^3 p.s.i.	D638[b]		1800-1950	1700	2100	2400	3000	3600
12. Compressive modulus, 10^3 p.s.i.	D695							
13. Flexural modulus, 10^3 p.s.i. 73° F.	D790	2020	1400-2000	1500	1900	2100	2600	2700
200° F.	D790		489					
250° F.	D790							
300° F.	D790		320					
14. Izod impact, ft.-lb./in. of notch (1/8-in. thick specimen)	D256A	1.6	0.9-2.4	4.0	5.0	6.2	8.0	1.5
15. Hardness Rockwell	D785	M79	R118, R119					
Shore/Barcol	D2240/ D2583							
Thermal								
16. Coef. of linear thermal expansion, 10^{-6} in./in./°F.	D696		18-21					
17. Deflection temperature under flexural load, °F. 264 p.s.i.	D648	429	396-440	470	475	480	480	430
66 p.s.i.	D648	466	420-480					470
18. Thermal conductivity, 10^{-4} cal.-cm./ sec.-cm.²-°C.	C177		10.0					
Physical								
19. Specific gravity	D792	1.78	1.58-1.74	1.61	1.70	1.85	1.91	1.42
20. Water absorption (1/8-in. thick specimen), % 24 hr.	D570		0.04-0.05					0.05
Saturation	D570							
21. Dielectric strength (1/8-in. thick specimen), short time, v./mil	D149	399	550-687					
SUPPLIERS[a]		Eastman	Albis; AlliedSignal; Bayer Corp.; ComAlloy; DSM; DuPont; Hoechst Celanese; M.A. Polymers; MRC; RTP; Thermofil	Hoechst Celanese	Hoechst Celanese	Hoechst Celanese	Hoechst Celanese	ComAlloy; DSM; Ferro Eng. Therm.; RTP

Polyester, thermoplastic (Cont'd)

	Polyethylene terephthalate (Cont'd)			PETG	PCTG	Polyester/polycarbonate blends			Wholly aromatic (liquid crystal)			
	Recycled content, 30% glass fiber	Recycled content, 45% glass fiber	Recycled content, 35% glass/mineral-reinforced	Unfilled	Unfilled	High-impact	30% glass fiber-reinforced	Flame retarded	Unfilled medium melting point	Unfilled high melting point	30% carbon fiber-reinforced	40% glass fiber-filled
1a.												
1.									280-421	400	280	
				81-91								
2.	I: 530-550	I: 530-550	I: 530-550	I: 480-520 E: 490-550	I: 530-560	I: 460-630	I: 470-560	I: 480-550	I: 540-770	I: 700-850	555-600	I: 660-770
3.				1-20	1-20	8-18	10-18	10-20	1-16	5-18	1-14	5-14
4.	2-3	2-3	2-3	2.4-3	2.4-3				2.5-4	2.5-3	2.5-4	2.5-3
5.				0.002-0.005	0.002-0.005	0.0005-0.019	0.003-0.009	0.005-0.007	0.001-0.008	0-0.002	0-0.002	
6.	24,000	28,500	14,000-15,000	4100	7600	4500-9000	12,000-13,300	8900	15,900-27,000	12,500	35,000	13,600
7.	2.0	2.0	2.1-2.2	110	330	100-175		130	1.3-4.5	2	1.0	1.8
8.				7300	6500	5000-8100		7400				
9.						8600-10,000	10,860-11,600		6200-19,000	10,000	34,500	10,400
10.	35,500	45,000	21,500-22,000	10,200	9600	8500-12,500	20,000	14,100	19,000-35,500	19,000	46,000	20,500
11.						240-325			1400-2800	1750	5400	1870
12.									400-900	309	4800	420
13.	1400	2100	1400	300	260	310	780-850	380,000	1770-2700[e]	1860	4600	1320
									1500-1700[e]			
									1300-1500[e]	450F: 870		
									1200-1450[e]	575F: 450		
14.	1.5	2.0	1.1	1.9	NB	12-19-No break	3.1-3.2	13	1.7-10	1.2	1.4	1.6
15.				R106	R105	R112-116	R109-110	R122	M76; R60-66	R97	M99	79
16.						58-150	25	0.64	5-7	8.9	-2.65	14.9
17.	435	445	395-420	147	149	140-250	300-330	212	356-671	606	440	606
				158	165	210-265	400-415	239				
18.						5.2			2			
19.	1.58	1.70	1.60	1.27	1.23	1.20-1.26	1.44-1.51	1.3	1.35-1.84	1.79	1.49	1.70
20.				0.13	0.13	0.13-0.80	0.09-0.10	0.07	0-<0.1	0	<0.1	<0.1
						0.30-0.62		0.22	<0.1	0.02		
21.	565	540	450-550			396-500		660	600-980	470		510
	Hoechst Celanese	Hoechst Celanese	Hoechst Celanese	Eastman	Eastman	Bayer Corp.; ComAlloy; Eastman; Ferro Eng. Therm.; GE Plastics; MRC; Progressive Polymers	ComAlloy; Ferro Eng. Therm.; GE Plastics; MRC; RTP	Bayer Corp.	Amoco Polymers; Hoechst Celanese	Amoco Polymers	Hoechst Celanese; RTP	Amoco Polymers; RTP

Materials	Properties	ASTM test method	Polyester, thermoplastic (Cont'd)					Polyester, thermoset and alkyd	
			Wholly aromatic (liquid crystal) (Cont'd)					Cast	
			40% glass plus 10% mineral-filled	30-50% mineral-filled	30% glass fiber-reinforced	30% glass-reinforced, high HDT	Unfilled platable grade	Rigid	Flexible
Processing	1a. Melt flow (gm./10 min.)	D1238							
	1. Melting temperature, °C. T_m (crystalline)			327	280	355		Thermoset	Thermoset
	T_g (amorphous)								
	2. Processing temperature range, °F. (C = compression; T = transfer; I = injection; E = extrusion)		I: 660-770	I: 605-770	I: 555-770	I: 625-730	I: 600-620		
	3. Molding pressure range, 10^3 p.s.i.		5-14	1-14	1-14	4-8			
	4. Compression ratio		2.5-3	2.5-4	2.5-4	2.5-4	3-4		
	5. Mold (linear) shrinkage, in./in.	D955		0.003	0.001-0.09				
Mechanical	6. Tensile strength at break, p.s.i.	D638[b]	14,200	10,400-16,500	16,900-30,000	18,000-21,000	13,500	600-13,000	500-3000
	7. Elongation at break, %	D638[b]	2.3	1.1-4.0	1.7-2.7	1.7-2.2	2.9	<2.6	40-310
	8. Tensile yield strength, p.s.i.	D638[b]							
	9. Compressive strength (rupture or yield), p.s.i.	D695	9700	6900-7500	9900-21,000	9800-12,500		13,000-30,000	
	10. Flexural strength (rupture or yield), p.s.i.	D790	19,700	14,200-23,500	21,700-24,600	24,000-26,000		8500-23,000	
	11. Tensile modulus, 10^3 p.s.i.	D638[b]	1870	1500-2700	700-3000	2330-2600	1500	300-640	
	12. Compressive modulus, 10^3 p.s.i.	D695	473	490-2016	470-1000	447-700			
	13. Flexural modulus, 10^3 p.s.i. 73° F.	D790	1730	1250-2500	1680-2100	1800-2050	1500	490-610	
	200° F.	D790							
	250° F.	D790			900	1100			
	300° F.	D790			800	1100			
	14. Izod impact, ft.-lb./in. of notch ($^1/_8$-in. thick specimen)	D256A	1.9	0.8-3.0	2.0-3.0	2.0-2.5	0.6	0.2-0.4	>7
	15. Hardness Rockwell	D785		82	61-87	63			
	Shore/Barcol	D2240/ D2583						Barcol 35-75	Shore D84-94
Thermal	16. Coef. of linear thermal expansion, 10^{-6} in./in./°F.	D696	12.9-52.8	9-65	4.9-77.7	14-36		55-100	
	17. Deflection temperature under flexural load, °F. 264 p.s.i.	D648	493	429-554	485-655	518-568	410	140-400	
	66 p.s.i.	D648			489-530				
	18. Thermal conductivity, 10^{-4} cal.-cm./ sec.-cm.^{2}-°C.	C177	2.5	2.57	1.73	1.52			
Physical	19. Specific gravity	D792	1.78	1.63-1.89	1.60-1.67	1.6-1.66		1.04-1.46	1.01-1.20
	20. Water absorption ($^1/_8$-in. thick specimen), % 24 hr.	D570	<0.1	<0.1	<0.1	<0.1	0.03	0.15-0.6	0.5-2.5
	Saturation	D570							
	21. Dielectric strength ($^1/_8$-in. thick specimen), short time, v./mil	D149	1145	900-955	640-900	900-1050	600	380-500	250-400
	SUPPLIERS[a]		Amoco Polymers; RTP	Amoco Polymers; DuPont; Hoechst Celanese; RTP	Amoco Polymers; DuPont; Hoechst Celanese; RTP	Amoco Polymers; DuPont; Hoechst Celanese; RTP	Hoechst Celanese	Alpha/Owens-Corning; Aristech Chem; AZS; Cargill; Cook C&P; ICI Americas; Reichhold	Alpha/Owens-Corning; Aristech Chem.; Cargill; Cook C&P; ICI Americas; Reichhold

Polyester, thermoset and alkyd (Cont'd)

| | Glass fiber-reinforced | | | | | | | | EMI shielding (conductive) | |
	Preformed, chopped roving	Premix, chopped glass	Woven cloth	SMC	SMC, BMC low-density	SMC low-pressure	SMC low-shrink	BMC, TMC	SMC, TMC	BMC
1a.										
1.	Thermoset	Thermoset	Thermoset		Thermoset	Thermoset	Thermoset	Thermoset	Thermoset	Thermoset
2.	C: 170-320	C: 280-350	C: 73-250	C: 270-380 I: 280-310 T: 280-310	C: 270-330 I: 270-350 T: 270-350	C: 270-330	C: 270-330 I: 270-380	C: 280-380 I: 280-370 T: 280-320	C: 270-380 I: 270-370 T: 280-320	C: 310-380 I: 300-370 T: 280-320
3.	0.25-2	0.5-2	0.3	0.3-2	0.5-2	0.25-0.8	0.5-2	0.4-1.0	0.5-2	
4.	1.0	1.0		1.0	1.0	1.0	1.0	1.0	1.0	
5.	0.0002-0.002	0.001-0.012	0.0002-0.002	0.0005-0.004	0.0002-0.001	0.0002-0.001	0.0002-0.001	0.0003-0.004	0.0002-0.001	0.0005-0.004
6.	15,000-30,000	3000-10,000	30,000-50,000	7000-25,000	4000-20,000	7000-25,000	4500-20,000	3000-10,000	7000-8000	4000-4500
7.	1-5	<1	1-2	3	2-5	3	3-5			
8.										
9.	15,000-30,000	20,000-30,000	25,000-50,000	15,000-30,000	15,000-30,000	15,000-30,000	15,000-30,000	14,000-30,000	20,000-24,000	18,000
10.	10,000-40,000	7000-20,000	40,000-80,000	10,000-36,000	10,000-35,000	10,000-36,000	9000-35,000	11,000-24,000	18,000-20,000	12,000
11.	800-2000	1000-2500	1500-4500	1400-2500	1400-2500	1400-2500	1000-2500	1500-2500		
12.										
13.	1000-3000	1000-2000	1000-3000	1000-2200	1000-2500	1000-22,000	1000-2500		1400-1500	1400-1500
			660							
			430							
			270							
14.	2-20	1.5-16	5-30	7-22	2.5-18	7-24	2.5-15	2-13	10-12	5-7
15.	Barcol 50-80	Barcol 50-80	Barcol 60-80	Barcol 50-70		Barcol 40-70	Barcol 40-70	Barcol 50-65	Barcol 45-50	Barcol 50
16.	20-50	20-33	15-30	13.5-20	6-30	6-30	6-30			
17.	>400	>400	>400	375-500	>375	375-500	375-500	320-400	395-400+	400+
18.							18-22			
19.	1.35-2.30	1.65-2.30	1.50-2.10	1.65-2.6	1.0-1.5	1.65-2.30	1.6-2.4	1.72-2.10	1.75-1.80	1.80-1.85
20.	0.01-1.0	0.06-0.28	0.05-0.5	0.1-0.25	0.4-0.25	0.1-0.25	0.01-0.25	0.1-0.45		
									0.5	0.5
21.	350-500	345-420	350-500	380-500	300-400	380-500	380-450	300-390		
	Eagle-Picher; Glastic; Haysite; Jet Moulding; Plumb; Premix; Reichhold; Rostone	Applied Components; Bulk Molding Compounds; CYTEC; Eagle-Picher; Glastic; Haysite; Plumb; Premix; Reichhold; Rostone	Eagle-Picher; Glastic; Haysite; Plumb; Premix; Reichhold; Rostone	Applied Components; Budd; Eagle-Picher; Haysite; Jet Moulding; Plastics Mfg.; Polyply; Premix; Rostone	Rostone	Rostone	Applied Components; Budd; Eagle-Picher; Haysite; Jet Moulding; Polyply; Premix; Rostone	BP Chemicals; CYTEC; Eagle-Picher; Epic Resins; Glastic; Haysite; Jet Moulding; Plumb; Polyply; Premix; Rostone	Applied Components; Jet Moulding; Polyply; Premix; Rostone	Applied Components; Bulk Molding Compounds; Jet Moulding; Polyply; Premix; Rostone

	Properties	ASTM test method	Polyester, thermoset and alkyd (Cont'd) Alkyd molding compounds — Granular and putty, mineral-filled	Glass fiber-reinforced	Polyetherimide Unfilled	30% glass fiber-reinforced	EMI shielding (conductive); 30% carbon fiber	Polyethersulfone Extrusion/injection molding grade	10% glass fiber-reinforced
Processing	1a. Melt flow (gm./10 min.)	D1238							
	1. Melting temperature, °C. T_m (crystalline)		Thermoset	Thermoset					
	T_g (amorphous)				215-217	215	215		
	2. Processing temperature range, °F. (C = compression; T = transfer; I = injection; E = extrusion)		C: 270-350 I: 280-390 T: 320-360	C: 290-350 I: 280-380	I: 640-800	I: 620-800	I: 600-780	I: 635-735	I: 680-715
	3. Molding pressure range, 10^3 p.s.i.		2-20	2-25	10-20	10-20	10-30		
	4. Compression ratio		1.8-2.5	1-11	1.5-3	1.5-3	1.5-3		
	5. Mold (linear) shrinkage, in./in.	D955	0.003-0.010	0.001-0.010	0.005-0.007	0.001-0.004	0.0005-0.002	0.007-0.010	0.005-0.006
Mechanical	6. Tensile strength at break, p.s.i.	D638[b]	3000-9000	4000-9500	14,000	23,200-28,500	29,000-34,000	13,000	16,500
	7. Elongation at break, %	D638[b]			60	2-5	1-3	15-40	4.3
	8. Tensile yield strength, p.s.i.	D638[b]			15,200	24,500			
	9. Compressive strength (rupture or yield), p.s.i.	D695	12,000-38,000	15,000-36,000	21,900	23,500-30,700	32,000		
	10. Flexural strength (rupture or yield), p.s.i.	D790	6000-17,000	8500-26,000	22,000	33,000	37,000-45,000	18,500	24,500
	11. Tensile modulus, 10^3 p.s.i.	D638[b]	500-3000	2000-2800	430	1300-1600	2600-3300	410	740
	12. Compressive modulus, 10^3 p.s.i.	D695	2000-3000		480	550-938			
	13. Flexural modulus, 10^3 p.s.i. 73° F.	D790	2000	2000	480	1200-1300	2500-2600	370	650
	200° F.	D790			370	1100			
	250° F.	D790			360	1060			
	300° F.	D790			350	1040			
	14. Izod impact, ft.-lb./in. of notch ($1/8$-in. thick specimen)	D256A	0.3-0.5	0.5-16	1.0-1.2	1.7-2.0	1.2-1.6	1.5-1.6	1.3
	15. Hardness Rockwell	D785	E98	E95	M109-110	M114, M125, R123	M127	M85	M94
	Shore/Barcol	D2240/ D2583							
Thermal	16. Coef. of linear thermal expansion, 10^{-6} in./in./°F.	D696	20-50	15-33	47-56	20-21		31	19
	17. Deflection temperature under flexural load, °F. 264 p.s.i.	D648	350-500	400-500	387-392	408-420	405-420	383	414
	66 p.s.i.	D648			405-410	412-415	410-425	406	419
	18. Thermal conductivity, 10^{-4} cal.-cm./ sec.-cm.^{2}.-°C.	C177	12-25	15-25	1.6	6.0-9.3	17.6		
Physical	19. Specific gravity	D792	1.6-2.3	2.0-2.3	1.27	1.49-1.51	1.39-1.42	1.37	1.45
	20. Water absorption ($1/8$-in. thick specimen), % 24 hr.	D570	0.05-0.5	0.03-0.5	0.25	0.16-0.20	0.18-0.2		
	Saturation	D570			1.25	0.9		2.1	1.9
	21. Dielectric strength ($1/8$-in. thick specimen), short time, v./mil	D149	350-450	259-530	500	495-630			
	SUPPLIERS[a]		CYTEC; OxyChem; Plastics Eng.; Plumb; Premix	Cosmic Plastics; CYTEC; OxyChem; Plastics Eng.; Plumb; Premix; Rogers	GE Plastics	ComAlloy; DSM; Ferro Eng. Therm.; GE Plastics; LNP; Polymer Resources; RTP; Thermofil	ComAlloy; DSM; Ferro Eng. Therm.; LNP; RTP; Thermofil	BASF	BASF; RTP

	Polyethersulfone (Cont'd)			Polyethylene and ethylene copolymers							
				Low and medium density					High density		
						LDPE copolymers					Copolymers
	20-30% glass fiber-reinforced	20% mineral-filled	30% carbon-filled	Branched homo-polymer	Linear copolymer	Ethylene-vinyl acetate	Ethylene-ethyl acrylate	Ethylene-methyl acrylate	Polyethylene homopolymer	Rubber-modified	Low and medium molecular weight
1a.				0.25-27.0		1.4-2.0			5-18		
1.				98-115	122-124	103-110		83	130-137	122-127	125-132
				−25							
2.	I: 660-765	I: 662-716	I: 680-734	I: 300-450 E: 250-450	I: 350-500 E: 450-600	C: 200-300 I: 350-430 E: 300-380	C: 200-300 I: 250-500	E: 200-620	I: 350-500 E: 350-525	E: 360-450	I: 375-500 E: 300-500
3.				5-15	5-15	1-20	1-20		12-15		5-20
4.				1.8-3.6	3				2		2
5.	0.004-0.006	0.007-0.008		0.015-0.050	0.020-0.022	0.007-0.035	0.015-0.035		0.015-0.040		0.012-0.040
6.	20,000-22,000	7975	15,800	1200-4550	1900-4000	2200-4000	1600-2100	1650	3200-4500	2300-2900	3000-6500
7.	2.1-2.8	5	1.4	100-650	100-965	200-750	700-750	740	10-1200	600-700	10-1300
8.				1300-2100	1400-2800	1200-6000		1650	3800-4800	1400-2600	2600-4200
9.							3000-3600		2700-3600		2700-3600
10.	26,500-28,500										
11.	1150-1550	522	1740	25-41	38-75	7-29	4-7.5	12	155-158		90-130
12.											
13.	980-1300			35-48	40-105	7.7			145-225		120-180
14.	1.4-1.7	1.5		No break	1.0-No break	No break	No break		0.4-4.0		0.35-6.0
15.	M96-M97			Shore D44-50	Shore D55-56	Shore D17-45	Shore D27-38		Shore D66-73	Shore D55-60	Shore D58-70
16.	12-14	47	12	100-220		160-200	160-250		59-110		70-110
17.	419	399	433								
	430	414	440	104-112					175-196		149-176
18.				8					11-12		10
19.	1.53-1.60	1.52	1.53	0.917-0.932	0.918-0.940	0.922-0.943	0.93	0.942-0.945	0.952-0.965	0.932-0.939	0.939-0.960
20.				<0.01		0.005-0.13	0.04	0.0	<0.01		<0.01
	1.5-1.7	1.7									
21.				450-1000		620-760	450-550		450-500		450-500
	BASF; RTP	BASF; RTP	BASF; RTP	American Polymers; Bamberger Polymers; Chevron; Dow Plastics; DuPont; Eastman; Exxon; Mobil; Monmouth; Novacor; Quantum; RSG Polymers; Rexene; A. Schulman; Union Carbide; Wash. Penn; Westlake	Bamberger Polymers; Chevron; Dow Plastics; DuPont; Eastman; Exxon; Mobil; Monmouth; Montell Polyolefins; Novacor; Quantum; RSG Polymers; A. Schulman; Solvay Polymers; Union Carbide	AT Plastics; Chevron; DuPont; Exxon; Federal Plastics; Mobil; Quantum; Rexene; A. Schulman; Union Carbide; Westlake	Union Carbide	Chevron; Exxon; A. Schulman	American Polymers; Bamberger Polymers; Chevron; Dow Plastics; Eastman; Exxon; Federal Plastics; Hoechst Celanese; M.A. Polymers; Mobil; Monmouth; Novacor; Paxon; Phillips; Quantum; RSG Polymers; A. Schulman; Shuman; Solvay Polymers; Union Carbide	M.A. Polymers; Paxon	American Polymers; Bamberger Polymers; Chevron; Dow Plastics; Eastman; Exxon; Hoechst Celanese; Mobil; Monmouth; Novacor; Paxon; Phillips; Quantum; RSG Polymers; A. Schulman; Shuman; Union Carbide

		Polyethylene and ethylene copolymers (Cont'd)						**Polyimide**
		High density (Cont'd)				**Crosslinked**		**Thermoplastic**
		Copolymers (Cont'd)						
Properties	ASTM test method	High molecular weight	Ultra high molecular weight	30% glass fiber-reinforced	20-30% long glass fiber-reinforced	Molding grade	Wire and cable grade	Unfilled
1a. Melt flow (gm./10 min.)	D1238	5.4-6.8						4.5-7.5
1. Melting temperature, °C. T_m (crystalline)		125-135	125-138	120-140	120-140			388
T_g (amorphous)								250-365
2. Processing temperature range, °F. (C = compression; T = transfer; I = injection; E = extrusion)		I: 375-500 E: 375-475	C: 400-500	I: 350-600	I: 525-600	C: 240-450 I: 250-300	E: 250-400	C: 625-690 I: 734-740 E: 734-740
3. Molding pressure range, 10^3 p.s.i.				1-2	10-20			3-20
4. Compression ratio								1.7-4
5. Mold (linear) shrinkage, in./in.	D955	0.015-0.040	0.040	0.002-0.006	0.002-0.004	0.007-0.090	0.020-0.050	0.0083
6. Tensile strength at break, p.s.i.	D638[b]	2500-4300	5600-7000	7500-9000	7000-8500	1600-4600	1500-3100	10,500-17,100
7. Elongation at break, %	D638[b]	170-800	350-525	1.5-2.5	2.0-2.5	10-440	180-600	7.5-90
8. Tensile yield strength, p.s.i.	D638[b]	2800-3900	3100-4000				1200-2000	12,500-13,000
9. Compressive strength (rupture or yield), p.s.i.	D695			6000-7000	5000-6000	2000-5500		17,500-40,000
10. Flexural strength (rupture or yield), p.s.i.	D790			11,000-12,000	8000-9500	2000-6500		10,000-28,800
11. Tensile modulus, 10^3 p.s.i.	D638[b]	136		700-900	800-900	50-500		300-400
12. Compressive modulus, 10^3 p.s.i.	D695					50-150		315-350
13. Flexural modulus, 10^3 p.s.i. 73° F.	D790	125-175	130-140	700-800	600-800	70-350	8-14	360-500
200° F.	D790							
250° F.	D790							
300° F.	D790							210
14. Izod impact, ft.-lb./in. of notch (1/8-in. thick specimen)	D256A	3.2-4.5	No break	1.1-1.5	2.5-3.5	1-20		1.5-1.7
15. Hardness Rockwell	D785		R50	R75-90	R75-90			E52-99,R129,M95
Shore/Barcol	D2240/ D2583	Shore D63-65	Shore D61-63			Shore D55-80	Shore D30-65	
16. Coef. of linear thermal expansion, 10^{-6} in./in./°F.	D696	70-110	130-200	48		100	100	45-56
17. Deflection temperature under flexural load, °F. 264 p.s.i.	D648		110-120	250	240-250	105-145	100-173	460-680
66 p.s.i.	D648	154-158	155-180	260-265	250-260	130-225		
18. Thermal conductivity, 10^{-4} cal.-cm./ sec.-cm.^{2}-°C.	C177			8.6-11				2.3-4.2
19. Specific gravity	D792	0.947-0.955	0.94	1.18-1.28	1.09-1.18	0.95-1.45	0.91-1.40	1.33-1.43
20. Water absorption (1/8-in. thick specimen), % 24 hr.	D570		<0.01	0.02-0.06	0.05-0.06	0.01-0.06	0.01-0.06	0.24-0.34
Saturation	D570							1.2
21. Dielectric strength (1/8-in. thick specimen), short time, v./mil	D149		710	500-550		230-550	620-760	415-560
SUPPLIERS[a]		Amoco Chemical; Bamberger Polymers; BASF; Chevron; Dow Plastics; Exxon; Hoechst Celanese; Mobil; Novacor; Paxon; Phillips; Quantum; Solvay Polymers	Hoechst Celanese; Montell Polyolefins	ComAlloy; DSM; Ferro Corp.; Ferro Eng. Therm.; LNP; M.A. Polymers; RTP; A. Schulman; Thermofil	DSM	Mobil; Monmouth; Phillips; Quantum; A. Schulman	AT Plastics; Quantum; Synergistics Industries; A. Schulman; Union Carbide	Ciba-Geigy; DuPont; Fluorocarbon; Mitsui Toatsu; Monsanto; Rhone-Poulenc

Materials   Processing   Mechanical   Thermal   Physical

| | Polyimide (Cont'd) | | | | | | | Polymethylpentene | | |
| | Thermoplastic (Cont'd) | | | | | Thermoset | | | | |
	30% glass fiber-reinforced	30% carbon fiber-reinforced	30% carbon fiber-reinforced, crystallized	15% graphite-filled	40% graphite-filled	Unfilled	50% glass fiber-reinforced	Unfilled	Filled	Paper coating
1a.								26	30	180
1.	388	388		388		Thermoset	Thermoset	230-240	240	240
	250	250		250	365					
2.	I: 734-788	I: 734-788	I: 734-788	I: 734-788	C: 690	460-485	C: 460 I: 390 T: 390	I: 510-610 E: 510-650	I: 510-610	E: 520-630
3.	10-30	10-30	10-30	10-30	3-5	7-29	3-10	1-10	1-10	
4.	1.7-2.3	1.7-2.3	1.7-2.3	1.7-2.3		1-1.2		2.0-3.5	2.0-3.5	2.5-3.0
5.	0.0044	0.0021		0.006		0.001-0.01	0.002	0.016-0.021	0.014-0.017	
6.	24,000	33,400	31,700	8000-8400	7600	4300-22,900	6400	2300-2500	2400	2500
7.	3	2	1	3.5	3	1		20-120	25	20
8.						4300-22,900		2200-3400	3400	3250
9.	27,500	30,200		25,000		19,300-32,900	34,000			
10.	35,200	46,700	43,600	11,000-14,100	18,000	6500-50,000	21,300	6300-8300		
11.	1720	3000			14,000	460-4650		160-280		
12.	458	573		330		421		114-171		
13.	1390	2780	3210	460-500		422-3000	1980	70-190	270	190
					700			36		
								26		
	1175	2450		260		1030-2690		17		
14.	2.2	2.0	2.4	1.1	0.7	0.65-15	5.6	2-3		
15.	R128. M104	R128, M105			E27	110M-120M	M118	R35-85	90	80
16.	17-53	6-47		41	38	15-50	13	65		65
17.	469	478	>572	680	680	572->575	660	120-130		
								180-190	230	
18.	8.9	11.7			41.4	5.5-12	8.5	4.0		4
19.	1.56	1.43	1.47	1.41	1.65	1.41-1.9	1.6-1.7	0.833-0.835	1.08	0.833
20.	0.23	0.23		0.19	0.14	0.45-1.25	0.7	0.01	0.11	0.01
					0.6					
21.	528			250		480-508	450	1096-1098		
	Mitsui Toatsu; RTP	Mitsui Toatsu; RTP	Mitsui Toatsu; RTP	DuPont; Mitsui Toatsu	DuPont; Fluorocarbon; Rhone-Poulenc	Ciba-Geigy; Rhone-Poulenc	Rhone-Poulenc	Mitsui Petrochemical	Mitsui Petrochemical	Mitsui Petrochemical

Polyphenylene oxide, modified

Alloy with polystyrene

	Properties	ASTM test method	Low glass transition	High glass transition	Impact-modified	15% glass fiber-reinforced	20% glass fiber-reinforced	30% glass fiber-reinforced	Mineral-filled
Processing	1a. Melt flow (gm./10 min.)	D1238							
	1. Melting temperature, °C. T_m (crystalline)								
	T_g (amorphous)		100-112	117-190	135	100-125	100-125	100-125	110-135
	2. Processing temperature range, °F. (C = compression; T = transfer; I = injection; E = extrusion)		I: 400-600 E: 420-500	I: 425-670 E: 460-525	I: 425-550	I: 400-630 E: 460-525	I: 400-630 E: 460-525	I: 400-630 E: 460-525	I: 540-590 E: 470-530
	3. Molding pressure range, 10^3 p.s.i.		12-20	12-20	10-15	10-40	10-40	10-40	12-20
	4. Compression ratio		1.3-3	1.3-3					2-3
	5. Mold (linear) shrinkage, in./in.	D955	0.005-0.008	0.006-0.008	0.006	0.002-0.004	0.002-0.004	0.001-0.004	0.005-0.007
Mechanical	6. Tensile strength at break, p.s.i.	D638[b]	6800-7800	9600	7000-8000	10,000-12,000	13,000-15,000	15,000-18,500	
	7. Elongation at break, %	D638[b]	48-50	60	35	5-8	8.0	2-5	25
	8. Tensile yield strength, p.s.i.	D638[b]	6500-7800	7000-9000				14,500	9500-11,000
	9. Compressive strength (rupture or yield), p.s.i.	D695	12,000-16,400	16,400	10,000			17,900	
	10. Flexural strength (rupture or yield), p.s.i.	D790	8300-12,800	9500-14,000	8200-11,000			20,000-23,000	
	11. Tensile modulus, 10^3 p.s.i.	D638[b]	310-380	355-380	345-360			1000-1300	
	12. Compressive modulus, 10^3 p.s.i.	D695							
	13. Flexural modulus, 10^3 p.s.i. 73° F.	D790	325-400	330-400	325-345	450-500	760	1100-1150	425-500
	200° F.	D790	260	305				1000	
	250° F.	D790							
	300° F.	D790							
	14. Izod impact, ft.-lb./in. of notch ($^1/_8$-in. thick specimen)	D256A	3-6	5	6.8	1.1-1.3	1.5	1.7-2.3	3-4
	15. Hardness Rockwell	D785	R115-116	R118-120	L108, M93	R106-110	R115	R115-116	R121
	Shore/Barcol	D2240/ D2583							
Thermal	16. Coef. of linear thermal expansion, 10^{-6} in./in./°F.	D696	38-70	33-77				14-25	
	17. Deflection temperature under flexural load, °F. 264 p.s.i.	D648	176-215	225-300	190-275	252-260	262-275	275-317	190-230
	66 p.s.i.	D648	230	279	205-245	273-280	280-290	285-320	
	18. Thermal conductivity, 10^{-4} cal.-cm./ sec.-cm.^{2}-°C.	C177	3.8	5.2				3.8-4.1	
Physical	19. Specific gravity	D792	1.04-1.10	1.04-1.09	1.27-1.36			1.27-1.36	1.24-1.25
	20. Water absorption ($^1/_8$-in. thick specimen), % 24 hr.	D570	0.06-0.1	0.06-0.12	0.1-0.07			0.06	0.07
	Saturation	D570							
	21. Dielectric strength ($^1/_8$-in. thick specimen), short time, v./mil	D149	400-665	500-700	530			550-630	490
	SUPPLIERS[a]		Ashley Polymers; GE Plastics; Polymer Resources; Shuman	GE Plastics; Hüls America; Polymer Resins; Polymer Resources	GE Plastics	Ashley Polymers	Ashley Polymers	Ashley Polymers; ComAlloy; Ferro Eng. Therm.; GE Plastics; LNP; Polymer Resources; RTP; Thermofil	ComAlloy; GE Plastics; Polymer Resources; RTP

	Polyphenylene oxide, modified (Cont'd)			Polyphenylene sulfide							
	Alloy with polystyrene (Cont'd)		Alloy with nylon								
	EMI shielding (conductive)										
	30% graphite fiber	40% aluminum flake	Un-reinforced	Unfilled	10-20% glass fiber-reinforced	30% glass fiber-reinforced	40% glass fiber-reinforced	53% glass/mineral-reinforced, high elong. and impact	30% glass fiber, 15% PTFE	30% long glass fiber-reinforced	40% long glass fiber-reinforced
1a.							30-41	50	30		
1.				285-290	275-285	275-285	275-290	354	275-290	310	299
				88		90	88-90				
2.	I: 500-600	I: 500-600		I: 590-640	I: 600-675	I: 590-640	I: 600-675	I: 590-620	640-660	580-620	590-620
3.	10-20	10-20		5-15		8-12	5-20	7-15		8-10	7-20
4.				2-3		3	3	3-4		3	3-4
5.	0.001	0.001	0.013-0.016	0.006-0.014	0.002-0.005	0.003-0.005	0.002-0.005	0.001-0.003	0.002-0.004	0.001-0.007	0.001-0.003
6.	18,700	6500	8500	7000-12,500	7500-14,000	22,000	17,500-29,100	24,000	21,000	21,000	23,000
7.	2.5	3.0	60	1-6	1.0-1.5	1.5	0.9-4	1.5	1.8	1.2	1.1
8.											
9.	20,000	6000		16,000	17,000-20,000		21,000-31,200	30,000		32,400	32,000-34,000
10.	24,000	9500	315	14,000-21,000	9500-20,000	28,000	22,700-43,600	35,000	29,000	32,900	35,000-36,400
11.	1150	750		480	850-1200		1100-2100		1340	1800	2200
12.											
13.	1100	850	290-315	550-600	900-1200	1700	1700-2160	2300	1360	1700	2300
		35					1000				
							730				
14.	1.3	0.6	3.8	<0.5	0.7-1.2	1.3	1.1-2.0	1.5	1.6	4.6	4.8-5.3
15.	R111	R110		R123-125	R121	M102.7-M103	R123, M100-104	100M	R116		
16.	11	11		27-49	16-20 Trans. dir.: 15.36-45		Flow dir: 12.1-22 Trans. dir.: 14.4-45	Flow: 19-32 Trans.: 32-80			
17.	240	230	250	212-275	440-480	507	485-515	510	480	490	500
	265	250	350	390	500-520	534	536				
18.				2.0-6.9			6.9-10.7				
19.	1.25	1.45	1.10	1.35	1.39-1.47	1.38-1.58	1.60-1.67	1.80	1.60	1.52-1.62	1.62
20.	0.04	0.03	0.3	0.01-0.07	0.05	<0.03	0.004-0.05	0.02	0.005		
			1.0								
21.				380-450		10^6; 380	347-450	300			
	ComAlloy; LNP; RTP; Thermofil	ComAlloy; Thermofil	Ashley Polymers; GE Plastics	Hoechst Celanese; Phillips	Akzo; LNP; RTP	Ferro Eng. Therm.; GE Plastics; Hoechst Celanese; LNP; Phillips; RTP	Albis; DSM; Ferro Eng. Therm.; GE Plastics; Hoechst Celanese; LNP; Phillips; RTP; Thermofil	Hoechst Celanese	Ferro Eng. Therm.; LNP; RTP	Hoechst Celanese; RTP	Hoechst Celanese; RTP

Materials				Polyphenylene sulfide (Cont'd)					
	Properties	**ASTM test method**	**50% long glass fiber-reinforced**	**60% long glass fiber-reinforced**	**65% mineral-and glass-filled**	**60% mineral-and glass-filled**	**50% glass and mineral-reinforced**	**55% glass and mineral-reinforced**	**60% glass and mineral-reinforced**
Processing	1a. Melt flow (gm./10 min.)	D1238			20-35	38	97	85	63
	1. Melting temperature, °C. T_m (crystalline)		315	315	285-290	285-290	275-285	275-285	275-285
	T_g (amorphous)				88		90		
	2. Processing temperature range, °F. (C = compression; T = transfer; I = injection; E = extrusion)		I: 580-620	I: 580-620	I: 600-675	I: 600-675	I: 600-675	I: 600-675	I: 600-675
	3. Molding pressure range, 10^3 p.s.i.		8-10	8-10	5-20	6-12	4-12		
	4. Compression ratio		3	3	3		3		
	5. Mold (linear) shrinkage, in./in.	D955	0.001-0.005	0.001-0.005	0.001-0.004	0.003-0.004	0.002-0.004	0.003-0.005	0.007
Mechanical	6. Tensile strength at break, p.s.i.	D638[b]	23,200	22,500	13,000-23,100	16,900	18,000-22,000	12,700	8100
	7. Elongation at break, %	D638[b]	1.0	1.0	1.3; <1.4	3	1.0-3.5	1.6	2.0
	8. Tensile yield strength, p.s.i.	D638[b]			11,000				
	9. Compressive strength (rupture or yield), p.s.i.	D695	34,200-35,000	30,800	11,000-32,300				
	10. Flexural strength (rupture or yield), p.s.i.	D790	37,300-38,000	38,300	17,500-33,900	27,100	28,000-33,700	20,300	11,200
	11. Tensile modulus, 10^3 p.s.i.	D638[b]	2600	3000					
	12. Compressive modulus, 10^3 p.s.i.	D695							
	13. Flexural modulus, 10^3 p.s.i. 73° F.	D790	2400	3000	1800-2400	2970	2240-2400	1750	1441
	200° F.	D790							
	250° F.	D790			2000				
	300° F.	D790			1200				
	14. Izod impact, ft.-lb./in. of notch ($^1/_8$-in. thick specimen)	D256A	5.0-5.5	5.5	0.5-1.37	0.8	0.8-1.3	0.7	0.5
	15. Hardness Rockwell	D785			R121, M102	R121, M100	R120	M66	M85
	Shore/Barcol	D2240/ D2583							
Thermal	16. Coef. of linear thermal expansion, 10^{-6} in./in./°F.	D696			12.9-20 14.3	10.6-10.9	12.2-14.6		14.3-14.8
	17. Deflection temperature under flexural load, °F. 264 p.s.i.	D648	505	510	500-510	510	500-510	491	340
	66 p.s.i.	D648			534				
	18. Thermal conductivity, 10^{-4} cal.-cm./ sec.-cm.²-°C.	C177							
Physical	19. Specific gravity	D792	1.72	1.84	1.78-2.03	1.92	1.78-1.8	1.82	1.90
	20. Water absorption ($^1/_8$-in. thick specimen), % 24 hr.	D570			0.02-0.07	0.07	0.02-0.07	0.07	0.08
	Saturation	D570							
	21. Dielectric strength ($^1/_8$-in. thick specimen), short time, v./mil	D149			280-450	323	280-343		
	SUPPLIERS[a]		Hoechst Celanese; RTP	Hoechst Celanese	Ferro Eng. Therm.; Hoechst Celanese; LNP; Phillips; RTP	Albis; LNP; RTP	Albis; Hoechst Celanese; LNP; RTP	LNP; Phillips; RTP	Albis; LNP; Phillips; RTP

	Polyphenylene sulfide (Cont'd)						Polyphthalamide				
		EMI shielding (conductive)									
	Encapsulation grades	60% stainless steel	30% carbon fiber	20% PAN carbon fiber	40% PAN carbon fiber	30% PAN carbon fiber, 15% PTFE	Un-reinforced	Extra-tough	33% glass-reinforced	45% glass-reinforced	33% glass-reinforced, V-0
1a.	287										
1.	275-285	299	275-285	275-285	275-285	275-285	310	310	310	310	310
2.	I: 600-675	I: 590-620	I: 500-675	I: 600-675	I: 600-675	I: 600-675	I: 610-660	I: 610-660	I: 610-660	I: 610-660	I: 610-660
3.		7-20	5-20				5-15	5-15	5-15	5-15	5-15
4.		3.0-4.0					2.5-3	2.5-3	2.5-3	2.5-3	2.5-3
5.	0.006-0.007	0.005-0.007	0.005-0.003	0.0008	0.0005	0.0008	0.015-0.020	0.015-0.020	0.002-0.005	0.002-0.003	0.002-0.004
6.	9800	8400	20,000-27,000	20,000-22,000	26,000-29,000	25,500			32,000	38,000	26,000
7.	1.0		0.5-3	2.0	1.0-1.5	2.6		11	2.0	2.0	
8.							15,100	10,800			
9.			26,000	24,000	27,000				40,000	45,500	
10.	15,100	17,800	26,000-36,000	28,000-31,000	40,000	34,000	23,000	15,800	45,000	54,000	37,300
11.			2500-3700	2500-2600	4400-4800	3600					
12.											
13.	1574	590	2450-3300	2200-2300	3900-4100	3000	475	330	1650	2100	1900
14.	0.4	0.2	0.8-1.2	1.0	1.2-1.5	1.2	1.0	20	2.0-2.4	2.5	1.5
15.	M98		R123	R122	R123		125	120	125	125	125
16.	18.0-18.2		6-16	11	8			33.0	13	8	
17.	328	450	500-505	495-500	505	500	248	248	545	549	523
			>505	500	505						
18.			8.6-17.9				1.7		2.3	2.6	
19.	1.86	1.37	1.42-1.47	1.38-1.4	1.46-1.49	1.47	1.17	1.13	1.43	1.56	1.71
20.	0.03		0.01-0.02	0.03	0.02	0.06	0.81	0.65	0.21	0.12	0.18
21.	338								530	560	458
	Phillips	Hoechst Celanese	DSM; Ferro Corp.; Ferro Eng. Therm.; LNP; RTP; Thermofil	DSM; Ferro Eng. Therm.; LNP; RTP	DSM; Ferro Eng. Therm.; LNP; RTP	DSM; Ferro Eng. Therm.; LNP; RTP	Amoco Polymers	Amoco Polymers	Amoco Polymers; LNP; RTP	Amoco Polymers; LNP; RTP	Amoco Polymers; RTP

				Polyphthalamide					
Materials	Properties	ASTM test method	40% mineral-reinforced	40% glass/mineral-reinforced, heat-stabilized	50% glass/mineral reinforced, V0	51% glass/mineral reinforced	15% glass-reinforced	15% glass-reinforced, V0	45% glass-reinforced, V0
Processing	1a. Melt flow (gm./10 min.)	D1238							
	1. Melting temperature, °C. T_m (crystalline)		310	310	310	310			
	T_g (amorphous)								
	2. Processing temperature range, °F. (C = compression; T = transfer; I = injection; E = extrusion)		I: 610-660	I: 620-650	I: 610-640	I: 610-640	I: 610-640	I: 610-640	I: 610-640
	3. Molding pressure range, 10^3 p.s.i.		5-15						
	4. Compression ratio		2.5-3						
	5. Mold (linear) shrinkage, in./in.	D955	0.008		0.3		0.006-0.007	0.005	0.002
Mechanical	6. Tensile strength at break, p.s.i.	D638[b]	17,000						
	7. Elongation at break, %	D638[b]		2.5	1.3	2	2.0		
	8. Tensile yield strength, p.s.i.	D638[b]		23,500	21,000	16,200	19,000	18,400	29,500
	9. Compressive strength (rupture or yield), p.s.i.	D695	26,000	33,000			30,000		
	10. Flexural strength (rupture or yield), p.s.i.	D790	30,000	31,500	31,000	28,400	23,700	26,600	41,100
	11. Tensile modulus, 10^3 p.s.i.	D638[b]			2400	1530			
	12. Compressive modulus, 10^3 p.s.i.	D695							
	13. Flexural modulus, 10^3 p.s.i. 73° F.	D790	1300	1200	2040	1300	1090	1700	2600
	200° F.	D790							
	250° F.	D790							
	300° F.	D790							
	14. Izod impact, ft.-lb./in. of notch ($^1/_8$-in. thick specimen)	D256A	0.8	0.9	1.2	1.3	1.1	0.8	1.7
	15. Hardness Rockwell	D785	125	125			127		
	Shore/Barcol	D2240/ D2583							
Thermal	16. Coef. of linear thermal expansion, 10^{-6} in./in./°F.	D696	19	2.9-4.6			3.5		
	17. Deflection temperature under flexural load, °F. 264 p.s.i.	D648	361	527	505	482	531	504	527
	66 p.s.i.	D648							
	18. Thermal conductivity, 10^{-4} cal.-cm./ sec.-cm.^{2}.°C.	C177	2.6						
Physical	19. Specific gravity	D792	1.54	1.54	1.82	1.68	1.26	1.58	1.78
	20. Water absorption ($^1/_8$-in. thick specimen), % 24 hr.	D570	0.14	0.16	0.12		0.30	0.28	0.17
	Saturation	D570							
	21. Dielectric strength ($^1/_8$-in. thick specimen), short time, v./mil	D149	>560	505	635		480		
		SUPPLIERS[a]	Amoco Polymers; RTP	Amoco Polymers; RTP	Amoco Polymers	Amoco Polymers; RTP	Amoco Polymers; RTP	Amoco Polymers; RTP	Amoco Polymers; RTP

	Polyphthalamide (Cont'd)		Polypropylene							
			Homopolymer							
	40% glass/ mineral reinforced	40% mineral-reinforced, heat-stabilized	Unfilled	10-40% talc-filled	10-40% calcium carbonate-filled	10-30% glass fiber-reinforced	10-50% mica-filled	40% glass fiber-reinforced	20-30% long glass fiber-reinforced	40% long glass fiber-reinforced
1a.			0.4-38.0	0.1-30.0	0.1-30.0	1-20	4-10	1-20		
1.		310	160-175	158-168	168	168	168	168	168	163
			−20							
2.	I: 610-650	I: 610-650	I: 375-550 E: 400-500	I: 350-550	I: 375-525	I: 425-475		I: 450-550	I: 360-440	I: 370-410
3.	5-15		10-20	10-20	8-20			10-25		6-12
4.	2.5-3		2.0-2.4							3-4
5.	0.004-0.007		0.010-0.025	0.008-0.022	0.007-0.018	0.002-0.008	0.002-0.015	0.003-0.005	0.0025-0.004	0.001-0.003
6.			4500-6000	3545-5000	3400-4500	6500-13,000		8400-15,000	7500-14,100	10,500-15,600
7.	2.5	1.6	100-600	3-60	10-245	1.8-7	3-10	1.5-4	2.1-2.2	1.7
8.	23,500	17,000	4500-5400	3500-5000	3000-4600	7000-10,000	4700-6500			
9.	33,000	24,000	5500-8000	7500	3000-7200	6500-8400		8900-9800	6500-14,400	10,400-15,900
10.	31,500	28,000	6000-8000	7000-9200	5500-7000	7000-20,000		10,500-22,000	10,000-22,800	20,800-25,200
11.			165-225	450-575	375-500	700-1000		1100-1500	750-900	970-1120
12.			150-300							
13.	1200	1200	170-250	210-670	230-450	310-780	420-1150	950-1000	550-800	920-1000
			50	400	320					
			35							
14.	0.9	0.7	0.4-1.4	0.4-1.4	0.5-1.0	1.0-2.2	0.50-0.85	1.4-2.0	3.5-7.8	8.0-10.04
15.	125	125	R80-102	R85-110	R78-99	R92-115	R82-100	R102-111	R105-117	
							066-78			
16.	2.9-4.6	3.8-4.1	81-100	42-80	28-50	21-62		27-32		
17.	527	325	120-140	132-180	135-170	253-288		300-330	250-295	300
			225-250	210-290	200-270	290-320		330	305	
18.			2.8	7.6	6.9	5.5-6.2		8.4-8.8	2.35	
19.	1.54	1.57	0.900-0.910	0.97-1.27	0.97-1.25	0.97-1.14	0.99-1.35	1.22-1.23	1.04-1.17	1.21
20.	0.16	0.14	0.01-0.03	0.01-0.03	0.02-0.05	0.01-0.05	0.01-0.06	0.05-0.06	0.05	
								0.09-0.10		
21.	505	455	600	500	410-500			500-510		
	Amoco Polymers; RTP	Amoco Polymers; RTP	American Polymers; Amoco Chemical; Aristech Chem.; Bamberger Polymers; ComAlloy; Epsilon; Exxon; Federal Plastics; Ferro Corp.; Fina; M.A. Polymers; Monmouth; Montell Polyolefins; Phillips; Quantum; RSG Polymers; Rexene; A. Schulman; Shell; Shuman; Solvay Polymers; Thermofil; Wash. Penn	Adell; Albis; Amoco; Bamberger Polymers; ComAlloy; DSM; Exxon; Federal Plastics; Ferro Corp.; M.A. Polymers; Montell Polyolefins; MRC; Polifil; Polycom Huntsman; RSG Polymers; RTP; A. Schulman; Thermofil; Wash. Penn	Adell; Albis; Bamberger Polymers; ComAlloy; DSM; Federal Plastics; Ferro Corp.; M.A. Polymers; Montell Polyolefins; Polifil; Polycom Huntsman; RSG Polymers; RTP; A. Schulman; Thermofil; Wash. Penn	Adell; Albis; Bamberger Polymers; ComAlloy; DSM; Eastman; Federal Plastics; Ferro Corp.; LNP; M.A. Polymers; Montell Polyolefins; MRC; Polifil; Polycom Huntsman; RSG Polymers; RTP; A. Schulman; Wash. Penn	Polycom Huntsman; RTP; A. Schulman; Washington Penn	Adell; Albis; ComAlloy; DSM; Ferro Corp.; LNP; M.A. Polymers; Montell Polyolefins; MRC; Polifil; RTP; A. Schulman; Thermofil	DSM; Hoechst Celanese; LNP; Polycom Huntsman; RTP	Hoechst Celanese; LNP; RTP

	Properties	ASTM test method	30% random glass mat	40% random glass mat	42% directionalized glass mat Parallel	42% directionalized glass mat Transverse	Impact-modified, 40% mica-filled	EMI shielding (conductive); 30% PAN carbon fiber	Unfilled
					Polypropylene (Cont'd)				
					Homopolymer (Cont'd)				**Copolymer**
Processing	1a. Melt flow (gm./10 min.)	D1238							0.6-44.0
	1. Melting temperature, °C. T_m (crystalline)		168	168	168	168	168	168	150-175
	T_g (amorphous)								−20
	2. Processing temperature range, °F. (C = compression; T = transfer; I = injection; E = extrusion)		C: 420-440	C: 420-440	C: 420-440	C: 420-440	I: 350-470	I: 360-470	I: 375-550 E: 400-500
	3. Molding pressure range, 10^3 p.s.i.		1-2	1-2	1-2	1-2			10-20
	4. Compression ratio								2-2.4
	5. Mold (linear) shrinkage, in./in.	D955	0.002-0.003	0.001-0.002	0.0005-0.0015	0.0025-0.0035	0.007-0.008	0.001-0.003	0.010-0.025
Mechanical	6. Tensile strength at break, p.s.i.	D638[b]	12,000	14,000	32,200	10,000	4500	6800	4000-5500
	7. Elongation at break, %	D638[b]	3	2.1	2.1	2.4	4	0.5	200-500
	8. Tensile yield strength, p.s.i.	D638[b]	12,000	14,000	32,200	10,000			3000-4300
	9. Compressive strength (rupture or yield), p.s.i.	D695		9000					3500-8000
	10. Flexural strength (rupture or yield), p.s.i.	D790	20,000	24,000	43,180	22,785	7000	9000	5000-7000
	11. Tensile modulus, 10^3 p.s.i.	D638[b]	670	850	1400	705	700	1750	130-180
	12. Compressive modulus, 10^3 p.s.i.	D695							
	13. Flexural modulus, 10^3 p.s.i. 73° F.	D790	620	800	1375	740	600	1650	130-200
	200° F.	D790							40
	250° F.	D790							30
	300° F.	D790							
	14. Izod impact, ft.-lb./in. of notch (1/8-in. thick specimen)	D256A	12.2	14			0.7	1.1	1.1-14.0
	15. Hardness Rockwell	D785							R65-96
	Shore/Barcol	D2240/ D2583							Shore D70-73
Thermal	16. Coef. of linear thermal expansion, 10^{-6} in./in./°F.	D696	15	15	14	22			68-95
	17. Deflection temperature under flexural load, °F. 264 p.s.i.	D648	310	310	310	310	205	245	130-140
	66 p.s.i.	D648							185-220
	18. Thermal conductivity, 10^{-4} cal.-cm./ sec.-cm.^{2}-°C.	C177							3.5-4.0
Physical	19. Specific gravity	D792	1.1	1.19	1.21	1.21	1.23	1.04	0.890-0.905
	20. Water absorption (1/8-in. thick specimen), % 24 hr.	D570						0.12	0.03
	Saturation	D570							
	21. Dielectric strength (1/8-in. thick specimen), short time, v./mil	D149		360					600
	SUPPLIERS[a]		Azdel	Azdel	Azdel	Azdel	Albis; ComAlloy; DSM; Federal Plastics; Ferro Corp.; M.A. Polymers; Polifil; Polycom Huntsman; A. Schulman	ComAlloy; DSM; LNP; RTP	American Polymers; Amoco Chemical; Aristech Chem.; Bamberger Polymers; ComAlloy; Epsilon; Exxon; Federal Plastics; Ferro Corp.; Fina; M.A. Polymers; Monmouth; Montell Polyolefins; Novacor; Phillips; Polycom Huntsman; Quantum; RSG Polymers; Rexene; A. Schulman; Shell; Shuman; Solvay Polymers; Wash. Penn

	Polypropylene (Cont'd)					Polystyrene and styrene copolymers				
	Copolymer (Cont'd)					Polystyrene homopolymers				Rubber-modified
	Unfilled, impact-modified	10-20% glass fiber-reinforced	30-40% glass fiber-reinforced	10-40% talc-filled	10-40% calcium carbonate-filled	High and medium flow	Heat-resistant	30% long and short glass fiber-reinforced	20% long and short glass fiber-reinforced	Flame-retarded, UL-V0
1a.		0.1-20	0.1-20	0.1-30	0.1-30					
1.	150-168	160-168	160-168							
	-20					74-105	100-110	110-120	115	
2.	I: 390-500 E: 400-500	I: 350-480	I: 350-480	I: 350-470 E: 425-475	I: 350-470	C: 300-400 I: 350-500 E: 350-500	C: 300-400 I: 350-500 E: 350-500	I: 400-460	I: 400-550	I: 400-450 E: 375-425
3.	10-20			15-20	15-20	5-20	5-20		10-20	6-15
4.	2-2.4			2-2.5	2-2.5	3	3-5			3
5.	0.010-0.025	0.003-0.01	0.001-0.01	0.009-0.017	0.006-0.022	0.004-0.007	0.004-0.007	0.001-0.002	0.001-0.003	0.003-0.006
6.	3500-5000	5000-8000	6000-10,000	3000-3775	2500-3465	5200-7500	6440-8200	11,000-13,000	10,000-12,000	2650-4100
7.	200-700	3.0-4.0	2.2-3.0	20-50	40-50	1.2-2.5	2.0-3.6	1-1.2	1.0-1.3	30-50
8.	1600-4000			2800-4100	2000-3800		6440-8150			3100-4400
9.	3500-6000	5500-5600	5400-5700			12,000-13,000	13,000-14,000	16,500-17,500	16,000-17,000	
10.	4000-6000	7000-11,000	9000-15,000	4500-5100	4000-6500	10,000-14,600	13,000-14,000	14,000-20,000	14,000-18,000	4500-7500
11.	50-150				3500	330-475	450-485	1200-1300	900-1200	240-300
12.						480-490	495-500			
13.	60-160	355-510	600-960	160-400	140-750	380-490	450-500	1200	950-1100	280-330
14.	2.2-No break	0.95-2.7	0.9-3.0	0.6-4.0	0.7-10.7	0.35-0.45	0.4-0.45	0.9-3.0	0.9-2.5	1.9-3.3
15.	R50-60	R100-103	R104-105	R83-88	R81-89	M60-75	M75-84	M85-95	M80-95, R119	R38-65
	Shore D45-55		Shore D45-55							
16.	68-95					50-83	68-85	20	39.6-40	45
17.	115-135	260-280	280	100-165	102-155	169-202	194-217	215-220	200-220	180-205
	167-192	305	310	195-260	140-235	155-204	200-224	225-230	220-230	176-181
18.	3.5-4.0					3.0	3.0		5.9	
19.	0.880-0.905	0.98-1.04	1.11-1.21	0.97-1.24	0.97-1.24	1.04-1.05	1.04-1.05	1.29	1.20	1.15-1.17
20.	0.03	0.01	0.01	0.02	0.02	0.01-0.03	0.01	0.09	0.07-0.01	0.0
						0.01-0.03	0.01		0.3	
21.	500					500-575	500-525	450	425	550
	American Polymers; ComAlloy; Epsilon; Exxon; Federal Plastics; Huntsman; M.A. Polymers; Monmouth; Montell Polyolefins; Phillips; Polycom Huntsman; Quantum; Rexene; A. Schulman; Solvay Polymers; Wash. Penn	Adell; Albis; ComAlloy; DSM; Eastman; Federal Plastics; Ferro Corp.; LNP; M.A. Polymers; Montell Polyolefins; Polifil; Polycom Huntsman; RTP; A. Schulman; Wash. Penn	Adell; Albis; Amoco Chemical; ComAlloy; DSM; Eastman; Ferro Corp.; LNP; Montell Polyolefins; Polifil; Polycom Huntsman; RTP; A. Schulman; Wash. Penn	Adell; Albis; Bamberger Polymers; ComAlloy; DSM; Eastman; Federal Plastics; Ferro Corp.; M.A. Polymers; Montell Polyolefins; Polifil; Polycom Huntsman; RTP; A. Schulman; Wash. Penn	Adell; Albis; Bamberger Polymers; ComAlloy; DSM; Federal Plastics; Ferro Corp.; M.A. Polymers; Montell Polyolefins; Polifil; Polycom Huntsman; RTP; A. Schulman; Wash. Penn	A & E Plastics; American Polymers; Amoco Chemical; Bamberger Polymers; BASF; Chevron; Dow Plastics; Federal Plastics; Fina; Huntsman; Mobil; RSG Polymers; A. Schulman;	A & E Plastics; American Polymers; Amoco Chemical; BASF; Chevron; Dow Plastics; Federal Plastics; Fina; Huntsman; Ivex; Mobil; A. Schulman	DSM; LNP; M.A. Polymers; RTP; A. Schulman	DSM; Ferro Corp.; Ferro Eng. Therm.; LNP; M.A. Polymers; Mobil; RTP; Thermofil	BASF; Dow Plastics; Huntsman; Mobil; RTP; A. Schulman

Polystyrene and styrene copolymers (Cont'd)

| | | Rubber-modified (Cont'd) | Styrene copolymers | | | | Acrylate-styrene-acrylonitrile (ASA) | |
| | | | Styrene-acrylonitrile (SAN) | | | | | |
Properties	ASTM test method	High impact	Molding and extrusion	Olefin rubber-modified	20% glass fiber-reinforced	Acrylic or acrylate SAN graft copolymer	ASA extrusion, blow molding, injection molding grades	ASA/PVC
1a. Melt flow (gm./10 min.)	D1238	5.8	1.4					
1. Melting temperature, °C. T_m (crystalline)								
T_g (amorphous)		9.3-105	100-200	–55/110	120			
2. Processing temperature range, °F. (C = compression; T = transfer; I = injection; E = extrusion)		I: 350-525 E: 375-500	C: 300-400 I: 360-550 E: 360-450	I: 480-530 E: 435-460	I: 400-550	I: 465-535	E: 380-450 I: 400-470	E: 330-360 I: 340-380
3. Molding pressure range, 10^3 p.s.i.		10-20	5-20	1-2	10-20		9-15	12-17
4. Compression ratio		4	3	2.7-3.2			2.8-3.0	2.0-2.5
5. Mold (linear) shrinkage, in./in.	D955	0.004-0.007	0.003-0.005	0.005-0.007	0.001-0.003	0.004-0.007	0.004-0.006	0.003-0.005
6. Tensile strength at break, p.s.i.	D638[b]	1900-6500	10,000-11,900	5100	15,500-18,000	5000-5850	4000-7500	
7. Elongation at break, %	D638[b]	20-65	2-3	15-30	1.2-1.8	20-21	25-40	40-70
8. Tensile yield strength, p.s.i.	D638[b]	2100-6000	9920-12,000	5000-6000		6500-6800	5200-5600	6300-6700
9. Compressive strength (rupture or yield), p.s.i.	D695		14,000-15,000		17,000-21,000			
10. Flexural strength (rupture or yield), p.s.i.	D790	3300-10,000	11,000-19,000	7700-8900	20,000-22,700		6000-8000	9000-10,000
11. Tensile modulus, 10^3 p.s.i.	D638[b]	160-370	475-560	300	1200-1710	320-330		
12. Compressive modulus, 10^3 p.s.i.	D695		530-580					
13. Flexural modulus, 10^3 p.s.i. 73° F.	D790	160-390	500-610	280-300	1000-1280		220-341	280-315
200° F.	D790							
250° F.	D790							
300° F.	D790							
14. Izod impact, ft.-lb./in. of notch ($^1/_8$-in. thick specimen)	D256A	0.95-7.0	0.4-0.63	13-15	1.0-3.0	6.2-11.2	9-11	17-20
15. Hardness Rockwell	D785	R50-82; L-60	M80, R83, 75	R100-102	M89-100, R122		R85-90	R99-102
Shore/Barcol	D2240/ D2583							
16. Coef. of linear thermal expansion, 10^{-6} in./in./°F.	D696	44.2	65-68	80	23.4-41.4		59	84
17. Deflection temperature under flexural load, °F. 264 p.s.i.	D648	170-205	203-220	197-200	210-230	180-203	185-190	165-170
66 p.s.i.	D648	165-200	220-224		220	203-214	200-210	175-185
18. Thermal conductivity, 10^{-4} cal.-cm./ sec.-cm.^{2}-°C.	C177		3.0		6.6			
19. Specific gravity	D792	1.03-1.06	1.06-1.08	1.02	1.22-1.40	1.07	1.05-1.06	1.21
20. Water absorption ($^1/_8$-in. thick specimen), % 24 hr.	D570	0.05-0.07	0.15-0.25	0.09	0.1-0.2	0.45	0.2-0.3	0.11-0.16
Saturation	D570		0.5		0.7			
21. Dielectric strength ($^1/_8$-in. thick specimen), short time, v./mil	D149		425	420	500		490	417-476
SUPPLIERS[a]		American Polymers; Amoco Chemical; Bamberger Polymers; BASF; Chevron; Dow Plastics; Federal Plastics; Fina; Huntsman; Mobil; Monsanto; Novacor; RSG Polymers; RTP; A. Schulman; Shuman	Albis; American Polymers; BASF; Bamberger Polymers; Dow Plastics; Federal Plastics; Huntsman; Monsanto; Network Polymers; RSG Polymers	Dow Plastics; Huntsman	ComAlloy; DSM; Ferro Corp.; Ferro Eng. Therm.; LNP; RTP; Thermofil	GE Plastics	BASF; GE Plastics; Monsanto; Network Polymers	GE Plastics

Polystyrene and styrene copolymers (Cont'd)

	Styrene copolymers (Cont'd)										
	ASA (Cont'd)		Styrene-maleic anhydride (SMA)						High heat-resistant copolymers		
	ASA/PC	Clear styrene-butadiene copolymers	Molding and extrusion	Impact-modified	12% glass fiber-reinforced	16% glass fiber-reinforced	20% glass fiber-reinforced	42% long glass fiber-reinforced	Injection molding	Impact-modified	20% glass fiber-reinforced
1a.		7-15			0.4-0.7	0.3-0.7				1.0-1.8	
1.											
		108	114								
2.	I: 400-480	I: 380-450 / E: 380-440	I: 430-510 / E: 400-500	I: 450-550 / E: 425-525	I: 450-550	I: 450-550	I: 400-550	I: 520-580	I: 450-550	I: 425-540 / E: 400-500	I: 425-550
3.	9-15		12-17	12-17			10-20				
4.				2.5-2.7							
5.	0.003-0.005	0.004-0.010	0.004-0.006	0.004-0.006	0.003-0.005	0.0025-0.0045	0.002-0.003	0.0016	0.005	0.003-0.006	0.003-0.004
6.			8100	4500-5500	9000-10,000	10,800-11,300		19,000	7100-8100	4600-5800	10,000-14,000
7.	70	20-180	1.8-30	10-35	2-3	2.5-2.7	2-3	1.3	1.7-1.9	10-20	1.4-3.5
8.	6600-6800	1900-4400	5200-8100	4500-6400			8100-11,000				
9.									11,400-14,200		
10.	9000-9700	2700-6400	8000-14,200	7600-13,000	15,000-17,500	17,500-19,200	16,300-17,000	37,000		8500-10,500	16,300-22,000
11.			340-390	270-360			750-880	2000	440-490	280-330	850-900
12.											
13.	285-295	153-215	320-470	280-490	575-650	675-745	720-800	1800	450-490	320-370	800-1050
14.	13	0.4-1.4 NB	0.4-2.0	2.5-6	1.6	1.6	2.1-2.7	5	0.4-0.6	1.5-4.0	2.1-2.6
15.	R99		R106-109	R75-109	R110	R110	R73	R120		R75-95	
16.	80		80	47-88	71	56		17.9	65-67	67-79	20
17.	205-210	143-170	–214-245	198-235	234	245	229-245	265	226-249	230-260	231-247
	225-230										
18.								5.5			
19.	1.11	1.01	1.05-1.08	1.05-1.09	1.14	1.18	1.21-1.22	1.45	1.07-1.10	1.05-1.08	1.20-1.22
20.	0.29	0.08	0.1	0.1-0.5			0.1		0.1	0.1	0.1
21.		300	415-480								
	GE Plastics	Phillips 66	Arco; DSM; Monsanto	Arco; Monsanto	Arco	Arco	Arco; DSM; Monsanto; RTP	Arco	Arco	Arco; Monsanto	Arco; ComAlloy; DSM; LNP; M.A. Polymers; RTP

	Properties	ASTM test method	Styrene methyl methacrylate	EMI shielding (conductive); 20% PAN carbon fiber	Casting resins — Liquid	Casting resins — Unsaturated	50-65% mineral-filled potting and casting compounds	Unreinforced molding
			Polystyrene and styrene copolymers (Cont'd) — Styrene copolymers (Cont'd)		**Polyurethane** — Thermoset			Thermoplastic
Processing	1a. Melt flow (gm./10 min.)	D1238						
	1. Melting temperature, °C. T_m (crystalline)				Thermoset	Thermoset	Thermoset	75-137
	T_g (amorphous)		100-105					
	2. Processing temperature range, °F. (C = compression; T = transfer; I = injection; E = extrusion)		I: 375-475	I: 430-500	C: 43-250		25 (casting)	I: 370-500 E: 370-510
	3. Molding pressure range, 10^3 p.s.i.		5-20		0.1-5			6-15
	4. Compression ratio		2.5-3.5					3
	5. Mold (linear) shrinkage, in./in.	D955	0.002-0.006	0.0005-0.003	0.020		0.001-0.002	0.004-0.010
Mechanical	6. Tensile strength at break, p.s.i.	D638[b]	8100-10,100	14,000	175-10,000	10,000-11,000	1000-7000	4500-9000
	7. Elongation at break, %	D638[b]	2.1-5.0	1	100-1000	3-6	5-55	60-550
	8. Tensile yield strength, p.s.i.	D638[b]						7,800-11,000
	9. Compressive strength (rupture or yield), p.s.i.	D695			20,000			
	10. Flexural strength (rupture or yield), p.s.i.	D790	14,100-16,000	20,700	700-4500	19,000		10,200-15,000
	11. Tensile modulus, 10^3 p.s.i.	D638[b]	440-520	2000	10-100			190-300
	12. Compressive modulus, 10^3 p.s.i.	D695	440-480		10-100			
	13. Flexural modulus, 10^3 p.s.i. 73° F.	D790	450-460	1900	10-100	610		4-310
	200° F.	D790						
	250° F.	D790						
	300° F.	D790						
	14. Izod impact, ft.-lb./in. of notch ($^1/_8$-in. thick specimen)	D256A	0.3-0.4	0.7	25 to flexible	0.4		1.5-1.8-No break
	15. Hardness Rockwell	D785	M80-85					R >100; M48
	Shore/Barcol	D2240/ D2583			Shore A10-13,D90	Barcol 30-35	Shore A90, D52-85	Shore 75A-70D
Thermal	16. Coef. of linear thermal expansion, 10^{-6} in./in./°F.	D696	40-72		100-200		71-100	0.5-0.8
	17. Deflection temperature under flexural load, °F. 264 p.s.i.	D648	205-210	220	Varies over wide range	190-200		158-260
	66 p.s.i.	D648		230				115-275
	18. Thermal conductivity, 10^{-4} cal.-cm./ sec.-cm.^{2}-°C.	C177			5		6.8-10	
Physical	19. Specific gravity	D792	1.08-1.13	1.14	1.03-1.5	1.05	1.37-2.1	1.12-1.24
	20. Water absorption ($^1/_8$-in. thick specimen), % 24 hr.	D570	0.11-0.17	0.1	0.2-1.5	0.1-0.2	0.06-0.52	0.15-0.19
	Saturation	D570						0.5-0.6
	21. Dielectric strength ($^1/_8$-in. thick specimen), short time, v./mil	D149			300-500		500-750 @1/16 in.	400
	SUPPLIERS[a]		Network Polymers; Novacor	DSM; LNP; RTP	Bayer Corp.; Cabot; Conap; Dow Plastics; Emerson & Cuming; Hexcel; Hysol; ITW Devcon; Polyurethane Corp. of America; Polyurethane Specialties; Union Carbide	Dow Plastics; Emerson & Cuming; Hexcel; Hyson; Polyurethane Corp. of America; Polyurethane Specialties	Conap	Bayer Corp.; Dow Plastics

	Polyurethane (Cont'd)							Polyvinylidene chloride copolymers			Silicone
	Thermoplastic (Cont'd)								Barrier film resins		Casting resins
	10-20% glass fiber-reinforced molding compounds	30% long glass fiber-reinforced	40% long glass fiber-reinforced	50% long glass fiber-reinforced	60% long glass fiber-reinforced	Long glass-reinforced molding compound	EMI shielding (conductive); 30% PAN carbon fiber	Injection molding	Un-plasticized	Plasticized	Flexible (including RTV)
1a.											
1.		240	240	245	255	75		172	160	172	Thermoset
	120-160							-15	0.2	-15	
2.	I: 360-410	I: 440-500	I: 440-500	I: 450-510	I: 460-530	I: 450-500	I: 360-450	C: 260-350 I: 300-400 E: 300-400	E: 320-390	E: 340-400	
3.	8-11	10-20	10-20	10-20	10-20			5-30	3-30	5-30	
4.		3	3	3	3			2.5	2-2.5	2-2.5	
5.	0.004-0.010	0.002-0.006	0.001-0.005	0.001-0.004	0.001-0.003	0.001	0.001-0.002	0.005-0.025	0.005-0.025	0.005-0.025	0.0-0.006
6.	4800-7500	24,200	28,600	32,500	34,900	27,000-33,000	13,000	3500-5000	2800	3500	350-1000
7.	3-70	3.0	2.3-2.5	2.2-2.3	1.9	2	20	160-240	350-400	250-300	20-700
8.							27,000-33,000	2800-3800		4900	
9.	5000	24,600-25,700	26,800-30,400	28,000-33,000	29,300-36,900			2000-2700			
10.	1700-6200	35,300	44,400	48,800-50,400	57,200	45,000-57,000	9000	4200-6200			
11.	0.6-1.40	1100	1500	2100	2300-2460	1700-2500	500	50-80	50-80	50-80	
12.								55-95			
13.	40-90	1100	1360-1500	1810-2000	2400	1500-2200	500	55-95			
14.	10-14-No break	6.2-7.5	8.4-9.1	10.1	13.4	8-16	10	0.4-1.0	0.3-1.0	0.3-1.0	
15.	R45-55							M60-65	R98-106	R98-106	
											Shore A10-70
16.	34					0.6-0.8		190	190	190	10-19
17.	115-130	185	195	205	215	200-260	180	130-150	130-150	130-150	
	140-145					230-370					
18.								3	3	3	3.5-7.5
19.	1.22-1.36	1.43	1.52	1.63	1.76		1.33	1.65-1.72	1.65-1.70	1.68-1.72	0.97-2.5
20.	0.4-0.55							0.1	0.1	0.1	0.1
	1.5										
21.	600							400-600		400-600	400-550
	DSM; LNP; RTP; Thermofil; A. Schulman; Union Carbide	Hoechst Celanese; RTP	Hoechst Celanese; RTP	Hoechst Celanese; RTP	Hoechst Celanese; RTP	Dow Plastics; RTP	LNP; RTP	Dow Plastics	Dow Plastics	Dow Plastics	Bayer Corp.; Dow Corning; Emerson & Cuming; GE Silicones

			Silicone (Cont'd)				Silicone epoxy	Sulfone polymers	
Materials			Liquid injection molding	Molding and encapsulating compounds	Silicone/poly-amide pseudo-interpenetrating networks[f]			Polysulfone	
Properties	ASTM test method	Liquid silicone rubber	Mineral- and/or glass-filled	Silicone/ nylon 66	Silicone/ nylon 12	Molding and encapsulating compound	Injection molding, flame-retarded, extrusion	Mineral-filled
Processing									
1a. Melt flow (gm./10 min.)	D1238						3.5-9	7-8.5
1. Melting temperature, °C. T_m (crystalline)		Thermoset	Thermoset	Interpenetrating network	Interpenetrating network	Thermoset		
T_g (amorphous)							187-190	190
2. Processing temperature range, °F. (C = compression; T = transfer; I = injection; E = extrusion)		I: 360-420	C: 280-360 I: 330-370 T: 330-370	I: 460-525	I: 360-410	C: 350	I: 625-750 E: 600-700	I: 675-775
3. Molding pressure range, 10^3 p.s.i.		1-2	0.3-6	8-15	7-12	0.4-1.0	5-20	10-20
4. Compression ratio			2.0-8.0				2.5-3.5	2.5-3.5
5. Mold (linear) shrinkage, in./in.	D955	0.0-0.005	0.0-0.005	0.004-0.007	0.005-0.007	0.005-0.006	0.0058-0.007	0.004-0.005
Mechanical									
6. Tensile strength at break, p.s.i.	D638[b]	725-1305	500-1500	10,100-12,500	5200-7400	500-8000		9500-9800
7. Elongation at break, %	D638[b]	300-1000	80-800	5-20	10	60	50-100	2-5
8. Tensile yield strength, p.s.i.	D638[b]						10,200-11,600	
9. Compressive strength (rupture or yield), p.s.i.	D695					28,000	40,000	
10. Flexural strength (rupture or yield), p.s.i.	D790			14,000-15,900	8000-8300	17,000	15,400-17,500	14,300-15,400
11. Tensile modulus, 10^3 p.s.i.	D638[b]						360-390	550-650
12. Compressive modulus, 10^3 p.s.i.	D695						374	
13. Flexural modulus, 10^3 p.s.i. 73° F.	D790			360-410	200-216		390	600-750
200° F.	D790						370	570-710
250° F.	D790						350	550-690
300° F.	D790						310	510-650
14. Izod impact, ft.-lb./in. of notch ($^1/_8$-in. thick specimen)	D256A			0.8-0.9	0.6-0.7	0.3	1.0-1.3	0.65-1.0
15. Hardness Rockwell	D785						M69	M70-74
Shore/Barcol	D2240/ D2583	Shore A20-70	Shore A10-80			Shore A68-95		
Thermal									
16. Coef. of linear thermal expansion, 10^{-6} in./in./°F.	D696	10-20	20-50			30-200	56	34-39
17. Deflection temperature under flexural load, °F. 264 p.s.i.	D648		>500				345	345-354
66 p.s.i.	D648						358	
18. Thermal conductivity, 10^{-4} cal.-cm./ sec.-cm.^{2}.°C.	C177		7.18			16	6.2	
Physical									
19. Specific gravity	D792	1.08-1.14	1.80-2.05	1.12-1.13	1.01-1.02	1.2-1.84	1.24-1.25	1.48-1.61
20. Water absorption ($^1/_8$-in. thick specimen), % 24 hr.	D570		0.15	0.6-0.8	0.12-0.15	0.2	0.3	
Saturation	D570		0.15-0.40				0.8	0.5-0.6
21. Dielectric strength ($^1/_8$-in. thick specimen), short time, v./mil	D149		200-550			246-500	425	450
SUPPLIERS[a]		Bayer Corp.	Dow Corning; ICI Fiberite; GE Silicones	LNP	LNP	Dow Corning; Emerson & Cuming	Amoco Polymers; BASF	Amoco Polymers; RTP

Sulfone polymers (Cont'd)

	Polysulfone (Cont'd)							Polyethersulfone		
	Extrusion/ injection molding grade	10% glass fiber-reinforced	20%-30% glass fiber-reinforced	20% glass fiber-reinforced	30% glass fiber-reinforced	EMI shielding (conductive); 30% carbon fiber	Polyaryl-sulfone	Unfilled	10% glass fiber-reinforced	20% glass fiber-reinforced
1a.					7-8		10-30		12	10
1.										
					189-190	190	220	220-230		220-225
2.	I: 610-680	I: 660-715	I: 660-715	I: 660-715	I: 600-700	I: 550-700	I: 630-800 E: 620-750	C: 645-715 I: 590-750 E: 625-720	I: 660-715	C: 610-750 I: 630-735 E: 570-650
3.						10-20	5-20	6-20	50-100	6-100
4.						2.5-3.5		2-2.5	2:1	2.0-3.5
5.	0.005-0.007	0.003-0.005	0.003-0.005	2.8×10^{-3}	0.001-0.006	0.0005-0.001	0.007-0.008	0.006-0.007	0.5	0.002-0.005
6.		14,500	16,500-18,000	16,700	14,500-18,100	23,000-23,500	9000	9800-13,800	16,000	15,200-20,000
7.	40-80	4.2	1.8-2.4	2.4	1.5-1.8	1.5-2	40-60	6-80	4.1	2-3.5
8.	11,500						10,400-12,000	12,200-13,000		18,000-18,800
9.					19,000	25,000		11,800-15,600		19,500-24,000
10.	17,500	20,000	21,000-23,500	21,000	20,000-23,500	31,000-35,000	12,400-16,100	17,000-18,700	21,000; 24,500	23,500-27,600
11.	390	667-670	1000-1450	1020	1350-1450	2150-2800	310-385	350-410	555; 740	825-1130
12.										
13.	370	600	850-1250	850	1050-1250	1900-2300	330-420	348-380	590; 650	750-980
										812-850
								330		840
								280		580-842
14.	1.0-1.2	1.3	1.5	1.5	1.1-1.5	1.2-1.8	1.6-12	1.4-No break	0.9; 1.3	1.1-1.7
15.	M69	M79	M83-M87	M83	M87-100	M80		M85-88	M94	M96-99
16.	31	18-32	11-14	25	20-25	6	31-49	55	34	23-32
17.	340	361	363-365	363	350-365	360-365	400	383-397	437	408-426; 437
	360	367	369-372	369	360-372	365-380		410	423	410-430
18.								3.2-4.4		
19.	1.24	1.31	1.40-1.49	1.40	1.46-1.49	1.36-1.7	1.29-1.37	1.37-1.46	1.43; 1.45	1.51-1.53
20.					0.3	0.15-0.25		0.12-1.7		0.15-0.4
	0.8	0.6-0.7	0.5-0.6	0.6	0.5		1.1-1.85	2.5-2.5	1.9	1.65-2.1; 1.7
21.					≥1500		370-380	400	440	375-500
	BASF	BASF; RTP	BASF; ComAlloy; RTP	BASF; ComAlloy; RTP	BASF; ComAlloy; DSM; Ferro Corp.; Ferro Eng. Therm.; LNP; RTP; Thermofil	DSM; Ferro Corp.; Ferro Eng. Therm.; LNP; RTP; Thermofil	Amoco Chemicals Polymers Business Group; Amoco Polymers; LNP; RTP	Amoco Chemicals	BASF; ICI Americas	Amoco Polymers; BASF; DSM; ICI Americas; LNP; RTP; Thermofil

| | | | Sulfone polymers (Cont'd) | | | | | Thermoplastic elastomers |
| | | | Polyethersulfone (Cont'd) | | Modified polysulfone | | | Polyolefin |
Properties	ASTM test method	30% glass fiber-reinforced	EMI shielding (conductive); 30% carbon fiber	Poly-phenyl sulfone	Injection molding and platable grades	Mineral-filled	30% glass fiber-reinforced	Low and medium hardness
Processing								
1a. Melt flow (gm./10 min.)	D1238	10		14-19	12-18		16	0.4-20.0
1. Melting temperature, °C. T_m (crystalline)								
T_g (amorphous)			225	220				165
2. Processing temperature range, °F. (C = compression; T = transfer; I = injection; E = extrusion)		I: 660-735	I: 600-750	I: 680-735	I: 690-750	I: 575-650	I: 520-750	C: 350-450 I: 360-475 E: 380-450
3. Molding pressure range, 10^3 p.s.i.		50-100	10-20	10-20	5-100	5-20	0.05-20	4-19
4. Compression ratio		2:1	2.5-3.5	2.2	2	2.5-3.5	2	2-3.5
5. Mold (linear) shrinkage, in./in.	D955	5.7×10^{-3}	0.0005-0.002	0.007	0.0025	0.006-0.007	0.001-0.003	0.015-0.020
Mechanical								
6. Tensile strength at break, p.s.i.	D638[b]	20,300	26,000-30,000	10,100	1100		15,000-19,000	650-2500
7. Elongation at break, %	D638[b]	1.9	1.3-2.5	60	2.3	50-100	1.9-3.0	150-780
8. Tensile yield strength, p.s.i.	D638[b]					10,500		
9. Compressive strength (rupture or yield), p.s.i.	D695		22,000					
10. Flexural strength (rupture or yield), p.s.i.	D790	26,000; 28,500	36,000-38,000	13,200	7700	16,500	20,000-25,800	
11. Tensile modulus, 10^3 p.s.i.	D638[b]	1480	2120-2900	340		400	830-1100	1.1-16.4
12. Compressive modulus, 10^3 p.s.i.	D695							
13. Flexural modulus, 10^3 p.s.i. 73° F.	D790	1170; 1300	2000-2600	350	881	480	900-1260	1..5-30
200° F.	D790					370		
250° F.	D790					340		
300° F.	D790					320		
14. Izod impact, ft.-lb./in. of notch ($^1/_8$-in. thick specimen)	D256A	1.4; 1.7	1.2-1.6	13		1.1	1.0-2.0	No break
15. Hardness Rockwell	D785	M97	R123			M74	M80-85	
Shore/Barcol	D2240/ D2583							Shore A64-92
Thermal								
16. Coef. of linear thermal expansion, 10^{-6} in./in./°F.	D696	31	10	17		53	27-54	
17. Deflection temperature under flexural load, °F. 264 p.s.i.	D648	415; 419	415-420	405	410	325	320-394	
66 p.s.i.	D648	430	420-430					
18. Thermal conductivity, 10^{-4} cal.-cm./ sec.-cm.^{2}-°C.	C177							4.5-5.0
Physical								
19. Specific gravity	D792	1.58	1.47-1.48	1.29-1.3	1.68	1.30	1.52-1.54	0.88-0.98
20. Water absorption ($^1/_8$-in. thick specimen), % 24 hr.	D570		0.29-0.35	0.37			0.10-0.20	0.01
Saturation	D570	1.5				0.8	0.43	
21. Dielectric strength ($^1/_8$-in. thick specimen), short time, v./mil	D149	440		360	380	460	400-510	410-445
SUPPLIERS[a]		Amoco Polymers; BASF; RTP	DSM; LNP; RTP; Thermofil	Amoco Polymers	Amoco Polymers	Amoco Polymers; RTP	Amoco Polymers; DSM; RTP	DuPont; Exxon; M.A. Polymers; Monsanto; Montell Polyolefins; A. Schulman; Teknor Apex; Union Carbide; Vi-Chem

Thermoplastic elastomers (Cont'd)

	Polyolefin (Cont'd) High hardness	Copolyester	Copolyester ether	Polyester	Polyether/ amide block copolymers	Block copolymers of styrene and butadiene or styrene and isoprene	Block copolymers of styrene and ethylene and/or butylene	Silicone-based, pseudo-interpenetrating networks[l] Silicone/ polyamide	Silicone/ polyester	Silicone/ polyolefin
1a.	0.4-10.0			3-12 @190-240		0.5-20				
1.		180-210	195-215	148-230	148-209			Interpenetrating network	Interpenetrating network	Interpenetrating network
	163-165		-3							
2.	C: 370-450 I: 350-480 E: 380-475	350-500	E: 400-500 I: 435-500	I: 340-500 E: 340-500	C: 365-480 I: 340-482 E: 340-460	C: 250-325 I: 300-425 E: 370-400	C: 300-380 I: 350-480 E: 330-380	I: 360-410 I: 355-580 E: 320-365	I: 360-450 E: 380-450	I: 340-425
3.	6-10		Low	1-15	8-12	0.3-3	1.5-20	5-25	6-14	6-11
4.	2-3.5			3-3.5		2.0-4.0	2.5-5.0	3-5	2.5-4	2.5-3.5
5.	0.007-0.021	0.012-0.027	0.004-0.008	0.003-0.018		0.001-0.022	0.003-0.022	0.004-0.007	0.004-0.015	0.010-0.020
6.	950-4000	1830-6000	4980	1000-6800	2000-7000	100-4350	600-3000	5200-12,500	5000	475-1000
7.	20-600	12-600	618	170-900	350-680	20-1350	600-940	5-275	950-1050	120-1000
8.	1600-4000	1350-5070	1370	1350-3900	3000-3500	3700-4400				
9.	22-39									
10.					1100-1900	5100-6400	0.1-100	2200-15,900	1700	
11.	0.34-34.1		18	1.1-130	2-60	0.8-235			18	
12.				7.5-48		3.6-120				
13.	2.7-300	5-280	18	5-175	2.9-66	4-215	0.1-100	35-410	25	
14.	5.0-16.0 No break	2.15-No break	NB (-30C)	2.5-No break	4.3-No break	No break	No break	No break; 0.6-0.9	No break	No break
15.				104						
	Shore D40-70	Shore D32-75	55/95 (D/A)	Shore D35-72	Shore A75-D72	Shore A40-D63-75	Shore A5-95	Shore D60	Shore D49-51	Shore A57-65
16.	36-110		150	85-190	210-230	67-140				
17.					45-135	<0-170				
				111-284	158-257	<0-190				
18.			5-6	3.6-4.5		3.6				
19.	0.90-1.15	1.15-2.15	1.13	1.10-1.28	1.0-1.03	0.90-1.2	0.9-1.2	1.01-1.21	1.21-1.16	0.96-0.97
20.	0-0.28	0.18-0.72	0.4	0.2-3.6	1.01-1.36	0.009-0.39	0.1-0.42	0.5-0.15	0.3-0.8	0.01
			0.14-0.7							
21.	390-465	370-485	350	350-600		300-520	450-800			
	M.A. Polymers; Monsanto; Montell Polyolefins; Quantum; A. Schulman; Teknor Apex; Vi-Chem; Wash. Penn	GE Plastics	Eastman	DuPont; GE Plastics; Hoechst Celanese	Elf Atochem N.A.; EMS; Hüls America	Fina; Firestone Synthetic Rubber; A. Schulman; Shell	Consolidated Polymer; Dow Plastics; Shell; Teknor Apex	Hüls America; LNP	Hüls America; LNP	LNP

	Properties	ASTM test method	Silicone/ polystyrene ethylene butadiene- styrene	Aromatic and aliphatic polyether and polyester urethane, unfilled	Aromatic poly- ether urethane, 15% carbon fiber- reinforced	Polyester	Polyether
			Silicone-based, pseudo-interpenetrating networks[f] (Cont'd)			**Polyurethane**	
			Silicone/	**Silicone/polyurethane**		**Solution coating resins**	
Processing	1a. Melt flow (gm./10 min.)	D1238					
	1. Melting temperature, °C. T_m (crystalline)		Interpenetrating network	Interpenetrating network	Interpenetrating network		
	T_g (amorphous)					−20 to +16	−49
	2. Processing temperature range, °F. (C = compression; T = transfer; I = injection; E = extrusion)		I: 390-490 E: 325-460	I: 325-450 E: 325-475	I: 410-475		
	3. Molding pressure range, 10^3 p.s.i.		6-20	5-15	6-15		
	4. Compression ratio		2-3.5	2.5-3.5	2.5-3.5		
	5. Mold (linear) shrinkage, in./in.	D955	0.005-0.040	0.010-0.020	0.011-0.014		
Mechanical	6. Tensile strength at break, p.s.i.	D638[b]	1000-3300	2100-6500	18,500	4500-7900	5500
	7. Elongation at break, %	D638[b]	500-1000	400-1300	5	290-630	530
	8. Tensile yield strength, p.s.i.	D638[b]					
	9. Compressive strength (rupture or yield), p.s.i.	D695					
	10. Flexural strength (rupture or yield), p.s.i.	D790		7000-8000	18,000		
	11. Tensile modulus, 10^3 p.s.i.	D638[b]		9-19		0.33-1.45[e]	0.7[e]
	12. Compressive modulus, 10^3 p.s.i.	D695					
	13. Flexural modulus, 10^3 p.s.i. 73° F.	D790		2-27	800		
	200° F.	D790					
	250° F.	D790					
	300° F.	D790					
	14. Izod impact, ft.-lb./in. of notch ($\frac{1}{8}$-in. thick specimen)	D256A	No break	No break	1.6		
	15. Hardness Rockwell	D785					
	Shore/Barcol	D2240/ D2583	Shore A50-84	Shore A55-87; D55-60	Shore D70	Shore A70-D54	
Thermal	16. Coef. of linear thermal expansion, 10^{-6} in./in./°F.	D696					
	17. Deflection temperature under flexural load, °F. 264 p.s.i.	D648					
	66 p.s.i.	D648					
	18. Thermal conductivity, 10^{-4} cal.-cm./ sec.-cm.^{2}-°C.	C177					
Physical	19. Specific gravity	D792	0.90-0.97	1.04-1.19	1.24	1.19-1.22	1.11
	20. Water absorption ($\frac{1}{8}$-in. thick specimen), % 24 hr.	D570	0.05-0.3	0.3-0.6	0.4		
	Saturation	D570					
	21. Dielectric strength ($\frac{1}{8}$-in. thick specimen), short time, v./mil	D149					
	SUPPLIERS[a]		Hüls America; LNP	Hüls America	LNP	BFGoodrich; Polyurethane Specialties	BFGoodrich; Polyurethane Specialties

Thermoplastic elastomers (Cont'd)

Materials

	Thermoplastic elastomers (Cont'd)						Urea	Vinyl polymers and copolymers		
	Polyurethane (Cont'd)				Elastomeric alloys				PVC and PVC-acetate MC, sheets, rods, and tubes	
	Molding and extrusion compounds									
	Polyester		Polyether							
	Low and medium hardness	High hardness	Low and medium hardness	High hardness	Low and medium hardness	High hardness	Alpha cellulose-filled	PVC molding compound, 20% glass fiber-reinforced	Rigid	Rigid, lead-stabilized
1a.			Thermoset							
1.					165	165	Thermoset			
	120-160	120-160	120-160	120-160				75-105	75-105	
2.	I: 340-435 E: 340-410	I: 400-440 E: 370-410	I: 350-430 E: 340-410 C: 72-120	I: 400-435 E: 380-440	I: 350-450	I: 350-450	C: 275-350 I: 290-320 T: 270-300	I: 380-400 E: 390-400	C: 285-400 I: 300-415	I: 380-400
3.	0.8-1.4	0.8-1.4	0.6-1.2	1-1.4	6-10	6-10	2-20	5-15	10-40	
4.					2.5-3.5	2.5-3.5	2.2-3.0	1.5-2.5	2.0-2.3	2.0
5.	0.008-0.015	0.005-0.015	0.008-0.015	0.008-0.012	1.5-2.5	1.5-2.5	0.006-0.014	0.001	0.002-0.006	
6.	3300-8400	4000-11,000	158-6750	6000-7240	650-3200	2300-4000	5500-13,000	8600-12,800	5900-7500	
7.	410-620	110-550	475-1000	340-600	300-750	250-600	<1	2-5	40-80	15
8.						1600			5900-6500	6000
9.							25,000-45,000		8000-13,000	
10.							5000-18,000	14,200-22,500	10,000-16,000	10,000
11.					0.7-5.0	16-100	1000-1500	680-970	350-600	340
12.										
13.					1.5-6.6	15-50	1300-1600	680-970	300-500	370
14.					No fracture	No fracture	0.25-0.40	1.0-1.9	0.4-22	6.3
15.							M110-120	R108-119		
	Shore A55-94	Shore D46-78	Shore A13-92	Shore D55-75	Shore 55-95A	Shore 32-72D			Shore D85-89	Shore D65-85
16.					82	82	22-36	24-36	50-100	66.8
17.							260-290	165-174	140-170	156
									135-180	
18.							2-10		3.5-5.0	
19.	1.17-1.25	1.15-1.28	1.02	1.14-1.21	0.90-1.39	0.90-1.31	1.47-1.52	1.43-1.50	1.30-1.58	1.39
20.		0.3					0.4-0.8	0.01	0.04-0.4	
21.			470	470	400-600	400-600	300-400		350-500	
	BASF; Bayer Corp.; Dainippon; Dow Plastics; Eagle-Picher; BFGoodrich; A. Schulman	BASF; Bayer Corp.; Dow Plastics; Eagle-Picher; BFGoodrich; A. Schulman	BASF; Bayer Corp.; Cabot; Dow Plastics; Eagle-Picher; BFGoodrich; A. Schulman	BASF; Bayer Corp.; Dow Plastics; Eagle-Picher; BFGoodrich; A. Schulman	Goodyear; Monsanto; Montell Polyolefins; North Coast Compounders; Novatec; A. Schulman	Monsanto; Montell Polyolefins; North Coast Compounders; A. Schulman	Budd; CYTEC; Patent Plastics; Perstorp; Plastics Mfg.	Thermofil	AlphaGary; Borden; Colorite; Formosa; Georgia Gulf; Keysor-Century; Novatec; OxyChem; Rimtec; RSG Polymers; Shintech; Stauffer; Synergistics; Tenneco; Union Carbide; Vi-Chem; Vista Chemical	AlphaGary; Synergistics; Vista Chemical

	Properties	ASTM test method	PVC and PVC-acetate MC, sheets, rods, and tubes (Cont'd)			Molding and extrusion compounds			
			Rigid, tin-stabilized	Flexible, unfilled	Flexible, filled	Vinyl formal	Chlorinated polyvinyl chloride	Vinyl butyral, flexible	PVC/acrylic blends

Vinyl polymers and copolymers (Cont'd)

Materials

Processing

	Properties	ASTM test method	Rigid, tin-stabilized	Flexible, unfilled	Flexible, filled	Vinyl formal	Chlorinated polyvinyl chloride	Vinyl butyral, flexible	PVC/acrylic blends
1a.	Melt flow (gm./10 min.)	D1238							
1.	Melting temperature, °C. T_m (crystalline)								
	T_g (amorphous)			75-105	75-105	105	110	49	
2.	Processing temperature range, °F. (C = compression; T = transfer; I = injection; E = extrusion)		I: 380-400	C: 285-350 I: 320-385	C: 285-350 I: 320-385	C: 300-350 I: 300-400	C: 350-400 I: 395-440 E: 360-420	C: 280-320 I: 250-340	I: 360-390 E: 390-410
3.	Molding pressure range, 10^3 p.s.i.			8-25	1-2	10-30	15-40	0.5-3	2-3
4.	Compression ratio		2.0	2.0-2.3	2.0-2.3		1.5-2.5		2-2.5
5.	Mold (linear) shrinkage, in./in.	D955		0.010-0.050	0.008-0.035 0.002-0.008	0.001-0.003	0.003-0.007		0.003

Mechanical

	Properties	ASTM test method	Rigid, tin-stabilized	Flexible, unfilled	Flexible, filled	Vinyl formal	Chlorinated polyvinyl chloride	Vinyl butyral, flexible	PVC/acrylic blends
6.	Tensile strength at break, p.s.i.	D638[b]		1500-3500	1000-3500	10,000-12,000	6800-9000	500-3000	6400-7000
7.	Elongation at break, %	D638[b]		200-450	200-400	5-20	4-100	150-450	35-100
8.	Tensile yield strength, p.s.i.	D638[b]	5140-7560				6000-8000		
9.	Compressive strength (rupture or yield), p.s.i.	D695		900-1700	1000-1800		9000-22,000		6800-8500
10.	Flexural strength (rupture or yield), p.s.i.	D790	6300-13,600			17,000-18,000	14,500-17,000		10,300-11,000
11.	Tensile modulus, 10^3 p.s.i.	D638[b]	310-435			350-600	341-475		340-370
12.	Compressive modulus, 10^3 p.s.i.	D695					335-600		
13.	Flexural modulus, 10^3 p.s.i. 73° F.	D790	305-420				380-450		350-380
	200° F.	D790							
	250° F.	D790							
	300° F.	D790							
14.	Izod impact, ft.-lb./in. of notch ($1/8$-in. thick specimen)	D256A	2.1-20.0	Varies over wide range	Varies over wide range	0.8-1.4	1.0-5.6	Varies over wide range	1-12
15.	Hardness Rockwell	D785				M85	R117-112	A10-100	R106-110
	Shore/Barcol	D2240/ D2583	Shore D69-78	Shore A50-100	Shore A50-100				

Thermal

	Properties	ASTM test method	Rigid, tin-stabilized	Flexible, unfilled	Flexible, filled	Vinyl formal	Chlorinated polyvinyl chloride	Vinyl butyral, flexible	PVC/acrylic blends
16.	Coef. of linear thermal expansion, 10^{-6} in./in./°F.	D696	1.0-18.1	70-250		64	62-78		44-79
17.	Deflection temperature under flexural load, °F. 264 p.s.i.	D648	69-163			150-170	202-234		167-185
	66 p.s.i.	D648	161-166				215-247		172-189
18.	Thermal conductivity, 10^{-4} cal.-cm./ sec.-cm.^{2}-°C.	C177		3-4	3-4	3.7	3.3		

Physical

	Properties	ASTM test method	Rigid, tin-stabilized	Flexible, unfilled	Flexible, filled	Vinyl formal	Chlorinated polyvinyl chloride	Vinyl butyral, flexible	PVC/acrylic blends
19.	Specific gravity	D792	1.32-1.45	1.16-1.35	1.3-1.7	1.2-1.4	1.49-1.58	1.05	1.26-1.35
20.	Water absorption ($1/8$-in. thick specimen), % 24 hr.	D570		0.15-0.75	0.5-1.0	0.5-3.0	0.02-0.15	1.0-2.0	0.09-.016
	Saturation	D570							
21.	Dielectric strength ($1/8$-in. thick specimen), short time, v./mil	D149		300-400	250-300	490	600-625	350	480

SUPPLIERS[a]	AlphaGary; Colorite; Georgia Gulf; Rimtec; Shintech; Synergistics; Vista Chemical	AlphaGary; Borden; Colorite; Keysor-Century; Novatec; Rimtec; A. Schulman; Shintech; Shuman; Synergistics; Teknor Apex; Tenneco; Union Carbide; Vi-Chem; Vista Chemical	AlphaGary; Borden; Colorite; Keysor-Century; Novatec; Rimtec; Shintech; Stauffer; Synergistics; Teknor Apex; Tenneco; Union Carbide; Vi-Chem; Vista Chemical	Monsanto	Elf Atochem N.A.; Georgia Gulf	Monsanto; Union Carbide	AlphaGary; Sumitomo

Notes for Appendix C

a. See list below for addresses of suppliers.

b. Tensile test method varies with material; D638 is standard for thermoplastics; D651 for rigid thermosetting plastics; D412 for elastomeric plastics; D882 for thin plastics sheeting.

c. Dry, as molded (approximately 0.2 percent moisture content).

d. As conditioned to equilibrium with 50 percent relative humidity.

e. Test method in ASTM D4092.

f. *Pseudo* indicates that the thermosetting and thermoplastic components were in the form of pellets or powder prior to fabrication.

g. Dow Plastics samples are unannealed.

Names and Addresses of Suppliers Listed in Appendix C

Adell Plastics, Inc.
4530 Annapolis Rd.
Baltimore, MD 21227
800-638-5218, 410-789-7780
Fax: 410-789-2804

Ain Plastics, Inc.
249 E. Sandford Blvd.
P.O. Box 151-M,
Mt. Vernon, NY 10550
800-431-2451, 914-668-6800
Fax: 914-668-8820

Albis Corp.
445 Hwy. 36 N.
P.O. Box 711
Rosenberg, TX 77471
800-231-5911, 713-342-3311
Fax: 713-342-3058
Telex: 166-181

Alliedsignal Inc.
Alliedsignal Engineered Plastics
P.O. Box 2332, 101 Columbia Rd.
Morristown, NJ 07962-2332
201-455-5010
Fax: 201-455-3507

A L M Corp.
55 Haul Rd.
Wayne, NJ 07470
201-694-4141
Fax: 201-831-8327

Alpha/Owens-Corning
P.O. Box 610
Collierville, TN 38027-0610
901-854-2800
Fax: 901-854-1183

AlphaGary Corp.
170 Pioneer Dr.
P.O. Box 808
Leominster, MA 01453
800-232-9741, 508-537-8071
Fax: 508-534-3021

American Polymers
P.O. Box 366
53 Milbrook St.
Worcester, MA 01606
508-756-1010
Fax: 508-756-3611

Ametek, Inc.
Haveg Div.
900 Greenbank Rd.
Wilmington, DE 19808
302-995-0400
Fax: 302-995-0491

Amoco Chemical Co.
200 E. Randolph Dr.
Mail Code 7802
Chicago, IL 60601-7125
800-621-4590, 312-856-3200
Fax: 312-856-4151

Applied Composites Corp.
333 N. Sixth St.
St. Charles, IL 60174
708-584-3130
Fax: 708-584-0659

Applied Polymer Systems, Inc.
P.O. Box 56404
Flushing, NY 11356-4040
718-539-4425
Fax: 718-460-4159

Arco Chemical Co.
3801 West Chester Pike
Newtown Square, PA 19073
800-345-0252 (PA Only), 215-359-2000

Aristech Chemical Corp.
Acrylic Sheet Unit
7350 Empire Dr.
Florence, KY 41042
800-354-9858
Fax: 606-283-6492

Ashley Polymers
5114 Ft. Hamilton Pkwy.
Brooklyn, NY 11219
718-851-8111
Fax: 718-972-3256
Telex: 42-7884

AtoHaas North America Inc.
100 Independence Mall W.
Philadelphia, PA 19106
215-592-3000
Fax: 215-592-2445

Ausimont USA, Inc.
Crown Point Rd. & Leonards Lane
P.O. Box 26
Thorofare, NJ 08086
800-323-2874, 609-853-8119
Fax: 609-853-6405

Bamberger, Claude P., Molding
    Compounds Corp.
111 Paterson Plank Rd.
P.O. Box 67
Carlstadt, NJ 07072
201-933-6262
Fax: 201-933-8129

Bamberger Polymers, Inc.
1983 Marcus Ave.
Lake Success, NY 11042
800-888-8959, 516-328-2772
Fax: 516-326-1005
Telex: 6711357

BASF Corp., Plastic Materials
3000 Continental Dr. N.
Mount Olive, NJ 07828-1234
201-426-2600

Bayer Corp.
100 Bayer Rd.
Pittsburgh, PA 15205-9741
800-662-2927, 412-777-2000

BFGoodrich Adhesive Systems Div.
123 W. Bartges St.
Akron, OH 44311-1081
216-374-2900
Fax: 216-374-2860

BFGoodrich Specialty Chemicals
9911 Breckville Rd.
Cleveland, OH 44141-3247
800-331-1144, 216-447-5000
Fax: 216-447-5750

Boonton Plastic Molding Co.
30 Plain St.
Boonton, NJ 07005-0030
201-334-4400
Fax: 201-335-0620

Borden Packaging, Div.
Borden Inc.
One Clark St.
North Andover, MA 01845
508-686-9591

BP Chemicals (Hitco) Fibers and
    Materials
700 E. Dyer Rd.
Santa Ana, CA 92705
714-549-1101

BP Chemicals, Inc.
4440 Warrensville Center Rd.
Cleveland, OH 44128
800-272-4367, 216-586-5847
Fax: 216-586-5839

BP Performance Polymers Inc.
Phenolic Business
60 Walnut Ave.
Suite 100
Clark, NJ 07066
908-815-7843
Fax: 908-815-7844

Budd Chemical Co.
Pennsville-Auburn Rd.
Carneys Point, NJ 08069
609-299-1708
Fax: 609-299-2998

Budd Co.
Plastics Div.
32055 Edward Ave.
Madison Heights, MI 48071
810-588-3200
Fax: 810-588-0798

Bulk Molding Compounds Inc.
3N497 N. 17th St.
St. Charles, IL 60174
708-377-1065
Fax: 708-377-7395

Cadillac Plastic & Chemical Co.
143 Indusco Ct.
P.O. Box 7035
Troy, MI 48007-7035
800-488-1200, 810-583-1200
Fax: 810-583-4715

Cast Nylons Ltd.
4300 Hamann Pkwy.
Willoughby, OH 44092
800-543-3619, 216-269-2300
Fax: 216-269-2323

Chevron Chemical Co.
Olefin & Derivatives
P.O. Box 3766
Houston, TX 77253
800-231-3828, 713-754-2000

Ciba-Geigy Corp., Ciba
  Additives
540 White Plains Rd.
P.O. Box 20005
Tarrytown, NY 10591-9005
800-431-2360, 914-785-2000
Fax: 914-785-4244

Color-Art Plastics, Inc.
317 Cortlandt St.
Belleville, NJ 07109-3293
201-759-2400

ComAlloy International Co.
481 Allied Dr.
Nashville, TN 37211
615-333-3453
Fax: 615-834-9941

Conap, Inc.
1405 Buffalo St.
Olean, NY 14760
716-372-9650
Fax: 716-372-1594

Consolidated Polymer Technologies,
  Inc.
11811 62nd St. N.
Largo, FL 34643
800-541-6880, 813-531-4191
Fax: 813-530-5603

Cook Composites & Polymers
P.O. Box 419389
Kansas City, MO 64141-6389
800-821-3590, 816-391-6000
Fax: 816-391-6215

Cosmic Plastics, Inc.
27939 Beale Ct.
Valencia, CA 91355
800-423-5613, 805-257-3274
Fax: 805-257-3345

Custom Manufacturers
858 S. M-18
Gladwin, MI 48624
800-860-4594, 517-426-4591
Fax: 517-426-4049

Custom Molders Corp.
2470 Plainfield Ave.
Scotch Plains, NJ 07076
908-233-5880
Fax: 908-233-5949

Custom Plastic Injection Molding
  Co. Inc.
3 Spielman Rd.
Fairfield, NJ 07004
201-227-1155

Cyro Industries
100 Enterprise Dr., 7th fl.
Rockaway, NJ 07866
201-442-6000

CYTEC Industries Inc.
5 Garret Mt. Plaza
West Paterson, NJ 07424
800-438-5615, 201-357-3100
Fax: 201-357-3065

CYTEC Industries Inc.
12600 Eckel Rd.
P.O. Box 148
Perrysburg, OH 43551
800-537-3360, 419-874-7941
Fax: 419-874-0951

Diamond Polymers
1353 Exeter Rd.
Akron, OH 44306
216-773-2700
Fax: 216-773-2799

Dow Chemical Co.
Polyurethanes

2040 Willard H. Dow Center
Midland, MI 48674
800-441-4369

Dow Corning Corp.
P.O. Box 0994
Midland, MI 48686-0994
517-496-4000
Fax: 517-496-4586

Dow Corning STI
47799 Halyard Dr.
Suite 99
Plymouth, MI 48170
313-459-7792
Fax: 313-459-0204

Dow Plastics
P.O. Box 1206
Midland, MI 48641-1206
800-441-4369

DSM Copolymer, Inc.
P.O. Box 2591
Baton Rouge, LA 70821
800-535-9960, 504-355-5655
Fax: 504-357-9574

DSM Engineering Plastics
2267 W. Mill Rd.
P.O. Box 3333
Evansville, IN 47732
800-333-4237, 812-435-7500
Fax: 812-435-7702

DSM Thermoplastic Elastomers, Inc.
29 Fuller St.
Leominster, MA 01453-4451
800-524-0120, 508-534-1010
Fax: 508-534-1005

DuPont Engineering Polymers
1007 Market St.
Wilmington, DE 19898
800-441-7515, 302-999-4592
Fax: 302-999-4358

Eagle-Picher Industries, Inc.
C & Porter Sts.
Joplin, MO 64802
417-623-8000
Fax: 417-782-1923

Eastman Chemical Co.
P.O. Box 511
Kingsport, TN 37662
800-327-8626

Elf Atochem North America, Inc.
Fluoropolymers
2000 Market St.
Philadelphia, PA 19103
800-225-7788, 215-419-7000
Fax: 215-419-7497

Elf Atochem North America, Inc.,
    Organic Peroxides
2000 Market St.
Philadelphia, PA 19103
800-558-5575, 215-419-7000
Fax: 215-419-7591

Emerson & Cuming, Inc./Grace
    Specialty Polymers
77 Dragon Ct.
Woburn, MA 01888
800-832-4929, 617-938-8630
Fax: 617-935-0125

Epic, Inc.
654 Madison Ave.
New York, NY 10021
212-308-7039
Fax: 212-308-7266

Epsilon Products Co.
Post Rd. and Blueball Ave.
P.O. Box 432
Marcus Hook, PA 19061
610-497-8850
Fax: 610-497-4694

EVAL Co. of America
1001 Warrenville Rd.
Suite 201
Lisle, IL 60532-1359
800-423-9762, 708-719-4610
Fax: 708-719-4622

Exxon Chemical Co.
13501 Katy Freeway
Houston, TX 77079-1398
800-231-6633, 713-870-6000
Fax: 713-870-6970

Federal Plastics Corp.
715 South Ave. E.
Cranford, NJ 07016
800-541-4424, 908-272-5800
Fax: 908-272-9021

Ferro Corp.
Filled & Reinforced Plastics Div.
5001 O'Hara Dr.

Evansville, IN 47711
812-423-5218
Fax: 812-435-2113

Ferro Corp., World Headquarters
1000 Lakeside Ave.
P.O. Box 147000
Cleveland, OH 44114-7000
216-641-8580
Fax: 216-696-6958
Telex: 98-0165

Fina Oil & Chemical Co.
Chemical Div.
P.O. Box 2159
Dallas, TX 75221
800-344-3462, 214-750-2806
Fax: 214-821-1433

Firestone Canada
P.O. Box 486
Woodstock, Ontario, Canada N4S 7Y9
800-999-6231, 519-421-5649
Fax: 519-537-6235

Firestone Synthetic Rubber &
  Latex Co.
P.O. Box 26611
Akron, OH 44319-0006
800-282-0222
Fax: 216-379-7875
Telex: 67-16415

Formosa Plastics Corp. USA
9 Peach Tree Hill Rd.
Livingston, NJ 07039
201-992-2090
Fax: 201-716-7208

Franklin Polymers, Inc.
P.O. Box 481
521 Yale Ave.
Pitman, NJ 08071-0481
800-238-7659, 609-582-6115
Fax: 609-582-0525

FRP Supply Div.
Ashland Chemical Co.
P.O. Box 2219
Columbus, OH 43216
614-790-4272
Fax: 614-790-4012
Telex: 24-5385 ASHCHEM

GE Plastics
One Plastics Ave.

Pittsfield, MA 01201
800-845-0600, 413-448-7110

GE Silicones
260 Hudson River Rd.
Waterford, NY 12188
800-255-8886
Fax: 518-233-3931

General Polymers Div.
Ashland Chemical Co.
P.O. Box 2219
Columbus, OH 43216
800-828-7659
Fax: 614-889-3195

George, P. D., Co.
5200 N. Second St.
St. Louis, MO 63147
314-621-5700

Georgia Gulf Corp.
PVC Div.
P.O. Box 629
Plaquemine, LA 70765-0629
504-685-1200

Glastic Corp.
4321 Glenridge Rd.
Cleveland, OH 44121
216-486-0100
Fax: 216-486-1091

GLS Corp.
Thermoplastic Elastomers Div.
740B Industrial Dr.
Cary, IL 60013
800-457-8777 (not IL), 708-516-8300
Fax: 708-516-8361

Goodyear Tire & Rubber Co.
Chemical Div.
1485 E. Archwood Ave.
Akron, OH 44316-0001
800-522-7659, 216-796-6253
Fax: 216-796-2617

Grace Specialty Polymers
77 Dragon Ct.
Woburn, MA 01888
617-938-8630

Haysite Reinforced Plastics
5599 New Perry Hwy.
Erie, PA 16509
814-868-3691
Fax: 814-864-7803

Hercules Inc.
Hercules Plaza
Wilmington, DE 19894
800-235-0543, 302-594-5000
Fax: 412-384-4291

Hercules Moulded Products
R.R. 3
Maidstone, Ontario, Canada N0R 1K0
519-737-6693
Fax: 519-737-1747

Hoechst Celanese Corp., Advanced
   Materials Group
90 Morris Ave.
Summit, NJ 07901
800-526-4960, 908-598-4000
Fax: 908-598-4330
Telex: 13-6346

Hoechst Celanese Corp.
Hostalen GUR Business Unit
2520 S. Shore Blvd.
Suite 110
League City, TX 77573
713-334-8500

Huls America Inc.
80 Centennial Ave.
Piscataway, NJ 08855-0456
908-980-6800
Fax: 908-980-6970

Huntsman Chemical Corp.
2000 Eagle Gate Tower
Salt Lake City, UT 84111
800-421-2411, 801-536-1500
Fax: 801-536-1581

Huntsman Corp.
P.O. Box 27707
Houston, TX 77227-7707
713-961-3711
Fax: 713-235-6437
Telex: 227031 TEX UR

Hysol Engineering Adhesives
Dexter Distrib. Programs
One Dexter Dr.
Seabrook, NH 03874-4018
800-767-8786, 603-474-5541
Fax: 603-474-5545

ICI Acrylics Canada Inc.
7521 Tranmere Dr.
Mississauga, Ontario, Canada L5S 1L4
800-387-4880, 905-673-3345
Fax: 905-673-1459

ICI Acrylics Inc.
10091 Manchester Rd.
St. Louis, MO 63122
800-325-9577, 314-966-3111
Fax: 314-966-3117

ICI Americas Inc.
Tatnall Bldg.
P.O. Box 15391
3411 Silverside Rd.
Wilmington, DE 19850-5391
800-822-8215, 302-887-5536
Fax: 302-887-2089

ICI Fiberite
Molding Materials
501 W. Third St.
Winona, MN 55987-5468
507-454-3611
Fax: 507-452-8195
Telex: 507-454-3646

ICI Polyurethanes Group
286 Mantua Grove Rd.
West Deptford, NJ 08066-1732
800-257-5547, 609-423-8300
Fax: 609-423-8580

ITW Adhesives
37722 Enterprise Ct.
Farmington Hills, MI 48331
800-323-0451, 313-489-9344
Fax: 313-489-1545

Jet Composites Inc.
405 Fairall St.
Ajax, Ontario, Canada L1S 1R8
416-686-1707
Fax: 416-427-9403

Jet Plastics
941 N. Eastern Ave.
Los Angeles, CA 90063
213-268-6706
Fax: 213-268-8262

Keysor-Century Corp.
26000 Springbrook Rd.
P.O. Box 924
Santa Clarita, CA 91380-9024
805-259-2360
Fax: 805-259-7937

Kleerdex Co.
100 Gaither Dr.
Suite B
Mt. Laurel, NJ 08054

800-541-7232, 609-866-1700
Fax: 609-866-9728

Laird Plastics
1400 Centrepark
Suite 500
West Palm Beach, FL 33401
800-610-1016, 407-684-7000
Fax: 407-684-7088

LNP Engineering Plastics Inc.
475 Creamery Way
Exton, PA 19341
800-854-8774, 610-363-4500
Fax: 610-363-4749

Lockport Thermosets Inc.
157 Front St.
Lockport, LA 70374
800-259-8662, 504-532-2541
Fax: 504-532-6806

M.A. Industries Inc.
Polymer Div.
303 Dividend Dr.
P.O. Box 2322
Peachtree City, GA 30269
800-241-8250, 404-487-7761

Mitsui Petrochemical Industries Ltd.
3-2-5 Kasumigaseki, Chiyoda-ku
Tokyo, Japan 100
03-3593-1630
Fax: 03-3593-0979

Mitsui Plastics, Inc.
11 Martine Ave.
White Plains, NY 10606
914-287-6800
Fax: 914-287-6850

Mitsui Toatsu Chemicals, Inc.
2500 Westchester Ave.
Suite 110
Purchase, NY 10577
914-253-0777
Fax: 914-253-0790

Mobil Chemical Co.
1150 Pittsford-Victor Rd.
Pittsford, NY 14534
716-248-1193
Fax: 716-248-1075

Mobil Polymers
P.O. Box 5445
800 Connecticut Ave.
Norwalk, CT 06856

203-854-3808
Fax: 203-854-3840

Monmouth Plastics Co.
800 W. Main St.
Freehold, NJ 07728
800-526-2820, 908-866-0200
Fax: 908-866-0274

Monsanto Co.
800 N. Lindbergh Blvd.
St. Louis, MO 63167
314-694-1000
Fax: 314-694-7625
Telex: 650-397-7820

Montell Polyolefins
Three Little Falls Centre
2801 Centerville Rd.
Wilmington, DE 19850-5439
800-666-8355, 302-996-6000
Fax: 302-996-5587

Morton International Inc.
Morton Plastics Additives
150 Andover St.
Danvers, MA 01923
508-774-3100
Fax: 508-750-9511

Network Polymers, Inc.
1353 Exeter Rd.
Akron, OH 44306
216-773-2700
Fax: 216-773-2799

Nova Polymers, Inc.
P.O. Box 8466
Evansville, IN 47716-8466
812-476-0339
Fax: 812-476-0592

Novacor Chemicals Inc.
690 Mechanic St.
Leominster, MA 01453
800-225-8063, 508-537-1111
Fax: 508-537-5685

Novacor Chemicals Inc.
Clear Performance Plastics
690 Mechanic St.
Leominster, MA 01453
800-243-4750, 508-537-1111
Fax: 508-537-6410

Novatec Plastics & Chemicals Co. Inc.
P.O. Box 597
275 Industrial Way W.

Eatontown, NJ 07724
800-782-6682, 908-542-6600

Nylon Engineering
12800 University Dr.
Suite 275
Ft. Myers, FL 33907
813-482-1100
Fax: 813-482-4202

Occidental Chemical Corp.
5005 LBJ Freeway
Dallas, TX 75244
214-404-3800

Patent Plastics Inc.
638 Maryville Pike S.W.
P.O. Box 9246
Knoxville, TN 37920
800-340-7523, 615-573-5411

Paxon Polymer Co.
P.O. Box 53006
Baton Rouge, LA 70892
504-775-4330

Performance Polymers Inc.
803 Lancaster St.
Leominster, MA 01453
800-874-2992, 508-534-8000
Fax: 508-534-8590

Perstorp Compounds Inc.
238 Nonotuck St.
Florence, MA 01060
413-584-2472
Fax: 413-586-4089

Perstorp Xytec, Inc.
9350 47th Ave. S.W.
P.O. Box 99057
Tacoma, WA 98499
206-582-0644
Fax: 206-588-5539

Phillips Chemical Co.
101 ARB Plastics Technical Center
Bartlesville, OK 74004
918-661-9845
Fax: 918-662-2929

Plaskolite, Inc.
P.O. Box 1497
Columbus, OH 43216
800-848-9124, 614-294-3281
Fax: 614-297-7287

Plaskon Electronic Materials, Inc.
100 Independence Mall West

Philadelphia, PA 19106
800-537-3350, 215-592-2081
Fax: 215-592-2295
Telex: 845-247

Plaslok Corp.
3155 Broadway
Buffalo, NY 14227
800-828-7913, 716-681-7755
Fax: 716-681-9142

Plastic Engineering & Technical
    Services, Inc.
2961 Bond
Rochester Hills, MI 48309
313-299-8200
Fax: 313-299-8206

Plastics Mfg. Co.
2700 S. Westmoreland St.
Dallas, TX 75223
214-330-8671
Fax: 214-337-7428

Polymer Resources Ltd.
656 New Britain Ave.
Farmington, CT 06032
800-243-5176, 203-678-9088
Fax: 203-678-9299

Polymers International Inc.
P.O. Box 18367
Spartanburg, SC 29318
803-579-2729
Fax: 803-579-4476

Polyply Composites Inc.
1540 Marion
Grand Haven, MI 49417
616-842-6330
Fax: 616-842-5320

PPG Industries, Inc.
Chemicals Group
One PPG Place
Pittsburgh, PA 15272
412-434-3131
Fax: 412-434-2891

Premix, Inc.
Rte. 20 & Harmon Rd.
P.O. Box 281
North Kingsville, OH 44068
216-224-2181
Fax: 216-224-2766

Prime Alliance, Inc.
1803 Hull Ave.

Des Moines, IA 50302
800-247-8038, 515-264-4110
Fax: 515-264-4100

Prime Plastics, Inc.
2950 S. First St.
Clinton, OH 44216
216-825-3451

Progressive Polymers, Inc.
P.O. Box 280
4545 N. Jackson
Jacksonville, TX 75766
800-426-4009, 903-586-0583
Fax: 903-586-4063

Quantum Composites, Inc.
4702 James Savage Rd.
Midland, MI 48642
800-462-9318, 517-496-2884
Fax: 517-496-2333

Reichhold Chemicals, Inc.
P.O. Box 13582
Research Triangle Park, NC 27709
800-448-3482, 919-990-7500
Fax: 919-990-7711

Reichhold Chemicals, Inc.
Emulsion Polymer Div.
2400 Ellis Rd.
Durham, NC 27703-5543
919-990-7500
Fax: 919-990-7711

Resinoid Engineering Corp.
P.O. Box 2264
Newark, OH 43055
614-928-6115
Fax: 614-929-3165

Resinoid Engineering Corp.
Materials Div.
7557 N. St. Louis Ave.
Skokie, IL 60076
708-673-1050
Fax: 708-673-2160

Rhone-Poulenc
Rte. 8
Rouseville Rd.
P.O. Box 98
Oil City, PA 16301-0098
814-677-2028
Fax: 814-677-2936

Rimtec Corp.
1702 Beverly Rd.

Burlington, NJ 08016
800-272-0069, 609-387-0011
Fax: 609-387-0282

Rogers Corp.
One Technology Dr.
Rogers, CT 06263
203-774-9605
Fax: 203-774-9630

Rogers Corp.
Molding Materials Div.
Mill and Oakland Sts.
P.O. Box 550
Manchester, CT 06045
800-243-7158, 203-646-5500
Fax: 203-646-5503

Ronald Mark Associates, Inc.
P.O. Box 776
Hillside, NJ 07205
908-558-0011
Fax: 908-558-9366

Rostone Corp.
P.O. Box 7497
Lafayette, IN 47903
317-474-2421
Fax: 317-474-5870

Rotuba Extruders, Inc.
1401 Park Ave. S.
Linden, NJ 07036
908-486-1000
Fax: 908-486-0874

R.S.G. Polymers Corp.
P.O. Box 1677
Valrico, FL 33594
813-689-7558
Fax: 813-685-6685

RTP Co.
580 E. Front St.
P.O. Box 5439
Winona, MN 55987-0439
800-433-4787, 507-454-6900
Fax: 507-454-2041

Schulman, A., Inc.
3550 W. Market St.
Akron, OH 44333
800-547-3746, 216-668-3751
Fax: 216-668-7204
Telex: SCHN 6874 22

Shell Chemical Co.
One Shell Plaza

Rm. 1671
Houston, TX 77002
713-241-6161

Shell Chemical Co.
Polyester Div.
4040 Embassy Pkwy.
Suite 220
Akron, OH 44333
216-798-6400
Fax: 216-798-6400

Shintech Inc.
Weslayan Tower
24 Greenway Plaza
Suite 811
Houston, TX 77046
713-965-0713
Fax: 713-965-0629

Shuman Co.
3232 South Blvd.
Charlotte, NC 28209
704-525-9980
Fax: 704-525-0622

Solvay Polymers, Inc.
P.O. Box 27328
Houston, TX 77227-7328
800-231-6313, 713-525-4000
Fax: 713-522-2435

Sumitomo Plastics America, Inc.
900 Lafayette St.
Suite 510
Santa Clara, CA 95050-4967
408-243-8402
Fax: 408-243-8405

Synergistics Industries (NJ) Inc.
10 Ruckle Ave.
Farmingdale, NJ 07727
908-938-5980
Fax: 908-938-6933

Syracuse Plastics, Inc.
400 Clinton St.
Fayetteville, NY 13066
315-637-9881
Fax: 315-637-9260

Teknor Apex Co.
505 Central Ave.
Pawtucket, RI 02861
800-554-9892, 401-725-8000
Fax: 401-724-6250

Tetrafluor, Inc.
2051 E. Maple Ave.
El Segundo, CA 90245
310-322-8030
Fax: 310-640-0312

Texapol Corp.
177 Mikron Rd.
Lower Nazareth Comm. Park
Bethlehem, PA 18017
800-523-9242, 610-759-8222
Fax: 610-759-9460

Thermofil, Inc.
6150 Whitmore Lake Rd.
Brighton, MI 48116-1990
800-444-4408, 810-227-3500
Fax: 810-227-3824

Union Carbide Corp.
39 Old Ridgebury Rd.
Danbury, CT 06817-0001
800-335-8550, 203-794-5300

Vi-Chem Corp.
55 Cottage Grove St. S.W.
Grand Rapids, MI 49507
800-477-8501, 616-247-8501
Fax: 616-247-8703

Vista Chemical Co.
900 Threadneedle
P.O. Box 19029
Houston, TX 77079
713-588-3000

Washington Penn Plastic Co.
2080 N. Main St.
Washington, PA 15301
412-228-1260
Fax: 412-228-0962

Wellman Extrusion
P.O. Box 130
Ripon, WI 54971-0130
800-398-7876, 414-748-7421
Fax: 414-748-6093

Westlake Polymers Corp.
2801 Post Oak Blvd.
Suite 600
Houston, TX 77056
800-545-9477, 713-960-9111
Fax: 713-960-8761

APPENDIX D

ELECTRICAL PROPERTIES OF RESINS AND COMPOUNDS

	Volume resistivity, $\Omega \cdot cm$	Dielectric strength, V/mil	Dielectric constant at 1 MHz	Dissipation factor at 1 MHz
ABS	10^{16}	425	2.6	0.007
Acetal copolymer	10^{15}	450	3.7	0.005
Acetal homopolymer	10^{15}	450	3.7	0.005
Acrylic	10^{18}	500	2.2	0.3
Alkyds				
Glass-filled	10^{15}	375	4.6	0.02
Mineral-filled	10^{14}	400	4.7	0.02
Allyls, glass-filled (DAP/ DAIP)	10^{15}	400	3.5	0.01
Cellulose acetate	10^{13}	425	5.1	0.05
Chlorinated polyethylene	10^{16}	450	4.3	0.10
Elastomers				
COX	10^{15}	500	10.0	0.05
CR	10^{11}	700	8.0	0.03
CSM	10^{14}	700	8.0	0.07
EPDM	10^{16}	800	3.5	0.007
FPM	10^{13}	700	18.0	0.04
IIR	10^{17}	600	2.4	0.003
NR	10^{16}	800	3.0	0.003
SBR	10^{15}	800	3.5	0.003
SI	10^{15}	700	3.6	0.001
T	10^{12}	700	9.5	0.005
U	10^{12}	500	5.0	0.03
Epoxies				
Glass-filled	10^{16}	360	4.6	0.01
Mineral-filled	10^{16}	400	5.0	0.01
Fluoroplastics				
TFE	10^{18}	480	2.2	0.002
FEP	10^{18}	500	2.2	0.003
ETFE	10^{16}	490	1.7	0.007
ECTFE	10^{16}	490	1.7	0.007
Imides				
Polyamide imide	10^{17}	600	3.5	0.02
Polyetherimide	10^{17}	800	3.2	0.002
Polyimide	10^{16}	575	3.4	0.02
Liquid-crystal polymer	10^{14}	510	3.9	0.006
Melamine	10^{11}	300	7.9	0.13
Nylon	10^{14}	385	3.6	0.04

	Volume resistivity, $\Omega \cdot cm$	Dielectric strength, V/mil	Dielectric constant at 1 MHz	Dissipation factor at 1 MHz
Phenolics				
General-purpose	10^{13}	400	6.0	0.7
Glass-filled	10^{13}	350	6.0	0.8
Mineral-filled	10^{14}	400	6.0	0.10
Polyarylate	10^{16}	400	2.6	0.01
Polybutadiene	10^{16}	500	2.7	0.02
Polycarbonate	10^{15}	425	3.1	0.01
Polyesters				
PBT	10^{15}	425	3.1	0.03
PET	10^{15}	650	3.6	0.002
Thermoset	10^{14}	325	4.7	0.02
Polyetheretherketone	10^{16}	450	3.2	0.1
Polyethylene	10^{19}	475	2.3	0.0005
Polymethyl methacrylate	10^{17}	500	2.8	0.03
Polymethyl pentene	10^{16}	650	2.1	0.0001
Polyphenylene ether	10^{16}	500	2.9	0.001
Polyphenylene oxide	10^{17}	600	2.8	0.001
Polyphenylene sulfide	10^{16}	525	3.1	0.0007
Polypropylene	10^{17}	650	2.3	0.0004
Polystyrene	10^{16}	425	2.4	0.0004
Polyurethane	10^{13}	400	3.4	0.03
Polyvinyl chloride	10^{11}	400	6.0	0.1
Silicone	10^{16}	550	2.7	0.001
Sulfones				
Polyaryl	10^{16}	400	3.5	0.006
Polyether	10^{16}	400	3.5	0.006
Polyphenyl	10^{16}	375	3.4	0.007
Polysulfone	10^{16}	425	3.6	0.004

IMPORTANT NOTES

1. Since there are so many variations in possible plastic resin and compound formulations, the values given in this table are averages for each resin and compound type. *Each material supplier can provide the specific values for its individual products.*

2. These values are representative of values measured at room temperature, atmospheric pressure, and up to an electrical frequency of 1 MHz. Values for volume resistivity and dielectric strength decrease with increasing temperature, while values for dielectric constant and dissipation factor increase with increasing temperature. These changes are gradual (but increasingly significant) up to the glass-transition temperature of each material, at which temperature major changes occur. Increases in humidity and electrical frequency also have significant effects on electrical properties. Electrode configuration and material purity and homogeneity have significant effects on dielectric strength. (See Chap. 2 of the second edition of this Handbook for changes in values for selected materials as a function of temperature, frequency, humidity, and other system operational parameters.)

3. Voltage breakdown and the associated dielectric strength decrease at decreasing pressures (and hence at increasing altitudes) according to Paschen's law (see Chap. 12 of the second edition of C. A. Harper, *Electronic Packaging and Interconnection Handbook,* McGraw-Hill, New York, 1991).

4. These property values are representative of values obtained in standard tests of the American Society for Testing and Materials (ASTM). The ASTM test methods for measuring these values are D-257 for volume resistivity, D-149 for dielectric strength, and D-150 for dielectric constant and for dissipation factor. Dissipation factor is also sometimes called power factor, loss tangent, or tan δ. Dielectric constant is also sometimes called permittivity. (See also Chap. 2 of the second edition of this Handbook for further discussions.)

5. Due to the broad range of possible formulations for each class of materials, small differences in values shown in this appendix may not be significant. *For final design, supplier values for specific formulations and product application conditions should be used.*

APPENDIX E

SOURCES OF SPECIFICATIONS AND STANDARDS FOR PLASTICS AND COMPOSITES

As in other material categories, the reliable use of plastics and composites requires specifications to support procurement of these materials and standards to establish engineering and technical requirements for processes, procedures, practices, and methods for testing and using these materials. These specifications and standards may be general industry-wide, or they may be specific to one industry, such as the space-borne industry, wherein plastics and composites must exhibit special stability in the harsh environments of temperature and vacuum. Specifications and standards for this industry, for instance, would be controlled by documents from the National Aeronautics and Space Administration (NASA). Increasingly, the development and use of standards is becoming international. Since the total sum of these specifications and standards is voluminous, no attempt is made herein to itemize all of them. However, the names and addresses of organizations that are sources of these documents, and information concerning them, are listed below. Also, one comprehensive reference book that can be recommended to readers with interest in this area is given as Ref. 1 at the end of this listing. Another excellent reference book dealing with the major issues of flammability standards of plastics and composites is listed as Ref. 2.

NAMES AND ADDRESSES OF ORGANIZATIONAL SOURCES OF SPECIFICATIONS AND STANDARDS FOR PLASTICS AND COMPOSITES

1. Industry Standards

American National Standards Institute, Inc. (ANSI)
1430 Broadway
New York, NY 10018

Global Engineering Documents
2805 McGaw Avenue
P.O. Box 19539
Irvine, CA 92714

or
Global Engineering Documents
1990 M Street N.W., Suite 400
Washington, DC 20036

National Standards Association
1200 Quince Orchard Boulevard
Gaithersburg, MD 20878

2. Federal Standards and Specifications

General Services Administration (GSA)
Seventh and D Streets, S.W.
Washington, DC 20407

Global Engineering Documents
(See 1. Industry Standards, above)

E.1

National Standards Association
(See 1. Industry Standards, above)

Document Engineering Company, Inc.
15210 Stagg Street
Van Nuys, CA 91405

3. Military Specifications and Standards

DODSSP—Customer Service
Standardization Document Order
Desk
Building 4D
700 Robbins Avenue
Philadelphia, PA 19111-5094

Global Engineering Documents
(See 1. Industry Standards, above)

National Standards Association
(See 1. Industry Standards, above)

Document Engineering Company, Inc.
(See 2. Federal Standards and
Specifications, above)

4. Spaceborne Plastics and Composites Specifications and Standards (NASA)

For Plastics:

Individual NASA Space Centers

For Composites:

NASA
ACEE Composites Project Office
Langley Research Center
Hampton, VA 23665-5225

5. Foreign or International Standards

American National Standards
Institute, Inc. (See 1. Industry
Standards, above)

Global Engineering Documents (See
1. Industry Standards, above)

International Standards and Law
Information (ISLI)
P.O. Box 230
Accord, MA 02018

International Organization for
Standardization
1 Rue de Varembe
Case Postale 56
CH-1211 Geneva 20, Switzerland

6. National and International Electrical Standards

National Electrical Manufacturers
Association (NEMA)
2101 L Street, N.W.
Washington, DC 20037

International Electrotechnical
Commission (IEC)
3 Rue de Varembe
Case Postale 131
CH-1211 Geneva 20, Switzerland

National Standards Association
   (See 1. Industry Standards and
   22. Specifications, Standards, and
   Industry Approvals for Plastics and
   Testing of Plastics)

7. International Standards

International Standards Organization
   (ISO)
(Located in Geneva, Switzerland but
   documents and information available
   in the United States from American
   National Standards Institute [see
   1. Industry Standards, above])

8. Japanese Standards and Specifications

Japanese Standards Association
1-24, Akasaka 4-chome, Minatoku-ku
Tokyo 107, Japan

9. German (DIN) Standards

Beuth Verlag GmBH
Burggrafenstrasse 6
D-1000 Berlin 30, Germany

Deutsches Institut für Normung e.V.
   (DIN)
Postfach 1107
D-1000 Berlin 30, Germany

10. European Aerospace Specifications and Standards

European Association of Aerospace
   Manufacturers (AECMA)
88, Boulevard Malesherbes
F-75008 Paris, France

Generally available in U.S. from
   National Standards Association—see
   1. Industry Standards, above

11. European Plastics Manufacturing Standards

Association of Plastics Manufacturers
   in Europe (APME)
Avenue Louise 250
Box 73
B-1050 Brussels, Belgium

12. Plastics Standards in United Kingdom

British Standards Institute
2 Park Street
London W1A 2BS, U.K.

British Plastics Federation
5 Belgrave Square
London SWIX 8 PH, U.K.

13. European Plastics Machinery Standards

European Committee of Machinery
   Manufacturers for the Plastics and
   Rubber Industry (EUROMAP)
Kirchenweg 4
CH-8032 Zurich, Switzerland

14. Rubber and Plastic Standards in United Kingdom

RAPRA Technology, Ltd.
Shawbury, Shrewsbury
Shropshire SY4 4NR, U.K.

15. Plastics Standards in France

French Association for
   Standardization (AFNOR)
Tour Europe, Cedex 7
92080 Paril La Défense
Paris, France

16. Specifications and Standards for Aerospace Products

Aerospace Industries Association of
   America, Inc. (AIA)
1250 I Street, N.W.
Washington, DC 20005

17. National Standards for U.S. Standards Developing Bodies

American National Standards
   Institute, Inc. (ANSI)
(See 1. Industry Standards, above)

18. Standards for Testing

American Society for Testing and
   Materials (ASTM)
1916 Race Street
Philadelphia, PA 19103-1187

19. Specifications and Standards for the Plastics Industry

Society of the Plastics Industry (SPI)
355 Lexington Avenue
New York, NY 10017
or
1275 K Street, N.W., Suite 400
Washington, DC 20005

20. Specifications and Standards for Composites Materials Characterization

Composite Materials Characterization, Inc.
Attn: Cecil Schneider
Lockheed Martin Aeronautical Systems Company
Advanced Structures and Materials Division
Marietta, GA 30063

21. Standards in Measurements

National Institute of Standards and Technology (NIST)
Office of Standard Reference Materials
Room B311, Chemistry Building
Gaithersburg, MD 20899
(Note: Previously known as National Bureau of Standards [NBS])

22. Specifications, Standards, and Industry Approvals for Plastics and Testing of Plastics

Underwriters Laboratories, Inc. (UL)
333 Pfingston Road
Northbrook, IL 60062
   (Offices also in Melville, Long Island,
   NY; Santa Clara, CA; Tampa, FL)

23. Standardization of Composite Materials Test Methods

Suppliers of Advanced Composite Materials Association (SACMA)
1600 Wilson Boulevard, Suite 1008
Arlington, VA 22209

REFERENCES

1. Traceski, F. T., *Specifications and Standards for Plastics and Composites,* ASM International, Materials Park, OH, 1990.
2. Hilado, C. J., *Flammability Handbook for Plastics,* 4th ed., Technomic Publishing Co., Inc., Lancaster, PA, 1990.

INDEX

Abbreviations
  for plastics. *See Appendix B*
  for rubber, **1.**21–**1.**22, **9.**13
ABS plastics
  alloys, **1.**41, **1.**44
  flame retardants, **1.**20–**1.**22
  nature of, **1.**41, **1.**42
  platable, **1.**42
  properties, **1.**42–**1.**45. *See also* Tests and
    properties
  structure, **1.**41
  trade names, **1.**42
Acetals
  dimensional changes, **1.**46
  fiber-reinforced, **1.**49
  moisture effects, **1.**46, **1.**47
  nature of, **1.**45
  properties, **1.**47, **1.**48. *See also* Tests and
    properties
  radiation effects, **1.**47, **1.**48
  TFE fiber-filled, **1.**49
  trade names, **1.**42
Acrylics
  adhesives, **7.**96
  coatings, **6.**11
  dimensional stability, **1.**50–**1.**52
  liquid resins, **4.**60–**4.**62
  moisture effects, **1.**51
  nature of, **1.**49, **1.**50
  optical properties, **1.**50
  properties, **1.**50–**1.**52. *See also* Tests and
    properties
  structure, **1.**49
  trade names, **1.**42
Adhesives and bonding
  acrylics, **7.**96
  adhesive selection, **7.**80–**7.**84
    for acetals, **7.**94
    for elastomers, **7.**83
    for fluorocarbons, **7.**95
    for glass, **7.**84
    for metals, **7.**80
    for nylon, **7.**95

Adhesives and bonding, adhesive selection
    (*Cont.*):
    for plastics, **7.**94–**7.**97
    for polycarbonates, **7.**95
    for polyethylene, **7.**95–**7.**96
    for polypropylene, **7.**95
    for polystyrene, **4.**47, **7.**97
    for wood, **7.**83
  adhesive theory, **7.**5
  anaerobic, **7.**67
  application methods, **7.**109
  ASTM tests, **7.**20–**7.**29
  bond requirements, **7.**8
  cyanoacrylate, **7.**67
  design and test, **7.**13–**7.**29
  direct-heat welding, **7.**85
  effects of environments, **7.**97–**7.**107
    chemicals, **7.**105
    humidity, **7.**103
    radiation, **7.**107
    solvents, **7.**105
    temperature, **7.**98–**7.**103
    vacuum, **7.**107
    water, **7.**103
    weathering, **7.**104
  elastomeric, **7.**69
  electromagnetic adhesives, **7.**92
  epoxy-based, **7.**61
  equipment, **7.**83
  failure mechanisms, **7.**9
  films, **7.**111
  heat sealing, **7.**90
  high-temperature polyaromatics, **7.**100
  inspection, **7.**115
  joint stresses, **7.**13
  phenolic based, **7.**65
  phenoxy, **7.**67
  polybenzimidazole, **7.**101
  polyesters, **7.**66
  polyurethane, **7.**67
  primers, **7.**37
  quality control, **7.**114
  silicones, **7.**70

Adhesives and bonding (*Cont.*):
  solvent cementing, **7.**93
  surface preparation, **7.**38–**7.**56
    for aluminum, **7.**38
    for copper, **7.**38
    for elastomers, **7.**52
    for metals, **7.**38
    for plastics, **7.**46
    for steel, **7.**46
    for titanium, **7.**46
  test methods, **7.**20–**7.**29
Advanced composites
  adhesive bonding, **3.**60–**3.**63
  aramid fibers, **3.**1, **3.**8, **3.**41
  autoclave fabrication, **7.**31–**7.**35
  bismalemide, **3.**11, **3.**17, **3.**18
  boron-epoxy, **3.**2, **3.**9
  braiding, **3.**51
  carbon fibers, **3.**1
  ceramic fibers, **3.**9
  component design and fabrication,
    3.57–**3.**59
  damage tolerance, **3.**59, **3.**60
  definition, **3.**1
  design, **3.**54–**3.**65
  design allowables, **3.**53
  environmental effects, **3.**64, **3.**65
  epoxy, **3.**11–**3.**17
  fabrication techniques
    fiber placement, **3.**47, **3.**48
    filament winding, **3.**37–**3.**47
    lay-up, **3.**32–**3.**36
    methods and procedures, **3.**31–**3.**33
    pultrusion, **3.**48–**3.**52
    resin transfer molding, **3.**36
  fiber-matrix systems, **3.**20–**3.**22
  fibers, **3.**4–**3.**7
  flow diagram for manufacture, **3.**21
  fracture toughness, **3.**59
  glass fiber, **3.**4
  graphite fiber, **3.**5, **3.**9
  hazards, **3.**71–**3.**73
  health and safety, **3.**71–**3.**73
  hybrid, **3.**4
  intralaminar shear, **3.**7
  Kevlar, **3.**5, **3.**41–**3.**46
  lay-up fabrication, **3.**32, **3.**33
  macrocracking, **3.**56
  material systems, **3.**4
  materials testing, **3.**54–**3.**69
  matrix, **3.**1, **3.**7–**3.**20
  mechanics, **3.**52–**3.**54
  methods of analysis, **3.**28–**3.**31
  micromechanics, **3.**28
  nondestructive testing, **3.**71
  organic fibers, **3.**8, **3.**41
  orthotropic, **3.**1
  ply orientation, **3.**23–**3.**28

Advanced composites (*Cont.*):
  ply properties, **3.**22–**3.**28
  Poisson's ratio, **3.**29
  polyacrylonitrile (PAN), **3.**1
  polyester, **3.**11–**3.**13
  polyimide, **3.**11, **3.**17–**3.**19
  precursor, **3.**5
  prepreg, **3.**40
  pressure vessels, **3.**42
  product verification, **3.**69–**3.**71
  protective armor, **3.**6, **3.**42
  quasi-isotropic laminate, **3.**28
  repairs, **3.**60, **3.**61
  roving fibers, **3.**20, **3.**39
  safety and health, **3.**71–**3.**73
  stacking sequence, **3.**56
  tape fibers, **3.**20
  testing, **3.**65–**3.**71
  thermoplastic, **3.**8, **3.**11
  tow fibers, **3.**39
  unidirectional properties, **3.**7, **3.**22–**3.**25
  vacuum bag fabrication, **3.**31
  weaving, **3.**51
  Young's modulus, **3.**29
Alkyds
  coatings, **6.**11
  electrical data, **2.**41
  nature of, **1.**25
  properties, **1.**25. *See also* Tests and proper-
    ties
  shrinkage stability, **1.**25
  trade names, **1.**24
Allyls, **1.**26
Aminos
  coatings, **6.**25
  electrical data, **2.**21
  mechanical data, **2.**41
  nature of, **1.**31
  properties, **1.**31–**1.**35, **2.**16. *See also* Tests
    and properties
  structure, **1.**32
  trade names, **1.**24
Aramids
  nature of, **1.**72
  properties, **1.**74, **2.**4–**2.**13
  trade names, **1.**44
Asbestos, **2.**11
Automotive elastomers
  ACM, **9.**19
  AEM, **9.**18, **9.**19
  application functions, **9.**34
  ASTM, **9.**13, **9.**36
  body sealing, **9.**10
  butyl, **9.**18
  CM, **9.**20
  CO, **9.**20
  comparison with plastics, **9.**1, **9.**2, **9.**3
  compounding, **9.**21, **9.**22

Automotive elastomers (*Cont.*):
  cost, **9.**5
  cross-linking, **9.**3
  CSM, **9.**20
  damping, **9.**7
  devulcanization, **9.**5
  durability, **9.**4
  dynamic seals, **9.**10
  ECO, **9.**20
  energy management, **9.**7
  EPDM, **9.**15, **9.**16
  EVM, **9.**19
  extenders, **9.**24
  fillers, **9.**24
  fluid sealing, **9.**9
  fluoroelastomers, **9.**21
  functionality, **9.**4
  gaskets, **9.**9
  GECO, **9.**20
  introduction to, **9.**1, **9.**2
  isolation, **9.**7
  mechanical applications, **9.**10
  natural rubber, **9.**14, **9.**15
  neoprene (CR), **9.**20
  nitrile (NBR), **9.**16–**9.**18
  nomenclature, **9.**13
  ozone resistance, **9.**33
  plasticizers, **9.**24
  recycling, **9.**39. *See also* recycling
  reversion, **9.**5
  SAE, **9.**13, **9.**36
  safety considerations, **9.**5, **9.**6
  SBR, **9.**19
  selection, **9.**31–**9.**34
  silicone elastomers, **9.**21
  silicones, **9.**2
  specifications, **9.**35–**9.**39
  stabilizers, **9.**24
  structural applications, **9.**10
  thermal resistance, **9.**32
  types, **9.**13
  ultraviolet attack, **9.**2, **9.**9
  use environment, **9.**7
  uses of, **9.**3, **9.**6, **9.**8, **9.**10–**9.**12
  vulcanization, **9.**2, **9.**22–**9.**24
  See also Elastomers
Automotive thermoplastics
  ABS, **9.**26
  acetals, **9.**25
  acrylics, **9.**25, **9.**26
  engineering thermoplastics, **9.**25–**9.**29
  exterior applications, **9.**11
  glass-transition temperature, **9.**2
  interior decorative applications, **9.**11, **9.**12
  introduction to, **9.**1, **9.**2
  mechanical applications, **9.**10
  nylon, **9.**26, **9.**27
  PBT, **9.**27

Automotive thermoplastics (*Cont.*):
  PET, **9.**28, **9.**29
  polyamide, **9.**26, **9.**27
  polycarbonate, **9.**27
  polyesters, **9.**28, **9.**29
  polyether-imide, **9.**29
  polyphenylene sulfide, **9.**29
  polyphtalamide, **9.**29
  polypropylene, **9.**28
  polysulfone, **9.**29
  polyvinylchloride (PVC), **9.**28
  PTFE, **9.**30
  SAN, **9.**26
  selection, **9.**31–**9.**34
  specialty thermoplastics, **9.**29, **9.**30
  structural applications, **9.**10
  thermoplastic olefins, **9.**28
  thermoplastic polyesters, **9.**28, **9.**29
  thermoplastic polyurethane, **9.**30
  transparent applications, **9.**11
  ultraviolet attack, **9.**2
  See also Thermoplastics
Automotive thermosets
  introduction to, **9.**1, **9.**2
  glass-transition temperature, **9.**2
  polyurethanes, **9.**30
  RIM, **9.**31
  selection, **9.**31–**9.**34
  sheet molding compound, **9.**31
  ultraviolet attack, **9.**2
  urethanes, **9.**30
  See also Thermosets

Bismaleimides, **2.**30
Blow molding, **1.**7

Calendering, **1.**7
Casting, **4.**8. *See also* Liquid resins
Cellulosics
  coatings, **6.**25
  nature of, **1.**52
  properties, **1.**53
  structure, **1.**52
  trade names, **1.**42
Chemical and environmental resistance
  ASTM tests, **1.**15, **1.**16
  coatings, **6.**6, **6.**14, **6.**15
  laminates, **2.**6
  liquid epoxies, **4.**13
  moisture, **1.**34, **1.**37, **1.**47, **1.**48, **1.**52
  nylons, **4.**58–**4.**60
  silicones, **4.**40
  See also individual types
Coatings
  abrasion resistance, **6.**25
  acrylics, **6.**11
  alkyds, **6.**11
  application methods, **6.**33, **6.**38, **6.**39

Coatings (*Cont.*):
  cellulosics, **6.**25
  chlorinated rubber, **6.**27
  coefficients of friction, **6.**28
  curing, **6.**45
  ecology, **6.**3
  electrical properties, **6.**12, **6.**17, **6.**18, **6.**19,
    6.20, **6.**22
  epoxies, **6.**16
  fluidized, **6.**3, **6.**41
  fluorocarbons, **6.**27, **6.**30
  fungus resistance, **6.**14
  melamines, **6.**25
  moisture-vapor transmission, **6.**31
  nylons, **6.**29, **6.**30
  phenolic, **6.**28
  polyesters, **6.**16
  polyethylenes, **6.**29
  polyimides, **6.**31
  powder, **6.**41. *See also* Coatings, fluidized
  prepainted metal. *See* Coatings, application methods
  pretreatments, **6.**3–**6.**7. *See also* Coatings, surface preparation
  properties, **6.**6, **6.**12, **6.**13, **6.**14, **6.**15, **6.**17, **6.**19, **6.**20, **6.**22, **6.**25, **6.**26, **6.**28, **6.**30, **6.**31, **6.**32, **6.**36
  silicones, **6.**46
  solvent resistance, **6.**46
  specifications, **6.**10
  surface preparation
    by abrasive cleaning, **6.**5
    by alkaline cleaning, **6.**5
    by detergent cleaning, **6.**6
    by emulsion cleaning, **6.**6
    by solvent cleaning, **6.**6
    by steam cleaning, **6.**6
    by vapor degreasing, **6.**6
  surface pretreatment
    for aluminum, **6.**6
    for copper, **6.**7
    for galvanized steel, **6.**7
    for stainless steel, **6.**7
    for steel, **6.**7
    for titanium, **6.**7
    for zinc and cadmium, **6.**7
  thermal conductivity, **6.**36
  transfer efficiency, **6.**39
  types, **6.**11–**6.**33
  vinyls, **6.**23
  wire insulation, **6.**32, **6.**33
Copolyester thermoplastic elastomers, **5.**2, **5.**14
  chemistry, **5.**14
  designing with, **5.**14
  fluid resistance, **5.**15
  morphology, **5.**10
  properties, **5.**14, **5.**15
Cyanate esters, **4.**64, **4.**66

Cycle time, **7.**5, **7.**31
Cyclic thermoplastics, **4.**63

Decorative laminates, **2.**39
Definitions. *See Appendix A*
Depolymerized rubber, **4.**56–**4.**57
Diallyl phthalates
  liquid resins, **4.**46, **4.**48
  nature of, **1.**26, **1.**27
  trade names, **1.**24
Dielectric constant. *See* Electrical properties and tests
Dielectric strength. *See* Electrical properties and tests
Dissipation factor. *See* Electrical properties and tests
Dynamic vulcanization, **5.**9, **5.**21

Elastomeric alloys. *See* Thermoplastic vulcanizates
Elastomeric polyamides, **5.**17
  chemistry, **5.**17
  fluid resistance, **5.**18
  morphology, **5.**18
  processing, **5.**18
  properties, **5.**18
Elastomers
  acrylic, **1.**90, **1.**95
  as adhesives, **7.**69
  bonding of, **7.**83
  butadiene, **1.**90, **1.**96
  butyl, **1.**90, **1.**94
  costs, **1.**90
  epichlorohydrin, **1.**90, **1.**96
  EPM terpolymer, **1.**90, **1.**94
  EPT copolymer, **1.**90, **1.**94
  fluoroelastomers, **1.**90, **1.**96
  Hypalon, **1.**90, **1.**94
  isoprene, **1.**90, **1.**92, **1.**93
  natural rubber, **1.**89, **1.**90
  neoprene, **1.**91, **1.**93
  nitrile, **1.**91, **1.**94
  nomenclature, **1.**90–**1.**91, **9.**13
  physical properties, **1.**90, **1.**91
  properties, **1.**90–**1.**99
  radiation stability, **1.**99
  relative costs, **1.**90, **1.**91
  SBR, **1.**90, **1.**93
  silicones, **1.**91, **1.**96
  testing and evaluation, **1.**90, **1.**99, **3.**41
  thermal stability, **1.**90, **1.**91
  trade names, **1.**90, **1.**91
  types, **1.**90, **1.**91, **9.**13
  urethane, **1.**91, **1.**96
  See also Automotive elastomers; thermoplastic elastomers; thermoplastic vulcanizates
Electrical properties and tests
  aramids, **2.**4–**2.**13

Electrical properties and tests (*Cont.*):
  arc resistance, **2.**16, **2.**29
  epoxies, **2.**34
  fiber-reinforced thermoplastics, **2.**31
  fluorocarbons, **2.**31
  phenolics, **2.**34
  polyesters, **2.**25
  specifications, **1.**15
  tests, **1.**15, **1.**16
  See also Appendices C and D
Electroplating on plastics, **1.**83, **1.**84
Engineering thermoplastics, **1.**44, **9.**25–**9.**30
Epoxies
  adhesives, **7.**61
  coatings, **6.**16
  curing agents, **1.**29
  electrical data, **2.**16–**2.**34
  mechanical data, **2.**16–**2.**34
  properties, **2.**16–**2.**34. *See also* Tests and
    properties
  structures, **1.**27, **1.**28
  trade names, **1.**24
  types, **1.**27
  See also Liquid resins
Ethylene-propylene liquid polymers, **4.**58
Expanding monomers, **4.**64
Extrusion, **1.**8, **1.**10

Fiber-reinforced thermoplastics, properties
  of, **1.**38, **1.**39, **1.**79, **2.**45
Fibers, **2.**5–**2.**15. *See also* Fiber-reinforced
  thermoplastics; Laminated structures;
  Man-made fibers
Filament winding, **1.**8, **1.**39, **2.**58
Fillers
  effect in molding compounds, **1.**38
  fiber-reinforced thermoplastics, **1.**38, **1.**39,
    1.79, **1.**80
Films and tapes
  abrasion resistance, **6.**25
  moisture-vapor transmission, **6.**31
  nature of, **1.**80
  properties, **1.**50–**1.**82
  wire insulation, **6.**32, **6.**33
Finishing
  painting, **1.**82, **1.**83
  plating, **1.**83, **1.**84. *See also* Coatings
Fluorocarbons
  coatings, **6.**27, **6.**30
  electrical data, **2.**31
  hardness, **1.**55
  mechanical data, **2.**31
  nature of, **1.**53–**1.**57
  properties, **2.**31. *See also* Tests and proper-
    ties
  structures, **1.**54
  thermal expansion, **1.**55, **2.**31
  trade names, **1.**54
  types, **1.**53–**1.**57

Glass fibers, **2.**5–**2.**15. *See also* Fiber-reinforced
  thermoplastics; Man-made fibers
Glass-transition temperature, **9.**2
Glossary. *See Appendix A*

Hand lay-up, **1.**9. *See also* Processing
Health and safety, **3.**71–**3.**73
Heat welding, **4.**2, **4.**23, **4.**28, **4.**34, **4.**62

Ionomers
  nature of, **1.**57, **1.**58
  properties, **1.**57, **1.**58
  trade names, **1.**42

Joining, plastics, **7.**94–**7.**97. *See also* Adhesives and bonding; Product design

Kevlar, **2.**4, **2.**13

Laminated structures
  asbestos fiber, **2.**11
  boron composites, **2.**14
  copper-clad laminates, **2.**39, **2.**63
  copper peel strength, **2.**39
  definitions, **2.**1
  design criteria, **2.**2–**2.**5
  epoxies, **2.**37
  filament winding, **1.**8, **2.**58
  glass fibers, **2.**4
  graphite composites, **2.**13
  Kevlar composites, **2.**4–**2.**13
  laminate selector charts, **2.**1
  molded tubing, **2.**4
  NEMA, **2.**34
  phenolics, **2.**35
  polyesters, **2.**38
  polyimides, **2.**41
  processing, **2.**1
  pultrusion, **1.**13, **2.**56–**2.**58
  reinforced plastics, **2.**2
  rods, **2.**4, **2.**34
  rolled tubing, **2.**34
  sheet molding, **2.**26
  silicones, **2.**29, **2.**34
  thermoplastic laminates, **2.**47
  tolerances, **2.**5
  See also Advanced composites
Liquid resins
  acid curing agents, **4.**16
  acrylics, **4.**60–**4.**62
  advantages, **4.**1–**4.**2
  aliphatic amines, **4.**15–**4.**16
  allylics, **4.**46–**4.**52
  amines, **4.**15–**4.**16
  anhydride curing agents, **4.**16–**4.**17
  aromatic amines, **4.**15–**4.**16
  chemical resistance, **4.**13, **4.**40
  chlorinated epoxies, **4.**14
  curing agents, **4.**11, **4.**15–**4.**17, **4.**28

Liquid resins (*Cont.*):
  cycloaliphatic epoxies, **4.**10, **4.**12
  epoxies, **4.**9–**4.**23
  epoxy modifiers, **4.**17–**4.**18
  fillers, **4.**19–**4.**22, **4.**28
  flame-retardant, **4.**14, **4.**16
  flexibilizers, **4.**17, **4.**18
  flexibilizing curing agents, **4.**17
  hydrolytic stability, **4.**31, **4.**36–**4.**37
  latent curing agents, **4.**17
  MOCA, **4.**31, **4.**34
  monomers, **4.**24–**4.**27
  novolac epoxies, **4.**12–**4.**13
  nylons, **4.**58–**4.**60
  phenolics, **4.**42, **4.**46, **4.**47
  polybutadienes, **4.**52–**4.**56
  polyester applications, **4.**28
  polyester curing agents, **4.**28
  polyesters, **4.**23–**4.**29
    flexible, **4.**25
  polysulfides, **4.**57–**4.**58
  polyurethanes, **4.**30–**4.**37
  properties, **4.**3–**4.**5
  RTV, **4.**38–**4.**40
  selection criteria, **4.**6, **4.**7
  silicones, **4.**37–**4.**42, **4.**43–**4.**46
  styrene in polyesters, **4.**24, **4.**26, **4.**27
  thermal aging data, **4.**16, **4.**31, **4.**34
  vinyl plastisols, **4.**62–**4.**63
  viscosity data, **4.**12, **4.**18
Liquid silicone rubber, **4.**41, **4.**44–**4.**46
Low-permeability thermoplastics
  nature of, **1.**58
  properties, **1.**58
  trade names, **1.**42

Man-made fibers
  boron, **2.**14
  carbon, **2.**14
  fiber characteristics, **2.**5
  fiber forms, **2.**5
  filament, **2.**6
  finishing, **2.**70
  industry classifications, **2.**5
  Nomex, **2.**13
  nylons, **2.**13
Matched-die molding, **1.**8
Mechanical properties and tests
  ASTM tests, **1.**15
  creep resistance, ABS, **1.**42–**1.**45
  deflection of thermoplastics, **1.**45
  dimensional stability
    acetals, **1.**47, **1.**48
    acrylics, **1.**50, **1.**51
    nylons, **1.**60, **1.**62
    thermoplastics, **1.**80, **1.**81
  federal tests, **1.**16
  flexural modulus, ABS, **1.**45
  fluorocarbons, **1.**53–**1.**57, **4.**31

Mechanical properties and tests (*Cont.*):
  hardness comparisons, **1.**55, **7.**7
  hardness of fluorocarbons, **1.**55
  polypropylene, **2.**33
  tensile creep (ABS), **1.**44
  thermal expansion
    comparisons, **2.**3
    fluorocarbons, **1.**55
  thermosetting, **2.**40–**2.**43
  viscoelastic behavior. *See* Tests and properties; *specific polymers*
  See also Appendix C

Nylons
  aramids, **1.**62
  castable liquids, **4.**58–**4.**60
  coatings, **6.**29, **6.**30
  dimensional changes, **1.**59–**1.**62
  fiber-reinforced, **1.**60
  high-temperature, **1.**62
  man-made fibers, **2.**13
  moisture effects, **1.**60
  nature of, **1.**58–**1.**63
  structure, **1.**59
  trade names, **1.**42
  types, **1.**59

Packaging. *See* Plastics in packaging
Painting of plastics, **1.**82, **1.**83
Paper, **2.**13
Parylenes
  nature of, **1.**63
  process, **1.**53, **1.**64
  properties, **1.**64, **1.**65
  structure, **1.**63
  trade names, **1.**42
Phenolics
  coatings, **6.**28
  liquid resins, **4.**42, **4.**46, **4.**47
  nature of, **1.**33
  properties, **1.**26, **1.**33, **2.**15
  structure, **1.**36
  trade names, **1.**24
Photopolymers, **4.**67
Physical properties. *See Appendix C*
Plastics in packaging
  annual demand, **8.**3
  ethylene vinyl alcohols, **8.**13
  introduction, **8.**1–**8.**3
  mass transfer, **8.**30
    diffusion across single sheet,
      8.34–**8.**39
    diffusion across composite structure,
      8.39–**8.**43
    diffusion and sorption equations,
      8.32–**8.**34
    impermeable foil layer, **8.**50
    migration and sorption at equilibrium,
      8.59–**8.**61

Plastics in packaging, mass transfer (*Cont.*):
  migration and sorption in a semi-infinite
    polymer plane, **8.**49–**8.**59
  packaging interactions, **8.**30, **8.**31
  permanent transport and packaging
    shelf life, **8.**46–**8.**49
  permeability measurement, **8.**43–**8.**46
  types of interactions, **8.**31, **8.**32
 nylons, **8.**14, **8.**15
 polycarbonates, **8.**16
 polyethylene, **8.**3–**8.**8
  braided copolymers, **8.**4–**8.**6
  branched homopolymers, **8.**4
  categories, **8.**3
  linear, **8.**6–**8.**8
 polyethylene naphthalene, **8.**16
 polyethylene terephthalates, **8.**15, **8.**16
 polypropylene, **8.**8–**8.**10
 polystyrenes, **8.**12, **8.**13
 polyvinyl chlorides, **8.**10
 properties of, **8.**17–**8.**30
  barrier, **8.**26, **8.**27
  electrical, **8.**24, **8.**30
  mechanical, **8.**23–**8.**26
  morphology and density, **8.**17–**8.**21
  optical, **8.**28, **8.**29
  surface and adhesion, **8.**27, **8.**28
  thermophysical, **8.**21–**8.**23
 vinylidene chlorides, **8.**11
Polyallomers
 nature of, **1.**70
 properties, **1.**70. *See also* Tests and properties
 trade names, **1.**43
Polyaryl ether
 nature of, **1.**65
 properties, **1.**66
 trade names, **1.**42
Polyaryl sulfone
 nature of, **1.**65
 properties, **1.**66
 trade names, **1.**42
Polybutadienes
 liquid resins, **4.**52–**4.**56
 nature of, **1.**36
 properties, **1.**37
 structure, **1.**36
 trade names, **1.**24
Polycarbonates
 moisture effects, **1.**67, **1.**68
 nature of, **1.**67
 properties, **1.**68
 structure, **1.**67, **2.**2
 trade names, **1.**42
Polyesters
 electrical data. *See Appendices C and D*
 nature of, **1.**68
 properties, **1.**68, **2.**17
 thermosetting
  adhesives, **7.**66

Polyesters, thermosetting (*Cont.*):
 chemical resistance, **2.**17
 coatings, **6.**16
 electrical data, **2.**17
 exotherms, **1.**38
 liquid resins, **4.**23–**4.**29
 mechanical data, **2.**17–**2.**64
 nature of, **1.**38, **1.**39
 properties, **1.**26, **1.**38, **1.**39, **2.**40–**2.**42
 structure, **1.**3
 trade names, **1.**42
 See also Liquid resins
Polyethersulfone
 nature of, **1.**69
 properties, **1.**69, **1.**70
 structure, **1.**69
 trade names, **1.**42
Polyethylenes
 coatings, **6.**29
 cross-linked, **1.**72
 irradiated, **1.**72
 nature of, **1.**70–**1.**72
 structure, **1.**70
 trade names, **1.**43
Polyimides
 adhesives, **7.**101
 coatings, **6.**31
 electrical data, **2.**27
 mechanical data, **2.**27
 nature of, **1.**72
 properties, **1.**72–**1.**74, **2.**27. *See also* Tests
  and properties
 structure, **1.**73
 trade names, **1.**43
Polymers
 abbreviations. *See Appendix B*
 classes, **1.**2
 flame retardants, **1.**20
 flame test summary, **1.**21
 flammability, **1.**22, **1.**23, **4.**14, **4.**16, **6.**14
 nature of, **1.**1, **1.**2
 painting of, **1.**82, **1.**83
 plating of, **1.**83, **1.**84
 product design guidelines, **1.**5–**1.**14
 property tables, **1.**18, **1.**19, **4.**3, **4.**4. *See also*
  Tests and properties; *Appendix C*
 raw-material, **1.**5
 structures, **1.**1, **1.**2. *See specific polymer*
 symbols. *See Appendix B*
 thermoplastics, **1.**2, **1.**3, **1.**40–**1.**82. *See also*
  Automotive thermoplastics; *specific*
  type or specific subject
 thermosets, **1.**2, **1.**21–**1.**40. *See also* Automotive thermosets; *specific type or*
  specific subject
Polymethylpentene
 nature of, **1.**74
 properties, **1.**74, **1.**75. *See also* Tests and
  properties

Polymethylpentene (*Cont.*):
trade names, **1.**43
Polyolefins
cross-linked, **1.**72. *See also* Polyallomers;
Polyethylene; Polypropylene
Polyphenylene oxides
nature of, **1.**75
properties, **1.**66, **1.**68. *See also* Tests and
properties
structure, **1.**76
trade names, **1.**43
Polyphenylene sulfides
nature of, **1.**75, **1.**76
properties, **1.**76. *See also* Tests and properties
structure, **1.**75
trade names, **1.**43, **1.**44
Polypropylenes
mechanical data, **2.**33
nature of, **1.**72
properties, **1.**72, **2.**33. *See also* Tests and
properties
structure, **1.**70
trade names, **1.**43
Polystyrenes
cross-linked, **1.**77
nature of, **1.**76–**1.**78
properties, **1.**76, **1.**78. *See also* Tests and
properties
structure, **1.**76
thermosetting, **1.**77
trade names, **1.**43
Polysulfides
elastomers, **1.**91
liquid resins, **4.**57–**4.**58
Polysulfones
fiber-reinforced, **1.**78
nature of, **1.**70
properties, **1.**66. *See also* Tests and properties
trade names, **1.**43
Polyurethanes
adhesives, **9.**67
coatings, **6.**21
hydrolytic stability, **4.**31, **4.**36–**4.**37
liquid resins, **4.**30–**4.**37. *See also* Liquid
resins
Prepreg materials, **2.**2
Price per unit property of important plastics,
comparisons, **1.**18, **1.**19
Processing
basics of, **1.**5–**1.**14, **10.**1
blow molding, **1.**7, **10.**9, **10.**10
bulk molding compounds, **2.**46
calendering, **1.**7
casting, **1.**7, **4.**6, **4.**8, **10.**3, **10.**4
centrifugal casting, **4.**8
cold-forming, **1.**7
comparison molding, **1.**11, **10.**10–**10.**14
comparisons, **1.**7

Processing (*Cont.*):
costs, **1.**7
cross-linking, **1.**2, **1.**3
descriptions, **1.**7
extrusion, **1.**8, **10.**7
filament winding, **1.**8, **2.**58, **10.**5
hand lay-up, **10.**4, **10.**5
injection molding, **1.**8, **10.**14–**10.**18
laminating, **1.**8, **2.**1
liquid resins, **4.**2, **4.**6, **4.**8–**4.**9, **10.**3, **10.**4
machining, **10.**19–**10.**23
methods, **1.**5–**1.**14
molding equipment, **10.**3–**10.**18
painting, **1.**82, **1.**83
parylene, **1.**63–**1.**65
plating, **1.**83, **1.**84
properties. *See Appendix C*
pultrusion, **1.**13, **2.**56, **10.**7, **10.**8
reaction injection molding, **2.**55, **10.**18
resin transfer molding, **10.**5
rotational molding, **1.**9, **10.**4
sheet molding, **2.**26
slush molding, **1.**9
thermoforming, **1.**9, **1.**14, **10.**6
transfer molding, **1.**9, **10.**10–**10.**14. *See also*
Liquid resins
vapor phase, **1.**82, **1.**83
wet lay-up, **1.**9, **2.**51
See also Appendix C; Laminated structures; Product design; *specific product*
Product design
adhesive joints, **7.**13–**7.**29, **10.**19–**10.**21. *See
also* Adhesives and bonding
deflashing, **10.**23, **10.**24
design procedure, **10.**1
drilling and tapping, **10.**22, **10.**23
flow lines, **10.**26–**10.**28
gates, **10.**26–**10.**28
guidelines, **1.**7–**1.**10
inserts, **10.**33
laminates, **4.**2
lead forming, **10.**24
materials selection, **2.**61–**2.**70
mechanical finishing, **10.**19–**10.**23
mold construction, **10.**31–**10.**33
molding methods, **1.**7, **1.**8
parting lines, **10.**30
process comparisons, **1.**5–**1.**14
process descriptions, **1.**5–**1.**14
prototyping, **10.**2
reaming, **10.**22
sink marks, **10.**29
taps, **10.**22
tolerances, **10.**28, **10.**29
ultrasonic welding, **10.**19
welding, **10.**19, **10.**20
See also specific material or process
Pultrusion, **1.**3, **3.**48–**3.**52

Radiation resistance, **1.**86
Reaction injection molding, **4.**9
Reactions
  addition, **1.**3
  bulk, **1.**4
  condensation, **1.**31
  cross-linking, **1.**2
  emulsion, **1.**4
  irradiation, **1.**72
  nylons, **1.**58
  parylene, **1.**63
  phenolics, **1.**33, **2.**12
  polyester, **1.**38
  polysulfone, **1.**78
  raw-material sources, **1.**5
  solution, **1.**4
  suspension, **1.**4
Recycling
  ABS, **11.**39
  automotive plastics, **11.**36–**11.**38
  benefits, **11.**6
  bottle-deposit systems, **11.**11
  collection of materials, **11.**9, **11.**11–**11.**13
  commercial plastics, **11.**32–**11.**34
    markets for recycled materials, **11.**33
    processes, **11.**32
  contamination issues, **11.**7
  contamination with nonplastics, **11.**8
  contamination with other plastics, **11.**8
  design for, **11.**46
  design issues, **11.**45
  drop-off systems, **11.**11
  environmental certification, **11.**47
  future of, **11.**48–**11.**50
  green marketing, **11.**47
  high-density polyethylene, **11.**21–**11.**24
    markets for recycled material, **11.**23
    milk bottles, 11,21
    pigmented materials, **11.**22
    properties of recycled material, **11.**24
  introduction, **11.**1
  low-density polyethylene, **11.**24–**11.**26
    stretch film, **11.**24
  mixed waste processing, **11.**13
  municipal solid waste, **11.**4–**11.**6
  nylon, **11.**38
  polycarbonate, **11.**38
  polyethylene terephthalate, **11.**14–**11.**21
    films, **11.**18
    markets for recycled material, **11.**19
    properties of recycled material, **11.**19
    recycling processes, **11.**5, **11.**16
  polypropylene, **11.**28–**11.**30
    automotive battery cases, **11.**28
  polystyrene, **11.**26–**11.**28
    expanded PS, **11.**26
    markets for recycled material, **11.**28
  polyvinyl chloride, **11.**30–**11.**32

Recycling, polyvinyl chloride (*Cont.*):
  bottles, **11.**30
  materials for recycled material, **11.**32
  quality issues, **11.**9
  by resin type, **11.**6
  safety concerns, **11.**7
  soft-drink bottles, **11.**14
  solid waste concerns, **11.**2–**11.**5
  sorting and separation, **11.**39–**11.**44
  thermal, **11.**44–**11.**45
  thermosets, **11.**34–**11.**36
Reinforced plastics. *See* Laminated struc-
    tures; Advanced composites
Resin transfer molding, **4.**9
Resistivity. *See* Electrical properties and tests
Rotational molding, **1.**9, **4.**8

Scrap, **5.**1, **5.**6, **5.**43. *See also* Recycling
Screw extrusion, **1.**8, **1.**10
Silicones
  adhesives, **7.**70
  elastomers, **1.**84–**1.**89
  electrical data, **2.**26
  flexible, **1.**95
  liquid resins, **4.**37–**4.**42
  mechanical data, **4.**43–**4.**46
  nature of, **1.**95
  properties, **1.**91, **2.**26
  rigid, **2.**26
  trade names. *See* Liquid resins
Slush molding, **1.**9
Standards and specifications
  ANSI, **2.**61
  ASTM, **1.**15, **2.**61, **6.**10
  copper-clad laminates, **2.**34, **2.**38, **2.**43
  DOD, **2.**65
  federal specifications, **2.**65
  federal standard test methods, **6.**10
  federal standards, **1.**16
  guide to insulation specifications, **2.**61
  IPC, **2.**63
  NEMA, **2.**26, **2.**34, **6.**33
  UL, **2.**66
  See also Appendix E
Styrene-diene TPEs, **5.**2, **5.**11
  applications, **5.**13
  chemistry, **5.**12
  markets, **5.**13
  morphology, **5.**10
  properties, **5.**13

Terms and definitions. *See Appendix A*
Tests and properties
  abrasion resistance, **6.**25
  adhesives, **7.**20–**7.**29
  analytical, **1.**15
  ASTM and federal tests, cross reference
    of, **1.**16

Tests and properties (*Cont.*):
  ASTM tests, significance of, **1.**15
  chemical, **1.**15
  coatings, **6.**8–**6.**10, **6.**12–**6.**15, **6.**17–**6.**22,
    6.24–**6.**26, **6.**28, **6.**30–**6.**33, **6.**36–**6.**37
  elastomers, **1.**97–**1.**99
  electrical. *See Appendix D*
  federal, **1.**16
  flammability, **1.**20, **1.**21, **4.**14, **4.**16
  hardness comparisons, **1.**17
  liquid resins, **4.**3–**4.**5
  mechanical. *See Appendix C*
  optical, **1.**15
  permanence, **1.**15
  testing and evaluation, **1.**16
  thermal, **1.**15. *See also Appendix C*
  thermal-conductivity data, **2.**3
  thermal expansion comparisons, **2.**3
  thermoplastics, **1.**40–**1.**82
  thermosets. *See specific polymer or test*
  water-absorption comparisons, **1.**15. *See
    also* Water absorption
  See also Appendix C
Testing standards. *See Appendix E*
Thermal properties. *See Appendix C*
Thermoplastic elastomers, **5.**1
  advantages, **5.**5
  applications, **5.**33
  automotive uses, **5.**38
  block morphology, **5.**9
  blow molding, **5.**36
  chemistry, **5.**40
  cost/performance, **5.**11
  disadvantages, **5.**7
  electrical applications, **5.**20
  EPDM/polypropylene, **5.**19, **5.**21
  extrusion, **5.**32
  food contact applications, **5.**39
  glass-transition temperature, **5.**8
  growth, **5.**3
  history, **5.**3
  injection molding, **5.**34
  markets, **5.**38
  mechanical goods, **5.**39
  melting point, **5.**8
  morphology, **5.**8, **5.**9
  NBR/polypropylene, **5.**21
  NBR/PVC, **5.**19, **5.**27
  NR/polypropylene, **5.**27
  processing, **5.**6, **5.**27
  production volumes, **5.**3
  properties, **5.**9, **5.**12, **5.**15, **5.**17, **5.**18, **5.**19,
    5.20, **5.**23
  reactor TPEs, **5.**27
  recycle, **5.**39. *See also* Recycling
  usage, **5.**3
  See also Elastomers

Thermoplastic polyurethanes, **5.**2, **5.**10, **5.**11,
    5.15
  abrasion of, **5.**17
  chemistry, **5.**16
  morphology, **5.**16
  properties, **5.**16
Thermoplastics. *See* Polymers, thermoplas-
    tics; *specific polymer*
Thermoplastic vulcanizates
  anisotropy, **5.**23
  applications, **5.**27
  chemistry, **5.**21
  cross-linking, **5.**20
  EPDM/polypropylene, **5.**21
  fluid resistance, **5.**25
  mechanical properties, **5.**22, **5.**23, **5.**25
  morphology, **5.**21
  NBR/polypropylene, **5.**25
  service temperatures, **5.**25
  vulcanizates, **5.**2, **5.**8, **5.**11, **5.**20
Thermosets. *See* Polymers, thermosets; *spe-
    cific types*
Tires, **5.**41
Trade names
  elastomers, **1.**90, **1.**91
  thermoplastics, **1.**42–**1.**44
  thermosets, **1.**24

Ultrasonic welding, **7.**92

Vacuum and space, **1.**47
Vapor-phase polymerization, parylene,
    1.63
Vinyl esters, **4.**64, **4.**65
Vinyls
  coatings, **6.**23, **6.**26
  nature of, **1.**78, **1.**79
  organosol, **6.**23, **6.**26
  plastisols, **4.**23, **4.**26, **4.**62, **6.**63
  properties, **1.**78, **1.**79. *See also* Tests and
    properties
  solutions, **6.**23, **6.**26
  trade names, **1.**43
Voltage breakdown. *See* Electrical properties
    and tests
Vulcanized rubber, **2.**48

Water-absorption
  acetals, **1.**48
  acrylics, **1.**51
  nylons, **1.**60–**1.**62
  polycarbonates, **1.**67, **1.**68
  polyurethanes, **2.**13, **4.**36–**4.**37
  testing, **1.**15, **1.**16
Water-vapor transmission rates of plastic
    coatings and films, **6.**31